# Hypersonic Shock Wave Turbulent Boundary Layers

## Direct Numerical Simulation, Large Eddy Simulation and Experiment

**Online at: https://doi.org/10.1088/978-0-7503-5002-0**

# Hypersonic Shock Wave Turbulent Boundary Layers

Direct Numerical Simulation, Large Eddy Simulation and Experiment

**Doyle Knight and Nadia Kianvashrad**
*Department of Mechanical and Aerospace Engineering*
*Rutgers, The State University of New Jersey,*
*New Brunswick, New Jersey 08903, USA*

**IOP** Publishing, Bristol, UK

ISBN 978-0-7503-5002-0 (ebook)
ISBN 978-0-7503-5000-6 (print)
ISBN 978-0-7503-5003-7 (myPrint)
ISBN 978-0-7503-5001-3 (mobi)

DOI 10.1088/978-0-7503-5002-0

Version: 20230601

IOP ebooks

British Library Cataloguing-in-Publication Data: A catalogue record for this book is available from the British Library.

Published by IOP Publishing, wholly owned by The Institute of Physics, London

IOP Publishing, No.2 The Distillery, Glassfields, Avon Street, Bristol, BS2 0GR, UK

US Office: IOP Publishing, Inc., 190 North Independence Mall West, Suite 601, Philadelphia, PA 19106, USA

*To Michael S Holden*

# Contents

# Preface

This monograph is motivated by the resurgence of interest in hypersonic aerothermodynamics. We have somewhat arbitrarily chosen to define hypersonic as freestream Mach number greater than or equal to five. The scope is limited in two specific areas. First, we only consider hypersonic turbulent boundary layers, in both zero and modest pressure gradients (i.e., in the absence of boundary layer separation), and shock wave boundary layer interactions. In all cases we omit any reference to hypersonic boundary layer transition, including transitional shock wave boundary layer interactions, as well as hypersonic free shear layers. Second, we examine both experiment and simulation, but we restrict our attention in the latter to direct numerical simulation (DNS) and large eddy simulation (LES), omitting any consideration of Reynolds-averaged Navier–Stokes methods.

The book is organized as follows. Chapter 1 provides an introduction to hypersonic aerothermodynamics with emphasis on recent developments. Chapter 2 describes the fundamental mathematical theories of hypersonic turbulent flows. Chapter 3 discusses the structure of equilibrium hypersonic turbulent boundary layers. Chapters 4 and 5 present experimental data for hypersonic turbulent boundary layers and hypersonic shock wave turbulent boundary layer interactions, respectively. Chapters 6 and 7 present DNS and LES results for hypersonic turbulent boundary layers and shock wave turbulent boundary layer interactions, respectively. Chapter 8 summarizes the structure of equilibrium hypersonic turbulent boundary layers and suggests areas for future research. The appendix provides short summaries of hypersonic wind tunnels.

The experimental and computational chapters are organized in chronological order by date of publication. This organization achieves two main purposes of the book. First, it emphasizes the contributions of the researchers. Second, it provides a history of the research. We have made every attempt to be as comprehensive as possible, and we apologize for any unforeseen omissions. Selected results are included for each paper, and the reader is directed to the original source for further information. Data have been digitized from the original publication, and every effort has been made to include all data points. In those cases where the density of data points and the resolution of the figure preclude clear identification of each point, we have used our best judgment to include as many data points as possible. The sizes of the symbols are chosen for clarity and do not necessarily indicate the uncertainty in the measurements.

Special thanks to Guillaume Grossir (von Kármán Institute for Fluid Dynamics), David Mie (University of Queensland), Erich Schülein (German Aerospace Center), and Owen Williams (University of Washington) for their assistance. Special thanks to Rutgers University Library staff for their assistance in this research.

Doyle Knight and Nadia Kianvashrad
New Brunswick, New Jersey
September 2022

# Author biographies

## Doyle Knight

Doyle Knight received his PhD in Aeronautics from the California Institute of Technology in 1974. Following two years' service in the United States Air Force as an Aeronautical Engineer and one year's Postdoctoral Fellowship in Applied Mathematics at the California Institute of Technology, he joined the faculty of the Department of Mechanical and Aerospace Engineering at Rutgers, The State University of New Jersey. He is Distinguished Professor of Aerospace and Mechanical Engineering. His research interests include gas dynamics and design optimization. His research in gas dynamics includes supersonic and hypersonic shock wave boundary layer interaction, incipient separation on pitching airfoils, turbulence model development, high-speed inlet unstart, and effects of unsteady energy deposition in supersonic flows. His research activity in design optimization focuses on the application of computational fluid dynamics to the automated optimal design of high-speed air vehicles. He has published more than 100 journal papers and more than 200 refereed conference papers. He is the author of the books *Elements of Numerical Methods for Compressible Flows* and *Energy Deposition for High Speed Flow Control* published by Cambridge University Press. He is the author or co-author of seven book chapters. He is a member of the Board of Directors (Bureau) of the European Conference for AeroSpace Sciences.

## Nadia Kianvashrad

Nadia Kianvashrad received her MS in Aerospace Engineering at Sharif University of Technology, Tehran, Iran, and her PhD in Mechanical and Aerospace Engineering at Rutgers, The State University of New Jersey. She is a Postdoctoral Associate in the Department of Mechanical and Aerospace Engineering at Rutgers University. Her research interests include numerical simulation of hypersonic flows including non-equilibrium effects and turbulence, energy deposition for high-speed flow and flight control, automated design optimization using computational fluid dynamics, development of higher-order methods in computational fluid dynamics, and reduced-order modeling of high-speed aerodynamics. She is the author of more than 20 papers including conference and journal publications. She is currently a member of the AIAA Fluid Dynamics Technical Committee.

# Nomenclature

*Roman*

| | |
|---|---|
| $c_{\mathrm{f}}$ | skin friction coefficient |
| $c_{\mathrm{p}}$ | specific heat at constant pressure |
| $c_v$ | specific heat at constant volume |
| $e$ | total energy per unit mass |
| $e_i$ | internal energy per unit mass |
| $h$ | static enthalpy per unit mass |
| $H$ | total enthalpy per unit mass |
| $\widetilde{H}$ | Favre-averaged total enthalpy per unit mass |
| $k$ | molecular thermal conductivity, Boltzmann's constant |
| $\widetilde{k}$ | turbulence kinetic energy (Favre averaging) |
| $k_{\mathrm{t}}$ | turbulent thermal conductivity |
| $K$ | Knudsen number |
| $m$ | molecular mass |
| $M$ | Mach number |
| $M_k$ | root-mean-square fluctuating Mach number |
| $p$ | static pressure |
| $Pr$ | molecular Prandtl number |
| $Pr_{\mathrm{t}}$ | turbulent Prandtl number |
| $q_i$ | molecular heat flux vector $(q_1, q_2, q_3)$ |
| $R$ | gas constant for species |
| $Re$ | Reynolds number |
| $Re_\delta$ | Reynolds number based on boundary layer thickness |
| $Re_\theta$ | Reynolds number based on momentum thickness |
| $St$ | Stanton number |
| $T$ | static temperature |
| $\widetilde{T}$ | Favre-averaged static temperature |
| $T_{\mathrm{t}}$ | total temperature |
| $T_{\mathrm{aw}}$ | adiabatic wall temperature |
| $T_{\mathrm{r}}$ | recovery temperature (same as $T_{\mathrm{aw}}$) |
| $\widetilde{T}_{\mathrm{t}}$ | Favre-averaged total temperature |
| $u_i$ | Cartesian velocity components $(u_1, u_2, u_3) = (u, v, w)$ |
| $\overline{u}_i$ | Reynolds-averaged Cartesian velocity component |
| $\widetilde{u}_i$ | Favre-averaged Cartesian velocity component |
| $u_i'$ | fluctuating Cartesian velocity component (Reynolds averaging) |
| $u_i''$ | fluctuating Cartesian velocity component (Favre averaging) |
| $u_\tau$ | friction velocity $(\sqrt{\tau_{\mathrm{w}}/\rho_{\mathrm{w}}})$ |
| $u_{\mathrm{VD}}$ | Van Driest transformed velocity |
| $x_i$ | Cartesian coordinates $(x_1, x_2, x_3) = (x, y, z)$ |
| $y^+$ | dimensionless wall distance $(yu_\tau/\nu_{\mathrm{w}})$ |

*Greek*

| | |
|---|---|
| $\gamma$ | ratio of specific heats $(c_{\mathrm{p}}/c_v)$ |
| $\delta$ | boundary layer thickness |
| $\delta_{ij}$ | Kronecker delta |
| $\theta$ | momentum thickness |

$\kappa$ von Kármán's constant ($\kappa = 0.41$)
$\lambda$ second molecular viscosity coefficient
$\mu$ dynamic molecular viscosity
$\mu_t$ turbulent eddy viscosity
$\nu$ kinematic molecular viscosity
$\rho$ density
$\tau_{ij}$ molecular viscous stress tensor
$\Pi$ wake factor

*Subscripts*

$e$ conditions at edge of boundary layer
$\infty$ freestream conditions in test section
$w$ evaluated at the wall

*Overbars and Tildes*

$-$ Reynolds average
$\sim$ Favre average

# Chapter 1

## Introduction

Hypersonic is herein arbitrarily defined as flow at Mach Numbers in excess of 5.0.

Lobb *et al* (1955)

Almost everyone has their own definition of the term hypersonic. If we were to conduct something like a public opinion poll among those present, and asked everyone to name a Mach number above which the flow of a gas should properly be described as hypersonic flow there would be a majority of answers round about five or six, but it would be quite possible for someone to advocate, and defend, numbers as small as three, or as high as twelve.

Philip Roe
Lecture at von Kármán Institute for Fluid Dynamics,
January 1970

Hypersonic flow regions of viscous/inviscid interaction in laminar and turbulent flows represent one of the most taxing group of aerothermal problems for the designer and certainly are the most difficult to predict in detail with any accuracy.

Holden (1986)

The beginning of the 21st century has been marked by a resurgence in development of hypersonic vehicles for civilian and military applications. Space Exploration Technologies Corporation (SpaceX), a private company founded in 2002 by Elon Musk, successfully sent astronauts to the International Space Station in 2020.

doi:10.1088/978-0-7503-5002-0ch1  

The Russian Federation successfully tested the Mach 7 scramjet-powered missile 3M22 Tsirkon in 2021 (Raskin 2021). Additional hypersonic vehicle design studies include the Boeing hypersonic airliner (Adams 2018) and the Reaction Engines Limited LAPCAT A2 co-funded by the European Union (Steelant 2010).

Holden (1986) emphasized that the accurate prediction of the aerothermodynamic loading (i.e., surface heat transfer, pressure, and skin friction) due to the boundary layers on the surface of hypersonic vehicles is essential to effective design. The vehicle surface experiences laminar, transitional, and turbulent boundary layer flows which may be complicated by non-equilibrium effects including thermochemical reaction, vibrational–translational–electronic energy exchange, radiation, and surface reactions. Within this panorama of complexity, the accurate prediction of turbulent boundary layers at hypersonic speeds, including the effects of the interaction of shock waves with turbulent boundary layers, is paramount (Babinsky and Harvey 2011).

The conventional approach to prediction of hypersonic turbulent boundary layers is based upon the Reynolds-averaged Navier–Stokes (RANS) equations (section 2.2) with the effects of turbulence represented by a set of additional equations for turbulence variables such as the turbulence kinetic energy and rate of dissipation of turbulence kinetic energy (Wilcox 2006). However, RANS methods have shown limited success in prediction of aerothermodynamic loading at hypersonic speeds (Roy and Blottner 2006).

Direct Numerical Simulation (DNS) and Large Eddy Simulation (LES) (section 2.4) simulate the dynamic unsteady behavior of turbulent boundary layers and offer the potential for more accurate representation of the turbulence. DNS resolves the entire range of unsteady turbulent behavior from the large-scale energy-containing eddies (comparable in size to the boundary layer thickness) to the Kolmogorov scale where the turbulence energy is dissipated into heat (Tennekes and Lumley 1983). LES resolves the limited range of unsteady turbulent behavior from the large-scale energy-containing eddies to the inertial subrange where the turbulent motion is considered isotropic and driven by the nonlinear energy cascade from the larger scales (Tennekes and Lumley 1983). DNS and LES have been successfully applied to simulate turbulent boundary layers subject to strong pressure gradients from incompressible to supersonic flows over approximately the past 50 years (Smagorinsky 1963, Deardorff 1970, Galperin and Orszag 1993, Métais and Ferziger 1997, Moin and Mahesh 1998, Sandham 2005, Sagaut *et al* 2006). However, applications to hypersonic turbulent boundary layers (including shock wave boundary layer interactions) have appeared only in the past two decades.

The book is organized as follows. Chapter 2 presents the mathematical theory for perfect gas hypersonic turbulent boundary layers and introduces the three methods for simulation, namely, DNS, wall-resolved LES, and wall-modeled LES. Chapter 3 describes the structure of the equilibrium flat plate zero pressure gradient compressible turbulent boundary layer. Chapter 4 presents a survey of experimental data on hypersonic turbulent boundary layers for both zero and nonzero pressure gradients. Chapter 5 presents experimental data for hypersonic shock wave turbulent boundary layer interactions. Chapters 6 and 7 describe DNS and LES simulations of hypersonic turbulent boundary layers and shock boundary layer interaction. Chapter 8 provides a summary of the state of the art and proposes areas for future research. Practical limitations on length have precluded the inclusion of a number of important phenomena in hypersonic turbulent boundary layer physics including gas–surface interactions, radiation, rarefaction, and transition.

## References

Adams E 2018 *Boeing's Proposed Hypersonic Plane Is Really, Really Fast* https://www.wired.com/story/boeing-hypersonic-mach-5-jet-concept/

Babinsky H and Harvey J 2011 *Shock Wave–Boundary-Layer Interactions* (New York: Cambridge University Press)

Deardorff J 1970 A numerical study of three dimensional turbulent channel flow at large Reynolds numbers *J. Fluid Mech.* **41** 453–80

Galperin B and Orszag S (ed) 1993 *Large Eddy Simulation of Complex Engineering and Geophysical Flows* (Cambridge: Cambridge University Press)

Holden M 1986 A review of aerothermal problems associated with hypersonic flight *24th Aerospace Sciences Meeting* AIAA-86-0267

Lobb R, Winkler E and Persh J 1955 Experimental investigation of turbulent boundary layers in hypersonic flow *J. Aeronaut. Sci.* **22** 1–9

Métais O and Ferziger J (ed) 1997 *New Tools in Turbulence Modeling* (New York: Springer)

Moin P and Mahesh K 1998 Direct numerical simulation: a tool in turbulence research *Annu. Rev. Fluid Mech.* **30** 539–78

Raskin S 2021 *Russian Successfully Tests State-of-the-Art Cruise Missile* https://nypost.com/2021/07/19/russia-successfully-tests-tsirkon-cruise-missile-i/

Roy C and Blottner F 2006 Review and assessment of turbulence models for hypersonic fows *Prog. Aerosp. Sci.* **42** 469–530

Sagaut P, Deck S and Terracol M 2006 *Multiscale and Multiresolution Approaches to Turbulence* (London: Imperial College Press)

Sandham N 2005 Turbulence simulation *Prediction of Turbulent Flows* ed G Hewitt and J Vassilicos (Cambridge: Cambridge University Press) pp 207–35

Smagorinsky J 1963 General circulation experiments with the primitive equations, I. The basic experiment *Mon. Weather Rev.* **91** 99–164

Steelant J 2010 *Hypersonic Technology Developments with EU Co-Funded Projects* EN-AVT-185 NATO Research and Technology Organization

Tennekes H and Lumley J 1983 *A First Course in Turbulence* (Cambridge, MA: MIT Press)

Wilcox D 2006 *Turbulence Modeling for CFD* 3rd edn (La Canada, CA: DCW Industries, Inc.)

**IOP** Publishing

Hypersonic Shock Wave Turbulent Boundary Layers
Direct Numerical Simulation, Large Eddy Simulation and Experiment
**Doyle Knight and Nadia Kianvashrad**

# Chapter 2

## Mathematical theory

Again, the internal motion of water assumes one or other of two broadly distinguishable forms—either the elements of the fluid follow one another along lines of motion which lead in the most direct manner to their destination, or they eddy about in sinuous paths the most indirect possible.

Osborne Reynolds (1883)

The computation of turbulent flows has been a problem of major concern since the time of Osborne Reynolds.

William Reynolds (1976)

With ILES [implicit large eddy simulation], no SGS [subgrid scale] model is used; instead, the numerical dissipation associated with the computational scheme is used as an implicit SGS model. There is some justification for this because the main function of an SGS model is to provide the correct dissipation rate of turbulent kinetic energy, which is passed down to the small scales through the nonlinear cascade process.

Gregory Blaisdell (2008)


### Abstract

This chapter introduces the mathematical theory of compressible turbulent flows. The description is limited to calorically and thermally perfect gases for simplicity, although in practical applications in hypersonic flight non-equilibrium effects may be present depending upon the range of temperatures. These effects include dissociation, ionization, thermochemical reaction, and radiation. Additionally, the discussion is limited to continuum flows wherein the mean free path of the gas molecules is small compared to the characteristic flowfield scale. The mathematical theory of non-equilibrium continuum gas flows is described in Brun (2009) and Nagnibeda and Kustova (2009), and rarefied gas dynamics in Kogan (1969) and Boyd and Schwartzentruber (2017).

## 2.1 Navier–Stokes equations

The governing equations for a calorically and thermally perfect homogeneous gas are the Navier–Stokes equations derivable as moments of Boltzmann's equation (Chapman and Cowling 1970). Using the Einstein summation convention,[1]

$$\frac{\partial \rho}{\partial t} + \frac{\partial \rho u_j}{\partial x_j} = 0 \tag{2.1}$$

$$\frac{\partial \rho u_i}{\partial t} + \frac{\partial \rho u_i u_j}{\partial x_j} = -\frac{\partial p}{\partial x_i} + \frac{\partial \tau_{ij}}{\partial x_j} \tag{2.2}$$

$$\frac{\partial \rho e}{\partial t} + \frac{\partial (\rho e + p) u_j}{\partial x_j} = \frac{\partial}{\partial x_j}(u_i \tau_{ij} - q_j) \tag{2.3}$$

$$p = \rho R T \tag{2.4}$$

where $\rho$ is the density, $u_i$ are the Cartesian components of velocity, $p$ is the static pressure, $R$ is the gas constant, and $e$ is the total energy per unit mass defined as

$$e = e_i + \frac{1}{2} u_i^2 \tag{2.5}$$

where $e_i = c_v T$ is the internal energy per unit mass and $T$ is the static temperature. The static enthalpy per unit mass is

$$h = e_i + \frac{p}{\rho} \tag{2.6}$$

[1] The Einstein summation convention implies summation of all values of integer indices when the same index appears twice. Consider the Cartesian velocity vector $\boldsymbol{u} = (u_1, u_2, u_3)$. The divergence of the velocity is

$$\text{div}\, \boldsymbol{u} = \sum_{i=1}^{i=3} \frac{\partial u_i}{\partial x_i}$$

Written using the Einstein summation convention,

$$\text{div}\, \boldsymbol{u} = \frac{\partial u_i}{\partial x_i}.$$

The molecular viscous stress tensor is

$$\tau_{ij} = \mu\left(\frac{\partial u_i}{\partial x_j} + \frac{\partial u_j}{\partial x_i}\right) + \lambda\frac{\partial u_k}{\partial x_k}\delta_{ij} \tag{2.7}$$

where $\mu$ is the dynamic molecular viscosity and $\lambda$ is the second molecular viscosity coefficient. The molecular heat flux is

$$q_j = -k\frac{\partial T}{\partial x_j} \tag{2.8}$$

where $k$ is the molecular thermal conductivity. A molecular Prandtl number is defined as

$$Pr = \frac{\mu c_p}{k} \tag{2.9}$$

where $c_p$ is the specific heat at constant pressure.

Application of the calorically and thermally perfect Navier–Stokes equations requires that the intermolecular spacing is small compared to the characteristic length scale of the simulation, and the range of temperatures is small and below the level for thermochemical reactions to occur. The first condition may be stated in terms of the Knudsen number defined as

$$K = \frac{\lambda}{L} \tag{2.10}$$

where $\lambda$ is the mean free path of the gas and $L$ is the flowfield characteristic length scale (e.g., the boundary layer thickness). Typically, $K < 0.01$ for the description of the gas as a continuum[2] (Karniadakis *et al* 2000) (i.e., application of the Navier–Stokes equations). The second condition in the case of air requires that the stagnation temperature be significantly lower than the level where substantial oxygen dissociation occurs, i.e., $T_t \underset{\approx}{<} 2000$ K (Bauer 1990).

[2] The Knudsen number is related to the Mach and Reynolds numbers as follows. The mean free path of a gas is (Vincenti and Kruger 1965)

$$\lambda = \frac{\mu}{\rho}\sqrt{\frac{\pi m}{2kT}}$$

for a Maxwell–Boltzmann distribution where $k$ is Boltzmann's constant and $m$ is molecular mass. Thus

$$K = \sqrt{\frac{\gamma\pi}{2}}\frac{M}{Re}$$

assuming an ideal gas where $M$ is the Mach number, $Re$ is the Reynolds number, and $\gamma = c_p/c_v$ is the ratio of specific heats.

$$M = \frac{U}{a} \quad \text{and} \quad Re = \frac{\rho UL}{\mu}$$

Note that $U$, $a$, $\rho$, and $\mu$ are local values of the bulk velocity, speed of sound, density, and dynamic molecular viscosity, respectively.

## 2.2 Reynolds-averaged Navier–Stokes equations

As discussed in the reviews cited in section 2.6, the computational resources for Direct Numerical Simulation (DNS) of the Navier–Stokes equations increases rapidly with Reynolds number due to the requirement for resolution of all relevant physical and temporal scales. The computational resource requirement is impractical for most engineering applications. Consequently, a form of averaging over the smallest physical and temporal scales is performed.

The concept of suitably averaging the Navier–Stokes equations was introduced by Reynolds (1885), and the resultant equations are denoted the Reynolds-averaged Navier–Stokes (RANS) equations. The *ensemble average* of a property $f$ at a particular location $\boldsymbol{x}$ and time $t$ in the flowfield is defined as

$$\bar{f}(\boldsymbol{x}, t) = \lim_{N \to \infty} \frac{1}{N} \sum_{\nu=1}^{\nu=N} f^{(\nu)}(\boldsymbol{x}, t) \tag{2.11}$$

where $f^{(\nu)}(\boldsymbol{x}, t)$ is the $\nu$th realization of the measurement (e.g., repetition of the same experiment). For statistically stationary turbulent flows, the ergodic hypothesis (Monin and Yaglom 1971) states that the ensemble average can be replaced by a suitable time average,[3]

$$\bar{f}(\boldsymbol{x}) = \lim_{\Delta t \to \infty} \frac{1}{\Delta t} \int_{t}^{t+\Delta t} f(\boldsymbol{x})\, dt \tag{2.12}$$

The ensemble and conventional time average satisfy the *Reynolds conditions,*

$$\overline{f + g} = \bar{f} + \bar{g} \tag{2.13}$$

$$\overline{af} = a\bar{f} \quad \text{where } a \text{ is a constant} \tag{2.14}$$

$$\bar{a} = a \tag{2.15}$$

$$\overline{\frac{\partial f}{\partial \xi}} = \frac{\partial \bar{f}}{\partial \xi} \quad \text{where } \xi \text{ is spatial or temporal} \tag{2.16}$$

$$\overline{\bar{f} g} = \bar{f}\, \bar{g} \tag{2.17}$$

### 2.2.1 Reynolds averaging

Reynolds (1885) introduced a decomposition of an instantaneous variable $f(\boldsymbol{x}, t)$ as

$$f(\boldsymbol{x}, t) = \bar{f}(\boldsymbol{x}, t) + f'(\boldsymbol{x}, t) \tag{2.18}$$

[3] The requirement of statistically stationary turbulent flow is unnecessarily strict. For example, consider a turbulent boundary layer on a pitching airfoil. If the frequency of the pitching motion is small compared to the smallest frequencies of the turbulence (i.e., $\bar{u}/\delta$ corresponding to the large eddies in the boundary layer), then the Fourier spectrum of the unsteady motion is bimodal, and therefore the time interval for the averaging would be restricted to the frequency range of the turbulent fluctuations (Monin and Yaglom 1971).

where by definition

$$\overline{f'} = 0 \tag{2.19}$$

On the basis of the Reynolds conditions (2.13) to (2.17), the ensemble-averaged Navier–Stokes equations become

$$\frac{\partial \bar{\rho}}{\partial t} + \frac{\partial \overline{\rho u_j}}{\partial x_j} = 0 \tag{2.20}$$

$$\frac{\partial \overline{\rho u_i}}{\partial t} + \frac{\partial \overline{\rho u_i u_j}}{\partial x_j} = -\frac{\partial \bar{p}}{\partial x_i} + \frac{\partial \bar{\tau}_{ij}}{\partial x_j} \tag{2.21}$$

$$\frac{\partial \overline{\rho e}}{\partial t} + \frac{\partial \overline{(\rho e + p) u_j}}{\partial x_j} = \frac{\partial}{\partial x_j}\left(\overline{u_i \tau_{ij}} - \bar{q}_j\right) \tag{2.22}$$

$$\bar{p} = \overline{\rho T} R \tag{2.23}$$

where

$$\overline{\rho u_i} = \bar{\rho}\bar{u}_i + \overline{\rho' u_i'} \tag{2.24}$$

$$\overline{\rho u_i u_j} = \bar{\rho}\bar{u}_i\bar{u}_j + \bar{\rho}\overline{u_i' u_j'} + \overline{\rho' u_j'}\,\bar{u}_i + \overline{\rho' u_i'}\,\bar{u}_j + \overline{\rho' u_i' u_j'} \tag{2.25}$$

$$\overline{\rho e} = \bar{\rho}\bar{e} + \overline{\rho' e'} \tag{2.26}$$

$$\bar{e} = c_v \bar{T} + \frac{1}{2}(\bar{u}_i\bar{u}_i + \overline{u_i' u_i'}) \tag{2.27}$$

$$\overline{(\rho e + p) u_j} = \overline{\rho h u_j} + \frac{1}{2}\overline{\rho u_i u_i u_j} \tag{2.28}$$

$$\overline{\rho h u_j} = \bar{\rho}\bar{h}\bar{u}_j + \bar{\rho}\overline{h' u_j'} + \overline{\rho' u_j'}\bar{h} + \overline{\rho' h'}\,\bar{u}_j + \overline{\rho' h' u_j'} \tag{2.29}$$

$$\overline{\rho u_i u_i u_j} = \bar{\rho}\left(\bar{u}_i\bar{u}_i\bar{u}_j + \bar{u}_j\overline{u_i' u_i'} + 2\bar{u}_i\overline{u_i' u_j'} + \overline{u_i' u_i' u_j'}\right) + \tag{2.30}$$

$$2\bar{u}_i\bar{u}_j\overline{\rho' u_i'} + \bar{u}_j\overline{\rho' u_i' u_i'} + \bar{u}_i\bar{u}_i\overline{\rho' u_j'} + 2\bar{u}_i\overline{\rho' u_i' u_j'} + \overline{\rho' u_i' u_i' u_j'} \tag{2.31}$$

$$\overline{u_i \tau_{ij}} = \bar{u}\bar{\tau}_{ij} + \overline{u_i' \tau_{ij}'} \tag{2.32}$$

$$\overline{\rho T} = \bar{\rho}\,\bar{T} + \overline{\rho' T'} \tag{2.33}$$

It is evident that the Reynolds averaging of the Navier–Stokes equations introduces a significant number of correlations such as $\overline{\rho' u_i'}$ which would need to be modeled. An example of modeling is presented in Van Driest (1951). A more compact form of the ensemble-averaged Navier–Stokes equations is described in the next section.

### 2.2.2 Favre averaging

Favre (1965) introduced a density weighted average for compressible turbulent flows[4]

$$\tilde{f} = \frac{\overline{\rho f}}{\bar{\rho}} \tag{2.34}$$

with corresponding decomposition

$$f(\boldsymbol{x}, t) = \tilde{f}(\boldsymbol{x}, t) + f''(\boldsymbol{x}, t) \tag{2.35}$$

where by definition

$$\overline{\rho f''} = 0 \quad \text{or equivalently} \quad \widetilde{f''} = 0 \tag{2.36}$$

The utility of Favre averaging in simplifying notation can be seen by considering $\overline{\rho u v}$. Using conventional averaging

$$\overline{\rho u v} = \bar{\rho}\bar{u}\bar{v} + \bar{\rho}\,\overline{u'v'} + \overline{\rho' u'}\,\bar{v} + \overline{\rho' v'}\,\bar{u} + \overline{\rho' u' v'} \tag{2.37}$$

while for Favre averaging

$$\overline{\rho u v} = \bar{\rho}\tilde{u}\tilde{v} + \overline{\rho u'' v''} \tag{2.38}$$

The Favre-averaged Navier–Stokes equations for a perfect gas are

$$\frac{\partial \bar{\rho}}{\partial t} + \frac{\partial \bar{\rho}\tilde{u}_j}{\partial x_j} = 0 \tag{2.39}$$

$$\frac{\partial \bar{\rho}\tilde{u}_i}{\partial t} + \frac{\partial \bar{\rho}\tilde{u}_i\tilde{u}_j}{\partial x_j} = -\frac{\partial \bar{p}}{\partial x_i} + \frac{\partial}{\partial x_j}\left(-\overline{\rho u_i'' u_j''} + \bar{\tau}_{ij}\right) \tag{2.40}$$

$$\begin{aligned} \frac{\partial \bar{\rho}\tilde{e}}{\partial t} + \frac{\partial(\bar{\rho}\tilde{e} + \bar{p})\tilde{u}_j}{\partial x_j} &= \frac{\partial}{\partial x_j}\left(-c_p\,\overline{\rho T'' u_j''} - \bar{q}_j\right) \\ &+ \frac{\partial}{\partial x_j}\left(-\overline{\rho u_i'' u_j''}\,\tilde{u}_i + \bar{\tau}_{ij}\tilde{u}_i\right) \end{aligned} \tag{2.41}$$

$$\bar{p} = \bar{\rho} R \tilde{T} \tag{2.42}$$

where the Favre-averaged total energy per unit mass is

$$\tilde{e} = c_v \tilde{T} + \frac{1}{2}\tilde{u}_i\tilde{u}_i + \tilde{k} \tag{2.43}$$

[4] See also van Miegham (1949).

where $\tilde{k}$ is the turbulence kinetic energy per unit mass defined by

$$\bar{\rho}\tilde{k} = \frac{1}{2}\overline{\rho u_i'' u_i''} \tag{2.44}$$

The Favre-averaged total enthalpy per unit mass is

$$\tilde{H} = \tilde{e} + \frac{\bar{p}}{\bar{\rho}} \tag{2.45}$$

and thus from (2.42)

$$\tilde{H} = c_\mathrm{p}\tilde{T} + \frac{1}{2}\tilde{u}_i\tilde{u}_i + \tilde{k} \tag{2.46}$$

or

$$\tilde{H} = c_\mathrm{p}\tilde{T}\left[1 + \frac{\gamma - 1}{2}\left(\widetilde{M}^2 + M_k^2\right)\right] \tag{2.47}$$

where the mean Mach number $\widetilde{M}$ is defined as

$$\widetilde{M} = \frac{\sqrt{\tilde{u}_i^2}}{\sqrt{\gamma R \tilde{T}}} \tag{2.48}$$

where $\tilde{a} = \sqrt{\gamma R \tilde{T}}$ is the mean speed of sound, and the root-mean-square fluctuating Mach number $M_k$ is defined as

$$M_k = \frac{\sqrt{2\tilde{k}}}{\sqrt{\gamma R \tilde{T}}} \tag{2.49}$$

The total temperature is defined as

$$\tilde{T}_\mathrm{t} = \tilde{T}\left[1 + \frac{\gamma - 1}{2}\left(\tilde{M}^2 + M_k^2\right)\right] \tag{2.50}$$

and thus $\tilde{H} = c_\mathrm{p}\tilde{T}_\mathrm{t}$.

In deriving the Favre-averaged energy equation, the following term is neglected:

$$\frac{\partial}{\partial x_j}\left(-\frac{1}{2}\overline{\rho u_i'' u_i'' u_j''} + \overline{u_i'' \tau_{ij}}\right) \tag{2.51}$$

## 2.3 Turbulent Prandtl and Lewis numbers

Closure of the Favre-averaged Navier–Stokes equations requires models for

| | |
|---|---|
| $-\overline{\rho u_i'' u_j''}$ | Reynolds stress |
| $-c_\mathrm{p}\overline{\rho T'' u_j''}$ | turbulent heat flux |

The gradient transport hypothesis for the Reynolds stress is[5]

$$-\overline{\rho u_i'' u_j''} = \mu_t\left(\frac{\partial \tilde{u}_i}{\partial x_j} + \frac{\partial \tilde{u}_j}{\partial x_i} - \frac{2}{3}\frac{\partial \tilde{u}_k}{\partial x_k}\delta_{ij}\right) - \frac{2}{3}\bar{\rho}\tilde{k}\delta_{ij} \tag{2.52}$$

where $\mu_t$ is the turbulent eddy viscosity and $\delta_{ij}$ is the Kronecker delta. The corresponding hypothesis for the turbulent heat flux is

$$-c_p\,\overline{\rho T'' u_j''} = k_t\frac{\partial \tilde{T}}{\partial x_j} \tag{2.53}$$

The turbulent Prandtl number is defined as

$$Pr_t = \frac{\mu_t c_p}{k_t} \tag{2.54}$$

and is typically taken to be a constant (Wilcox 2006). For a turbulent boundary layer on a flat plate with Cartesian velocity components $u$ and $v$ in the streamwise ($x$) and wall normal ($y$) directions, respectively, equations (2.52), (2.53), and (2.54) yield

$$Pr_t = \frac{\overline{\rho u'' v''}\ \partial \tilde{T}/\partial y}{\overline{\rho T'' v''}\ \partial \tilde{u}/\partial y} \tag{2.55}$$

Using Reynolds averaging, the equivalent form of the turbulent Prandtl number is

$$Pr_t = \frac{\overline{\rho u' v'}\ \partial \bar{T}/\partial y}{\overline{\rho T' v'}\ \partial \bar{u}/\partial y} \tag{2.56}$$

## 2.4 Large Eddy Simulation

The governing equations for Large Eddy Simulation (LES) of a compressible perfect gas are obtained by spatially filtering the compressible Navier–Stokes equations. The effect of spatial filtering is to remove the small scale (high frequency) components of the flow while retaining the unsteadiness of the larger dynamic turbulent motion. The filtering operation introduces the subgrid scale (SGS) stresses and heat flux analogous to the Reynolds stress and turbulent heat flux in the RANS equations.

The conventional spatial average of an arbitrary variable $\mathcal{F}(x_i, t)$ is defined as

$$\overline{\mathcal{F}} = \int_{\mathcal{V}} G(x_i - \xi_i, \Delta)\mathcal{F}(\xi_i, t)\, d\xi_i \tag{2.57}$$

[5] Note that the sum of the diagonal terms is

$$\sum_{i=1}^{i=3} -\overline{\rho u_i'' u_i''} = -2\bar{\rho}\tilde{k}$$

which agrees with (2.44).

where $G(x_i - \xi_i, \Delta)$ is the filter function, $\Delta$ is a measure of the filter width and is related to the computational mesh size, and $\mathcal{V}$ is the filter volume (Ferziger 1993). The Favre filter is used in compressible flows for simplicity. The Favre filter of an arbitrary variable $\mathcal{F}(x_i, t)$ is defined as

$$\widetilde{\mathcal{F}} = \frac{\overline{\rho \mathcal{F}}}{\bar{\rho}} \tag{2.58}$$

A variable is decomposed into its Favre-filtered component and fluctuating component as

$$\mathcal{F}(x_i, t) = \widetilde{\mathcal{F}}(x_i, t) + \mathcal{F}''(x_i, t) \tag{2.59}$$

The spatial filtering of the compressible Navier–Stokes equations for a perfect gas yield the following (using the Einstein notation):

$$\frac{\partial \bar{\rho}}{\partial t} + \frac{\partial \bar{\rho} \tilde{u}_j}{\partial x_j} = 0 \tag{2.60}$$

$$\frac{\partial \bar{\rho} \tilde{u}_i}{\partial t} + \frac{\partial \bar{\rho} \tilde{u}_i \tilde{u}_j}{\partial x_j} = -\frac{\partial \bar{p}}{\partial x_i} + \frac{\partial \mathcal{T}_{ij}}{\partial x_j} \tag{2.61}$$

$$\frac{\partial \bar{\rho} \tilde{e}}{\partial t} + \frac{\partial (\bar{\rho} \tilde{e} + \bar{p}) \tilde{u}_j}{\partial x_j} = \frac{\partial \mathcal{H}_j}{\partial x_j} \tag{2.62}$$

$$\bar{p} = \bar{\rho} R \tilde{T} \tag{2.63}$$

where $x_i$ represents the Cartesian coordinates ($i = 1, 2, 3$), $\bar{\rho}$ is the mean density, $\tilde{u}_i$ are the Cartesian components of the filtered velocity, $\bar{p}$ is mean pressure, $\mathcal{T}_{ij}$ is the total stress, $\mathcal{H}_j$ is the energy flux due to heat transfer and work done by the total stress, and $\tilde{e}$ is the filtered total energy per unit mass,

$$\bar{\rho} \tilde{e} = \bar{\rho} c_v \tilde{T} + \frac{1}{2} \bar{\rho} \tilde{u}_i \tilde{u}_i + \bar{\rho} k \tag{2.64}$$

where $\bar{\rho} k$ is the SGS turbulence kinetic energy per unit volume,

$$\bar{\rho} k = \frac{1}{2} \bar{\rho} (\widetilde{u_i u_i} - \tilde{u}_i \tilde{u}_i) \tag{2.65}$$

The total stress is

$$\mathcal{T}_{ij} = \tau_{ij} + \overline{\sigma}_{ij} \tag{2.66}$$

where $\tau_{ij}$ is the SGS stress tensor,

$$\tau_{ij} = -\bar{\rho}\left(\widetilde{u_i u_j} - \tilde{u}_i \tilde{u}_j\right) \tag{2.67}$$

and hence $\tau_{ii} = -2\bar{\rho}k$. The molecular viscous stress tensor $\bar{\sigma}_{ij}$ is approximated by (Moin *et al* 1991)

$$\bar{\sigma}_{ij} = \mu(\tilde{T})\left(\frac{\partial \tilde{u}_i}{\partial x_j} + \frac{\partial \tilde{u}_j}{\partial x_i} - \frac{2}{3}\frac{\partial \tilde{u}_k}{\partial x_k}\delta_{ij}\right) \tag{2.68}$$

where $\mu(\tilde{T})$ is the molecular viscosity based on the Favre-filtered static temperature $\tilde{T}$. The total heat transfer is

$$\mathcal{Q}_j = Q_j - \bar{q}_j \tag{2.69}$$

where $Q_j$ is the SGS heat flux

$$Q_j = -c_p\bar{\rho}\left(\widetilde{u_jT} - \tilde{u}_j\tilde{T}\right) \tag{2.70}$$

and $\bar{q}_j$ is the molecular heat flux

$$\bar{q}_j = -\kappa(\tilde{T})\frac{\partial \tilde{T}}{\partial x_j} \tag{2.71}$$

where $\kappa(\tilde{T})$ the molecular thermal conductivity. The energy flux $\mathcal{H}_j$ is

$$\mathcal{H}_j = \mathcal{Q}_j + \mathcal{T}_{ij}\tilde{u}_i \tag{2.72}$$

Note that the exact form of the filtered energy equation, assuming the derivative and filtering operations commute, is

$$\frac{\partial \bar{\rho}\tilde{e}}{\partial t} + \frac{\partial}{\partial x_j}(\bar{\rho}\tilde{e} + \bar{p})\tilde{u}_j = \frac{\partial}{\partial x_j}\left(\mathcal{Q}_j + \mathcal{W}_j\right) \tag{2.73}$$

where

$$\mathcal{W}_j = -\frac{1}{2}\bar{\rho}\left(\widetilde{u_iu_iu_j} - \tilde{u}_i\tilde{u}_i\tilde{u}_j - 2k\tilde{u}_j\right) + \overline{\sigma_{ij}u_i} \tag{2.74}$$

By analogy to the RANS equations, we assume

$$-\frac{1}{2}\bar{\rho}\left(\widetilde{u_iu_iu_j} - \tilde{u}_i\tilde{u}_i\tilde{u}_j - 2k\tilde{u}_j\right) \approx \tau_{ij}\tilde{u}_i \tag{2.75}$$

and

$$\overline{\sigma_{ij}u_j} \approx \bar{\sigma}_{ij}\tilde{u}_i \tag{2.76}$$

and thus

$$\mathcal{W}_j = \mathcal{T}_{ij}\tilde{u}_i \tag{2.77}$$

equation (2.75) can be argued on the following basis. For analysis of equation (2.74), if it is assumed that the contribution due to the variation of $\rho$ on the subgrid scale is neglected, and furthermore if it is assumed that $\tilde{\tilde{u}}_i \approx \tilde{u}_i$ in (2.74), then the error in (2.75) is $-\frac{1}{2}\bar{\rho}\overline{u_i''u_i''u_k''}$ which is typically negligible compared to $\tau_{ij}\tilde{u}_i$ since $u_i'' \ll \tilde{u}_i$.

The closure of the system of equations (2.60) to (2.63) requires a model for the SGS stress $\tau_{ij}$ and heat flux $Q_j$ and the specification of appropriate boundary conditions for the flow variables as described below. Numerical simulation of the LES governing equations throughout the entire boundary layer is denoted *wall resolved LES.*

### 2.4.1 Explicit Large Eddy Simulation

Explicit LES defines formal mathematical models for $\tau_{ij}$ and $Q_j$. The earliest model was developed by Smagorinsky (1963) for a general circulation model of the weather. The SGS stress $\tau_{ij}$ is modeled using the concept of an eddy viscosity. By analogy to kinetic theory (Vincenti and Kruger 1965), the eddy viscosity $\mu_{\text{SGS}}$ is assumed to be of the form

$$\mu_{\text{SGS}} = \bar{\rho} u \ell \tag{2.78}$$

where $u$ and $\ell$ are the characteristic velocity and length scales, respectively, of the SGS motion, and the local density $\bar{\rho}$ is included in accordance with Morkovin's hypothesis (section 3.5). The length scale is assumed to be the filter width $\Delta$, and the velocity scale is a tensor invariant velocity gradient $\sqrt{\tilde{S}_{mn}\tilde{S}_{mn}}$ multiplied by the filter width $\Delta$. Incorporating a dimensionless coefficient $2C_R$, the SGS eddy viscosity becomes (Okong'o and Knight 1998)

$$\mu_{\text{SGS}} = 2C_R\, \bar{\rho}\, \Delta^2 D \sqrt{\tilde{S}_{mn}\tilde{S}_{mn}} \tag{2.79}$$

where $D$ is the Van Driest damping factor (Van Driest 1956) included for wall bounded flows

$$D = 1 - \exp\left(-\frac{y u_\tau}{\nu_{\text{w}} A}\right) \tag{2.80}$$

with $A = 26$. The SGS stress is

$$\tau_{ij} = \underbrace{2C_R\, \bar{\rho}\, \Delta^2 D \sqrt{\tilde{S}_{mn}\tilde{S}_{mn}}}_{\mu_{\text{SGS}}} \left(\tilde{S}_{ij} - \frac{1}{3}\tilde{S}_{kk}\delta_{ij}\right) \tag{2.81}$$

where $\tilde{S}_{ij}$ is the rate-of-strain tensor based on the resolved velocity field

$$\tilde{S}_{ij} = \frac{1}{2}\left(\frac{\partial \tilde{u}_i}{\partial x_j} + \frac{\partial \tilde{u}_j}{\partial x_i}\right) \tag{2.82}$$

The SGS heat flux is modeled assuming an eddy viscosity with turbulent Prandtl number $Pr_{t_{\text{SGS}}}$

$$Q_j = \bar{\rho}\, c_{\text{p}} \frac{C_R}{Pr_{t_{\text{SGS}}}} \Delta^2 D \sqrt{\tilde{S}_{mn}\tilde{S}_{mn}}\, \frac{\partial \tilde{T}}{\partial x_j} \tag{2.83}$$

Typical values are $C_R = 0.0042$ and $Pr_{t_{SGS}} = 0.4$ (Okong'o and Knight 1998) for a compressible turbulent boundary layer. However, a variety of different values for $C_R$ have been found necessary for different turbulent flows (Lesieur and Métais 1996). Germano *et al* (1991) developed a dynamic SGS incompressible eddy viscosity model which was extended to compressible flows by Moin *et al* (1991). The dynamic SGS model determines a local[6] time-dependent value of the Smagorinsky constant $C_R$ by introducing a second filtering operation of the filtered equations. Additional details are presented in Meneveau and Katz (2000). The structure function model of Métais and Lesieur (1992) determines the eddy viscosity by assuming the subgrid scales are close to isotropic and follow a Kolmogorov cascade, thereby defining the eddy viscosity for incompressible flow by

$$\mu_t = \frac{2}{3}\bar{\rho}C_K^{-\frac{2}{3}}\left[\frac{E(k_c)}{k_c}\right]^{\frac{1}{2}} \tag{2.84}$$

where $C_K$ is the Kolmogorov constant and $E(k_c)$ is the local kinetic energy spectrum at $k_c = \pi/\Delta$ of the resolved velocity field. The filtered structure function model of Ducros *et al* (1996) applies a filter to the large-scale fluctuations of the velocity field prior to computing the structure function.

A variety of explicit LES models have been developed that do not assume an eddy viscosity model for the SGS stress and heat transfer. Bardina *et al* (1980) developed the scale similarity model which introduces a second filtering of the resolved flowfield at a scale $\hat{\Delta} > \Delta$. The SGS stress tensor is related to the filtered resolved field according to

$$\tau_{ij} = C_s\bar{\rho}\left(\overline{\tilde{u}_i\tilde{u}_j} - \bar{\tilde{u}}_i\bar{\tilde{u}}_j\right) \tag{2.85}$$

where the overline $\bar{\tilde{u}}_j$ indicates the filtering of the instantaneous resolved velocity field $\tilde{u}_i$ using the second filter width $\hat{\Delta}$ and $C_s$ is a constant. However, the model was found to dissipate an insufficient amount of energy to the unresolved scales, and therefore an additional dissipative Smagorinsky term was added (Bardina *et al* 1980). Other types of scale similarity models include those of Liu *et al* (1994), Vreman *et al* (1997), Layton and Lewandowski (2003), and Kobayashi (2018).

Reviews of explicit LES models for incompressible and compressible flows include Rogallo and Moin (1984), Meneveau and Katz (2000), Lu and Rutland (2016), and Moser *et al* (2021).

### 2.4.2 Implicit large eddy simulation

The implicit SGS models, also known as monotone integrated LES or implicit LES (ILES), originated with the work of Boris (1990) and Boris *et al* (1992). ILES is based on the concept that the dissipation of energy from the resolved scales to the subgrid scales is achieved by the inviscid flux algorithm, and thus no formal mathematical model is prescribed for the SGS stresses and heat flux. In other words,

[6] The procedure uses an averaging in the statistically homogeneous direction.

$$\tau_{ij} = 0 \tag{2.86}$$

$$Q_j = 0 \tag{2.87}$$

Boris *et al* (1992) stated that

> Numerical experience at NRL [Naval Research Laboratory] and elsewhere ··· suggests that the nonlinear filter built into monotone computational fluid dynamics (CFD) algorithms really serves the same purposes as a subgrid scale model.

The original work of Boris *et al* (1992) was performed using the flux-corrected transport algorithm (Boris and Book 1973, Book *et al* 1975, Boris and Book 1976, Zalesak 1979). Among the monotone algorithms cited by Boris *et al* (1992) are those of Leer (1979), Colella and Woodward (1984), and Woodward and Colella (1984).

The concept of a monotone flux algorithm was introduced by Godunov (1959). A numerical scheme is considered to be monotone if it does not create unphysical oscillations. Consider the one-dimensional time-dependent Euler equations

$$\frac{\partial \mathcal{Q}}{\partial t} + \frac{\partial \mathcal{F}}{\partial x} = 0 \tag{2.88}$$

where the dependent vector is

$$\mathcal{Q} = \begin{Bmatrix} \rho \\ \rho u \\ \rho e \end{Bmatrix} \tag{2.89}$$

and the flux vector is

$$\mathcal{F} = \begin{Bmatrix} \rho u \\ \rho u^2 + p \\ \rho e u + p u \end{Bmatrix} \tag{2.90}$$

Consider a grid of control volumes shown in figure 2.1. Integration of (2.88) over a volume $V_i$ yields

$$\frac{dQ_i}{dt} + \frac{\left(F_{i+\frac{1}{2}} - F_{i-\frac{1}{2}}\right)}{\Delta x} = 0 \tag{2.91}$$

where $Q_i$ is the volume-averaged vector of dependent variables

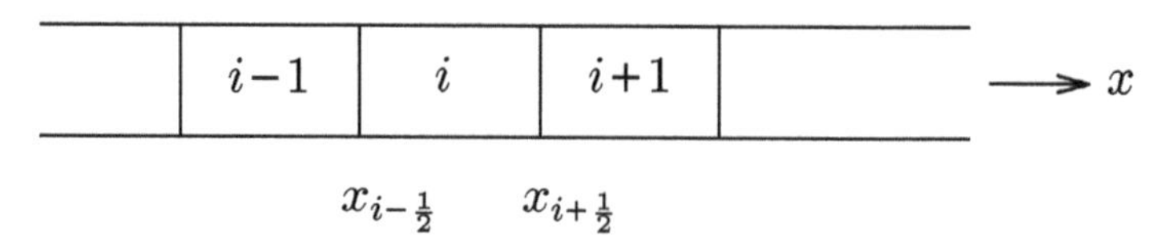


**Figure 2.1.** Grid of control volumes $V_{i-1}$, $V_i$, $V_{i+1}$.

$$Q_i(t) = \frac{1}{V_i} \int_{V_i} Q \, dx \, dy \tag{2.92}$$

and $F_{i+\frac{1}{2}}$ is the flux at the cell face $x_{i+\frac{1}{2}}$

$$F_{i+\frac{1}{2}} = \frac{1}{A_{i+\frac{1}{2}}} \int_{x_{i+\frac{1}{2}}} \mathcal{F} \, dy \tag{2.93}$$

Consider the numerical solution of (2.91) using the following expression for the inviscid flux:

$$\begin{aligned} F_{i+\frac{1}{2}} &= \mathcal{F}(Q_{i+\frac{1}{2}}) \\ Q_{i+\frac{1}{2}} &= \frac{1}{2}(Q_i + Q_{i+1}) \end{aligned} \tag{2.94}$$

It can be shown that the flux algorithm (2.94) is numerically stable (Knight 2006). However, the algorithm introduces unphysical oscillations in the vicinity of a shock wave. Figure 2.2 shows the initial condition and later evolution of the velocity $u(x, t)$ with a periodic disturbance initial condition

$$u(x, 0) = \epsilon a_o \sin \kappa x \quad \text{for} \quad 0 \leqslant \kappa x \leqslant 2\pi \tag{2.95}$$

where $\kappa$ is the wavenumber and $a_o$ is the speed of sound based upon the stagnation temperature. A shock wave initially forms at

$$\kappa a_o t = \frac{2}{(\gamma + 1)\epsilon} \tag{2.96}$$

The computed solution displays unphysical oscillations analogous to the Gibbs phenomenon in the Fourier series representation of a discontinuous function (Greenberg 1998). The flux algorithm (2.94) is not monotone.

Next consider the solution of (2.91) using Roe's method (Roe 1981). The inviscid flux vector is defined by

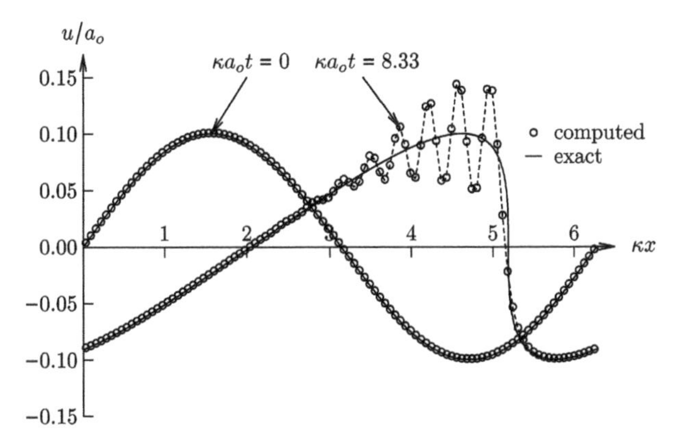


**Figure 2.2.** Unphysical oscillations at formation of shock wave for $\gamma = 1.4$ and $\epsilon = 0.1$.

$$F_{i+\frac{1}{2}} = \frac{1}{2}\left[F_l + F_r + \tilde{S}|\tilde{\Lambda}|\tilde{S}^{-1}\left(Q^l_{i+\frac{1}{2}} - Q^r_{i+\frac{1}{2}}\right)\right] \tag{2.97}$$

where $F_l = \mathcal{F}(Q^l_{i+\frac{1}{2}})$ and $F_r = \mathcal{F}(Q^r_{i+\frac{1}{2}})$ with $Q^l_{i+\frac{1}{2}}$ and $Q^r_{i+\frac{1}{2}}$ the left-face and right-face reconstructed expressions for $Q$ at $i + \frac{1}{2}$. The matrix $\tilde{S}$ is the right eigenvectors of the Jacobian of $\mathcal{F}$ evaluated using Roe variables (Knight 2006)

$$\tilde{S} = \begin{Bmatrix} 1 & 1 & 1 \\ \tilde{u} & \tilde{u}+\tilde{a} & \tilde{u}-\tilde{a} \\ \frac{1}{2}\tilde{u}^2 & \tilde{H}+\tilde{u}\tilde{a} & \tilde{H}-\tilde{u}\tilde{a} \end{Bmatrix} \tag{2.98}$$

and $|\tilde{\Lambda}|$ is the diagonal matrix of eigenvalues

$$|\tilde{\Lambda}| = \begin{Bmatrix} |\tilde{u}| & 0 & 0 \\ 0 & |\tilde{u}+\tilde{a}| & 0 \\ 0 & 0 & |\tilde{u}-\tilde{a}| \end{Bmatrix} \tag{2.99}$$

where $\tilde{a}$, $\tilde{H}$, and $\tilde{u}$ are the Roe-averaged speed of sound, total enthalpy per unit mass, and velocity, respectively (Roe 1981). Figure 2.3 shows the initial and subsequent evolution of the velocity $u(x, t)$ with the periodic initial condition (2.95). No oscillations appear at the moment of the shock formation. The flux algorithm (2.97) is monotone.

Since ILES assumes that the SGS stresses and heat flux are determined by the inviscid flux algorithm, it is possible to determine the mathematical form of the effective SGS stresses and heat flux by Taylor series expansion of the governing equations in finite volume form. The exact mathematical form depends on the flux algorithm. Examples are provided in Fureby *et al* (1997) and Fureby and Grinstein (1999). Detailed reviews of ILES are presented in Grinstein and Fureby (2002), Grinstein and Fureby (2004), and Grinstein *et al* (2007).

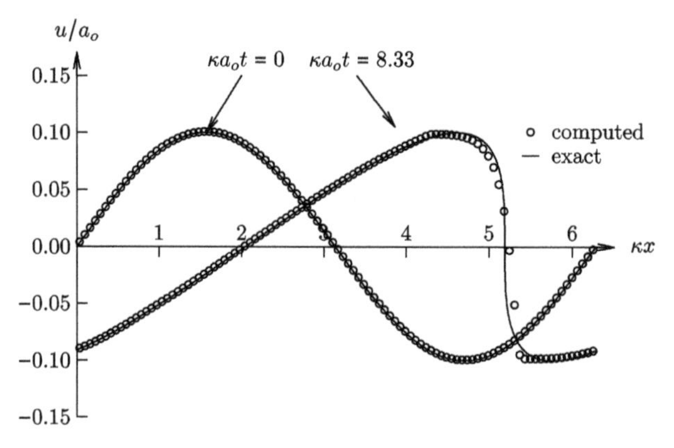


**Figure 2.3.** Results for Roe's flux algorithm with $\gamma = 1.4$ and $\epsilon = 0.1$.

### 2.4.3 Wall modeled Large Eddy Simulation

The required computer resources for LES increase rapidly with Reynolds number due to the requirement for adequate resolution of the near wall region responsible for turbulence production (section 2.6). A variety of methods have been developed to mitigate the Reynolds number effect on computer resources by substituting an alternate model for the fluid motion in the near wall region. These techniques are denoted *wall modeled LES* and may be divided into two main categories as discussed below. Recent reviews include Piomelli and Balaras (2002), Piomelli (2008), Spalart (2009), Kawai and Larsson (2012), Larsson *et al* (2016), Bose and Park (2018), and Durbin (2018).

#### *2.4.3.1 Wall shear stress and heat flux models*

The wall shear stress and heat flux models solve the LES equations extending to the wall but using a grid capable of resolving only the flow structures in the outer portion of the boundary layer. Thus a separate model is required for determining the wall shear stress and heat flux. A simple model[7] is based on the concept of a statistically equilibrium near wall region where convection is negligible and the mean flow is steady. Consider a flat plate zero pressure gradient boundary layer with Cartesian coordinates $x$, $y$, and $z$ in the streamwise, wall normal, and spanwise directions, respectively, with corresponding velocity components $u$, $v$, and $w$. In the boundary layer approximation, the Favre-averaged momentum (2.40) and energy equations (2.41) become

$$0 = \frac{\partial}{\partial y}\left( -\overline{\rho u'' v''} + \mu \frac{\partial \tilde{u}}{\partial y} \right) \tag{2.100}$$

$$0 = \frac{\partial}{\partial y}\left( -\overline{\rho v'' w''} + \mu \frac{\partial \tilde{w}}{\partial y} \right) \tag{2.101}$$

$$0 = \frac{\partial}{\partial y}\left( -c_p \overline{\rho T'' v''} + k \frac{\partial \tilde{T}}{\partial y} - \overline{\rho u'' v''}\, \tilde{u} - \overline{\rho v'' w''}\, \tilde{w} + \mu \left( \frac{\partial \tilde{u}}{\partial y} \tilde{u} + \frac{\partial \tilde{w}}{\partial y} \tilde{w} \right) \right) \tag{2.102}$$

A simple mixing length model is assumed for the Reynolds stress and heat flux:

$$-\overline{\rho u'' v''} = \mu_t \frac{\partial \tilde{u}}{\partial y} \tag{2.103}$$

$$-\overline{\rho v'' w''} = \mu_t \frac{\partial \tilde{w}}{\partial y} \tag{2.104}$$

[7] Denoted the equilibrium stress balance model by Bose and Park (2018).

with

$$\mu_t = \kappa \bar{\rho} y u_\tau D \tag{2.105}$$

where $\kappa$ is von Kármán's constant and $D$ is the Van Driest damping factor (2.80). Assuming a constant turbulent Prandtl number $Pr_{t_{wm}}$, the governing equations become

$$0 = \frac{d}{dy}\left[(\mu + \mu_t)\frac{d\tilde{u}}{dy}\right] \tag{2.106}$$

$$0 = \frac{d}{dy}\left[(\mu + \mu_t)\frac{d\tilde{w}}{dy}\right] \tag{2.107}$$

$$0 = \frac{d}{dy}\left[c_p\left(\frac{\mu}{c_p} + \frac{\mu_t}{Pr_{t_{wm}}}\right)\frac{d\tilde{T}}{dy} + (\mu + \mu_t)\left(\tilde{u}\frac{d\tilde{u}}{dy} + \tilde{w}\frac{d\tilde{w}}{dy}\right)\right] \tag{2.108}$$

Equations (2.106) to (2.108) are ordinary differential equations, with no-slip and adiabatic or isothermal boundary conditions at the wall $y = 0$, and matching to the LES solution at some level above the wall (Kawai and Larsson 2012) at each timestep of the LES computation. The equations may be efficiently solved using a fine grid in the wall normal direction to determine the surface shear stress and heat flux for the LES computation. A similar method was developed by Knight (1984) for the compressible RANS equations. Streamwise and spanwise pressure gradients can also be introduced into the model by assuming the pressure gradient across the modeled sublayer is negligible. The assumption of a statistically equilibrium near wall region can be relaxed by considering the unsteady and convective terms in the momentum and energy equations and by incorporating an eddy viscosity model for the SGS stress and heat flux. Examples include Balaras *et al* (1996), Cabot and Moin (2000), and Wang and Moin (2002).

#### *2.4.3.2 Reynolds-averaged Navier–Stokes models*

The second category comprises a combination of a RANS model in the near wall region and LES in the outer region of the boundary layer and in large regions of separated flow. Perhaps the best known is the Detached Eddy Simulation (DES) method originally developed by Spalart *et al* (1997). The model is based on the observation that RANS models are least accurate in regions of massive separation such as occur in airfoil stall and downstream of bluff trailing edges or afterbodies. The DES model was first incorporated into the Spalart–Allmaras (SA) RANS model (Spalart and Allmaras 1994). The SA model is based upon a turbulent eddy viscosity $\mu_t$ defined by

$$\mu_t = \rho\tilde{\nu} f_{v1} \quad \text{where} \quad f_{v1} = \frac{\chi^3}{\chi^3 + C_{v1}^3} \quad \text{and} \quad \chi = \frac{\tilde{\nu}}{\nu} \tag{2.109}$$

where $\nu$ is the kinematic molecular viscosity, $C_{v1}$ is a constant, and $\tilde{\nu}$ is defined by

$$\frac{\partial \tilde{\nu}}{\partial t} + u_j \frac{\partial \tilde{\nu}}{\partial x_j} = \mathcal{P}(\tilde{\nu}) - \mathcal{D}(\tilde{\nu}) + \mathcal{I}(\tilde{\nu}) \tag{2.110}$$

where $\mathcal{P}(\tilde{\nu})$, $\mathcal{D}(\tilde{\nu})$, and $\mathcal{I}(\tilde{\nu})$ are empirical models of the production, dissipation, and diffusion of $\tilde{\nu}$ (Spalart and Allmaras 1994). In particular, the dissipation term is

$$\mathcal{D}(\tilde{\nu}) = f(\tilde{\nu})\left(\frac{\tilde{\nu}}{d}\right)^2 \tag{2.111}$$

where $f(\tilde{\nu})$ is a nonlinear function of $\tilde{\nu}$ and $d$ is the distance to the wall. The SA DES model replaces $d$ by $\tilde{d}$ defined by

$$\tilde{d} = \min\left(d,\ C_{\mathrm{DES}}\Delta\right) \tag{2.112}$$

where

$$\Delta = \max\left(\Delta_x,\ \Delta_y,\ \Delta_z\right) \tag{2.113}$$

Within the boundary layer, $d \ll \Delta$ typically, and thus the conventional RANS SA model applies. Within large separated regions (e.g., at airfoil stall), $d \gg \Delta$, and thus a Smagorinsky-type LES model determines $\tilde{d}$. A variety of subsequent modifications to DES have been developed, including delayed DES and improved delayed DES. A review is presented in Spalart (2009).

A second example is the approach of Boles *et al* (2009). The unsteady Favre-averaged Navier–Stokes equations are coupled with the shear stress transport model (Menter 1994). The Reynolds stress and turbulent heat flux are modeled using an eddy viscosity defined as a weighted sum of a RANS description and a Smagorinsky model

$$\mu_{\mathrm{t}} = \rho\left[\Gamma \frac{k}{\omega} + (1 - \Gamma)\nu_{t_{SGS}}\right] \tag{2.114}$$

where $k/\omega$ is the RANS kinematic eddy viscosity defined by the shear stress transport model and $\nu_{t_{SGS}}$ is the kinematic LES eddy viscosity defined by the Smagorinsky model

$$\nu_{t_{SGS}} = C_{\mathrm{s}}\Delta^2 S \tag{2.115}$$

where $C_{\mathrm{s}} = 0.01$ and $\Delta$ is the geometric average of the local grid spacing

$$\Delta = \left(\Delta_x \Delta_y \Delta_z\right)^{\frac{1}{3}} \tag{2.116}$$

and $S$ is the norm of a modified rate-of-strain tensor

$$S = \left[\frac{\partial \tilde{u}_i}{\partial x_j}\frac{\partial \tilde{u}_j}{\partial x_i} + \frac{\partial \tilde{u}_i}{\partial x_j}\frac{\partial \tilde{u}_i}{\partial x_j} - \frac{2}{3}\left(\frac{\partial \tilde{u}_i}{\partial x_i}\right)^2\right]^{\frac{1}{2}} \tag{2.117}$$

The blending function $\Gamma$ is based on the distance from the wall $y$ and a modeled form of the Taylor microscale,

$$\Gamma = \frac{1}{2}\left\{1 - \tanh\left[5\left(\frac{\kappa}{\sqrt{C_\mu}}\eta^2 - 1\right) - \phi\right]\right\} \tag{2.118}$$

where

$$\eta = \frac{y}{\alpha_1 \lambda} \quad \text{and} \quad \lambda = \sqrt{\frac{\nu}{C_\mu \omega}} \tag{2.119}$$

The value of $\phi = \tanh^{-1} 0.98$ fixes the location where $\kappa\eta^2/\sqrt{C_\mu} = 1$ to $\Gamma = 0.99$. The constant $\alpha_1$ is selected to require the average LES and RANS transition location where $\Gamma = 0.99$ to a location where the Law of the Wall and Wake begins to depart from the Law of the Wall. The inflow boundary condition is obtained through a recycling/rescaling procedure. A subsequent modification of the approach is presented in Gieseking *et al* (2010).

## 2.5 Inflow boundary conditions

A fundamental issue with DNS and LES is the specification of boundary conditions. Figure 2.4 shows an example problem. A supersonic boundary layer develops on a flat plate with transition from laminar to turbulent flow assumed to occur upstream of the corner where the flat plate and ramp intersect. Simulation of the flowfield using DNS or LES requires the specification of a computational domain with boundary conditions applied on the external surfaces. Two methods are shown in figure 2.4. The first domain extends from the leading edge of the flat plate. The boundary conditions on the inflow surface are fixed values for the pressure, temperature, and velocity defined by the test conditions in the wind tunnel. The boundary conditions on the plate are no-slip and either adiabatic or isothermal wall. The transition of the boundary layer from laminar to turbulent flow is included in the simulation either by natural instabilities or by forcing as discussed below.

The second domain is smaller with the inflow surface located downstream of transition. In this case the inflow surface includes the turbulent boundary layer, and consequently dynamic (i.e., time-varying) boundary conditions for the pressure, temperature, and velocity are specified. This requires an inflow turbulence generation method. Several of these approaches are discussed below together with selected examples. Recent reviews of turbulence inflow boundary conditions are

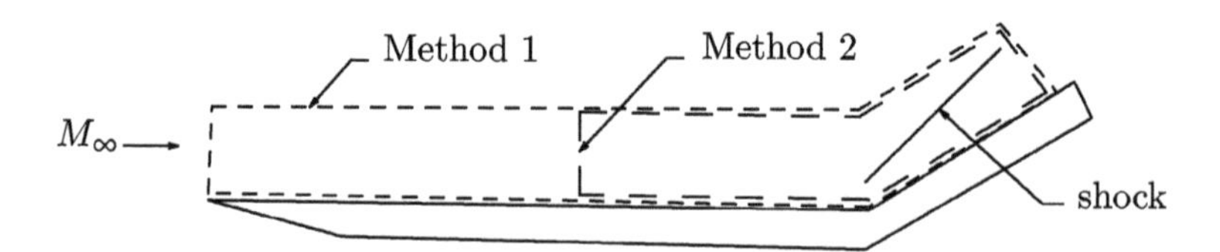


**Figure 2.4.** Computational domains for supersonic flow past compression corner.

Keating *et al* (2004), Xu and Martín (2004), Georgiadis *et al* (2010), Tabor and Baba-Ahmadi (2010), Wu (2017), and Dhamankar *et al* (2018).

### 2.5.1 Forcing transition

Several approaches to forcing transition of the boundary layer from laminar to turbulent have been implemented. Consider a boundary layer developing on a flat plate where $x$, $y$, and $z$ are the streamwise, wall normal, and spanwise directions, respectively. Sayadi *et al* (2013) imposed a time-dependent surface normal velocity over a strip $x_1 \leqslant x \leqslant x_2$ to stimulate transition for a Mach 0.2 boundary layer. The wall normal velocity is

$$v(x, z, t) = A_1 f(x) \sin \omega_1 t + A_2 f(x) g(z) \sin \omega_2 t \tag{2.120}$$

where $\omega_1$ and $\omega_2$ are the frequencies of two-dimensional Tollmien–Schlichting (Schlichting 1968) and oblique waves (Reed and Saric 1989), respectively. The streamwise profile of Fasel and Konzelmann (1990) is specified as

$$f(x) = 15.187\,5\xi^5 - 35.437\,5\xi^4 + 20.24\xi^3 \tag{2.121}$$

where

$$\xi = \begin{cases} \dfrac{x - x_1}{x_m - x_1} & x_1 \leqslant x \leqslant x_m \\ \dfrac{x_2 - x}{x_2 - x_m} & x_m \leqslant x \leqslant x_2 \end{cases} \tag{2.122}$$

where $x_m = \frac{1}{2}(x_1 + x_2)$ and $g(z) = \cos 2\pi z/\lambda$ where $\lambda$ is the spanwise wavelength. A blowing and suction approach was similarly implemented by Spyropoulos and Blaisdell (1998), Pirozzoli *et al* (2004), and Pirozzoli and Grasso (2006).

Boundary layer transition can be forced to develop more rapidly through the use of geometric perturbations to the surface. Gloerfelt and Berland (2013) incorporated a small spanwise step to stimulate rapid transition in a Mach 0.5 boundary layer. Step heights of $0.14\delta$ and $0.26\delta$ and streamwise locations $11\delta \leqslant x \leqslant 15\delta$ and $5.7\delta \leqslant x \leqslant 9\delta$ were used for coarse- and fine-grain LES, respectively, where $\delta$ is the inflow laminar boundary layer thickness.

Another approach to forcing boundary layer transition is the inclusion of a body force term in the governing equations for a selected region. Mullenix and Visbal (2013) incorporated a constant counterflow simulated dielectric barrier discharge to create a separation bubble. The shear layer instability associated with the separated flow causes transition of the boundary layer.

### 2.5.2 Auxiliary simulation

The dynamic inflow boundary conditions can also be obtained from auxiliary simulation(s). Rizzetta and Visbal (2001) and Rizzetta and Visbal (2002) utilized two separate simulations to provide the inflow boundary conditions for a supersonic compression ramp (figure 2.5). A DNS is performed in a half channel domain with

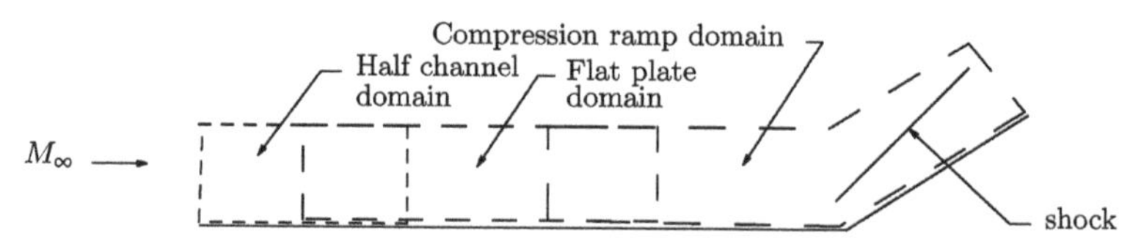


**Figure 2.5.** Computational domains.

streamwise and spanwise periodic boundary conditions together with symmetry boundary condition on the upper surface. The turbulent fluctuations at a midplane in the half channel domain were recorded and rescaled in the vertical direction to correspond to the boundary layer thickness for the inflow laminar Blasius boundary layer profile in the flat plate domain. The fluctuations caused transition within the flat plate domain, and an equilibrium turbulent boundary developed on the flat plate. The instantaneous flowfield variables were recorded at a downstream plane within the flat plate domain and further processed to provide the dynamical inflow boundary conditions for the compression ramp domain (Rizzetta and Visbal 2001). The method is similar to the auxiliary simulation technique of Adams (2000).

### 2.5.3 Recycling and rescaling

The recycling and rescaling method utilizes the DNS or LES simulation itself for determining the dynamic inflow boundary conditions. The concept is illustrated in figure 2.6. An equilibrium turbulent boundary layer develops on a flat plate. At each timestep the instantaneous flowfield at a downstream location denoted the recycle station is used to generate the boundary condition at the inflow surface using the known structure of an equilibrium turbulent boundary layer.

Lund *et al* (1998) developed a recycling and rescaling method for LES of incompressible turbulent boundary layers inspired by the method of Spalart and Leonard (1985). The method was extended to supersonic adiabatic or near adiabatic turbulent boundary layers by Urbin *et al* (1999) and Urbin and Knight (2001) and is described here in detail. The analysis is presented here assuming Reynolds averaging for LES, although the method can also be implemented for Favre averaging. For simplicity, overbars are omitted for the mean density, static temperature, and velocity.

The mean streamwise velocity $u$ in a zero pressure gradient boundary layer exhibits a two-layer behavior (Smits and Dussauge 1996):

$$u_{\text{vd}} = u_\tau f_i(y^+) \qquad \text{inner region} \tag{2.123}$$

$$u_{\text{vd}_e} - u_{\text{vd}} = u_\tau f_o(\eta) \qquad \text{outer region} \tag{2.124}$$

where $f_i(y^+)$ and $f_o(\eta)$ are universal functions[8] and

[8] In the outer region, alternatives to the length scale $\delta$ include the momentum thickness and Clauser or Rotta thickness (Smits and Dussauge 1996).

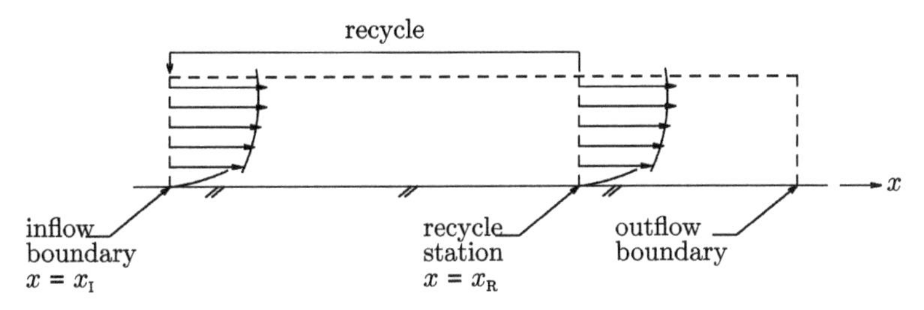


**Figure 2.6.** Inflow and recycle locations.

$$y^+ = \frac{y u_\tau}{\nu_w} \tag{2.125}$$

$$\eta = \frac{y}{\delta} \tag{2.126}$$

where $\tau_w$ is the local mean wall shear stress, $u_\tau = \sqrt{\tau_w/\rho_w}$ is the local friction velocity, $\nu_w = \mu_w/\rho_w$ is the kinematic viscosity evaluated at the wall, $\delta$ is the local boundary layer thickness, and $u_{vd}$ is the Van Driest transformed velocity (section 3.2.1)

$$u_{vd} = \frac{u_e}{A}\left\{\sin^{-1}\left[\frac{2A^2 u/u_e - B}{\sqrt{B^2 + 4A^2}}\right] + \sin^{-1}\left[\frac{B}{\sqrt{B^2 + 4A^2}}\right]\right\} \tag{2.127}$$

where $u_e$ is the streamwise velocity at the edge of the boundary layer and

$$A = \sqrt{\frac{(\gamma - 1)}{2} Pr_t M_e^2 \frac{T_e}{T_w}} \tag{2.128}$$

$$B = \left[1 + Pr_t^{1/2}\frac{(\gamma - 1)}{2} M_e^2\right]\frac{T_e}{T_w} - 1 \tag{2.129}$$

where $Pr_t$ is the turbulent Prandtl number.

Denote the inflow and recycle stations by I and R, respectively. From (2.123) and (2.124),

$$u_{VD_R} = u_{\tau_R} f_i(y^+) \tag{2.130}$$

$$u_{vde} - u_{VD_R} = u_{\tau_R} f_o(\eta) \tag{2.131}$$

Since $f_i(y^+)$ and $f_o(\eta)$ are universal functions,

$$u_{VD_I}(y_I^+) = \beta\, u_{VD_R}(y_I^+) \quad \text{inner region} \tag{2.132}$$

$$u_{VD_I}(\eta_I) = \beta\, u_{VD_R}(\eta_I) + (1 - \beta) u_{VD_e} \quad \text{outer region} \tag{2.133}$$

where

$$\beta = \frac{u_{\tau_{\mathrm{I}}}}{u_{\tau_{\mathrm{R}}}} \tag{2.134}$$

For a given location $y_{\mathrm{I}}$, the corresponding location $y_{\mathrm{R}}$ is obtained from $y_{\mathrm{R}}^{+} = y_{\mathrm{I}}^{+}$ in the inner region and $\eta_{\mathrm{I}} = \eta_{\mathrm{R}}$ in the outer region; thus

$$y_{\mathrm{R}} = \begin{cases} \beta y_{\mathrm{I}} & \text{inner region} \\ \alpha y_{\mathrm{I}} & \text{outer region} \end{cases} \tag{2.135}$$

where $\alpha = \delta_{\mathrm{I}}/\delta_{\mathrm{R}}$.

The mean velocity at the inflow $u_{\mathrm{I}}$ in the inner and outer regions is obtained from (2.127). The mean velocity profile at the inflow combines the inner and outer mean velocity profiles using a smoothing function (Lund *et al* 1998)

$$u_{\mathrm{I}} = [1 - S(\eta_{\mathrm{I}})]u_{\mathrm{I}}\big|_{\mathrm{inner}} + S(\eta_{\mathrm{I}})u_{\mathrm{I}}\big|_{\mathrm{outer}} \tag{2.136}$$

where

$$S(\eta) = \begin{cases} \frac{1}{2}\left[1 + \tanh\left[\frac{a(\eta - b)}{(1 - 2b)\eta + b}\right][\tanh a]^{-1}\right] & \text{for } \eta \leqslant 1 \\ 1 & \text{for } \eta > 1 \end{cases} \tag{2.137}$$

where $a = 4$ and $b = 0.2$ are constants. The value of $a$ determines the width of the region where $S$ transitions from zero to one, and $b$ determines the location of the transition. A value $b = 0.2$ provides a smooth transition at $\eta \approx 0.2$. The mean streamwise velocity profile at the inflow boundary is thus obtained from the mean streamwise velocity profile at the recycle station assuming an equilibrium turbulent boundary layer. In other words, the LES itself provides the mean inflow streamwise velocity profile. The wall normal mean velocity $v$ is similarly rescaled in terms of inner and outer regions

$$v_{\mathrm{I}}(y_{\mathrm{I}}^{+}) = v_{\mathrm{R}}(y_{\mathrm{I}}^{+}) \quad \text{inner region} \tag{2.138}$$

$$v_{\mathrm{I}}(\eta_{\mathrm{I}}) = v_{\mathrm{R}}(\eta_{\mathrm{I}}) \quad \text{outer region} \tag{2.139}$$

and the wall normal mean velocity at the inflow combines the inner and outer profiles similarly to (2.136). The spanwise mean velocity at the inflow boundary is zero. Assuming a Crocco–Busemann relation (3.147), the mean static temperature $T$ is similarly rescaled in terms of inner and outer regions

$$T_{\mathrm{I}}(y_{\mathrm{I}}^{+}) = T_{\mathrm{R}}(y_{\mathrm{I}}^{+}) \quad \text{inner region} \tag{2.140}$$

$$T_{\mathrm{I}}(\eta_{\mathrm{I}}) = T_{\mathrm{R}}(\eta_{\mathrm{I}}) \quad \text{outer region} \tag{2.141}$$

and the mean static temperature at the inflow combines the inner and outer profiles similarly to (2.136). The mean density $\rho$ at the inflow is obtained from the ideal gas equation assuming constant mean static pressure. Thus, the mean inflow boundary conditions are determined.

The streamwise root-mean-square fluctuating velocity shows a two-layer behavior when scaled by the local mean density (Smits and Dussauge 1996) in accordance with Morkovin's hypothesis (section 3.5):

$$\rho\,\overline{u'^2} = \tau_w\, g_i(y^+) \quad \text{inner region} \tag{2.142}$$

$$\rho\,\overline{u'^2} = \tau_w\, g_o(\eta) \quad \text{outer region} \tag{2.143}$$

where $g_i(y^+)$ and $g_o(\eta)$ are universal functions. Therefore,

$$\overline{u'^2} = \frac{\rho_w}{\rho} u_\tau^2 g_i(y^+) \quad \text{inner region} \tag{2.144}$$

$$\overline{u'^2} = \frac{\rho_w}{\rho} u_\tau^2 g_o(\eta) \quad \text{outer region} \tag{2.145}$$

Since the mean pressure is constant across the boundary layer,

$$\frac{\rho_w}{\rho} = \frac{T}{T_e}\frac{T_e}{T_w} \tag{2.146}$$

and therefore, from (2.138) and (2.139) for fixed $T_w/T_e$,

$$\overline{u'^2} = u_\tau^2 h_i(y^+, u_\tau/u_e) \quad \text{inner region} \tag{2.147}$$

$$\overline{u'^2} = u_\tau^2 h_o(\eta, u_\tau/u_e) \quad \text{outer region} \tag{2.148}$$

where $h_i$ and $h_o$ are universal functions. Therefore, the streamwise velocity fluctuations are rescaled as

$$u_I'(y_I^+, z, t) = \beta u_R'(y_I^+, z, t) \quad \text{inner region} \tag{2.149}$$

$$u_I'(\eta_I, z, t) = \beta u_R'(\eta_I, z, t) \quad \text{outer region} \tag{2.150}$$

A similar scaling is used for $v'$ and $w'$. The streamwise fluctuating velocities at the inflow are obtained similar to (2.136). The inner and outer static temperature fluctuations are similarly scaled and combined without the factor $\beta$.

A variety of different implementations of the recycling and rescaling method have been developed for compressible turbulent boundary layers. Examples include Schröder *et al* (2001), Stolz and Adams (2003), Xiao *et al* (2003), Sagaut *et al* (2004), Xu and Martín (2004), Simens *et al* (2009), Pirozzoli *et al* (2010), Lagha *et al* (2011), Morgan *et al* (2011), and Pirozzoli and Bernardini (2013).

### 2.5.4 Synthetic inflow turbulence

The simplest approach to imitating a flat plate turbulent boundary layer inflow condition is to add random white noise perturbations in pressure, temperature, and velocity to known mean velocity and temperature profiles while assuming constant static pressure. The mean velocity (section 3.2) and temperature (section 3.3.1) are determined by the chosen Reynolds number $Re_\delta$ based upon the inflow boundary layer thickness $\delta$ and wall temperature ratio $T_w/T_\infty$. The distribution of fluctuations across the inflow boundary can be scaled to reflect experimental turbulence statistics. However, this approach lacks any phase information between the fluctuating variables and thus does not properly characterize the turbulent eddies (Keating *et al* 2004). Also, the skewness of the fluctuating velocity derivatives are zero, and therefore the inflow condition lacks nonlinear energy transfer (Lund *et al* 1998). Consequently, the flowfield downstream of the inflow boundary must readjust to achieve an equilibrium turbulent boundary layer.

Smirnov *et al* (2001) developed a random flow generation method for generating inflow conditions for an incompressible turbulent boundary layer based upon the method of Kraichnan (1970). Using a specified kinematic Reynolds stress tensor $R_{ij} = \overline{u_i' u_j'}$, an orthogonal transformation tensor is determined to diagonalize $R_{ij}$. A time-dependent velocity field is defined using the method of Kraichnan (1970) with specified length and time scales of the turbulence, and modeled turbulence spectrum. The resultant fluctuating velocities are rescaled and transformed to the original coordinate system. The resultant time-dependent velocities yield the specified Reynolds stress tensor, length, and time scales. Modifications of the random flow generation method have been developed by Batten *et al* (2004), Huang *et al* (2010), and Yu and Bai (2014).

A variety of other methods for generating inflow boundary conditions have been developed, principally for incompressible flows. A detailed summary is presented in Wu (2017) and Dhamankar *et al* (2018). These include approaches based upon proper orthogonal decomposition, digital filters, synthetic vortices, and volumetric forcing.

## 2.6 Reynolds number dependence

Several reviews have estimated the Reynolds number dependence of DNS and LES. These include Chapman (1979), Reynolds (1990), Choi and Moin (2012), and Griffin *et al* (2021).

## References

Adams N 2000 Direct simulation of the turbulent boundary layer along a compression ramp at M = 3 and $Re_\theta = 1685$ *J. Fluid Mech.* **420** 47–83

Balaras E, Benocci C and Piomelli U 1996 Two-layer approximate boundary conditions for large eddy simulations *AIAA J.* **34** 1111–9

Bardina J, Ferziger J and Reynolds W 1980 Improved subgrid scale models for large eddy simulation *13th Fluid and Plasma Dynamics Conference* 80–1357

Batten P, Goldberg U and Chakravarthy S 2004 Interfacing statistical turbulence closures with large eddy simulation *AIAA J.* **42** 485–92

Bauer E 1990 *Physics of High-Temperature Air. Part 1. Basics* D-487 Institute for Defense Analyses

Blaisdell G 2008 Implicit large eddy simulation: computing turbulent fluid dynamics *AIAA J.* **46** 3168–70

Boles J, Choi J-I, Edwards J and Baurle R 2009 Simulations of high speed internal flows using LES/RANS models *47th AIAA Aerospace Sciences Meeting* AIAA Paper 2009-1324 (Reston, VA: American Institute of Aeronautics and Astronautics)

Book D, Boris J and Hain K 1975 Flux-corrected transport II: generalization of the method *J. Comput. Phys.* **18** 248–83

Boris J 1990 On large eddy simulation using subgrid turbulence models *Whither Turbulence? Turbulence at the Crossroads: Workshop Held at Cornell University* J Lumley (Berlin: Springer) 344–53

Boris J and Book D 1973 Flux-corrected transport I: SHASTA–a fluid transport algorithm that works *J. Comput. Phys.* **11** 38–69

Boris J and Book D 1976 Flux-corrected transport III: minimal error FCT algorithms *J. Comput. Phys.* **20** 397–431

Boris J, Grinstein F, Oran E and Kolbe R 1992 New insights into large eddy simulation *Fluid Dyn. Res.* **10** 199–228

Bose S and Park G 2018 Wall-modeled large-eddy simulation for complex turbulent flows *Annu. Rev. Fluid Mech.* **50** 535–61

Boyd I and Schwartzentruber T 2017 *Nonequilibrium Gas Dynamics and Molecular Simulation* (Cambridge: Cambridge University Press)

Brun R 2009 *Introduction to Reactive Gas Dynamics* (New York: Oxford University Press)

Cabot W and Moin P 2000 Approximate wall boundary conditions in the large eddy simulation of high Reynolds number flow *Flow Turbul. Combust.* **63** 269–91

Chapman D 1979 Computational aerodynamics development and outlook *AIAA J.* **17** 1293–313

Chapman S and Cowling T 1970 *The Mathematical Theory of Non-Uniform Gases* 3rd edn (Cambridge: Cambridge University Press)

Choi H and Moin P 2012 Grid-point requirements for large eddy simulation: Chapman's estimates revisited *Phys. Fluids* **24** 011702

Colella P and Woodward P 1984 The piecewise parabolic method (PPM) for gas-dynamical simulations *J. Comput. Phys.* **54** 174–201

Dhamankar N, Blaisdell G and Lyrintzis A 2018 Overview of turbulent inflow boundary conditions for large eddy simulation *AIAA J.* **56** 1317–34

Ducros F, Comte P and Lesieur M 1996 Large eddy simulation of transition to turbulence in a boundary layer developing spatially over a flat plate *J. Fluid Mech.* **326** 1–36

Durbin P 2018 Some recent developments in turbulence closure modeling *Annu. Rev. Fluid Mech.* **50** 77–103

Fasel H and Konzelmann U 1990 Non-parallel stability of a flat plate boundary layer using the complete Navier–Stokes equations *J. Fluid Mech.* **221** 311–47

Favre A 1965 Equations des Gaz turbulents compressibles *J. Méc.* **4** 361–421

Ferziger J 1993 Subgrid-scale modeling *Large Eddy Simulation of Complex Engineering and Geophysical Flows* ed B Galperin and S Orszag (Cambridge: Cambridge University Press) pp 37–54

Fureby C and Grinstein F 1999 Monotonically integrated large eddy simulation of free shear flows *AIAA J.* **37** 544–56

Fureby C, Tabor G, Weller H and Gosman A 1997 A comparitive study of subgrid scale models in homogeneous isotropic turbulence *Phys. Fluids* **9** 1416–29

Georgiadis N, Rizzetta D and Fureby C 2010 Large eddy simulations: current capabilities, recommended practices, and future research *AIAA J.* **48** 1772–84

Germano M, Piomelli U, Moin P and Cabot W 1991 A dynamic subgrid-scale eddy viscosity model *Phys. Fluids* A **3** 1760–5

Gieseking D, Choi J-I, Edwards J and Hassan H 2010 Simulation of shock/boundary layer interactions using improved LES/RANS models *48th AIAA Aerospace Sciences Meeting* AIAA Paper 2010-111 (Reston, VA: American Institute of Aeronautics and Astronautics)

Gloerfelt X and Berland J 2013 Turbulent boundary layer noise: direct radiation at Mach number 0.5 *J. Fluid Mech.* **723** 318–51

Godunov S 1959 A finite difference method for the computation of discontinuous solutions of the equations of fluid dynamics *Mat. Sb.* **47** 357–93

Greenberg M 1998 *Advanced Engineering Mathematics* 2nd edn (Englewood, NJ: Prentice-Hall)

Griffin K, Fu L and Moin P 2021 *The Effect of Compressibility on Grid-Point and Time-Step Requirements for Simulations of Wall-Bounded Turbulent Flows* Center for Turbulence Research

Grinstein F and Fureby C 2002 Recent progress on MILES for high Reynolds number flows *J. Fluids Eng. Trans. ASME* **124** 848–61

Grinstein F and Fureby C 2004 From canonical to complex flows: recent progress on monotonically integrated LES *Comput. Sci. Eng.* **6** 36–49

Grinstein F, Margolin L and Rider W (ed) 2007 *Implicit Large Eddy Simulation–Computing Turbulent Fluid Dynamics* (Cambridge: Cambridge University Press)

Huang S, Li Q and Wu J 2010 A general inflow turbulence generator for large eddy simulation *J. Wind Eng. Ind. Aerodyn.* **98** 600–17

Karniadakis G, Beskok A and Aluru N 2000 *Microflows and Nanoflows: Fundamentals and Simulation* (Berlin: Springer)

Kawai S and Larsson J 2012 Wall modeling in large eddy simulation: length scales, grid resolution, and accuracy *Phys. Fluids* **24** 015105

Keating A, Piomelli U, Balaras E and Kaltenbach H-J 2004 A priori and a posteriori tests of inflow conditions for large eddy simulation *Phys. Fluids* **16** 4696–712

Knight D 1984 A hybrid explicit-implicit numerical algorithm for the three dimensional compressible Navier-Stokes equations *AIAA J.* **22** 1056–63

Knight D 2006 *Elements of Numerical Methods for Compressible Flows* (Cambridge: Cambridge University Press)

Kobayashi H 2018 Improvement of the SGS model by using a scale similarity model based on the analysis of SGS force and SGS energy transfer *Int. J. Heat Fluid Flow* **72** 329–36

Kogan M 1969 *Rarefied Gas Dynamics* (New York: Plenum Press)

Kraichnan R 1970 Diffusion by a random velocity field *Phys. Fluids* **11** 22–31

Lagha M, Kim J, Eldredge J and Zhong X 2011 A numerical study of compressible turbulent boundary layers *Phys. Fluids* **23** 015106

Larsson J, Kawai S, Bodart J and Bermejo-Moreno I 2016 Large eddy simulation with modeled wall stress: recent progress and future directions *Mech. Eng. Rev.* **3** 1–23

Layton W and Lewandowski R 2003 A simple and stable scale-similarity model for large eddy simulation: energy balance and existence of weak solutions *Appl. Math. Lett.* **16** 1205–9

Leer B V 1979 Towards the ultimate conservative difference scheme. V. A second-order sequel to Godunov's method *J. Comput. Phys.* **32** 101–36

Lesieur M and Métais O 1996 New trends in large eddy simulations of turbulence *Annu. Rev. Fluid Mech.* **28** 45–82

Liu S, Meneveau C and Katz J 1994 On the properties of similarity subgrid scale models as deduced from measurements in a turbulent jet *J. Fluid Mech.* **275** 83–119

Lu H and Rutland C 2016 Structural subgrid scale modeling for large eddy simulation: a review *Acta Mech. Sin* **32** 567–78

Lund T, Wu X and Squires K 1998 Generation of turbulent inflow data for spatially-developing boundary layer simulations *J. Comput. Phys.* 233–58

Meneveau C and Katz J 2000 Scale-invariance and turbulence models for large-eddy simulation *Annu. Rev. Fluid Mech.* **32** 1–32

Menter F 1994 Two equation eddy viscosity turbulence models for engineering applications *AIAA J.* **32** 1598–605

Métais O and Lesieur M 1992 Spectral large eddy simulation of isotropic and stably stratified turbulence *J. Fluid Mech.* **239** 157–94

Moin P, Squires K, Cabot W and Lee S 1991 A dynamic subgrid-scale model for compressible turbulence and scalar transport *Phys. Fluids* A **11** 2746–57

Monin A and Yaglom A 1971 *Statistical Fluid Mechanics: Mechanics of Turbulence* vol 1 (Cambridge, MA: MIT Press)

Morgan B, Larsson J, Kawai S and Lele S 2011 Improving low frequency characteristics of recycling/rescaling inflow turbulence generation *AIAA J.* **49** 582–97

Moser R, Haering S and Yalla G 2021 Statistical properties of subgrid-scale turbulence models *Annu. Rev. Fluid Mech.* **53** 255–86

Mullenix N and Visbal D G M 2013 Spatially developing supersonic turbulent boundary layer with a body-force-based method *AIAA J.* **51** 1805–19

Nagnibeda E and Kustova E 2009 *Non-Equilibrium Reacting Gas Flows* (Berlin: Springer)

Okong'o N and Knight D 1998 Compressible large eddy simulation using unstructured grids: channel and boundary flow layers *34th AIAA/ASME/SAE/ASEE Joint Propulsion Conference and Exhibit* AIAA Paper 98-3315 (Reston, VA: American Institute of Aeronautics and Astronautics)

Piomelli U 2008 Wall layer models for large eddy simulation *Prog. Aerosp. Sci.* **44** 437–46

Piomelli U and Balaras E 2002 Wall-layer models for large-eddy simulations *Annu. Rev. Fluid Mech.* **34** 349–74

Pirozzoli S and Bernardini M 2013 Probing high Reynolds number effects in numerical boundary layers *Phys. Fluids* **25** 02174

Pirozzoli S, Bernardini M and Grasso F 2010 Direct numerical simulation of transonic shock/boundary layer interaction under conditions of incipient separation *J. Fluid Mech.* **657** 361–93

Pirozzoli S and Grasso F 2006 Direct numerical simulation of impinging shock wave/turbulent boundary layer interaction at $M = 2.25$ *Phys. Fluids* **18** 065113

Pirozzoli S, Grasso F and Gatski T 2004 Direct numerical simulation and analysis of a spatially evolving supersonic turbulent boundary layer at $M = 2.25$ *Phys. Fluids* **16** 530–45

Reed H and Saric W 1989 Stability of three-dimensional boundary layers *Annu. Rev. Fluid Mech.* **21** 235–84

Reynolds O 1883 An experimental investigation of the circumstances which determine whether the motion of water shall be direct or sinuous, and of the law of resistance in parallel channels *Phil. Trans. R. Soc.* **174** 935–82

Reynolds O 1885 On the dynamical theory of incompressible viscous fluids and the determination of the criterion *Phil. Trans. R. Soc.* A **186** 123–64

Reynolds W 1976 Computation of turbulent flows *Annu. Rev. Fluid Mech.* **8** 183–208

Reynolds W 1990 The potential and limitations of direct and large eddy simulation *Whither Turbulence? Turbulence at the Crossroads: Workshop Held at Cornell University* J Lumley (Berlin: Springer) pp 313–42

Rizzetta D and Visbal M 2001 Large-eddy simulation of supersonic compression-ramp flows *15th AIAA Computational Fluid Dynamics Conference* 2001-2858

Rizzetta D and Visbal M 2002 Application of large eddy simulation to supersonic compression ramps *AIAA J.* **40** 1574–81

Roe P 1981 Approximate Reimann solvers, parameter vectors, and difference schemes *J. Comput. Phys.* **43** 357–72

Rogallo R and Moin P 1984 Numerical simulation of turbulent flows *Annu. Rev. Fluid Dynamics* **16** 99–137

Sagaut P, Garnier E, Tromeur E, Larchevêque L and Labourasse E 2004 Turbulent inflow conditions for large eddy simulation of wall-bounded flows *AIAA J.* **42** 469–77

Sayadi T, Hamman C and Moin P 2013 Direct numerical simulation of complete H-type and K-type transitions with implications for the dynamics of turbulent boundary layers *J. Fluid Mech.* **724** 480–509

Schlichting H 1968 *Boundary-Layer Theory* 6th edn (New York: McGraw-Hill)

Schröder W, Meinke M, Ewert R and El-Askary W 2001 LES of a turbulent flow around a sharp leading edge *Direct and Large Eddy Simulation IV* ed B Geurts, R Friedrich and O Métais (Norwell, MA: Kluwer Academic) pp 353–63

Simens M, Jiménez J, Hoyas S and Mizuno Y 2009 A high-resolution code for turbulent boundary layers *J. Comput. Phys.* **228** 4218–31

Smagorinsky J 1963 General circulation experiments with the primitive equations, I. The basic experiment *Mon. Weather Rev.* **91** 99–164

Smirnov A, Shi S and Celik I 2001 Random flow generation technique for large eddy simulations and particle dynamics modeling *J. Fluids Eng.* **123** 359–71

Smits A and Dussauge J-P 1996 *Turbulent Shear Layers in Supersonic Flow* (Woodbury, NY: American Institute of Physics)

Spalart P 2009 Detached-eddy simulation *Annu. Rev. Fluid Mech.* **41** 181–202

Spalart P and Allmaras S 1994 A one-equation turbulence model for aerodynamic flows *Rech. Aérosp.* **1** 5–21

Spalart P, Jou W-H, Strelets M and Allmaras S 1997 Comments on the feasibility of LES for wings, and on a hybrid RANS/LES approach *Advances in DNS/LES* ed C Liu and Z Liu (Columbus, OH: Greyden Press) pp 137–47

Spalart P and Leonard A 1985 Direct numerical simulation of equilibrium turbulent boundary layers *Turbulent Shear Flows 5* ed F Durst, B Launder, J Lumley, F Schmidt and J Whitelaw (Berlin: Springer) pp 234–252

Spyropoulos E and Blaisdell G 1998 Large eddy simulation of a spatially evolving supersonic turbulent boundary layer flow *AIAA J.* **36** 1983–90

Stolz S and Adams N 2003 Large-eddy simulation of high-Reynolds-number supersonic boundary layers using the approximate deconvolution method and a rescaling and recycling technique *Phys. Fluids* **15** 2398–412

Tabor G and Baba-Ahmadi M 2010 Inlet conditions for large eddy simulation: a review *Comput. Fluids* **39** 553–67

Urbin G and Knight D 2001 Large eddy simulation of a supersonic boundary layer using an unstructured grid *AIAA J.* **39** 1288–95

Urbin G, Knight D and Zheltovodov A 1999 Compressible large eddy simulation using unstructured grid: supersonic turbulent boundary layer and compression corner *37th Aerospace Sciences Meeting and Exhibit* AIAA Paper 1999-427 (Reston, VA: American Institute of Aeronautics and Astronautics)

Van Driest E 1951 Turbulent boundary layer in compressible fluids *J. Aeronaut. Sci.* **18** 145–60

Van Driest E 1956 On turbulent flow near a wall *J. Aeronaut. Sci.* **23** 1007–11

van Miegham J 1949 *Les Équations Générales de la Mécanique et de l'Énergétique des Milleux Turbulens en Vue des Applications à la météorologie* Technical Report 34 Institut Royal Météorologie de Belgique

Vincenti W and Kruger C 1965 *Introduction to Physical Gas Dynamics* (Malabar, FL: Krieger Publishing Company)

Vreman B, Geurts B and Kuerten H 1997 Large eddy simulation of the turbulent mixing layer *J. Fluid Mech.* **339** 357–90

Wang M and Moin P 2002 Dynamic wall modeling for large eddy simulation of complex turbulent flows *Phys. Fluids* **14** 2043–51

Wilcox D 2006 *Turbulence Modeling for CFD* 3rd edn (La Canada, CA: DCW Industries, Inc.)

Woodward P and Colella P 1984 The numerical simulation of two-dimensional fluid flow with strong shocks *J. Comput. Phys.* **54** 115–73

Wu X 2017 Inflow turbulence generation methods *Annu. Rev. Fluid Mech.* **49** 23–49

Xiao X, Edwards J, Hassan H and Baurle R 2003 Inflow boundary conditions for hybrid large eddy/Reynolds averaged Navier–Stokes simulations *AIAA J.* **41** 1481–9

Xu S and Martín M P 2004 Assessment of inflow boundary conditions for compressible turbulent boundary layers *Phys. Fluids* **16** 2623–39

Yu R and Bai X-S 2014 A fully divergence-free method for generation of inhomogeneous and anisotropic turbulence with large spatial variation *J. Comput. Phys.* **256** 234–53

Zalesak S 1979 Fully multidimensional flux-corrected transport algorithm for fluids *J. Comput. Phys.* **31** 335–62

**IOP** Publishing



# Chapter 3

# Equilibrium turbulent boundary layers

It remains to call attention to the chief outstanding difficulty of our subject.

Horace Lamb (1945)


## Abstract

An important concept in turbulence is an *equilibrium turbulent boundary layer*. The term *equilibrium* indicates a state wherein the mean and root-mean-square fluctuating properties can be scaled based upon local properties such as the wall shear stress and boundary layer thickness, together perhaps with inviscid parameters such as the pressure gradient. In particular, the concept of the equilibrium flat plate zero pressure gradient turbulent boundary is important since it provides the basis for comparison of experimental results from different facilities. In this chapter, we present the structure of an equilibrium hypersonic turbulent boundary layer in terms of the mean velocity and mean temperature (or mean total temperature), and the statistics of the fluctuating turbulent variables.


## 3.1 Incompressible Law of the Wall and Wake

Consider an incompressible zero pressure gradient flat plate turbulent boundary layer. In the boundary layer approximation the Reynolds-averaged conservation of momentum becomes

$$\frac{\partial \rho \bar{u}\bar{u}}{\partial x} + \frac{\partial \rho \bar{u}\bar{v}}{\partial y} = \frac{\partial}{\partial y}\left(-\rho \overline{u'v'} + \mu \frac{\partial \bar{u}}{\partial y}\right) \tag{3.1}$$

doi:10.1088/978-0-7503-5002-0ch3  

where $\bar{u}$ and $\bar{v}$ are the Cartesian velocity components in the streamwise $x$- and wall normal $y$-directions. Assume there exists a region adjacent to the wall where the convective terms are negligible. Equation (3.1) becomes

$$\frac{\partial}{\partial y}\left(-\rho\overline{u'v'} + \mu\frac{\partial \bar{u}}{\partial y}\right) = 0 \tag{3.2}$$

and integrating from the wall

$$-\rho\overline{u'v'} + \mu\frac{\partial \bar{u}}{\partial y} = \tau_{\mathrm{w}} \tag{3.3}$$

where $\tau_{\mathrm{w}}$ is the wall shear stress

$$\tau_{\mathrm{w}} = \mu\frac{\partial \bar{u}}{\partial y}\bigg|_{\mathrm{w}} \tag{3.4}$$

Outside the viscous sublayer (where molecular shear is significant)

$$-\rho\overline{u'v'} = \tau_{\mathrm{w}} \tag{3.5}$$

Boussinesq (1877) introduced the concept of a turbulent eddy viscosity $\mu_{\mathrm{t}}$ by analogy to molecular viscosity

$$-\rho\overline{u'v'} = \mu_{\mathrm{t}}\frac{\partial \bar{u}}{\partial y} \tag{3.6}$$

Prandtl (1925) introduced the concept of the mixing length $\ell$ by analogy to kinetic theory

$$-\rho\overline{u'v'} = \underbrace{\rho\ell^2\left|\frac{\partial \bar{u}}{\partial y}\right|}_{\mu_{\mathrm{t}}}\frac{\partial \bar{u}}{\partial y} \tag{3.7}$$

and postulated

$$\ell = \kappa y \tag{3.8}$$

where $\kappa$ is a constant. Integrating (3.5) and (3.7),

$$\frac{\bar{u}}{u_\tau} = \frac{1}{\kappa}\log\left(\frac{y u_\tau}{\nu}\right) + C \tag{3.9}$$

over the portion of the boundary layer above the viscous sublayer (where molecular viscosity is significant) and below the outer region of the boundary layer (where convection is significant) where $u_\tau$ is the friction velocity defined as

$$u_\tau = \sqrt{\frac{\tau_{\mathrm{w}}}{\rho}} \tag{3.10}$$

with $\nu = \mu/\rho$. The friction velocity $u_\tau$ is related to the skin friction coefficient according to

$$c_\mathrm{f} = 2\left(\frac{u_\tau}{u_\infty}\right)^2 \tag{3.11}$$

where $c_\mathrm{f} = \tau_\mathrm{w}/\frac{1}{2}\rho u_\infty^2$. The constant $\kappa$ is von Kármán's constant and equal to $0.40 \pm 0.01$, and $C = 5.1 \pm 0.2$ for a smooth wall (Coles 1953). Coles (1956, 1968) showed that the profile can be extended to the outer edge of the boundary layer by incorporating a wake contribution

$$\frac{\bar{u}}{u_\tau} = \frac{1}{\kappa}\log\left(\frac{yu_\tau}{\nu}\right) + C + \frac{2\Pi}{\kappa}\sin^2\left(\frac{\pi}{2}\frac{y}{\delta}\right) \tag{3.12}$$

where $\delta$ is the boundary layer thickness. The parameter $\Pi$ is a function of the Reynolds number $Re_\delta = \rho u_\infty \delta/\mu$, or equivalently, $\Pi$ is a function of $Re_\theta$ where $\theta$ is the momentum thickness, and also Mach number (Baronti and Libby 1966). Examples of experimental results are shown in figure 3.1 together with the asymptotic expression of Coles (1962). The above expression (3.12) holds for $yu_\tau/\nu \gtrsim 50$ and is denoted the Law of the Wall and Wake.

Within the viscous sublayer where molecular shear stress dominates turbulent shear stress, (3.3) becomes

$$\mu\frac{\partial \bar{u}}{\partial y} = \tau_\mathrm{w} \tag{3.13}$$

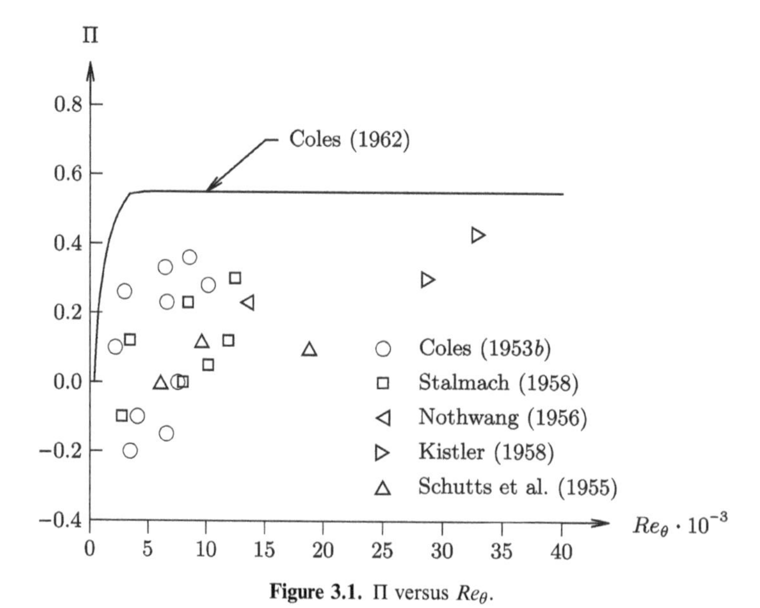


**Figure 3.1.** $\Pi$ versus $Re_\theta$.

and thus

$$\frac{\bar{u}}{u_\tau} = \frac{yu_\tau}{\nu} \tag{3.14}$$

The above expression holds for $yu_\tau/\nu \lesssim 10$. Equations (3.12) and (3.14) are connected by a buffer region $10 \lesssim yu_\tau/\nu \lesssim 50$ wherein the mean velocity profile changes smoothly. Equation (3.12) provides a relationship between the nondimensional friction velocity $u_\tau/u_\infty$ and Reynolds number $Re_\delta$, noting that $\Pi$ is a weak function of $Re_\delta$,

$$\frac{u_\infty}{u_\tau} = \frac{1}{\kappa}\log\left(Re_\delta \frac{u_\tau}{u_\infty}\right) + C + 2\frac{\Pi}{\kappa} \tag{3.15}$$

In comparing computational and experimental mean velocity profile data, typically (3.14) and (3.12) are plotted to the value of $yu_\tau/\nu$ where they match, thereby recognizing that the buffer layer profile is omitted. An example is presented in figure 3.2 for $Re_\delta = 10^5$ and $\Pi = 0.55$ with the matching point $y^+ = 11.06$ where the dashed line indicates the actual experimental velocity profile in the buffer region.

The incompressible law of the wall and wake for a rough wall is

$$\frac{\bar{u}}{u_\tau} = \frac{1}{\kappa}\log\left(\frac{yu_\tau}{\nu}\right) + C - H\left(\frac{ku_\tau}{\nu}\right) + \frac{2\Pi}{\kappa}\sin^2\left(\frac{\pi}{2}\frac{y}{\delta}\right) \tag{3.16}$$

where $H\left(\frac{ku_\tau}{\nu}\right)$ is Hama's roughness function (Hama 1954, Clauser 1954, 1956) where $k$ is the effective roughness height.

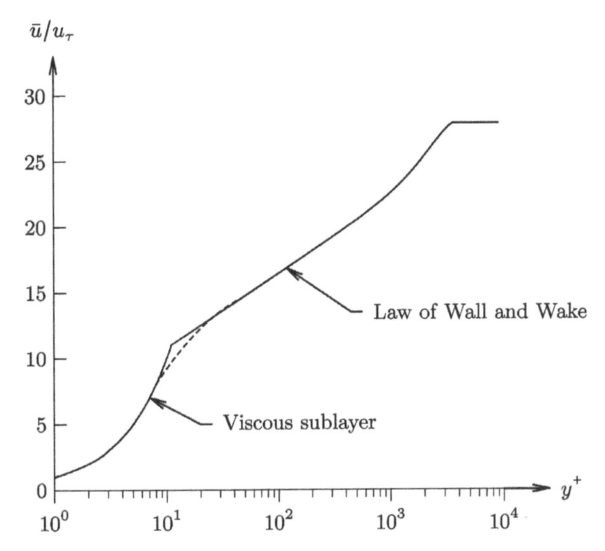


**Figure 3.2.** Combined viscous sublayer and law of wall and wake.

An expression for the velocity defect in the outer portion of the boundary layer can be obtained from (3.12) or (3.16) as

$$\frac{\bar{u} - u_\infty}{u_\tau} = \frac{1}{\kappa}\log\left(\frac{y}{\delta}\right) - \frac{2\Pi}{\kappa}\left[1 - \sin^2\left(\frac{\pi}{2}\frac{y}{\delta}\right)\right] \tag{3.17}$$

The *wake strength* is defined as

$$\Delta\left(\frac{\bar{u}}{u_\tau}\right) = \max\left(\frac{\bar{u}}{u_\tau} - \frac{1}{\kappa}\log\frac{yu_\tau}{\nu} - C\right) \tag{3.18}$$

and thus according to (3.12) or (3.16)

$$\Delta\left(\frac{\bar{u}}{u_\tau}\right) = \frac{2\Pi}{\kappa} \tag{3.19}$$

Experimental data indicate that $\Delta(\bar{u}/u_\tau)$ is an increasing function of $Re_\theta$ asymptoting to 2.75 for $Re_\theta > 6000$ (Fernholz and Finley 1980). This corresponds to an asymptotic value for $\Pi$ equal to 0.55.

## 3.2 Compressible velocity transformations

A variety of methods have been proposed to transform the mean streamwise velocity profile in a zero pressure gradient compressible turbulent boundary layer to the incompressible law of the wall and wake (3.12) including in some methods also the viscous sublayer. Several methods are discussed below. Additional methods for transforming mean velocity and/or turbulence statistics to the incompressible forms include those by Brun *et al* (2008), Zhang *et al* (2014), Patel *et al* (2016), Wu *et al* (2017), and Modesti and Pirozzoli (2019).

### 3.2.1 Van Driest transformation with no mass transfer at wall

Van Driest (1951) showed that a transformation of the mean streamwise velocity in a compressible zero pressure gradient turbulent boundary layer satisfied the Law of the Wall. The result was extended by Matthews *et al* (1970) and Sun and Childs (1976) to include the Law of the Wake (Coles 1968).

We present the derivation of Van Driest (1951) based upon Favre-averaged variables. Consider a zero pressure gradient flat plate turbulent boundary layer. In the boundary layer approximation the conservation of momentum (2.40) in the streamwise $x$-direction and conservation of energy (2.41) become

$$\frac{\partial\bar{\rho}\,\tilde{u}\tilde{u}}{\partial x} + \frac{\partial\bar{\rho}\,\tilde{u}\tilde{v}}{\partial y} = \frac{\partial}{\partial y}\left(-\overline{\rho u''v''} + \mu\frac{\partial\tilde{u}}{\partial y}\right) \tag{3.20}$$

$$\frac{\partial(\bar{\rho}\,\tilde{e}+\bar{p})\tilde{u}}{\partial x} + \frac{\partial(\bar{\rho}\,\tilde{e}+\bar{p})\tilde{v}}{\partial y} = \frac{\partial}{\partial y}\left[-c_p\overline{\rho T''v''} - \bar{q}_y + \left(-\overline{\rho u''v''} + \mu\frac{\partial\tilde{u}}{\partial y}\right)\tilde{u}\right] \tag{3.21}$$

Assuming there is a region adjacent to the wall where the convective terms are negligible, (3.20) and (3.21) become

$$\frac{\partial}{\partial y}\left(-\overline{\rho u''v''} + \mu\frac{\partial \tilde{u}}{\partial y}\right) = 0 \tag{3.22}$$

$$\frac{\partial}{\partial y}\left[-c_p\overline{\rho T''v''} - \bar{q}_y + \left(-\overline{\rho u''v''} + \mu\frac{\partial \tilde{u}}{\partial y}\right)\tilde{u}\right] = 0 \tag{3.23}$$

Integrating (3.22) and (3.23) from the wall,

$$-\overline{\rho u''v''} + \mu\frac{\partial \tilde{u}}{\partial y} = \tau_w \tag{3.24}$$

$$-c_p\overline{\rho T''v''} - \bar{q}_y + \left(-\overline{\rho u''v''} + \mu\frac{\partial \tilde{u}}{\partial y}\right)\tilde{u} = -q_w \tag{3.25}$$

where $\tau_w$ and $q_w$ are the mean shear stress and heat flux at the wall defined by

$$\tau_w = \mu_w\frac{\partial \tilde{u}}{\partial y}\bigg|_w$$

$$q_w = -k_w\frac{\partial \tilde{T}}{\partial y}\bigg|_w$$

At sufficient distance from the wall, the viscous shear stress and molecular heat flux are negligible compared to their turbulent values, and thus

$$-\overline{\rho u''v''} = \tau_w \tag{3.26}$$

$$-c_p\overline{\rho T''v''} + (-\overline{\rho u''v''})\tilde{u} = -q_w \tag{3.27}$$

Assuming an eddy viscosity model for the Reynolds shear stress and turbulent heat flux,

$$-\overline{\rho u''v''} = \mu_t\frac{\partial \tilde{u}}{\partial y} \tag{3.28}$$

$$-c_p\overline{\rho T''v''} = k_t\frac{\partial \tilde{T}}{\partial y} \tag{3.29}$$

where $\mu_t$ and $k_t$ are the turbulent eddy viscosity and conductivity, respectively. The turbulent Prandtl number $Pr_t$ is defined as

$$Pr_t = \frac{\mu_t c_p}{k_t} \tag{3.30}$$

and is assumed to be constant. Assuming a mixing length model for the turbulent eddy viscosity,

$$\mu_t = \bar{\rho}\ell^2 \frac{\partial \tilde{u}}{\partial y} \tag{3.31}$$

where the mixing length $\ell$ is

$$\ell = \kappa y \tag{3.32}$$

where $\kappa$ is a constant. Thus (3.28) becomes

$$\bar{\rho}(\kappa y)^2 \left( \frac{\partial \tilde{u}}{\partial y} \right)^2 = \tau_w \tag{3.33}$$

Define the friction velocity $u_\tau$ according to

$$\tau_w = \rho_w u_\tau^2 \tag{3.34}$$

and thus

$$\sqrt{\frac{\bar{\rho}}{\rho_w}} \kappa y \frac{\partial \tilde{u}}{\partial y} = u_\tau \tag{3.35}$$

Formally integrating and non-dimensionalizing the $y$ coordinate by the viscous length scale $\nu_w/u_\tau$ where $\nu_w$ is the kinematic viscosity at the wall,

$$\int_{\tilde{u}_1}^{\tilde{u}} \sqrt{\frac{\bar{\rho}}{\rho_w}}\, d\tilde{u} = \frac{u_\tau}{\kappa} \log \frac{y u_\tau}{\nu_w} + C u_\tau \tag{3.36}$$

where the subscript $_1$ indicates the lower limit of (3.26) and (3.27) and $C$ is a constant. The expression on the left side is denoted the *Van Driest transformed velocity*,[1]

$$u_{VD} = \int_{\tilde{u}_1}^{\tilde{u}} \sqrt{\frac{\bar{\rho}}{\rho_w}}\, d\tilde{u} \tag{3.37}$$

It is important to note that the integral in (3.37) does *not* extend to the wall since the analysis from (3.26) to (3.37) is applicable only in the region where molecular viscosity is insignificant (i.e., outside the viscous sublayer). Furthermore, it is evident that the constant $C$ appearing in (3.36) depends upon the lower limit $\tilde{u}_1$ of integration in (3.36) and hence upon the details of the flow in the viscous sublayer including the wall temperature (e.g., adiabatic or isothermal). This is discussed in more detail in section 3.2.4.

Since the mean pressure is constant across the boundary layer, $\bar{\rho}/\rho_w = T_w/\tilde{T}$, and thus

$$u_{VD} = \int_{\tilde{u}_1}^{\tilde{u}} \sqrt{\frac{T_w}{\tilde{T}}}\, d\tilde{u} \tag{3.38}$$

[1] White (1974) denotes the expression the *Van Driest effective velocity*.

Assuming a Crocco–Busemann-type relationship (3.147) for the temperature

$$\tilde{T} = T_{\mathrm{w}}\left[1 + B\left(\frac{\tilde{u}}{u_\infty}\right) - A^2\left(\frac{\tilde{u}}{u_\infty}\right)^2\right] \tag{3.39}$$

the energy equation (3.27) yields

$$\underbrace{\frac{c_{\mathrm{p}}}{Pr_{\mathrm{t}}}\mu_{\mathrm{t}}\frac{\partial \tilde{u}}{\partial y}}_{\tau_{\mathrm{w}}} T_{\mathrm{w}}\left(\frac{B}{u_\infty} - 2\frac{A^2}{u_\infty^2}\tilde{u}\right) + \tau_{\mathrm{w}}\tilde{u} = -q_{\mathrm{w}} \tag{3.40}$$

and equating terms

$$A = \sqrt{\frac{(\gamma - 1)}{2} Pr_{\mathrm{t}} M_\infty^2 \frac{T_\infty}{T_{\mathrm{w}}}} \tag{3.41}$$

$$B = -\frac{Pr_{\mathrm{t}}\, q_{\mathrm{w}} u_\infty}{c_{\mathrm{p}} \tau_{\mathrm{w}} T_{\mathrm{w}}} \tag{3.42}$$

Assuming (3.39) is valid throughout the boundary layer, evaluation at the wall and freestream yields

$$Pr = Pr_{\mathrm{t}} \tag{3.43}$$

$$B = \left[1 + \frac{(\gamma - 1)}{2} Pr_{\mathrm{t}} M_\infty^2\right]\frac{T_\infty}{T_{\mathrm{w}}} - 1 \tag{3.44}$$

where $Pr = \mu c_{\mathrm{p}}/k$ is the molecular Prandtl number. Equation (3.43) is reasonably satisfied for air since $Pr = 0.72$ and $Pr_{\mathrm{t}} \approx 0.89$ according to measurements (chapter 4). Furthermore, applying (3.39) at $\tilde{u} = u_\infty$ yields

$$1 + B - A^2 = \frac{T_\infty}{T_{\mathrm{w}}} \tag{3.45}$$

and applying (3.44) for an adiabatic wall implies $B = 0$ and hence

$$T_{\mathrm{aw}} = T_\infty\left[1 + \frac{(\gamma - 1)}{2} Pr_{\mathrm{t}} M_\infty^2\right] \tag{3.46}$$

and therefore

$$A^2 = \frac{T_{\mathrm{aw}} - T_\infty}{T_{\mathrm{w}}} \tag{3.47}$$

$$B = \frac{T_{\mathrm{aw}} - T_{\mathrm{w}}}{T_{\mathrm{w}}} \tag{3.48}$$

Substituting (3.39) into (3.38),

$$\frac{u_\infty}{A}\sin^{-1}\left(\frac{2A^2\tilde{u}/u_\infty - B}{\sqrt{B^2 + 4A^2}}\right) + \text{constant} = \frac{u_\tau}{\kappa}\log\frac{yu_\tau}{\nu_w} + Cu_\tau \tag{3.49}$$

We now show that the constant on the left side of (3.49) is

$$\text{constant} = \frac{u_\infty}{A}\sin^{-1}\left(\frac{B}{\sqrt{B^2 + 4A^2}}\right) \tag{3.50}$$

Since (3.49) must be valid for $T_w = T_\infty$ in the limit $M_\infty \to 0$, we set $T_w = T_\infty$, and thus from (3.44), $B = A^2$. Since $A^2 \propto M_\infty^2$, the leading terms of the left side of (3.49) are

$$\frac{u_\infty}{A}\sin^{-1}\left(\frac{2A^2\tilde{u}/u_\infty - B}{\sqrt{B^2 + 4A^2}}\right) = \tilde{u} - \frac{1}{2}u_\infty + \mathcal{O}(M_\infty^3) \tag{3.51}$$

$$\frac{u_\infty}{A}\sin^{-1}\left(\frac{B}{\sqrt{B^2 + 4A^2}}\right) = \frac{1}{2}u_\infty + \mathcal{O}(M_\infty^3) \tag{3.52}$$

and thus in the limit $M_\infty \to 0$, equations (3.49) and (3.50) yield

$$\tilde{u} = \frac{u_\tau}{\kappa}\log\frac{yu_\tau}{\nu} + Cu_\tau \tag{3.53}$$

The compressible Law of the Wall is therefore

$$u_{VD} = \frac{u_\tau}{\kappa}\log\frac{yu_\tau}{\nu_w} + Cu_\tau \tag{3.54}$$

where the Van Driest transformed velocity is

$$u_{VD} = \frac{u_\infty}{A}\left[\sin^{-1}\left(\frac{2A^2(\tilde{u}/u_\infty) - B}{\sqrt{B^2 + 4A^2}}\right) + \sin^{-1}\left(\frac{B}{\sqrt{B^2 + 4A^2}}\right)\right] \tag{3.55}$$

The compressible Law of the Wall and Wake for a smooth wall is therefore

$$\frac{u_{VD}}{u_\tau} = \frac{1}{\kappa}\log\left(\frac{yu_\tau}{\nu_w}\right) + C + \frac{2\Pi}{\kappa}\sin^2\left(\frac{\pi}{2}\frac{y}{\delta}\right) \tag{3.56}$$

where $\kappa = 0.40 \pm 0.01$ is von Kármán's constant and $u_\tau$ is the friction velocity defined by (3.10) with $\rho$ replaced by $\rho_w$.

The constant $C$ arises from the lower limit $\tilde{u}_l$ in (3.36), in other words, the mean velocity at which molecular viscous stress is negligible compared to the turbulent Reynolds shear stress. For specificity, we can assume $u_l$ is the velocity at the height where $\mu\partial\tilde{u}/\partial y$ equals one percent of the Reynolds shear stress $-\overline{\rho u''v''}$ at the same height. It is therefore evident that the constant $C$ will depend upon the temperature boundary condition at the wall (i.e., isothermal or adiabatic) and hence depend upon a dimensionless wall temperature difference $(T_{aw} - T_w)/T_{aw}$ or similar dimensionless measure of the wall heat flux. This is confirmed by Danberg (1964) (see figure 4.16). A range of values for the constant $C$ is quoted in the literature, from $C = 5.2$ for adiabatic walls (Bradshaw 1977) to $C = 6.82$ for cold walls ($T_w < T_{aw}$) (Duan *et al* 2010, Duan and Martín 2011, Brooks *et al* 2017).

Within the viscous sublayer (where the turbulent shear stress is negligible), equation (3.24) yields

$$\mu \frac{\partial \tilde{u}}{\partial y} = \tau_w \tag{3.57}$$

For adiabatic and near adiabatic walls, the mean temperature variation is small across the viscous sublayer, and thus the molecular viscosity may be approximated by its value at the wall and (3.57) becomes

$$\mu_w \frac{\partial \tilde{u}}{\partial y} = \tau_w \tag{3.58}$$

which yields

$$\frac{\tilde{u}}{u_\tau} = \frac{y u_\tau}{\nu_w} \tag{3.59}$$

where $\nu_w = \mu_w / \rho_w$.

The corresponding compressible Law of the Wall and Wake for a rough wall is[2]

$$\frac{u_{VD}}{u_\tau} = \frac{1}{\kappa} \log\left(\frac{y u_\tau}{\nu_w}\right) + C - H\left(\frac{k u_\tau}{\nu_w}\right) + \frac{2\Pi}{\kappa} \sin^2\left(\frac{\pi}{2}\frac{y}{\delta}\right) \tag{3.60}$$

An expression for the velocity defect in the outer portion of the boundary layer can be obtained from (3.56) or (3.60) as

$$\frac{u_{VD} - u_{VD\infty}}{u_\tau} = \frac{1}{\kappa} \log\left(\frac{y}{\delta}\right) - \frac{2\Pi}{\kappa}\left[1 - \sin^2\left(\frac{\pi}{2}\frac{y}{\delta}\right)\right] \tag{3.61}$$

where $u_{VD\infty}$ is the value of $u_{VD}$ evaluated at the edge of the boundary layer $y = \delta$.

The compressible *wake strength* is defined as

$$\Delta\left(\frac{u_{VD}}{u_\tau}\right) = \max\left(\frac{u_{VD}}{u_\tau} - \frac{1}{\kappa} \log \frac{y u_\tau}{\nu} - C\right) \tag{3.62}$$

and thus according to (3.56) or (3.60)

$$\Delta\left(\frac{u_{VD}}{u_\tau}\right) = \frac{2\Pi}{\kappa} \tag{3.63}$$

For an equilibrium (i.e., free from tripping or transition effects) compressible turbulent boundary layer, $\Delta(u_{VD}/u_\tau)$ increases from a value of approximately zero

[2] The function $H(\frac{k u_\tau}{\nu_w})$ is not necessarily the same as for incompressible rough walls.

at $Re_\theta = 500$ to 2.75 for $Re_\theta > 6000$ (Fernholz and Finley 1980) which corresponds to an asymptotic value for $\Pi$ equal 0.55.

### 3.2.2 Van Driest transformation with mass transfer at wall

The compressible law of the wall was extended to include mass transfer at the wall (see also Dorrance and Dore 1954) by Rubesin (1954). Consider a zero pressure gradient flat plate turbulent boundary layer. In the boundary layer approximation the governing equations are

$$\frac{\partial \bar{\rho}\,\tilde{u}}{\partial x} + \frac{\partial \bar{\rho}\,\tilde{v}}{\partial y} = 0 \tag{3.64}$$

$$\frac{\partial \bar{\rho}\,\tilde{u}\tilde{u}}{\partial x} + \frac{\partial \bar{\rho}\,\tilde{u}\tilde{v}}{\partial y} = \frac{\partial}{\partial y}\left( -\overline{\rho u''v''} + \mu\frac{\partial \tilde{u}}{\partial y}\right) \tag{3.65}$$

$$\frac{\partial(\bar{\rho}\,\tilde{e}+\bar{p})\tilde{u}}{\partial x} + \frac{\partial(\bar{\rho}\,\tilde{e}+\bar{p})\tilde{v}}{\partial y} = \frac{\partial}{\partial y}\left[ -c_p\overline{\rho T''v''} - \bar{q}_y + \left( -\overline{\rho u''v''} + \mu\frac{\partial \tilde{u}}{\partial y}\right)\tilde{u}\right] \tag{3.66}$$

Similar to section 3.2.1, we neglect the streamwise convection of mass, momentum, and energy; however, the normal convection cannot be ignored due to the injection (or removal) of mass at the surface. The boundary layer equations therefore become

$$\frac{\partial \bar{\rho}\,\tilde{v}}{\partial y} = 0 \tag{3.67}$$

$$\frac{\partial \bar{\rho}\,\tilde{u}\tilde{v}}{\partial y} = \frac{\partial}{\partial y}\left( -\overline{\rho u''v''} + \mu\frac{\partial \tilde{u}}{\partial y}\right) \tag{3.68}$$

$$\frac{\partial(\bar{\rho}\,\tilde{e}+\bar{p})\tilde{v}}{\partial y} = \frac{\partial}{\partial y}\left[ -c_p\overline{\rho T''v''} - \bar{q}_y + \left( -\overline{\rho u''v''} + \mu\frac{\partial \tilde{u}}{\partial y}\right)\tilde{u}\right] \tag{3.69}$$

The integral of (3.67) is

$$\bar{\rho}\tilde{v} = \rho_w v_w \tag{3.70}$$

and therefore (3.68) and (3.69) become

$$\rho_w v_w\frac{\partial \tilde{u}}{\partial y} = \frac{\partial}{\partial y}\left( -\overline{\rho u''v''} + \mu\frac{\partial \tilde{u}}{\partial y}\right) \tag{3.71}$$

$$\rho_w v_w\frac{\partial \tilde{H}}{\partial y} = \frac{\partial}{\partial y}\left[ -c_p\overline{\rho T''v''} - \bar{q}_y + \left( -\overline{\rho u''v''} + \mu\frac{\partial \tilde{u}}{\partial y}\right)\tilde{u}\right] \tag{3.72}$$

where from (2.46) in the boundary layer approximation

$$\tilde{H} = c_p \tilde{T} + \frac{1}{2} \tilde{u}^2 \tag{3.73}$$

where $\tilde{H}$ is the total enthalpy per unit mass and assuming $M_k \ll M$ where $M_k$ is defined in (2.49). The total shear stress $\tau$ is defined as

$$\tau = -\overline{\rho u'' v''} + \mu \frac{\partial \tilde{u}}{\partial y} \tag{3.74}$$

and thus from (3.71)

$$\tau = \tau_w + \rho_w v_w \tilde{u} \tag{3.75}$$

where $\tau_w$ is the wall shear stress.

In the fully turbulent region, $\tau = -\overline{\rho u'' v''}$, and assuming an eddy viscosity model for the Reynolds shear stress and a constant turbulent Prandtl number in accordance with (3.28) to (3.32), equation (3.71) becomes

$$\frac{\partial \tilde{u}^+}{\partial y^+} = \sqrt{\frac{\rho_w}{\bar{\rho}}} \frac{1}{\kappa y^+} \sqrt{1 + 2 \frac{c_q}{c_f} \frac{\tilde{u}^+}{u_\infty^+}} \tag{3.76}$$

where

$$c_f = \frac{\tau_w}{\frac{1}{2} \rho_\infty u_\infty^2} \tag{3.77}$$

$$c_q = \frac{\rho_w v_w}{\rho_\infty u_\infty} \tag{3.78}$$

$$\tilde{u}^+ = \frac{\tilde{u}}{u_\tau} \tag{3.79}$$

$$y^+ = \frac{y u_\tau}{\nu_w} \tag{3.80}$$

Define

$$du^{++} = \sqrt{\frac{\bar{\rho}}{\rho_w}} \left[ 1 + 2 \frac{c_q}{c_f} \frac{\tilde{u}^+}{u_\infty^+} \right]^{-\frac{1}{2}} d\tilde{u}^+ \tag{3.81}$$

Integrating from some value of $y_1^+$ within the fully turbulent region,

$$u^{++}(y^+) - C_1 = \frac{1}{\kappa} \log y^+ - C_2 \tag{3.82}$$

where $C_1$ and $C_2$ are constants depending on the choice of $y_1^+$. Now assume formally that the integration is to the wall and therefore

$$u^{++} = \frac{1}{\kappa}\log y^{+} + C \tag{3.83}$$

where $C$ is a constant which incorporates the physics of the viscous sublayer, and

$$u^{++} = \int_{0}^{\tilde{u}^{+}} \sqrt{\frac{\bar{\rho}}{\rho_{w}}}\left[1 + 2\frac{c_{q}}{c_{f}}\frac{\tilde{u}^{+}}{u_{\infty}^{+}}\right]^{-\frac{1}{2}} d\tilde{u}^{+} \tag{3.84}$$

Integration of (3.84) requires a Crocco–Busemann expression for $\bar{\rho}/\rho_{w} = T_{w}/\tilde{T}$. Integrating (3.72),

$$\rho_{w}v_{w}\left(c_{p}\tilde{T}+\frac{1}{2}\tilde{u}^{2}\right) - \rho_{w}v_{w}c_{p}T_{w} = \left(\frac{\mu}{Pr} + \frac{\mu_{t}}{Pr_{t}}\right)\frac{\partial c_{p}\tilde{T}}{\partial y} + q_{w} + (\mu + \mu_{t})\frac{\partial \frac{1}{2}\tilde{u}^{2}}{\partial y} \tag{3.85}$$

Assuming the Crocco relationship for the temperature (3.39), substituting into (3.85) and evaluating at $y = 0$ yield

$$B = -\frac{Pr\, q_{w}u_{\infty}}{c_{p}\tau_{w}T_{w}} \tag{3.86}$$

Evaluating (3.85) in the fully turbulent region and equating powers in $\tilde{u}$ yields two equations for $A$, namely,

$$A = \sqrt{\frac{(\gamma - 1)}{2}Pr_{t}M_{\infty}^{2}\frac{T_{\infty}}{T_{w}}\frac{1}{(2 - Pr_{t})}} \tag{3.87}$$

$$A = \sqrt{\frac{(\gamma - 1)}{2}Pr_{t}M_{\infty}^{2}\frac{T_{\infty}}{T_{w}} - \frac{\rho_{w}v_{w}(Pr_{t}u_{\infty})^{2}q_{w}}{2c_{p}T_{w}\tau_{w}^{2}}\left(\frac{1}{Pr_{t}} - 1\right)} \tag{3.88}$$

and therefore (3.39) holds provided $Pr_{t} = Pr = 1$. Therefore,

$$A = \sqrt{\frac{(\gamma - 1)}{2}M_{\infty}^{2}\frac{T_{\infty}}{T_{w}}} \tag{3.89}$$

$$B = -\frac{q_{w}u_{\infty}}{c_{p}\tau_{w}T_{w}} \tag{3.90}$$

Equation (3.84) becomes

$$u^{++} = \frac{u_{\infty}}{u_{\tau}}\sqrt{C_{fq}}\int_{0}^{(\tilde{u}/u_{\infty})}\left\{\left[1 + B\left(\frac{\tilde{u}}{u_{\infty}}\right) - A^{2}\left(\frac{\tilde{u}}{u_{\infty}}\right)^{2}\right]\left[\frac{1}{2}\frac{c_{f}}{c_{q}} + \left(\frac{\tilde{u}}{u_{\infty}}\right)\right]\right\}^{-\frac{1}{2}} d\left(\frac{\tilde{u}}{u_{\infty}}\right) \tag{3.91}$$

The solution is

$$u^{++} = i\frac{u_{\infty}}{u_{\tau}}\sqrt{C_{fq}}\left\{\mathcal{A}\left(\frac{\tilde{u}}{u_{\infty}}\right)F\left[m\left(\frac{\tilde{u}}{u_{\infty}}\right)\,\middle|\,k^{2}\right] - \mathcal{A}(0)F[m(0)\mid k^{2}]\right\} \tag{3.92}$$

where

$$\mathcal{A} = \left(C_{fq} + \frac{\tilde{u}}{u_\infty}\right)^{\frac{3}{2}} \cdot$$
$$\left[2 - 4\left[A^2 C_{fq}^2 + BC_{fq} - 1\right]\left[\left(C_{fq} + \frac{\tilde{u}}{u_\infty}\right)\left(-\sqrt{4A^2 + B^2} + 2A^2 C_{fq} + B\right)\right]^{-1}\right]^{\frac{1}{2}} \cdot$$
$$\left[1 - 2\left(A^2 C_{fq}^2 + BC_{fq} - 1\right)\left[\left(C_{fq} + \frac{\tilde{u}}{u_\infty}\right)\left(\sqrt{4A^2 + B^2} + 2A^2 C_{fq} + B\right)\right]^{-1}\right]^{\frac{1}{2}} \tag{3.93}$$
$$\left[\left(A^2 C_{fq}^2 + BC_{fq} - 1\right)\left(\sqrt{4A^2 + B^2} - 2A^2 C_{fq} - B\right)^{-1}\right]^{-\frac{1}{2}} \cdot$$
$$\left[-\left(C_{fq} + \frac{\tilde{u}}{u_\infty}\right)\left(A^2\left(\frac{\tilde{u}}{u_\infty}\right)^2 - B\frac{\tilde{u}}{u_\infty} - 1\right)\right]^{-\frac{1}{2}}$$

$$\mathcal{B} = \left[2\left(A^2 C_{fq}^2 + BC_{fq} - 1\right)\right]^{\frac{1}{2}}$$
$$\left[\left(C_{fq} + \frac{\tilde{u}}{u_\infty}\right)\left(-2C_{fq}A^2 - B + \sqrt{4A^2 + B^2}\right)\right]^{-\frac{1}{2}} \tag{3.94}$$

$$k^2 = \left[2C_{fq}A^2 + B - \sqrt{4A^2 + B^2}\right]\left[2C_{fq}A^2 + B + \sqrt{4A^2 + B^2}\right]^{-1} \tag{3.95}$$

$$m\left(\frac{\tilde{u}}{u_\infty}\right) = i\,\sinh^{-1}\left(\mathcal{B}\left(\frac{\tilde{u}}{u_\infty}\right)\right) \tag{3.96}$$

$$C_{fq} = \frac{1}{2}\frac{c_f}{c_q} \tag{3.97}$$

where $F\left[m\left(\frac{\tilde{u}}{u_\infty}\right)|k^2\right]$ is the elliptic integral of the first kind and $i = \sqrt{-1}$.

### 3.2.3 Validity of Van Driest transformation

The Van Driest velocity transformation (3.36) for a flat plate zero pressure gradient turbulent boundary layer is

$$\int_{u_1}^{u} \sqrt{\frac{\overline{\rho}}{\rho_w}}\, du = \frac{u_\tau}{\kappa}\log\frac{yu_\tau}{\nu_w} + Cu_\tau$$

It is important to recall that the Van Driest transformation arises from two assumptions:

1. Constant stress and constant energy flux layer in the region where molecular viscosity and thermal conductivity are negligible (i.e., above the viscous sublayer). These are (3.26) and (3.27).
2. Mixing length model for turbulent eddy viscosity (3.31) and (3.32).

Five important observations may be made:

1. The intent of the Van Driest transformation is to collapse the mean streamwise velocity profile *outside the viscous sublayer* to a single expression (3.56) valid for a range of Mach numbers and surface heat flux.
2. There is no assumption in (3.36) that the wall is adiabatic. Indeed, the analysis of Van Driest included wall heat transfer resulting in the expression (3.55) where the constant $B$ defined in (3.42) specifically incorporates wall heat transfer. However, the inclusion of heat transfer in (3.55) required the additional assumption that a Crocco–Busemann relationship (3.39) holds across the *entire* boundary layer.
3. The mixing length model (3.31) and (3.32) used by Van Driest 1951 embodies the concept (later explicitly stated by Morkovin 1962) that the effects of compressibility in turbulent boundary layers is principally due to the variation in mean density, i.e., the incompressible mixing length model can be extended to compressible flow by simply allowing for the local density $\overline{\rho}$ in (3.7).
4. It is the mixing length eddy viscosity model equations (3.31) and (3.32) that results in the $\sqrt{\overline{\rho}/\rho_w}$ factor in (3.36).
5. The integral in (3.36) does *not* extend to the wall. To use the definition (3.36) to calculate the Van Driest transformed velocity from experiment, DNS or LES would require the identification of $u_l$ (e.g., the velocity at the location where, say, $\mu\partial\tilde{u}/\partial y$ is one percent of the Reynolds shear stress $-\overline{\rho u''v''}$ at the same location).
6. The constant $C$ depends upon the value of $u_l$ in (3.36) and hence upon the structure of the viscous sublayer. In other words, $C$ depends upon the wall temperature (e.g., adiabatic or isothermal). This is discussed in more detail in section 3.2.4.

Assessment of the accuracy of the Van Driest transformation (i.e., the capability of collapsing the mean streamwise velocity profile $\tilde{u}$ to the expression (3.56) where $u_{VD}$ is defined by (3.55) for a range of Mach numbers and wall heat transfer) can be performed using experiment, DNS, and LES. Details are presented in chapters 4 and 6.

### 3.2.4 The other Van Driest transformation

With the advent of DNS and LES, it has become commonplace to evaluate a *modified* Van Driest transformed velocity

$$u_{VD} = \int_{o}^{u} \sqrt{\frac{\overline{\rho}}{\rho_w}}\, du \tag{3.98}$$

where the lower limit of the integral in (3.36) is *arbitrarily* taken to be $u_l = 0$. Recent examples include Lagha *et al* (2011), Patel *et al* (2016), Trettel and Larsson (2016), Zhang *et al* (2017), Huang *et al* (2020), and Volpiani *et al* (2020). Recall that the factor $\sqrt{\bar{\rho}/\rho_w}$ arises from assumptions (1) and (2) in section 3.2.3. In other words, the factor $\sqrt{\bar{\rho}/\rho_w}$ arises due to the restriction of the integral to the layer where the molecular viscous shear stress is negligible, and the assumption of a mixing layer eddy viscosity model. Consequently, the arbitrary assumption that $u_l = 0$ in (3.37) is not justified.

### 3.2.5 Trettel and Larsson (2016)

Trettel and Larsson (2016) developed a generalization of the compressible law of the wall. Consider a compressible zero pressure gradient turbulent boundary layer with mean velocity $\bar{u}$, density $\bar{\rho}$, and dynamic molecular viscosity $\bar{\mu}$. All three mean quantities depend upon the wall normal coordinate $y$. Consider a transformed ('incompressible') zero pressure gradient turbulent boundary layer with mean velocity $\overline{U}$ depending upon a transformed coordinate $Y$. The transformed velocity has density $\rho_w$ and dynamic molecular viscosity $\mu_w$ identical to the wall values of the compressible boundary layer. The compressible and transformed velocity profiles are assumed to have the same local mean wall shear stress $\tau_w$.

From dimensional analysis (Bradshaw 1994) and assuming a logarithmic law exists in a region above the viscous sublayer,

$$\frac{d\bar{u}}{dy} = \frac{1}{\kappa}\frac{1}{y}\sqrt{\frac{\tau_w}{\bar{\rho}}} \tag{3.99}$$

$$\frac{d\overline{U}}{dY} = \frac{1}{\kappa}\frac{1}{Y}\sqrt{\frac{\tau_w}{\rho_w}} \tag{3.100}$$

where $\kappa$ is von Kármán's constant. Dividing (3.100) by (3.99) yields

$$\frac{d\overline{U}}{d\bar{u}} = \frac{y}{Y}\sqrt{\frac{\bar{\rho}}{\rho_w}}\frac{dY}{dy} \tag{3.101}$$

Equations (3.99) to (3.101) include as a special case of the Van Driest transformation if it is assumed that $Y = y$. From (3.101)

$$\overline{U} = \int_{\bar{u}_l}^{\bar{u}} \sqrt{\frac{\bar{\rho}}{\rho_w}}\, d\bar{u} \tag{3.102}$$

which is identical to (3.37). Integrating (3.100),

$$\int_{\bar{u}_l}^{\bar{u}} \sqrt{\frac{\bar{\rho}}{\rho_w}}\, d\bar{u} = \frac{u_\tau}{\kappa}\log\frac{yu_\tau}{\nu_w} + Cu_\tau \tag{3.103}$$

which is (3.36).

Trettel and Larsson (2016) furthermore assume that total shear stress in the constant stress layer extending from the wall to the logarithmic region is the same for the transformed and incompressible cases

$$\bar{\mu}\frac{d\bar{u}}{dy} - \overline{\rho u''v''} = \mu_w \frac{d\bar{U}}{dY} - \rho_w \overline{u'v'} \tag{3.104}$$

and apply Morkovin scaling to the turbulent shear stress

$$\overline{\rho u''v''} = \rho_w \overline{u'v'} \tag{3.105}$$

and thus from (3.104)

$$\bar{\mu}\frac{d\bar{u}}{dy} = \mu_w \frac{d\bar{U}}{dY} \tag{3.106}$$

and therefore

$$\frac{d\bar{U}}{dY} = \frac{\bar{\mu}}{\mu_w}\frac{d\bar{u}}{dy} \tag{3.107}$$

Furthermore from (3.101)

$$\frac{d\bar{U}}{dY} = \frac{y}{Y}\sqrt{\frac{\bar{\rho}}{\rho_w}}\frac{d\bar{u}}{dy} \tag{3.108}$$

and using (3.107)

$$Y = \frac{\mu_w}{\bar{\mu}}\sqrt{\frac{\bar{\rho}}{\rho_w}}\, y \tag{3.109}$$

and therefore

$$\frac{dY}{dy} = \frac{\mu_w}{\bar{\mu}}\sqrt{\frac{\bar{\rho}}{\rho_w}}\left[1 + \frac{1}{2}\frac{1}{\bar{\rho}}\frac{d\bar{\rho}}{dy}y - \frac{1}{\bar{\mu}}\frac{d\bar{\mu}}{dy}y\right] \tag{3.110}$$

Using (3.107)

$$\frac{d\bar{U}}{d\bar{u}} = \sqrt{\frac{\bar{\rho}}{\rho_w}}\left[1 + \frac{1}{2}\frac{1}{\bar{\rho}}\frac{d\bar{\rho}}{dy}y - \frac{1}{\bar{\mu}}\frac{d\bar{\mu}}{dy}y\right] \tag{3.111}$$

Denote

$$u_{TL} = \bar{U} \tag{3.112}$$

then

$$u_{TL} = \int_0^{\bar{u}} \sqrt{\frac{\bar{\rho}}{\rho_w}}\left[1 + \frac{1}{2}\frac{1}{\bar{\rho}}\frac{d\bar{\rho}}{dy}y - \frac{1}{\bar{\mu}}\frac{d\bar{\mu}}{dy}y\right] d\bar{u} \tag{3.113}$$

Note that the lower limit of the integral is the wall in contrast to (3.37). The nondimensional transformed velocity $u_{\mathrm{TL}}/u_\tau$ is a function of $Y^+$ where

$$Y^+ \equiv \frac{Yu_\tau}{\nu_\mathrm{w}} \tag{3.114}$$

and thus

$$Y^+ = \frac{\mu_\mathrm{w}}{\bar{\mu}}\sqrt{\frac{\bar{\rho}}{\rho_\mathrm{w}}}\, y^+ \tag{3.115}$$

where $y^+$ is the dimensionless wall distance in the Van Driest transformed velocity

$$y^+ \equiv \frac{yu_\tau}{\nu_\mathrm{w}} \tag{3.116}$$

The transformation was examined for supersonic channel and boundary layers at a range of Reynolds numbers and rates of wall heat transfer and produced excellent collapse of the mean velocity profiles (Trettel and Larsson 2016). Huang *et al* (2020) observed that the transformation yielded better collapse of the mean velocity profile in the viscous sublayer for hypersonic cold wall boundary layers compared to the Van Driest transformation but exhibited similar variation in the value of $C$.

### 3.2.6 Volpiani *et al* (2020)

Volpiani *et al* (2020) developed a model for transforming the mean velocity profile in a zero pressure gradient flat plate compressible turbulent boundary layer to the incompressible profile (3.12) with $C = 5.2$, and also matching the viscous sublayer profile (3.14). A transformed wall normal distance $y_\mathrm{V}$ and streamwise velocity $\tilde{u}_\mathrm{V}$ were defined by

$$y_\mathrm{V} = \int_\mathrm{o}^{y} \underbrace{\left(\frac{\bar{\rho}}{\rho_\mathrm{w}}\right)^{\frac{1}{2}}\left(\frac{\bar{\mu}}{\mu_\mathrm{w}}\right)^{-\frac{3}{2}}}_{f} dy \tag{3.117}$$

$$\tilde{u}_\mathrm{V} = \int_\mathrm{o}^{u} \underbrace{\left(\frac{\bar{\rho}}{\rho_\mathrm{w}}\right)^{\frac{1}{2}}\left(\frac{\bar{\mu}}{\mu_\mathrm{w}}\right)^{-\frac{1}{2}}}_{g} du \tag{3.118}$$

The scaling functions were determined as follows. First, assuming universality of the viscous sublayer or Morkovin-scaled Reynolds shear stress (Trettel and Larsson 2016) yields

$$\left(\frac{\bar{\mu}}{\mu_\mathrm{w}}\right) f = g \tag{3.119}$$

Second, the scaling functions were assumed to satisfy

$$f = \left(\frac{\overline{\rho}}{\rho_w}\right)^b \left(\frac{\overline{\mu}}{\mu_w}\right)^{-a} \tag{3.120}$$

$$g = \left(\frac{\overline{\rho}}{\rho_w}\right)^b \left(\frac{\overline{\mu}}{\mu_w}\right)^{1-a} \tag{3.121}$$

which satisfies (3.119). Third, the DNS results of four flat plate compressible turbulent boundary layers with Mach numbers between 2.3 and 5.84 and wall temperature ratios $T_w/T_{aw} = 0.25$ to 1.9 were used to determine the values of $a$ and $b$ which yield the lowest error in comparison with the incompressible law of the wall in the region $40 \leqslant y^+ \leqslant 100$. The resultant values are $a = \frac{3}{2}$ and $b = \frac{1}{2}$.

The transformed mean velocity profile for three DNS cases at Mach 2.3 with $T_w/T_{aw} = 0.5$ to 1.9 display close agreement with the incompressible profile in the viscous sublayer, logarithmic, and wake regions (Volpiani *et al* 2020). Similar agreement was observed for other DNS cases with Mach number up to 13.64 (Volpiani *et al* 2020). The dimensionless streamwise $\overline{\rho}\widetilde{u''u''}/\tau_w$, wall normal $\overline{\rho}\widetilde{v''v''}/\tau_w$, and spanwise $\overline{\rho}\widetilde{w''w''}/\tau_w$ turbulent stresses versus $y_V$ display close agreement with the corresponding incompressible profiles (Volpiani *et al* 2020). However, the dimensionless Reynolds shear stress $-\overline{\rho}\widetilde{u''v''}/\tau_w$ versus $y_V$ shows significant Mach number dependence for the cases considered.

### 3.2.7 Griffin *et al* (2021)

Griffin *et al* (2021) developed a model for transforming the mean velocity profile in a zero pressure gradient flat plate compressible turbulent boundary layer to the incompressible profile. Following Huang *et al* (1995), the semilocal length $\ell_l$ and velocity $u_{\tau l}$ scales are introduced:

$$\ell_l = \frac{\nu(y)}{u_{\tau l}} \tag{3.122}$$

$$u_{\tau l} = \sqrt{\frac{\tau_w}{\overline{\rho}(y)}} \tag{3.123}$$

where $\nu(y)$ and $\overline{\rho}(y)$ are the local mean molecular kinematic viscosity and density, respectively. The dimensionless total shear stress $\tau^+$ is the sum of the viscous (V) and Reynolds (R) shear stresses

$$\tau^+ = \tau_V^+ + \tau_R^+ \tag{3.124}$$

where

$$\tau_V^+ = \frac{1}{\tau_w}\mu\frac{\partial\tilde{u}}{\partial y} \tag{3.125}$$

$$\tau_R^+ = \frac{1}{\tau_W}(-\overline{\rho u''v''}) \tag{3.126}$$

Note that

$$\tau_V^+ = \mu^+ \frac{\partial \tilde{u}^+}{\partial y^+} \tag{3.127}$$

where $\mu^+ = \mu/\mu_W$, $\tilde{u}^+ = \tilde{u}/u_\tau$, and $y^+ = yu_\tau/\nu_W$. A scaled wall normal distance $y^*$ is defined by

$$y^* = \frac{y}{\ell_l} \tag{3.128}$$

Note that $y^* = y^+/\ell_l^+$ where

$$\ell_l^+ = \frac{\mu}{\mu_W}\sqrt{\frac{\rho_W}{\bar{\rho}}} \tag{3.129}$$

A dimensionless 'total' mean shear $S_T^+$ is defined by

$$\tau^+ = S_T^+\left(1 + \frac{\tau_R^+}{S_R^+}\right) \tag{3.130}$$

where

$$S_R^+ = \sqrt{\frac{\rho_W}{\bar{\rho}}}\frac{\partial \tilde{u}^+}{\partial y^+} \tag{3.131}$$

Using (3.123),

$$S_T = \frac{\tau^+ S_R^+}{\tau^+ + S_R^+ - \tau_V^+} \tag{3.132}$$

The transformed mean velocity $u_G$ is defined as

$$u_G(y^*) = \int_0^{y^*} S_T \, dy^* \tag{3.133}$$

In the constant stress layer region between the surface and the outer limit of the logarithmic region, $\tau^+ \approx 1$, and thus

$$S_T = \frac{S_R^+}{1 + S_R^+ - \tau_V^+} \tag{3.134}$$

where $S_R^+$ and $\tau_V^+$ are defined by (3.131) and (3.127), respectively. The transformation (3.133) was found to map the mean velocity profiles for both adiabatic and diabatic boundary layers to the incompressible profile in both the viscous sublayer and the law of the wall with $C \approx 5.2$ in the latter.

## 3.3 Mean velocity–mean temperature relations

### 3.3.1 Crocco–Busemann relation

A relationship between the mean total enthalpy (or total temperature) and mean velocity in a zero pressure gradient turbulent boundary layer can be developed from the Favre-averaged Navier–Stokes equations (section 2.2.2) under certain restrictions (Walz 1962, Crocco 1963, Rotta 1965). In the boundary layer approximation, the conservation of streamwise momentum and the energy equation become

$$\bar{\rho}\tilde{u}\frac{\partial\tilde{u}}{\partial x} + \bar{\rho}\tilde{v}\frac{\partial\tilde{u}}{\partial y} = -\frac{d\bar{p}}{dx} + \frac{\partial}{\partial y}\left(-\overline{\rho u''v''} + \tau_{xy}\right) \tag{3.135}$$

$$\bar{\rho}\tilde{u}\frac{\partial\tilde{H}}{\partial x} + \bar{\rho}\tilde{v}\frac{\partial\tilde{H}}{\partial y} = \frac{\partial}{\partial y}\left[-c_{\mathrm{p}}\overline{\rho T''v''} - \bar{q}_y - \overline{\rho u''v''}\tilde{u}+\tau_{xy}\tilde{u}\right] \tag{3.136}$$

Assuming an eddy viscosity model for the Reynolds stress (2.52) and turbulent heat flux (2.53),

$$\bar{\rho}\tilde{u}\frac{\partial\tilde{u}}{\partial x} + \bar{\rho}\tilde{v}\frac{\partial\tilde{u}}{\partial y} = -\frac{d\bar{p}}{dx} + \frac{\partial}{\partial y}\left[(\mu_{\mathrm{t}} + \mu)\frac{\partial\tilde{u}}{\partial y}\right] \tag{3.137}$$

$$\bar{\rho}\tilde{u}\frac{\partial\tilde{H}}{\partial x} + \bar{\rho}\tilde{v}\frac{\partial\tilde{H}}{\partial y} = \frac{\partial}{\partial y}\left[(k_{\mathrm{t}} + k)\frac{\partial\tilde{T}}{\partial y} + (\mu_{\mathrm{t}} + \mu)\frac{\partial\tilde{u}}{\partial y}\tilde{u}\right] \tag{3.138}$$

Since $\tilde{v} \ll \tilde{u}$ and assuming $M_k \ll \tilde{M}$,

$$\tilde{H} \approx c_{\mathrm{p}}\tilde{T}+\frac{1}{2}\tilde{u}^2 \tag{3.139}$$

the energy equation becomes

$$\bar{\rho}\tilde{u}\frac{\partial\tilde{H}}{\partial x} + \bar{\rho}\tilde{v}\frac{\partial\tilde{H}}{\partial y} = \frac{\partial}{\partial y}\left\{\left(\frac{\mu_{\mathrm{t}}}{Pr_{\mathrm{t}}} + \frac{\mu}{Pr}\right)\frac{\partial\tilde{H}}{\partial y} + \left[\mu_{\mathrm{t}}\left(1 - \frac{1}{Pr_{\mathrm{t}}}\right) + \mu\left(1 - \frac{1}{Pr}\right)\right]\frac{\partial\tilde{u}}{\partial y}\tilde{u}\right\} \tag{3.140}$$

Hence if $d\bar{p}/dx = 0$ and assuming $Pr_{\mathrm{t}} = Pr = 1$, the equations become

$$\bar{\rho}\tilde{u}\frac{\partial\tilde{u}}{\partial x} + \bar{\rho}\tilde{v}\frac{\partial\tilde{u}}{\partial y} = \frac{\partial}{\partial y}\left[(\mu_{\mathrm{t}} + \mu)\frac{\partial\tilde{u}}{\partial y}\right] \tag{3.141}$$

$$\bar{\rho}\tilde{u}\frac{\partial\tilde{H}}{\partial x} + \bar{\rho}\tilde{v}\frac{\partial\tilde{H}}{\partial y} = \frac{\partial}{\partial y}\left[(\mu_{\mathrm{t}} + \mu)\frac{\partial\tilde{H}}{\partial y}\right] \tag{3.142}$$

The equation for $\tilde{H}$ is the same as the equation for $\tilde{u}$. The solution is

$$\frac{\tilde{H}-H_{\mathrm{w}}}{H_{\infty}-H_{\mathrm{w}}}=\frac{\tilde{u}}{u_{\infty}} \qquad dp/dx=0 \text{ and } Pr_{\mathrm{t}}=Pr=1 \tag{3.143}$$

or equivalently, for constant specific heats,

$$\frac{\tilde{T}_{\mathrm{t}}-T_{\mathrm{w}}}{T_{t_{\infty}}-T_{\mathrm{w}}}=\frac{\tilde{u}}{u_{\infty}} \qquad dp/dx=0 \text{ and } Pr_{\mathrm{t}}=Pr=1 \tag{3.144}$$

where $_{\infty}$ and $_{\mathrm{w}}$ indicate the edge of the boundary layer and wall, respectively. This is known as the *Crocco–Busemann relation* in recognition of the analogous result obtained for compressible laminar boundary layers by Busemann (1935) and Crocco (1932). Furthermore,

$$\tilde{T}=T_{\mathrm{w}}+\left(T_{\infty}+\frac{1}{2c_{\mathrm{p}}}u_{\infty}^{2}-T_{\mathrm{w}}\right)\frac{\tilde{u}}{u_{\infty}}-\frac{\tilde{u}^{2}}{2c_{\mathrm{p}}} \qquad dp/dx=0 \text{ and } Pr_{\mathrm{t}}=Pr=1 \tag{3.145}$$

and thus the adiabatic wall temperature is[3]

$$T_{\mathrm{aw}}=T_{\infty}\left[1+\frac{(\gamma-1)}{2}M_{\infty}^{2}\right] \qquad dp/dx=0 \text{ and } Pr_{\mathrm{t}}=Pr=1 \tag{3.146}$$

and hence an equivalent form of the Crocco–Busemann relation is

$$\tilde{T}=T_{\mathrm{w}}+(T_{\mathrm{aw}}-T_{\mathrm{w}})\left(\frac{\tilde{u}}{u_{\infty}}\right)+(T_{\infty}-T_{\mathrm{aw}})\left(\frac{\tilde{u}}{u_{\infty}}\right)^{2} \qquad dp/dx=0 \text{ and } Pr_{\mathrm{t}}=Pr=1 \tag{3.147}$$

### 3.3.2 Walz's relation

Walz (1969) developed a more general temperature–velocity relation based upon the method of Van Driest (1959) for laminar flow. Consider the Favre-averaged zero pressure gradient compressible turbulent boundary layer equations

$$\frac{\partial\bar{\rho}\tilde{u}}{\partial x}+\frac{\partial\bar{\rho}\tilde{v}}{\partial y}=0 \tag{3.148}$$

$$\bar{\rho}\tilde{u}\frac{\partial\tilde{u}}{\partial x}+\bar{\rho}\tilde{v}\frac{\partial\tilde{u}}{\partial y}=\frac{\partial\tau}{\partial y} \tag{3.149}$$

$$\bar{\rho}\tilde{u}c_{\mathrm{p}}\frac{\partial\tilde{T}}{\partial x}+\bar{\rho}\tilde{v}c_{\mathrm{p}}\frac{\partial\tilde{T}}{\partial y}=\tau\frac{\partial\tilde{u}}{\partial y}+\frac{\partial q}{\partial y} \tag{3.150}$$

[3] Differentiate (3.145) with respect to $y$ and set equal to zero at the wall assuming $\partial\tilde{u}/\partial y\neq0$ at the wall.

where $\tau$ and $q$ are the total shear stress and total heat transfer assuming an eddy viscosity model for the turbulent shear stress and heat flux

$$\tau = (\mu + \mu_t)\frac{\partial \tilde{u}}{\partial y} \tag{3.151}$$

$$q = (k + k_t)\frac{\partial \tilde{T}}{\partial y} \tag{3.152}$$

Consider a transformation from independent variables $(x, y)$ to independent variables $(\tilde{x}, \tilde{u})$ where $\tilde{x} = x$. Define a 'total' Prandtl number

$$Pr_t = \frac{(\mu + \mu_t)c_p}{(k + k_t)} \tag{3.153}$$

Assuming $\partial\tilde{u}/\partial x \approx 0$ and $\partial\tilde{T}/\partial x \approx 0$, the energy equation becomes

$$\frac{1}{Pr_t}\frac{\partial^2 \hat{T}}{\partial \hat{u}^2} + \left(\frac{1}{Pr_t} - 1\right)\frac{1}{\tau}\frac{\partial \tau}{\partial \hat{u}}\frac{\partial \hat{T}}{\partial \hat{u}} + \frac{u_\infty^2}{c_p T_\infty} = 0 \tag{3.154}$$

where

$$\hat{T} = \frac{\tilde{T}}{T_\infty} \tag{3.155}$$

$$\hat{u} = \frac{\tilde{u}}{u_\infty} \tag{3.156}$$

Since (3.154) is linear in $\hat{T}$, a formal integration yields

$$\hat{T} = 1 + \frac{(T_r - T_w)}{T_\infty}(f_1 - 1) + \frac{(\gamma - 1)}{2} r M_\infty^2 (1 - f_2) \tag{3.157}$$

where $T_r$ is the recovery temperature (also denoted the adiabatic wall temperatures $T_{aw}$)

$$T_r = T_\infty\left[1 + \frac{(\gamma - 1)}{2} r M_\infty^2\right] \tag{3.158}$$

and the recovery factor $r$ is

$$r = 2Pr_t \int_o^1 \left(\frac{\tau}{\tau_w}\right)^{Pr_t - 1}\left[\int_o^{\hat{u}} \left(\frac{\tau}{\tau_w}\right)^{1 - Pr_t} d\hat{u}\right] d\hat{u} \tag{3.159}$$

Define the dimensionless total shear stress

$$\hat{\tau} = \frac{\tau}{\tau_w} \tag{3.160}$$

Also

$$f_1 = \left[\int_0^{\hat{u}} \hat{\tau}^{Pr_t-1} d\hat{u}\right]\left[\int_0^1 \hat{\tau}^{Pr_t-1} d\hat{u}\right]^{-1} \tag{3.161}$$

$$f_2 = \left\{\int_0^{\hat{u}} \hat{\tau}^{Pr_t-1}\left[\int_0^{\hat{u}} \hat{\tau}^{1-Pr_t} d\hat{u}\right] d\hat{u}\right\} \cdot \left\{\int_0^1 \hat{\tau}^{Pr_t-1}\left[\int_0^{\hat{u}} \hat{\tau}^{1-Pr_t} d\hat{u}\right] d\hat{u}\right\}^{-1} \tag{3.162}$$

Note that if $Pr_t = 1$, then

$$r = 1 \tag{3.163}$$

$$f_1 = \hat{u} \tag{3.164}$$

$$f_2 = \hat{u}^2 \tag{3.165}$$

and therefore

$$\hat{T} = 1 + \frac{(T_r - T_w)}{T_\infty}(\hat{u} - 1) + \frac{(\gamma - 1)}{2} M_\infty^2 (1 - \hat{u}^2) \qquad \text{for } Pr_t = 1 \tag{3.166}$$

or equivalently

$$\tilde{T} = T_w + (T_r - T_w)\frac{\tilde{u}}{u_\infty} - \frac{\tilde{u}^2}{2c_p} \qquad \text{for } Pr_t = 1 \tag{3.167}$$

which is identical to (3.145) using (3.158) and (3.163). Furthermore,

$$T_r = T_\infty\left[1 + \frac{(\gamma - 1)}{2} M_\infty^2\right] \qquad \text{for } Pr_t = 1 \tag{3.168}$$

and

$$\frac{T_t - T_w}{T_{t_\infty} - T_w} = \frac{\tilde{u}}{u_\infty} \tag{3.169}$$

Van Driest (1959) found that (3.164) and (3.165), obtained assuming $Pr_t = 1$, are reasonable approximations for $f_1$ and $f_2$. Arbitrarily assuming $Pr_t = 1$ in the integrand for the recovery factor $r$ in (3.159) but not in the factor outside the integral yields

$$r = Pr_t \tag{3.170}$$

Walz's relation becomes

$$\hat{T} = 1 + \frac{(T_r - T_w)}{T_\infty}(\hat{u} - 1) + \frac{(\gamma - 1)}{2} r M_\infty^2 (1 - \hat{u}^2) \qquad \text{where } r = Pr_t \tag{3.171}$$

or equivalently

$$\tilde{T} = T_{\rm w} + (T_{\rm r} - T_{\rm w})\left(\frac{\tilde{u}}{u_\infty}\right) + (T_\infty - T_{\rm r})\left(\frac{\tilde{u}}{u_\infty}\right)^2 \tag{3.172}$$

Furthermore

$$T_{\rm r} = T_\infty\left[1 + \frac{(\gamma - 1)}{2}Pr_{\rm t}M_\infty^2\right] \tag{3.173}$$

and

$$\frac{T_{\rm t} - T_{\rm w}}{T_{t_\infty} - T_{\rm w}} = A\left(\frac{\tilde{u}}{u_\infty}\right) + B\left(\frac{\tilde{u}}{u_\infty}\right)^2 \tag{3.174}$$

where

$$A = \left[1 + \frac{(\gamma - 1)}{2}rM_\infty^2 - \frac{T_{\rm w}}{T_\infty}\right]\left[1 + \frac{(\gamma - 1)}{2}M_\infty^2 - \frac{T_{\rm w}}{T_\infty}\right]^{-1} \tag{3.175}$$

$$B = \left[\frac{(\gamma - 1)}{2}M_\infty^2(1 - r)\right]\left[1 + \frac{(\gamma - 1)}{2}M_\infty^2 - \frac{T_{\rm w}}{T_\infty}\right]^{-1} \tag{3.176}$$

Note that (3.172), (3.173), and (3.174) differ from the corresponding Crocco–Busemann relations (3.147), (3.146), and (3.144), respectively, by the presence of the recovery factor $r = Pr_{\rm t}$ and reduce to (3.147), (3.146), and (3.144) when $r = 1$.

### 3.3.3 Zhang *et al* relation

Zhang *et al* (2014) developed a mean velocity–mean temperature relation as part of their study on a generalized Reynolds analogy for compressible turbulent flows. Consider a flat plate zero pressure gradient turbulent boundary layer. Assuming $Pr = Pr_{\rm t} = 1$, and in the boundary layer approximation, equation (3.143) yields

$$\tilde{H} - H_{\rm w} = U_{\rm w}\tilde{u} \tag{3.177}$$

where

$$\tilde{H} = c_{\rm p}\tilde{T} + \frac{1}{2}\tilde{u}^2 \tag{3.178}$$

to a first approximation and

$$U_{\rm w} = \frac{H_\infty - H_{\rm w}}{u_\infty} \tag{3.179}$$

Also

$$U_{\rm w} = c_{\rm p}\frac{\partial\tilde{T}}{\partial\tilde{u}}\bigg|_{\rm w} \tag{3.180}$$

obtained by differentiating (3.177) with respect to $\tilde{u}$ and evaluating at the wall. In deriving (3.177) and (3.178), it is assumed that $\tilde{v} \ll \tilde{u}$ and $M_k \ll \overline{M}$. To the same approximation, $\tilde{H} \approx \overline{H}, \tilde{u} \approx \bar{u}$ and thus (3.177), (3.178), and (3.180) can be replaced by

$$\overline{H} - H_w = U_w \bar{u} \tag{3.181}$$

$$\overline{H} = c_p \overline{T} + \frac{1}{2} \bar{u}^2 \tag{3.182}$$

$$U_w = c_p \frac{\partial \overline{T}}{\partial \bar{u}} \bigg|_w \tag{3.183}$$

An equivalent expression for $U_w$ is

$$U_w = -Pr \frac{q_w}{\tau_w} \tag{3.184}$$

Zhang *et al* (2014) generalize (3.181) as

$$\overline{H}_g - H_w = U_w \bar{u} \tag{3.185}$$

where $H_g$ is denoted the *general recovery total enthalpy* and defined by

$$H_g = c_p T + \frac{1}{2} r_g u^2 \tag{3.186}$$

and $r_g$ is the *general recovery factor*. Note that $H_g$, $T$, and $u$ in (3.186) are the *instantaneous* values of the general recovery total enthalpy, static temperature, and streamwise velocity, respectively. To a first approximation,

$$\overline{H}_g = c_p \overline{T} + \frac{1}{2} r_g \bar{u}^2 \tag{3.187}$$

$$H_g' = c_p T' + r_g \bar{u} u' \tag{3.188}$$

Zhang *et al* (2014) further assume

$$H_g' + c_p \phi' = U_w u' \tag{3.189}$$

where $c_p \phi'$ represents the instantaneous difference between $H_g'$ and $U_w u'$. Combining (3.188) and (3.189)

$$c_p T' + r_g \bar{u} u' + c_p \phi' = U_w u' \tag{3.190}$$

Multiplying (3.190) by $(\rho v)'$ and averaging yield

$$r_g = \frac{c_p}{\bar{u}} \left( \frac{\partial \overline{T}}{\partial \bar{u}} \bigg|_w - \frac{1}{Pr_e} \frac{\partial \overline{T}}{\partial \bar{u}} \right) \tag{3.191}$$

where $Pr_e$ is the *effective turbulent Prandtl number* defined by

$$Pr_e = \frac{Pr_t}{1 + \epsilon} \tag{3.192}$$

where $Pr_t$ is *turbulent Prandtl number*

$$Pr_t = \frac{\overline{(\rho v)'u'}}{\overline{(\rho v)'T'}} \frac{\partial \overline{T}/\partial y}{\partial \overline{u}/\partial y} \tag{3.193}$$

and

$$\epsilon = \frac{\overline{(\rho v)'\phi'}}{\overline{(\rho v)'T'}} \tag{3.194}$$

Substitution of (3.191) in (3.185) yields

$$\overline{T} - \frac{\overline{u}}{2}\left( \left.\frac{\partial \overline{T}}{\partial \overline{u}}\right|_{\mathrm{w}} + \frac{1}{Pr_e} \frac{\partial \overline{T}}{\partial \overline{u}} \right) - T_{\mathrm{w}} = 0 \tag{3.195}$$

Zhang *et al* (2014) observed that $Pr_e \approx 1$ for the DNS of Pirozzoli and Bernardini (2011) and their own DNS results. Therefore, (3.195) becomes

$$\frac{\overline{u}}{2} \frac{\partial \overline{T}}{\partial \overline{u}} - T = -\frac{\overline{u}}{2} \left.\frac{\partial \overline{T}}{\partial \overline{u}}\right|_{\mathrm{w}} - T_{\mathrm{w}} \tag{3.196}$$

Assuming $\overline{T}$ is a function of $\overline{u}$ only and writing

$$\overline{T} = T_{\mathrm{w}} + A\overline{u} + B\overline{u}^2 \tag{3.197}$$

thus

$$A = \left.\frac{d\overline{T}}{d\overline{u}}\right|_{\mathrm{w}} \tag{3.198}$$

$$B = \left[ T_\infty - T_{\mathrm{w}} - \left.\frac{d\overline{T}}{d\overline{u}}\right|_{\mathrm{w}} u_\infty \right] / u_\infty^2 \tag{3.199}$$

Therefore

$$\overline{T} = T_{\mathrm{w}} + \left.\frac{d\overline{T}}{d\overline{u}}\right|_{\mathrm{w}} \overline{u} + \left( T_\infty - T_{\mathrm{w}} - u_\infty \left.\frac{d\overline{T}}{d\overline{u}}\right|_{\mathrm{w}} \right) \left( \frac{\overline{u}}{u_\infty} \right)^2 \tag{3.200}$$

Using (3.191) and (3.200)

$$r_g = \frac{2c_{\mathrm{p}}}{u_\infty^2}(T_{\mathrm{w}} - T_\infty) - 2Pr\frac{q_{\mathrm{w}}}{u_\infty \tau_{\mathrm{w}}} \tag{3.201}$$

where

$$\left.\frac{d\overline{T}}{d\overline{u}}\right|_{w} = -\frac{Pr\, q_w}{c_p \tau_w} \tag{3.202}$$

Defining a generalized mean recovery temperature

$$T_{rg} = T_\infty + \frac{1}{2} r_g \frac{u_\infty^2}{c_p} \tag{3.203}$$

the mean temperature—mean velocity relation (3.200) becomes

$$\overline{T} = T_w + \left(T_{rg} - T_w\right)\left(\frac{\overline{u}}{u_\infty}\right) + \left(T_\infty - T_{rg}\right)\left(\frac{\overline{u}}{u_\infty}\right)^2 \tag{3.204}$$

The relationships between the generalized recovery temperature $T_{rg}$ and recovery factor $r_g$ and the corresponding recovery temperature $T_r$ and recovery factor $r$ (3.158) can be derived as follows. The recovery (adiabatic wall) temperature $T_r$ and recovery factor $r$ are related by (3.158) and written in terms of Reynolds averages:

$$T_r = T_\infty\left[1 + \frac{(\gamma - 1)}{2} r M_\infty^2\right] \tag{3.205}$$

The Reynolds Analogy Factor $s$ (section 3.4) is defined as

$$s = \frac{2St}{c_f} \tag{3.206}$$

where $St$ is the Stanton number

$$St = \frac{q_w}{\rho_e u_\infty c_p (T_w - T_r)} \tag{3.207}$$

and $c_f$ is the skin friction coefficient

$$c_f = \frac{\overline{\tau}_w}{\frac{1}{2}\rho_\infty u_\infty^2} \tag{3.208}$$

From (3.201),

$$r_g = r[sPr + (1 - sPr)\Theta] \tag{3.209}$$

where

$$\Theta = \frac{T_w - T_\infty}{T_r - T_\infty} \tag{3.210}$$

and Pr is the molecular Prandtl number. Note that for an adiabatic wall ($T_w = T_r$), $r_g = r$.

From (3.203) and (3.205), the relationship between $T_r$ and $T_{rg}$ is

$$T_{rg} = T_r + \left(\frac{r_g}{r} - 1\right) r \frac{u_\infty^2}{2c_p} \tag{3.211}$$

or

$$T_{rg} = T_r + (sPr - 1)(T_r - T_w) \tag{3.212}$$

Substituting into (3.204),

$$\overline{T} = T_w + (T_r - T_w)\left[(1 - sPr)\left(\frac{\overline{u}}{u_\infty}\right)^2 + sPr\left(\frac{\overline{u}}{u_\infty}\right)\right] + (T_\infty - T_r)\left(\frac{\overline{u}}{u_\infty}\right)^2 \tag{3.213}$$

The above expression is identical to the result obtained by Duan and Martín (2011)

$$\overline{T} = T_w + (T_r - T_w) f\left(\frac{\overline{u}}{u_\infty}\right) + (T_\infty - T_r)\left(\frac{\overline{u}}{u_\infty}\right)^2 \tag{3.214}$$

where

$$f\left(\frac{\overline{u}}{u_\infty}\right) = 0.8259\left(\frac{\overline{u}}{u_\infty}\right) + 0.1741\left(\frac{\overline{u}}{u_\infty}\right)^2 \tag{3.215}$$

with $sPr = 0.8259$ which implies $s = 1.147$ for $Pr = 0.72$.

Figure 3.3 presents a comparison of the Crocco–Busemann, Walz, and Zhang *et al* relations for three DNS cases of Zhang *et al* (2014) at Mach 4.5 corresponding to $T_w/T_{aw} = 0.215$, 0.537, and 0.945. Best agreement is achieved with the Zhang *et al* relation.

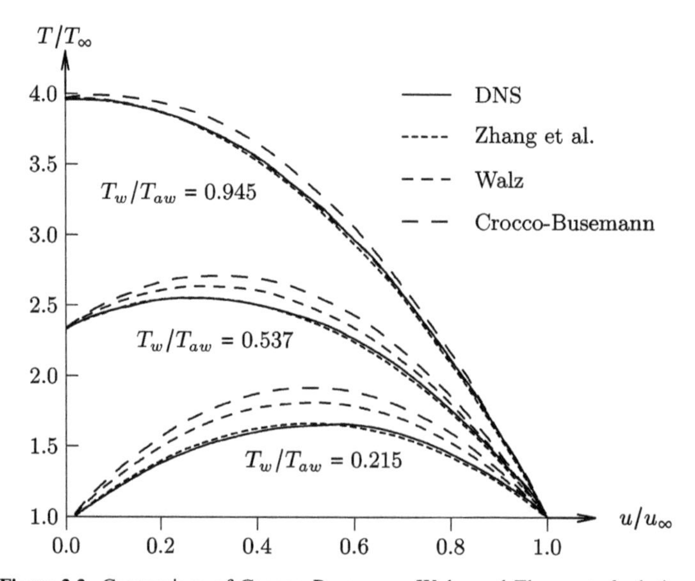


**Figure 3.3.** Comparison of Crocco–Busemann, Walz, and Zhang *et al* relations.

### 3.3.4 Hypersonic velocity–total temperature relation

Assuming a perfect gas,

$$\left(\frac{u}{u_\infty}\right)^2 = \frac{M^2 T}{M_\infty^2 T_\infty} = \frac{M^2}{M_\infty^2}\frac{T_t}{T_{t_\infty}}\frac{\left[1+\frac{1}{2}(\gamma-1)M_\infty^2\right]}{\left[1+\frac{1}{2}(\gamma-1)M^2\right]} \tag{3.216}$$

For $M \gg 1$

$$\left(\frac{u}{u_\infty}\right)^2 \approx \frac{T_t}{T_{t_\infty}} \tag{3.217}$$

and thus

$$\frac{H - H_w}{H_\infty - H_w} = \frac{(u/u_\infty)^2 - T_w/T_{t_\infty}}{1 - T_w/T_{t_\infty}} \tag{3.218}$$

For a cold wall ($T_w \ll T_{t_\infty}$) hypersonic turbulent boundary layer,

$$\frac{H - H_w}{H_\infty - H_w} \approx \left(\frac{u}{u_\infty}\right)^2 \tag{3.219}$$

Since the above relation is derived assuming perfect gas, equivalent expressions are

$$\frac{T_t - T_w}{T_{t_\infty} - T_w} = \left(\frac{u}{u_\infty}\right)^2 \tag{3.220}$$

and

$$\tilde{T} = T_w + (T_\infty - T_w)\left(\frac{\tilde{u}}{u_\infty}\right)^2 \tag{3.221}$$

Figure 3.4 shows the measurements of Holden (1991) for the boundary layer on a 6° half-angle cone for Mach numbers from 11.0 to 15.43 where $x$ is the distance of the profile upstream of the cone-flare junction in cm. The experimental data support (3.219).

## 3.4 Reynolds Analogy Factor

The Reynolds Analogy Factor represents a relationship between the Stanton number[4]

$$St = \frac{q_w}{\rho_\infty u_\infty c_p (T_w - T_{aw})} \tag{3.222}$$

[4] Note that $q_w = -\hat{k}\partial T/\partial y$. Therefore, if $T_w < T_{aw}$, then $\partial T/\partial y > 0$ and thus $q_w < 0$. Hence the Stanton number $St$ is positive.

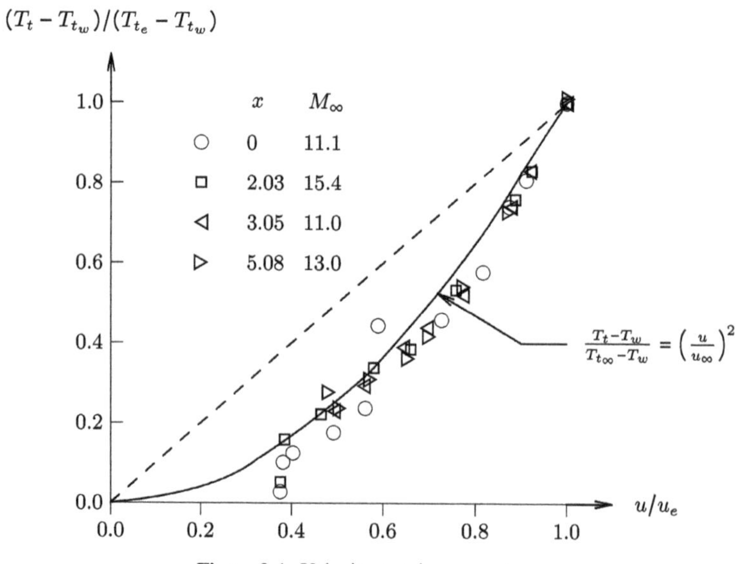


**Figure 3.4.** Velocity–total temperature.

where $T_{aw}$ is the *adiabatic wall temperature* (also denoted the *recovery temperature*) and the skin friction coefficient

$$c_f = \frac{\tau_w}{\frac{1}{2}\rho_\infty u_\infty^2} \tag{3.223}$$

where $\tau_w$ is the wall shear stress. The Reynolds Analogy Factor $s$ is defined as

$$s = \frac{2St}{c_f} \tag{3.224}$$

### 3.4.1 Osborne Reynolds

Cary (1970) attributes to Reynolds (1875) the expression

$$\frac{2St}{c_f} = 1 \tag{3.225}$$

for incompressible flow. However, this is not entirely deserved. Reynolds (1875) considered the heat transfer to an incompressible fluid moving through a length of pipe. The fluid enters at a uniform temperature $T_1$. The steady volume flow rate is $Q$, and the cross-sectional area is $A$. The mean velocity $V = Q/A$. The surface temperature of the pipe is kept fixed at $T_w \neq T_1$ throughout its length $L$. Reynolds (1875) stated that the heat transferred to the fluid per unit area per unit time is

$$H = A\Delta T + B\rho V\Delta T \tag{3.226}$$

where $\Delta T$ is 'the difference of temperature between the surface and the fluid' and '$A$ and $B$ are constants depending on the nature of the fluid'. The terms $A\Delta T$ and $B\rho V \Delta T$ correspond to heat transfer by conduction and convection, respectively.

Reynolds (1875) also stated that

> If, therefore, a fluid were forced along a fixed length of pipe which was maintained at a uniform temperature greater or less than the initial temperature of the gas, we should expect the following results.
>
> (1) Starting with a velocity zero, the gas would then acquire the same temperature as the tube
> (2) As the velocity increased the temperature at which the gas would emerge would gradually diminish, rapidly at first, but in a decreasing ratio until it would become sensibly constant and independent of the velocity. The velocity after which the temperature of the emerging gas would be sensibly constant can only be found for each particular gas by experiment; but it would seem reasonable to suppose that it would be the same as that at which the resistance offered by friction to the motion of the fluid would be sensibly proportional to the square of the velocity. It having been found both theoretically and by experiment that this resistance is connected with th diffusion of the gas by a formula:
>
> $$R = A'V + B'\rho V^2 \text{ (2)}$$
>
> And various considerations lead to the supposition that $A$ and $B$ in (1) are proportional to $A'$ and $B'$ in (2).

In the above, (1) refers to (3.226).The observation that at sufficiently large velocity the temperature of exiting gas would become independent of velocity was demonstrated by Stanton (1897).

The resistance $R$ may be interpreted as the pressure drop per unit length in fully developed pipe flow. The first term corresponds to laminar flow, and the second term to turbulent flow.

Consider incompressible fully developed turbulent flow in the pipe of diameter $D$. The pressure drop $\Delta p$ per length $L$ is

$$\Delta p = f \frac{L}{D} \frac{1}{2} \rho V^2 \tag{3.227}$$

where $f$ is the friction factor and depends upon the Reynolds number $Re = \rho VD/\mu$ and relative surface roughness $\epsilon$. Neglecting the conduction contribution to $H$ and the laminar contribution to $R$,

$$H \approx B\rho V \Delta T$$
$$R \approx B'\rho V^2$$

Thus

$$B' = \frac{1}{2}\frac{f}{D} \tag{3.228}$$

The wall shear stress is

$$\tau_w = \frac{D}{4}\frac{\Delta p}{L} \tag{3.229}$$

and defining the friction coefficient

$$c_f = \frac{\tau_w}{\frac{1}{2}\rho V^2} \tag{3.230}$$

therefore

$$c_f = \frac{1}{2}DB' \tag{3.231}$$

Define the Stanton number

$$St = \frac{H}{\rho c V \Delta T} \tag{3.232}$$

where $c$ is the specific heat of the fluid. Thus

$$\frac{2St}{c_f} = \frac{4}{cD}\frac{B}{B'} \tag{3.233}$$

Based upon the supposition that $B$ and $B'$ are proportional, the statements in Reynolds (1875) imply

$$\frac{2St}{c_f} = \text{constant} \tag{3.234}$$

but do not provide a value for the constant. Moreover, the result applies to incompressible pipe flow only.

### 3.4.2 Colburn (1933)

Colburn (1933) developed the empirical relationship for turbulent flow,

$$\frac{2St}{c_f} = Pr^{-\frac{2}{3}} \tag{3.235}$$

where $Pr$ is the laminar Prandtl number.

### 3.4.3 Van Driest

The analysis of Van Driest (section 3.2.1) also provides an expression for the Reynolds Analogy Factor. From equation (3.42),

$$B = -\frac{Pr_t q_w u_\infty}{c_p \tau_w T_w} \tag{3.236}$$

Using equations (3.44) and (3.46),

$$B = \frac{T_{aw}}{T_w} - 1 \tag{3.237}$$

From equation (3.222),

$$St = \frac{1}{Pr_t} \frac{\tau_w}{\rho_\infty u_\infty^2} \tag{3.238}$$

and thus

$$\frac{2St}{c_f} = \frac{1}{Pr_t} \tag{3.239}$$

### 3.4.4 Turbulence model

An expression for the Reynolds Analogy Factor for a compressible, zero pressure gradient boundary layer can be deduced from an asymptotic analysis of the two-equation $k - \epsilon$ turbulence model. The analysis follows the method of Saffman and Wilcox (1974) which deduced a similar result for the $k - \omega$ turbulence model. The standard $k - \epsilon$ model was developed by Jones and Launder (1972) and Launder and Sharma (1974) for incompressible flows. In the boundary layer approximation for an incompressible, flat plate, two-dimensional, steady, zero pressure gradient flow, the Reynolds-averaged equations are

$$\frac{\partial \overline{u}}{\partial x} + \frac{\partial \overline{v}}{\partial y} = 0 \tag{3.240}$$

$$\rho \overline{u} \frac{\partial \overline{u}}{\partial x} + \rho \overline{v} \frac{\partial \overline{u}}{\partial y} = \frac{\partial}{\partial y}\left(-\rho \overline{u'v'} + \overline{\tau}_{xy}\right) \tag{3.241}$$

$$\rho \overline{u} \frac{\partial \overline{k}}{\partial x} + \rho \overline{v} \frac{\partial \overline{k}}{\partial y} = -\rho \overline{u'v'} \frac{\partial \overline{u}}{\partial y} - \rho \overline{\epsilon} + \frac{\partial}{\partial y}\left[\left(\mu + \frac{\mu_T}{\sigma_k}\right)\frac{\partial \overline{k}}{\partial y}\right] \tag{3.242}$$

$$\rho \overline{u} \frac{\partial \overline{\epsilon}}{\partial x} + \rho \overline{v} \frac{\partial \overline{\epsilon}}{\partial y} = -C_{\epsilon 1}\frac{\overline{\epsilon}}{\overline{k}} \rho \overline{u'v'} \frac{\partial \overline{u}}{\partial y} - C_{\epsilon 2}\frac{\rho \overline{\epsilon}^2}{\overline{k}} + \frac{\partial}{\partial y}\left[\left(\mu + \frac{\mu_T}{\sigma_\epsilon}\right)\frac{\partial \overline{\epsilon}}{\partial y}\right] \tag{3.243}$$

where $\overline{k}$ and $\overline{\epsilon}$ are the turbulence kinetic energy per unit mass and the rate of dissipation of turbulence kinetic energy per unit mass, respectively, and the molecular and Reynolds shear stresses are

$$\overline{\tau}_{xy} = \mu \frac{\partial \overline{u}}{\partial y} \tag{3.244}$$

$$-\rho\overline{u'v'} = \mu_T \frac{\partial \overline{u}}{\partial y} \tag{3.245}$$

where the turbulent eddy viscosity is

$$\mu_T = C_\mu \, \rho \, \frac{\overline{k}^2}{\overline{\epsilon}} \tag{3.246}$$

The dimensionless constants are (Launder and Sharma 1974, Wilcox 2006)

$$C_{\epsilon 1} = 1.44, \quad C_{\epsilon 2} = 1.92, \quad C_\mu = 0.09, \quad \sigma_k = 1.0, \quad \sigma_\epsilon = 1.3 \tag{3.247}$$

There is a neighborhood of the boundary layer adjacent to the wall where the convective terms are negligible. Equation (3.241) can be integrated from the wall to obtain

$$-\rho\overline{u'v'} + \overline{\tau}_{xy} = \tau_{\rm w} \tag{3.248}$$

where $\tau_{\rm w}$ is the wall shear stress. Within this neighborhood, but sufficiently far from the wall that $\mu_T \gg \mu$,

$$-\rho\overline{u'v'} = \tau_{\rm w} \tag{3.249}$$

and the governing equations for $k$ and $\epsilon$ become

$$0 = -\rho\overline{u'v'}\frac{\partial \overline{u}}{\partial y} - \rho\overline{\epsilon} + \frac{\partial}{\partial y}\left[\frac{\mu_T}{\sigma_k}\frac{\partial \overline{k}}{\partial y}\right] \tag{3.250}$$

$$0 = -C_{\epsilon 1}\frac{\overline{\epsilon}}{\overline{k}}\rho\overline{u'v'}\frac{\partial \overline{u}}{\partial y} - C_{\epsilon 2}\frac{\rho\overline{\epsilon}^2}{\overline{k}} + \frac{\partial}{\partial y}\left[\frac{\mu_T}{\sigma_\epsilon}\frac{\partial \overline{\epsilon}}{\partial y}\right] \tag{3.251}$$

Equations (3.250) and (3.251) admit a solution

$$\overline{u} = \frac{u_\tau}{\kappa}\log\frac{y u_\tau}{\nu} + C u_\tau \tag{3.252}$$

$$\overline{k} = C_\mu^{-\frac{1}{2}} u_\tau^2 \tag{3.253}$$

$$\overline{\epsilon} = \frac{u_\tau^3}{\kappa y} \tag{3.254}$$

where $C$ is a constant, $\nu = \mu/\rho$, and the friction velocity $u_\tau$ is defined by

$$\tau_{\rm w} = \rho u_\tau^2 \tag{3.255}$$

and

$$\kappa = \sqrt{(C_{\epsilon_2} - C_{\epsilon 1})\sigma_\epsilon\sqrt{C_\mu}} \tag{3.256}$$

Equation (3.252) is the Law of the Wall (Schlichting 1968). Using equation (3.247), $\kappa = 0.43$ which is close to the experimental value $\kappa = 0.40$ (Schlichting 1968).

Following Morkovin (section 3.5), we postulate a compressible extension of the $k - \epsilon$ model for a compressible, flat plate, two-dimensional, steady, zero pressure gradient flow using Favre-averaged variables:

$$\frac{\partial \bar{\rho}\tilde{u}}{\partial x} + \frac{\partial \bar{\rho}\tilde{v}}{\partial y} = 0 \tag{3.257}$$

$$\bar{\rho}\tilde{u}\frac{\partial \tilde{u}}{\partial x} + \bar{\rho}\tilde{v}\frac{\partial \tilde{u}}{\partial y} = \frac{\partial}{\partial y}\left(-\overline{\rho u''v''} + \bar{\tau}_{xy}\right) \tag{3.258}$$

$$\frac{\partial(\bar{\rho}\tilde{e}+\bar{p})\tilde{u}}{\partial x} + \frac{\partial(\bar{\rho}\tilde{e}+\bar{p})\tilde{v}}{\partial x} = \frac{\partial}{\partial y}\left(-c_p\overline{\rho T''v''} - \bar{q}_y - \overline{\rho u''v''}\;\tilde{u}+\bar{\tau}_{xy}\tilde{u}\right) \tag{3.259}$$

$$\bar{\rho}\tilde{u}\frac{\partial \tilde{k}}{\partial x} + \bar{\rho}\tilde{v}\frac{\partial \tilde{k}}{\partial y} = -\overline{\rho u''v''}\frac{\partial \tilde{u}}{\partial y} - \bar{\rho}\tilde{\epsilon} + \frac{\partial}{\partial y}\left[\left(\mu + \frac{\mu_T}{\sigma_k}\right)\frac{1}{\bar{\rho}}\frac{\partial \bar{\rho}\tilde{k}}{\partial y}\right] \tag{3.260}$$

$$\bar{\rho}\tilde{u}\frac{\partial \tilde{\epsilon}}{\partial x} + \bar{\rho}\tilde{v}\frac{\partial \tilde{\epsilon}}{\partial y} = -C_{\epsilon 1}\frac{\tilde{\epsilon}}{\tilde{k}}\overline{\rho u''v''}\frac{\partial \tilde{u}}{\partial y} - C_{\epsilon 2}\frac{\bar{\rho}\tilde{\epsilon}^2}{\tilde{k}} + \frac{\partial}{\partial y}\left[\left(\mu + \frac{\mu_T}{\sigma_\epsilon}\right)\frac{\partial \tilde{\epsilon}}{\partial y}\right] \tag{3.261}$$

where

$$\tilde{e} = c_v\tilde{T} + \frac{1}{2}\tilde{u}^2 + \tilde{k} \tag{3.262}$$

$$\bar{q}_y = -\hat{k}\frac{\partial \tilde{T}}{\partial y} \tag{3.263}$$

where $\hat{k}$ is the molecular thermal conductivity.

Note the assumed form of the diffusion term for the turbulence kinetic energy equation. The expression

$$\frac{1}{\bar{\rho}}\frac{\partial \bar{\rho}\tilde{k}}{\partial y}$$

is chosen instead of

$$\frac{\partial \tilde{k}}{\partial y}$$

so that a solution of the form

$$\bar{\rho}\tilde{k} = C_\mu^{-\frac{1}{2}}\tau_w$$

is obtained. This is similar to the incompressible result equation (3.253)

$$\bar{\rho}\bar{k} = C_\mu^{-\frac{1}{2}}\bar{\rho}u_\tau^2 = C_\mu^{-\frac{1}{2}}\tau_w$$

By analogy to the derivation of Van Driest (section 3.2.1), we consider a neighborhood of the boundary layer adjacent to the wall where the convection of momentum, energy, turbulence kinetic energy $\tilde{k}$, and the rate of turbulence kinetic energy dissipation $\tilde{\epsilon}$ are neglected. The governing equations become (Wilcox 2006)

$$0 = \frac{\partial}{\partial y}\left(-\overline{\rho u''v''} + \overline{\tau}_{xy}\right) \tag{3.264}$$

$$0 = \frac{\partial}{\partial y}\left(-c_p\overline{\rho T''v''} - \overline{q}_y - \overline{\rho u''v''}\,\tilde{u} + \tau_{xy}\tilde{u}\right) \tag{3.265}$$

$$0 = -\overline{\rho u''v''}\frac{\partial \tilde{u}}{\partial y} - \overline{\rho}\tilde{\epsilon} + \frac{\partial}{\partial y}\left(\frac{1}{\sigma_k}\frac{\mu_T}{\overline{\rho}}\frac{\partial \overline{\rho}\tilde{k}}{\partial y}\right) \tag{3.266}$$

$$0 = -C_{\epsilon 1}\frac{\tilde{\epsilon}}{\tilde{k}}\overline{\rho u''v''}\frac{\partial \tilde{u}}{\partial y} - C_{\epsilon 2}\frac{\overline{\rho}\tilde{\epsilon}^2}{\tilde{k}} + \frac{\partial}{\partial y}\left(\frac{\mu_T}{\sigma_\epsilon}\frac{\partial \tilde{\epsilon}}{\partial y}\right) \tag{3.267}$$

Integration of equations (3.264) and (3.265) from the wall to the region outside the viscous sublayer where molecular shear stress and heat flux are negligible yields

$$-\overline{\rho u''v''} = \tau_w \tag{3.268}$$

$$-c_p\overline{\rho T''v''} - \overline{\rho u''v''}\,\tilde{u} = -q_w \tag{3.269}$$

Using equation (3.268), a solution to equation (3.266) is

$$\overline{\rho}\tilde{k} = C_\mu^{-\frac{1}{2}}\tau_w \tag{3.270}$$

$$\frac{\partial \tilde{u}}{\partial y} = \frac{\overline{\rho}\tilde{\epsilon}}{\tau_w} \tag{3.271}$$

From equation (2.53) the energy equation (3.269) becomes

$$c_p\frac{\mu_T}{Pr_t}\frac{\partial \tilde{T}}{\partial y} + \tau_w\tilde{u} = -q_w \tag{3.272}$$

Assuming that the mean temperature is a function of $\tilde{u}$, equation (3.272) becomes

$$\frac{c_p}{Pr_t}\frac{d\tilde{T}}{d\tilde{u}} + \tilde{u} = -\frac{q_w}{\tau_w} \tag{3.273}$$

Assuming the static pressure $\overline{p}$ is constant across the boundary layer,

$$\frac{\tilde{T}}{T_w} = \frac{\rho_w}{\overline{\rho}} \tag{3.274}$$

and integrating equation (3.273),

$$\frac{\rho_w}{\bar{\rho}} = -\frac{1}{2}\frac{Pr_t}{c_p T_w}\tilde{u}^2 - \frac{q_w Pr_t}{c_p T_w \tau_w}\tilde{u} + \text{constant} \tag{3.275}$$

The above equation is valid within the portion of the constant stress layer where $\mu_T \gg \mu$. Assuming an adiabatic or near adiabatic wall, the temperature variation within the viscous sublayer (i.e., from $y = 0$ to the height where $\mu_T \gg \mu$) is small, and thus taking the limit of equation (3.275) as $y \to 0$, the value of the constant is unity. Therefore,

$$\frac{\rho_w}{\bar{\rho}} = 1 + Bv - A^2 v^2 \tag{3.276}$$

where

$$A^2 = \frac{(\gamma - 1)}{2} Pr_t \frac{T_\infty}{T_w} M_\infty^2 \tag{3.277}$$

$$B = -\frac{Pr_t q_w u_\infty}{c_p T_w \tau_w} \tag{3.278}$$

$$v = \frac{u}{u_\infty} \tag{3.279}$$

The equation for $\tilde{\epsilon}$ becomes

$$\frac{d^2\tilde{\epsilon}}{dv^2} - R^2(1 + Bv - A^2v^2)^{-1}\tilde{\epsilon} = 0 \tag{3.280}$$

where

$$R = \sqrt{\sigma_\epsilon (C_{\epsilon 2} - C_{\epsilon 1})\sqrt{C_\mu}}\,\frac{u_\infty}{u_\tau} = \kappa\,\frac{u_\infty}{u_\tau} \tag{3.281}$$

The asymptotic solution to equation (3.280) is obtained by the Wentzel, Kramers, Brillouin and Jeffreys (WKBJ) method[5] (Carrier *et al* 1966)

$$\tilde{\epsilon} \sim E(1 + Bv - A^2v^2)^{\frac{1}{4}} \exp\left[-\frac{R}{A}\sin^{-1}\left(\frac{2A^2v - B}{\sqrt{B^2 + 4A^2}}\right)\right] \tag{3.282}$$

where $E$ is a constant. Substituting into equation (3.271),

$$(1 + Bv - A^2v^2)^{\frac{3}{4}} \exp\left[\frac{R}{A}\sin^{-1}\left(\frac{2A^2v - B}{\sqrt{B^2 + 4A^2}}\right)\right] dv = \frac{E\rho_w}{u_\infty \tau_w} dy \tag{3.283}$$

[5] Two solutions are obtained by the WKBJ method; however, the second solution corresponds to an exponential growth in $v$ which is unphysical.

Define a change of variable

$$\zeta = \sin^{-1}\left(\frac{2A^2 v - B}{\sqrt{B^2 + 4A^2}}\right) \tag{3.284}$$

and integrate from a point outside the viscous sublayer to an arbitrary point within the constant stress layer

$$\frac{1}{A}\left(\frac{B^2 + 4A^2}{4A^2}\right)^{\frac{5}{4}} \int_{\zeta_m}^{\zeta} (\cos\zeta)^{\frac{5}{2}} \exp\left(\frac{R}{A}\zeta\right) d\zeta = \frac{E\rho_w}{u_\infty \tau_w} y + \text{constant} \tag{3.285}$$

Integrating by parts and taking the lowest order term in $R^{-1}$,

$$\frac{1}{R}\left(\frac{B^2 + 4A^2}{4A^2}\right)^{\frac{5}{4}} (\cos\zeta)^{\frac{5}{2}} \exp\left(\frac{R}{A}\zeta\right) = \frac{E\rho_w}{u_\infty \tau_w} y + \text{constant} \tag{3.286}$$

Inverting,

$$\frac{u_\infty}{u_\tau}\frac{1}{A}\sin^{-1}\left[\frac{2A^2 v - B}{\sqrt{B^2 + 4A^2}}\right] = \frac{1}{\kappa}\log\frac{y u_\tau}{\nu_w} + \text{constant} \tag{3.287}$$

By similar arguments as in section 3.2.1

$$\underbrace{\frac{u_\infty}{A}\left[\sin^{-1}\left(\frac{2A^2(u/u_\infty) - B}{\sqrt{B^2 + 4A^2}}\right) + \sin^{-1}\left(\frac{B}{\sqrt{B^2 + 4A^2}}\right)\right]}_{u_{VD}} = \frac{u_\tau}{\kappa}\log\frac{y u_\tau}{\nu_w} + C u_\tau \tag{3.288}$$

which is the compressible Law of the Wall equation (3.54).

The rate of turbulence dissipation is obtained from equation (3.271):

$$\tilde{\epsilon} = \text{constant}\left(\frac{\rho_w}{\bar{\rho}}\right)^{\frac{1}{4}} \frac{u_\tau^{\frac{3}{2}}}{\kappa y} \tag{3.289}$$

and since the solution must be valid in the limit of incompressible flow,

$$\tilde{\epsilon} = \left(\frac{\rho_w}{\bar{\rho}}\right)^{\frac{1}{4}} \frac{u_\tau^{\frac{3}{2}}}{\kappa y} \tag{3.290}$$

which is valid in the region where $\mu_T \gg \mu$ and convective terms are negligible.

From equation (3.276),

$$\frac{T_\infty}{T_w} = 1 + Bv - A^2 v^2 \tag{3.291}$$

Next assume that the above equation holds across the entire boundary layer[6] and apply at the edge of the boundary layer to obtain

$$\frac{T_\infty}{T_w} = 1 + B - A^2 \tag{3.292}$$

For an adiabatic wall $B = 0$ from equation (3.278), and thus

$$T_{aw} = T_\infty\left(1 + \frac{(\gamma - 1)}{2} Pr_t M_\infty^2\right) \tag{3.293}$$

where $T_{aw}$ is the adiabatic wall temperature (recovery temperature). For a non-adiabatic wall

$$T_\infty = T_w(1 - B + A^2) \tag{3.294}$$

and thus

$$B = \left[1 + \frac{(\gamma - 1)}{2} Pr_t M_\infty^2\right]\frac{T_\infty}{T_w} - 1 \tag{3.295}$$

which is the same as equation (3.44). From equation (3.278)

$$-\frac{Pr_t q_w u_\infty}{c_p T_w \tau_w} = \left[1 + \frac{(\gamma - 1)}{2} Pr_t M_\infty^2\right]\frac{T_\infty}{T_w} - 1 \tag{3.296}$$

and using equation (3.292)

$$St = \frac{q_w}{\rho_\infty c_p u_\infty (T_w - T_{aw})} = \frac{1}{Pr_t}\frac{\tau_w}{\rho_\infty u_\infty^2} \tag{3.297}$$

and thus

$$\frac{2St}{c_f} = \frac{1}{Pr_t} \tag{3.298}$$

which is identical to equation (3.239).

## 3.5 Morkovin's hypothesis

Morkovin (1962) postulated that, at moderate freestream Mach numbers ($M \leqslant 5$), the effects of compressibility in turbulent shear flows can be attributed to the mean variations of fluid properties (e.g., density and temperature). We reproduce his argument herein using Favre-averaged variables.

Multiplying the Favre-averaged momentum equation (2.40) by $\tilde{u}_i$ and summing over $i$ and using (2.39) yields

[6] This is not precisely correct, of course, since equation (3.291) is based upon the assumption that convective terms are negligible which is not correct in the outer portion of the boundary layer.

$$\bar{\rho}\frac{D}{Dt}\frac{1}{2}\tilde{u}_i^2 = -\tilde{u}_i\frac{\partial \bar{p}}{\partial x_i} + \tilde{u}_i\frac{\partial}{\partial x_j}\left(\bar{\tau}_{ij} - \overline{\rho u_i'' u_j''}\right) \tag{3.299}$$

where the convective derivative following the mean motion is

$$\frac{D}{Dt} = \frac{\partial}{\partial t} + \tilde{u}_j\frac{\partial}{\partial x_j} \tag{3.300}$$

Equation (3.299) can be rewritten as

$$\underbrace{\frac{\partial}{\partial x_j}\left[\tilde{u}_i\left(-\bar{p}\,\delta_{ij} + \bar{\tau}_{ij} - \overline{\rho u_i'' u_j''}\right)\right]}_{\text{I}} = \underbrace{\bar{\rho}\frac{D}{Dt}\frac{1}{2}\tilde{u}_i^2}_{\text{II}} + \underbrace{\left(\bar{\tau}_{ij} - \overline{\rho u_i'' u_j''}\right)\frac{\partial \tilde{u}_i}{\partial x_j}}_{\text{III}} - \underbrace{\bar{p}\frac{\partial \tilde{u}_i}{\partial x_i}}_{\text{IV}} \tag{3.301}$$

The term (I) represents the net rate of work done on a fluid element by the mean motion and is converted into (II) the change of the mean kinetic energy along the mean path, (III) removed through molecular dissipation and turbulence production, and (IV) stored as internal energy of the fluid element. The same equation holds for incompressible flows with the density $\rho$ being constant. Morkovin considered the corresponding Reynolds-averaged form of the previous equation:

$$\underbrace{\frac{\partial}{\partial x_j}\left[\bar{u}_i\left(-\bar{p}\,\delta_{ij} + \bar{\tau}_{ij} - \bar{\rho}\overline{u_i' u_j'}\right)\right]}_{\text{I}} = \underbrace{\left(\bar{\rho}\bar{u}_j + \overline{\rho' u_j'}\right)\frac{\partial}{\partial x_j}\left(\frac{1}{2}\bar{u}_i^2\right)}_{\text{II}} + \underbrace{\left(\bar{\tau}_{ij} - \bar{\rho}\overline{u_i' u_j'}\right)\frac{\partial \bar{u}_i}{\partial x_j}}_{\text{III}} - \underbrace{\bar{p}\frac{\partial \bar{u}_i}{\partial x_i}}_{\text{IV}} \tag{3.302}$$

On this basis, Morkovin argued that the dimensionless energy of the streamwise fluctuations $\bar{\rho}\,\overline{u'u'}/\tau_w$ and the dimensionless shear stress $-\bar{\rho}\,\overline{u'v'}/\tau_w$ would be insensitive to Mach number where $\tau_w$ is the mean local wall shear stress. Since $\tau_w = \rho_w u_\tau^2$, for an equilibrium flat plate boundary layer,

$$\frac{\bar{\rho}}{\rho_w}\frac{\overline{u'u'}}{u_\tau^2} = f\left(\frac{y}{\delta}\right) \tag{3.303}$$

$$\frac{\bar{\rho}}{\rho_w}\frac{\overline{u'v'}}{u_\tau^2} = g\left(\frac{y}{\delta}\right) \tag{3.304}$$

where the functions $f(y/\delta)$ and $g(y/\delta)$ are independent of Mach number. Figure 3.5 shows the streamwise turbulence intensity $(\bar{\rho}\overline{u'u'}/\rho_w u_\tau^2)^{\frac{1}{2}}$ versus $y/\delta$ for a limited set of experimental data including the incompressible turbulent boundary layer data of Klebanoff and Diehl (1952), the incompressible pipe flow data of Laufer (1955), and the supersonic turbulent boundary layer data of Morkovin and Phinney (1958) and Kistler (1959) which support the hypothesis (3.303).

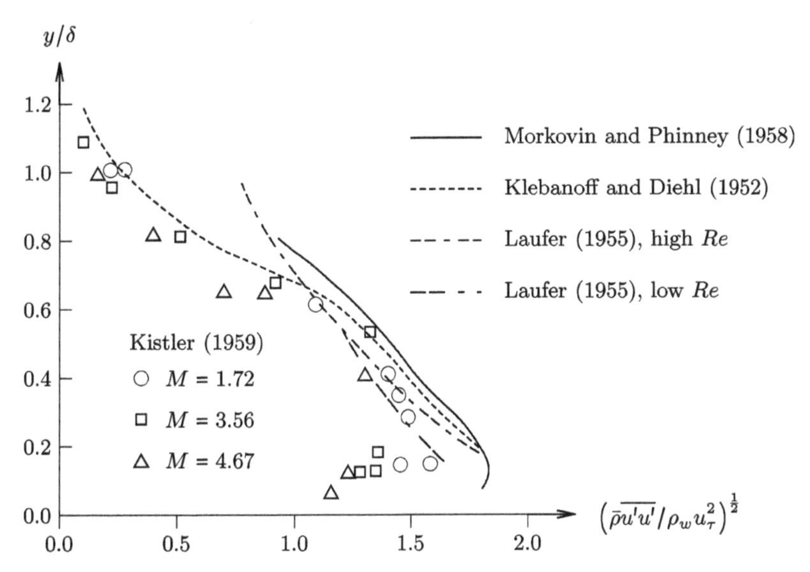


**Figure 3.5.** Streamwise turbulence intensity versus $y/\delta$ from Morkovin (1962).

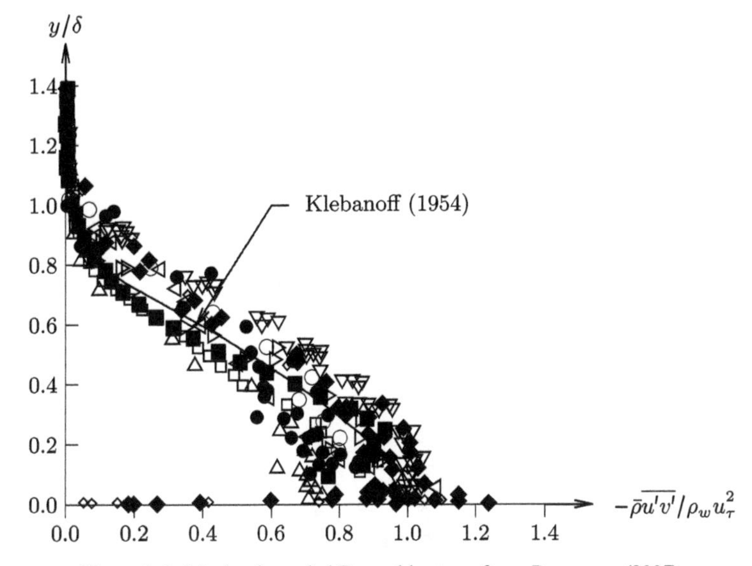


**Figure 3.6.** Morkovin-scaled Reynolds stress from Bowersox (2007).

Further evidence is shown in figures 3.6, 3.7, and 3.8 for the Morkovin-scaled Reynolds shear stress, streamwise, and wall normal turbulence intensities, respectively, from Bowersox (2007) with symbols defined in table 3.1 where $M = 0$ refers to incompressible flow. The results of Klebanoff (1954) for an incompressible turbulent boundary layer are shown as a solid line. Notwithstanding the scatter in the data, particularly in the outer portion of the boundary layer for the Morkovin-scaled wall normal turbulence intensity, there is general agreement among the profiles using Morkovin scaling.

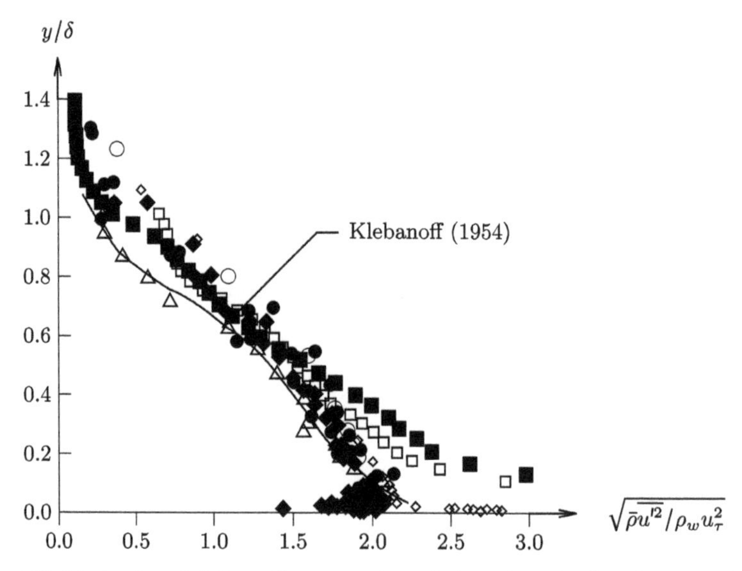


**Figure 3.7.** Morkovin-scaled Reynolds streamwise turbulence intensity from Bowersox (2007).

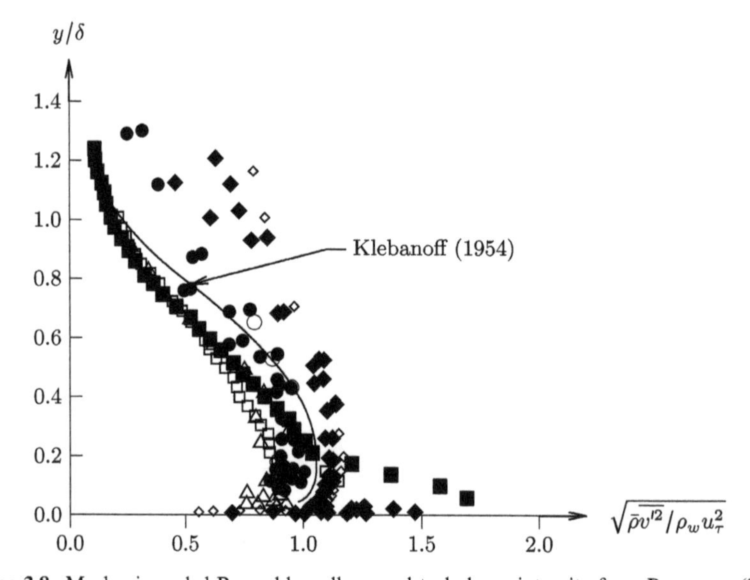


**Figure 3.8.** Morkovin-scaled Reynolds wall normal turbulence intensity from Bowersox (2007).

## 3.6 Morkovin's Strong Reynolds Analogies

Morkovin (1962) postulated relationships between the total temperature, static temperature, and velocity fluctuations in a zero pressure gradient compressible turbulent boundary layer. We reproduce the arguments here using conventional averaging.

**Table 3.1.** Symbols for figures 3.6, 3.7, and 3.8.

| Symbol | $M_\infty$ | Surface | Reference |
|---|---|---|---|
| ◇ | 0 | Smooth | George and Simpson (2000) |
| ▽ | 0 | Smooth | Ligrani and Moffat (1986) |
| △ | 2.3 | Smooth | Elena and LaCharme (1988) |
| ○ | 2.7 | Smooth | Latin and Bowersox (2000) |
| □ | 2.8 | Smooth | Ekoto *et al* (2008) |
| ▷ | 2.8 | Smooth | Luker *et al* (2000) |
| ◁ | 2.9 | Smooth | Robinson *et al* (1983) |
| ◆ | 0 | Rough | George and Simpson (2000) |
| ● | 2.7 | Rough | Latin and Bowersox (2000) |
| ■ | 2.8 | Rough | Ekoto *et al* (2008) |

The conventional average of the conservation of mass yields

$$\frac{\partial \bar{\rho}}{\partial t} + \frac{\partial}{\partial x_j}\left(\bar{\rho}\,\bar{u}_j + \overline{\rho' u_j'}\right) = 0 \tag{3.305}$$

The conventional average of the conservation of momentum is

$$\begin{aligned}\bar{\rho}\frac{\partial \bar{u}_i}{\partial t} + \left(\bar{\rho}\,\bar{u}_j + \overline{\rho' u_j'}\right)\frac{\partial \bar{u}_i}{\partial x_j} &= -\frac{\partial \bar{p}}{\partial x_i} + \frac{\partial}{\partial x_j}\left(\bar{\tau}_{ij} - \bar{\rho}\,\overline{u_i' u_j'}\right)\\ &\quad - \frac{\partial}{\partial x_j}\left(\overline{\rho' u_i'}\,\bar{u}_j + \overline{\rho' u_i' u_j'}\right)\end{aligned} \tag{3.306}$$

using (3.305). Neglecting the third term on the right-hand side,

$$\begin{aligned}\bar{\rho}\frac{\partial \bar{u}_i}{\partial t} + \left(\bar{\rho}\,\bar{u}_j + \overline{\rho' u_j'}\right)\frac{\partial \bar{u}_i}{\partial x_j} &= -\frac{\partial \bar{p}}{\partial x_i}\\ &\quad + \frac{\partial}{\partial x_j}\left(\bar{\tau}_{ij} - \bar{\rho}\,\overline{u_i' u_j'}\right)\end{aligned} \tag{3.307}$$

Defining

$$\bar{\rho}\frac{D}{Dt} = \bar{\rho}\frac{\partial \bar{u}_i}{\partial t} + \left(\bar{\rho}\,\bar{u}_j + \overline{\rho' u_j'}\right)\frac{\partial \bar{u}_i}{\partial x_j} \tag{3.308}$$

the momentum equation is

$$\bar{\rho}\frac{D\bar{u}_i}{Dt} = -\frac{\partial \bar{p}}{\partial x_i} + \frac{\partial}{\partial x_j}\left(\bar{\tau}_{ij} - \bar{\rho}\,\overline{u_i' u_j'}\right) \tag{3.309}$$

The conventional average of the energy equation is

$$\bar{\rho}\frac{D\bar{H}}{Dt} = \frac{\partial}{\partial x_j}\left[\frac{k}{c_p}\frac{\partial \bar{H}}{\partial x_j} - \bar{\rho}\,\overline{u_j' H'} + \mu\left(1 - \frac{1}{Pr}\right)\frac{\partial}{\partial x_j}\left(\frac{1}{2}\bar{u}_j^2 + \frac{1}{2}\overline{u_j' u_j'}\right)\right] + \frac{\partial \bar{p}}{\partial t} \tag{3.310}$$

where $H$ is the total enthalpy per unit mass,

$$H = h + \frac{1}{2}u_i u_i \tag{3.311}$$

and $h = c_p T$ is the static enthalpy per unit mass. Furthermore, $k$ and $\mu$ are the average molecular thermal conductivity and dynamic molecular viscosity, respectively. Note that there are additional terms neglected in (3.310) and assumed small. The molecular Prandtl number is

$$Pr = \frac{\mu c_p}{k} \tag{3.312}$$

The mean and fluctuating total enthalpy per unit mass are

$$\overline{H} = c_p \overline{T} + \frac{1}{2}(\overline{u}_i \overline{u}_i + \overline{u_i' u_i'}) \tag{3.313}$$

$$H' = c_p T' + \overline{u}_i u_i' \tag{3.314}$$

The total temperature $T_t$ is defined as

$$H = c_p T_t \tag{3.315}$$

where

$$\overline{H} = c_p \overline{T}_t \tag{3.316}$$

$$H' = c_p T_t' \tag{3.317}$$

and thus

$$\overline{T}_t = \overline{T} + \frac{1}{2c_p}(\overline{u}_i \overline{u}_i + \overline{u_i' u_i'}) \tag{3.318}$$

$$T_t' = T' + \frac{1}{2c_p}(2\overline{u}_i u_i' + u_i' u_i' - \overline{u_i' u_i'}) \tag{3.319}$$

Following Young (1951), Morkovin considers a zero pressure gradient boundary layer with steady mean flow and $Pr = 1$. In the usual boundary layer approximation, the streamwise momentum equation and energy equation are

$$\overline{\rho}\frac{D\overline{u}}{Dt} = \frac{\partial}{\partial y}\left(\mu\frac{\partial \overline{u}}{\partial y} - \overline{\rho}\ \overline{u'v'}\right) \tag{3.320}$$

$$\overline{\rho}\frac{D\overline{T}_t}{Dt} = \frac{\partial}{\partial y}\left(\mu\frac{\partial \overline{T}_t}{\partial y} - \overline{\rho}\ \overline{T_t' v'}\right) \tag{3.321}$$

where $u$ is the streamwise velocity and $y$ is the wall normal coordinate. Young (1951) noted that since $T_t - T_{t_w}$ and $u$ satisfy the same boundary condition at $y = 0$ where $T_{t_w}$ is the total temperature at the wall, (3.320) and (3.321) admit two solutions:

$$\overline{T}_t - \overline{T}_{t_w} = K\overline{u} \tag{3.322}$$

$$T'_t = Ku' \tag{3.323}$$

since $T'_{t_w} = 0$ where $K$ is a constant. Define

$$\overline{q}_w = k\frac{\partial \overline{T}}{\partial y}\bigg|_w \tag{3.324}$$

where $\overline{q}_w$ is the heat transfer *to* the wall. Therefore,

$$K = Pr_w\frac{\overline{q}_w}{\overline{\tau}_w} \tag{3.325}$$

where $Pr_w$ is the Prandtl number at the wall[7]. Therefore

$$c_p(\overline{T}_t - \overline{T}_{t_w}) = Pr_w\frac{\overline{q}_w}{\overline{\tau}_w}\overline{u} \tag{3.326}$$

$$c_p T'_t = Pr_w\frac{\overline{q}_w}{\overline{\tau}_w}u' \tag{3.327}$$

Equations (3.326) and (3.327) were denoted the *strong Reynolds analogy* by Morkovin. On the basis of these results, Morkovin deduced five results for an adiabatic flat plate turbulent boundary layer. In the following we consider a Cartesian coordinate system where $x$, $y$, and $z$ are the streamwise, wall normal, and spanwise coordinates with velocity components $u$, $v$, and $w$.

### 3.6.1 Strong Reynolds Analogy no. 1

For an adiabatic wall, (3.327) implies $T'_t = 0$. From (3.319) for a Cartesian coordinate system and neglecting higher order fluctuating quantities,

$$T'_t = T' + \frac{1}{c_p}(\overline{u}u' + \overline{v}v' + \overline{w}w') = 0 \tag{3.328}$$

For a nominally two-dimensional flat plate boundary layer, $\overline{v} \ll \overline{u}$ and $\overline{w} = 0$. Furthermore, $v' \ll u'$ as indicated for example in the measurements of Konrad (1993) for an adiabatic turbulent boundary layer at Mach 2.9 and shown in figures 3.9 and 3.10. Consequently,

$$c_p T' + \overline{u}u' = 0 \tag{3.329}$$

---

[7] Of course, Pr is assumed to be unity in the analysis.

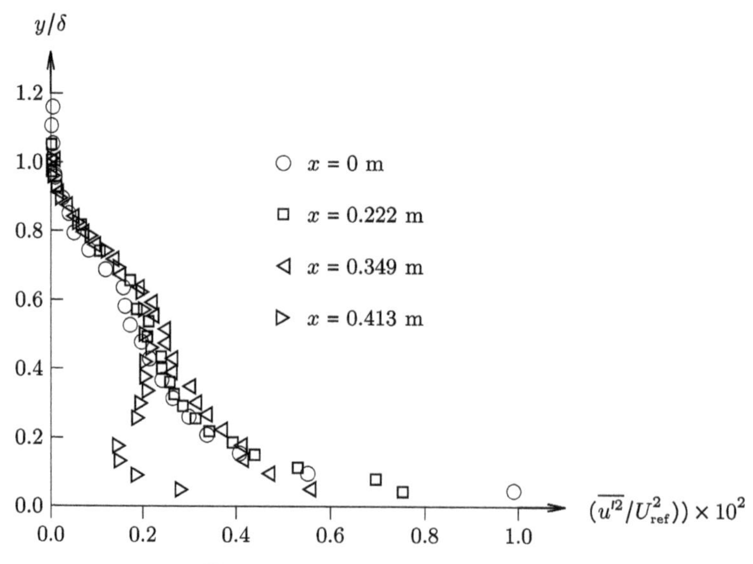


**Figure 3.9.** Streamwise Reynolds stress.

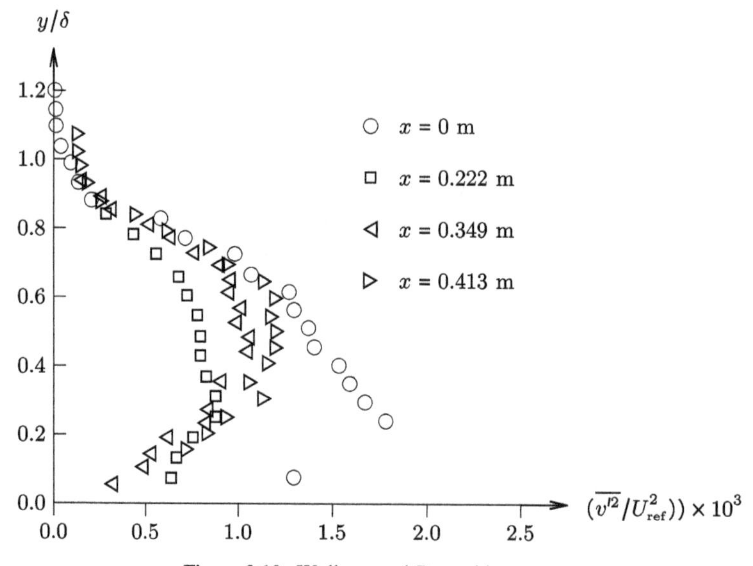


**Figure 3.10.** Wall normal Reynolds stresses.

and consequently

$$\frac{(T'/\overline{T})}{(\gamma - 1)\overline{M}^2 \, (u'/\overline{u})} = -1 \tag{3.330}$$

### 3.6.2 Strong Reynolds Analogy no. 2

From (3.329), it is straightforward to show

$$\frac{\sqrt{\overline{T'^2}}/\overline{T}}{(\gamma-1)\overline{M}^2\sqrt{\overline{u'^2}}/\overline{u}} = 1 \tag{3.331}$$

where $\overline{M}$ is the mean *local* Mach number defined by

$$\overline{M} = \frac{\overline{u}}{\gamma R\overline{T}} \tag{3.332}$$

### 3.6.3 Strong Reynolds Analogy no. 3

From (3.329), the velocity temperature correlation $R_{uT}$ defined by

$$R_{uT} = \frac{\overline{T'u'}}{\sqrt{\overline{T'^2}\,\overline{u'^2}}} \tag{3.333}$$

becomes

$$R_{uT} = -1 \tag{3.334}$$

### 3.6.4 Strong Reynolds Analogy no. 4

From (3.326) and (3.327)

$$\frac{\sqrt{\overline{T'^2}}}{\overline{T}_{t_w} - T_\infty} = 2\frac{\overline{u}}{U_\infty}\frac{\sqrt{\overline{u'^2}}}{U_\infty} \tag{3.335}$$

### 3.6.5 Strong Reynolds Analogy no. 5

From the definition of the turbulent eddy viscosity $\mu_t$ (2.52), turbulent thermal conductivity $k_t$ (2.53), and turbulent Prandtl number $Pr_t$ (2.54), it can be directly shown that

$$Pr_t = 1 \tag{3.336}$$

### 3.6.6 Morkovin's analysis of the five Strong Reynolds Analogies

Morkovin (1962) presented a critical assessment of the five results of the Strong Reynolds Analogy for adiabatic boundary layers. With regards to the Strong Reynolds Analogy no. 1, he concluded that the total temperature fluctuations $T_t'$ were not necessarily negligible and provided the experimental data[8] shown in figure 3.11.

With regards to the Strong Reynolds Analogy nos. 2 and 4, Morkovin (1962) stated that they are '… verified within 20% or less'. Concerning the Strong Reynolds

[8] Morkovin notes, 'Only the extreme measurements, as noted above, were selected from the Morkovin–Phinney set. There were many more measurements, generally in between the values shown here'.

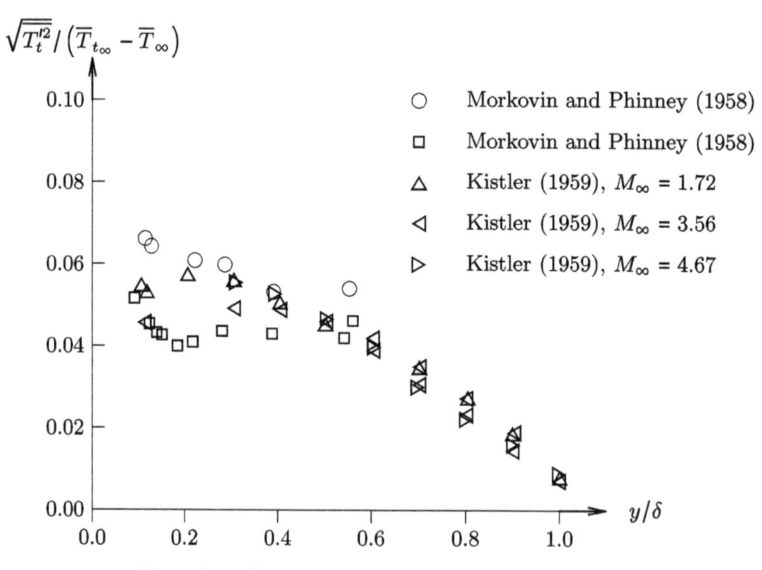


**Figure 3.11.** Total temperature fluctuations versus $y/\delta$.

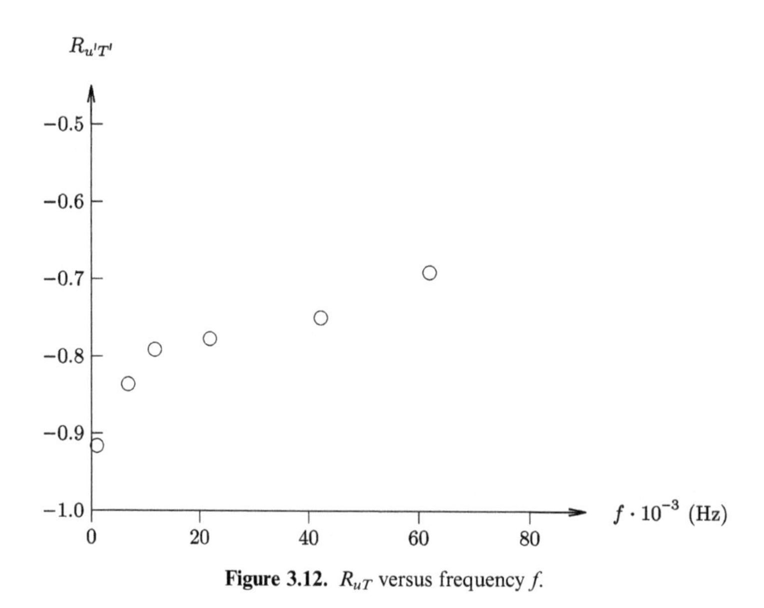


**Figure 3.12.** $R_{uT}$ versus frequency $f$.

Analogy no. 3, he concluded that the experimental data range from −0.7 to −0.9 depending on the hot wire anemometry material. Figure 3.12 displays data of Morkovin and Phinney (1958) for $R_{uT}$ versus frequency for turbulent boundary layer at Mach 1.77 at a location where $\overline{M}$ = 1.33. With regards to Strong Reynolds Analogy no. 5, Morkovin (1962) concluded that '... limited measurements of $Pr_t$ yielded results between 0.9 and 0.93'.

## 3.7 Cebeci and Smith Strong Reynolds Analogy

Cebeci and Smith (1974) proposed a modification to the Strong Reynolds Analogy no. 1 as follows. Assuming a zero pressure gradient $dp/dx = 0$ and $Pr = Pr_t = 1$ and using Reynolds averaging, the Crocco–Busemann relation (3.143) is

$$\frac{\overline{H} - \overline{H}_w}{\overline{H}_e - \overline{H}_w} = \frac{\overline{u}}{\overline{u}_e}$$

which can be written as

$$\overline{H} = A + B\overline{u} \tag{3.337}$$

where

$$A = \overline{H}_w \tag{3.338}$$

$$B = (\overline{H}_e - \overline{H}_w)/\overline{u}_e \tag{3.339}$$

Neglecting (a) wall normal and spanwise velocities compared to the streamwise velocity and (b) second order terms,

$$H' = h' + \overline{u}u' \tag{3.340}$$

Cebeci and Smith (1974) assume that (3.337) applies to $H'$ also, and thus for a perfect gas

$$\frac{T'}{\overline{T}} = -\left[(\gamma - 1)\overline{M}^2 + \frac{(T_w - T_e)}{\overline{T}}\frac{\overline{u}}{\overline{u}_e}\right]\frac{u'}{\overline{u}} \tag{3.341}$$

The first term on the right side corresponds to Morkovin's Strong Reynolds Analogy no. 1 equation (3.330). Squaring (3.341), averaging, and taking the square root yield

$$\frac{\sqrt{\overline{T'^2}}/\overline{T}}{(\gamma - 1)\overline{M}^2\sqrt{\overline{u'^2}}/\overline{u}} = 1 + \frac{(T_w - T_e)}{\overline{T}}\left(\frac{\overline{u}}{u_e}\right)\frac{1}{(\gamma - 1)\overline{M}^2} \tag{3.342}$$

## 3.8 Huang *et al* Strong Reynolds Analogy

Huang *et al* (1995) proposed a modification to the Strong Reynolds Analogy no. 3 as follows. Assuming a mixing length model for the streamwise velocity and static temperature fluctuations in the boundary layer,

$$u' = \ell_{u'}\frac{\partial\overline{u}}{\partial y} \tag{3.343}$$

$$T' = \ell_{T'}\frac{\partial\overline{T}}{\partial y} \tag{3.344}$$

where $\ell_{u'}$ and $\ell_{T'}$ are the mixing lengths for the turbulent transport of momentum and internal energy, respectively. Furthermore, assume that the length scales are proportional:

$$\frac{\ell_{u'}}{\ell_{T'}} = c \tag{3.345}$$

where $c$ is assumed independent of time. Thus

$$u' \frac{\partial \overline{T}}{\partial y} = c\, T' \frac{\partial \overline{u}}{\partial y} \tag{3.346}$$

Multiplying (3.346) by $\rho v'$ and averaging,

$$\overline{\rho v' u'}\, \frac{\partial \overline{T}}{\partial y} = c\, \overline{\rho v' T'}\, \frac{\partial \overline{u}}{\partial y} \tag{3.347}$$

For a turbulent boundary layer, equation (2.56) yields

$$c = \frac{\overline{\rho v' u'}\ \partial \overline{T} / \partial y}{\overline{\rho v' T'}\ \partial \overline{u} / \partial y} = Pr_{\mathrm{t}} \tag{3.348}$$

From (3.346)

$$\frac{T'}{u'} = \frac{1}{Pr_{\mathrm{t}}} \frac{\partial \overline{T} / \partial y}{\partial \overline{u} / \partial y} \tag{3.349}$$

From section 3.3.1, the mean temperature $\overline{T}$ is a function of the mean streamwise velocity $\overline{u}$ for both adiabatic and isothermal (non-adiabatic) boundary layers, and thus the obverse also holds. Hence,

$$\frac{\partial \overline{T} / \partial y}{\partial \overline{u} / \partial y} = \frac{1}{d\overline{u} / d\overline{T}} \tag{3.350}$$

From the definition of total temperature,

$$\overline{T}_{\mathrm{t}} = \overline{T} + \frac{1}{2c_{\mathrm{p}}}(\overline{u}^2 + \overline{v}^2 + \overline{w}^2) + \frac{1}{c_{\mathrm{p}}} k \tag{3.351}$$

where $k = \frac{1}{2}(\overline{u'^2} + \overline{v'^2} + \overline{w'^2})$ is the turbulence kinetic energy. For a two-dimensional turbulent boundary layer, $\overline{v} \ll \overline{u}$, $\overline{w} = 0$, and $k \ll \overline{u}^2$; thus

$$\overline{T}_{\mathrm{t}} \approx \overline{T} + \frac{1}{2c_{\mathrm{p}}} \overline{u}^2 \tag{3.352}$$

Thus

$$\frac{d\overline{u}}{d\overline{T}} = \frac{c_{\mathrm{p}}}{\overline{u}} \left( \frac{d\overline{T}_{\mathrm{t}}}{d\overline{T}} - 1 \right) \tag{3.353}$$

Therefore, equation (3.349) yields

$$\frac{(T'/\overline{T})}{(\gamma - 1)\overline{M}^2\,(u'/\overline{u})} = \frac{1}{Pr_t}\frac{1}{(d\overline{T}_t/d\overline{T} - 1)} \tag{3.354}$$

where $\overline{M} = \overline{u}/\sqrt{\gamma R\overline{T}}$. Therefore, equation (3.354) is an alternative to (3.330) (Strong Reynolds Analogy no. 1). Equation (3.354) is derived without (3.328) (i.e., total temperature fluctuations are negligible) and is therefore more applicable to non-adiabatic boundary layers. Taking the root mean square of (3.354) yields

$$\frac{\left(\sqrt{\overline{T'^2}}/\overline{T}\right)}{(\gamma - 1)\overline{M}^2\left(\sqrt{\overline{u'^2}}/\overline{u}\right)} = \frac{1}{Pr_t}\frac{1}{|\, d\overline{T}_t/d\overline{T} - 1\,|} \tag{3.355}$$

Note that the process of squaring (3.354), averaging, and then taking the square root implies the absolute value of the term $d\overline{T}_t/d\overline{T} - 1$ since this term may be either positive or negative within the boundary layer.

Gaviglio (1987) derives a similar result:

$$\frac{\left(\sqrt{\overline{T'^2}}/\overline{T}\right)}{(\gamma - 1)\overline{M}^2\left(\sqrt{\overline{u'^2}}/\overline{u}\right)} = \frac{1}{(1 - d\overline{T}_t/d\overline{T})} \tag{3.356}$$

The absence of the $Pr_t^{-1}$ term on the right side is due to the assumption by Gaviglio (1987) that $|\ell_{u'}| = |\ell_{T'}|$ in (3.343) and (3.344). The term $(1 - d\overline{T}_t/d\overline{T})$ appears in the denominator instead of $(d\overline{T}_t/d\overline{T} - 1)$ as in (3.355) due to the assumption $\ell_{u'} = -\ell_{T'}$ in equation (31) of Gaviglio (1987). However, an absolute value should also be used for the term $1 - d\overline{T}_t/d\overline{T}$ for reasons cited above. Figure 3.13 compares (3.354) with DNS data for a supersonic turbulent channel flow at $M = 1.5$ and 3 where the Mach number is based upon the bulk velocity and the speed of sound at the wall (Huang *et al* 1995). Close agreement is evident.

A similar result to (3.355) was obtained by Zhang *et al* (2014). Gaviglio (1987) and Brun *et al* (2008) developed an analogous result with the factor $Pr_t^{-1}$ replaced by 1 and $Pr_m^{-1}$, respectively, where $Pr_m$ is the mixed Prandtl number

$$Pr_m = \frac{(\mu + \mu_t)c_p}{k + k_t} \tag{3.357}$$

and thus

$$\frac{\left(\sqrt{\overline{T'^2}}/\overline{T}\right)}{(\gamma - 1)\overline{M}^2\left(\sqrt{\overline{u'^2}}/\overline{u}\right)} = \frac{1}{Pr_m}\frac{1}{|\, d\overline{T}_t/d\overline{T} - 1\,|} \tag{3.358}$$

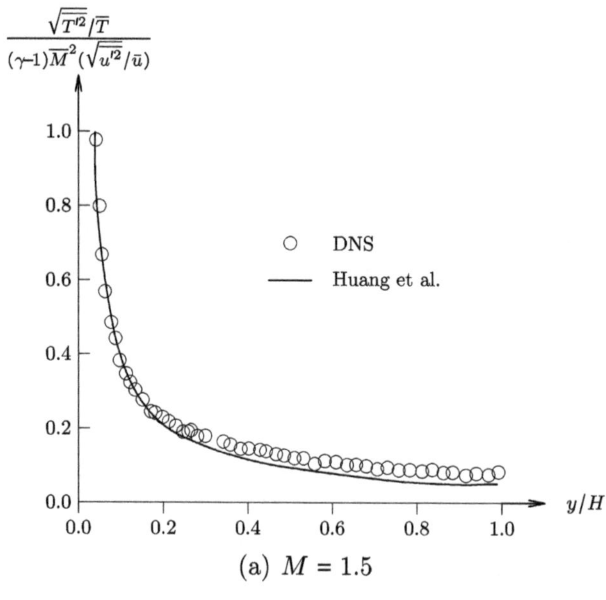


(a) $M = 1.5$

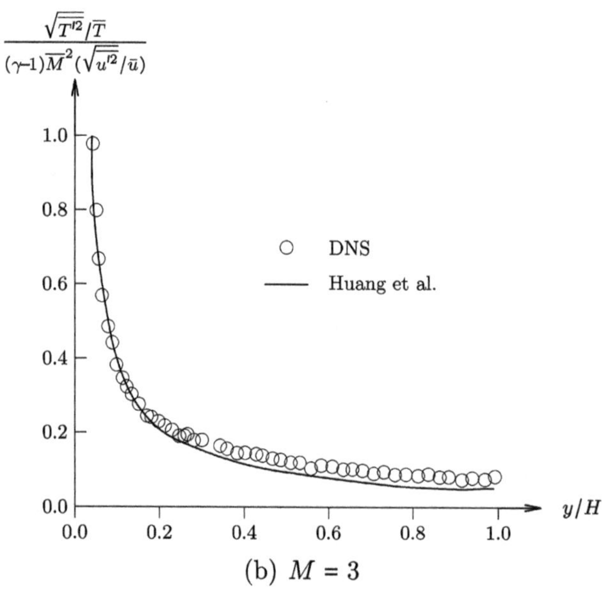


(b) $M = 3$

**Figure 3.13.** Huang *et al* Strong Reynolds Analogy.

Duan *et al* (2011) extended the above analysis to chemically reacting flows with variable heat capacities.[9] Assuming the mixing length models (3.343) and (3.344) with $\ell_{u'} = Pr_t \ell_{T'}$,

[9] The analysis was based on the Favre fluctuating variables instead of the Reynolds-averaged fluctuating variables. For consistency with the Huang *et al* Strong Reynolds Analogy, the analysis of Duan *et al* (2011) is presented with the former.

$$\sqrt{\overline{T'^2}} = -\frac{1}{Pr_t}\frac{\partial \overline{T}}{\partial \overline{u}}\sqrt{\overline{u'^2}} \tag{3.359}$$

where the negative sign is taken in the outer region of the boundary layer where $\partial \overline{T}/\partial y < 0$.

## References

Baronti P and Libby P 1966 Velocity profiles in turbulent compressible boundary layers *AIAA J.* **4** 193–202

Boussinesq J 1877 Théorie del'écoulement tourbillant *Mém. prés. Acad. Sci.* **23** 46–50

Bowersox R 2007 Survey of high-speed rough wall boundary layers: invited presentation *37th AIAA Fluid Dynamics Conference and Exhibit* AIAA Paper 2007-3998 (Reston, VA: American Institute of Aeronautics and Astronautics)

Bradshaw P 1977 Compressible turbulent shear layers *Annu. Rev. Fluid Mech.* **9** 33–54

Bradshaw P 1994 Turbulence: the chief outstanding difficulty of our subject *Exp. Fluids* **16** 203–16

Brooks J, Gupta A, Helm C, Martín M P, Smith M and Marineau E 2017 Mach 10 PIV flow field measurements of a turbulent boundary layer and shock turbulent boundary layer interaction *33rd AIAA Aerodynamic Measurement Technology and Ground Testing Conference* 2017-3325

Brun C, Boiarciuc M, Haberkorn M and Comte P 2008 Large eddy simulation of compressible channel flow *Theor. Comput. Fluid Dyn.* **22** 189–212

Busemann A 1935 Gasströmung mit laminarer grenzschicht entlang einer platte *Z. Angew. Math. Mech.* **15** 23–5

Carrier G, Krook M and Pearson C 1966 *Functions of a Complex Variable* (New York: McGraw-Hill)

Cary A 1970 *Summary of Available Information on Reynolds Analogy for Zero-Pressure-Gradient, Compressible, Turbulent-Boundary-Layer Flow* NASA Technical Note D-5560 (Washington, DC: National Aeronautics and Space Administration)

Cebeci T and Smith A 1974 *Analysis of Turbulent Boundary Layers* (New York: Academic)

Clauser F 1954 Turbulent boundary layers in adverse pressure gradients *J. Aeronaut. Sci.* **21** 91–108

Clauser F 1956 The turbulent boundary layer *Appl. Mech.* **4** 1–51

Colburn A 1933 A method of correlating forced convection heat transfer data and a comparison with fluid friction *Trans. Am. Inst. Chem. Eng.* **29** 174–211

Coles D 1953 Measurements in the Boundary Layer on a Smooth Flat Plate in Supersonic Flow. Part I: The Problem of the Turbulent Boundary Layer *Technical Report 20-69* Technical Report 20-69 Jet Propulsion Laboratory, California Institute of Technology

Coles D 1956 The law of the wake in the turbulent boundary layer *J. Fluid Mech.* **1** 191–226

Coles D 1962 *The Turbulent Boundary Layer in a Compressible Fluid* R-403-PR RAND Corporation

Coles D 1968 The Young Person's Guide to the Data *AFOSR-IFP-Stanford Conference on Computation of Turbulent Boundary Layers*

[9] The analysis was based on the Favre fluctuating variables instead of the Reynolds-averaged fluctuating variables. For consistency with the Huang *et al* Strong Reynolds Analogy, the analysis of Duan *et al* (2011) is presented with the former.

Crocco L 1932 Sulla trasmissione del calore da una lamina piana a un fluido scorrente ad alta velocita *L Aerotec* **12** 181–97

Crocco L 1963 Transformation of the compressible turbulent boundary layer with heat exchange *AIAA J.* **1** 2723–31

Danberg J 1964 *Characteristics of the Turbulent Boundary Layer with Heat and Mass Transfer at* $M = 6.7$ Technical Report NOLTR 64-99 US Naval Ordnance Laboratory

Dorrance W and Dore F 1954 The effect of mass transfer on the compressible turbulent boundary layer skin friction and heat transfer *J. Aeronaut. Sci.* **21** 404–10

Duan L, Beekman I and Martín M P 2010 Direct numerical simulation of hypersonic turbulent boundary layers. Part 2. Effect of wall temperature *J. Fluid Mech.* **655** 419–45

Duan L, Beekman I and Martín M P 2011 Direct numerical simulation of hypersonic turbulent boundary layers. Part 3. Effect of Mach number *J. Fluid Mech.* **672** 245–67

Duan L and Martín M P 2011 Direct numerical simulation of hypersonic turbulent boundary layers. Part 4. Effect of high enthalpy *J. Fluid Mech.* **684** 25–59

Ekoto I, Bowersox R, Beutner T and Goss I 2008 Supersonic boundary layers with periodic surface roughness *AIAA J.* **46** 486–97

Elena M and LaCharme J-P 1988 Experimental study of a supersonic turbulent boundary layer using a laser doppler anemometer *J. Theor. Appl. Mech.* **7** 175–90

Fernholz H and Finley P 1980 *A Critical Commentary on Mean Flow Data for Two-Dimensional Compressible Turbulent Boundary Layers* AGARDograph No. 253 Advisory Group on Aerospace Research and Development

Gaviglio J 1987 Reynolds analogies and experimental study of heat transfer in the supersonic boundary layer *Int. J. Heat Mass Transf.* **30** 911–26

George J and Simpson R 2000 Some effects of sparsely distributed three-dimensional roughness elements in two dimensional turbulent boundary layers *38th Aerospace Sciences Meeting and Exhibit* 2000-0915

Griffin K, Fu L and Moin P 2021 Velocity transformation for compressible wall-bounded turbulent flows with and without heat transfer *Proc. Natl Acad. Sci.* **118** e2111144118

Hama F 1954 Boundary layer characteristics for smooth and rough surfaces *Trans. - Soc. Nav. Archit. Mar. Eng.* **62** 333–58

Holden M 1991 Studies of the mean and unsteady structure of turbulent boundary layer separation in hypersonic flow *22nd Fluid Dynamics, Plasma Dynamics and Lasers Conference* AIAA Paper 91-1778 (Reston, VA: American Institute of Aeronautics and Astronautics)

Huang J, Nicholson G, Duan L, Choudhari M and Bowersox R 2020 Simulation and modeling of cold wall hypersonic turbulent boundary layers on flat plate *AIAA Scitech 2020 Forum* AIAA Paper 2020-0571 (Reston, VA: American Institute of Aeronautics and Astronautics)

Huang P, Coleman G and Bradshaw P 1995 Compressible turbulent channel flows: DNS results and modelling *J. Fluid Mech.* **305** 185–218

Jones W and Launder B 1972 The prediction of laminarization with a two equation model of turbulence *Int. J. Heat Mass Transf.* **15** 301–14

Kistler A 1959 Fluctuation measurements in a supersonic turbulent boundary layer *Phys. Fluids* **2** 290–6

Klebanoff P 1954 *Characteristics of Turbulence in a Boundary Layer with Zero Pressure Gradient* Technical Report 3178 (Washington, DC: National Advisory Committee on Aeronautics)

Klebanoff P and Diehl Z 1952 *Some Features of Artificially Thickened Fully Developed Turbulent Boundary Layers with Zero Pressure Gradient* Technical Report 1110 (Washington, DC: National Advisory Committee on Aeronautics)

Konrad W 1993 A three-dimensional supersonic turbulent boundary layer generated by an isentropic compression *PhD Thesis* Princeton University

Lagha M, Kim J, Eldredge J and Zhong X 2011 A numerical study of compressible turbulent boundary layers *Phys. Fluids* **23** 015106

Lamb H 1945 *Hydrodynamics* (New York: Dover)

Latin R and Bowersox R 2000 Flow properties of a supersonic boundary layer with wall roughness *AIAA J.* **38** 1804–21

Laufer J 1955 The structure of turbulence in fully developed pipe flow *Technical Report 1174* (Washington, DC: National Advisory Committee on Aeronautics)

Launder B and Sharma B 1974 Application of the energy dissipation model of turbulence to the calculation of flow near a spinning disc *Lett. Heat Mass Transf.* **1** 131–8

Ligrani P and Moffat R 1986 Structure of transitionally rough and fully rough turbulent boundary layers *J. Fluid Mech.* **162** 69–98

Luker J, Bowersox R and Buter T 2000 Influence of a curvature driven favorable pressure gradient on a supersonic boundary layer *AIAA J.* **38** 1351–9

Matthews D, Childs M and Paynter G 1970 Use of Coles' universal wake function for compressible turbulent boundary layers *J. Aircr.* **7** 137–40

Modesti D and Pirozzoli S 2019 Direct numerical simulation of supersonic pipe flow at moderate Reynolds number *Int. J. Heat Fluid Flow* **76** 100–12

Morkovin M 1962 Effects of compressibility on turbulent flows *Mécanique de la Turbulence, Colloques Internationaux du Centre National de la Recherche Scientifique* ed A Favre (Paris: Centre National de la Recherche Scientifique) pp 367–80

Morkovin M and Phinney R 1958 *Extended Applications of Hot-Wire Anemometry to High Speed Turbulent Boundary Layers*

Patel A, Boersma B and Pecnik R 2016 The influence of near-wall density and viscosity gradients on turbulence in channel flows *J. Fluid Mech.* **809** 793–820

Pirozzoli S and Bernardini M 2011 Direct numerical simulation database for impinging shock wave turbulent boundary layer interaction *AIAA J.* **49** 1307–12

Prandtl L 1925 Über die ausgebildete turbulenz *Z. Angew. Math. Mech.* **5** 136–9

Reynolds O 1875 On the extent and action of the heating surface for steam boilers *Proc Manch. Lit. Phil. Soc.* **14** 7–12

Robinson S, Seegmiller H and Kussoy M 1983 Hot-wire and laser doppler anemometer measurements in a supersonic boundary layer *16th Fluid and Plasmadynamics* Conf. AIAA Paper 83-1723 (Reston, VA: American Institute of Aeronautics and Astronautics)

Rotta J 1965 Heat transfer and temperature distribution in turbulent boundary layers at supersonic and hypersonic flow *Recent Developments in Boundary Layer Research* AGARDograph 97 Part I 35–63

Rubesin M 1954 *An Analytical Estimation of the Effect of Transpiration Cooling on the Heat-Transfer and Skin-Friction Characteristics of a Compressible, Turbulent Boundary Layer* Technical Note 3341 (Washington, DC: National Advisory Committee on Aeronautics)

Saffman P and Wilcox D 1974 Turbulence-model predictions for turbulent boundary layers *AIAA J.* **12** 541–6

Schlichting H 1968 *Boundary-Layer Theory* 6th edn (New York: McGraw-Hill)

Schutts W, Harting W and Weiler J 1955 *Turbulent Boundary Layer and Skin Friction Measurements on a Smooth, Thermally Insulated Flat Plate at Supersonic Speeds* Technical Report DRL-364, CM-823 Defense Research Laboratory, University of Texas, Austin, TX

Stanton T 1897 On the passage of heat between metal surfaces and liquids in contact with them *Phil. Trans.* A **190** 67–88

Sun C-C and Childs M 1976 Wall-wake velocity profile for compressible nonadiabatic flows *AIAA J.* **14** 820–2

Trettel A and Larsson J 2016 Mean velocity scaling for compressible wall turbulence with heat transfer *Phys. Fluids* **28** 026102

Van Driest E 1951 Turbulent boundary layer in compressible fluids *J. Aeronaut. Sci.* **18** 145–60 and 216

Van Driest E 1959 Convective heat transfer in gases *Turbulent Flows and Heat Transfer* ed C-C Lin (Princeton, NJ: Princeton University Press) 339–427

Volpiani P, Bernardini M and Larsson J 2020 Effects of a nonadiabatic wall on hypersonic shock/boundary-layer interactions *Phys. Rev. Fluids* **5** 014602-1–20

Volpiani P, Iyer P, Pirozzoli S and Larsson J 2020 Data-driven compressibility transformation for turbulent wall layers *Phys. Rev. Fluids* **5** 052602

Walz A 1962 *Mécanique de la Turbulence* ed A Favre (Paris: Centre National de la Recherche Scientifique), 299–350

Walz A 1969 *Boundary Layers of Flow and Temperature* (Cambridge, MA: MIT Press)

White F 1974 *Viscous Fluid Flow* (New York: McGraw-Hill)

Wilcox D 2006 *Turbulence Modeling for CFD* 3rd edn (CA: DCW Industries, Inc.)

Wu B, Bi W, Hussain F and She Z 2017 On the invariant mean velocity profile for compressible turbulent boundary layers *J. Turbul.* **18** 186–202

Young A 1951 *The Equations of Motion and Energy and the Velocity Profile of a Turbulent Boundary Layer in a Compressible Fluid* Techincal Report 42 College of Aeronautics, Cranfield

Zhang C, Duan L and Choudhari M 2017 Effect of wall cooling on boundary-layer-induced pressure fluctuations at Mach 6 *J. Fluid Mech.* **822** 5–30

Zhang Y, Bi W, Hussain F and She Z 2014 A generalized Reynolds analogy for compressible wall-bounded turbulent flows *J. Fluid Mech.* **739** 392–420

**IOP** Publishing

Hypersonic Shock Wave Turbulent Boundary Layers

Direct Numerical Simulation, Large Eddy Simulation and Experiment

**Doyle Knight and Nadia Kianvashrad**

# Chapter 4

# Experiments—hypersonic turbulent boundary layers

## Abstract

Experimental research on hypersonic turbulent boundary layers began in the 1950s due to the interest in space flight for both civilian and military applications. The present chapter examines a selected set of experimental data on hypersonic turbulent boundary layers with zero and modest streamwise pressure gradients. The following chapter examines hypersonic shock wave turbulent boundary layer interactions. The set of experiments described herein was chosen to both illustrate the history of research in hypersonic boundary layers and provide an understanding of the state of the art. A selected set of data is shown for each experiment. Typically it is the primary measurements that are shown. Mean quantities evaluated at the edge of the boundary layer are denoted by the subscript $_e$ as in $M_e$ for the Mach number, etc. Conditions in a wind tunnel test section or reservoir are denoted by the subscript $_\infty$ as in $M_\infty$. Note that for a flat plate at angle of attack the values of $M_e$ and $M_\infty$ are not the same.[1] The designation 'na' indicates 'not available'. Mean experimental quantities are shown without tilde ~ or overbar - (e.g., $u$, $T$, etc).

[1] The term 'stagnation pressure' was sometimes used to describe the pressure measured by a pitot tube in a supersonic or hypersonic boundary layer. The stagnation pressure at a point is the pressure obtained by bringing the fluid to rest adiabatically and reversibly (i.e., isentropically). In supersonic or hypersonic flow, the pressure measured by a pitot tube is not the stagnation pressure at that point, but the stagnation pressure downstream of a normal shock at that point in accordance with the Rayleigh pitot formula (Anderson 2006) and denoted the 'pitot pressure'. Nonetheless, there are examples of the term 'stagnation pressure' being used for 'pitot pressure', for example

> "Measurements of cold wall turbulent hypersonic boundary layer *stagnation pressure* and stagnation temperature profiles have been made on the wall of a conical nozzle in a hypersonic gun tunnel facility. A high speed traversing *pitot tube* and a short time response stagnation temperature probe were developed for the profile measurements (Perry and East 1968a)." (Italics added)

doi:10.1088/978-0-7503-5002-0ch4

Further detailed information is available in the reviews by Fernholz and Finley (1977, 1980), Fernholz *et al* (1981), Settles and Dodson (1993), and Roy and Blottner (2006).

## 4.1 Wegener *et al* (1953)

Wegener *et al* (1953) investigated a cold wall turbulent boundary layer at Mach 7. The experiment was performed at the United States Naval Ordnance Laboratory 12 cm by 12 cm hypersonic tunnel no. 4 (section A.23.1). The measurements were obtained on the centerline of one of the diverging walls of the nozzle where the Mach number was 7.0 and the freestream Mach number gradient was approximately 0.019 7$M$ cm$^{-1}$. The freestream conditions are shown in table 4.1. Experimental measurements include pitot pressure and total temperature profiles and surface pressure and heat transfer.

Figures 4.1, 4.2, and 4.3 display profiles of Mach number $M$, total temperature $T_t$, and velocity $u/u_e$, respectively. Figure 4.4 presents the Van Driest transformed velocity $u_{VD}/u_\tau$ and normalized velocity $u/u_\tau$ versus $y^+$. Also shown are the viscous sublayer $u/u_\tau = y^+$ and law of the wall and wake with $C = 5.1$ and $\Pi = 0.55$. The measured velocity profile agrees closely with the Law of the Wall and Wake. The measurements within the viscous sublayer may be in error due to probe wall interference (4.2).

## 4.2 Winkler and Persh (1954)

Winkler and Persh (1954) investigated the cold wall hypersonic turbulent boundary layer on the nozzle wall of the Naval Ordnance Laboratory 12 cm × 12 cm hypersonic tunnel no. 4. (section A.23.1). A steady nozzle wall temperature distribution was maintained by a coolant flow. Experimental diagnostics include surface pitot probes, freestream static probes, and boundary layer surveys of pitot, static, and stagnation temperature at a location $x$ = 50 cm from the nozzle throat. The flow conditions are listed in table 4.2.

**Table 4.1.** Freestream conditions.

| Quantity | Value |
|---|---|
| $M_e$ | 7.0 |
| $p_{t_e}$ (MPa) | 1.01 |
| $T_{t_e}$ (K) | 593 |
| $T_w/T_e$ | 5.7 |
| $T_w/T_{aw}$ | 0.59[a] |
| $H_e$ (MJ kg$^{-1}$) | 0.595 |
| $Re \times 10^{-6}$ (m$^{-1}$) | 4.50 |
| $\delta^*$ (cm) | 2.6 |

[a]Using (3.46) with $Pr_t = 0.89$.

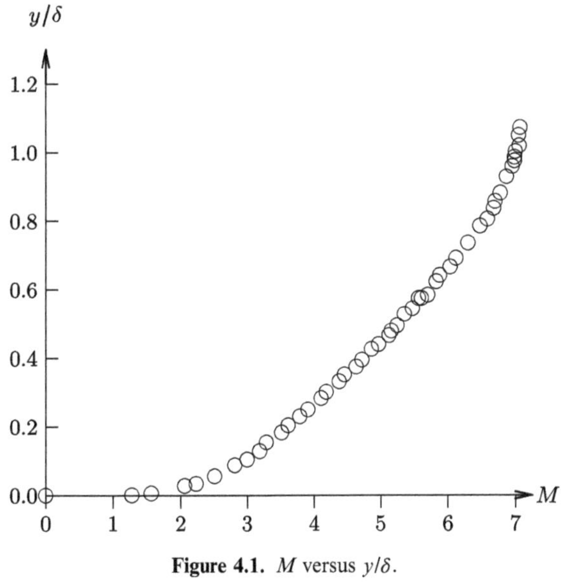


**Figure 4.1.** $M$ versus $y/\delta$.

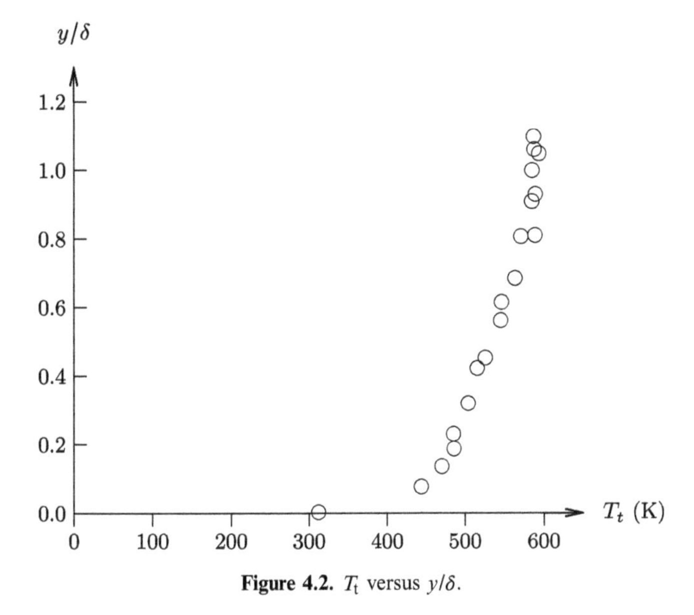


**Figure 4.2.** $T_t$ versus $y/\delta$.

Figures 4.5 and 4.6 display the dimensionless streamwise velocity $u/u_\infty$ and dimensionless static temperature $T/T_\infty$ versus $y/\delta$ for cases 1 and 2 listed in table 4.3. Also shown is the approximate fit $u/u_\infty = (y/\delta)^{1/8}$.

## 4.3 Lobb *et al* (1955)

Lobb *et al* (1955a, 1955b) conducted an experimental study of a cold wall zero pressure gradient turbulent boundary layer at Mach numbers from 5 to 7.

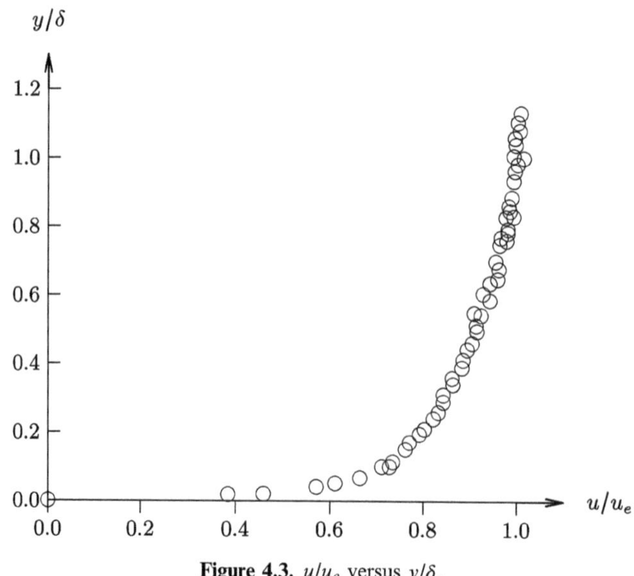


**Figure 4.3.** $u/u_e$ versus $y/\delta$.

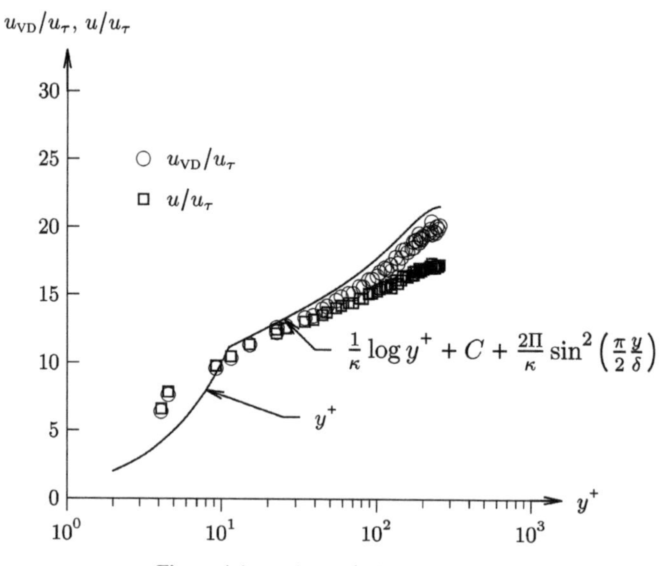


**Figure 4.4.** $u_{\mathrm{VD}}/u_\tau$ and $u/u_\tau$ versus $y^+$.

The experiments were performed in the United States Naval Ordnance Laboratory hypersonic wind tunnel no. 4 (section A.23.1). The measurements were obtained on the nozzle wall boundary layer. The freestream conditions are listed in table 4.4. Experimental measurements include boundary layer profiles of pitot and static pressure, total temperature, and wall measurements of temperature and heat transfer.[2].

[2] Note that Lobb *et al* (1955a, 1955b) define $T_e$ as the equilibrium (i.e., adiabatic) wall temperature.

**Table 4.2.** Freestream conditions.

| Quantity | Min | Max |
|---|---|---|
| $M_\infty$ | 5.0 | 5.0 |
| $p_{t_\infty}$ (MPa) | 0.10 | 0.81 |
| $T_{t_\infty}$ (K) | 430 | 430 |
| $H_\infty$ (MJ kg$^{-1}$) | 0.43 | 0.43 |
| $Re_\theta \times 10^{-3}$ | 0.988[a] | 10.68[a] |

[a]At $x = 50$ cm from throat.

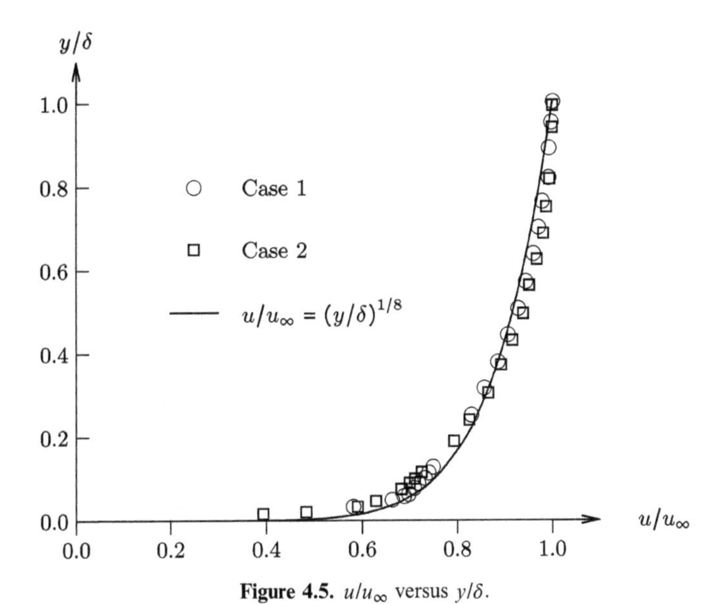


**Figure 4.5.** $u/u_\infty$ versus $y/\delta$.

Figures 4.7 and 4.8 display the normalized velocity $u/u_\tau$ and the normalized Van Driest transformed velocity $u_{VD}/u_\tau$ for case 1 ($(T_{aw} - T_w)/T_\infty = 0$ adiabatic wall) and case 4 ($(T_{aw} - T_w)/T_\infty = 2.36$, cold wall), respectively. Lobb *et al* (1955a) presented the former, and the latter was calculated from the experimental data using (3.41), (3.44) and (3.55). Also shown is the linear viscous sublayer $u/u_\tau = y^+$ and the Law of the Wall and Wake (3.56) using $\kappa = 0.4$, $C = 5.5$, and $\Pi = 0.55$ with $\delta^+ = 570.3$ and 820.9, respectively. The adiabatic wall experimental data show close agreement with the Law of the Wall and Wake using the Van Driest transformed velocity $u_{VD}/u_\tau$, while the normalized velocity $u/u_\tau$ shows significant differences. The cold wall data for both $u_{VD}/u_\tau$ and $u/u_\tau$ are essentially identical except in the outermost portion of the boundary layer, and they also are in good agreement with viscous sublayer profile and the Law of the Wall and Wake. The close agreement between $u_{VD}/u_\tau$ and $u/u_\tau$ in the cold wall case is in contrast with the significant difference between the two profiles for the adiabatic wall case (figure 4.7). In the former, the temperature ratio across the boundary layer $T_w/T_e = 3.27$, while in the latter case $T_w/T_e = 5.42$, and

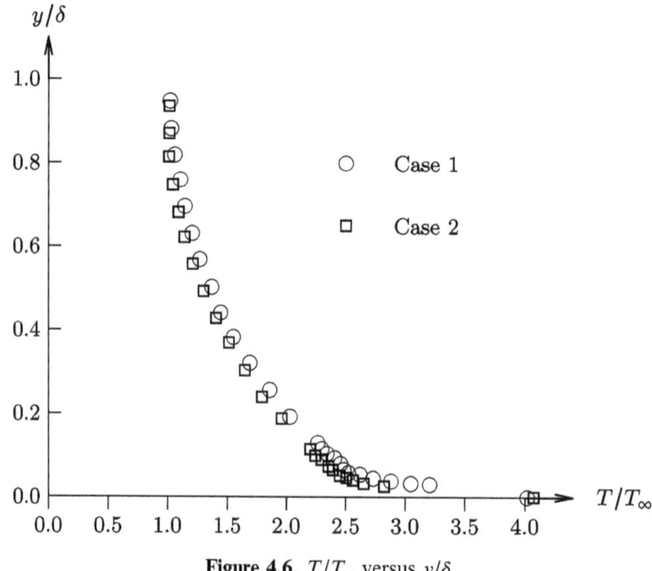


**Figure 4.6.** $T/T_\infty$ versus $y/\delta$.

**Table 4.3.** Boundary layer profiles.

| Quantity | Case 1 | Case 2 |
|---|---|---|
| $p_{t_\infty}$ (MPa) | 0.40 | 0.81 |
| $T_w/T_{aw}$ | 0.75 | 0.76 |
| $Re_\theta \times 10^{-3}$ | 4.23[a] | 10.65[a] |

[a]At $x$ = 50 cm from throat.

consequently from (3.38) there is a smaller difference between $u_{VD}$ and $u$ in the cold wall case.

## 4.4 Hill (1956)

Hill (1956) conducted an experimental investigation of a cold wall zero pressure gradient turbulent boundary layer in nitrogen at Mach 9. The experiments were performed in the hypersonic tunnel at the Applied Physics Laboratory, Johns Hopkins University (section A.4). The measurements were obtained on the nozzle wall boundary layer. The freestream conditions[3] are shown in table 4.5. Experimental measurements include boundary layer surveys of pitot pressure, static pressure, and total temperature. Figure 4.9 displays the velocity profile $u/u_\tau$ (not $u_{VD}/u_\tau$) versus $y^+$. The profiles show reasonably close resemblance to the viscous sublayer and Law of the Wall profiles.

[3] An additional experiment was performed at $M_e = 8.99$ and $p_{t_e} = 2.86$ MPa; however, the author indicates that the boundary layer was transitional, and therefore the data are omitted from the table.

**Table 4.4.** Freestream conditions.

| Quantity | Value Case 1[a] | Case 2[a] | Case 3[a] | Case 4[a] | Case 5[a] | Case 6[b] |
|---|---|---|---|---|---|---|
| $M_e$ | 4.93 | 5.01 | 5.03 | 5.06 | 5.75 | 5.79 |
| $p_{t_e}$ (MPa) | 0.312 | 0.508 | 0.795 | 0.869 | 1.36 | 1.70 |
| $T_{t_e}$ (K) | 325 | 399 | 513 | 562 | 401 | 477 |
| $T_w/T_e$ (K) | 5.42 | 4.29 | 3.49 | 3.27 | 6.19 | 5.35 |
| $(T_{aw} - T_w)/T_e$ | 0.0 | 1.23 | 2.07 | 2.36 | 1.05 | 1.98 |
| $T_w/T_{aw}$ | 1.0 | 0.78[c] | 0.63[c] | 0.59[c] | 0.90[c] | 0.77[c] |
| $H_e$ (MJ kg$^{-1}$) | 0.326 | 0.401 | 0.515 | 0.564 | 0.403 | 0.479 |
| $Re \times 10^{-6}$ (m$^{-1}$) | 8.57 | 9.15 | 9.75 | 8.46 | 18.0 | 16.6 |
| $\delta^*$ (cm) | 0.712 | 0.613 | 0.582 | 0.572 | 0.831 | 0.789 |
| $\theta$ (cm) | 0.0624 | 0.0708 | 0.0813 | 0.0871 | 0.0643 | 0.0745 |

| Quantity | Case 7[b] | Case 8[b] | Case 9[a] | Case 10[a] | Case 11[a] | Case 12[a] | Case 13[b] |
|---|---|---|---|---|---|---|---|
| $M_e$ | 5.82 | 6.78 | 6.78 | 6.83 | 6.83 | 7.67 | 8.18 |
| $p_{t_e}$ (MPa) | 1.72 | 2.14 | 2.17 | 1.55 | 2.89 | 2.45 | 3.22 |
| $T_{t_e}$ (K) | 551 | 586 | 639 | 468 | 586 | 643 | 655 |
| $T_w/T_e$ (K) | 4.41 | 5.22 | 4.64 | 6.34 | 5.24 | 5.94 | 6.6 |
| $(T_{aw} - T_w)/T_e$ | 2.98 | 4.05 | 4.63 | 3.05 | 4.16 | 5.66 | 7.03 |
| $T_w/T_{aw}$ | 0.63[c] | 0.57[c] | 0.51[c] | 0.68[c] | 0.56[c] | 0.52[c] | 0.51[c] |
| $H_e$ (MJ kg$^{-1}$) | 0.553 | 0.588 | 0.642 | 0.470 | 0.588 | 0.646 | 0.658 |
| $Re \times 10^{-6}$ (m$^{-1}$) | 13.2 | 10.2 | 9.02 | 10.9 | 14.0 | 7.73 | 8.37 |
| $\delta^*$ (cm) | 0.735 | 1.034 | 0.929 | 1.091 | 0.941 | 1.312 | 1.316 |
| $\theta$ (cm) | 0.0865 | 0.0825 | 0.0882 | 0.0784 | 0.0903 | 0.1052 | 0.0114 |

[a] Lobb *et al* (1955a).
[b] Lobb *et al* (1955b).
[c] Using 3.46 with $Pr_t = 0.89$.

## 4.5 Hill (1959)

Hill (1959b) investigated hypersonic turbulent boundary layers at Mach 8.25 to 10.06. The experiments were conducted in the hypersonic wind tunnel at the Johns Hopkins University Applied Physics Laboratory (section A.4) on the tunnel wall at 0.1524 cm upstream of the nozzle exit (station A) and 5.08 cm upstream of the nozzle exit (station B). The boundary layer was thus subject to a favorable pressure gradient. The freestream conditions are listed in table 4.6. Experimental measurements include boundary layer profiles of pitot pressure, static pressure, and total pressure and surface heat transfer and skin friction.

Figure 4.10 displays the Reynolds Analogy Factor $2St/c_f$ versus Mach number. The measurements fall slightly below $Pr_t^{-1}$.

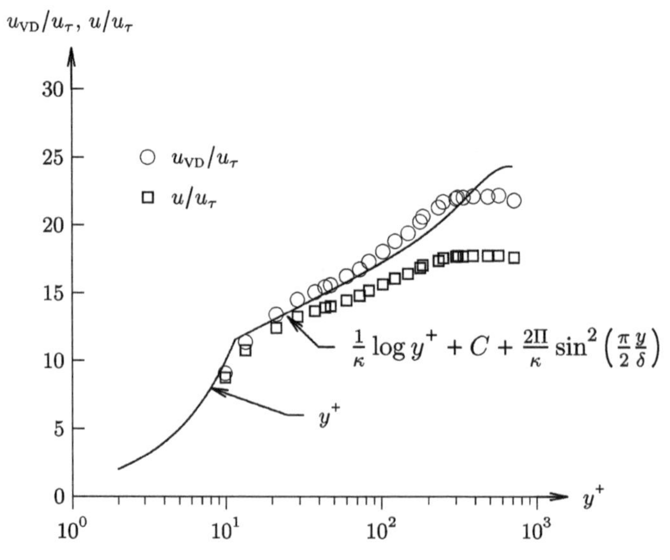


**Figure 4.7.** $u_{VD}/u_\tau$ and $u/u_\tau$ versus $y^+$ at $M_e = 5$ for $(T_{aw} - T_w)/T_\infty = 0$.

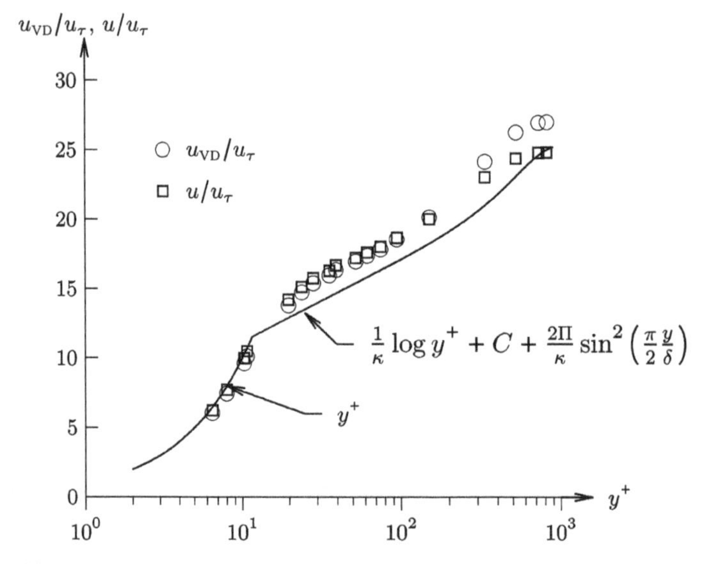


**Figure 4.8.** $u_{VD}/u_\tau$ and $u/u_\tau$ versus $y^+$ at $M_e = 5$ for $(T_{aw} - T_w)/T_\infty = 2.36$.

## 4.6 Winkler and Cha (1959)

Winkler and Cha (1959) and Winkler (1961) investigated a zero pressure gradient isothermal turbulent boundary layer at Mach 5.2. The experiments were performed in the Naval Ordnance Laboratory hypersonic tunnel no. 4 (section A.23.1) on an internally cooled flat plate. The freestream conditions are listed in table 4.7.

**Table 4.5.** Freestream conditions.

| Quantity | Value | | |
|---|---|---|---|
| | Case 1 | Case 2 | Case 3 |
| $M_e$ | 9.04 | 9.07 | 9.10 |
| $p_{t_e}$ (MPa) | 3.55 | 4.24 | 5.27 |
| $T_{t_e}$ (K) | 756 | 756 | 756 |
| $T_w$ (K) | 347 | 358 | 374 |
| $T_w/T_{t_e}$ | 0.49 | 0.50 | 0.52 |
| $T_w/T_{aw}$ | 0.55[a] | 0.56[a] | 0.58[a] |
| $H_e$ (MJ kg$^{-1}$) | 0.84 | 0.84 | 0.84 |
| $Re \times 10^{-6}$ (m$^{-1}$) | 5.1 | 6.1 | 7.5 |
| $\delta$ (cm) | 0.87 | 0.86 | 0.85 |
| $\theta$ (cm) | 0.031 | 0.031 | 0.031 |

[a] Using (3.46) with $Pr_t = 0.89$.

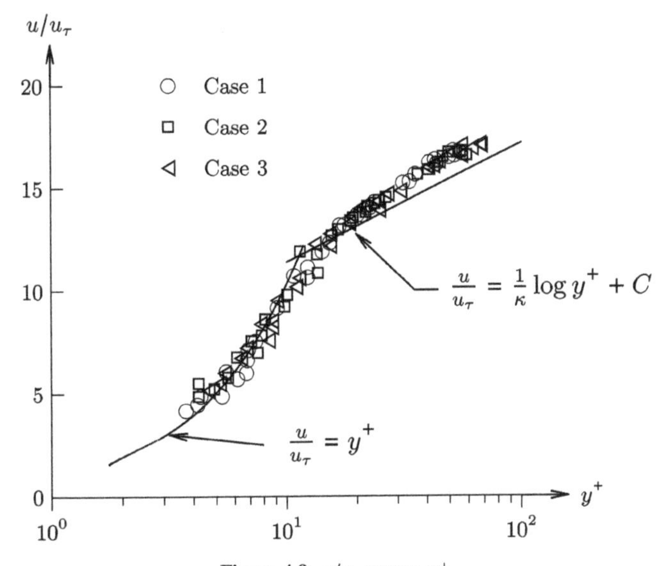


**Figure 4.9.** $u/u_\tau$ versus $y^+$.

Experimental measurements include boundary layer surveys of pitot pressure and total temperature, wall static pressure, and heat transfer.

Figures 4.11, 4.12, and 4.13 display the Van Driest transformed velocity $u_{VD}/u_\tau$ and normalized velocity $u/u_\tau$ for $T_w/T_{t_e} = 0.83$, 0.77, and 0.61, respectively. The value of $u_\tau$ is obtained from the measured skin friction based upon the velocity profile closest to the wall. Also shown is the viscous sublayer profile $u/u_\tau = y^+$ and Law of the Wall and Wake with $\Pi = 0.55$. The Van Driest transformed profile shows close agreement with the Law of the Wall and Wake for $T_w/T_{t_e} = 0.83$, but it

**Table 4.6.** Freestream conditions.

| Quantity | Min | Max |
|---|---|---|
| $M_e$ | 8.25 | 10.06 |
| $p_{t_e}$ (MPa) | 3.45 | 6.21 |
| $T_{t_e}$ (K) | 756 | 825 |
| $T_w$ (K) | 317 | 383 |
| $T_w/T_{aw}$ | 0.47 | 0.52 |
| $H_e$ (MJ kg$^{-1}$) | 0.76 | 0.83 |
| $Re_\theta$ | 1300 | 2965 |
| $\delta^*$ (cm) | 0.325 | 0.577 |
| $\theta$ (cm) | 0.0225 | 0.0315 |

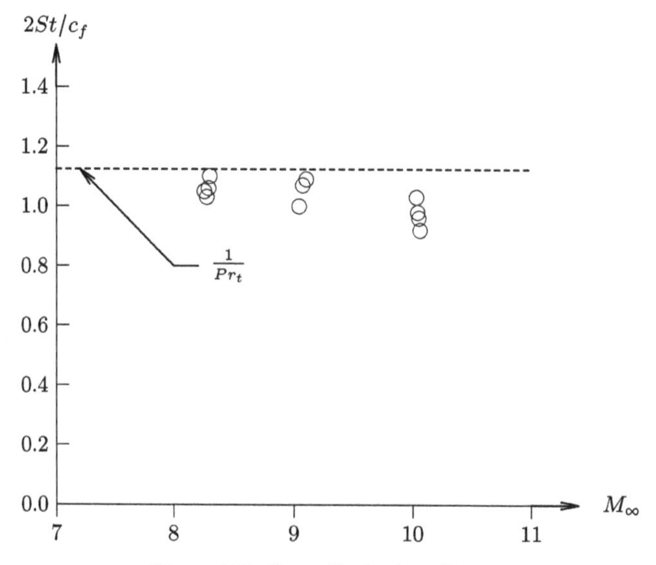


**Figure 4.10.** Reynolds Analogy Factor.

**Table 4.7.** Freestream conditions.

| Quantity | Min | Max |
|---|---|---|
| $M_e$ | 4.98 | 5.29 |
| $p_{t_e}$ (MPa) | 0.549 | 0.761 |
| $T_{t_e}$ (K) | 345 | 498 |
| $T_w/T_{t_e}$ | 0.6 | 0.85 |
| $T_w/T_{aw}$ | 0.66[a] | 0.94[a] |
| $H_e$ (MJ kg$^{-1}$) | 0.346 | 0.500 |
| $Re \times 10^{-6}$ (m$^{-1}$) | 8.8 | 15.5 |
| $\delta$ (cm) | 0.26 | 0.88 |
| $\delta^*$ (cm) | 0.122 | 0.400 |
| $\theta$ (cm) | 0.0114 | 0.0343 |

[a] Using (3.46) with $Pr_t = 0.89$.

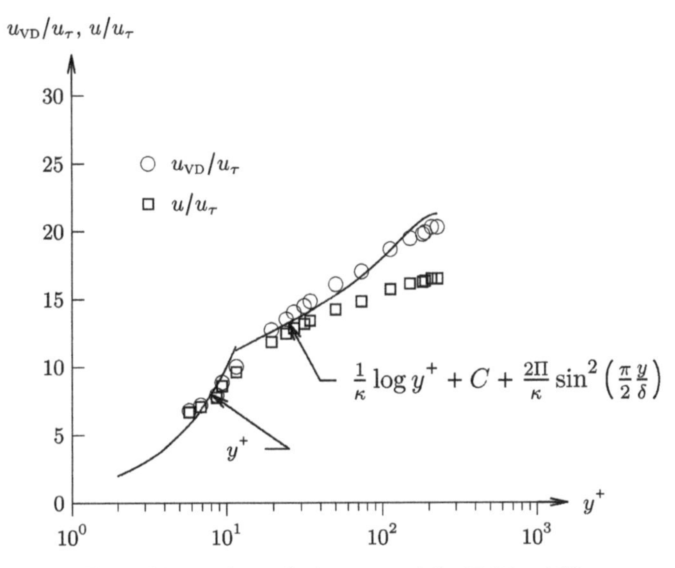


**Figure 4.11.** $u_{VD}/u_\tau$ and $u/u_\tau$ versus $y^+$ for $T_w/T_{t_e} = 0.83$.

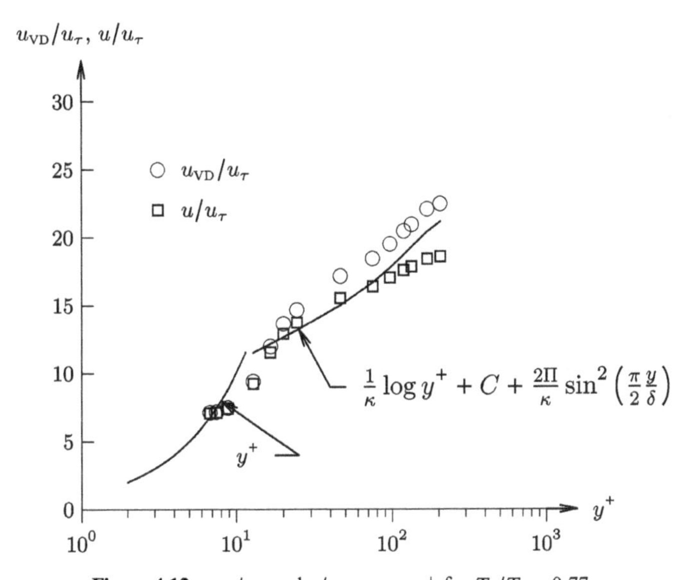


**Figure 4.12.** $u_{VD}/u_\tau$ and $u/u_\tau$ versus $y^+$ for $T_w/T_{t_e} = 0.77$.

increasingly differs for $T_w/T_{t_e} = 0.77$ and 0.61. However, the height of the first velocity measurement is at $y^+ = 5.78$, 6.79, and 18.83 for $T_w/T_{t_e} = 0.83$, 0.77, and 0.61, respectively, and consequently at least for $T_w/T_{t_e} = 0.61$ the measured velocity profile is not within the viscous sublayer.

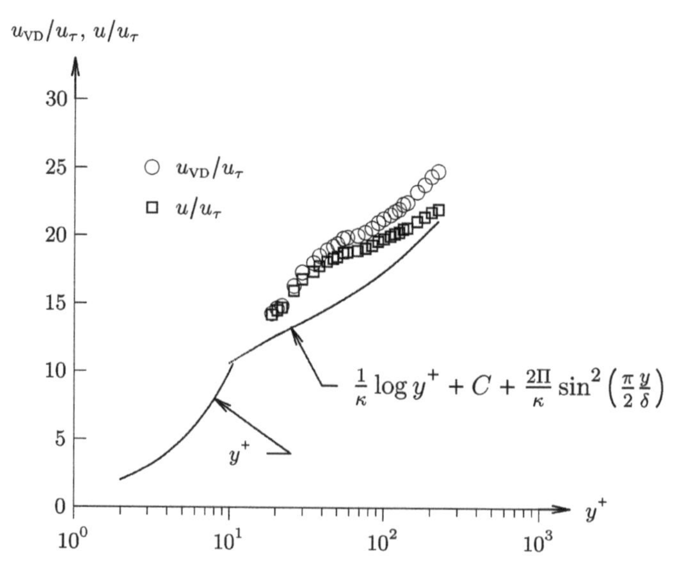


**Figure 4.13.** $u_{VD}/u_\tau$ and $u/u_\tau$ versus $y^+$ for $T_w/T_{t_e} = 0.61$.

## 4.7 Danberg (1964)

Danberg (1964) investigated a cold wall zero pressure gradient hypersonic turbulent boundary layer at Mach 6.7 with and without surface mass transfer.[4] The experiment was conducted in the Naval Ordnance Laboratory hypersonic tunnel no. 4 (section A.23.1) on a flat plate with a porous insert. The plate dimensions were 59.06 cm long and 25.4 cm wide. The porous insert was constructed of sintered stainless steel powder spheres (40 micron diameter) and was 48.85 cm long and 17.78 cm wide. The solidity of the porous insert was 42.7%. The freestream conditions are listed in table 4.8. Experimental measurements include boundary layer surveys of pitot pressure and total temperature, wall static pressure, and heat transfer. Profiles were obtained at four locations from the leading edge of the flat plate: 37.78 cm, 42.86 cm, 47.94 cm, and 53.02 cm. Measurements were obtained for four values of the mass transfer coefficient $c_q = \rho_w v_w / \rho_e u_e$ from $c_q = 0$ to $2.5 \times 10^{-3}$.

Figure 4.14 displays the Mach number versus distance from the wall for $c_q = 0$ to $2.5 \times 10^{-3}$ for $T_w/T_e = 4$ and $Re_x = 4 \times 10^6$. The boundary layer thickness increases significantly with the surface mass transfer coefficient with a near doubling between $c_q = 0$ and $c_q = 2.5 \times 10^{-3}$. Figure 4.15 presents the Stanton number $St$ versus skin friction coefficient $c_f$ for three different values of $T_w/T_e$ from 4.1 to 7.6 and multiple locations along the porous plate. The Reynolds analogy $St = \frac{1}{2}c_f$ holds approximately. Figure 4.16 shows the constant $C$ in the law of the wall versus $(T_{aw} - T_w)/T_{aw}$ to examine the effects of surface heat transfer. Data are also included from Monaghan and Cooke (1951), Lobb *et al* (1955b), and Winkler and Cha (1959). Also shown in the range of values for an adiabatic wall. The effect of heat

[4] Danberg (1967) contains a full tabulation of the experimental data.

**Table 4.8.** Freestream conditions.

| Quantity | Min | Max |
|---|---|---|
| $M_e$ | 6.7 | 6.7 |
| $p_{t_e}$ (MPa) | 1.54 | 3.85 |
| $T_{t_e}$ (K) | 550 | 550 |
| $T_w/T_e$ | 3.73 | 7.87 |
| $H_e$ (MJ kg$^{-1}$) | 0.50 | 0.50 |
| $T_w/T_{aw}$ | 0.41[a] | 0.87[a] |
| $Re \times 10^{-6}$ (m$^{-1}$) | 8.0 | 19.0 |
| $\delta$ (cm) | 0.53 | 2.40 |
| $\delta^*$ (cm) | 0.28 | 1.74 |
| $\theta$ (cm) | 0.0238 | 0.274 |

[a] Using (3.46) with $Pr_t = 0.89$.

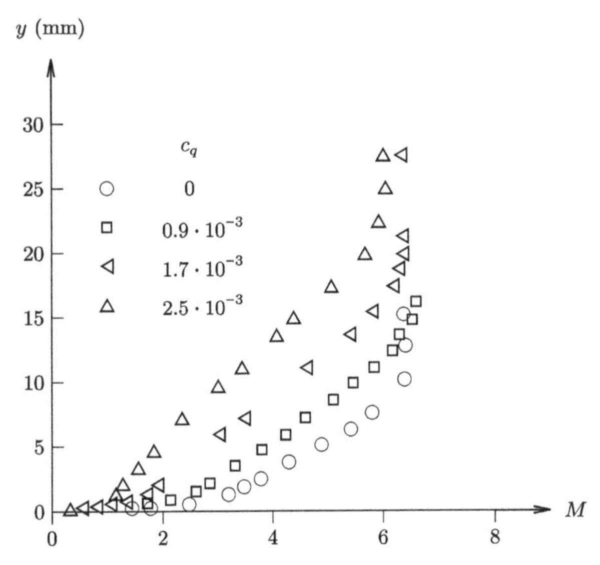


**Figure 4.14.** $M$ versus $y$ for $c_q = 0$ to $2.5 \times 10^{-3}$ at $T_w/T_e = 4$.

transfer to the wall is to increase the value of $C$, while the opposite holds for heat transfer to the gas. Figures 4.17 and 4.18 display the dimensionless total temperature ratio $(\tilde{T}_t - T_w)(T_{t_\infty} - T_w)$ versus $u/u_e$ for $T_w/T_e = 4.7$ and 7.07. Also shown are the Crocco relation (3.144) and the corresponding result using Walz's relation for temperature (3.174). With minor exceptions, the data correlate closely with (3.174).

Figure 4.19 displays the Van Driest transformed velocity $u_{VD}$ versus $y^+$, with $u_{VD}$ being computed using (3.38), for $T_w/T_e = 4.7$ and 7.07 at $c_q = 0$. The effect of heat transfer to the wall is to increase the value of $C$. Figure 4.20 shows $u^{++}$ versus $y^+$ for $c_q = 0$ to $1.21 \times 10^{-3}$ for $T_w/T_e \approx 4$, and figure 4.21 shows $u^{++}$ versus $y^+$ for $c_q = 0$ to

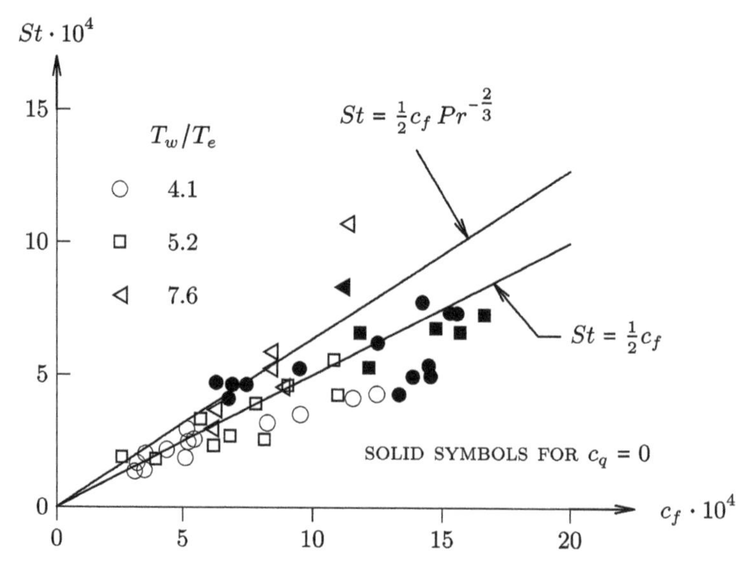


**Figure 4.15.** $St$ versus $c_f$ for $T_w/T_e = 4.1$ to 7.6.

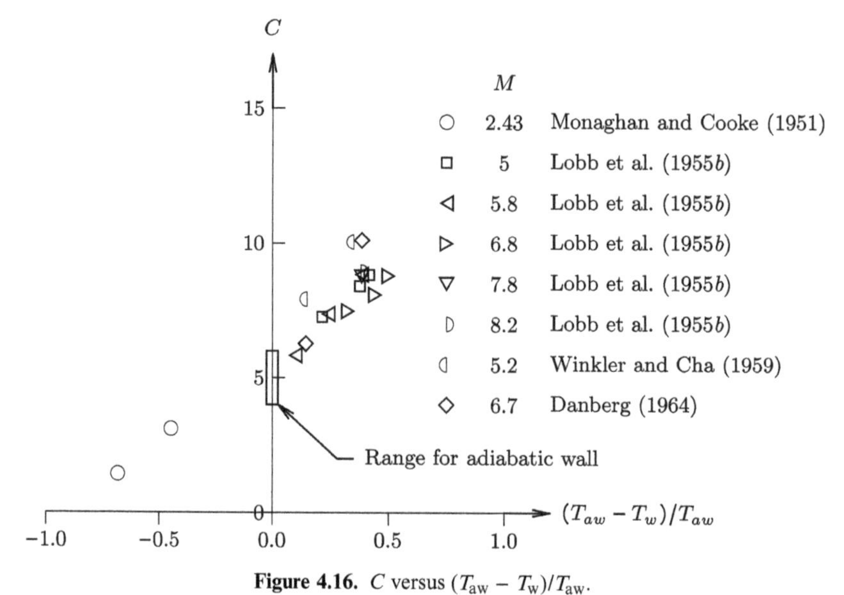


**Figure 4.16.** $C$ versus $(T_{aw} - T_w)/T_{aw}$.

$1.01 \times 10^{-3}$ for $T_w/T_e \approx 7$. In both cases the value of $u^{++}$ is obtained from (3.84). The effect of increasing mass transfer is to reduce the value of $C$.

## 4.8 Adcock *et al* (1965)

Adcock and Peterson (1965) conducted an experimental examination of a zero pressure gradient adiabatic turbulent boundary layer on the external surface of a cylinder at Mach 6. The model is shown in figure 4.22. The experiments were

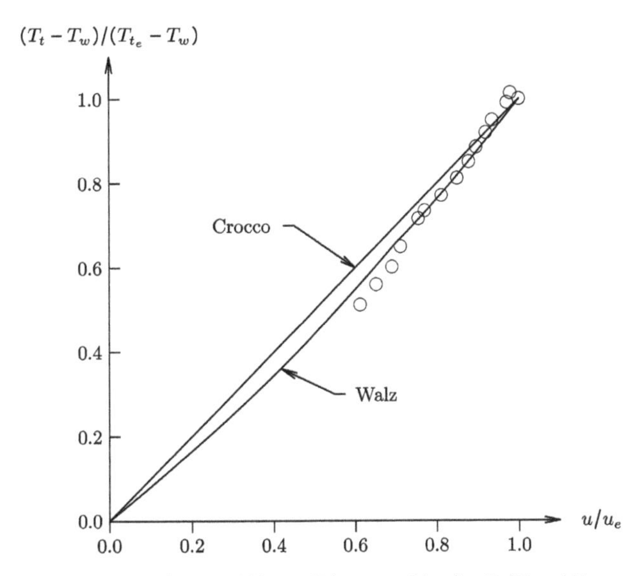


**Figure 4.17.** $(T_t - T_w)/(T_{t_e} - T_w)$ versus $\tilde{u}/u_e$ for $T_w/T_e = 4.7$.

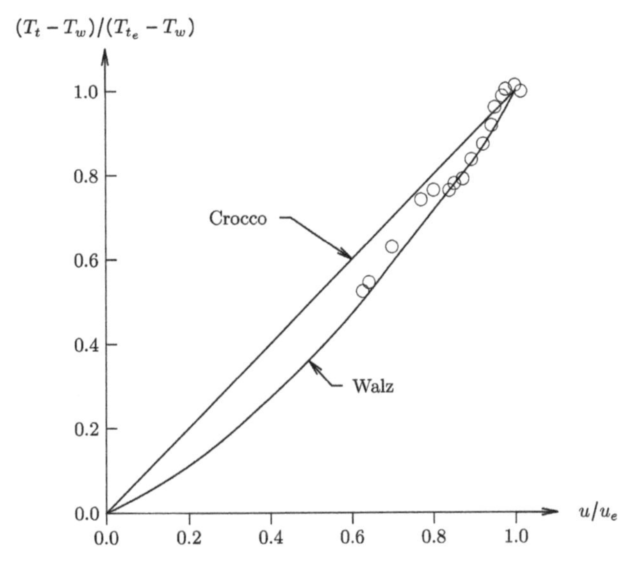


**Figure 4.18.** $(T_t - T_w)/(T_{t_e} - T_w)$ versus $\tilde{u}/u_e$ for $T_w/T_e = 7.07$.

performed at the NASA Langley 20 in Mach 6 tunnel. The measurements were obtained on the outer surface of a hollow cylinder at selected distances $x_{le}$ from the leading edge. The freestream conditions are listed in table 4.9. Experimental measurements include boundary layer surveys of pitot pressure, static pressure, and total temperature and surface pressure and temperature.

Figure 4.23 displays the normalized velocity $u/u_\tau$ and Van Driest transformed velocity $u_{VD}/u_\tau$ for case 9. Also shown is the viscous sublayer and Law of the Wall and Wake. In the latter, two different values of $\Pi$ are shown. The Van Driest

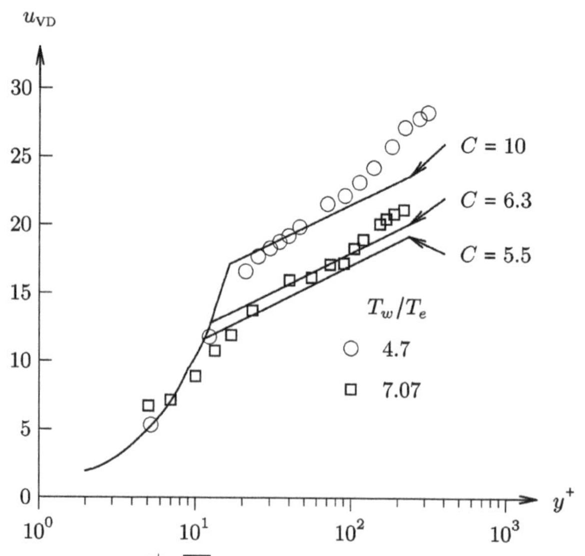


**Figure 4.19.** $\int_o^{u^+} \sqrt{\frac{T_W}{T}}\, du^+$ versus $y^+$ for $T_W/T_e = 4.7$ and 7.07.

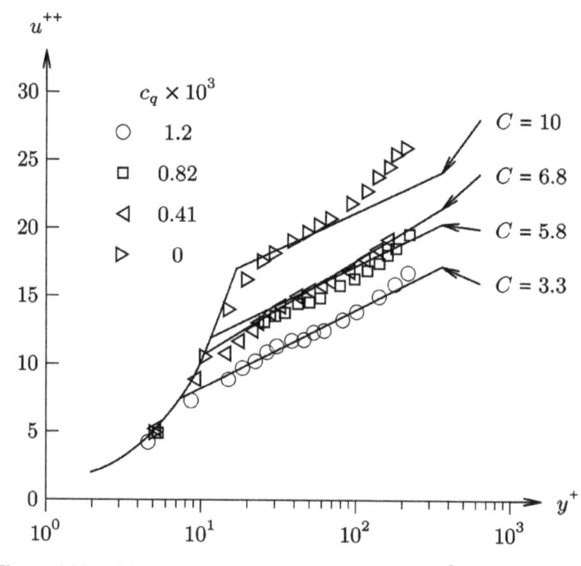


**Figure 4.20.** $u^{++}$ versus $y^+$ for $c_q = 0$ to $1.21 \times 10^{-3}$ for $T_W/T_e \approx 4.7$.

transformed velocity shows good agreement with the Law of the Wall and Wake, while the untransformed velocity $u/u_\tau$ shows significant disagreement. The lowest values of $u$ may be inaccurate due to interference between the pitot probe and the wall.[5]

[5] The pitot probe height is 0.178 mm, and the first through third data points are at a distance 0.134 mm from the wall.

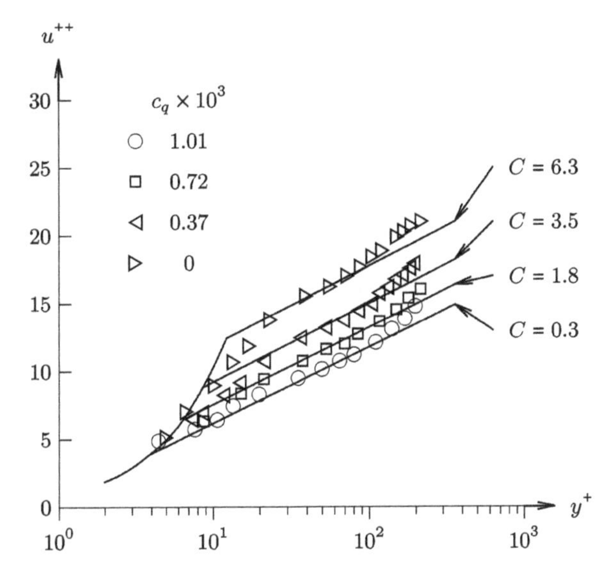


**Figure 4.21.** $u^{++}$ versus $y^+$ for $c_q = 0$ to $1.01 \times 10^{-3}$ for $T_w/T_e \approx 7$.

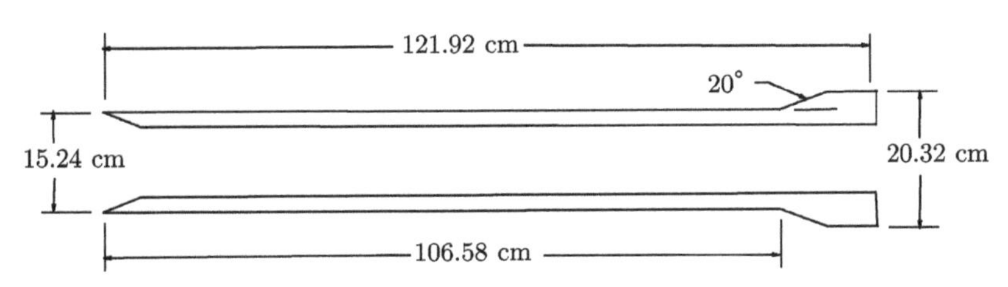


**Figure 4.22.** Model.

## 4.9 Young (1965)

Young (1965) conducted an experimental investigation of smooth and rough wall hypersonic turbulent boundary layers at Mach 5 for isothermal and adiabatic walls. The experiments were performed in the University of Texas Defense Research Laboratory Mach 5 tunnel. The measurements were made on a flat plate 48.77 cm in length and 15.19 cm in width for both smooth and rough surfaces. The surface roughness was milled spanwise V-grooves with 90° apex angle at the trough of each groove and with pitch values $P$ equal to 0.127 mm, 0.254 mm, and 0.762 mm and corresponding depths $H = \frac{1}{2}P$. Boundary layer profiles measurements were obtained at a location 31.75 cm from the leading edge. The freestream conditions are listed in table 4.10. Five separate experiments were performed for each of the four wall conditions (cases 1 to 4) with different values of $T_w/T_{aw}$. Experimental diagnostics include boundary layer profiles of pitot pressure and stagnation temperature and wall measurements of heat transfer, skin friction, static pressure, and temperature.

**Table 4.9.** Freestream conditions.

| Quantity | Value | | | | |
|---|---|---|---|---|---|
| | Case 1 | Case 2 | Case 3 | Case 4 | Case 5 |
| $M_e$ | 5.95 | 5.95 | 5.95 | 5.96 | 5.96 |
| $p_{t_e}$ (MPa) | 3.61 | 3.61 | 3.58 | 3.61 | 3.61 |
| $T_{t_e}$ (K) | 488 | 484 | 484 | 491 | 484 |
| $T_w$ (K)[a] | 465 | 461 | 461 | 466 | 462 |
| $H_e$ (MJ kg$^{-1}$) | 0.490 | 0.486 | 0.486 | 0.493 | 0.486 |
| $Re \times 10^{-6}$ (m$^{-1}$) | 32.4 | 32.9 | 32.9 | 32.4 | 32.9 |
| $\delta^*$ (cm) | 0.121 | 0.168 | 0.208 | 0.257 | 0.292 |
| $\theta$ (cm) | 0.00668 | 0.00998 | 0.0124 | 0.0153 | 0.0171 |
| $x_{le}$ (cm) | 12.7 | 15.2 | 20.3 | 27.9 | 27.9 |
| | Case 6 | Case 7 | Case 8 | Case 9 | Case 10 |
| $M_e$ | 6.02 | 6.02 | 6.02 | 6.02 | 6.02 |
| $p_{t_e}$ (MPa) | 3.61 | 3.61 | 3.61 | 3.61 | 3.61 |
| $T_{t_e}$ (K) | 484 | 482 | 484 | 478 | 477 |
| $T_w$ (K)[a] | 462 | 457 | 462 | 453 | 453 |
| $H_e$ (MJ kg$^{-1}$) | 0.486 | 0.484 | 0.486 | 0.480 | 0.479 |
| $Re \times 10^{-6}$ (m$^{-1}$) | 34.3 | 34.8 | 34.3 | 35.4 | 35.3 |
| $\delta^*$ (cm) | 0.714 | 0.724 | 0.772 | 0.754 | 0.714[b] |
| $\theta$ (cm) | 0.0417 | 0.0427 | 0.0457 | 0.0345 | 0.0417[b] |
| $x_{le}$ (cm) | 83.8 | 94.0 | 101.6 | 101.6 | 83.8 |

[a] $T_w = T_{aw}$.
[b] Assumed same as case 6.

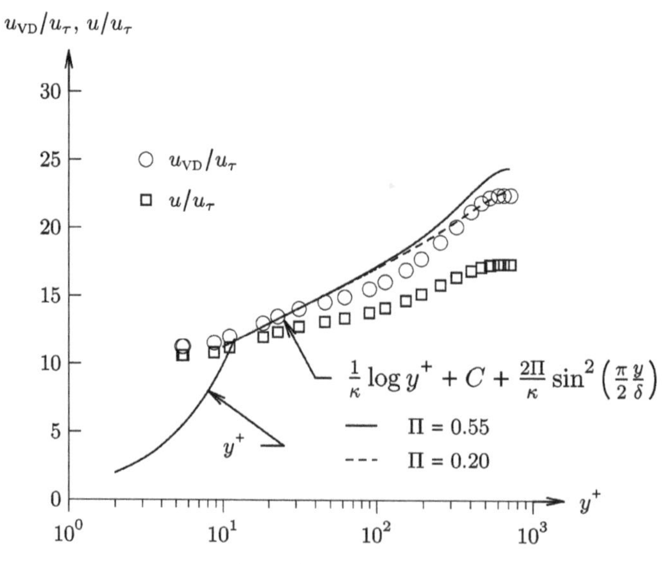


**Figure 4.23.** $u_{VD}/u_\tau$ and $u/u_\tau$ versus $y^+$.

**Table 4.10.** Freestream conditions.

| Quantity | Case 1 | | Case 2 | | Case 3 | | Case 4 | |
|---|---|---|---|---|---|---|---|---|
| | Smooth | | $P = 0.127$ *mm* | | $P = 0.254$ *mm* | | $P = 0.762$ *mm* | |
| | Min | Max | Min | Max | Min | Max | Min | Max |
| $M_e$ | 4.91 | 4.91 | 4.95 | 4.95 | 4.94 | 4.94 | 4.95 | 4.95 |
| $p_{t_e}$ (MPa) | 1.758 | 1.758 | 1.758 | 1.758 | 1.758 | 1.758 | 1.758 | 1.758 |
| $T_{t_e}$ (K) | 334 | 638 | 358 | 623 | 344 | 627 | 337 | 623 |
| $T_w$ (K) | 297.3 | 304.9 | 308.4 | 316.1 | 299.0 | 310.6 | 297.9 | 307.8 |
| $T_w/T_{aw}$ | 0.538 | 1.00 | 0.570 | 1.00 | 0.557 | 1.00 | 0.551 | 1.00 |
| $H_e$ (MJ kg$^{-1}$) | 0.335 | 0.641 | 0.359 | 0.625 | 0.345 | 0.630 | 0.338 | 0.625 |
| $Re \times 10^{-6}$ (m$^{-1}$) | 17.2 | 45.2 | 18.4 | 47.6 | 17.2 | 45.6 | 16.4 | 45.8 |
| $Hu_\tau/\nu_w$ | 0. | 0. | 5.2 | 6.3 | 11.5 | 13.6 | 40.4 | 47.1 |
| $\delta$ (cm) | 0.572 | 0.762 | 0.572 | 0.826 | 0.572 | 0.762 | 0.699 | 0.826 |
| $\theta$ (cm) | 0.0205 | 0.0281 | 0.0206 | 0.0258 | 0.0218 | 0.0269 | 0.0275 | 0.0359 |

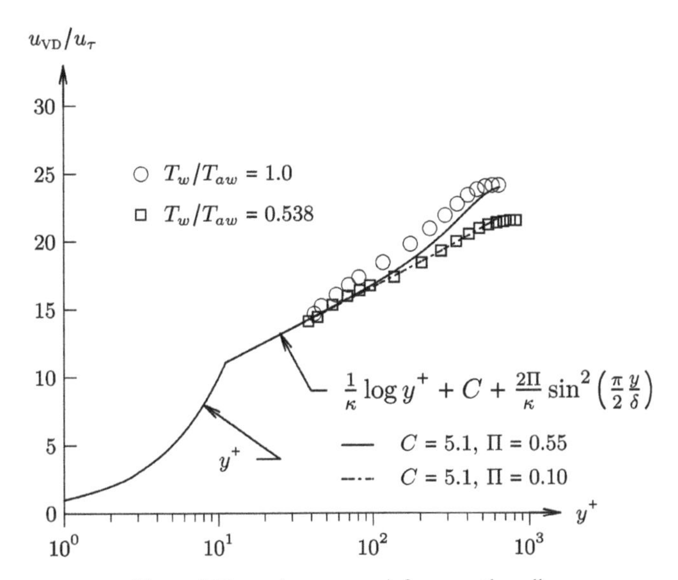


**Figure 4.24.** $u_{VD}/u_\tau$ versus $y^+$ for smooth wall.

Figures 4.24, 4.25, 4.26, and 4.27 display the Van Driest transformed velocity $u_{VD}/u_\tau$ versus $y^+$ for the four cases. Results are shown for adiabatic wall and the coldest wall condition. For case 1, the Van Driest transformed velocity matches closely with the Law of the Wall and Wake with $C = 5.1$ for both adiabatic wall and $T_w/T_{aw} = 0.538$. According to figure 4.16, $C \approx 8.5$ for $T_w/T_{aw} = 0.538$, thus indicating a significant difference between the two experiments for a cold smooth wall. Note that the edge Mach numbers differ with $M_e = 4.95$ for Young (1965) versus $M_e = 6.7$

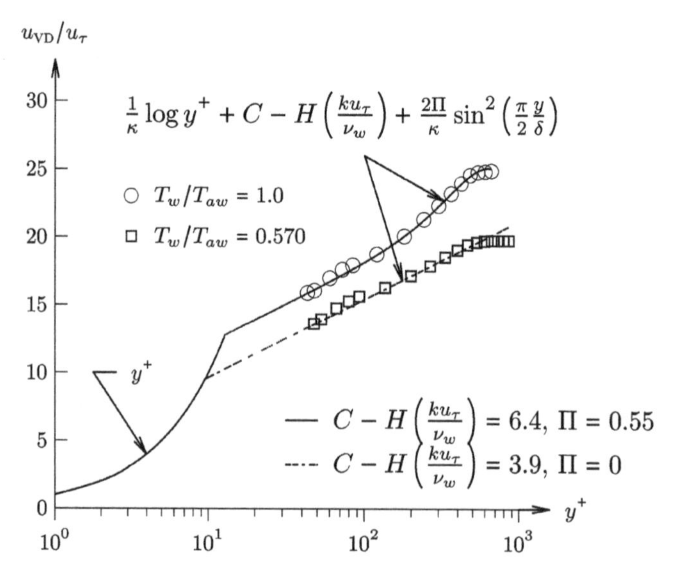


**Figure 4.25.** $u_{VD}/u_\tau$ versus $y^+$ for rough wall ($P$ = 0.127 mm).

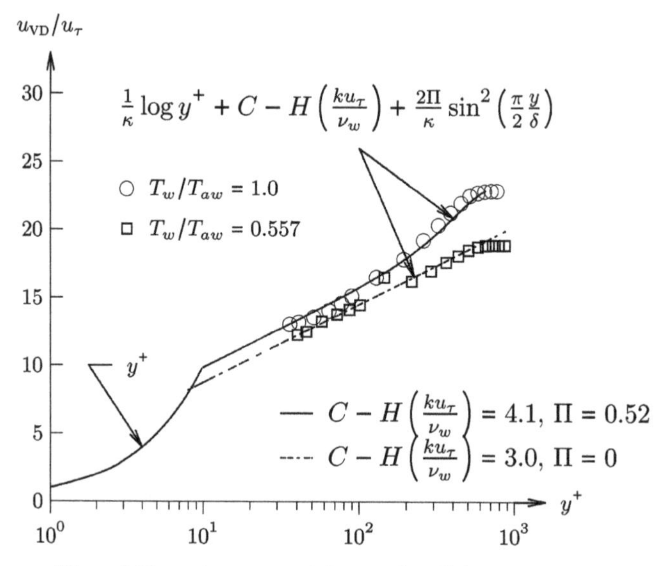


**Figure 4.26.** $u_{VD}/u_\tau$ versus $y^+$ for rough wall ($P$ = 0.254 mm).

for Danberg (1964). Figures 4.25, 4.26, and 4.27 show the Van Driest transformed velocity $u_{VD}/u_\tau$ versus $y^+$ for the three rough wall cases with results shown for adiabatic wall and the lowest values of $T_w/T_{aw}$. The resultant values of $C - H(ku_\tau/\nu_w)$ are shown in figure 4.28. With one exception, the results indicate $C - H(ku_\tau/\nu_w)$ decreases with increasing roughness.

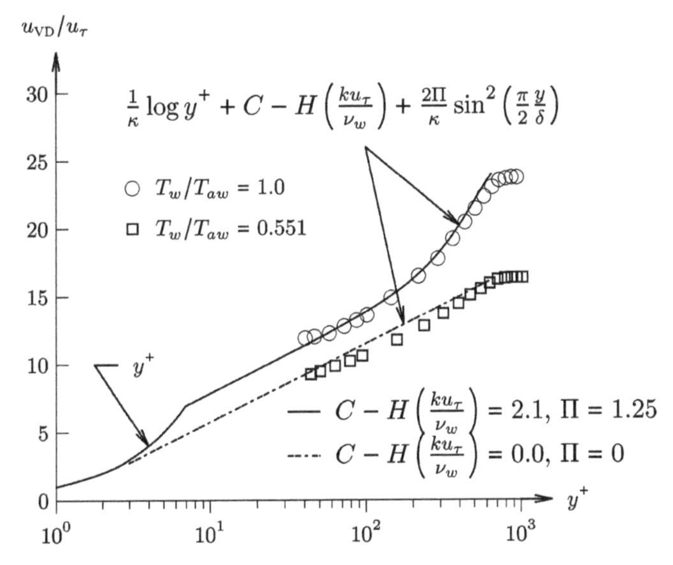


**Figure 4.27.** $u_{VD}/u_\tau$ versus $y^+$ for rough wall ($P = 0.762$ mm).

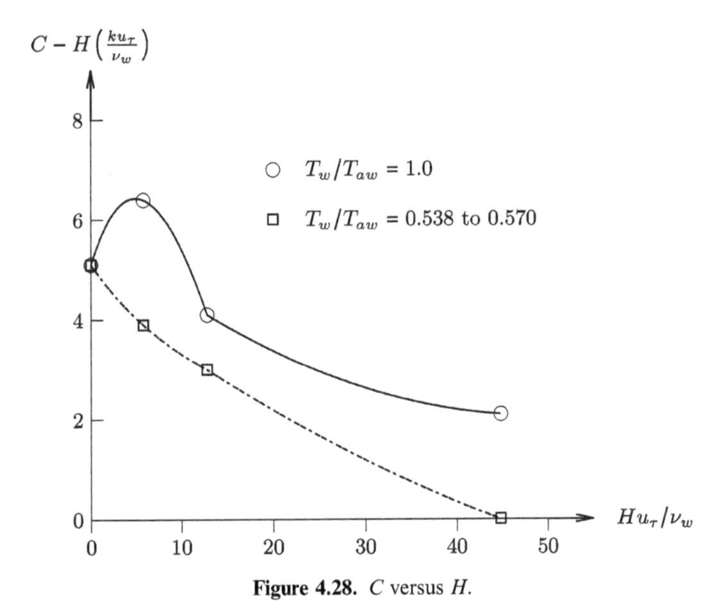


**Figure 4.28.** $C$ versus $H$.

## 4.10 Scaggs (1966)

Scaggs (1966) performed a series of experiments on a cold wall hypersonic turbulent boundary layer at Mach 6.5 and 11.5. The experiments were conducted in the Ohio State University 12 in hypersonic wind tunnel. The boundary layer on the wall of the nominal Mach 6 and Mach 12 nozzles was surveyed at the locations indicated in table 4.11 where $x_e$ is the $x$-coordinate of the nozzle exit and $x$ is the locations of the

**Table 4.11.** Location of surveys.

| | Survey locations | | | | | | | |
|---|---|---|---|---|---|---|---|---|
| | $M = 7$ | | | | $M = 12$ | | | |
| Quantity | 1 | 2 | 3 | 4 | 1 | 2 | 3 | 4 |
| $x_e - x$ (cm) | 8.89 | 8.89 | 26.67 | 26.67 | 11.43 | 11.43 | 30.48 | 30.48 |
| Pitot pressure surveys | | | | | | | | |
| $M_e$ | 6.65 | 6.72 | 6.46 | 6.54 | 11.52 | 11.66 | 11.30 | 11.40 |
| $Re_x \times 10^{-6}$ | 0.71 | 2.24 | 0.66 | 2.05 | 1.28 | 2.82 | 1.18 | 2.68 |
| $T_w/T_{t_\infty}$ | 0.3118 | 0.4082 | 0.3167 | 0.4473 | 0.2684 | 0.2913 | 0.3752 | 0.3010 |
| $\delta$ (cm) | 3.87 | 3.60 | 3.29 | 2.77 | 6.13 | 4.85 | 5.61 | 4.60 |
| Total temperature surveys | | | | | | | | |
| $M_e$ | 6.65 | 6.72 | 6.46 | 6.54 | 11.52 | 11.66 | 11.30 | 11.40 |
| $Re_x \times 10^{-6}$ | 0.71 | 2.24 | 0.61 | 2.05 | 1.28 | 2.82 | 1.18 | 2.68 |
| $T_w/T_{t_\infty}$ | 0.3118 | 0.4082 | 0.3167 | 0.4473 | 0.2684 | 0.2913 | 0.2752 | 0.3010 |
| $\delta$ (cm) | 3.87 | 3.56 | 3.29 | 2.79 | 6.13 | 4.85 | 5.61 | 4.60 |

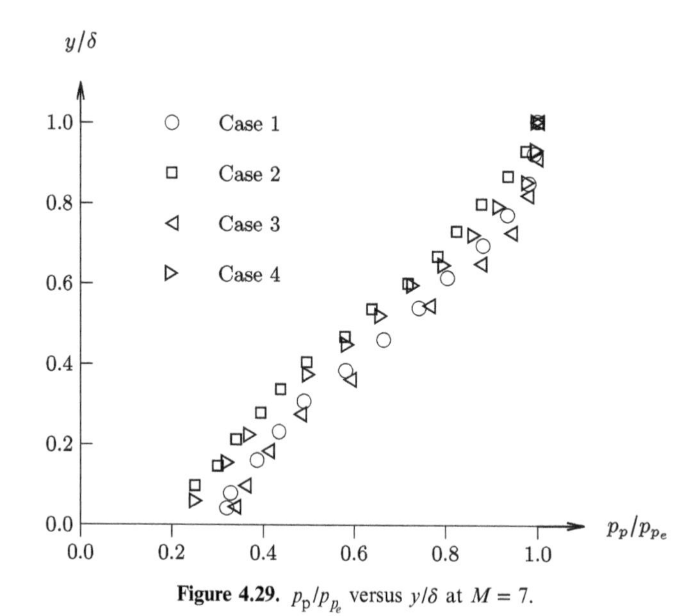


**Figure 4.29.** $p_p/p_{p_e}$ versus $y/\delta$ at $M = 7$.

measurements. Experimental diagnostics include boundary layer surveys of pitot pressure and total temperature and wall pressure and temperature.

Figures 4.29 and 4.30 display the normalized pitot pressure $p_p/p_{p_e}$ versus $y/\delta$ at nominal Mach numbers of 7 and 12, respectively, for cases 1 to 4 for each Mach number (table 4.11). All profiles at Mach 7 indicate a turbulent boundary layer; however, only case 2 at Mach 12 indicates a turbulent flow. Figures 4.31 and 4.32

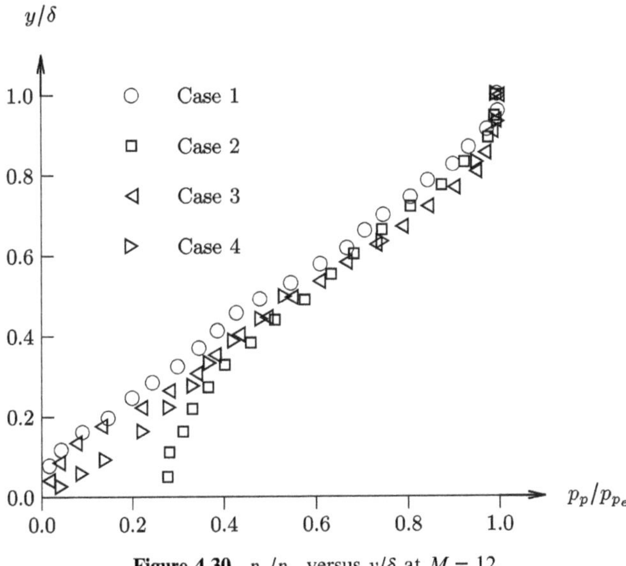


**Figure 4.30.** $p_p/p_{p_e}$ versus $y/\delta$ at $M = 12$.

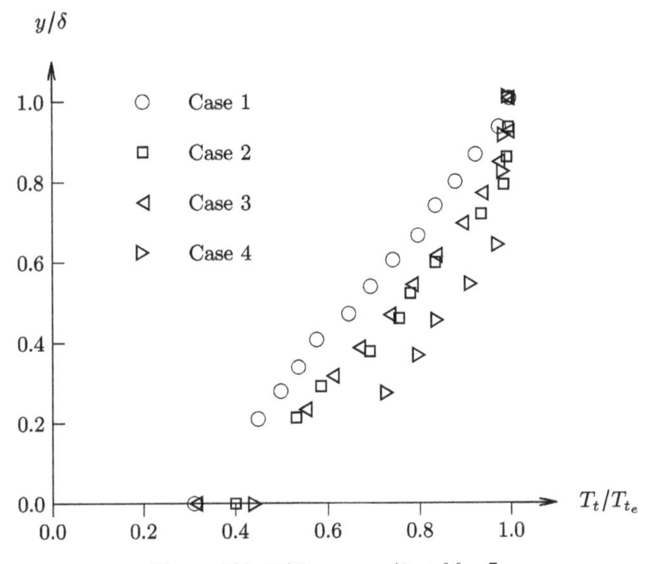


**Figure 4.31.** $T_t/T_{t_e}$ versus $y/\delta$ at $M = 7$.

show the normalized total temperature profiles $T_t/T_{t_e}$ versus $y/\delta$ for nominal Mach numbers of 7 and 12, respectively, for cases 1 to 4 for each (table 4.11).

## 4.11 Samuels *et al* (1967)

Samuels *et al* (1967) investigated a cold wall zero pressure gradient turbulent boundary layer in air at Mach 6. The experiments were conducted in the NASA Langley 20 in Mach 6 tunnel. The measurements were made on the outer surface of

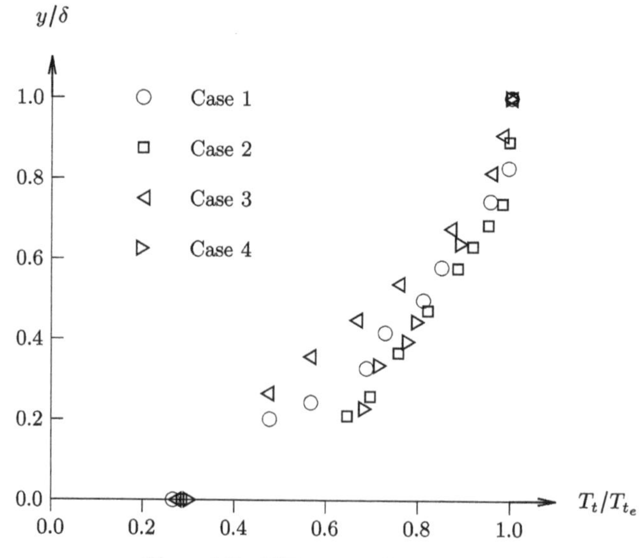


**Figure 4.32.** $T_t/T_{t_e}$ versus $y/\delta$ at $M = 12$.

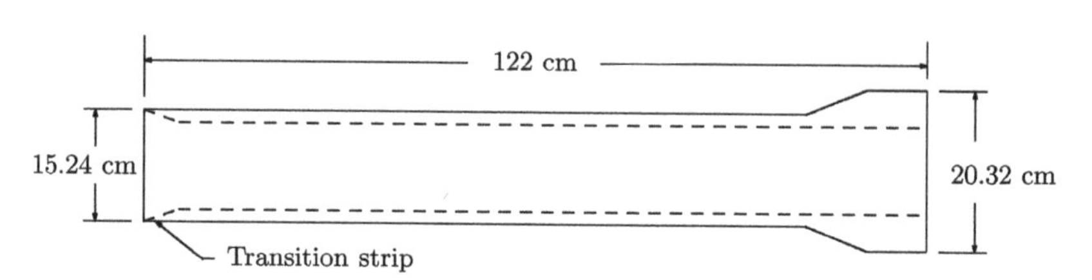


**Figure 4.33.** Model.

a hollow cylinder flare model (figure 4.33). The outer diameter of the model is 15.24 cm for the first 107 cm followed by a 20° flare to an outer diameter of 20.32 cm. The total length is 122 cm. Boundary layer profile measurements were obtained at five stations from 20 cm to 102 cm from the leading edge. The cylinder walls were cooled using liquid freon pumped through the hollow walls of the cylinder. A boundary layer trip was located at a distance of 2.92 cm from the leading edge. The freestream conditions are shown in table 4.12. Experimental measurements include boundary layer profiles of pitot pressure and total temperature and wall measurements of heat transfer, skin friction, static pressure, and temperature.

Figure 4.34 displays the normalized velocity $u/u_\tau$ and Van Driest transformed velocity $u_{\text{VD}}/u_\tau$ at $x = 101.6$ cm for case 1. Also shown is the viscous sublayer and Law of the Wall and Wake using $\Pi = 0.55$. There is no significant difference between the two normalized velocities in the viscous sublayer or Law of the Wall region due to the cold wall condition. The Van Driest transformed velocity shows better agreement in the wake region.

**Table 4.12.** Freestream conditions.

| Quantity | Value | | | |
|---|---|---|---|---|
| | Case 1 | | Case 2 | |
| $M_e$ | 6 | | 6 | |
| $p_{t_e}$ (MPa) | 3.62 | | 3.62 | |
| $T_{t_e}$ (K) | 485 | | 542 | |
| $T_w$ (K) | 217 | | 211 | |
| $T_w/T_{t_e}$ | 0.50 | | 0.44 | |
| $T_w/T_{aw}$ | 0.55 | | 0.49 | |
| $H_e$ (MJ kg$^{-1}$) | 0.49 | | 0.54 | |
| $Re \times 10^{-6}$ (m$^{-1}$) | 32.6 | | 28.6 | |
| | Min | Max | Min | Max |
| $\delta$ (cm) | 0.39 | 1.29 | 0.40 | 1.25 |
| $\theta$ (cm) | 0.011 | 0.044 | 0.010 | 0.040 |

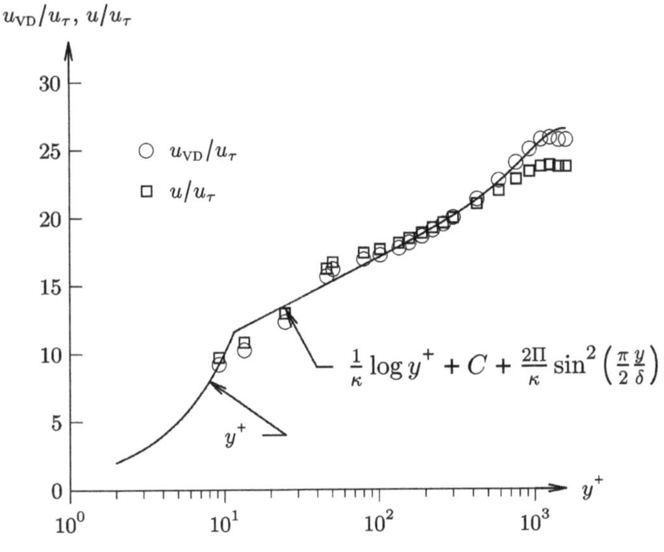


**Figure 4.34.** $u_{VD}/u_\tau$ and $u/u_\tau$ versus $y^+$.

## 4.12 Wallace (1967)

Wallace (1967) examined a cold wall hypersonic turbulent boundary layer in air on a flat plate and tunnel wall. The experiments were performed at the Cornell Aeronautical Laboratory in the 8 ft hypersonic shock tunnel. The range of freestream conditions are listed in tables 4.13 and 4.14. Experiments were performed on both blunted and sharp leading edge flat plates with lengths 60.96 cm and 67.81 cm, respectively, and on the tunnel contoured wall. Experimental measurements include surface heat transfer, pressure, and skin friction.

**Table 4.13.** Freestream conditions—flat plate.

| | Case | | | |
|---|---|---|---|---|
| Quantity | 1 | 2 | 3 | 4 |
| $M_e$ | 7.4 | 8.1 | 10.7 | 7.4 |
| $p_{t_\infty}$ (MPa) | 131 | 138 | 128 | 31 |
| $T_{t_\infty}$ (K) | 2111 | 1000 | 1556 | 1333 |
| $T_\infty$ (K) | 197.3 | 69.4 | 69.4 | 116.7 |
| $T_w/T_{t_\infty}$ | 0.14 | 0.30 | 0.19 | 0.22 |
| $T_w/T_{aw}$ | 0.16 | 0.33 | 0.21 | 0.25 |
| $H_\infty$ (MJ kg$^{-1}$) | 2.12 | 1.00 | 1.56 | 1.33 |
| $Re/m \times 10^{-6}$ | 65.6 | 265.7 | 39.4 | 29.5 |
| $\alpha$ (deg) | 0 to 10 | 0 to 10 | 0 to 20 | 0 |

$T_w/T_{aw} = (T_w/T_{t_e})\left[1 + \frac{(\gamma-1)}{2}M_e^2\right]\left[1 + \frac{(\gamma-1)}{2}Pr_t M_e^2\right]^{-1}$

**Table 4.14.** Freestream conditions—nozzle wall.

| | Case | | | | | | |
|---|---|---|---|---|---|---|---|
| Quantity | 1 | 2 | 3 | 4 | 5 | 6 | 7 |
| $M_e$ | 8.0 | 7.8 | 6.6 | 6.8 | 7.8 | 8.0 | 7.6 |
| $p_{t_\infty}$ (MPa) | 29.0 | 6.89 | 6.89 | 6.41 | 15.9 | 39.3 | 3.59 |
| $T_{t_\infty}$ (K) | 1028 | 1000 | 3194 | 2778 | 1278 | 1333 | 1222 |
| $T_\infty$ (K) | 77.7 | 83.3 | 411 | 328 | 106 | 100 | 106 |
| $T_w/T_{t_\infty}$ | 0.29 | 0.30 | 0.093 | 0.11 | 0.23 | 0.22 | 0.24 |
| $T_w/T_{aw}$ | 0.32 | 0.33 | 0.10 | 0.12 | 0.26 | 0.25 | 0.27 |
| $H_\infty$ (MJ kg$^{-1}$) | 1.03 | 1.00 | 3.20 | 2.78 | 1.28 | 1.33 | 1.22 |
| $Re/m \times 10^{-6}$ | 36.1 | 8.53 | 1.21 | 1.44 | 13.5 | 31.8 | 3.28 |

$T_w/T_{aw} = (T_w/T_{t_e})\left[1 + \frac{(\gamma-1)}{2}M_e^2\right]\left[1 + \frac{(\gamma-1)}{2}Pr_t M_e^2\right]^{-1}$

Figure 4.35 displays the Reynolds Analogy Factor for $H_w/H_\infty$ from 0.07 to 0.50. The results indicate $2St/c_f = 1.0 \pm 0.2$.

## 4.13 Wallace (1968, 1969)

Wallace (1968, 1969) performed an experimental investigation of a cold wall zero pressure gradient turbulent boundary layer in air at Mach 8.2 to 8.8. The experiments were performed in the Cornell Aeronautical Laboratory 48 in hypersonic shock tunnel. The measurements were obtained on the nozzle wall boundary layer. The freestream conditions are listed in table 4.15. Experimental measurements include boundary layer surveys of pitot pressure and density using electron beams.

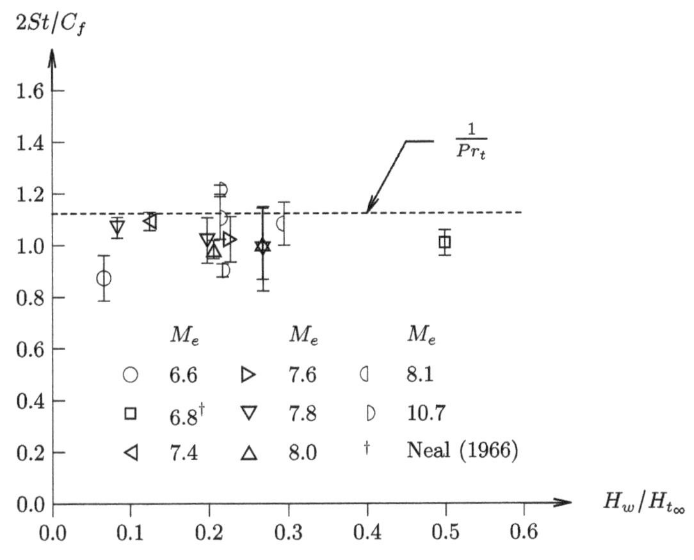


**Figure 4.35.** Reynolds Analogy Factor.

**Table 4.15.** Freestream conditions.

| Quantity | Case 1 | Case 2 | Case 3 | Case 4 |
|---|---|---|---|---|
| $M_e$ | 8.80 | 8.87 | 8.81 | 8.82 |
| $p_{t_e}$ (MPa) | 2.58 | 2.69 | 2.55 | 2.65 |
| $T_{t_e}$ (K) | 1989 | 1972 | 1967 | 1994 |
| $T_w$ (K) | 292 | 292 | 292 | 292 |
| $T_w/T_{t_e}$ | 0.15 | 0.15 | 0.15 | 0.15 |
| $T_w/T_{aw}$ | 0.16 | 0.17 | 0.17 | 0.16 |
| $H_e$ (MJ kg$^{-1}$) | 2.26 | 2.24 | 2.23 | 2.27 |
| $Re \times 10^{-6}$ (m$^{-1}$) | 0.60 | 0.62 | 0.61 | 0.60 |
| $\delta$ (cm) | 14.0 | 14.0 | 14.0 | 14.0 |
| | **Case 5** | **Case 6** | **Case 7** | **Case 8** |
| $M_e$ | 8.22 | 8.25 | 8.62 | 8.89 |
| $p_{t_e}$ (MPa) | 5.82 | 5.82 | 1.65 | 2.00 |
| $T_{t_e}$ (K) | 3231 | 3225 | 2108 | 1622 |
| $T_w$ (K) | 292 | 292 | 292 | 292 |
| $T_w/T_{t_e}$ | 0.09 | 0.09 | 0.14 | 0.18 |
| $T_w/T_{aw}$ | 0.10 | 0.10 | 0.15 | 0.20 |
| $H_e$ (MJ kg$^{-1}$) | 4.10 | 4.09 | 2.42 | 1.78 |
| $Re \times 10^{-6}$ (m$^{-1}$) | 0.51 | 0.50 | 0.36 | 0.70 |
| $\delta$ (cm) | 13.2 | 13.2 | 15.0 | 15.0 |

$T_{aw}$ from 3.46 using $Pr_t = 0.89$.

Figure 4.36 displays the normalized mean pitot pressure $p_p/p_{p_\infty}$ for four cases plus the data from run 580 of Samuels *et al* (1967) at Mach 6. Figure 4.37 shows the root-mean-square pitot pressure for six cases. All cases show an inflection point at $y/\delta \approx 0.4$ to 0.6 depending on the case, with the location of the local maximum for the first case (symbol ○) identified by the arrow. Figure 4.38 shows the normalized density fluctuations. Peak values of $\sqrt{\rho'^2}/\bar{\rho}$ as large as 0.09 are measured.

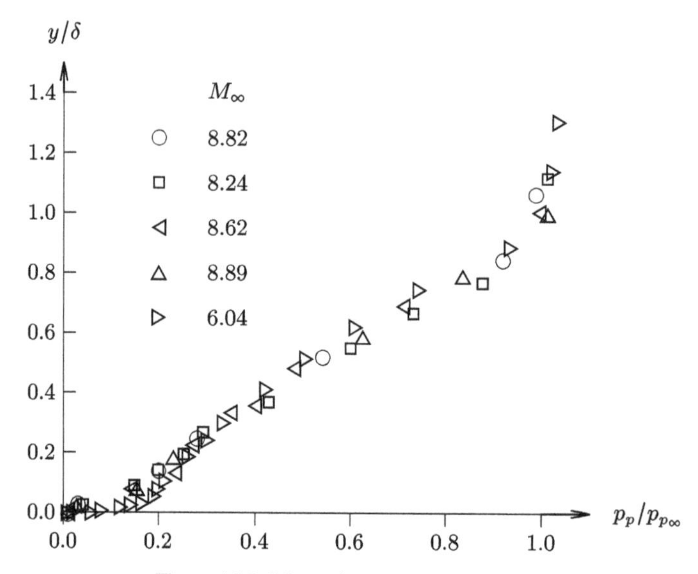


**Figure 4.36.** Mean pitot pressure versus $y/\delta$.

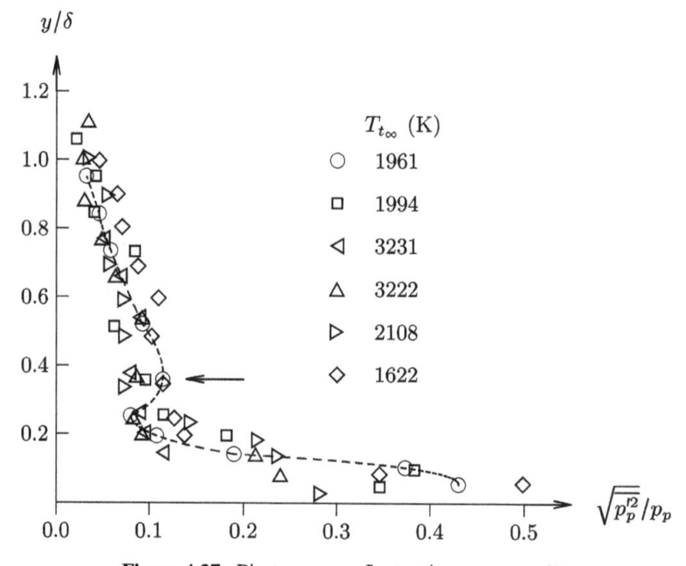


**Figure 4.37.** Pitot pressure fluctuations versus $y/\delta$.

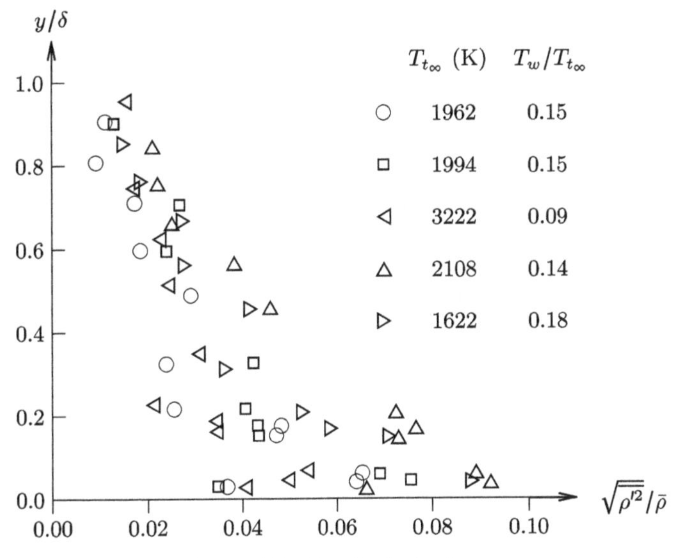


**Figure 4.38.** Density fluctuations.

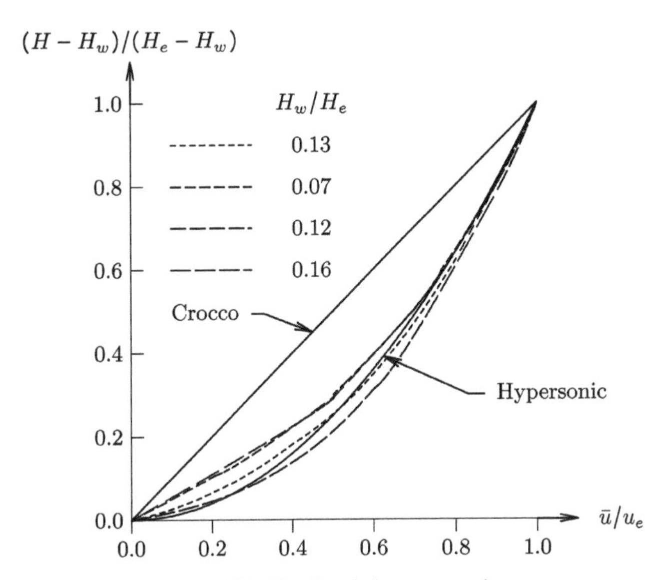


**Figure 4.39.** Total enthalpy versus $u/u_e$.

Interestingly, the lowest peak value among the five cases corresponds to the lowest value of $T_W/T_{t_\infty}$. Figure 4.39 displays the normalized stagnation enthalpy versus normalized velocity, together with the Crocco relation (3.143) and hypersonic velocity–total temperature relation (3.219). The experimental data are closely approximated by the latter. See also Harvey and Bushnell (1969a, 1969b).

## 4.14 Bushnell *et al* (1969)

Bushnell *et al* (1969) presented a compilation of experimental data to assess the velocity–total temperature relation for nominally 'flat plate' and nozzle wall configurations. The former include hollow cylinder and cone configurations in addition to flat plates. Figure 4.40 shows the experimental data for the 'flat plate' cases with the details of the experiments listed in table 4.16. Bushnell *et al* (1969) note that although the amount of scatter is appreciable, the mean of the data is approximated by the Crocco–Busemann relation (3.144). Figure 4.41 shows the experimental data for the nozzle wall cases with the details of the experiments listed

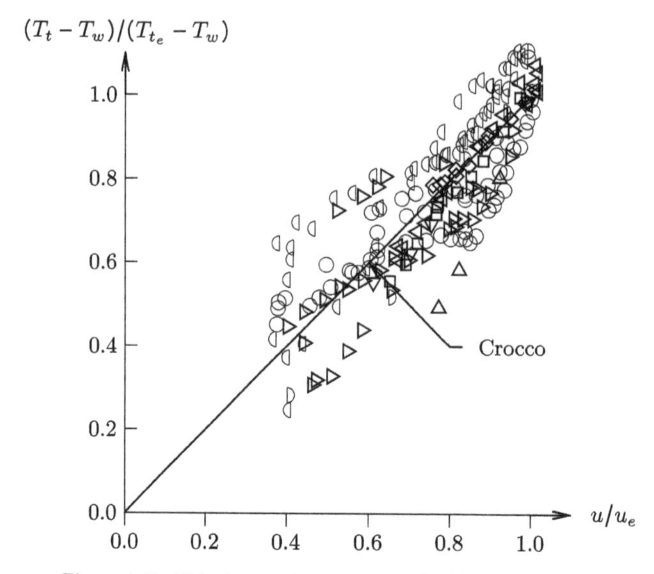


**Figure 4.40.** Velocity–total temperature for 'flat plate' flows.

**Table 4.16.** Symbols for figure 4.40.

| Symbol | $M_\infty$ | $Re_\theta \times 10^{-3}$ | $T_w/T_{t_\infty}$ | Model | Reference |
|---|---|---|---|---|---|
| ○ | 5.2 | 1.7 to 3.5 | 0.56 to 0.84 | FP | Winkler and Cha (1959) |
| □ | 6.4 | 6.0 | 0.52 | FP | Danberg (1967) |
| ◁ | 6.0 | 11 | 0.38, 0.49 | HC | Samuels *et al* (1967) |
| ▷ | 5.1 | 3.1 to 4.0 | 0.72 to 0.83 | FP | Danberg (1959) |
| △ | 10.2 | 2.3 | 0.28 | C | Softley and Sullivan (1968) |
| ▽ | 5.75 | 38 | 0.63 | HC | Hoydysh and Zakkay (1969) |
| ◗ | 10.5 | 1.3 | 0.3 | FP | Watson *et al* (1966) |
| ◖ | 5, 6 and 8 | 2.0 to 13.0 | 0.4 to 0.7 | HC | Stroud and Miller (1965) |
| ◇ | 6.5 | 2.2 to 5.9 | 0.30 to 0.38 | FP | Bushnell *et al* (1969)[a] |

C, cone; FP, flat plate; HC, hollow cylinder.
[a] Unpublished data from NASA Ames cited by Bushnell *et al* (1969).

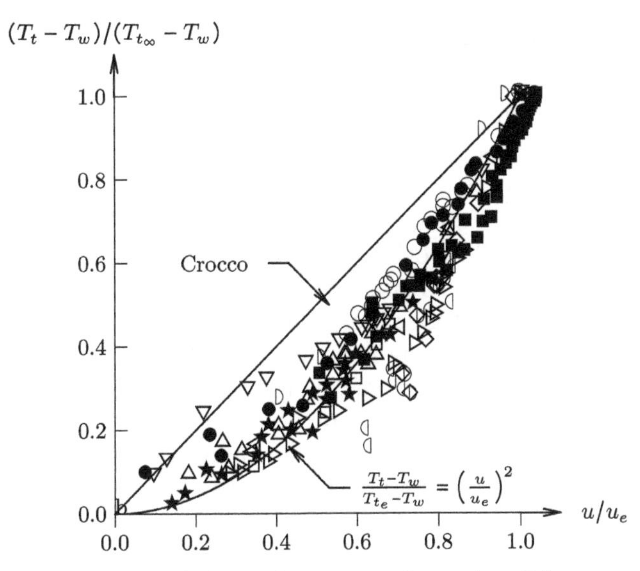


**Figure 4.41.** Velocity–total temperature for nozzle wall flows.

**Table 4.17.** Symbols for figure 4.41.

| Symbol | $M_\infty$ | $Re_\theta \times 10^{-3}$ | $T_W/T_{t_\infty}$ | Model | Reference |
|---|---|---|---|---|---|
| ○ | 6 | 19 to 49 | 0.63 | NW | Bertram and Neal (1965) |
| □ | 11.5 | 16 | 0.28 | NW | Perry and East (1968a) |
| ◁ | 6.8 | 13 | 0.5 | NW | Bertram and Neal (1965) |
| ▷ | 9.1 | 1.9 | 0.48 | NW | Hill (1959a) |
| △ | 5.1 to 8.2 | 7.4 to 12.6 | 0.46 to 0.53 | NW | Stroud and Miller (1965) |
| ▽ | 4.7 | 6 | 0.6 | NW | Lee *et al* (1968) |
| D | ≈14 | ≈2 | 0.2 | C | Sheer and Nagamatsu (1968) |
| ᗡ | 6.5 | 2.1 to 4.5 | 0.31 and 0.43 | NW | Scaggs (1966) |
| ◇ | 11.7 | 2.6 to 4.5 | 0.27 and 0.30 | NW | Scaggs (1966) |
| ★ | 8.2 to 8.9 | 3.7 to 9.4 | 0.07 to 0.16 | NW | Wallace (1968) |
| ● | 19.47 | 3.4 to 5.1 | 0.17 | NW | Bushnell *et al* (1969) |
| ■ | 7.9 | 11 to 31 | 0.44 | NW | Bushnell *et al* (1969) |

C, cone with $dp/dx$; NW, nozzle wall.

in table 4.17. Bushnell *et al* (1969) indicate that the mean of the data is approximated by the quadratic relation

$$\frac{T_t - T_W}{T_{t_e} - T_W} = \left(\frac{u}{u_e}\right)^2 \tag{4.1}$$

and concluded that a distance of approximately $60\delta$ was required for the velocity–total temperature relation to relax to (3.143).

## 4.15 Hoydysh and Zakkay (1969)

Hoydysh and Zakkay (1969) investigated the effects of an adverse pressure gradient on a hypersonic cold wall turbulent boundary layer at Mach 5.75. The experiments were performed in the Mach 6 hypersonic blowdown facility in the Guggenheim Aerospace Laboratory of New York University. The measurements were obtained on three different models comprising axisymmetric forebodies with flares. The forebody shapes are a hollow cylinder, a right circular cone with 7.5° half angle blended into a cylinder, and a streamlined cylindrical surface. Flare shapes include a circular arc and a fourth-degree polynomial. One of the models is shown in figure 4.42, and the corresponding freestream Mach number versus $x$ on the flare is shown in figure 4.43. The freestream conditions for the hollow cylinder flare model with the fourth-degree polynomial flare are listed in table 4.18. Experimental measurements include surface heat transfer and pressure and boundary layer profiles of pitot pressure and total temperature at up to 13 stations on the flare. A subsequent report by Zakkay and Wang (1972) describes a similar experiment with the adverse pressure gradient generated by a cowl. Figure 4.44 displays the Van Driest

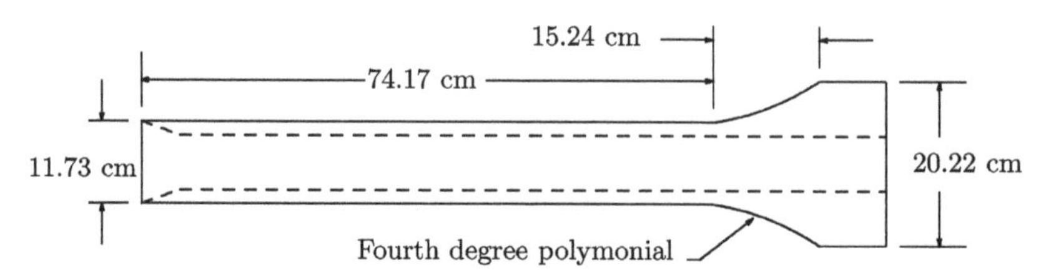


**Figure 4.42.** Hollow cylinder flare.

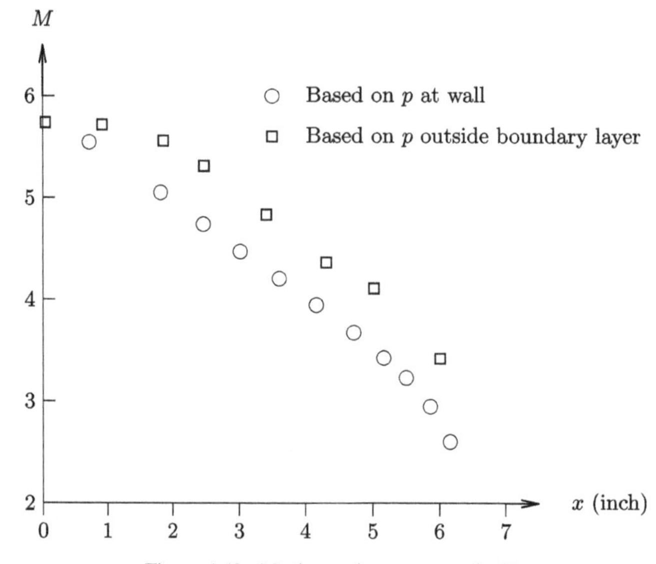


**Figure 4.43.** Mach number versus $x$ (inch).

**Table 4.18.** Freestream conditions.

| Quantity | Value |
|---|---|
| $M_e$ | 5.75[a] |
| $p_{t_e}$ (MPa) | 12.8[b] |
| $T_{t_e}$ (K) | 500 |
| $T_w/T_{aw}$ | 0.634 |
| $H_e$ (MJ kg$^{-1}$) | 0.50 |
| $Re \times 10^{-6}$ (m$^{-1}$) | 118.1 |

[a] At start of flare.
[b] Calculated using $Re$ m$^{-1}$.

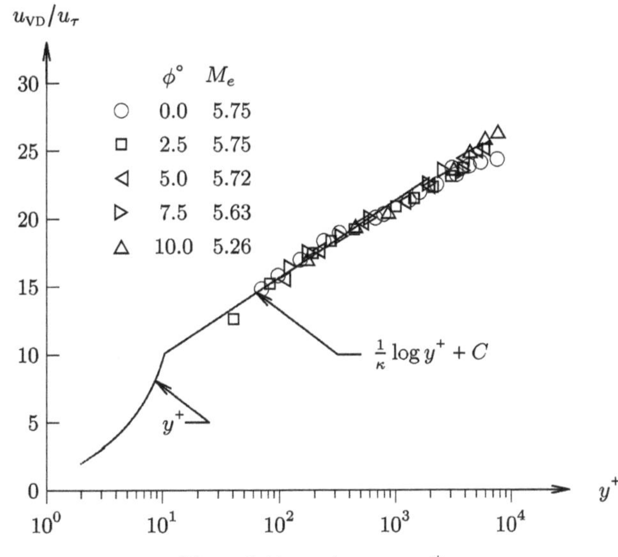


**Figure 4.44.** $u_{VD}/u_\tau$ versus $y^+$.

transformed velocity profile $u_{VD}/u_\tau$ versus $y^+$ at five stations on the flare for the model shown in figure 4.42. The profiles show close agreement with the Law of the Wall.

## 4.16 Lee *et al* (1969)

Lee *et al* (1969) conducted an investigation of a cold wall zero pressure gradient hypersonic turbulent boundary layer at Mach 5. The experiments were performed at the United States Naval Ordnance Laboratory tunnel no. 7 (section A.23.2). The measurements were obtained on the nozzle wall boundary layer at distances of 1.15 m, 1.45 m, 1.76 m, and 2.32 m from the nozzle throat. The freestream conditions are listed in table 4.19. Experimental measurements include pitot pressure and total temperature profiles and surface pressure, skin friction, and heat transfer.

**Table 4.19.** Freestream conditions.

| | Value | | | | |
|---|---|---|---|---|---|
| Quantity | Case 1 | Case 2 | Case 3 | Case 4 | Case 5 |
| Run | 12 196 | 12 091 | 12 086 | 12 292 | 12 201 |
| $M_e$ | 4.69 | 4.70 | 4.64 | 4.70 | 4.59 |
| $p_{t_e}$ (MPa) | 0.10 | 0.10 | 0.10 | 0.10 | 0.10 |
| $T_{t_e}$ (K) | 435.5 | 430.1 | 423.3 | 337.8 | 429.2 |
| $T_w/T_{aw}$ | 0.73 | 0.74 | 0.74 | 0.95 | 0.73 |
| $H_e$ (MJ kg$^{-1}$) | 0.44 | 0.43 | 0.43 | 0.34 | 0.43 |
| $Re \times 10^{-6}$ (m$^{-1}$) | 1.84 | 1.83 | 1.96 | 2.64 | 2.00 |
| $\delta^*$ (cm) | 1.56 | 1.89 | 2.21 | 2.36 | 3.05 |
| $\theta$ (cm) | 0.264 | 0.277 | 0.307 | 0.249 | 0.447 |
| Re $_\theta$ | 4855 | 5080 | 6030 | 6580 | 8938 |
| | Case 6 | Case 7 | Case 8 | Case 9 | Case 10 |
| Run | 3132 | 3142 | 3131 | 12 195 | 3141 |
| $M_e$ | 4.91 | 4.70 | 4.78 | 4.60 | 4.73 |
| $p_{t_e}$ (MPa) | 0.52 | 0.52 | 0.51 | 0.52 | 0.52 |
| $T_{t_e}$ (K) | 607.9 | 653.2 | 558.1 | 435.1 | 568.2 |
| $T_w/T_{aw}$ | 0.52 | 0.49 | 0.56 | 0.72 | 0.56 |
| $H_e$ (MJ kg$^{-1}$) | 0.61 | 0.66 | 0.56 | 0.44 | 0.57 |
| $Re \times 10^{-6}$ (m$^{-1}$) | 5.06 | 5.00 | 6.09 | 9.81 | 6.10 |
| $\delta^*$ (cm) | 1.73 | 1.67 | 1.67 | 1.18 | 1.67 |
| $\theta$ (cm) | 0.298 | 0.340 | 0.312 | 0.196 | 0.315 |
| Re $_\theta$ | 150 83 | 170 04 | 190 40 | 192 11 | 192 13 |
| | Case 11 | Case 12 | Case 13 | Case 14 | Case 15 |
| Run | 12 902 | 12 085 | 11 221 | 12 198 | 12 194 |
| $M_e$ | 4.77 | 4.69 | 4.63 | 4.75 | 4.71 |
| $p_{t_e}$ (MPa) | 0.52 | 0.52 | 0.51 | 0.51 | 1.03 |
| $T_{t_e}$ (K) | 434.2 | 430.8 | 331.5 | 435.6 | 416.4 |
| $T_w/T_{aw}$ | 0.73 | 0.74 | 0.98 | 0.73 | 0.76 |
| $H_e$ (MJ kg$^{-1}$) | 0.44 | 0.43 | 0.33 | 0.44 | 0.42 |
| $Re \times 10^{-6}$ (m$^{-1}$) | 9.18 | 9.54 | 14.88 | 9.12 | 19.93 |
| $\delta^*$ (cm) | 1.71 | 1.72 | 1.80 | 1.93 | 1.04 |
| $\theta$ (cm) | 0.237 | 0.250 | 0.188 | 0.307 | 0.184 |
| Re $_\theta$ | 21 798 | 23 881 | 28 040 | 28 040 | 36 760 |
| | Case 16 | Case 17 | Case 18 | Case 19 | |
| Run | 2021 | 11 283 | 12 197 | 6211 | |
| $M_e$ | 4.86 | 4.67 | 4.74 | 4.67 | |
| $p_{t_e}$ (MPa) | 1.04 | 1.02 | 1.04 | 1.03 | |
| $T_{t_e}$ (K) | 416.3 | 318.6 | 408.4 | 411.9 | |
| $T_w/T_{aw}$ | 0.76 | 1.02 | 0.77 | 0.79 | |

| | | | | |
|---|---|---|---|---|
| $H_e$ (MJ kg$^{-1}$) | 0.42 | 0.32 | 0.41 | 0.41 |
| $Re \times 10^{-6}$ (m$^{-1}$) | 18.84 | 31.14 | 18.27 | 20.68 |
| $\delta^*$ (cm) | 1.48 | 1.59 | 2.03 | 1.79 |
| $\theta$ (cm) | 0.198 | 0.159 | 0.282 | 0.274 |
| Re $_\theta$ | 37 367 | 49 430 | 51 518 | 56 737 |

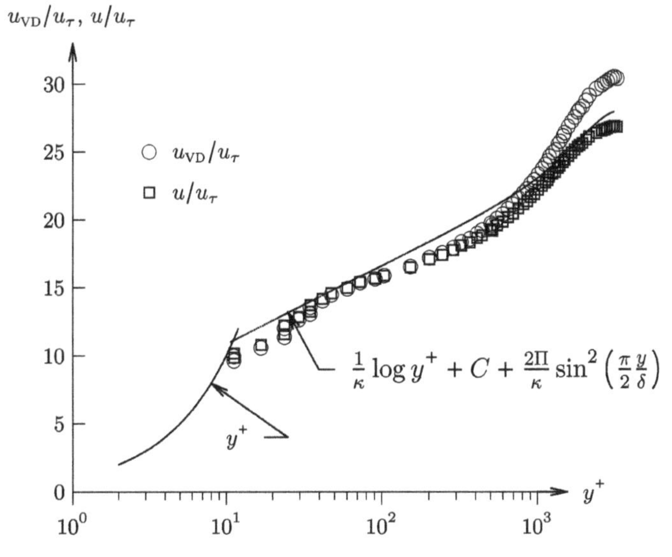


**Figure 4.45.** $u_{VD}/u_\tau$ and $u/u_\tau$ versus $y^+$.

Figure 4.45 displays the Van Driest transformed velocity $u_{VD}/u_\tau$ and normalized velocity $u/u_\tau$ versus $y^+$ for run 12 197 ($Re_\theta$ = 51 518, $T_w/T_{aw}$ = 0.77). Also shown is the linear and Law of the Wall and Wake with $\Pi$ = 0.55. The value of $u_\tau$ is obtained from table 3 in Lee *et al* (1969) for a separate run at $Re_\theta$ = 50 200 and $T_w/T_{aw}$ = 0.74. The profiles for $u_{VD}/u_\tau$ and $u/u_\tau$ are essentially identical in the logarithmic region and differ only in the wake region $y^+ > 10^3$ corresponding to $y/\delta > 0.34$ where $\delta$ = 5.36 cm. Figure 4.46 shows the Reynolds Analogy Factor 2 $St/c_f$ versus $Re_\theta$. A general trend of decreasing 2 $St/c_f$ with increasing $Re_\theta$ is observed with values approaching approximately unity.

## 4.17 Matthews and Trimmer (1969)

Matthews and Trimmer (1969) investigated a zero pressure gradient turbulent boundary layer in air at Mach 6 to 10. The experiments were performed in the von Kármán Gas Dynamics Facility at the Arnold Engineering Development Center (section A.5). The measurements were obtained on the nozzle wall boundary layer. The freestream conditions are listed in table 4.20 in three groups based upon

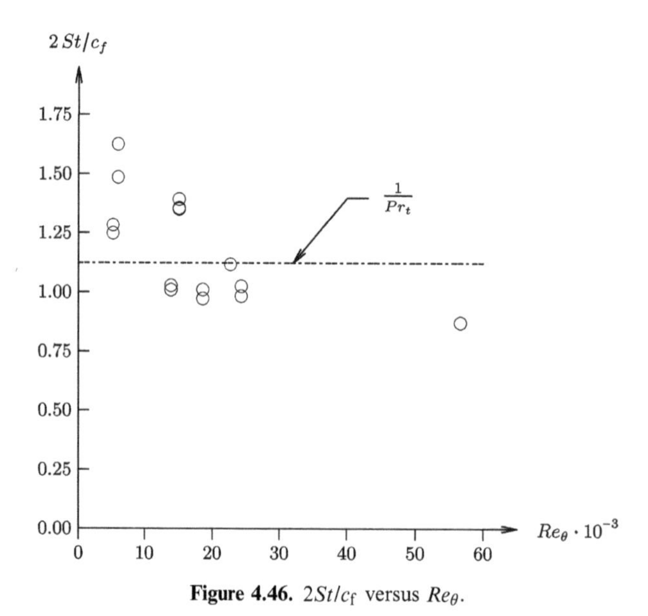


**Figure 4.46.** $2St/c_f$ versus $Re_\theta$.

**Table 4.20.** Freestream conditions.

| | Value | | | | | |
|---|---|---|---|---|---|---|
| | Group 1 | | Group 2 | | Group 3 | |
| Quantity | Min | Max | Min | Max | Min | Max |
| $M_e$ | 5.93 | 5.95 | 7.90 | 8.04 | 9.86 | 10.18 |
| $p_{t_e}$ (MPa) | 0.346 | 1.379 | 0.674 | 5.474 | 1.372 | 12.41 |
| $T_{t_e}$ (K) | 445 | 460 | 734 | 750 | 929 | 1094 |
| $H_e$ (MJ kg$^{-1}$) | 0.447 | 0.462 | 0.737 | 0.753 | 0.933 | 1.098 |
| $Re \times 10^{-6}$ (m$^{-1}$) | 3.41 | 12.8 | 1.44 | 11.3 | 1.05 | 6.66 |
| | Max | Min | Max | Min | Max | Min |
| $T_w/T_{t_e}$ | 0.662 | 0.640 | 0.401 | 0.393 | 0.317 | 0.269 |
| $T_w/T_{aw}$ | 0.73 | 0.71 | 0.45 | 0.44 | 0.35 | 0.30 |
| $\delta$ (cm) | 13.3 | 11.2 | 19.6 | 16.0 | 28.2 | 25.1 |
| $\delta^*$ (cm) | 4.93 | 3.84 | 7.47 | 5.72 | 9.88 | 7.90 |
| $\theta$ (cm) | 0.505 | 0.460 | 0.864 | 0.655 | 1.275 | 0.861 |

Notes:

1. Values for $T_w/T_{t_e}$ are typical.
2. Range of values for $\delta$, $\delta^*$, and $\theta$ correspond to range of variables listed above.
3. $T_w = 294$ K.
4. $T_{aw}$ using (3.46) with $Pr_t = 0.89$.

the nominal Mach number. Experimental measurements include boundary layer surveys of pitot pressure and total temperature.

Figure 4.47 displays the dimensionless total temperature ratio $(T_t - T_w)/(T_{t_e} - T_w)$ versus $u/u_e$. The data for one experiment in each of the three groups are shown.

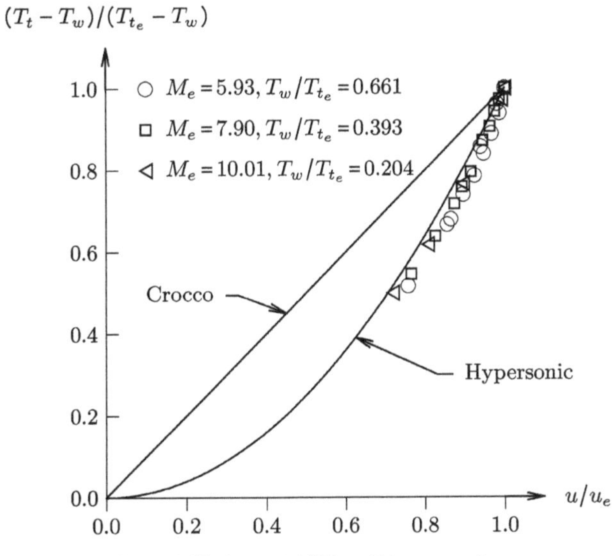


**Figure 4.47.** $(T_t - T_w)/(T_{t_e} - T_w)$ versus $u/u_e$.

**Table 4.21.** Flow conditions.

| Quantity | Value | |
|---|---|---|
| | Min | Max |
| $M_e$ | 4.6 | 11.7 |
| $H_w/H_\infty$ | 0.09 | 0.30 |
| $T_w/T_{aw}$ | 0.10 | 0.34 |
| $Re_x \times 10^{-6}$ | 1.0 | 222.0 |
| $T_w/T_{aw} = (H_w/H_\infty)\left[1 + \frac{(\gamma-1)}{2}M_e^2\right]\left[1 + \frac{(\gamma-1)}{2}Pr_t M_e^2\right]^{-1}$ | | |

The data show close agreement with the hypersonic velocity–total temperature relation (3.219).

## 4.18 Cary (1970)

Cary (1970) compiled experimental data on the Reynolds Analogy Factor for zero pressure gradient hypersonic turbulent boundary layers on flat plates and nozzle walls. The data for flat plates were obtained from Hironimus (1966) and for nozzle walls from Lee *et al* (1968). The range of Mach numbers, $H_w/H_\infty$, and Reynolds number $Re_x = \rho_\infty U_\infty x/\mu_\infty$ for the flat plate data are shown in table 4.21.

Figures 4.48 and 4.49 present the Reynolds Analogy Factor $2St/c_f$ versus $H_w/H_\infty$ and $Re_x$, respectively, for the flat plate experiments. The Mach number for each measurement is also shown. The theoretical result (3.298) is a reasonable

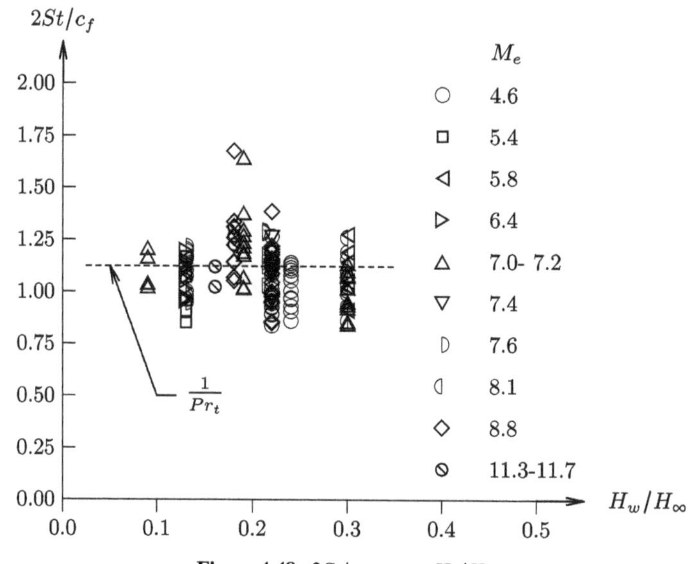


**Figure 4.48.** $2St/c_f$ versus $H_w/H_\infty$.

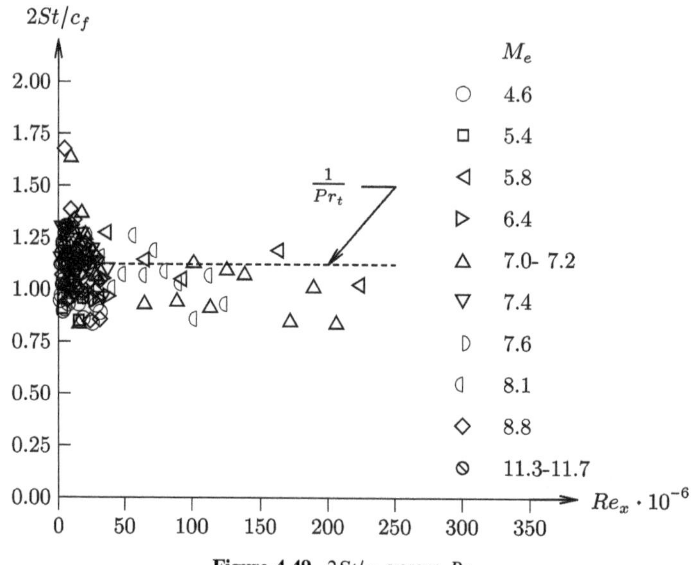


**Figure 4.49.** $2St/c_f$ versus $Re_x$.

approximation with minimum and maximum values for $2St/c_f = 0.838$ and 1.677, respectively. No apparent trend with either $H_w/H_\infty$ or $Re_x$ is apparent.

## 4.19 Jones and Feller (1970)

Jones and Feller (1970) examined a zero pressure gradient turbulent boundary layer in air at Mach 6. The experiments were conducted in the NASA Langley Mach 6 high Reynolds number tunnel. The measurements were obtained at four streamwise

**Table 4.22.** Freestream conditions.

| | Value | | | | | | | |
|---|---|---|---|---|---|---|---|---|
| | Station 94 | | Station 94 | | Station 172 | | Station 215 | |
| Quantity | Min | Max | Min | Max | Min | Max | Min | Max |
| $M_e$ | 5.866 | 5.978 | 5.764 | 5.821 | 5.581 | 5.778 | 5.450 | 5.692 |
| $p_{t_e}$ (MPa) | 0.45 | 4.24 | 0.45 | 4.14 | 0.45 | 3.89 | 0.45 | 3.55 |
| $T_{t_e}$ (K) | 441 | 518 | 434 | 514 | 445 | 512 | 452 | 506 |
| $T_w/T_{t_e}$ | 0.40 | 0.69 | 0.40 | 0.69 | 0.40 | 0.69 | 0.40 | 0.69 |
| $T_w/T_{aw}$ | 0.44 | 0.76 | 0.44 | 0.76 | 0.44 | 0.76 | 0.44 | 0.76 |
| $H_e$ (MJ kg$^{-1}$) | 0.44 | 0.52 | 0.43 | 0.51 | 0.45 | 0.51 | 0.45 | 0.51 |
| $Re \times 10^{-6}$ (m$^{-1}$) | 4.57 | 34.6 | 4.80 | 36.56 | 5.02 | 37.1 | 5.35 | 34.1 |
| | Max | Min | Max | Min | Max | Min | Max | Min |
| $\delta$ (cm) | 5.44 | 4.14 | 6.74 | 4.79 | 8.54 | 6.32 | 10.69 | 7.28 |
| $\delta^*$ (cm) | 2.01 | 1.39 | 2.45 | 1.73 | 3.08 | 2.22 | 3.81 | 2.58 |
| $\theta$ (cm) | 0.176 | 0.132 | 0.207 | 0.163 | 0.273 | 0.208 | 0.368 | 0.234 |

Notes:
1. Values for $T_w/T_{t_e}$ are typical.
2. Range of values for $\delta$, $\delta^*$, and $\theta$ correspond to range of variables listed above.
3. $T_{aw}$ calculated using (3.46) with $Pr_t = 089$.

locations on the nozzle wall boundary layer. The freestream conditions are listed in table 4.22. At each streamwise station measurements were obtained at four different stagnation pressures with two different wall temperatures for each stagnation pressure. Experimental measurements include boundary layer surveys of pitot pressure and total temperature and wall static pressure.

Figure 4.50 shows the normalized velocity $u/u_\tau$ and the Van Driest transformed velocity $u_{VD}/u_\tau$ for station 94 at $M_e = 5.953$, $p_{t_e} = 0.217$ MPa, $T_{t_e} \approx 490$ K, and $T_w/T_{t_e} = 0.69$. Also shown is the viscous sublayer $u/u_{tau} = y^+$ and the Law of the Wall and Wake with $C = 5.5$ and $\Pi = 0.2$ and 0.55. It is evident that the experimental Van Driest transformed velocity is in closer agreement with the Law of the Wall and Wake. Note that $c_f$ is estimated from the momentum equation. The Law of the Wall lies 1.78 above the experimental data (○). The value $C = 5.5 - 1.8 = 3.7$ yields close agreement.

## 4.20 Beckwith *et al* (1971)

Beckwith *et al* (1971) conducted an experimental investigation of a cold wall zero pressure gradient turbulent boundary layer in nitrogen at Mach 19.28 to 19.65. The experiments were performed in the NASA Langley hypersonic nitrogen tunnel. The measurements were obtained on the nozzle wall boundary layer at a distance of 2.083 m from the nozzle throat. The freestream conditions are listed in table 4.23. Experimental measurements include boundary layer surveys of pitot and static pressure and total temperature.

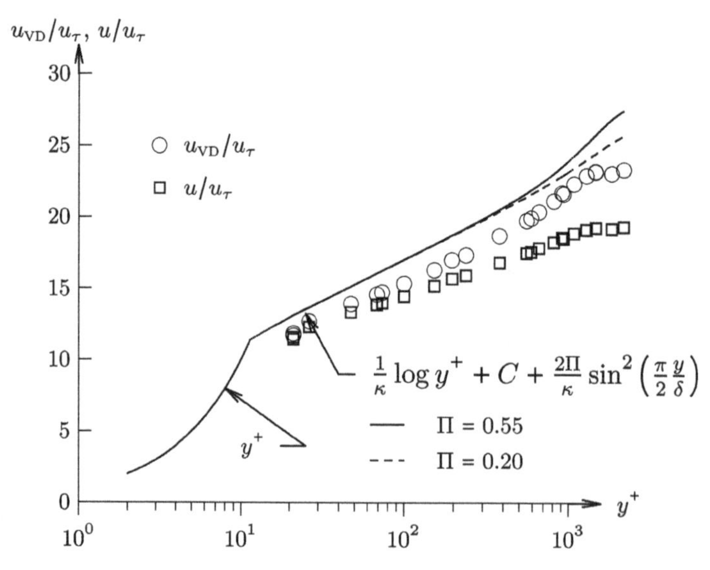


**Figure 4.50.** $u_{VD}/u_\tau$ and $u/u_\tau$ versus $y^+$.

**Table 4.23.** Freestream conditions.

| Quantity | Value | | | | | |
|---|---|---|---|---|---|---|
| | Case 1a | Case 1b | Case 2a | Case 2b | Case 3a | Case 3b |
| $M_e$ | 19.28 | 19.28 | 19.42 | 19.42 | 19.65 | 19.65 |
| $p_{t_e}$ (MPa) | 29.7 | 28.3 | 43.1 | 43.8 | 55.5 | 55.9 |
| $T_{t_e}$ (K) | 1680 | 1695 | 1780 | 1695 | 1835 | 1835 |
| $T_w$ (K) | 300 | 300 | 300 | 300 | 300 | 300 |
| $T_w/T_{t_e}$ | 0.18 | 0.18 | 0.17 | 0.18 | 0.16 | 0.16 |
| $T_w/T_{aw}$ | 0.20[a] | 0.20[a] | 0.19[a] | 0.20[a] | 0.18[a] | 0.18[a] |
| $H_e$ (MJ kg$^{-1}$) | 1.74 | 1.76 | 1.85 | 1.76 | 1.90 | 1.90 |
| $Re \times 10^{-6}$ (m$^{-1}$) | 2.02 | 1.90 | 2.62 | 2.90 | 2.98 | 3.00 |
| $\delta$ (cm) | 11.2 | 11.2 | 9.4 | 9.4 | 8.65 | 8.65 |
| $\delta^*$ (cm) | 5.92 | 5.92 | 4.62 | 4.62 | 4.31 | 4.31 |
| $\theta$ (cm) | 0.178 | 0.178 | 0.173 | 0.173 | 0.155 | 0.155 |

[a]Using (3.46) with $Pr_t = 0.89$.

Figure 4.51 shows the normalized mean pitot pressure profile in the boundary layer.[6]. The pitot pressure in the lower 20% of the boundary layer is considerably lower than in figure 4.36. Figure 4.52 presents the normalized mean total temperature profile. The darkened symbol is the measured wall temperature normalized by $T_{t_\infty}$. Figure 4.53 shows the normalized mean static pressure distribution in the

[6]Experimental data for $y/\delta > 1.4$ have been omitted.

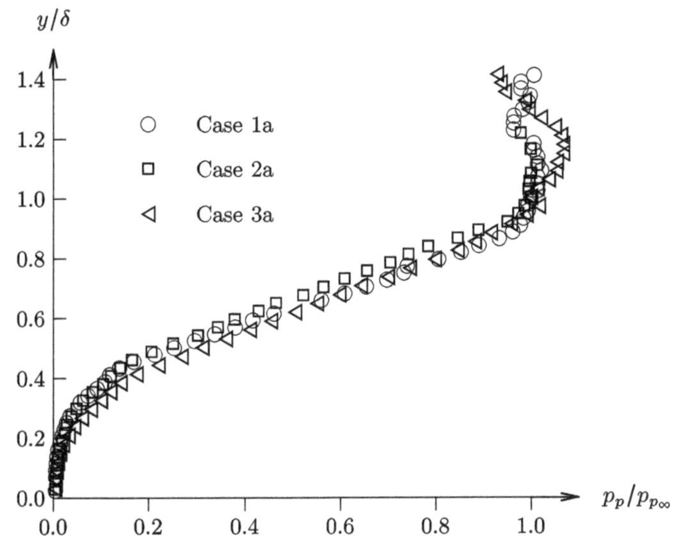


**Figure 4.51.** Mean pitot pressure versus $y/\delta$.

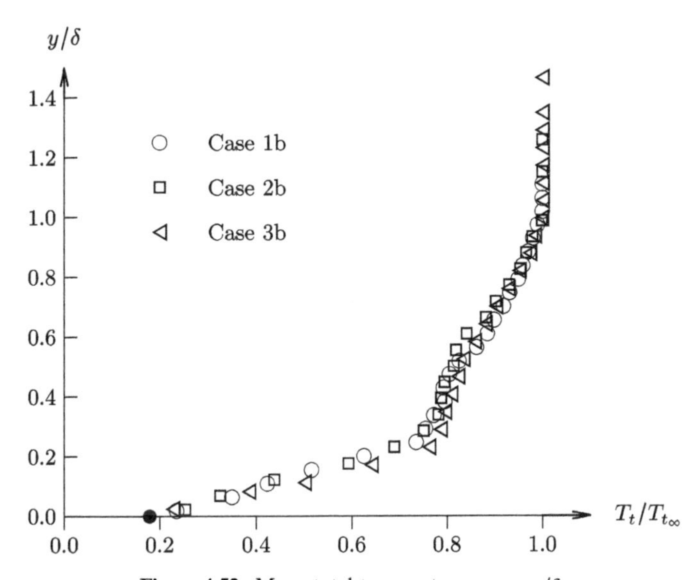


**Figure 4.52.** Mean total temperature versus $y/\delta$.

boundary layer for case 2. A significant gradient normal to the wall is evident. Figure 4.54 presents the normalized mean total temperature $(T_t - T_w)/(T_{t_e} - T_w)$ versus $u/u_e$. With the exception of case 1b, the data in the lower portion of the boundary layer $u/u_e < 0.8$ (corresponding to $y/\delta < 0.3$) satisfy (3.144) while in the outer portion of the boundary layer the data are more closely approximated by (3.220).

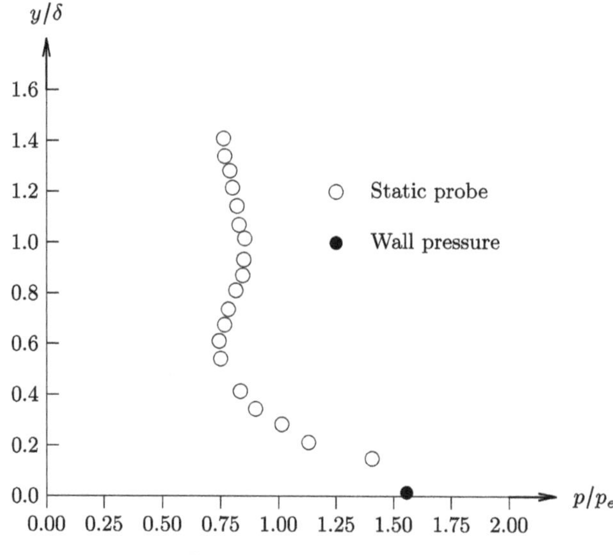


**Figure 4.53.** $p/p_e$ versus $y/\delta$.

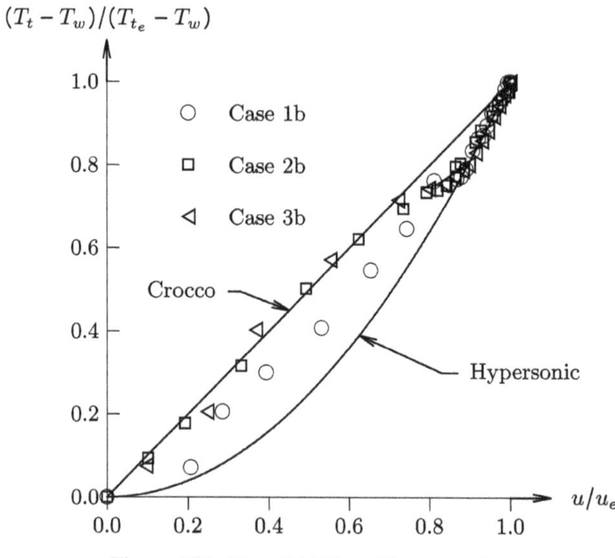


**Figure 4.54.** $(T_t - T_w)/(T_{t_e} - T_w)$ versus $u/u_e$.

## 4.21 Fischer *et al* (1971)

Fischer *et al* (1970) and Fischer *et al* (1971) conducted an experimental investigation of an adiabatic zero pressure gradient turbulent boundary layer at $M_\infty \approx 20$. The experiments were performed at the NASA Langley Mach 20 hypersonic helium tunnel. The measurements were taken on the nozzle wall boundary layer at several streamwise locations. Detailed measurements were obtained at station 139 at

**Table 4.24.** Freestream conditions.

| Quantity | Value |
|---|---|
| $M_e$ | 16.1–22.3 |
| $p_{t_\infty}$ (MPa) | 0.52–20.8 |
| $T_{t_\infty}$ (K) | 300 |
| $T_w$ (K) | 300 |
| $T_w/T_{aw}$ | ≈1 |
| $\delta^*$ (cm)[a] | 21.1–14.0 |
| $\theta$ (cm)[a] | 0.0607–0.0201 |
| $Re \times 10^{-6}$ (m$^{-1}$) | 2.0–55.6 |

[a]Limits correspond to range in *Re*.

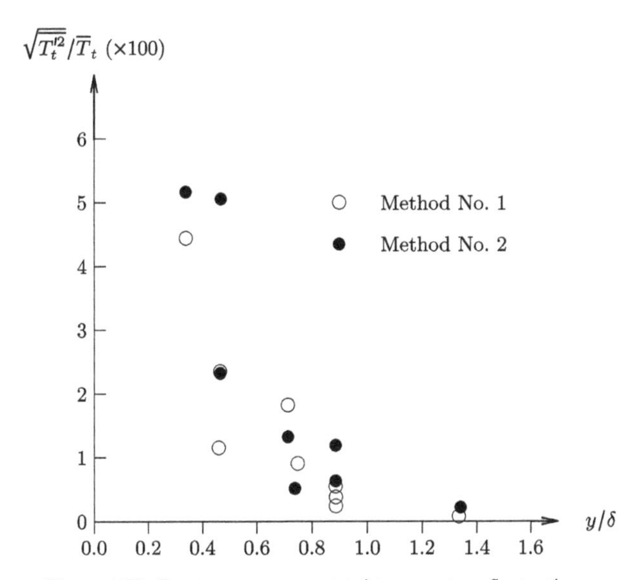


**Figure 4.55.** Root-mean-square total temperature fluctuations.

stagnation pressures from 0.52 MPa to 20.8 covering transitional and fully turbulent boundary layers and a range of $M_e$ from 16.1 to 22.3. The freestream conditions at station 139 are listed in table 4.24. Experimental measurements include boundary layer surveys of pitot pressure, static pressure, and total temperature, hot wire surveys, and surface pressure.

Figure 4.55 shows the root-mean-square total temperature fluctuations for $M_e = 20.4$ ($p_{t_\infty} = 7.0$ MPa) at station 139. Results are shown for two different calibration methods used with the hot wire anemometry. Root-mean-square fluctuations up to five percent are seen near the wall. Figure 4.56 displays the mean pressure distribution across the boundary layer at station 139 for $M_e = 21.6$ ($p_{t_\infty} = 14.5$ MPa). A significant variation in mean pressure is evident across the boundary layer with the wall pressure 35% above the edge pressure. Fischer *et al* (1971) attribute the variation in mean pressure to the presence of transitional or

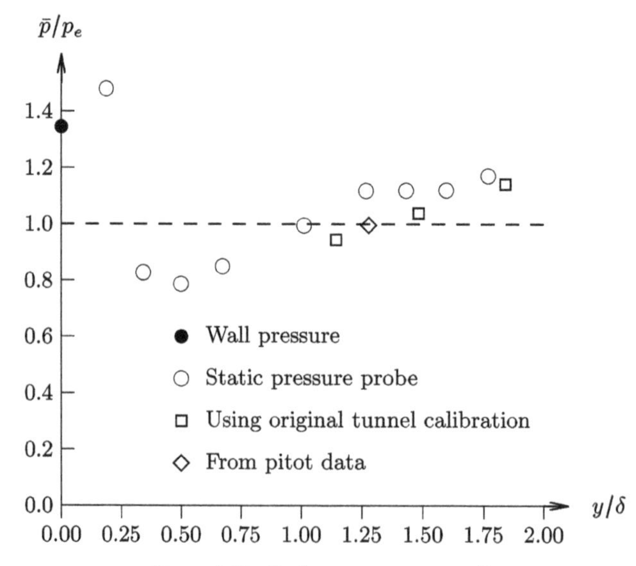


**Figure 4.56.** Static pressure versus $y/\delta$.

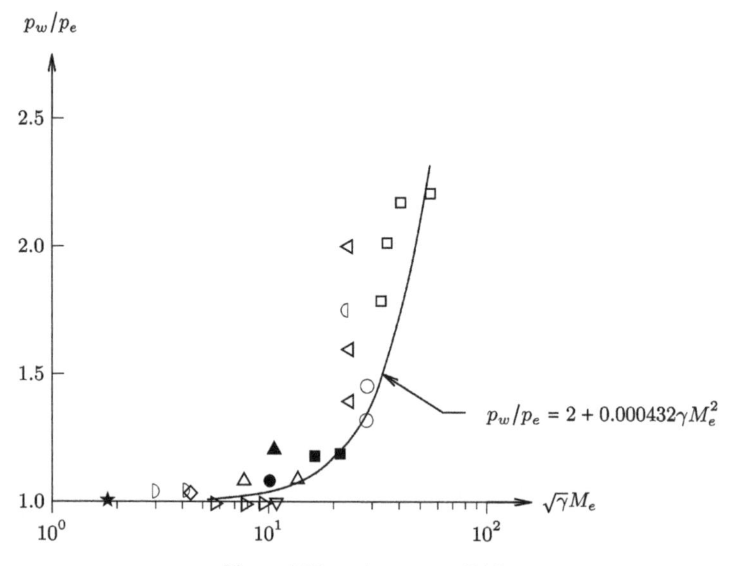


**Figure 4.57.** $p_w/p_e$ versus $\sqrt{\gamma}M_e$.

turbulence effects. Figure 4.57 presents a detailed compilation of experimental data for wall pressure ratios $p_w/p_e$ versus $\sqrt{\gamma}M_e$ for zero pressure gradient turbulent boundary layers. Symbols are listed in table 4.25. For $\sqrt{\gamma}M_e < 6$ the wall pressure ratio is essentially equal to one. The solid curve representing the general trend of the experimental data is

$$\frac{p_w}{p_e} = 1 + 0.000\,432\gamma M_e^2 \tag{4.2}$$

**Table 4.25.** Symbols for figure 4.57.

| Symbol | Gas | Configuration | Reference |
|---|---|---|---|
| ○ | Helium | NW | Fischer *et al* (1971) |
| □ | Helium | NW | Kemp and Sreekanth (1969) |
| ◁ | na | NW | Harvey and Clark[a] |
| ▷ | Air | NW | Lobb *et al* (1955b) |
| △ | Air | NW | Scaggs (1966) |
| ▽ | Air | NW | Perry and East (1968b) |
| D | Air | NW | Speaker and Ailman (1966) |
| ◖ | na | NW | Creel[a] |
| ◇ | Air | FP | Coles (1953a) |
| ★ | Helium | W | Wagner *et al* (1970) |
| ● | Air | FP | Cary[a] |
| ■ | Helium | C | Maddalon and Henderson (1968) |
| ▲ | Air | C | Everhart and Hamilton (1967) |

NW, nozzle wall; FP, flat plate; W, wedge; C, cone.
[a] Unpublished, cited in Fischer *et al* (1971).

## 4.22 Hopkins and Inouye (1971)

Hopkins and Inouye (1971) evaluated several theories for prediction of skin friction and heat transfer in supersonic and hypersonic flat plate zero pressure gradient turbulent boundary layers. The theories include those of Coles (1964), Sommer and Short (1955), Spaulding and Chi (1964), and Van Driest (1956). Hopkins and Inouye (1971) recommended the theory of Van Driest (1956) for prediction of the skin friction coefficient, with a Reynolds Analogy Factor of 1.0 for hypersonic Mach numbers and 1.2 for supersonic and lower Mach numbers.

The method of Van Driest (1956) is based upon a transformation from compressible to an equivalent incompressible flow. The Kármán–Schoenherr equation relating the local incompressible skin friction coefficient $\bar{c}_f$ and Reynolds number based upon the incompressible momentum thickness $\bar{Re}_\theta$ is

$$\frac{1}{\bar{c}_f} = 17.08\left(\log_{10}\bar{Re}_\theta\right)^2 + 25.11\log_{10}\bar{Re}_\theta + 6.102 \tag{4.3}$$

The relationships between the compressible and incompressible variables are

$$\bar{c}_f = F_c c_f \tag{4.4}$$

$$\bar{Re}_\theta = F_\theta Re_\theta \tag{4.5}$$

where $c_f$ is the compressible skin friction coefficient and $Re_\theta$ is the Reynolds number based upon the compressible momentum thickness

$$\theta = \int_0^\delta \frac{\rho u}{\rho_e u_e}\left(1 - \frac{u}{u_e}\right)dy \tag{4.6}$$

where $\delta$ is the local boundary layer thickness.

The transformation factors are

$$F_c = \frac{(\gamma - 1)}{2} r \left[ \sin^{-1}\left( \frac{2A^2 - B}{\sqrt{4A^2 + B^2}} \right) + \sin^{-1}\left( \frac{B}{\sqrt{4A^2 + B^2}} \right) \right]^{-2} \tag{4.7}$$

$$F_\theta = \frac{\mu_e}{\mu_w} \tag{4.8}$$

$$A = \left[ \frac{(\gamma - 1)}{2} r \frac{T_e}{T_w} \right]^{\frac{1}{2}} \tag{4.9}$$

$$B = \left[ 1 + \frac{(\gamma - 1)}{2} r - \frac{T_w}{T_e} \right] \frac{T_e}{T_w} \tag{4.10}$$

where $r$ is the recovery factor defined by

$$T_{aw} = T_e \left( 1 + \frac{(\gamma - 1)}{2} r M_e^2 \right) \tag{4.11}$$

For a given value of $Re_\theta$, $\bar{R}e_\theta$ is calculated using equations (4.8) and (4.5), and $\bar{c}_f$ is obtained from equation (4.3). The local compressible skin friction coefficient is then obtained from equation (4.4) using (4.7).

The accuracy of the Van Driest II method is summarized in figures 4.58, 4.59, 4.60, and 4.61. Figure 4.58 displays the percent relative error between the predicted

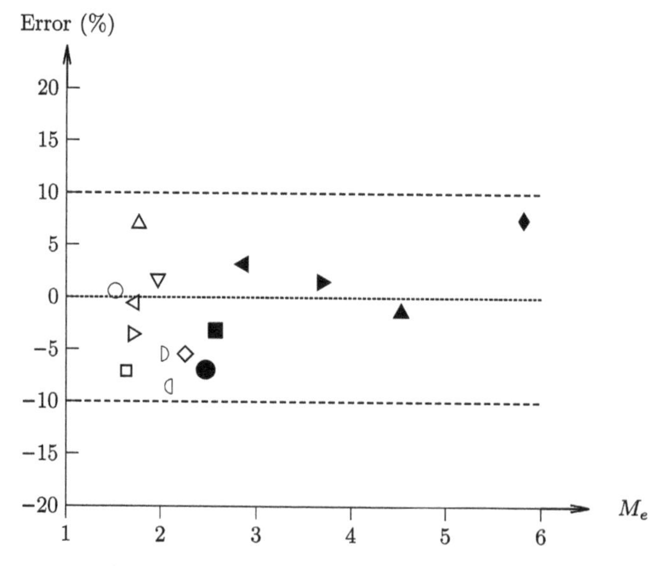


**Figure 4.58.** Error versus $M_e$ for adiabatic flat plate.

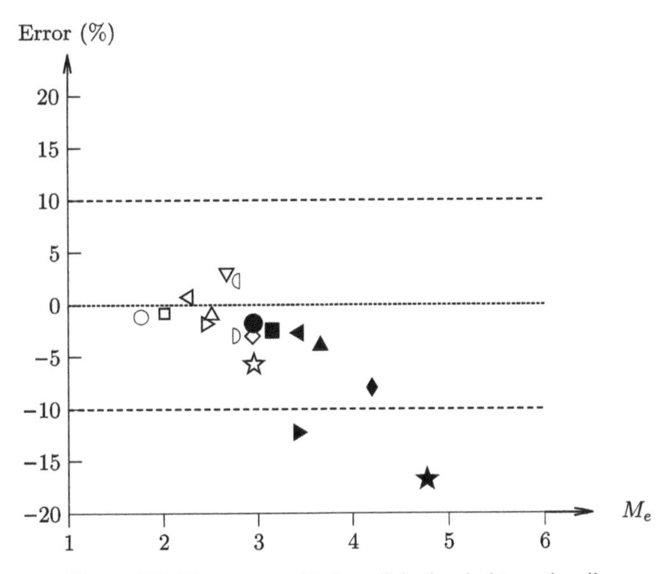


**Figure 4.59.** Error versus $M_e$ for adiabatic wind tunnel wall.

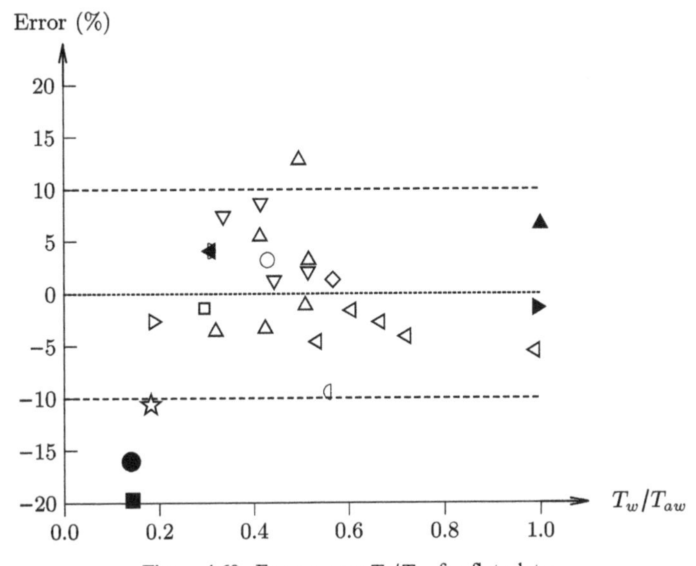


**Figure 4.60.** Error versus $T_w/T_{aw}$ for flat plates.

$c_{f_{VD}}$ and experimental $c_{f_{exp}}$ for adiabatic flat plates with $c_{f_{exp}}$ and $Re_\theta$ directly measured. The symbols are defined in table 4.26. The error is defined as

$$\text{Error} = \frac{(c_{f_{exp}} - c_{f_{VD}})}{c_{f_{VD}}} \times 100 \tag{4.12}$$

The Van Driest II theory is seen to be accurate to within ±10%. Figure 4.59 shows the percent relative error for adiabatic tunnel walls. The symbols are defined in

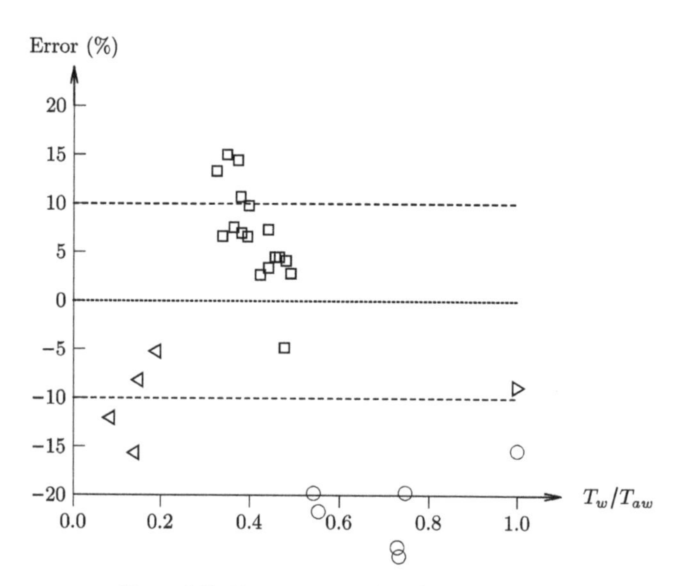


**Figure 4.61.** Error versus $T_w/T_{aw}$ for tunnel walls.

**Table 4.26.** Symbols for figure 4.58.

| Symbol | $M_e$ | $Re_\theta \times 10^{-3}$ | Reference |
|---|---|---|---|
| ○ | 1.5 | 2 | Hakkinen (1955) |
| □ | 1.63 | 12 | Shutts *et al* (1955) |
| ◁ | 1.68 | 20 | Shutts *et al* (1955) |
| ▷ | 1.73 | 13 | Shutts *et al* (1955) |
| △ | 1.75 | 2 | Hakkinen (1955) |
| ▽ | 1.97 | 6 | Coles (1953b) |
| D | 2.00 | 12 | Shutts *et al* (1955) |
| ◖ | 2.11 | 14 | Shutts *et al* (1955) |
| ◇ | 2.25 | 7 | Shutts *et al* (1955) |
| ● | 2.46 | 12 | Shutts *et al* (1955) |
| ■ | 2.56 | 6 | Coles (1953b) |
| ◀ | 2.80 | 81 | Moore and Harkness (1965) |
| ▶ | 3.69 | 5 | Coles (1953b) |
| ▲ | 4.53 | 5 | Coles (1953b) |
| ◆ | 5.79 | 3 | Korkegi (1956) |

table 4.27. With two exceptions the Van Driest II theory is seen to be accurate to within ±10%. Figure 4.60 displays the percent relative error for isothermal flat plates. The symbols are defined in table 4.28. Similarly, with few exceptions the Van Driest II theory is accurate to within ±10%. Figure 4.61 shows the percent relative

**Table 4.27.** Symbols for figure 4.59.

| Symbol | $M_e$ | $Re_\theta \times 10^{-3}$ | Reference |
|---|---|---|---|
| ○ | 1.75 | 8 | Stalmach (1958) |
| □ | 2.01 | 8 | Stalmach (1958) |
| ◁ | 2.23 | 7 | Stalmach (1958) |
| ▷ | 2.46 | 68 | Hopkins and Keener (1966) |
| △ | 2.49 | 7 | Stalmach (1958) |
| ▽ | 2.67 | 690 | Moore and Harkness (1965) |
| D | 2.72 | 7 | Stalmach (1958) |
| ◖ | 2.80 | 360 | Moore and Harkness (1965) |
| ◇ | 2.95 | 7 | Stalmach (1958) |
| ☆ | 2.95 | 15 | Matting *et al* (1961) |
| ● | 2.96 | 61 | Hopkins and Keener (1966) |
| ■ | 3.16 | 7 | Stalmach (1958) |
| ◀ | 3.39 | 7 | Stalmach (1958) |
| ▶ | 3.45 | 56 | Hopkins and Keener (1966) |
| ▲ | 3.67 | 6 | Stalmach (1958) |
| ◆ | 4.20 | 13 | Matting *et al* (1961) |
| ★ | 4.75 | 28 | Lee *et al* (1968) |

**Table 4.28.** Symbols for figure 4.60.

| Symbol | $M_e$ | BL Trips | $Re_\theta \times 10^{-3}$ | $Re_x \times 10^{-6}$ | $T_w/T_{aw}$ | Reference |
|---|---|---|---|---|---|---|
| ○ | 2.8 | Yes | | 2 | 0.43 | Sommer and Short (1955) |
| □ | 3.8 | Yes | | 2 | 0.29 | Sommer and Short (1955) |
| ◁ | 4.9 | No | 6 | | 0.60–0.99 | Young (1965) |
| ▷ | 5.6 | Yes | | 3 | 0.20 | Sommer and Short (1955) |
| △ | 6.5 | Yes | 5 | | 0.32–0.41 | Hopkins *et al* (1970) |
| ▽ | 6.5 | No | 4 | | 0.34–0.41 | Hopkins *et al* (1970) |
| D | 6.6 | No | 4 | | 0.31 | Hopkins *et al* (1969) |
| ◖ | 6.8 | No | | 3 | 0.57 | Neal (1966) |
| ◇ | 6.8 | No | | 2 | 0.57 | Neal (1966) |
| ☆ | 7.0 | Yes | | 1 | 0.18 | Sommer and Short (1955) |
| ● | 7.4 | No | | 14 | 0.14 | Wallace and McLaughlin (1966) |
| ■ | 7.4 | No | | 12 | 0.14 | Wallace and McLaughlin (1966) |
| ◀ | 7.4 | No | 4 | | 0.31 | Hopkins *et al* (1969) |
| ▶ | 4.53 | Yes | 5 | | 1.0 | Coles (1953b) |
| ▲ | 5.79 | Yes | 3 | | 1.0 | Korkegi (1956) |

**Table 4.29.** Symbols for figure 4.61.

| Symbol | $M_e$ | $Re_\theta \times 10^{-3}$ | $T_w/T_{aw}$ | Reference |
|---|---|---|---|---|
| ○ | 4.7 | 24.5 | 0.54–0.75 | Lee *et al* (1968) |
| □ | 7.4 | 36.5 | 0.32–0.49 | Hopkins *et al* (1969) |
| ◁ | 8.6 | 6.3 | 0.08–0.19 | Wallace (1969) |
| ▷ | 4.2 | 13 | 1.0 | Matting *et al* (1961) |

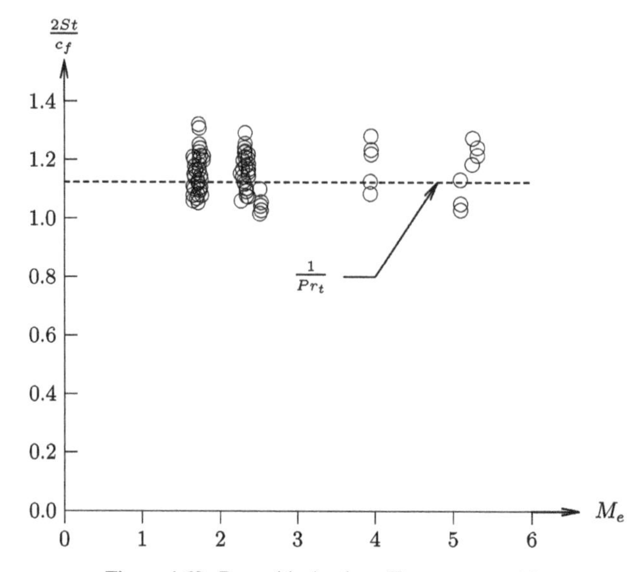


**Figure 4.62.** Reynolds Analogy Factor versus $M_e$.

error for isothermal wind tunnel walls. The symbols are defined in table 4.29. A similar conclusion holds for the Van Driest II theory. Figure 4.62 presents experimental data for the Reynolds Analogy Factor versus $M_e$ compiled by Chi and Spaulding (1966). The mean value is slightly above $Pr_t^{-1}$.

## 4.23 Voisinet *et al* (1971)

Voisinet *et al* (1971) conducted an experimental investigation of hypersonic turbulent boundary layers in an adverse pressure gradient at Mach 4.9 with adiabatic and cold walls. The experiments were performed in the Naval Ordnance Laboratory boundary layer channel (section A.23.2). The measurements were obtained on the tunnel wall whose shape downstream of achieving Mach 4.9 was curved to create the adverse pressure gradient. The flow conditions, listed in table 4.30, correspond to the flow in the test section immediately upstream of the beginning of the adverse pressure gradient. Experimental diagnostics include boundary layer surveys of pitot pressure, total temperature, and static pressure at

**Table 4.30.** Freestream conditions.

| Quantity | Min | Max |
|---|---|---|
| $M_e$ | 4.9 | 4.9 |
| $p_{t_e}$ (MPa) | 0.10 | 1.0 |
| $T_{t_e}$ (K) | 336 | 423 |
| $T_w/T_{aw}$ | 0.8 | 1.0 |
| $H_e$ (MJ kg$^{-1}$) | 0.34 | 0.42 |
| $Re \times 10^{-6}$ (m$^{-1}$) | 2.0 | 24.6 |
| $\delta$ (cm) | 3.8 | 8.9 |

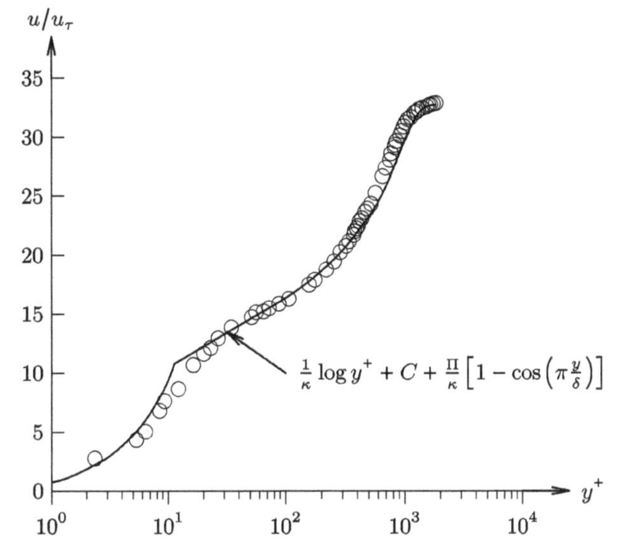


**Figure 4.63.** $u/u_\tau$ versus $y^+$.

six streamwise locations, surface heat transfer, and skin friction. The stations are numbered based upon their distance in inches from the nozzle throat.

Figure 4.63 displays $u/u_\tau$ versus $y^+$ for $p_{t_e}$ = 0.52 MPa, $T_{t_e}$ = 336 K, and $T_w$ = 294 K. Note that the mean velocity (not the Van Driest transformed velocity) is shown. The mean velocity agrees closely with the Law of the Wall and Wake of the form

$$\frac{u}{u_\tau} = \frac{1}{\kappa}\log y^+ + C + \frac{\Pi}{\kappa}\left[1 - \cos\left(\pi\frac{y}{\delta}\right)\right] \tag{4.13}$$

with $C = 4.78$ and $\Pi = 1.837$ with the wake function as suggested by Hinze (1959).

Figure 4.64 shows the velocity–total temperature relation at station 90 at the same freestream conditions as figure 4.63. A significant difference between the experimental data and the Crocco–Busemann relation (3.144) is evident. Close agreement with the following form of Walz's relation

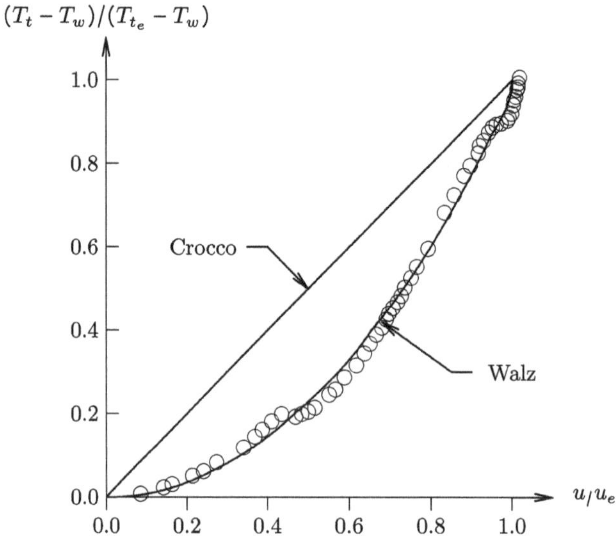


**Figure 4.64.** $u/u_e$ versus $(T_t - T_w)/(T_{t_e} - T_w)$.

$$\frac{T_t - T_w}{T_{t_e} - T_w} = \left\{ T_e \left[ \frac{T}{T_e} + \frac{(\gamma - 1)}{2} M_e^2 \left( \frac{u}{u_e} \right)^2 \right] - T_w \right\} \{ T_{t_e} - T_w \}^{-1} \tag{4.14}$$

where

$$\frac{T}{T_e} = \frac{T_w}{T_e} + \left( \frac{T_{aw} - T_w}{T_e} \right) \frac{u}{u_e} + \left( 1 - \frac{T_w}{T_e} \right) \left( \frac{u}{u_e} \right)^2 \tag{4.15}$$

is apparent. However, significant differences between experiment and both Crocco–Busemann and Walz appear at earlier stations. Additionally, the cooling of the nozzle throat had a significant effect on the velocity–total temperature relation.

Figure 4.65 shows the velocity–total temperature relation at station 90 for $p_{t_e} = 0.52$ MPa, $T_{t_e} = 440$ K, and $T_w = 294$ K using experimental data from two different probes. The difference between the experiment and (4.14) is more pronounced than in figure 4.64.

Figure 4.66 shows the Reynolds Analogy Factor versus skin friction coefficient $c_f$ at five stations for different freestream conditions. The data are generally clustered around $2St/c_f = Pr_t^{-1}$ with $Pr_t = 0.89$.

## 4.24 Holden (1972)

Holden (1972) conducted an extensive study of cold wall hypersonic turbulent boundary layers on a flat plate. The experiments were conducted at Mach numbers from 7 to 12 and wall-to-freestream stagnation temperature ratios from 0.14 to 0.3.

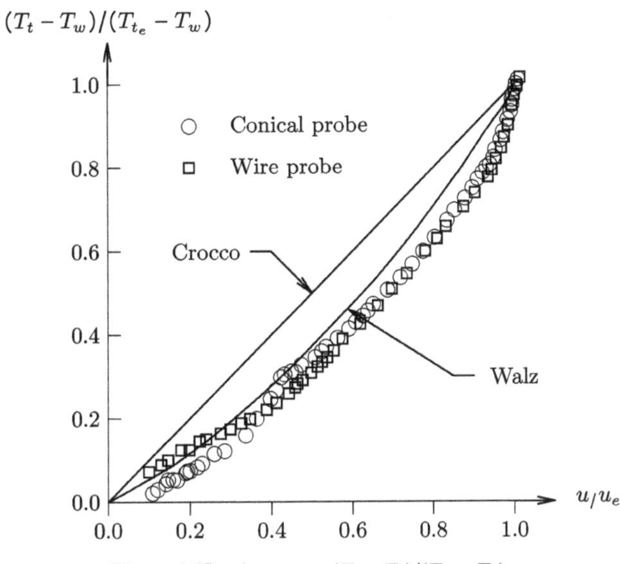


**Figure 4.65.** $u/u_e$ versus $(T_t - T_W)/(T_{t_e} - T_W)$.

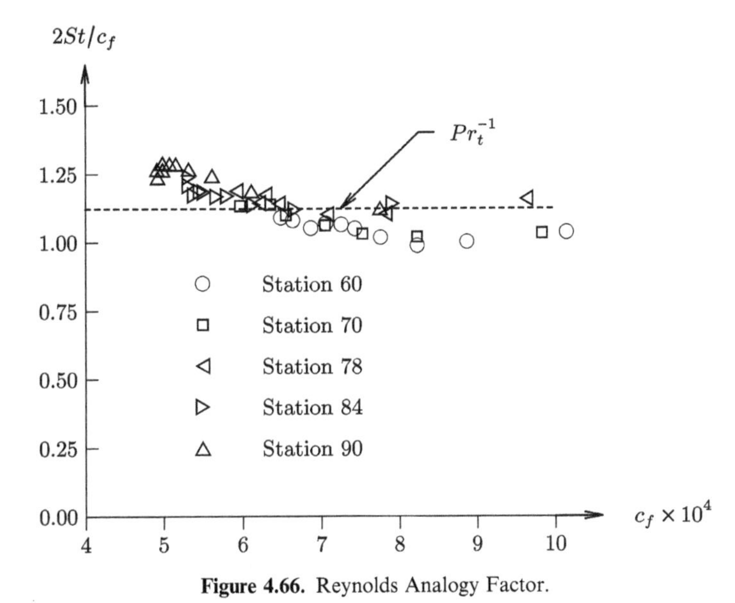


**Figure 4.66.** Reynolds Analogy Factor.

The experiments were performed in the Cornell Aeronautical Laboratories[7] 48 in and 96 in hypersonic shock tunnels. The freestream conditions are listed in table 4.31. Two test model flat plates with lengths of 99.06 cm and 132.08 cm were used. Experimental diagnostics include surface heat transfer, pressure, and skin friction.

[7] The Cornell Aeronautical Laboratories were later reorganized as the Calspan University of Buffalo Research Center.

**Table 4.31.** Freestream conditions.

| Quantity | Min | Max |
|---|---|---|
| $M_e$ | 7.36 | 12.04 |
| $H_e$ (MJ kg$^{-1}$) | 0.99 | 2.07 |
| $T_{t_e}$ (K) | 948 | 1871 |
| $H_w/H_e$ | 0.143 | 0.30 |
| $T_w/T_{aw}$ | 0.16 | 0.34 |
| $Re \times 10^{-6}$ (m$^{-1}$) | 15.1 | 191.8 |

$T_w/T_{aw} = (T_w/T_{t_e})\left[1 + \frac{(\gamma-1)}{2}M_e^2\right]\left[1 + \frac{(\gamma-1))}{2}Pr_t M_e^2\right]^{-1}$

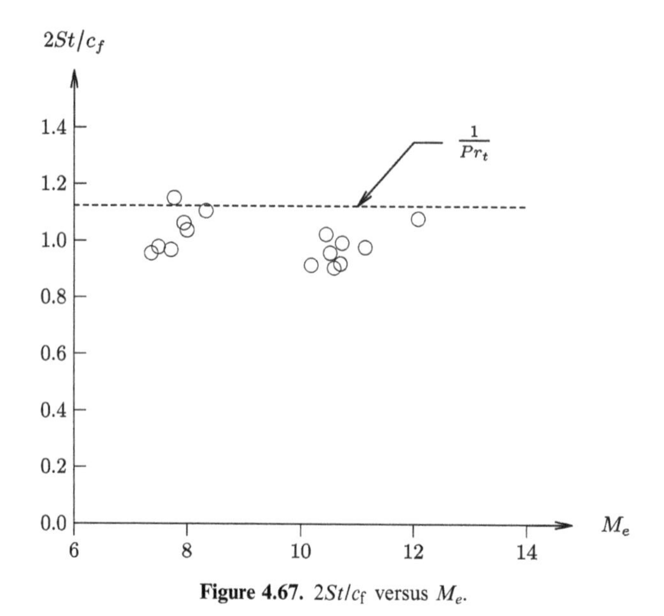


**Figure 4.67.** $2St/c_f$ versus $M_e$.

Figures 4.67 and 4.68 display the Reynolds Analogy Factor $2St/c_f$ versus Mach number $M_e$ and wall-to-freestream total enthalpy ratio $H_w/H_\infty$, respectively. The data indicate $2St/c_f = 1.02 \pm 0.07$. Additional extensive skin friction and heat transfer data are presented in Holden (1972).

## 4.25 Hopkins and Keener (1972)

Hopkins and Keener (1972a, 1972b) examined the effects of pressure gradient on the velocity–total temperature relation at Mach 7.4. The experiments were conducted in the NASA Ames hypersonic wind tunnel (section A.20.2) on the nozzle wall at a distance of 10 m from the nozzle throat. The flow conditions are listed in table 4.32. Experimental diagnostics include boundary layer profiles of pitot pressure and total

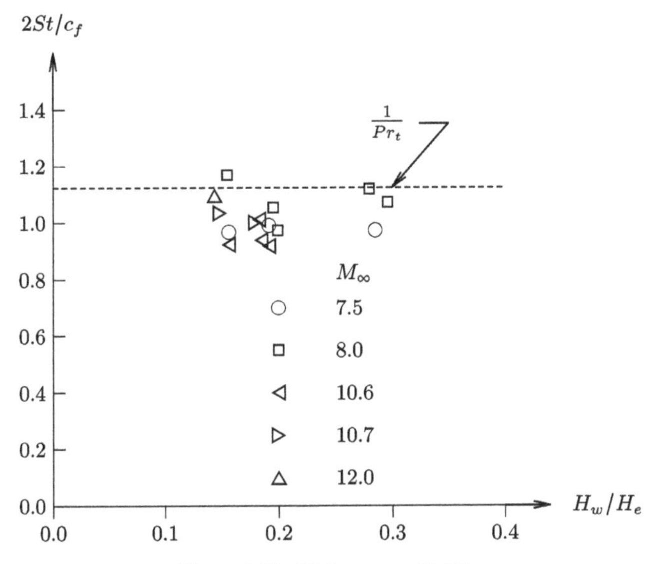


**Figure 4.68.** $2St/c_f$ versus $H_w/H_e$.

**Table 4.32.** Freestream conditions.

| Quantity | Min | Max |
|---|---|---|
| $M_\infty$ | 7.41 | 7.48 |
| $p_{t_\infty}$ (MPa) | 1.23 | 5.49 |
| $T_{t_\infty}$ (K) | 707 | 1044 |
| $T_w$ (K) | 302 | 316 |
| $T_w/T_{aw}$ | 0.48[a] | 0.34[a] |
| $H_\infty$ (MJ kg$^{-1}$) | 0.71 | 1.04 |
| $\delta^*$ (cm) | 5.43 | 7.24 |
| $\theta$ (mm) | 4.45 | 9.05 |

[a]Using (3.46).

temperature, and surface skin friction using a floating element balance. Previous experiments for a flat plate at Mach 6.5 (Hopkins *et al* 1970) were also analyzed.

Figure 4.69 shows the normalized total temperature difference $(T_t - T_w)/(T_{t_e} - T_w)$ versus $u/u_e$. The experimental data at the lower Mach number (Hopkins *et al* 1970) follow the Crocco relation, while the experimental data at the higher Mach number are closer to (3.220).

## 4.26 Horstman and Owen (1972)

Horstman and Owen (1972) and Owen and Horstman (1972) performed an experimental investigation of a cold wall hypersonic turbulent boundary layer on an ogive cylinder at Mach 7.2. The experiments were conducted in the NASA Ames 3.5 ft hypersonic wind tunnel (section A.20.2). The model is displayed in figure 4.70

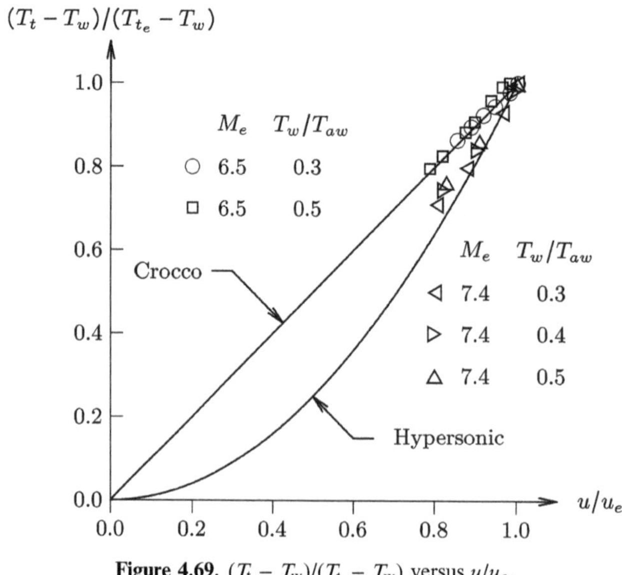


**Figure 4.69.** $(T_t - T_w)/(T_{t_e} - T_w)$ versus $u/u_e$.

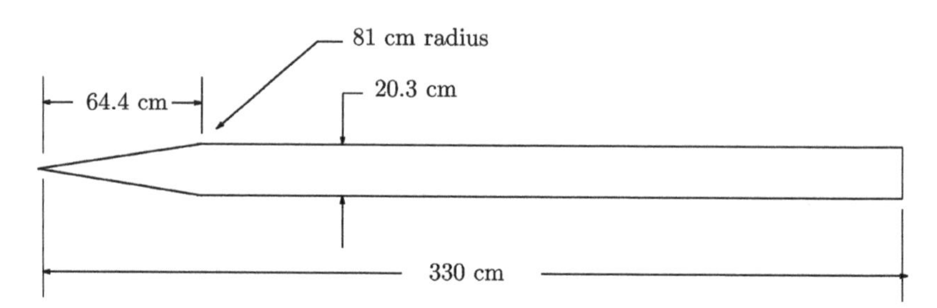


**Figure 4.70.** Model.

and the freestream conditions are indicated in table 4.33. Natural transition occurred between $x$ = 37 and 80 cm. Experimental diagnostics include surface pressure, skin friction, and heat transfer, boundary layer profiles of pitot pressure, and total temperature at several streamwise locations. Constant-current hot wire measurements were also obtained through the boundary layer at three streamwise locations. The static pressure was assumed constant across the boundary layer.

Figures 4.71, 4.72, and 4.73 display mean profiles of streamwise velocity, Mach number, and total temperature at three different streamwise locations. The total temperature profiles were measured directly. The Mach number profile was calculated from the measured pitot pressure and total temperature profiles, and the velocity profile from the measured total temperature and calculated Mach number. Figure 4.74 shows the turbulent shear stress computed from the profiles of velocity and density taking into account the small streamwise pressure gradient on the cylinder. Figure 4.75 displays the turbulent heat flux computed from the profiles of velocity and density. The turbulent Prandtl number profiles are shown in

**Table 4.33.** Freestream conditions.

| Quantity | Value |
|---|---|
| $M_\infty$ | 7.2 |
| $p_{t_\infty}$ (MPa) | 3.45 |
| $T_{t_\infty}$ (K) | 667 |
| $T_w$ (K) | 310 |
| $T_w/T_{aw}$ | 0.49 |
| $H_\infty$ (MJ kg$^{-1}$) | 0.67 |
| $\delta$ (cm) | 1.7–3.3 |
| $\theta$ (cm) | 0.053 5–0.118 8 |
| $Re \times 10^{-6}$ (m$^{-1}$) | 10.9 |

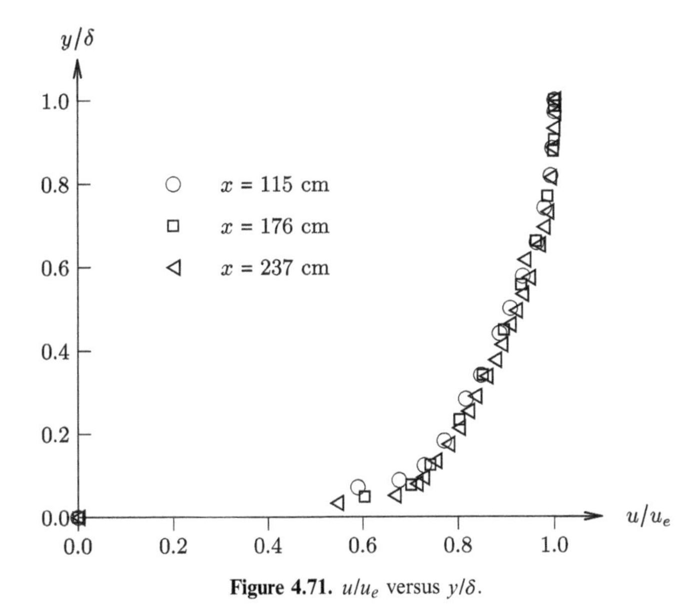


**Figure 4.71.** $u/u_e$ versus $y/\delta$.

figure 4.76 and display a non-uniform structure across the boundary layer. Further details regarding the turbulence structure including auto- and space-time correlation measurements and energy spectra are presented in Horstman and Owen (1972) and Owen and Horstman (1972).

Figure 4.77 displays the Van Driest transformed velocity versus the distance from the wall. A logarithmic region is evident. Figure 4.78 shows the ratio $(T_t - T_w)/(T_{t_e} - T_w)$ versus $u/u_e$. Neither a linear nor a quadratic behavior is observed.

## 4.27 Keener and Hopkins (1972)

Keener and Hopkins (1972) examined a zero pressure gradient cold wall hypersonic boundary layer on a flat plate at Mach 6.5. The experiments were performed in the

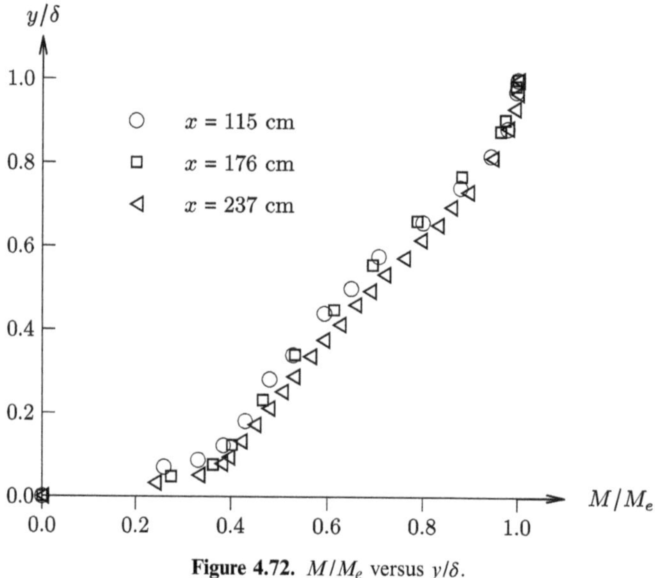


**Figure 4.72.** $M/M_e$ versus $y/\delta$.

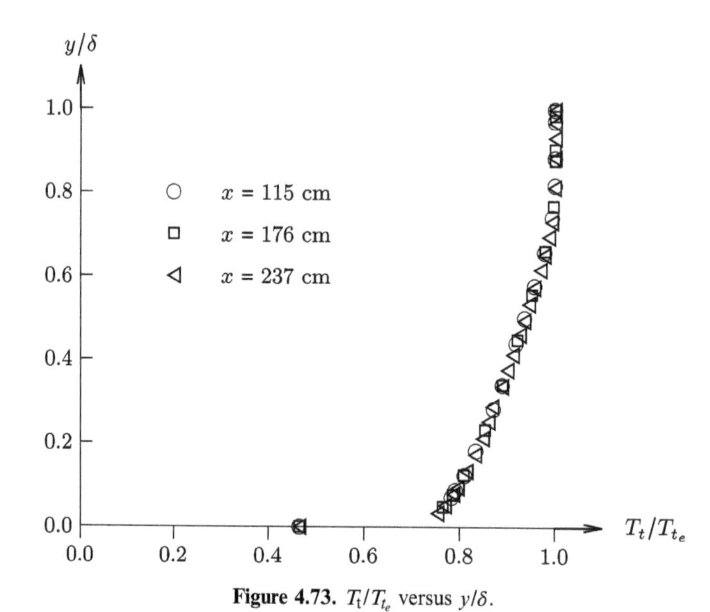


**Figure 4.73.** $T_t/T_{t_e}$ versus $y/\delta$.

NASA Ames 3.5 ft hypersonic wind tunnel (section A.20.2). The flat plate is 122 cm in length and 45.7 cm in width. Measurements were obtained at a station 99.6 cm from the leading edge. The experimental diagnostics include boundary layer surveys of pitot pressure, total temperature and static pressure, and wall shear stress. The freestream conditions at the measurement station are listed in table 4.34.

Figure 4.79 displays the Van Driest transformed velocity $u_{\text{VD}}$ versus $y^+$ for cases 1, 3, and 5. The transformed velocity shows close agreement with the Law of the Wall

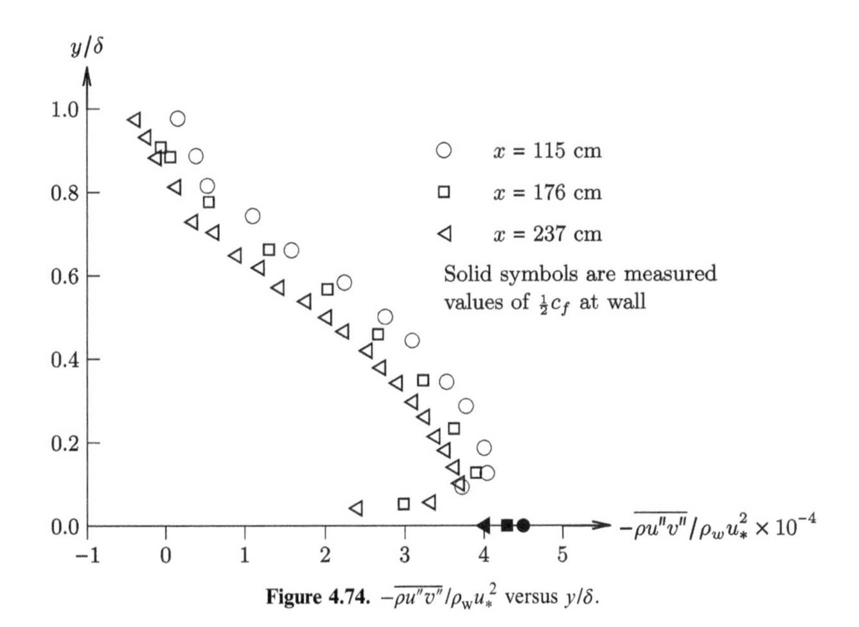


**Figure 4.74.** $-\overline{\rho u''v''}/\rho_w u_*^2$ versus $y/\delta$.

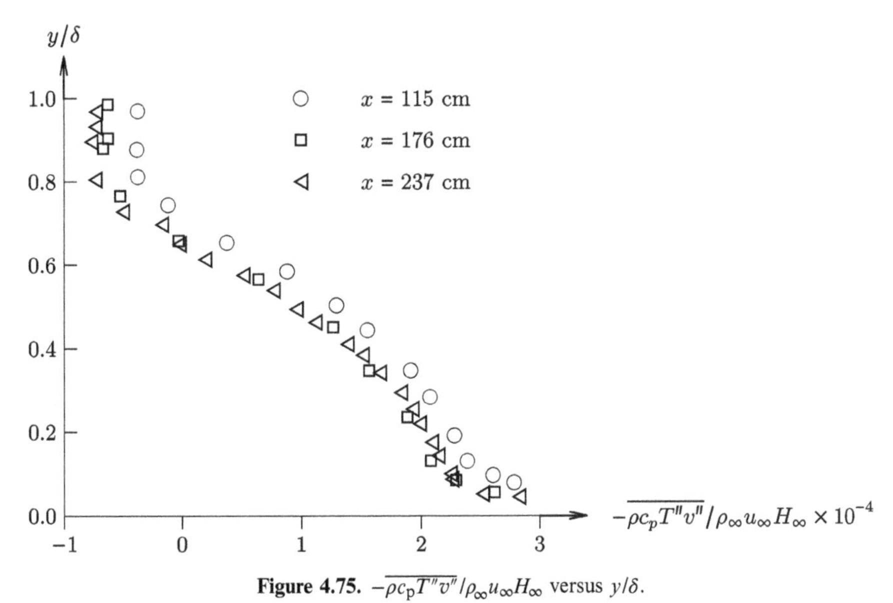


**Figure 4.75.** $-\overline{\rho c_p T''v''}/\rho_\infty u_\infty H_\infty$ versus $y/\delta$.

for $C = 5.1$. Figure 4.80 shows the velocity defect $(u_{VD} - u_{VDe})/u_\tau$ versus $y/\delta$. Also shown is the velocity defect law 3.61 with $\Pi = 0.55$. Good agreement is observed in the outer portion of the boundary layer.

## 4.28 Keener and Polek (1972)

Keener and Polek (1972) investigated the Reynolds Analogy Factor for cold wall zero pressure gradient turbulent boundary layers at Mach numbers from 5.9 to 7.7.

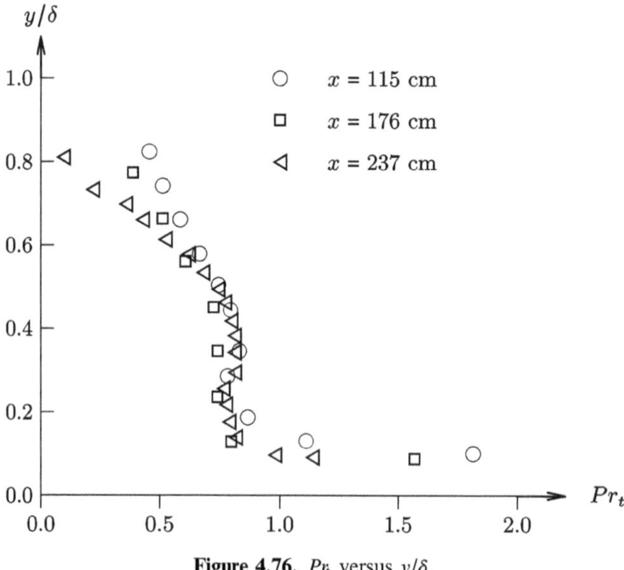


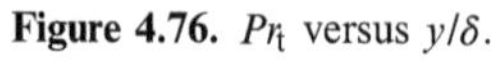
**Figure 4.76.** $Pr_t$ versus $y/\delta$.

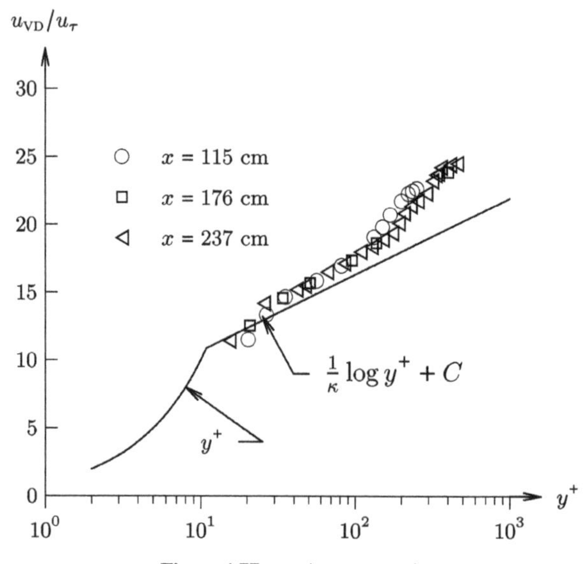


**Figure 4.77.** $u_{VD}/u_\tau$ versus $y^+$.

The experiments were performed in the NASA Ames 3.5 ft hypersonic wind tunnel (section A.20.2). Measurements of surface heat transfer and skin friction were obtained at a location 100 cm from the leading edge of a sharp-edged flat plate of dimensions 119 cm in length and 43.8 cm in width. The estimated uncertainty in measurement of heat transfer and skin friction is ±5%. The model was injected into the airstream at angles of attack from 0° to 9.3° resulting in edge Mach numbers

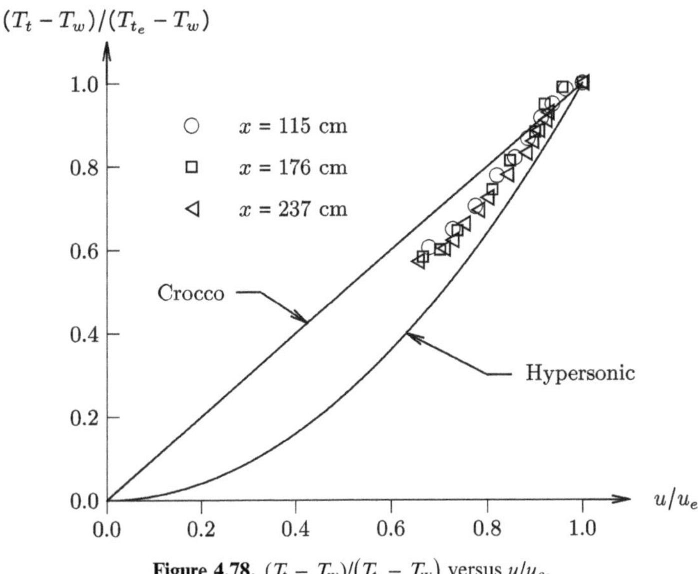


**Figure 4.78.** $(T_t - T_w)/(T_{t_e} - T_w)$ versus $u/u_e$.

**Table 4.34.** Freestream conditions.

| | Value | | | | | |
|---|---|---|---|---|---|---|
| Quantity | Case 1 | Case 2 | Case 3 | Case 4 | Case 5 | Case 6 |
| $M_e$ | 6.21 | 6.39 | 6.42 | 6.42 | 6.50 | 6.50 |
| $p_{t_e}$ (MPa) | 3.94 | 6.66 | 3.85 | 2.32 | 4.52 | 4.52 |
| $T_{t_e}$ (K) | 1028 | 1089 | 764 | 784 | 684 | 689 |
| $T_w$ (K) | 324 | 326 | 312 | 306 | 318 | 318 |
| $T_w/T_{aw}$ | 0.34 | 0.32 | 0.45 | 0.43 | 0.51 | 0.51 |
| $H_e$ (MJ kg$^{-1}$) | 1.03 | 1.09 | 0.77 | 0.79 | 0.69 | 0.69 |
| $\delta$ (cm) | 1.02 | 1.37 | 1.22 | 1.75 | 1.35 | 1.65 |
| $\theta$ (cm) | 0.049 8 | 0.062 5 | 0.051 8 | 0.074 4 | 0.051 8 | 0.067 6 |
| $Re \times 10^{-6}$ (m$^{-1}$) | 5.20 | 8.45 | 9.11 | 5.32 | 13.3 | 13.1 |

from 5.9 to 7.8. The range of boundary layer edge conditions are shown in table 4.35.

Figure 4.81 displays the Reynolds Analogy Factor versus $Re_\theta$. The adiabatic wall temperature was calculated from (3.158) assuming $r = 0.88$. The maximum scatter is ±9% with most results falling within ±4% of $2St/C_f = 1.0$.

## 4.29 Kemp and Owen (1972)

Kemp and Owen (1971, 1972a, 1972b) measured mean and fluctuating flow properties in a cold wall hypersonic boundary layer formed on the nozzle wall of the NASA Ames Mach 50 helium tunnel (section A.20.3). Measurements were

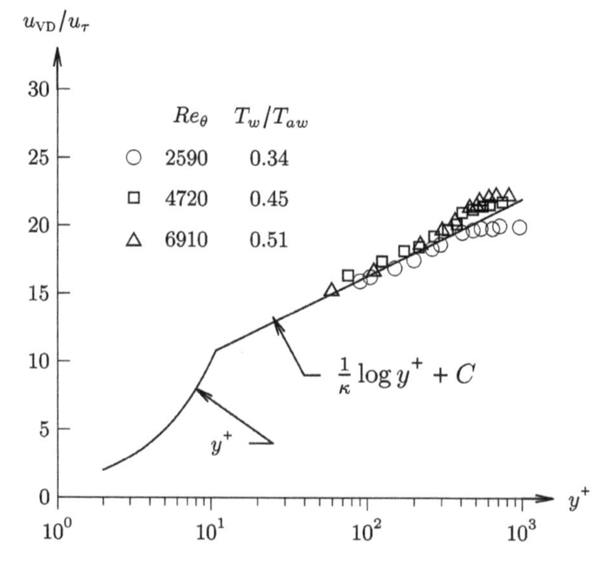


**Figure 4.79.** $u_{VD}/u_\tau$ versus $y^+$.

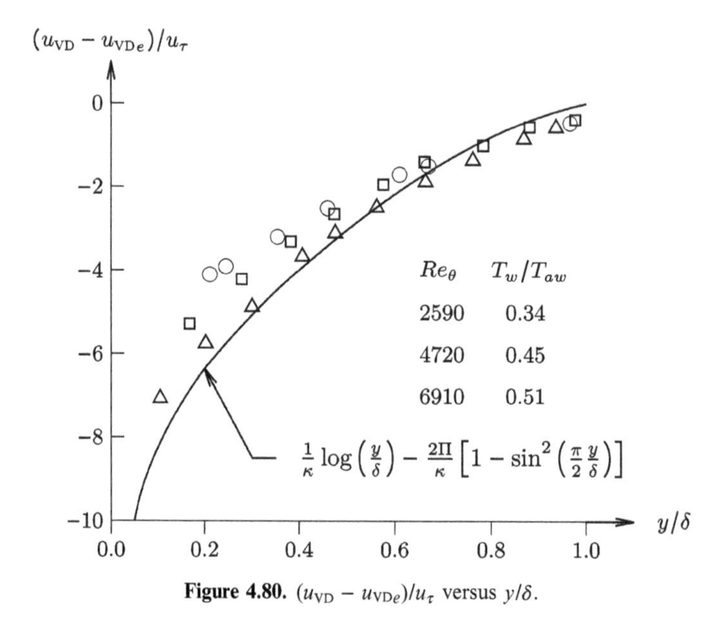


**Figure 4.80.** $(u_{VD} - u_{VDe})/u_\tau$ versus $y/\delta$.

obtained at four locations downstream of the nozzle throat. The range of boundary layer edge conditions are shown in table 4.36.

Figure 4.82 displays the mean velocity at the location $x = 2.793$ m downstream of the nozzle throat. At this location the edge Mach number $M_e = 42.1$. The velocity was computed from the measured pitot pressure and stagnation temperature using three different assumptions for the static pressure in the boundary layer, namely, (a) $p = p_e$, (b) $p = p_w$, and (c) $p(y) = p_w + (y/\delta)(p_e - p_w)$. It is interesting to compare

**Table 4.35.** Freestream conditions.

| Quantity | Value |
|---|---|
| $M_e$ | 5.9–7.8 |
| $T_w/T_{aw}$ | 0.32–0.50 |
| $Re_\theta \times 10^{-4}$ | 0.2–1.8 |

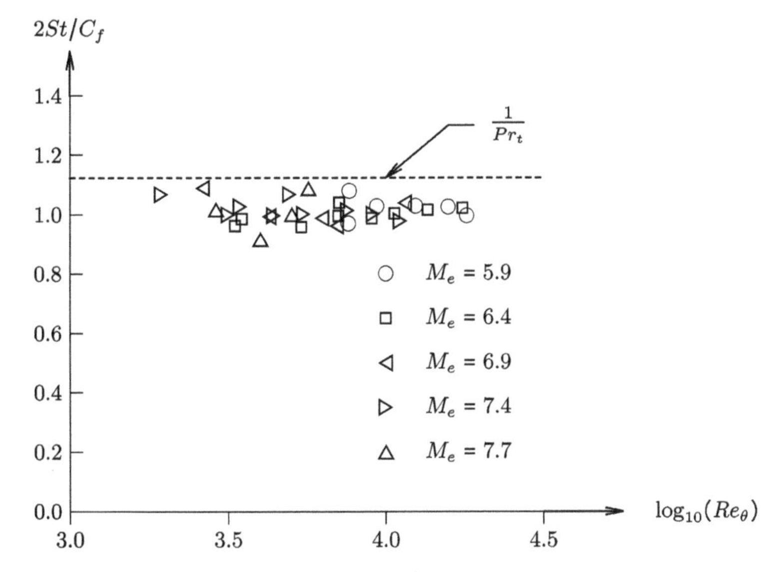


**Figure 4.81.** Reynolds Analogy Factor versus $Re_\theta$.

**Table 4.36.** Freestream conditions.

| Quantity | Min | Max |
|---|---|---|
| $M_e$ | 19.2 | 44.6 |
| $p_{t_e}$ (MPa) | 6.6 | 27.6 |
| $T_{t_e}$ (K) | 315 | 988 |
| $T_w$ (K) | 300 | 409 |
| $T_w/T_{t_e}$ | 0.35 | 1.0 |
| $T_w/T_{aw}$ | 0.35[a] | 1.19[a] |
| $\delta$ (cm) | 2.36 | 24.1 |
| $\theta$ (cm) | 0.011 7 | −0.085 5 |
| $Re \times 10^{-6}$ ($m^{-1}$) | 2.0 | 33.4 |

[a] Using (3.46) with $Pr_t = 0.89$.

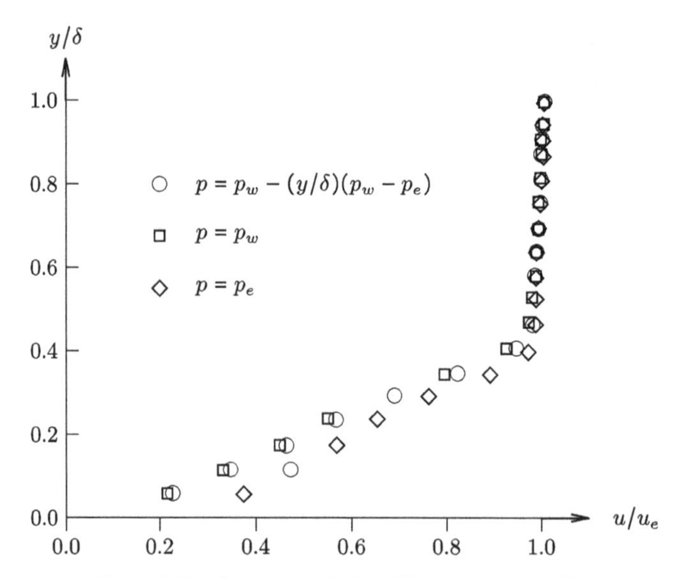


**Figure 4.82.** $u/u_e$ versus $y/\delta$ for different $p(y/\delta)$ profiles.

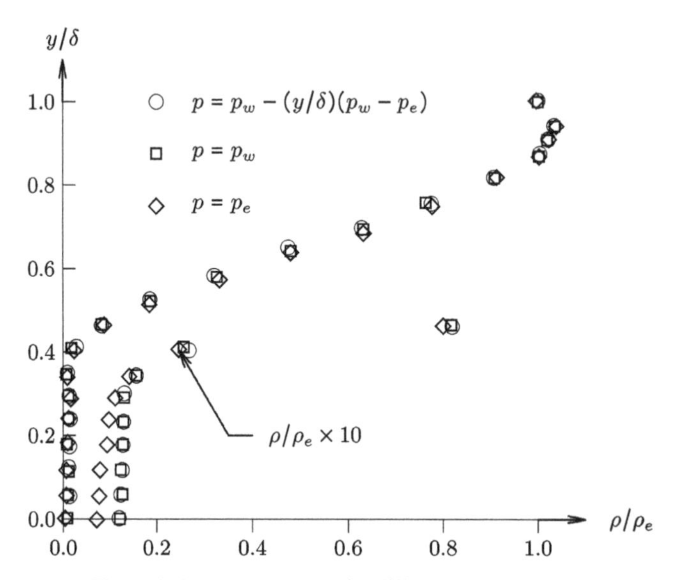


**Figure 4.83.** $\rho/\rho_e$ versus $y/\delta$ for different $p(y/\delta)$ profiles.

the velocity profile with the results of Horstman and Owen (1972) in figure 4.71. In the present case, $u/u_e = 0.6$ at $y/\delta = 0.2$, while in Horstman and Owen (1972) $u/u_e = 0.6$ at $y/\delta = 0.05$. Figure 4.83 shows the density at the same location. The large density gradient across the boundary layer is evident.

Figure 4.84 presents $(T_t - T_w)/(T_{t_e} - T_w)$ versus $u/u_e$ at $x = 1.067$ m for $T_w/T_{t_e} = 0.36$ and 0.72. At this location the edge Mach number $M_e = 27$. It is evident that the quadratic behavior (3.220) holds in the outer portion of the

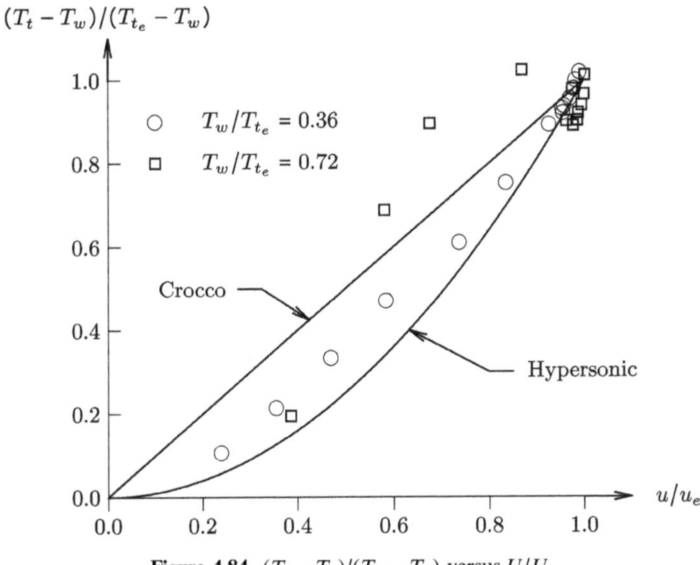


**Figure 4.84.** $(T_t - T_w)/(T_{t_e} - T_w)$ versus $U/U_e$.

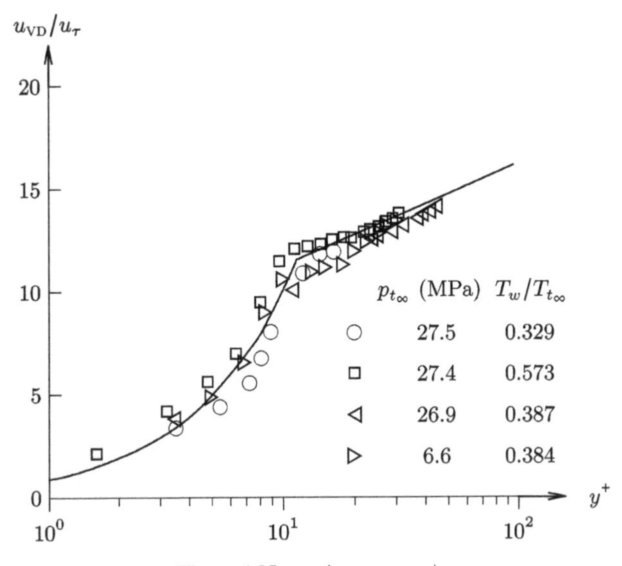


**Figure 4.85.** $u_{VD}/u_\tau$ versus $y^+$.

boundary layer. The actual extent of agreement covers approximately 75% of the boundary layer thickness.

Figure 4.85 displays the Van Driest transformed velocity $u_{VD}/u_\tau$ versus $y^+$. A logarithmic region is evident for $10 < y^+ < 100$; however, the slope yields a value for von Kármán's constant $\kappa = 0.46 \pm 0.01$. Figure 4.86 shows the mass flow fluctuations normalized by the local mean mass flow $\sqrt{\overline{(\rho u')^2}}/\rho u$ versus $y/\delta$. Peak values of

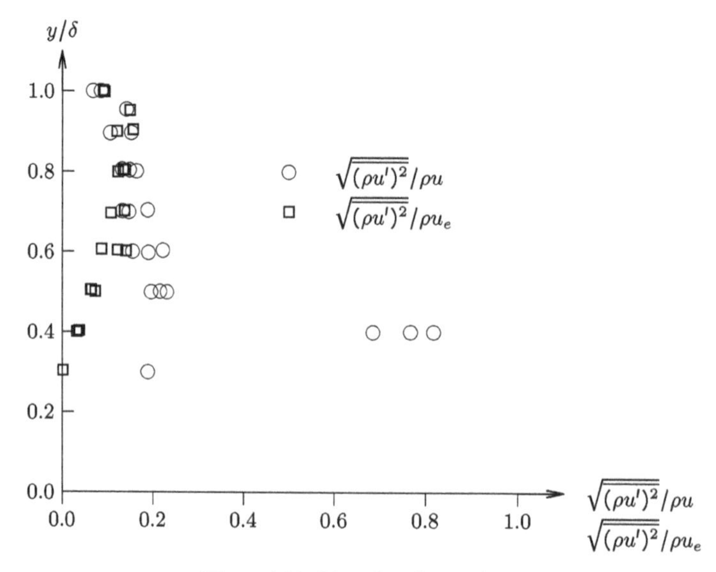


**Figure 4.86.** Mass flow fluctuations.

0.8 are measured. Results are also shown normalized by mean freestream mass flow $\sqrt{\overline{(\rho u')^2}}/\rho u_e$.

## 4.30 Laderman and Demetriades (1971, 1973, 1974)

Laderman and Demetriades (1971, 1973, 1974) conducted an experimental investigation of a cold wall hypersonic turbulent boundary layer on a flat plate at zero streamwise pressure gradient and Mach 9.4. The freestream conditions are shown in table 4.37. Experimental diagnostics include pitot and static pressure probes, total temperature thermocouple, and constant-current hot-wire anemometer.

Figure 4.87 displays the static pressure across the boundary layer. The symbol ○ represents the static probe measurements, while the symbol □ is the measured static pressure corrected for viscous effects using the polynomial fit of the static probe measurements. The latter yields close agreement with the freestream static pressure as calculated using measured freestream pitot pressure and stagnation pressure; however, the corrected static pressure profile within the boundary layer increases by approximately 50% from the freestream to the wall. Similar behavior was observed by Kemp and Owen (1972a).

Figure 4.88 shows the Van Driest transformed velocity $u_{\rm VD}/u_\tau$ versus the dimensionless wall distance $y^+$. A logarithmic region is evident for $10 < y^+ < 100$. The wake parameter $\Pi = 1.4$ which is significantly above the expected value $\Pi = 0.55$. The viscous sublayer velocity does not scale as $u_{\rm VD} = y^+$; however, there is no rationale for assuming this to be correct for a cold wall hypersonic turbulent boundary layer.

Figure 4.89 examines the validity of the Strong Reynolds Analogy (3.331)

**Table 4.37.** Freestream conditions (Laderman and Demetriades 1974).

| Quantity | Value |
|---|---|
| $M_\infty$ | 9.4 |
| $p_{t_\infty}$ (MPa) | 4.27 |
| $T_{t_\infty}$ (K) | 810 |
| $T_w$ (K) | 304 |
| $T_w/T_{aw}$ | 0.404 |
| $H_\infty$ (MJ kg$^{-1}$) | 0.81 |
| $\delta$ (cm) | 10 |
| $\delta^*$ (cm) | 5.7 |
| $\theta$ (cm) | 0.72 |
| $Re \times 10^{-6}$ (m$^{-1}$) | 12.7 |

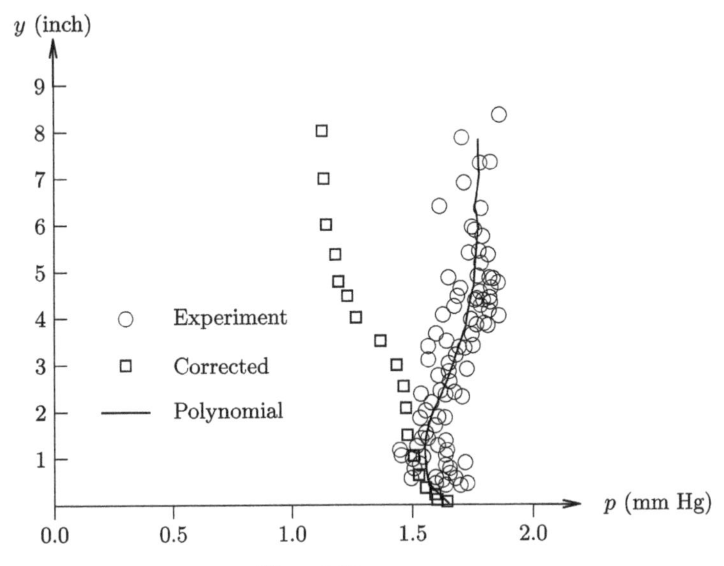


**Figure 4.87.** $p$ versus $y$.

$$\frac{\sqrt{\overline{T'^2}}/\bar{T}}{(\gamma - 1)\bar{M}^2\sqrt{\overline{u'^2}}/\bar{u}} = 1$$

It is evident that the Strong Reynolds Analogy does not hold. Figure 4.90 displays the root-mean-square mass flux fluctuations and total temperature fluctuations each normalized by the local mean values (Laderman and Demetriades 1973). Also shown are the corresponding measurements of Kistler (1959) at Mach 4.67. Results are shown for three different wires. The peak normalized total temperature fluctuations are 2%, and mass flow fluctuations are approximately 13%.

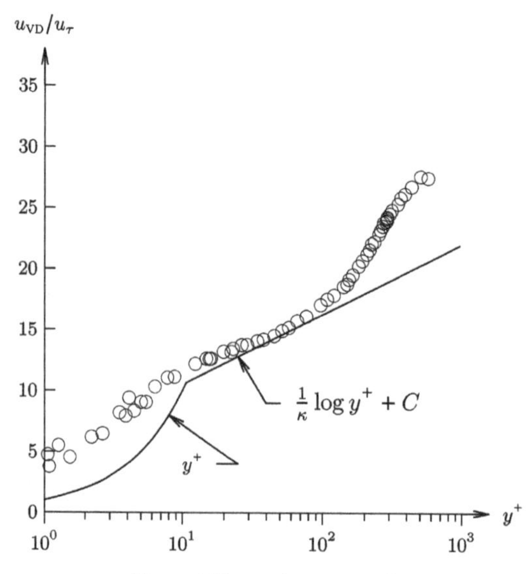


**Figure 4.88.** $u_{VD}/u_\tau$ versus $y^+$.

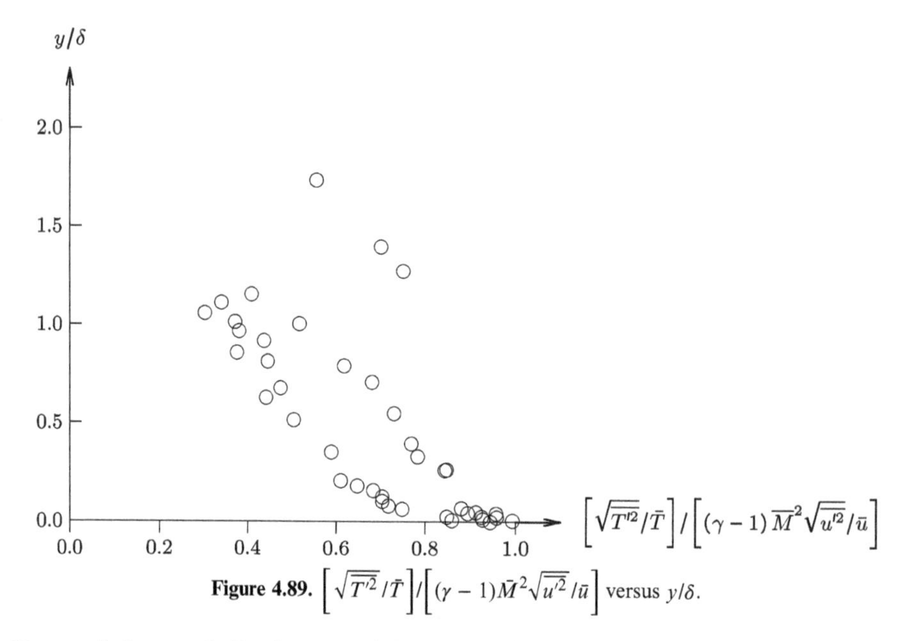


**Figure 4.89.** $\left[\sqrt{\overline{T'^2}}/\bar{T}\right]/\left[(\gamma-1)\bar{M}^2\sqrt{\overline{u'^2}}/\bar{u}\right]$ versus $y/\delta$.

Demetriades and Laderman (1973) provide experimental measurements of Reynolds streamwise, wall normal, and shear stress in a hypersonic turbulent boundary layer at $M_\infty = 9.4$.

## 4.31 Stone and Cary (1972)

Stone and Cary (1972) examined the capability of sonic wall jets to promote transition to turbulence in hypersonic boundary layers as an alternate to discrete

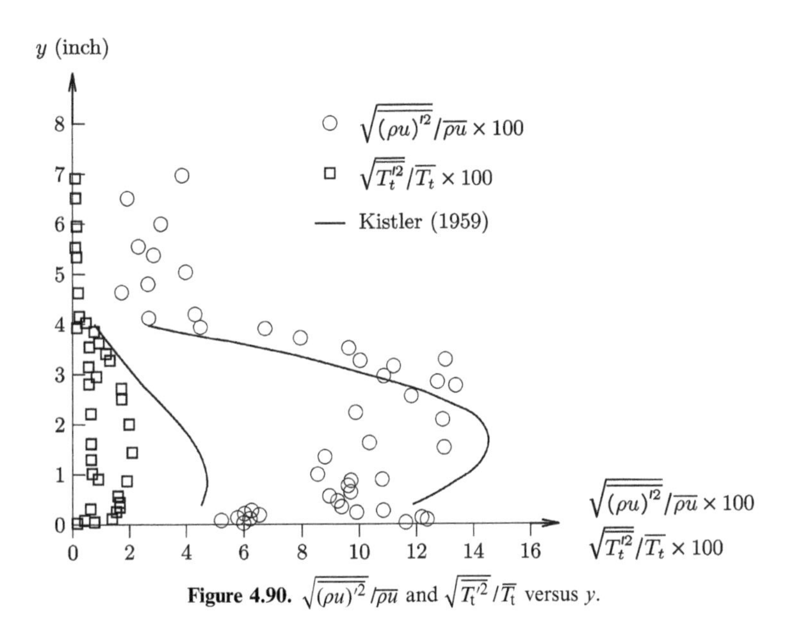


**Figure 4.90.** $\sqrt{\overline{(\rho u)'^2}}/\overline{\rho u}$ and $\sqrt{\overline{T_t'^2}}/\overline{T_t}$ versus $y$.

**Table 4.38.** Freestream conditions.

| Quantity | Min | Max |
|---|---|---|
| $M_e$ | 7.54 | 7.87 |
| $p_{t_e}$ (MPa) | 13.51 | 13.82 |
| $T_{t_e}$ (K) | 833 | 844 |
| $T_w/T_{aw}$ | ≈1 | |
| $H_e$ (MJ kg$^{-1}$) | 0.84 | 0.85 |
| $Re \times 10^{-6}$ (m$^{-1}$) | 2.6 | 2.9 |

surface roughness elements. The experiments were performed on a flat plate at Mach 6 and 8.5 for cold wall conditions. The experiments were conducted in the NASA Langley 20 in Mach 6 and Mach 8.5 hypersonic wind tunnels. The freestream conditions for the Mach 8.5 tests are listed in table 4.38. Two spanwise rows of 17 wall jets were located at 2.5 cm and 7.6 cm from the leading edge of the flat plate. Additional tests were performed using spherical roughness elements. The test model for the Mach 6 experiments had a single spanwise row of nine wall jets located 2.5 cm from the leading edge. Experimental diagnostics include surface heat transfer and pressure for both Mach numbers, and boundary layer pitot pressure and total temperature profiles at two locations downstream of the sonic wall jets for the Mach 8.5 case. Figure 4.91 displays the temperature–velocity relationship for the nominal Mach 8.5 configuration. Results are shown for no wall jets and for two survey locations downstream of the wall jets. The experimental data show closer agreement

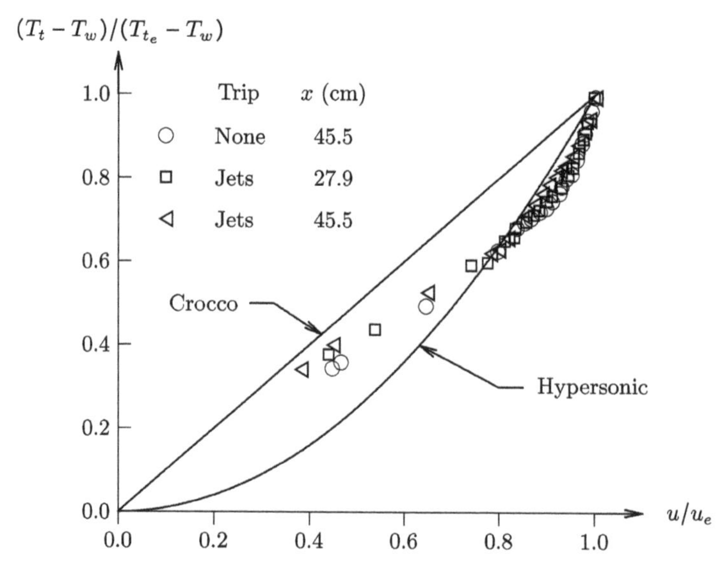


**Figure 4.91.** Effect of jet trips on total temperature–velocity relationship.

**Table 4.39.** Freestream conditions.

| Quantity | Min | Max | Run 812175 | Run 901302 | Run 002162 |
|---|---|---|---|---|---|
| $M_e$ | 4.742 | 4.902 | 4.823 | 4.89 | 4.857 |
| $p_{t_e}$ (MPa) | 0.099 3 | 1.078 | 0.514 | 0.519 | 0.514 |
| $T_{t_e}$ (K) | 346.6 | 451.4 | 352.5 | 415.3 | 422.1 |
| $T_w/T_{aw}$ | 0.25 | 1.0 | 1.0 | 0.8 | 0.25 |
| $H_e$ (MJ kg$^{-1}$) | 0.348 | 0.453 | 0.354 | 0.417 | 0.424 |
| $Re \times 10^{-6}$ (m$^{-1}$) | 1.756 | 24.68 | 12.13 | 9.307 | 9.079 |
| $\delta$ (cm) | 4.873 | 8.970 | 5.806 | 5.671 | 6.574 |
| $\delta^*$ (cm) | 1.626 | 3.018 | 2.236 | 2.165 | 2.007 |
| $\theta$ (cm) | 0.187 8 | 0.554 8 | 0.252 | 0.288 6 | 0.368 3 |

with the quadratic temperature–velocity relation in the outer portion of the boundary layer.

## 4.32 Voisinet and Lee (1972)

Voisinet and Lee (1972) conducted an extensive investigation of a zero pressure gradient hypersonic turbulent boundary layer at Mach 4.9 for adiabatic and cold wall conditions. The experiments were performed at the Naval Ordnance Laboratory tunnel no. 7 on the flat tunnel wall. The freestream conditions are listed in table 4.39. A total of 62 boundary layer profiles were measured at five locations from 1.524 m to 2.286 m from the tunnel throat. Measurements were obtained for $T_w/T_{aw} = 0.25$, 0.8, and 1.0. Experimental diagnostics include boundary layer profiles of pitot pressure and stagnation temperature and wall shear stress and heat transfer.

Figures 4.92, 4.93, and 4.94 display the Van Driest transformed velocity $u_{VD}/u_\tau$ versus $y^+$ for $T_w/T_{aw} = 1.0$, 0.8, and 0.25, respectively. All three profiles show agreement with the Law of the Wall where the value of $C = 4.0$ is chosen for $T_w/T_{aw} = 0.25$ based upon figure 4.16. The profiles are best fit in the outer part of the boundary layer by $\Pi = 1.8$, 1.65, and 0.55, respectively.

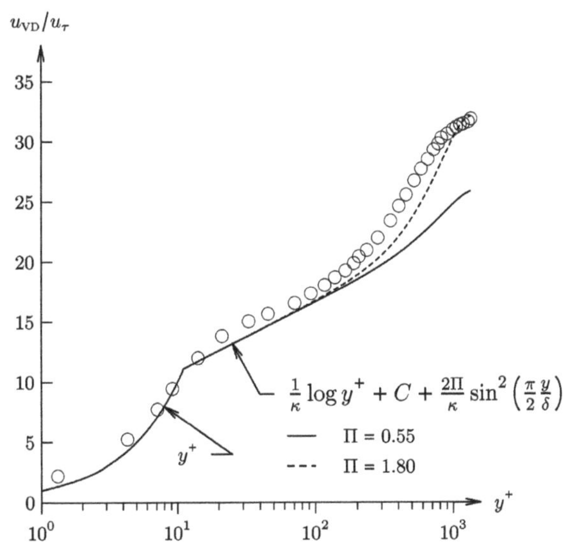


**Figure 4.92.** $u_{VD}/u_\tau$ versus $y^+$ for $T_w/T_{aw} = 1.0$ (run 812175).

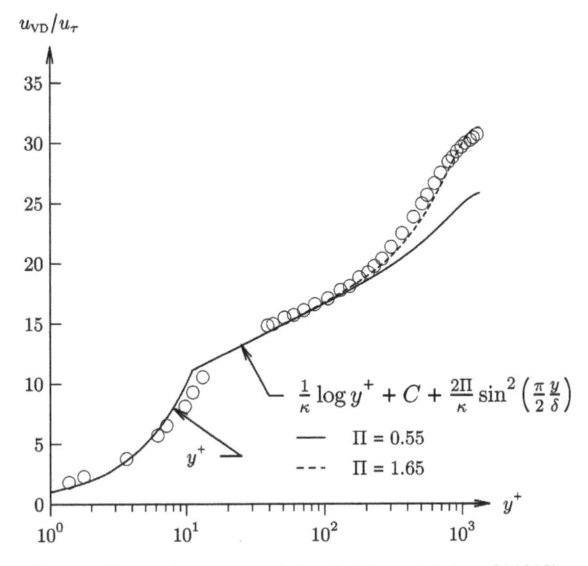


**Figure 4.93.** $u_{VD}/u_\tau$ versus $y^+$ for $T_w/T_{aw} = 0.8$ (run 901302).

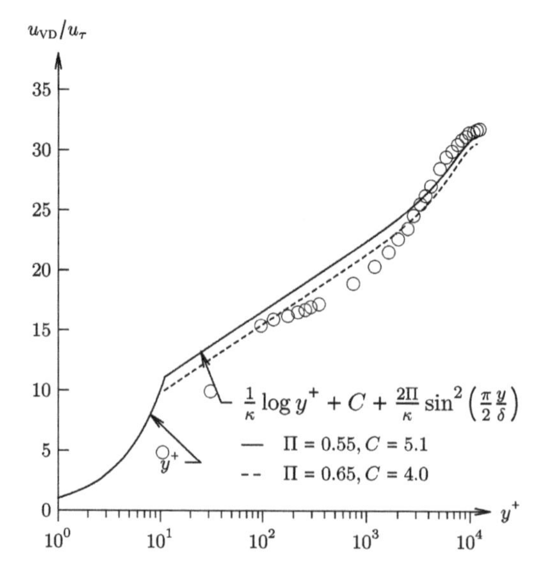


**Figure 4.94.** $u_{VD}/u_\tau$ versus $y^+$ for $T_w/T_{aw} = 0.25$ (run 002162).

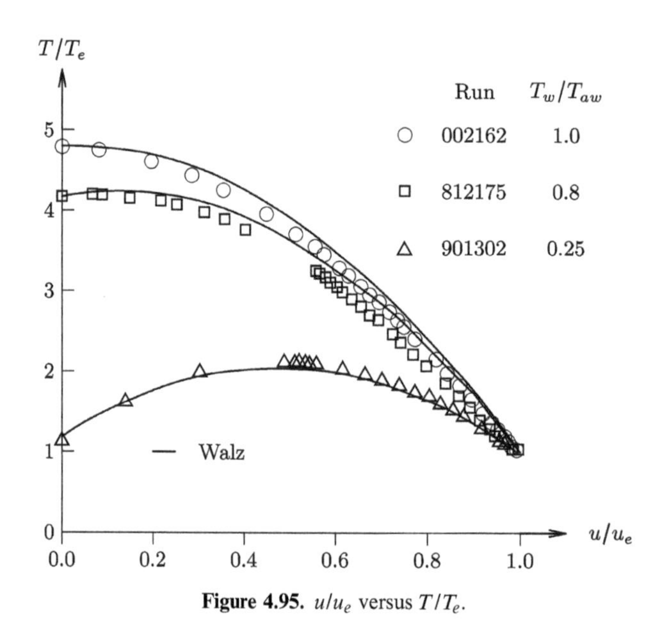


**Figure 4.95.** $u/u_e$ versus $T/T_e$.

Figure 4.95 displays the static temperature ratio $T/T_e$ versus the velocity ratio $u/u_e$ for $T_w/T_{aw} = 0.25$, 0.8, and 1.0. Also shown is Walz's relation (3.171). Close agreement is observed between experiment and (3.171). Figure 4.96 shows the Stanton number $St$ versus skin friction coefficient $c_f$. The experimental results fall below both (3.225) and (3.235). Note that Voisinet and Lee (1972) cautions that a recovery factor $r = 0.89$ was used to determine $T_{aw}$ (and thus $St$) which may not be

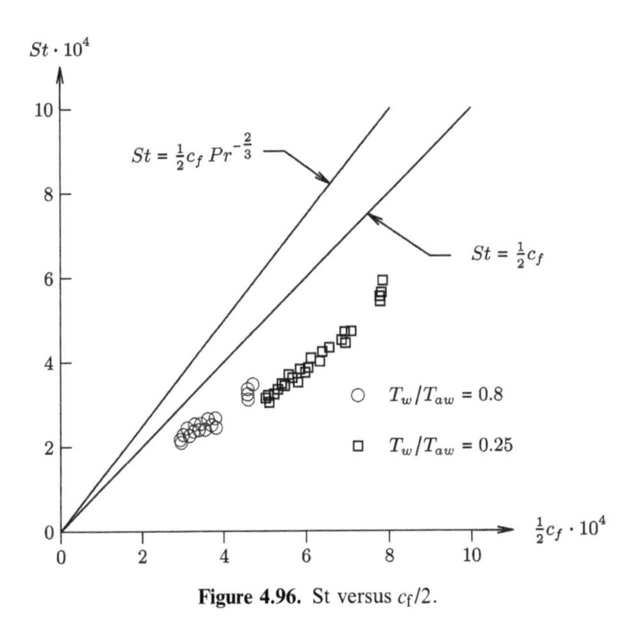


**Figure 4.96.** St versus $c_f/2$.

appropriate due to the upstream temperature history including the effect of energy removal at the nozzle throat.

## 4.33 Backx (1973–1976)

Backx (1973, 1974, 1975) and Backx and Richards (1976) performed a detailed study of a hypersonic turbulent boundary at Mach 15 and 19.8. The measurements were conducted in the tunnel wall boundary layer at the von Kármán Institute for Fluid Dynamics longshot tunnel (section A.48.1). The flow conditions are shown in table 4.40. Experimental diagnostics include boundary layer profiles of mass flow, pitot pressure, and stagnation temperature and wall heat transfer and static pressure.

Figure 4.97 shows the boundary layer profiles of pitot pressure $p_{\mathrm{p}}$ normalized by the pitot pressure at the boundary layer edge $p_{p_e}$ versus $y/\delta$ for $M_\infty = 15$ and 19.8. The wall static pressure and edge static pressures differ significantly, with $p_{\mathrm{w}}/p_e = 1.62$ and 1.8 for $M_\infty = 15$ and 19.8, respectively. Figure 4.98 displays the total temperature $T_{\mathrm{t}}$ normalized by the total temperature at the edge of the boundary layer $T_{t_e}$.

## 4.34 Feller (1973)

Feller (1973) investigated the effect of the upstream wall temperature history on the velocity–total temperature relationship at Mach 6. The experiments were performed in the NASA Langley Mach 6 high Reynolds number tunnel (table A.57). Boundary layer profile measurements of pitot pressure and total temperature were obtained in the nozzle wall boundary layer at a station 2.39 m downstream of the throat. Electrical heaters were installed on the outside of the piping, settling chamber and throat block of the tunnel. Four cases were considered, namely (1) unheated settling

**Table 4.40.** Freestream conditions.

| Quantity | Case 1 | Case 2 |
|---|---|---|
| Gas | $N_2$ | $N_2$ |
| $M_\infty$ | 15 | 19.8 |
| $p_{t_\infty}$ (MPa) | 231[a] | 328[a] |
| $T_{t_\infty}$ (K) | 1600[a] | 2050[a] |
| $T t_e$ (K) | 1775[b] | 2300[b] |
| $Re \times 10^{-6}$ ($m^{-1}$) | 27.9 | 11.2 |
| $T_w/T_{t_\infty}$ (K) | 0.15[c] | 0.11[c] |
| $\delta$ (mm) | 55 | 63 |
| $\delta^*$ (mm) | 23.4[d], 22.5[e] | 27.8[d, e] |
| $\theta$ (mm) | 1.24[d], 1.51[e] | 2.17[d], 2.39 |

[a] Reservoir condition, Backx (1975) table 6.
[b] At boundary layer edge, Backx (1975) table 6.
[c] Backx (1975) table 1.
[d] Assuming $p=$ constant in boundary layer.
[e] Assuming $p = p_w + (p_e - p_w)y/\delta$.

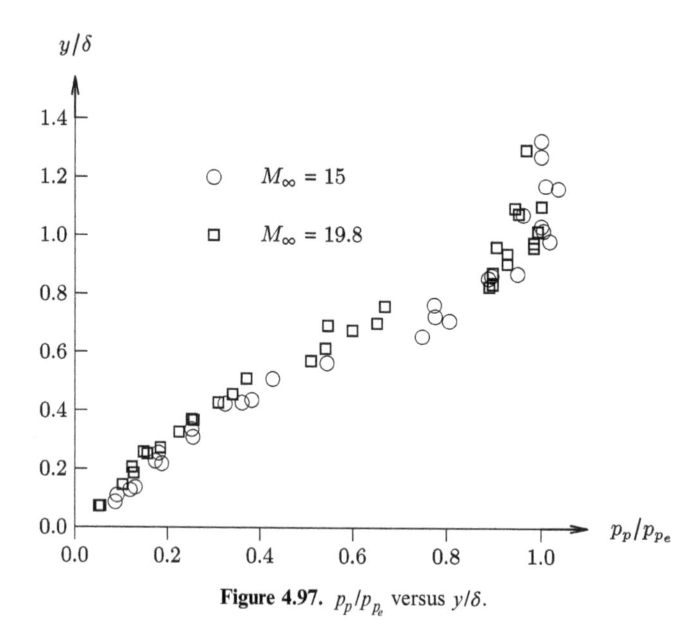


**Figure 4.97.** $p_p/p_{p_e}$ versus $y/\delta$.

chamber and throat, (2) settling chamber unheated, throat heated to approximate air stagnation temperature, (3) settling chamber heated to air stagnation chamber, throat unheated, and (4) both settling chamber and throat heated above air stagnation temperature. The wall temperature distributions are shown in figure 4.99. The flow conditions are listed in table 4.41.

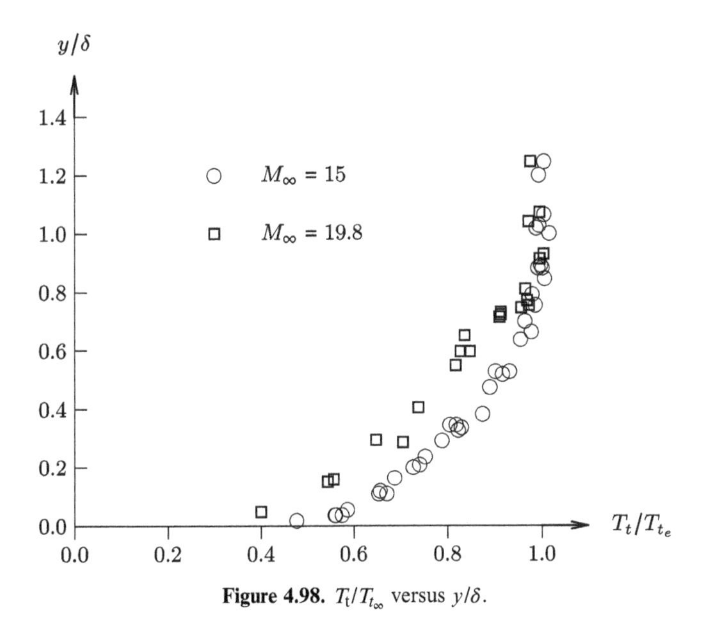


**Figure 4.98.** $T_t/T_{t_\infty}$ versus $y/\delta$.

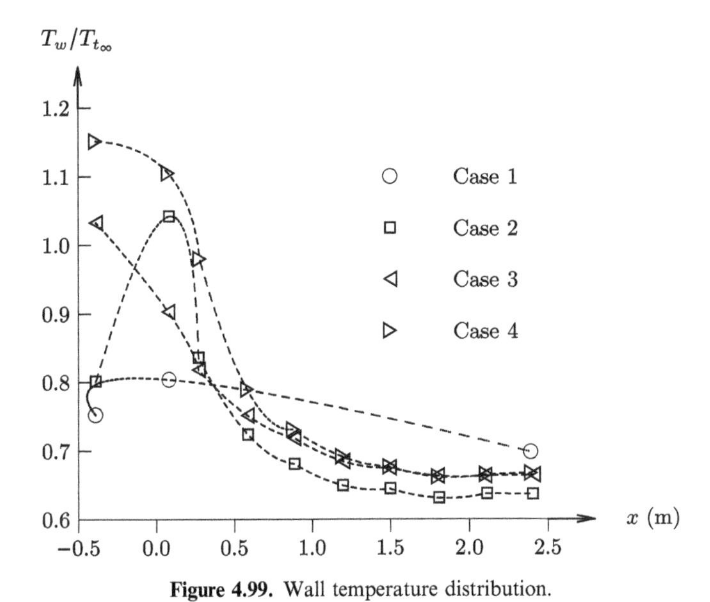


**Figure 4.99.** Wall temperature distribution.

Figure 4.100 displays the velocity–total temperature profiles for the four cases. Feller (1973) notes that the dimensionless velocity profiles $u/u_e$ versus $y/\delta$ are essentially identical based upon the power law fit (table 4.41)

$$\frac{u}{u_e} = \left(\frac{y}{\delta}\right)^n \tag{4.16}$$

**Table 4.41.** Freestream conditions.

| Quantity | Case 1 | Case 2 | Case 3 | Case 4 |
|---|---|---|---|---|
| $M_e$ | 5.98 | 6.02 | 6.0 | 6.03 |
| $p_{t_\infty}$ (MPa) | 4.24 | 4.03 | 4.08 | 4.07 |
| $T_{t_\infty}$ (K) | 512 | 480 | 486 | 478 |
| $T_w/T_{t_\infty}$ SC | 0.75 | 0.80 | 1.03 | 1.15 |
| $T_w/T_{t_\infty}$ NT | 0.80 | 1.04 | 0.90 | 1.10 |
| $T_w/T_{t_\infty}$ SS | 0.69 | 0.64 | 0.66 | 0.68 |
| $T_w/T_{aw}$ SS | 0.76[a] | 0.71[a] | 0.73[a] | 0.75[a] |
| $H_\infty$ (MJ kg) | 0.51 | 0.48 | 0.49 | 0.48 |
| $Re \times 10^{-6}$ ($m^{-1}$) | 32.3 | 31.0 | 30.9 | 33.8 |
| $\delta$ (cm) | 3.63 | 3.56 | 3.63 | 3.38 |
| $\delta^*$ (cm) | 1.39 | 1.45 | 1.44 | 1.45 |
| $\theta$ (mm) | 1.32 | 1.26 | 1.23 | 1.09 |
| $n$ | 9.30 | 9.64 | 9.62 | 9.66 |

NT, nozzle throat; SC, settling chamber; SS, survey station.
[a]Using (3.46) with $Pr_t = 0.89$.

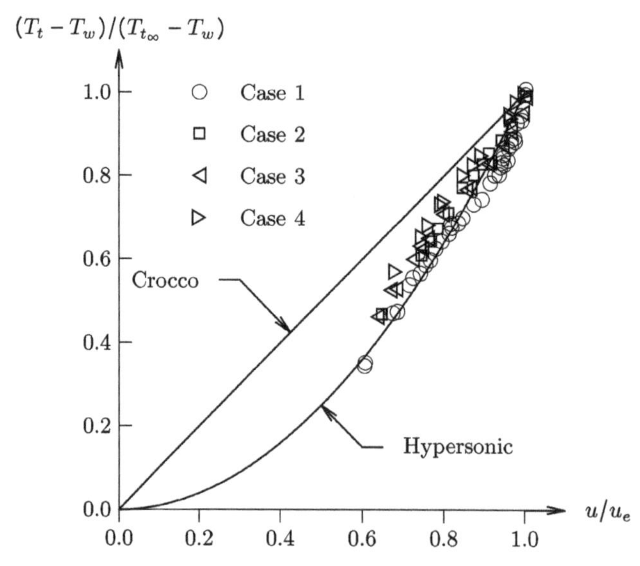


**Figure 4.100.** Velocity–total temperature profiles.

and therefore concludes that the differences in the velocity–total temperature profiles are due almost entirely to the effects of the upstream wall temperature distributions.

## 4.35 Watson *et al* (1973)

Watson *et al* (1973) investigated transitional and turbulent zero pressure gradient boundary layers at Mach 10 and near adiabatic wall conditions. The experiments were conducted at the NASA Langley Mach 20 high Reynolds number helium tunnel (section A.22.13). The model comprised a flat plate 236 cm long and 101.5 cm wide set at a 5° angle of attack. The freestream conditions are shown in table 4.42. Boundary layer profiles were measured at 74.2 cm, 99.6 cm, 125 cm, and 211 cm from the leading edge. Experimental diagnostics include boundary layer profiles of pitot pressure and stagnation temperature and wall shear stress, heat transfer, and pressure.

Figure 4.101 shows the Van Driest transformed velocity $u_{VD}/u_\tau$ versus $y^+$. The measurements show close agreement with the Law of the Wall and Wake.

**Table 4.42.** Freestream conditions.

| Quantity | Min | Max |
|---|---|---|
| $M_e$ | 8.6 | 10.09 |
| $p_{t_e}$ (MPa) | 1.434 | 13.21 |
| $T_{t_e}$ (K) | ≈300 | ≈300 |
| $T_w/T_{aw}$ | ≈1 | ≈1 |
| $H_e$ (MJ kg$^{-1}$) | ≈1.56 | ≈1.56 |
| $Re_\infty \times 10^{-6}$ (m$^{-1}$) | 7.769 | 34.544 |
| $\delta$ (cm) | 2.032 | 5.08 |
| $\delta^*$ (cm)[a] | 0.806 | 2.63 |
| $\theta$ (cm)[a] | 0.00 956 | 0.038 7 |

[a] Assuming $T_t$ constant across boundary layer.

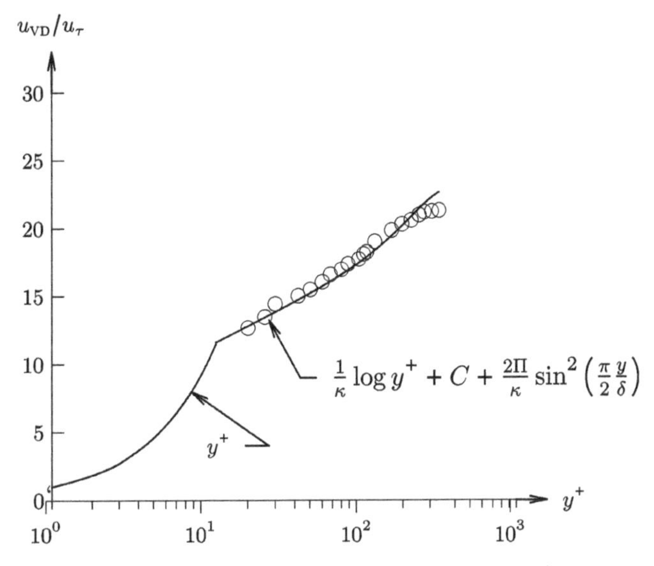


**Figure 4.101.** $u_{VD}/u_\tau$ versus $y^+$ for $\delta^+$=318 with $C$ = 5.5 and $\Pi$ = 0.55.

## 4.36 Raman (1974)

Raman (1974) investigated the surface pressure fluctuations on a flat plate turbulent boundary layer at Mach numbers from 5.2 to 10.4. The experiments were performed in the NASA Ames 3.5 ft hypersonic wind tunnel (section A.20.2). The flat plate model was 1.9 cm thick aluminum with a length of 120 cm and width of 45 cm. Measurements were obtained at a distance of 73 cm from the sharp leading edge. The surface temperature increased during the typical one minute duration of the experiment. A total of 60 experiments were performed. The freestream conditions are listed in table 4.43. Experimental measurements include surface pressure and temperature.

Figure 4.102 shows the normalized root-mean-square pressure fluctuations $(\sqrt{\overline{p'^2}}/\frac{1}{2}\rho_\infty u_\infty^2) \times 10^3$ versus Reynolds number $Re_{\delta^*}$ for three different Mach numbers. A definite trend of decreasing root-mean-square pressure fluctuations with increasing Reynolds number is evident. Figure 4.103 displays the normalized root-mean-square pressure fluctuations $(\sqrt{\overline{p'^2}}/\tau_w)$ versus $Re_\theta$. An increasing trend with Reynolds number is seen. Also shown are the measurements of Kistler and Chen (1962) at $M_\infty = 1.33$ to 3.99 which are consistent with the trend, and the results of Parrott *et al* (1989) which show an opposite trend with increasing $Re_\theta$.

## 4.37 Bloy (1975)

Bloy (1975) examined the effect of an expansion on a hypersonic boundary layer at Mach 16. The experiments were performed in the von Kármán Institute longshot tunnel (section A.48.1). The model is shown in figure 4.104. The model comprised a sharp-edged flat plate at angle of incidence of 10.6° connected to an adjustable trailing edge flap with angle of expansion $\alpha$ from $-5°$ to $-20°$. The lengths of the flat plate and flap are 30 cm and 18 cm, respectively, and the width is

**Table 4.43.** Freestream conditions.

| Quantity | Value | |
|---|---|---|
| | Min | Max |
| $M_e$ | 5.2 | 10.4 |
| $p_{t_e}$ (MPa) | 0.689 | 11.0 |
| $T_{t_e}$ (K) | 727 | 1192 |
| $T_w$ (K) | na | |
| $H_e$ (MJ kg$^{-1}$) | 0.73 | 1.12 |
| $Re \times 10^{-6}$ (m$^{-1}$) | 1.27 | 20.1 |
| $\delta$ (cm) | 0.919 | 1.747 |
| $\delta^*$ (cm) | 0.554 | 1.245 |
| $\theta$ (cm) | 0.019 5 | 0.058 5 |

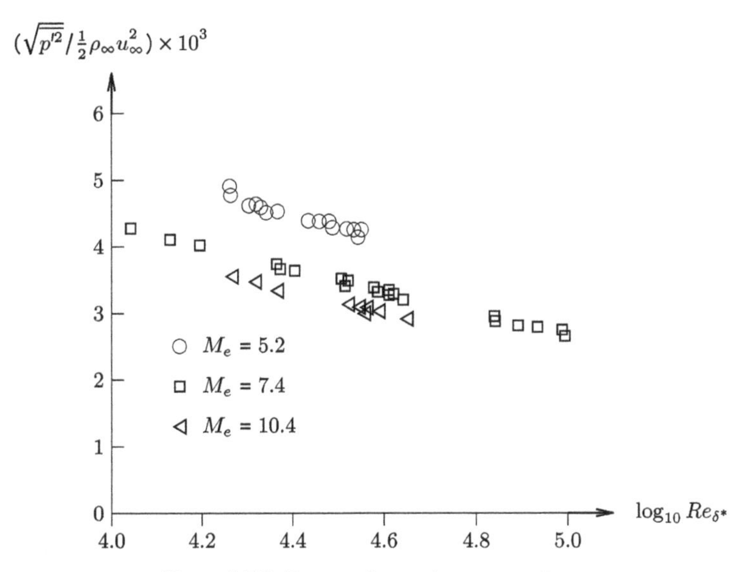


**Figure 4.102.** Pressure fluctuations versus $Re_{\delta^*}$.

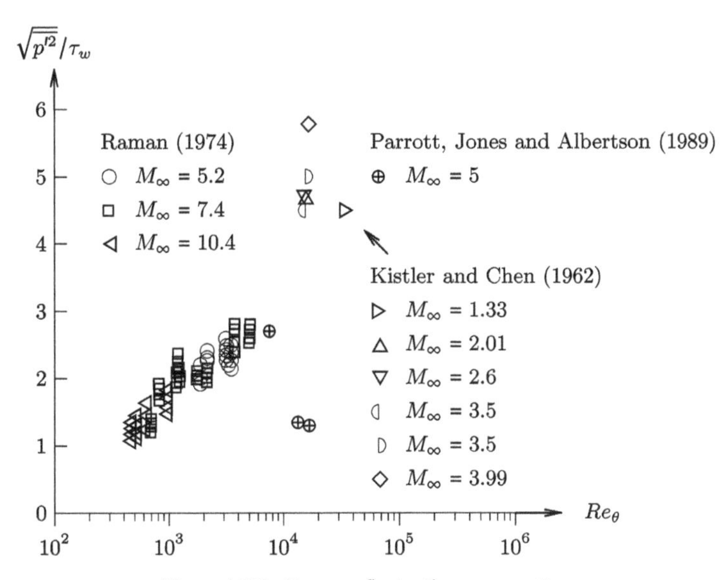


**Figure 4.103.** Pressure fluctuations versus $Re_\theta$.

18 cm. Experimental measurements include boundary layer pitot profiles and surface heat transfer and pressure. The freestream conditions are listed in table 4.44. Boundary layer pitot pressure profiles immediately upstream of the flap indicate that the boundary layer entering the expansion was transitional.

Figure 4.105 displays the experimental surface pressure on the flat plate and flap for $\alpha = 0$ to $-15°$. Figure 4.106 shows the experimental heat transfer on the flat plate

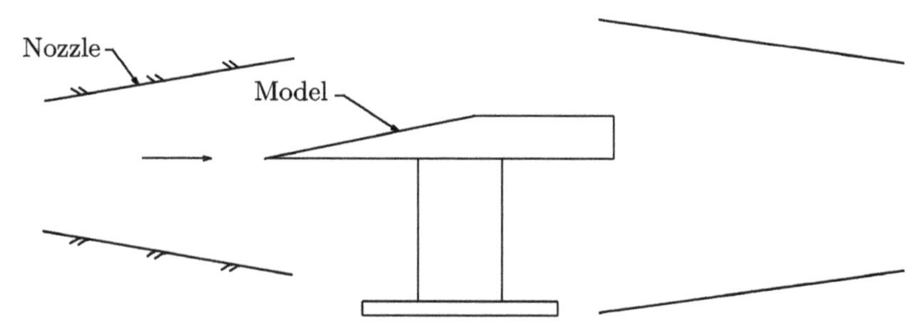


**Figure 4.104.** Model.

**Table 4.44.** Freestream conditions.

| Quantity | Value |
|---|---|
| Gas | $N_2$ |
| $M_\infty$ | 16 |
| $M_e$ | 7.35[a] |
| $p_{t_\infty}$ (MPa) | 499.5 |
| $T_{t_\infty}$ (K) | 2160 |
| $H_w/H_e$ | 0.13 |
| $Re \times 10^{-6}$ ($m^{-1}$) | 29.0 |

[a] At $x = 29.5$ cm.

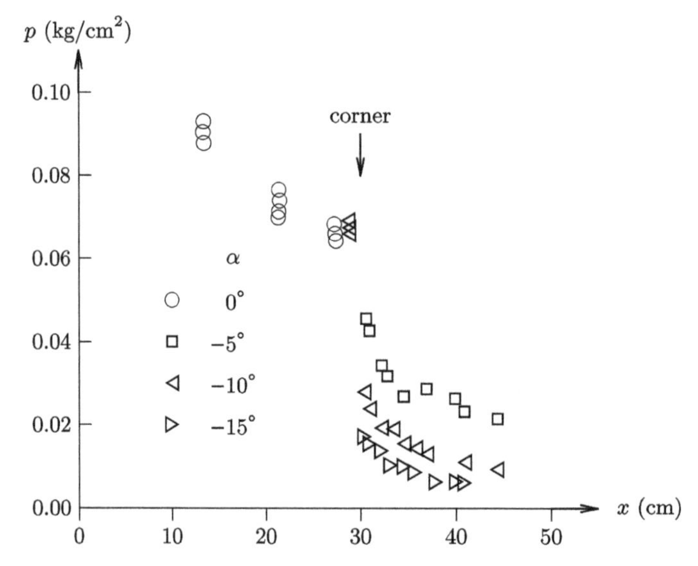


**Figure 4.105.** Surface pressure versus $x$.

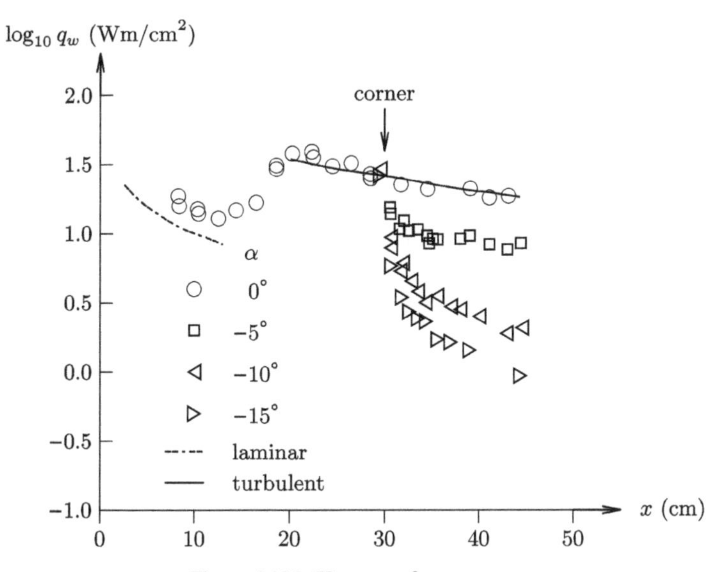


**Figure 4.106.** Heat transfer versus $x$.

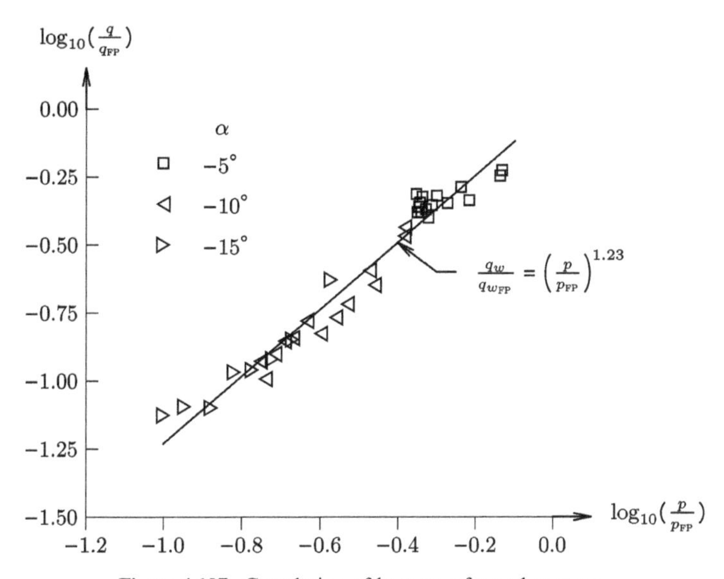


**Figure 4.107.** Correlation of heat transfer and pressure.

and flap for $\alpha = 0$ to $-15°$. The rise in heat transfer on the flat plate indicates transition from laminar to turbulent flow. Increasing the magnitude of the expansion angle $\alpha$ reduces the heat transfer on the flap. Figure 4.107 presents a correlation of the heat transfer on the flap with the corresponding surface pressure. The experimental data closely approximate the expression

$$\frac{q_{\rm w}}{q_{w_{\rm FP}}} = \left(\frac{p}{p_{\rm FP}}\right)^{1.23} \tag{4.17}$$

## 4.38 Mikulla and Horstman (1975)

Mikulla and Horstman (1975) performed an experimental investigation of a hypersonic turbulent boundary layer on a cone-ogive-cylinder at Mach 7.2. The experiments were conducted in the NASA Ames 3.5 ft hypersonic wind tunnel (section A.20.2). The model is shown in figure 4.70. Experimental measurements include boundary layer surveys using constant temperature anemometry with dual and triple wire probes at a location 175 cm from the model tip. The freestream conditions and edge conditions at the location of the boundary layer survey are shown in table 4.45.

Figure 4.108 shows the normalized mass flux–vertical velocity and mass flux–lateral velocity correlations. Surprisingly the lateral velocity correlation is comparable to the vertical velocity correlation within a factor of 1.8 or less in contrast to incompressible (Klebanoff 1955) and subsonic (Gibbings and Mikulla 1973) two-dimensional and axisymmetric boundary layers. Similarly, figure 4.109 shows nearly identical profiles for $\sqrt{\overline{v'^2}}$ and $\sqrt{\overline{w'^2}}$.

Figure 4.110 displays the normalized Reynolds shear stress. The calculation of the Reynolds stress from the hot wire anemometry data used the total temperature—vertical velocity correlation $\overline{T_{\rm t}'v'}$ in two methods, namely, (1) the measured $\overline{T_{\rm t}'v'}$ and (2) the calculated $\overline{T_{\rm t}'v'}$ from integration of the mean flow boundary layer equations. The two methods yielded very similar results. Figure 4.111 shows the intermittency profiles for mass flux $(\rho u)'$ including data from Owen and Horstman (1974), vertical $v'$ and lateral $w'$ fluctuations, and the intermittency distribution for an incompressible turbulent boundary layer obtained by Klebanoff (1954) and Klebanoff (1955). A significant difference in the outer half of the boundary layer is evident.

**Table 4.45.** Freestream conditions.

| Quantity | Value |
|---|---|
| $M_\infty$ | 7.2 |
| $M_e$ | 6.85 |
| $p_{t_\infty}$ (MPa) | 3.45 |
| $T_{t_\infty}$ (K) | 698 |
| $T_{\rm w}$ (K) | 310 |
| $T_{\rm w}/T_{\rm aw}$ | 0.47 |
| $H_\infty$ (MJ kg$^{-1}$) | 0.70 |
| $\delta$ (cm) | 2.7 |
| $Re \times 10^{-6}$ (m$^{-1}$) | 7.4 |

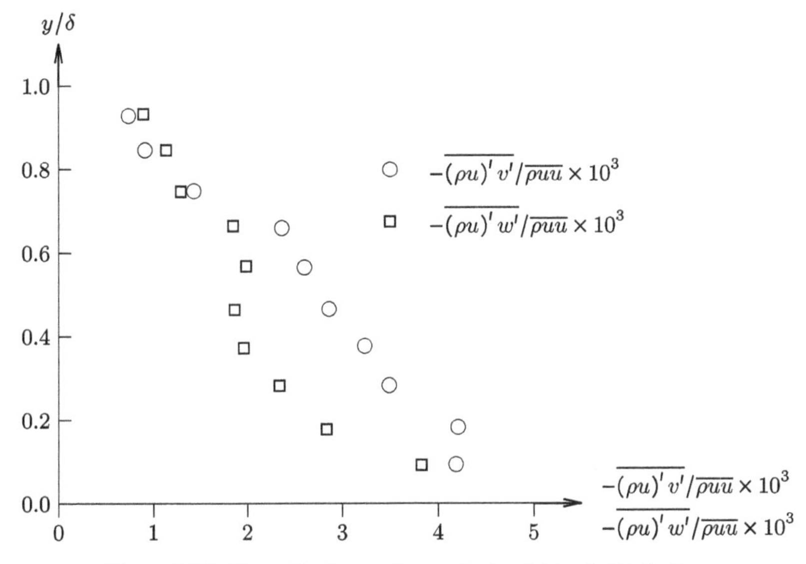


**Figure 4.108.** Normalized mass flux vertical and lateral distribution.

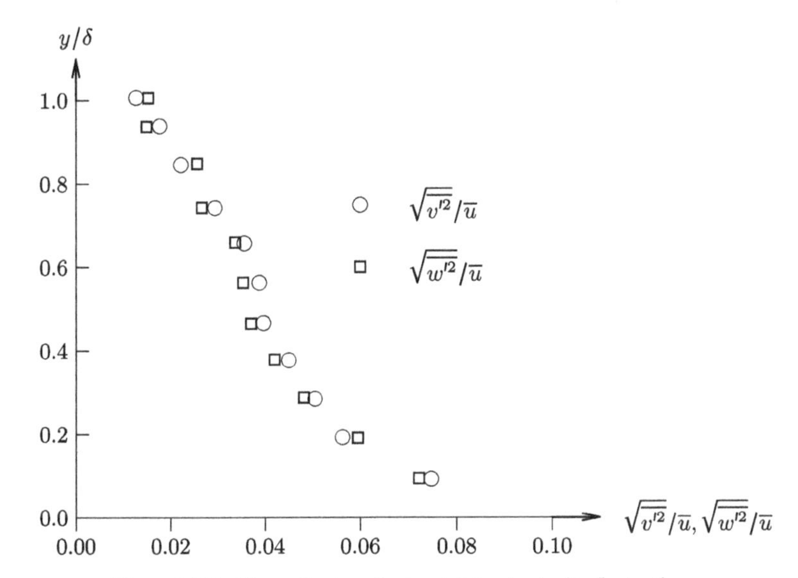


**Figure 4.109.** Normalized vertical and lateral velocity fluctuations.

## 4.39 Owen *et al* (1975)

Owen *et al* (1975) conducted an experimental investigation of a cold wall hypersonic turbulent boundary layer on an ogive cylinder at Mach 7.0. The model is shown in figure 4.70 and the freestream conditions in table 4.46. Natural transition occurred between $x = 37$ and 80 cm. Experimental diagnostics include boundary layer profiles of pitot pressure, static pressure, and total temperature. Constant current hot wire

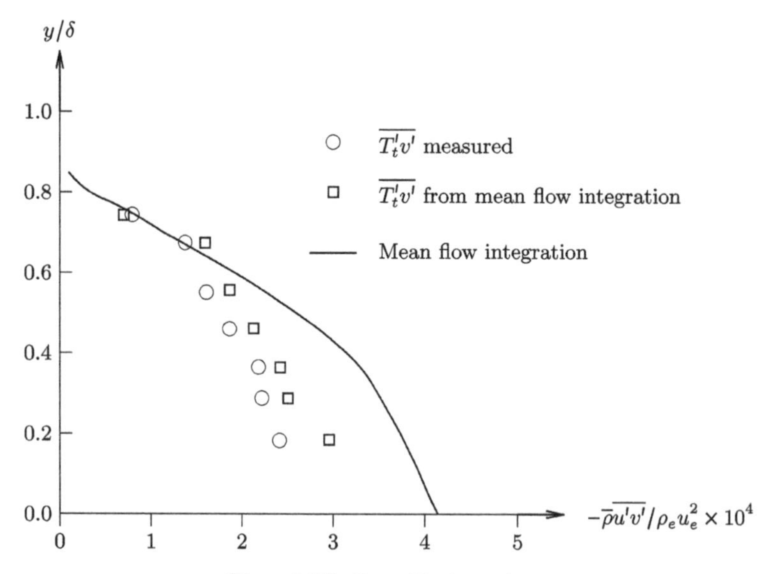


**Figure 4.110.** Reynolds shear stress.

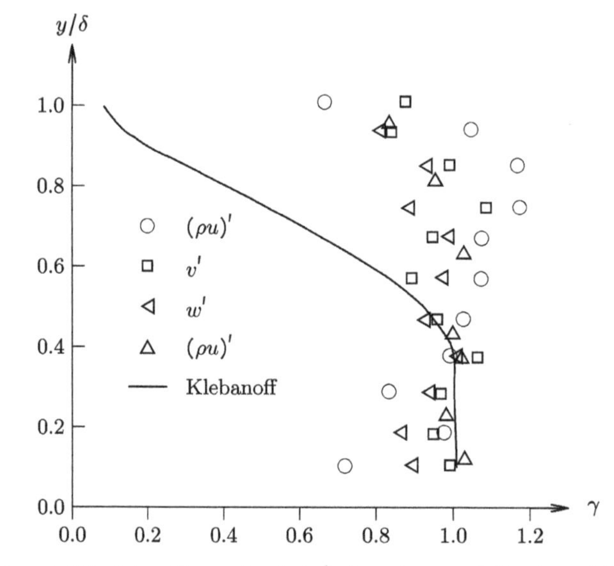


**Figure 4.111.** Intermittency. Symbol △ from Owen and Horstman (1974).

measurements were also performed through the boundary layer. In contrast to the static pressure measurements of Laderman and Demetriades (1974), the measured static pressure profiles showed no variation across the boundary layer within the measurement uncertainty.

Mean boundary layer profiles for velocity, Mach number, and density are displayed in figure 4.112. The dimensionless total temperature ratio

**Table 4.46.** Freestream conditions.

| Quantity | Value |
|---|---|
| $M_\infty$ | 7.0 |
| $p_{t_\infty}$ (MPa) | 3.45 |
| $T_{t_\infty}$ (K) | 695 |
| $T_w$ (K) | 300 |
| $T_w/T_{aw}$ | 0.46 |
| $H_\infty$ (MJ kg$^{-1}$) | 0.70 |
| $\delta$ (cm) | 3.3 |
| $\theta$ (cm) | 0.128 |
| $Re \times 10^{-6}$ (m$^{-1}$) | 10.9 |

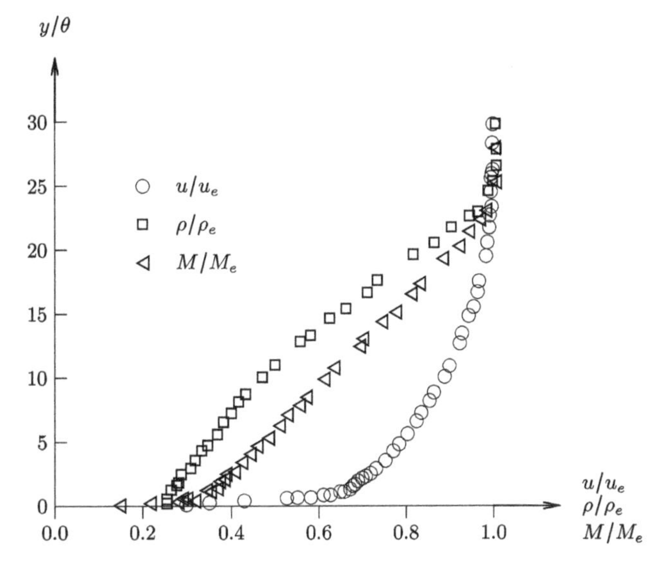


**Figure 4.112.** Mean measurements.

$(T_t - T_w)/(T_{t_e} - T_w)$ versus $u/u_e$ is presented in figure 4.113 and displays a linear relationship in the outer portion of the boundary layer.

The Van Driest velocity profile at $x$ = 225 cm is displayed in figure 4.114. The measured profile shows close agreement with the law of the wall and transitions to a linear profile to $y^+\approx 8$. The measured velocity defect shows close agreement with the incompressible defect law (figure 4.115). The turbulent Prandtl number profile is shown in figure 4.116 and displays a nonuniform shape. Figure 4.117 shows the scaled root-mean-square streamwise velocity fluctuations using Morkovin's hypothesis (section 3.5). The measurements of Kistler (1959) at near adiabatic wall conditions ($T_w/T_{t_e}$ = 1.0) at Mach 1.72, 3.56, and 4.67 are shown as separate dashed lines, and the range of data of Laderman and Demetriades (1974) at Mach 9.4 for a

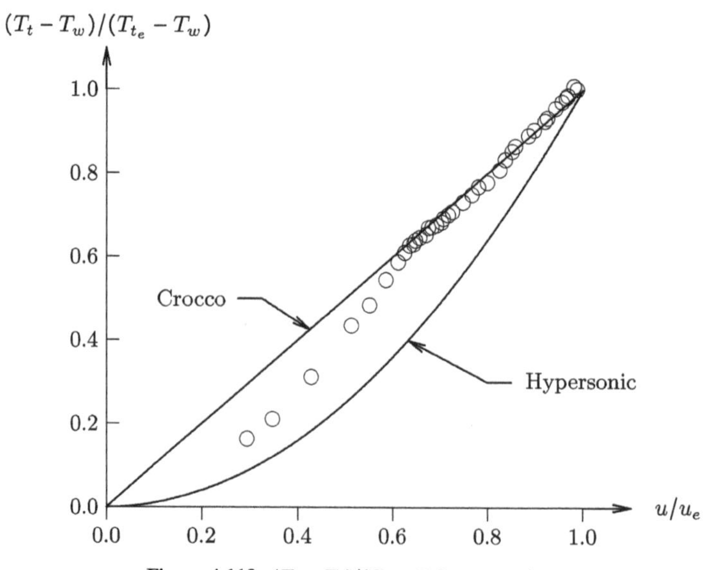


**Figure 4.113.** $(T_t - T_w)/(T_{t_e} - T_w)$ versus $u/u_e$.

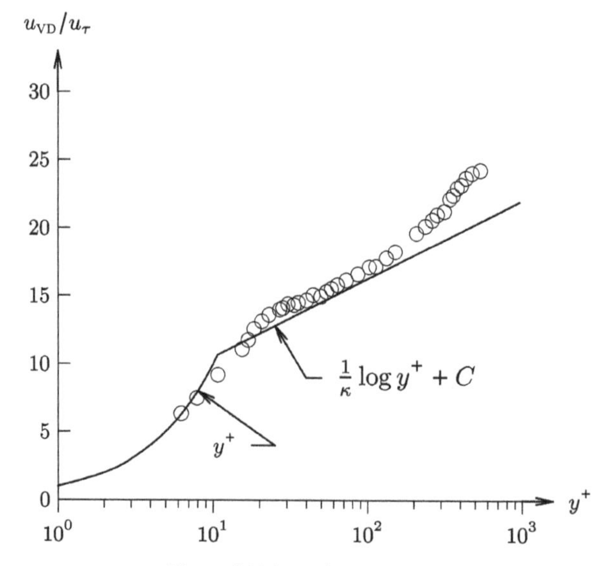


**Figure 4.114.** $u_{VD}/u_\tau$ versus $y^+$.

cold wall ($T_w/T_{t_e} = 0.38$) is shown. The results of Owen *et al* (1975) are in agreement with Laderman and Demetriades (1974) within experimental uncertainty and show significant differences with the data of Kistler (1959) at near adiabatic wall conditions.

## 4.40 Smith and Driscoll (1975)

Smith and Driscoll (1975) conducted an experimental investigation of a hypersonic adiabatic turbulent boundary layer at Mach 16 in the helium facility at the Princeton

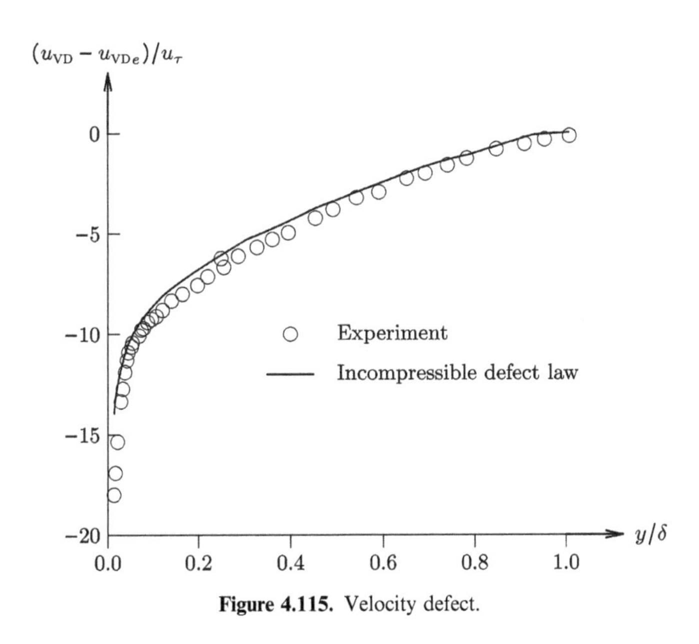


**Figure 4.115.** Velocity defect.

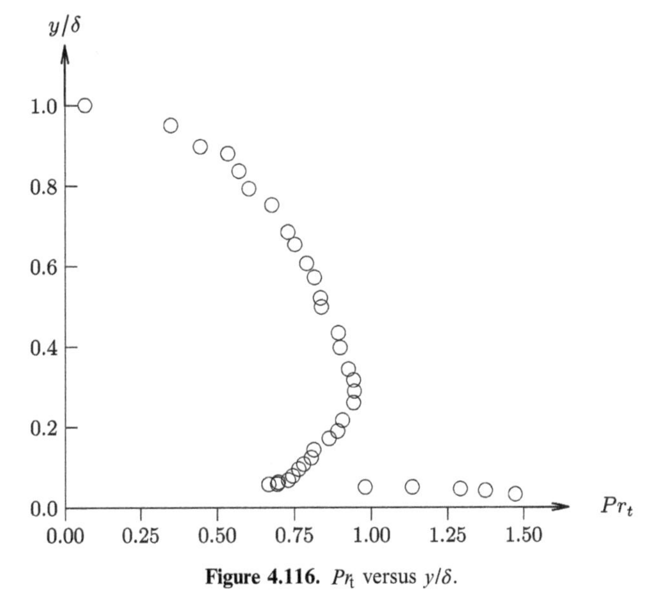


**Figure 4.116.** $Pr_t$ versus $y/\delta$.

Gas Dynamics Laboratory (section A.28.2). Measurements were performed on the tunnel wall boundary layer at a distance of 82 cm from the throat.[8] The freestream conditions are listed in table 4.47. Experimental measurements include boundary layer surveys of pitot pressure and density and temperature using electron beam fluorescence.

[8] The authors note '··· a compression region exists near the outer edge of the boundary layer, probably resulting from the compression corner created by the intersection of the axisymmetric nozzle with the flat test-section wall. Thus, we are *not* dealing with a canonical $\partial p/\partial x = 0$ (or $\partial p/\partial y = 0$, for that matter) boundary layer.'

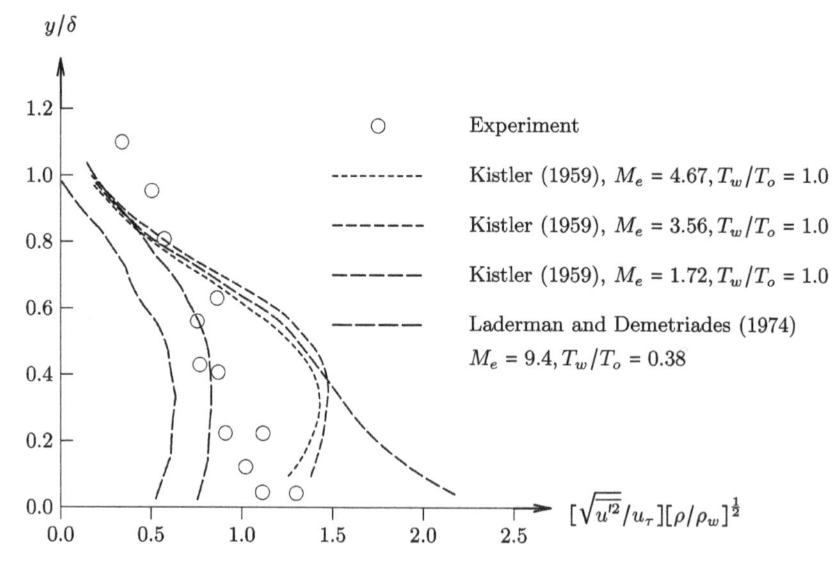


**Figure 4.117.** Scaled velocity fluctuations.

**Table 4.47.** Freestream conditions.

| Quantity | Value |
|---|---|
| Gas | He |
| $M_\infty$ | 16.3 |
| $p_{t_\infty}$ (MPa) | 1.55 |
| $p_\infty$ (Pa) | 20.5 |
| $T_\infty$ (K) | 3.5 |
| $n_\infty \times 10^{-17}$ ($cm^{-3}$) | 4.41 |
| $T_w$ (K) | 300 |

Figure 4.118 displays the mean number density distribution in the boundary layer. Both pitot pressure (●) and electron beam measurements (○, □, ◁, △) are indicated. Figure 4.119 shows the mean static temperature distribution using both measurement techniques. Reasonable agreement is evident for the mean density and somewhat less so for the temperature. Figure 4.120 presents the mean pressure distribution. A significant variation in pressure across the boundary layer is evident.

Figures 4.121, 4.122, and 4.123 display the distributions of root-mean-square fluctuations of number density, temperature, and pressure, respectively, normalized by their local mean values. Large fluctuations are apparent for all three variables. In particular, the normalized root-mean-square pressure and temperature fluctuations are comparable to their mean values.

Figures 4.124 and 4.125 show the probability density distributions for density and temperature, respectively. The probability distribution for density is highly skewed

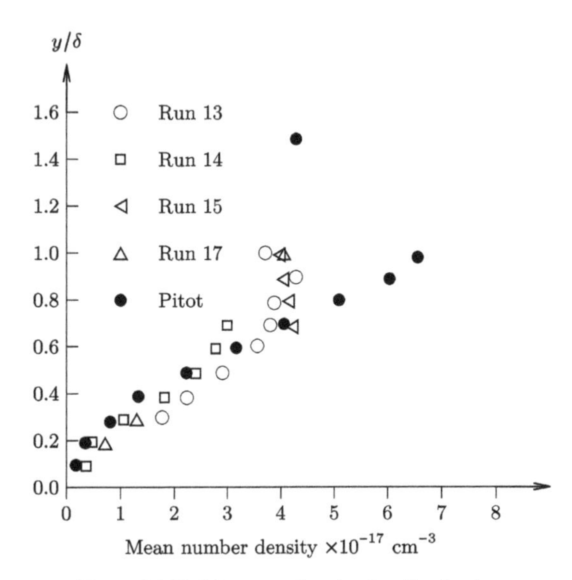


**Figure 4.118.** Mean number density distribution.

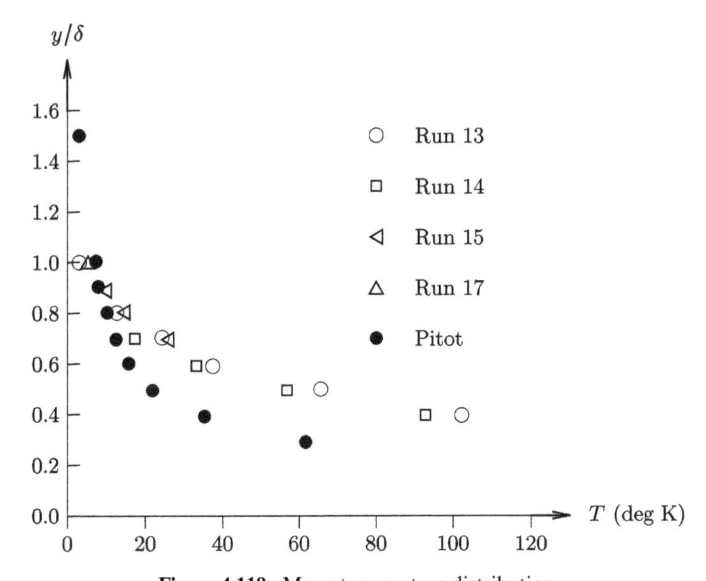


**Figure 4.119.** Mean temperature distribution.

for $y/\delta < 0.5$ but symmetric for $y/\delta > 0.5$. The probability density for temperature shows a similar behavior.

## 4.41 Laderman (1976)

Laderman (1976) examined the structure of a hypersonic turbulent boundary layer on a 4° half-angle cone at Mach 8 in the Arnold Engineering Development Center tunnel B (section A.5.1). The freestream conditions are listed in table 4.48. The

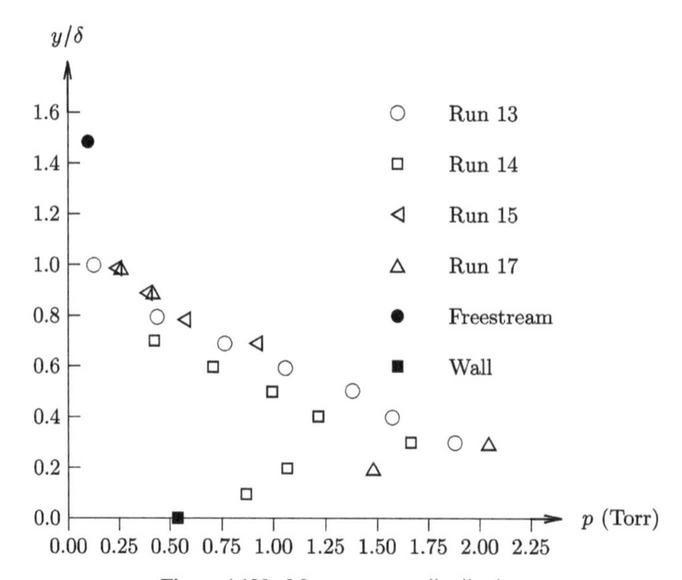


**Figure 4.120.** Mean pressure distribution.

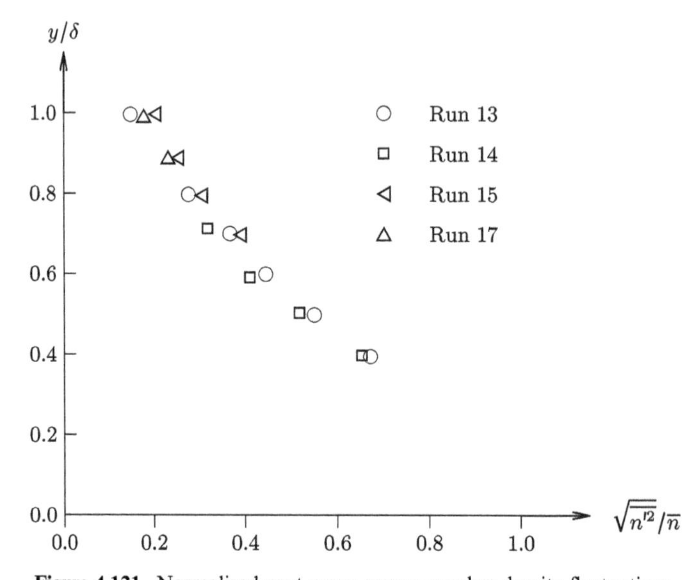


**Figure 4.121.** Normalized root-mean-square number density fluctuations.

boundary layer was surveyed at a distance 121.91 cm downstream from the cone apex. Experimental measurements include pitot pressure, total temperature, and X-wire constant current hot wire anemometry.

Figures 4.126 and 4.127 present the profiles of root-mean-square static temperature fluctuations for the cold wall and near adiabatic wall, respectively. Results are presented for several hot wire surveys. Comparison of the cold versus near adiabatic

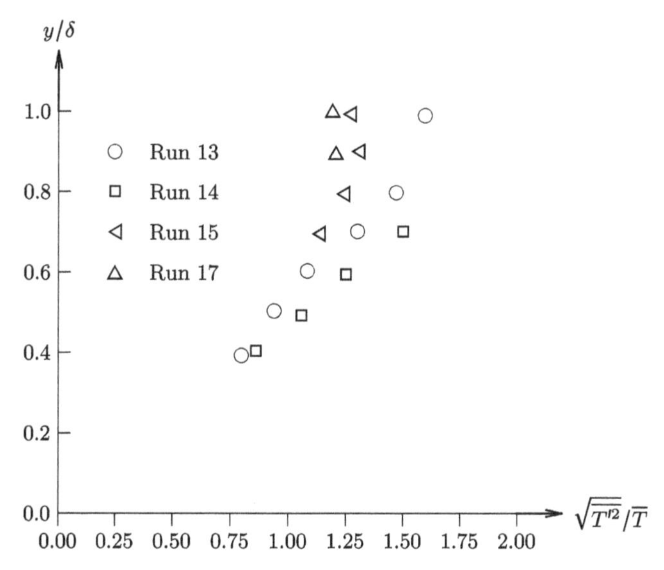


**Figure 4.122.** Normalized root-mean-square temperature fluctuations.

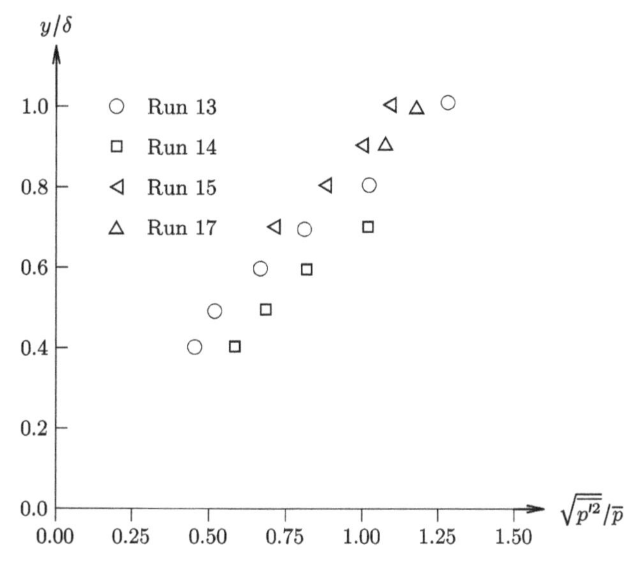


**Figure 4.123.** Normalized root-mean-square pressure fluctuations.

wall profiles indicates the reduction in peak static temperature fluctuations due to the cold wall.

Figures 4.128 and 4.129 display the root-mean-square velocity fluctuations $\sqrt{\overline{u'^2}}/\bar{u}$ and $\sqrt{\overline{v'^2}}/\bar{u}$ and correlation $\overline{u'v'}/\bar{u}^2$ for the cold wall and near adiabatic wall cases, respectively. The cold wall results show decreases in $\sqrt{\overline{u'^2}}/\bar{u}$ and $\overline{u'v'}/\bar{u}^2$ of approximately 20% and 40% to 50% compared to the near adiabatic wall case.

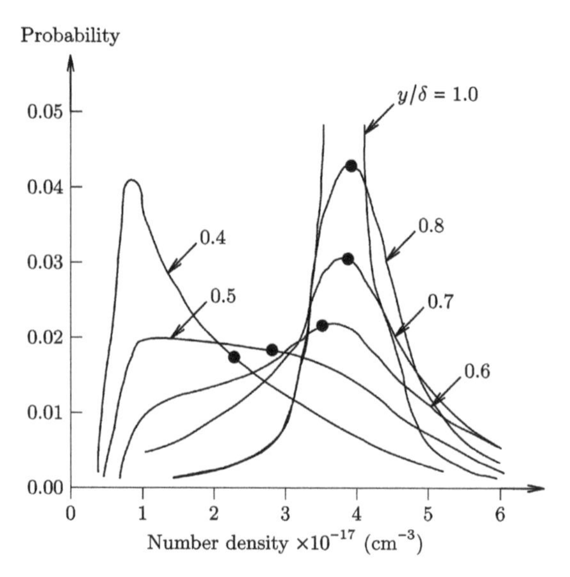


**Figure 4.124.** Probability density distribution for number density. ● indicates $\bar{n}$ at $y/\delta$ = 0.4, 0.5, 0.6, 0.7, 0.8.

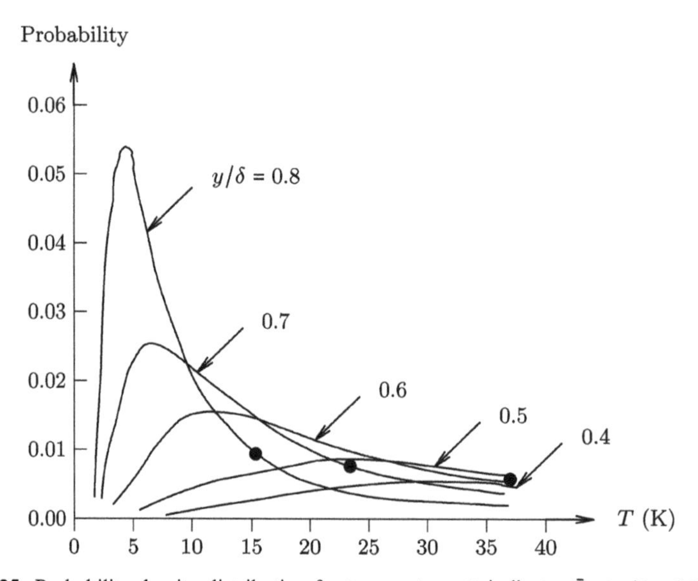


**Figure 4.125.** Probability density distribution for temperature. ● indicates $\bar{T}$ at $y/\delta$ = 0.8, 0.7, 0.6.

Figure 4.130 shows the velocity temperature correlation coefficients. The streamwise velocity temperature correlation coefficient is $R_{uT} = -1$ across the boundary layer and virtually identical for both cold wall and adiabatic wall cases (section 3.6.3). The normal velocity temperature correlation coefficient $R_{vT}$ varies significantly across the boundary layer.

Figure 4.131 displays the dimensionless Reynolds shear stress $-\bar{\rho}\overline{u'v'}/\rho_e u_e^2$ for both cases, together with published results at other Mach number and wall temperature conditions. Symbols are listed in table 4.49. The effect of increasing

**Table 4.48.** Freestream conditions.

| Quantity | Value Case 1 | Case 2 |
|---|---|---|
| $M_\infty$ | 8.0 | 8.0 |
| $M_e$ | 7.1 | 7.1 |
| $p_{t_e}$ (MPa) | 3.44 | 3,44 |
| $T_{t_e}$ (K) | 756 | 756 |
| $T_w$ (K) | 311 | 600 |
| $T_w/T_{aw}$ | 0.43 | 0.84 |
| $H_e$ (MJ kg$^{-1}$) | 0.76 | 0.76 |
| $\delta$ (cm) | 0.89 | 0.76 |
| $\theta$ (cm) | 0.028 | 0.020 |
| $Re \times 10^{-6}$ (m$^{-1}$) | 7.09 | 7.09 |

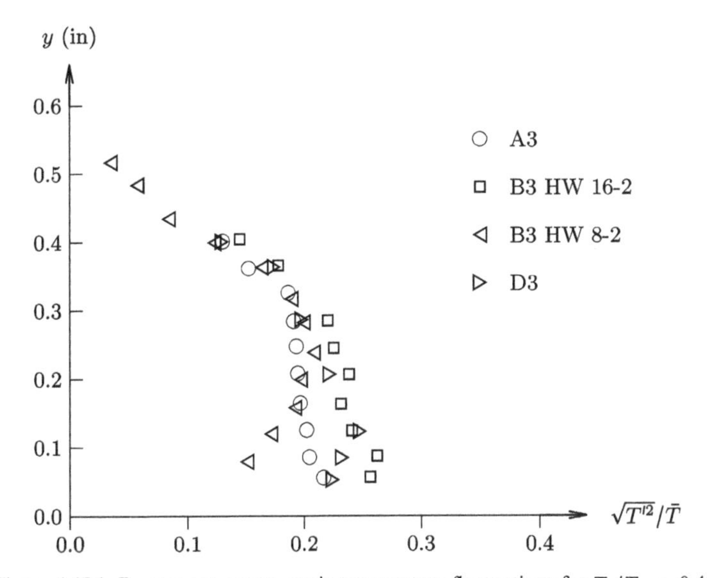


**Figure 4.126.** Root-mean-square static temperature fluctuations for $T_w/T_{aw} = 0.43$.

Mach number is to reduce the Reynolds shear stress, while the reduction in wall temperature at fixed Mach number has a similar effect.

## 4.42 Berg (1977)

Berg (1977) performed an experimental study of the effect of surface roughness on a hypersonic turbulent boundary layer. The experiments were conducted in the California Institute of Technology hypersonic wind tunnel (section A.7.1). The measurements were obtained on the nozzle wall boundary layer. Surface roughness

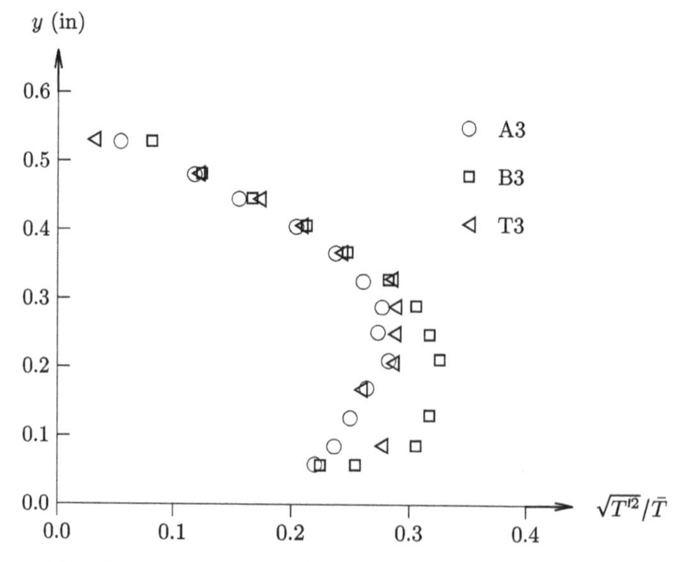


**Figure 4.127.** Root-mean-square static temperature fluctuations for $T_w/T_{aw} = 0.84$.

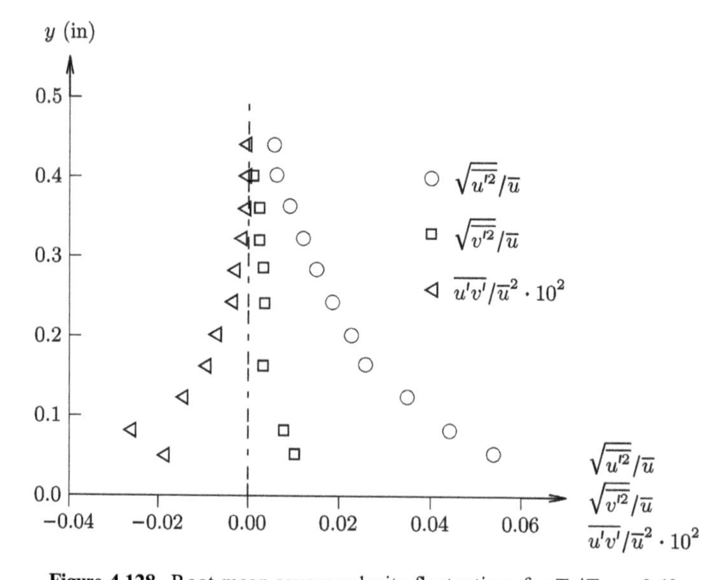


**Figure 4.128.** Root-mean-square velocity fluctuations for $T_w/T_{aw} = 0.43$.

comprised a series of spanwise rectangular groves with separation streamwise length $\ell = 4k$ and square cross-section of size $k$ as shown in figure 4.132. Three different roughness heights $k$ were examined as listed in table 4.50 where $k_s$ is the equivalent incompressible sand grain roughness (Nikuradse 1950) and $k_s^+ = k_s u_\tau / \nu_w$. Both smooth to rough and rough to smooth transitions were examined. The freestream conditions are listed in table 4.51. The extensive set of experimental measurements

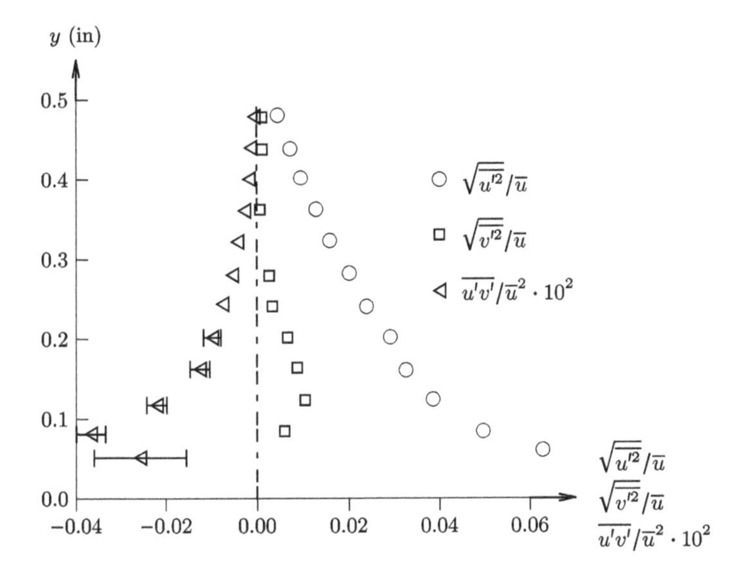


**Figure 4.129.** Root-mean-square velocity fluctuations for $T_w/T_{aw} = 0.84$.

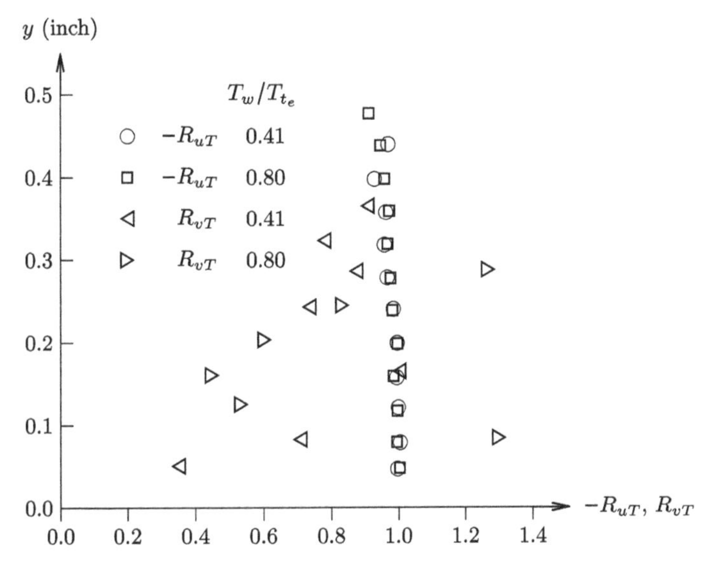


**Figure 4.130.** Velocity temperature correlation coefficients.

include boundary layer profiles of pitot pressure and total temperature, surface skin friction, and root-mean-square mass flux and total temperature fluctuations.

Figure 4.133 shows the Van Driest transformed velocity $u_{VD}/u_\tau$ versus $y^+$ for the smooth wall and a smooth to rough transition with $k = 1.27$ mm. The offset $H(k^+)$ for the rough wall is evident. The profiles are accurately represented by the compressible Law of the Wall and Wake (3.56) and (3.60).

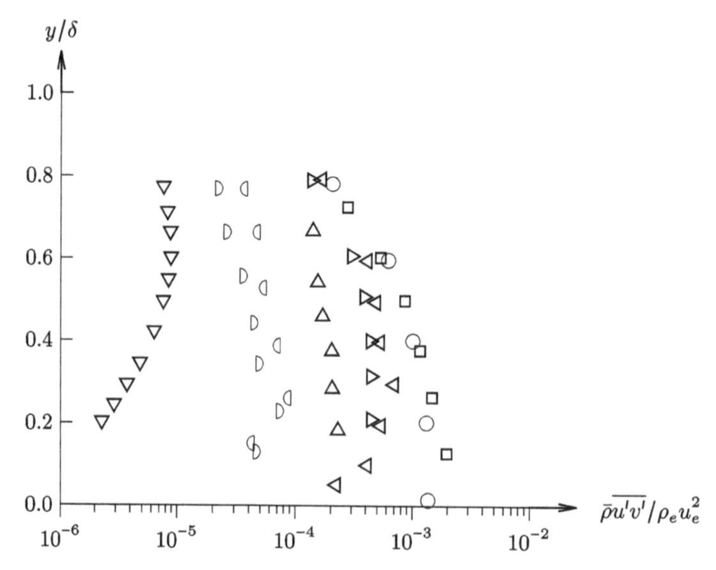


**Figure 4.131.** Normalized Reynolds shear stress.

**Table 4.49.** Symbols for figure 4.131.

| Symbol | $M_e$ | $T_w/T_{t_e}$ | $Re \times 10^{-3}$ | Reference |
|---|---|---|---|---|
| ○ | $\approx 0$ | 1 | 4.4 to 7.2 | Klebanoff (1955) |
| □ | $\approx 0$ | 1 | 0.5 to 0.7 | Townsend (1956) |
| ◁ | 2.9 | 1 | 74 | Johnson and Rose (1975) |
| ▷ | 3.0 | 1 | 10 to 14 | Yanta and Lee (1974) |
| △ | 7 | 0.46 | 9 | Laderman (1974) |
| ▽ | 9.4 | 0.38 | 36 | Demetriades and Laderman (1973) |
| ◖ | 7.1 | 0.8 | 1.4 | Laderman (1976) |
| D | 7.1 | 0.41 | 2.0 | Laderman (1976) |

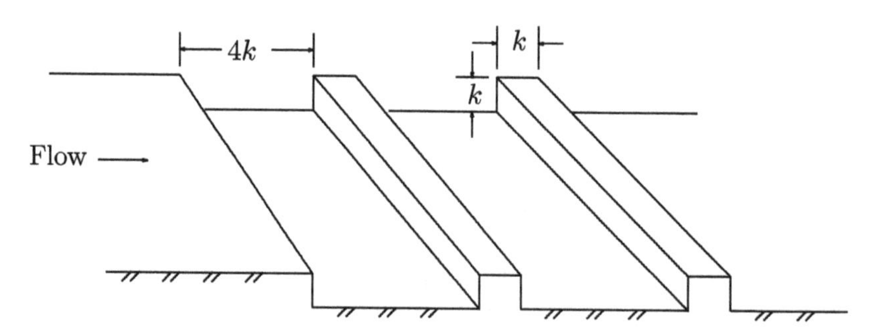


**Figure 4.132.** Roughness model.

**Table 4.50.** Roughness models.

| *Model* | $k$ (mm) | $k_s$ (mm) | $k_s^+$ |
|---|---|---|---|
| Smooth | 0 | 0 | 0 |
| 1 | 0.317 5 | 1.016 | 19 |
| 2 | 0.635 0 | 2.286 | 40 |
| 3 | 1.270 0 | 4.572 | 85 |

**Table 4.51.** Freestream conditions.

| Quantity | Value |
|---|---|
| $M_\infty$ | 6.01 ± 1% |
| $p_{t_\infty}$ (MPa) | 1.55 |
| $T_{t_\infty}$ (K) | 425 |
| $T_w$ (K) | 345 ± 2 |
| $T_w/T_{aw}$ | 0.89 |
| $H_\infty$ (MJ kg$^{-1}$) | 0.43 |
| $Re \times 10^{-6}$ (m$^{-1}$) | 15.7 |
| $\delta$ (mm) | 22.9[a] |
| $\delta^*$ (mm) | 10.9[a] |
| $\theta$ | 1.02[a] |

[a]Measured at 100.1 cm from tunnel throat.

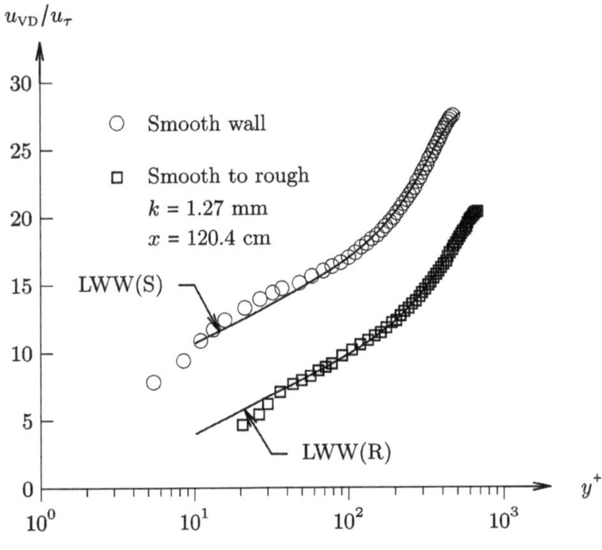


**Figure 4.133.** $u_{VD}/u_\tau$ versus $y^+$. LWW(S), law of the wall and wake (smooth); LWW(R), law of the wall and wake (rough).

Figure 4.134 displays the normalized total temperature fluctuations $\sqrt{\overline{T_t'^2}}/\bar{T}_t$ versus $y/\delta$, together with the data of Kistler (1959) and Owen *et al* (1975). Maximum relative root-mean-square values are 3% to 5%.

## 4.43 Materna (1977)

Materna (1977) investigated an adiabatic zero pressure gradient turbulent boundary layer in helium at Mach 16 at the Princeton University Gas Dynamics Laboratory hypersonic tunnel (section A.28.2). The measurements were obtained in the nozzle boundary layer. The freestream conditions are shown in table 4.52. Experimental measurements include pitot pressure, total temperature, wall pressure, and hot wire surveys.

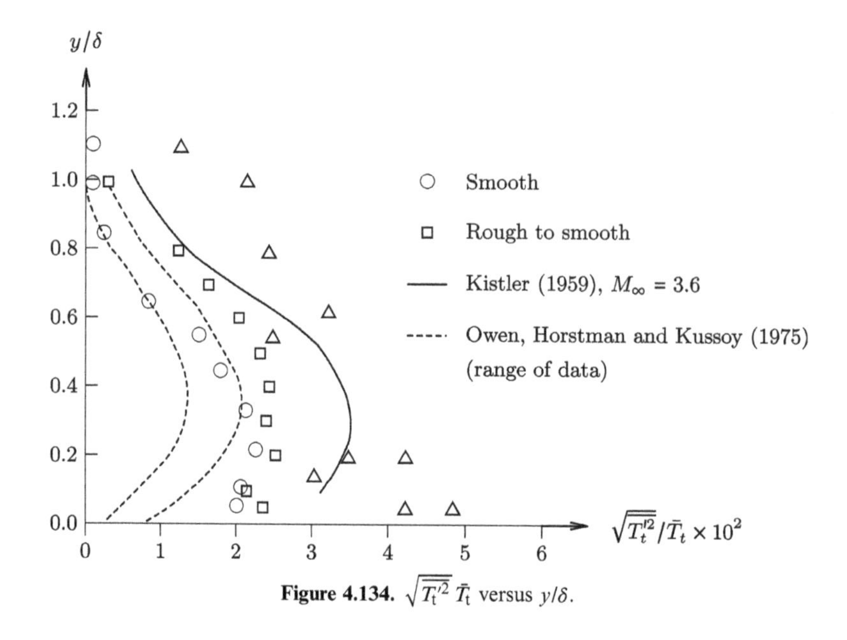


**Figure 4.134.** $\sqrt{\overline{T_t'^2}}\ \bar{T}_t$ versus $y/\delta$.

**Table 4.52.** Freestream conditions.

| Quantity | Value |
|---|---|
| Gas | He |
| $M_e$ | 16 |
| $p_{t_e}$ (MPa) | 1.55 |
| $T_{t_e}$ (K) | 300 |
| $T_w$ (K) | ≈300 |
| $T_w/T_{aw}$ | ≈1 |
| $\delta$ (cm) | 2.5 |
| $Re \times 10^{-6}$ (m$^{-1}$) | 6.8 |

Figure 4.135 shows the normalized root-mean-square mass flux fluctuations versus $y/\delta$. Normalization is shown for both the local mean mass flux $\overline{\rho u}$ and the freestream mass flux $\overline{\rho u}_e$. Within the boundary layer, the locally normalized root-mean-square mass flux reaches 50% at $y/\delta = 0.4$ which is the closest measurement to the wall. The corresponding value normalized by the freestream mass flux is 0.02. The difference is attributable to the low values of the density near the wall. Figure 4.136 displays the locally normalized root-mean-square total temperature

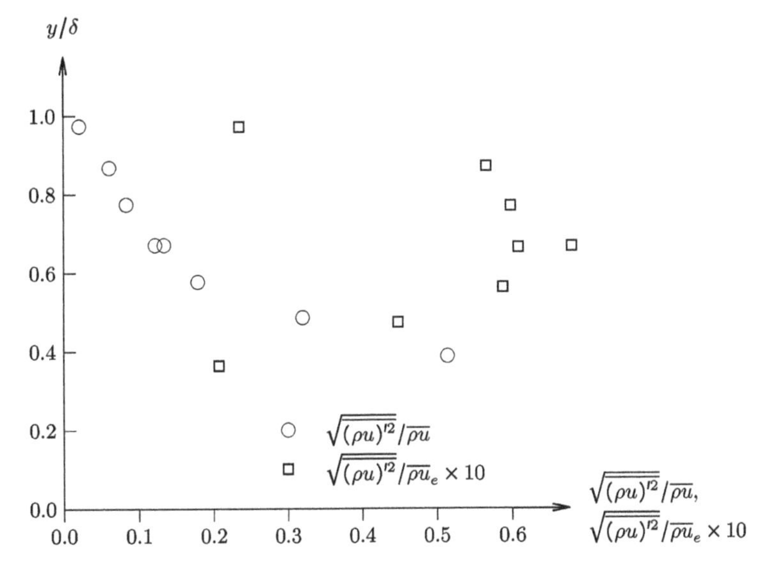


**Figure 4.135.** Normalized mass flow fluctuations versus $y/\delta$.

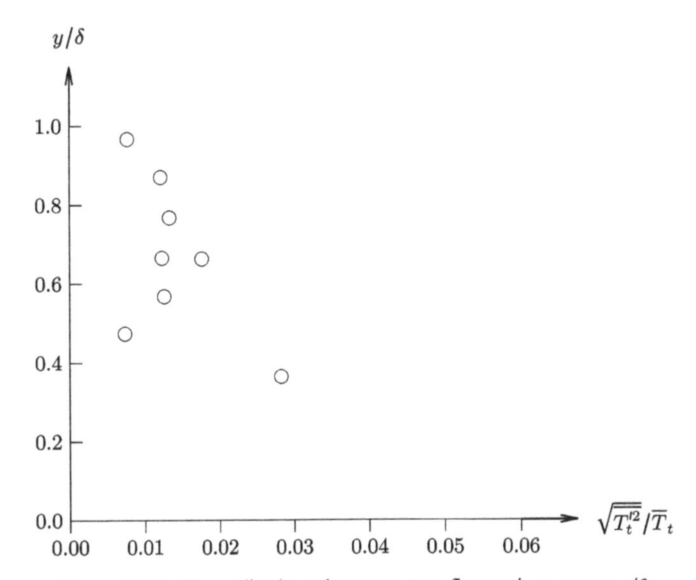


**Figure 4.136.** Normalized total temperature fluctuations versus $y/\delta$.

fluctuations. The maximum value is approximately 3% which is attributable to the nearly adiabatic wall boundary condition. Figure 4.137 shows the correlation coefficient $R_{\rho u T_t}$.

## 4.44 Owen and Calarese (1987)

Owen and Calarese (1987) conducted an experimental investigation of a zero pressure gradient turbulent flat plate boundary layer. The experiments were conducted in the Air Force Wright Aeronautical Laboratories Mach 6 high Reynolds number wind tunnel (section A.2.3). The freestream conditions are listed in table 4.53.

Figure 4.138 displays the Van Driest transformed velocity $u_{\mathrm{VD}}/u_\tau$ versus $y^+$. Close agreement is observed with the Law of the Wall. Figure 4.139 presents the streamwise turbulence intensity $\sqrt{\overline{u'^2}}/u_\tau$ versus $y/\delta$ together with the experimental

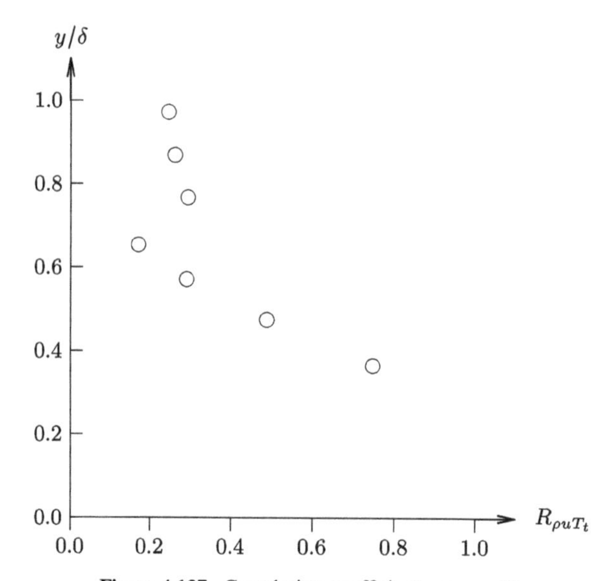


**Figure 4.137.** Correlation coefficient versus $y/\delta$.

**Table 4.53.** Freestream conditions.

| Quantity | Value |
|---|---|
| $M_\infty$ | 6.0 |
| $p_{t_\infty}$ (MPa) | 4.82–14.48 |
| $T_{t_\infty}$ (K) | 611 |
| $T_{\mathrm{w}}$ (K) | na |
| $\delta$ (cm) | na |

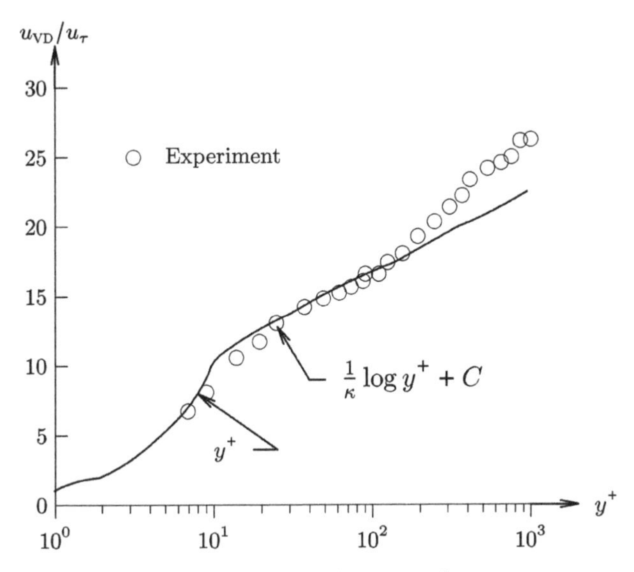


**Figure 4.138.** $u_{VD}/u_\tau$ versus $y^+$.

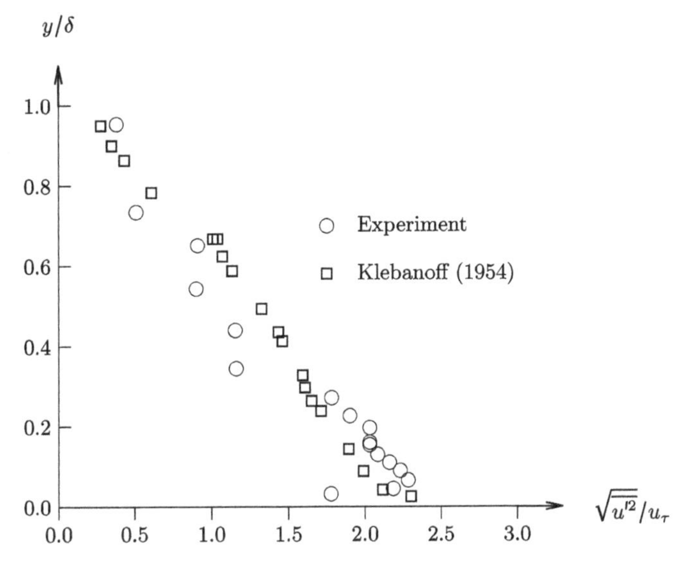


**Figure 4.139.** Streamwise turbulence intensity.

data of Klebanoff (1954) for an incompressible turbulent boundary layer. The experimental data at Mach 6 show close agreement with the incompressible results without invoking Morkovin's hypothesis (section 3.5).

## 4.45 McGinley *et al* (1994)

McGinley *et al* (1994) performed an experimental investigation of a hypersonic turbulent boundary layer in helium at Mach 11 in the Mach 20 high Reynolds

**Table 4.54.** Freestream conditions.

| Quantity | Value | |
|---|---|---|
| | Case 1 | Case 2 |
| $M_e$ | 10.5 | 11.0 |
| $p_{t_e}$ (MPa) | 5.5 | 11.0 |
| $T_{t_e}$ (K) | 312 | 316 |
| $T_w$ (K) | 302 | 302 |
| $T_w/T_{aw}$ | 0.98 | 0.99 |
| $\delta$ (cm) | 4.95 | 4.80 |
| $\delta^*$ (cm) | 2.47 | 2.36 |
| $\theta$ (cm) | 0.031 | 0.027 |
| $Re \times 10^{-6}$ (m$^{-1}$) | 21.3 | 44.1 |

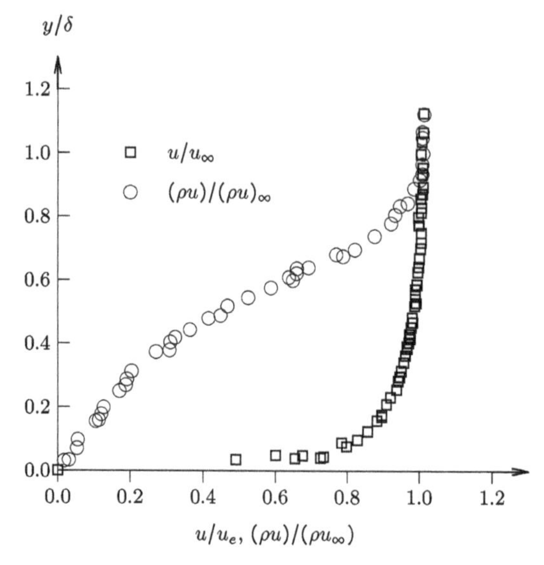


**Figure 4.140.** Mass flux and velocity profiles versus $y/\delta$.

number helium tunnel at NASA Langley Research Center (section A.22.13). The model is a 4° wedge, and the boundary layer edge conditions downstream of the wedge leading edge are shown in table 4.54 for the two different stagnation pressures. The boundary layer was examined at a location 2.16 m downstream of the wedge leading edge corresponding to $Re_x = 46.0 \times 10^6$ and $95.3 \times 10^6$ for the two cases. Experimental measurements include pitot pressure, wall pressure and temperature, and constant temperature hot wire anemometry.

Figure 4.140 displays the mean mass flux and velocity profiles for case 1. The mean velocity profile looks similar to an incompressible turbulent boundary layer

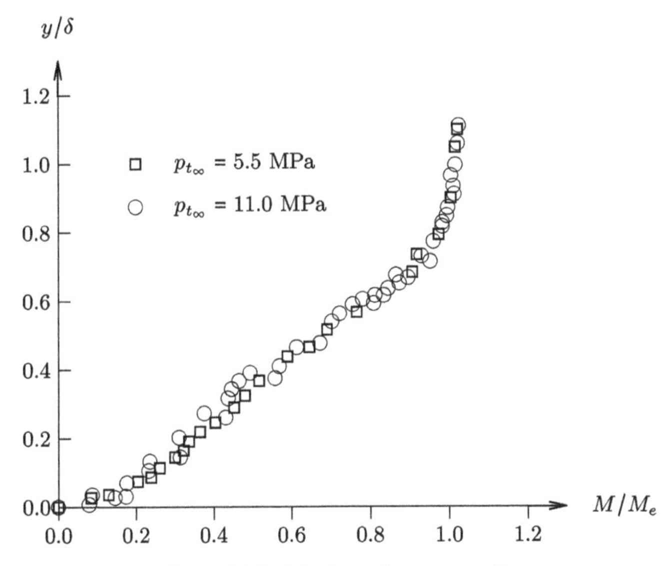


**Figure 4.141.** Mach number versus $y/\delta$.

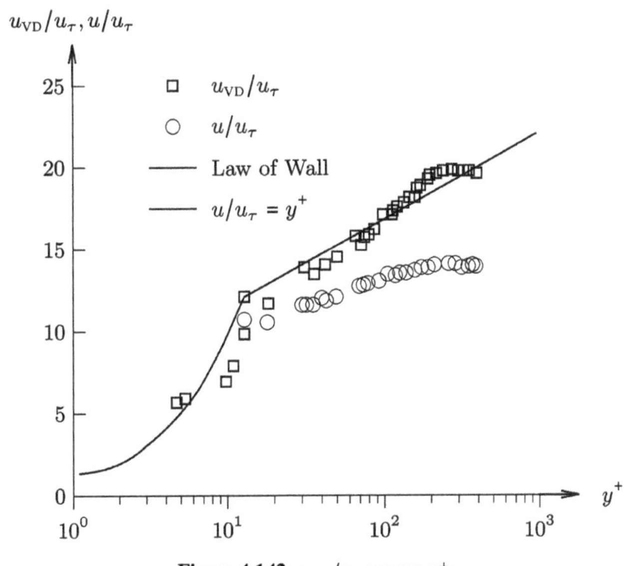


**Figure 4.142.** $u_{VD}/u_\tau$ versus $y^+$.

while the mass flux profile shows the effect of the large density variation of approximately a factor of 33 across the boundary layer. Figure 4.141 shows the mean Mach number profiles for cases 1 and 2. The profiles are essentially identical. Figure 4.142 presents the mean velocity versus the wall coordinate $y^+$ for case 2. The symbol (circle) ○ denotes the untransformed velocity $u/u_\tau$, and the symbol (square) denotes □ the Van Driest transformed velocity $u_{VD}/u_\tau$. The latter agrees closely with the Law of the Wall, while the lack of agreement of the former with the Law of the Wall shows the effect of the density gradient across the boundary layer. Figure 4.143

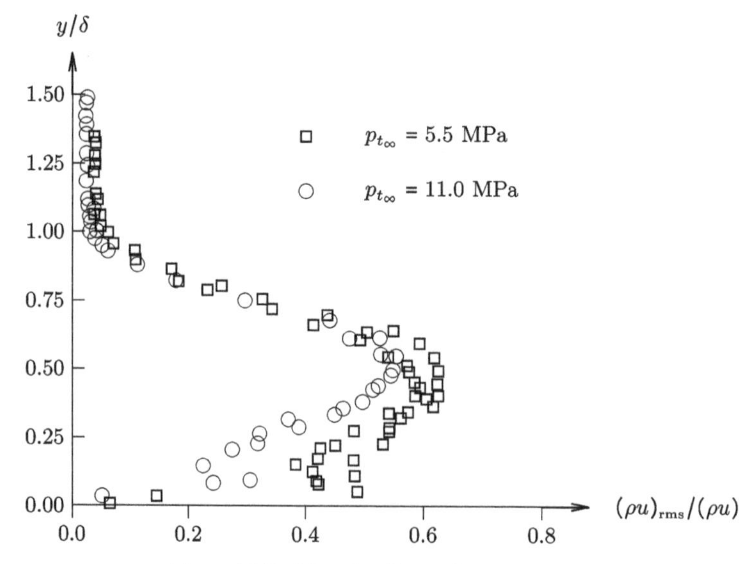


**Figure 4.143.** Mass flux fluctuations versus $y/\delta$.

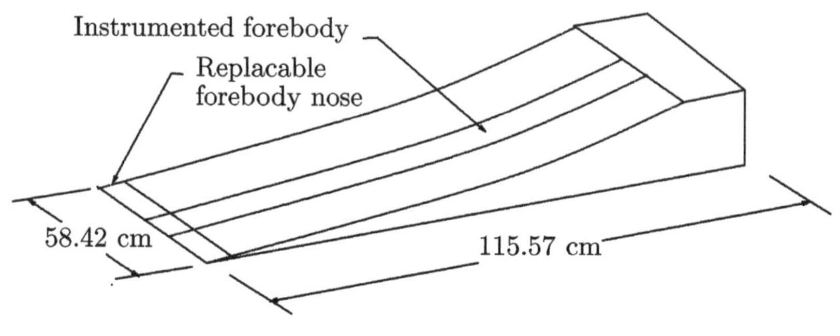


**Figure 4.144.** Model.

displays the profile of root-mean-square streamwise mass flux fluctuations normalized by the local mean mass flux for cases 1 and 2. The peak normalized root-mean-square fluctuations are 60% of the local mean mass flux indicating the strong effect of compressibility. Additional details regarding the statistics of the mass flux fluctuations are presented in McGinley *et al* (1994).

## 4.46 Holden and Chadwick (1995)

Holden and Chadwick (1995) conducted an experimental study of hypersonic transitional boundary layer interaction on a curved compression surface at Mach 10 to 12. The experiments were performed in the Calspan 48 in shock tunnel (section A.8.1). The model is shown in figure 4.144. A total of 18 tests were conducted with Reynolds numbers based upon the overall length $L$ = 1.156 m from $1.2 \times 10^6$ to $11.1 \times 10^6$, thereby achieving a range of locations of natural transition locations on the curved surface. The flow conditions are listed in table 4.55. Experimental

**Table 4.55.** Freestream conditions.

| Quantity | Case 1 | 2 | 3 | 4 | 5 | 6 | 7 | 8 | 9 |
|---|---|---|---|---|---|---|---|---|---|
| Run No. | 5 | 8 | 9 | 14 | 19 | 20 | 21 | 22 | 23 |
| $M_\infty$ | 10.0 | 10.0 | 10.0 | 10.1 | 10.0 | 10.0 | 10.6 | 11.9 | 10.6 |
| $p_{t_\infty}$ (MPa) | 27.3 | 22.2 | 16.4 | 6.6 | 9.3 | 16.2 | 25.4 | 24.3 | 20.0 |
| $T_{t_\infty}$ (K) | 1058 | 1116 | 1091 | 1290 | 1048 | 1085 | 1261 | 1693 | 1204 |
| $T_w$ (K) | 295 | 294 | 294 | 292 | 293 | 294 | 296 | 292 | 292 |
| $Re \times 10^{-6}$ (m$^{-1}$) | 9.6 | 7.5 | 5.7 | 1.9 | 3.4 | 5.6 | 6.2 | 2.8 | 5.1 |
| | 10 | 11 | 12 | 13 | 14 | 15 | 16 | 17 | 18 |
| Run No. | 24 | 25 | 26 | 27 | 28 | 31 | 36 | 38 | 39 |
| $M_\infty$ | 10.5 | 11.9 | 11.9 | 11.9 | 11.8 | 11.9 | 10.5 | 11.8 | 11.8 |
| $p_{t_\infty}$ (MPa) | 15.9 | 19.1 | 15.2 | 12.6 | 8.2 | 8.4 | 12.8 | 8.3 | 8.3 |
| $T_{t_\infty}$ (K) | 1202 | 1593 | 1536 | 1608 | 1573 | 1530 | 1218 | 1555 | 1560 |
| $T_w$ (K) | 294 | 295 | 293 | 293 | 294 | 293 | 293 | 293 | 294 |
| $Re \times 10^{-6}$ (m$^{-1}$) | 4.1 | 2.3 | 1.9 | 1.5 | 1.0 | 1.1 | 3.3 | 1.1 | 1.1 |

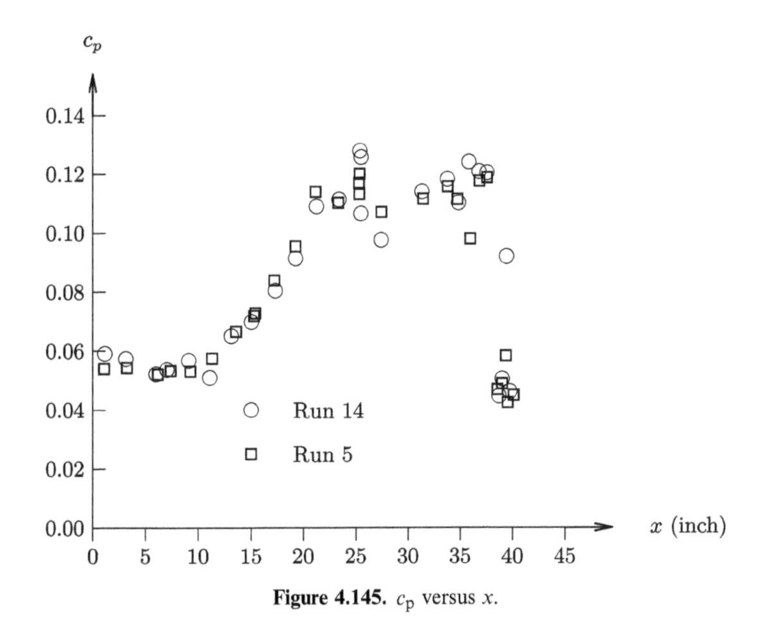


**Figure 4.145.** $c_p$ versus $x$.

diagnostics include 24 surface pressure gages and 58 surface heat transfer gages distributed off- and on-centerline.

Figure 4.145 shows the pressure coefficient $c_p$ versus $x$ for run 14 ($Re = 1.9 \times 10^6$ m$^{-1}$, $M_\infty = 10.1$) and run 5 ($Re = 9.6 \times 10^6$ m$^{-1}$, $M_\infty = 10.0$). Figure 4.146 shows the

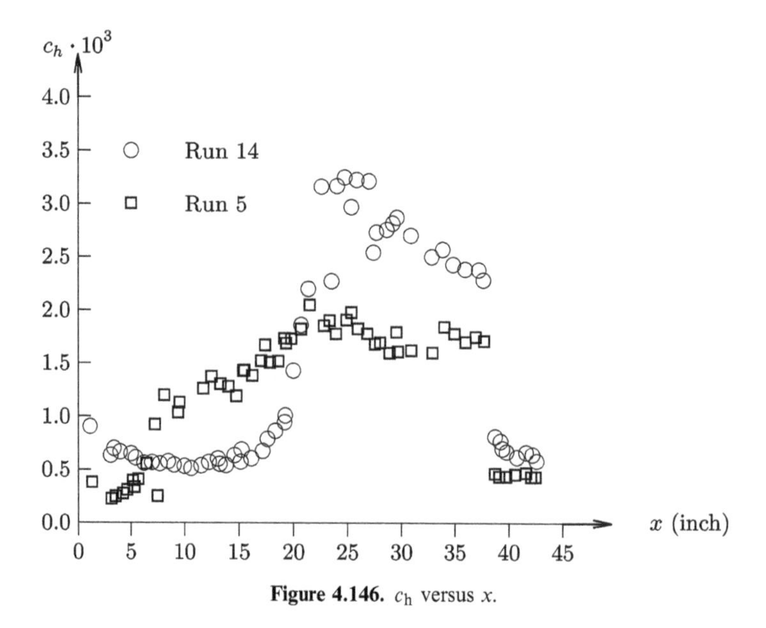


**Figure 4.146.** $c_h$ versus $x$.

heat transfer coefficient $c_h$ versus $x$ for the same cases. The pressure coefficient distribution is essentially the same for the two different Reynolds numbers. The heat transfer coefficient distribution shows marked differences, with the lower Reynolds number case (run 14) achieving significantly higher peak heat transfer than the higher Reynolds number case (run 5). The lower Reynolds number case transitions from laminar to turbulent within the adverse pressure gradient, while the higher Reynolds case transitions prior to the beginning of the adverse pressure gradient.

## 4.47 Auvity *et al* (2001)

Auvity *et al* (2001) examined the effect of transverse helium injection on a hypersonic zero pressure gradient boundary layer at Mach 8. The experiments were conducted in the Princeton Gas Dynamics Laboratory hypersonic wind tunnel (section A.28.1). This is one of few experimental investigations of mass injection in hypersonic boundary layers. The freestream conditions are listed in table 4.56. Helium gas was injected into a hypersonic boundary layer through a spanwise slot on the flat plate. The plate is 470 mm in length and 152 mm in width. The stagnation temperature of the injected gas was 530 K with a range of stagnation pressures up to 30 kPa, yielding a ratio $J$ of the jet momentum to the freestream momentum from zero to 0.20. Two different locations for the slot were examined at $x = 300$ mm (case 1) and $x = 220$ mm (case 2) from the leading edge of the flat plate. In the latter case the boundary layer was tripped. Filtered Rayleigh scattering (FRS) using $CO_2$ nanoparticles was used to visualize the instantaneous structure of the turbulent boundary layer downstream of the injection. Figure 4.147 (a) (no injection) and (b) (helium injection with $J = 0.13$) show ensemble planform averages of 40 instantaneous FRS images at a fixed height with the boundary layer. The vertical black line in the center of each image is the injection slot. The effect of injection is to impose a

**Table 4.56.** Freestream conditions.

| | Value | |
|---|---|---|
| Quantity | Case 1 | Case 2 |
| $M_\infty$ | 8.0 | 8.0 |
| $p_{t_\infty}$ (MPa) | 8.3 | 8.3 |
| $T_{t_\infty}$ (K) | 850 | 850 |
| $H_\infty$ (MJ kg$^{-1}$) | 0.85 | 0.85 |
| $Re \times 10^{-6}$ (m$^{-1}$) | 13.6 | 13.6 |

Note: Cases 1 and 2 are conditions 2 and 3 in Auvity *et al* (2001).

greater spanwise organization of the flow. Additional FRS images were obtained for a lower Reynolds number with a transitional boundary layer.

## 4.48 Goyne *et al* (2003)

Goyne *et al* (2003) conducted an investigation of high enthalpy turbulent flat plate boundary layers in air at cold wall conditions for Mach numbers from 5.3 to 6.7. The experiments were conducted in the free piston shock tunnel T4 at the University of Queensland (section A.45.1). The model is shown in figure 4.148. The flat plate comprised one surface of a test duct with length 1500 mm and width 146 mm. The freestream conditions are listed in table 4.57. The experimental diagnostics include surface heat transfer, pressure, and skin friction.

Figure 4.149 displays the Reynolds Analogy Factor (section 3.4) versus $H_w/H_\infty$. Results are presented for both the flat plate surface of the duct and a from a prior study (Goyne *et al* 1999) in a supersonic combustion duct with no combustion. The data show a decrease in the Reynolds Analogy Factor with decreasing $H_w/H_\infty$. Since the unit Reynolds number decreased with decreasing $H_w/H_\infty$, Goyne *et al* (2003) concluded that values of $2St/c_f$ below unity at low $H_w/H_\infty$ may be attributable to a transitional state of the boundary layer. The mean value of the test duct data is 0.86 ± 0.19 and of the supersonic combustion duct data is 1.1 ± 0.3. Also shown are the theoretical values $2St/c_f = Pr^{-\frac{2}{3}}$ and $Pr_t^{-1}$ for $Pr = 0.72$ and $Pr_t = 0.89$, respectively. Figure 4.150 shows the Reynolds Analogy Factor versus skin friction coefficient $c_f$. The trend of decreasing $2St/c_f$ with increasing $c_f$ was not explained by the authors.

## 4.49 Suraweera *et al* (2006)

Suraweera *et al* (2006) conducted an experimental study of hypersonic flat plate turbulent boundary layers at Mach numbers from 4.2 to 5.9. The experiments were performed in the T4 free piston reflected shock tunnel at the University of Queensland (section A.45.1). Surface skin friction and heat transfer measurements were obtained on one inner wall of a 1745 mm long duct with entrance cross section 60 mm × 100 mm or 57 mm × 100 mm depending on the test. The model is similar to figure 4.148. The freestream conditions are indicated in table 4.58. The Reynolds number, based upon the freestream conditions and distance from the entrance of the

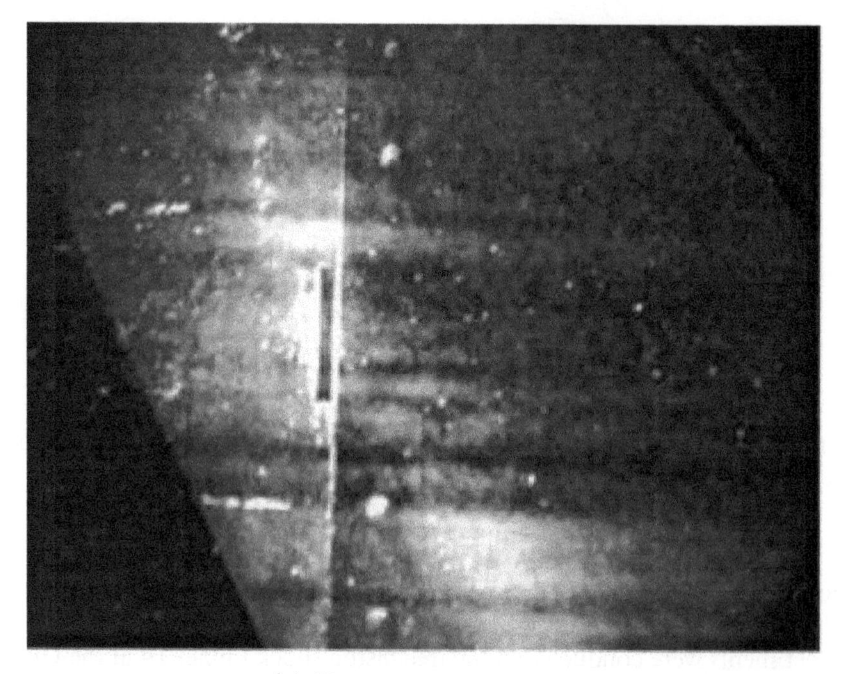

(a) No injection, $y$ = 7 mm

(b) $J$ = 0.13, $y$ = 8 mm

**Figure 4.147.** Ensemble average of 40 planform views of the boundary layer showing the effect of helium injection on turbulence structure. Reproduced from Autivy *et al* (2001). © 2001 American Institute of Physics.

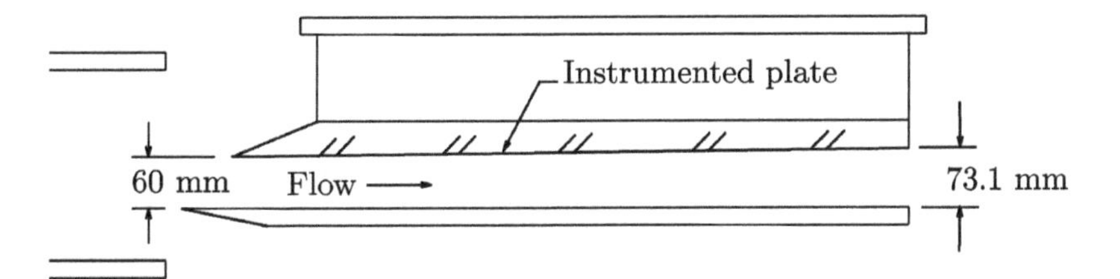


**Figure 4.148.** Model.

**Table 4.57.** Freestream conditions.

| Quantity | Value | |
|---|---|---|
| | Min | Max |
| $M_\infty$ | 5.3 | 6.7 |
| $p_\infty$ (kPa) | 0.87 | 17.9 |
| $T_\infty$ (K) | 326 | 1790 |
| $T_w/T_{aw}$ | 0.022 | 0.11 |
| $H_w/H_\infty$ | 0.02 | 0.10 |
| $H_\infty$ (MJ kg$^{-1}$) | 3.2 | 13.1 |
| $Re \times 10^{-6}$ (m$^{-1}$) | 0.43 | 16.6 |

$$T_w/T_{aw} = (T_w/T_{t_e})\left[1 + \frac{(\gamma-1)}{2}M_e^2\right]\left[1 + \frac{(\gamma-1))}{2}Pr_t M_e^2\right]^{-1}$$

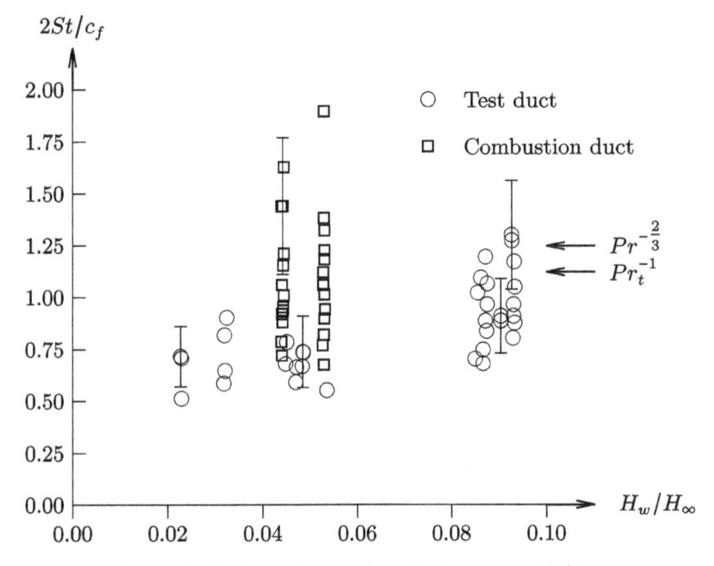


**Figure 4.149.** Reynolds Analogy Factor versus $H_w/H_\infty$.

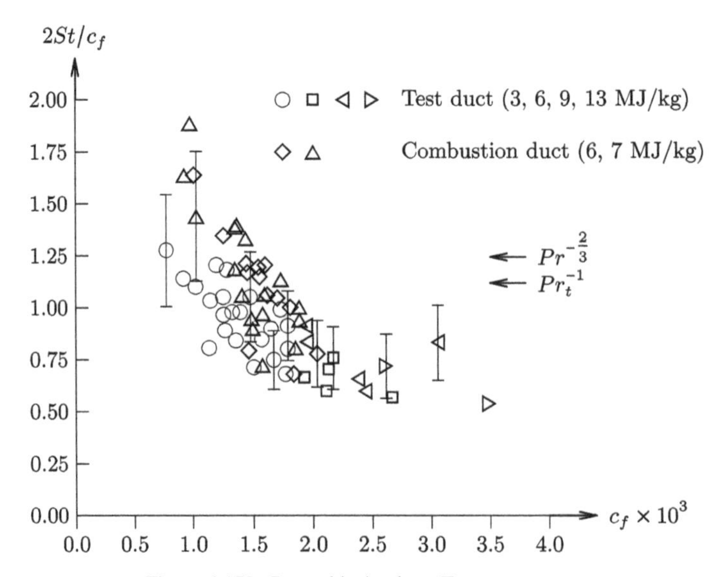


**Figure 4.150.** Reynolds Analogy Factor versus $c_f$.

**Table 4.58.** Freestream conditions.

| Quantity | Value |
|---|---|
| Gas | Air, $N_2$ |
| $M_\infty$ | 4.2–5.9 |
| $p_\infty$ (kPa) | 14.0–86.0 |
| $H_\infty$ (MJ kg$^{-1}$) | 4.8–9.5 |

duct, at the first measurement station exceeded $2 \times 10^6$ for all tests, thereby indicating a fully turbulent boundary layer (Suraweera *et al* 2006).

Figure 4.151 presents the Reynolds Analogy Factor $2St/c_f$ versus skin friction coefficient $c_f$ for the experiments, together with the experimental data of Hironimus (1966) and Goyne *et al* (2003). A distinct decrease in Reynolds Analogy Factor for $c_f > 2 \times 10^{-3}$ is evident. Tests using both $N_2$ and air showed essentially the same results, thus eliminating any effect of thermochemical reaction.

## 4.50 Maslov *et al* (2008)

Maslov *et al* (2008) performed an experimental study of the effects of ultrasonic absorptive coating on a transitional and turbulent boundary layer on a 7° half-angle cone at Mach 6. The experiments were performed in the Institute of Theoretical and Applied Mechanics Transit-M impulse tunnel in Novosibsk, Russian Federation (section A.18.5). The model is shown in figure 4.152 from Federov *et al* (2003). The cone length $L = 0.5$ m. The surface of the cone was covered with a felt-metal coating

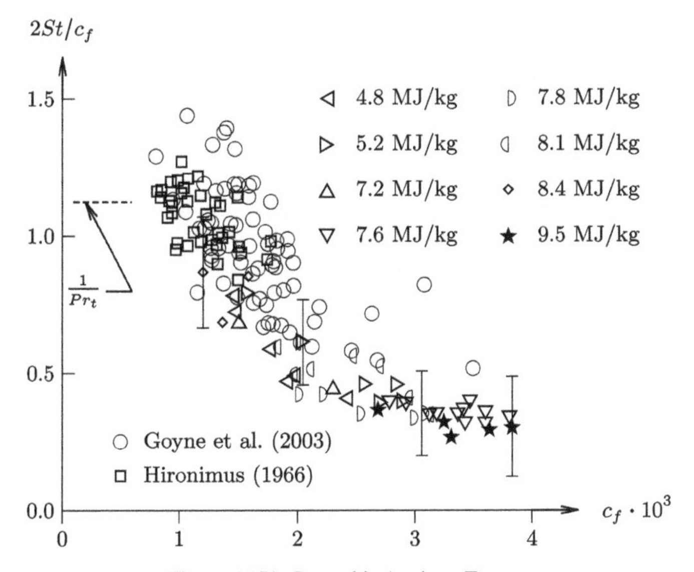


**Figure 4.151.** Reynolds Analogy Factor.

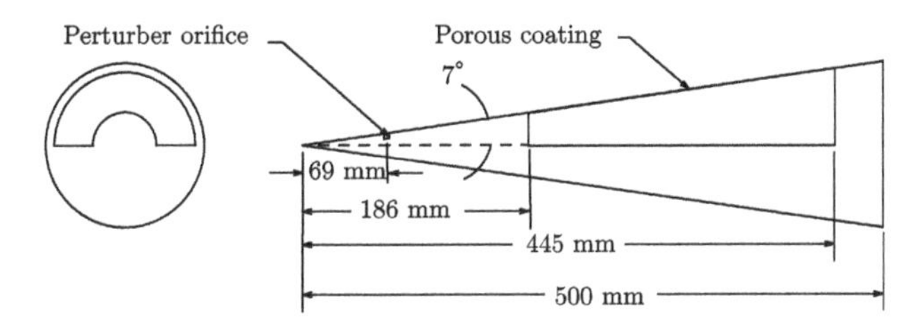


**Figure 4.152.** Model (dimensions in mm).

for 186 mm $< x <$ 445 mm. A magnified view of the upper surface of the felt-metal coating is shown in figure 4.153 from Federov *et al* (2003). The freestream conditions are listed in table 4.59. The experimental diagnostic was surface heat transfer using atomic layer thermal pile gauges.

Figure 4.154 displays the surface heat flux disturbance amplitude versus frequency for solid and porous surfaces under conditions of transitional and turbulent boundary layers. The porous coating suppresses the high frequency (>100 kHz) fluctuations for both the transitional and turbulent boundary layers.

## 4.51 Vaganov (2008)

Vaganov and Stolyarov (2008) measured the turbulent pressure fluctuations on a wind tunnel nozzle wall at Mach 7.5. The axisymmetric nozzle diameter $D = 1$ m, and the length is $L = 5$ m. The freestream conditions in the test section are listed in table 4.60. Experimental measurements were performed at $x/D = 2.9$ to 4.85 within the nozzle where $x$ is measured from the nozzle throat. Experimental diagnostics include surface pressure using Endevco piezoresistive differential transducers (type 8506-15) with diameter 2.3 mm and natural frequency approximately 135 kHz.

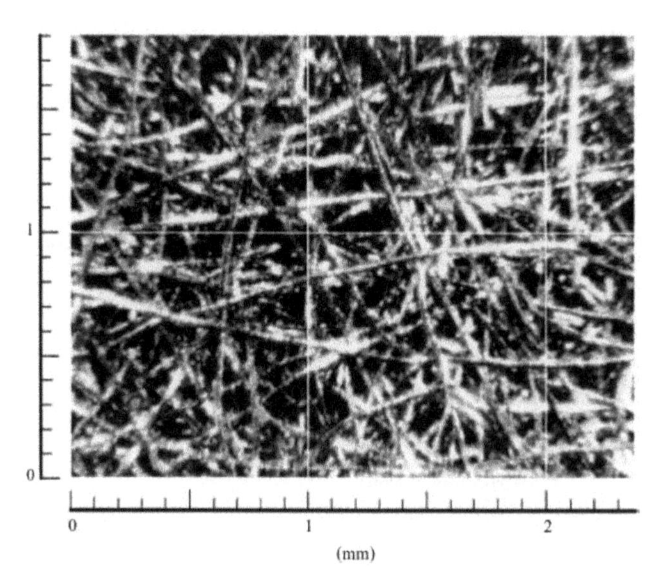


**Figure 4.153.** Upper surface of felt-metal coating 1 mm × 1 mm. Reproduced from Federov *et al* (2003). © Anatoly Alexandrovich Maslov.

**Table 4.59.** Freestream conditions.

| Quantity | Value |
|---|---|
| Gas | Air |
| $M_\infty$ | 6.0 |
| $p_{t_\infty}$ (MPa) | 0.22–0.99 |
| $T_{t_\infty}$ (K) | 397–485 |
| $T_w$ (K) | 289–292 |
| $T_w/T_{aw}$ | 0.6–0.74 |
| $Re \times 10^{-6}$ (m$^{-1}$) | 2.3–10 |

The normalized spectral density of the pressure fluctuations is shown in figure 4.155 versus the dimensionless frequency $fD/U_\infty$ at locations $x/D = 2.9, 4.26, 4.765,$ and 4.85. All locations are within the nozzle since $L/D = 5$. Also shown is the measurement channel noise. No significant periodic components are observed. The peak spectral density occurs at $fD/U_\infty \approx 0.1$, and approximately one-half of the spectral energy is within the range $0.02 < fD/U_\infty < 0.4$.

The statistical moments $M_n$ of the pressure fluctuations are defined by

$$M_n = \frac{1}{N_r}\sum_{k=1}^{N_r}\frac{1}{N_{d,k}}\sum_{j=1}^{N_{d,k}}\frac{1}{N-1}\sum_{i=1}^{N}[p(t_i, x) - \overline{p(t_i, x)}]^n \tag{4.18}$$

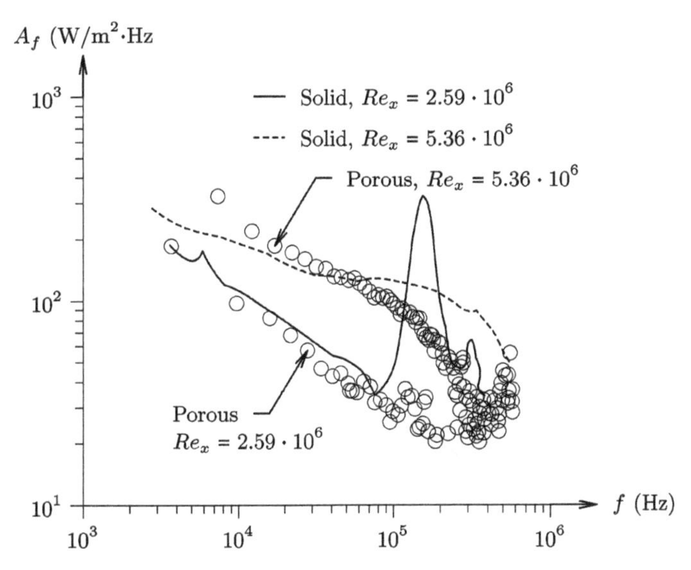


**Figure 4.154.** Disturbance spectra on solid and porous surfaces for transitional (solid lines) and turbulent (dashed lines) boundary layers.

**Table 4.60.** Freestream conditions.

| | Regime I | | Regime II | |
|---|---|---|---|---|
| Quantity | Min | Max | Min | Max |
| $M_\infty$ | 7.46 | 7.48 | 7.5 | 7.5 |
| $p_{t_\infty}$ (MPa) | 1.29 | 1.39 | 2.45 | 2.67 |
| $T_{t_\infty}$ (K) | 810 | 935 | 683 | 700 |
| $U_\infty$ (m s$^{-1}$) | 1240 | 1336 | 1132 | 1136 |
| $u_\tau$ (m s$^{-1}$) | 53.8 | 56 | 39.3 | 41.5 |
| $T_w$ (K) | 290 | 290 | 290 | 290 |
| $T_w/T_{aw}$ | 0.40[a] | 0.35[a] | 0.47[a] | 0.46[a] |
| Duration[b] (s) | 9.2 | 27 | 8.1 | 11.5 |

[a] Using (3.46).
[b] Recording interval.

where $N$ is the number of samples in a selected time series, $N_{d,k}$ is the number of selected time series, and $N_r$ is the number of realizations. The skewness and kurtosis (flatness) are defined as

$$\text{skewness} = M_3\, M_2^{-\frac{3}{2}}$$
$$\text{kurtosis} = M_4\, M_2^{-2} - 3$$

The results are shown in table 4.61. The statistical distribution of pressure fluctuations is strongly non-Gaussian as indicated by the skewness and kurtosis.

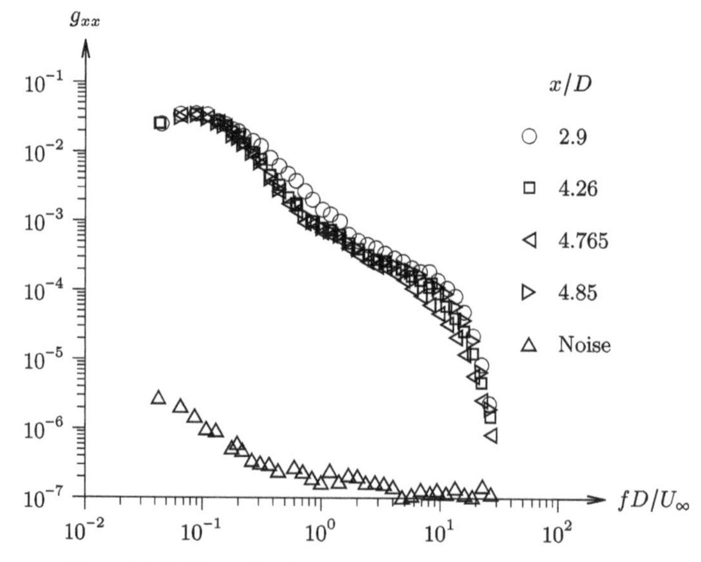


**Figure 4.155.** Normalized spectral density of fluctuating pressure.

**Table 4.61.** Statistics.

| Regime | $x/D$ | Skewness | Kurtosis |
|---|---|---|---|
| I | 2.900 | 0.252 | 0.44 |
| | 4.265 | 0.260 | 0.32 |
| | 4.765 | 0.262 | 0.32 |
| | 4.850 | 0.285 | 0.34 |
| II | 2.900 | 0.393 | 0.79 |
| | 4.265 | 0.352 | 0.53 |
| | 4.765 | 0.346 | 0.51 |
| | 4.850 | 0.360 | 0.50 |
| Noise | 4.765 | −0.003 | 0.04 |

## 4.52 Sahoo *et al* (2009)

Sahoo *et al* (2009) investigated the effect of surface roughness on a hypersonic flat plate turbulent boundary layer in air at Mach 7.3. The experiments were performed in the Mach 8 hypersonic boundary layer facility at the Princeton Gas Dynamics Laboratory (section A.28.1). The model is shown in figure 4.156, and the two different types of roughness models are shown in figure 4.157. The freestream conditions are shown in table 4.62. Experimental measurements include mean and fluctuating velocity using particle image velocimetry (PIV).

Figure 4.158 displays Van Driest transformed velocity for the smooth and rough walls and the corresponding Law of the Wall (3.56) and (3.60) with the omission of

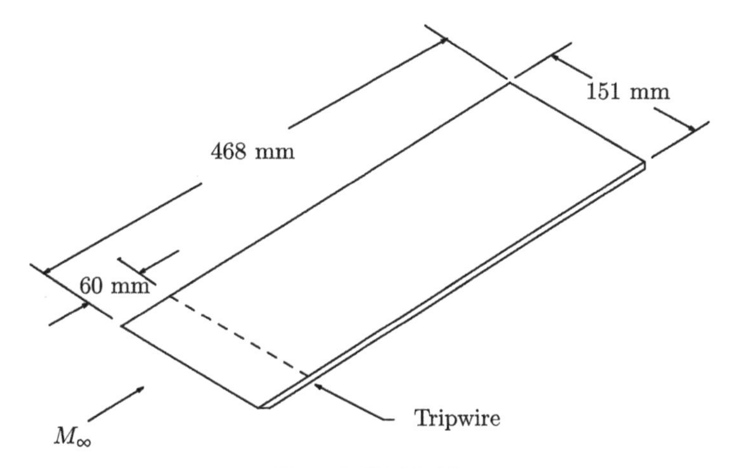


**Figure 4.156.** Model.

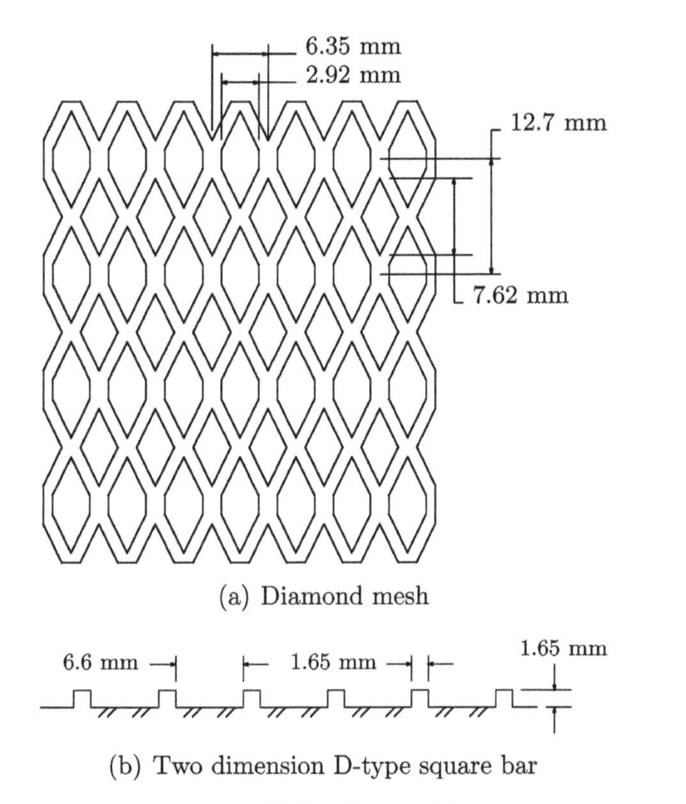


**Figure 4.157.** Roughness models.

Coles' wake function. The velocity profiles show close agreement with the Law of the Wall. Figure 4.159 shows the Van Driest transformed velocity defect. Close agreement is seen with the Coles' wake function in the region $0.6 \leqslant y/\delta \leqslant 1.0$.

**Table 4.62.** Freestream conditions.

| Quantity | Value | | |
|---|---|---|---|
| | Smooth | Diamond | Square |
| Gas | Air | Air | Air |
| $M_\infty$ | 7.3 | 7.3 | 7.3 |
| $p_{t_\infty}$ (MPa) | 7.14 | 7.07 | 7.10 |
| $T_{t_\infty}$ (K) | 756 | 747 | 760 |
| $T_w$ (k) | 352 | 338 | 337 |
| $T_w/T_{aw}$ | 0.49 | 0.48 | 0.47 |
| $H_\infty$ (MJ kg$^{-1}$) | 0.66 | 0.70 | 0.69 |
| $u_\tau$ (m s$^{-1}$) | 55.8 | 153 | 132 |
| $\delta$ (mm) | 7.21 | 10.4 | 12.1 |
| $\theta$ (mm) | 0.19 | 0.27 | 0.25 |
| $Re \times 10^{-6}$ (m$^{-1}$) | 17.9 | 18.7 | 18.0 |

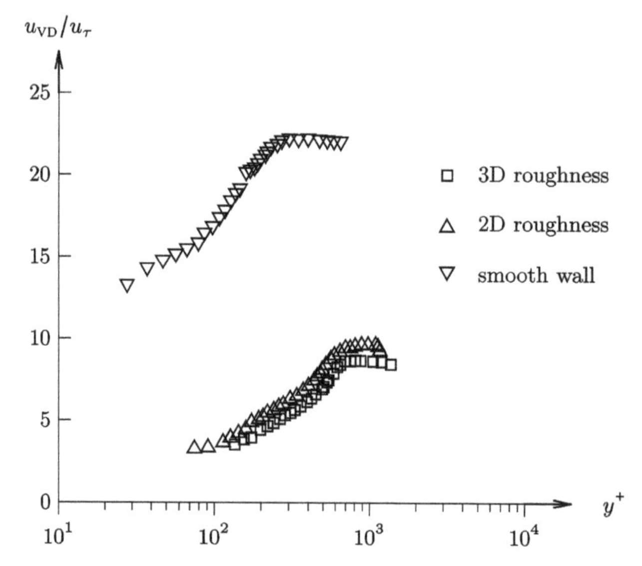


**Figure 4.158.** Van Driest transformed velocity.

Figure 4.160 displays the streamwise velocity fluctuations for the smooth wall scaled according to Morkovin, together with the measurements of Owen *et al* (1975) at $M_e = 6.7$ and the incompressible data of Klebanoff (1955). In contrast to the fluctuation data at supersonic speeds shown in figure 3.7, the Morkovin scaling does not collapse the data of Sahoo *et al* (2009) or Owen *et al* (1975) to the incompressible line.

Figures 4.161 and 4.162 show the density scaled streamwise and wall normal turbulence intensities. A significant effect of roughness is seen.

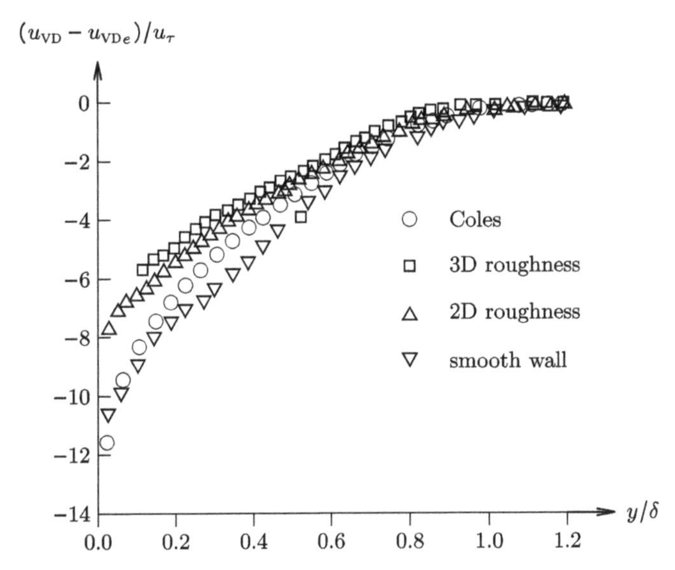


**Figure 4.159.** Van Driest transformed velocity defect.

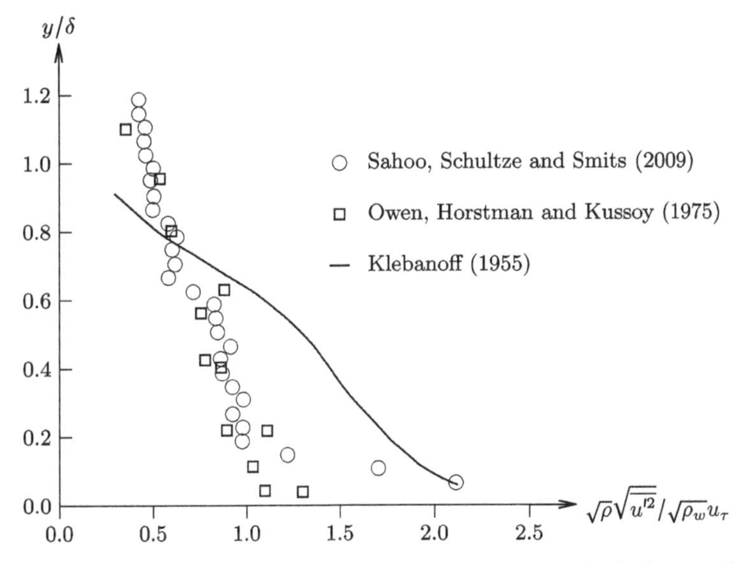


**Figure 4.160.** Streamwise velocity fluctuations scaled according to Morkovin for smooth plate.

## 4.53 Peltier *et al* (2011)

Peltier *et al* (2011) examined the effect of distributed surface roughness on a zero and favorable pressure gradient boundary layer at Mach 4.89. The experiments were performed in the National Aerothermodynamic Laboratory at Texas A&M University (section A.35). The experiments were obtained on the tunnel wall boundary layer at the measurement locations $x = 15.9$ cm and 29.8 cm (figure 4.163). A periodic diamond roughness pattern covered the tunnel wall

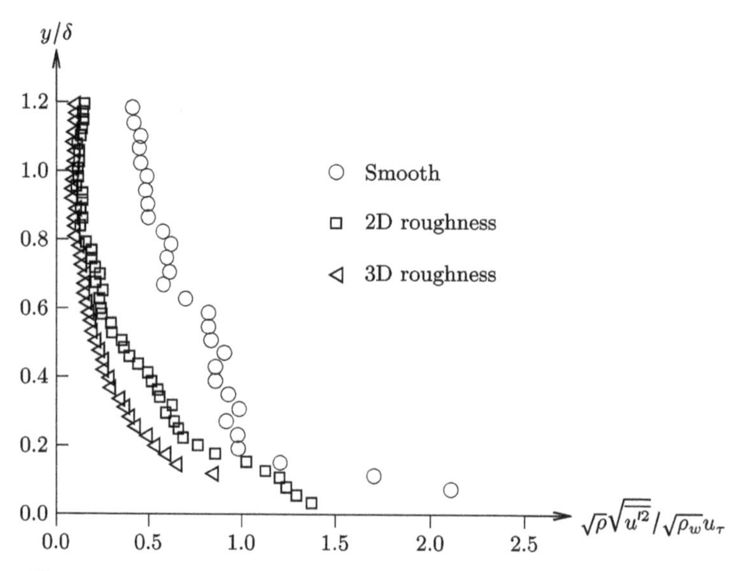


**Figure 4.161.** Streamwise velocity fluctuations scaled according to Morkovin.

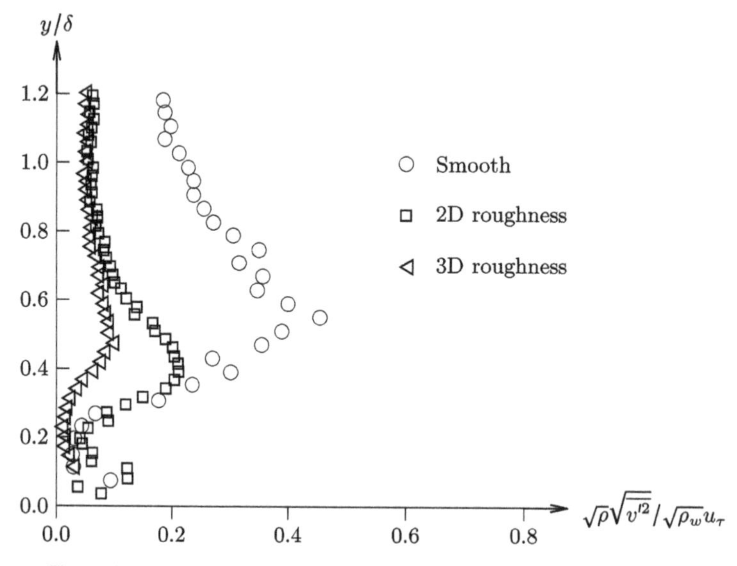


**Figure 4.162.** Normal velocity fluctuations scaled according to Morkovin.

(figure 4.164). The diamond elements were 1.59 mm wide and 9.0 mm long with a half-angle of 10° and height of 0.80 mm. Two different favorable pressure gradients were examined in addition to a zero pressure gradient. The flow conditions are indicated in table 4.63 where $\beta$ is the pressure gradient parameter defined as

$$\beta = \frac{\delta^*}{\tau_w}\frac{dp}{dx} \tag{4.19}$$

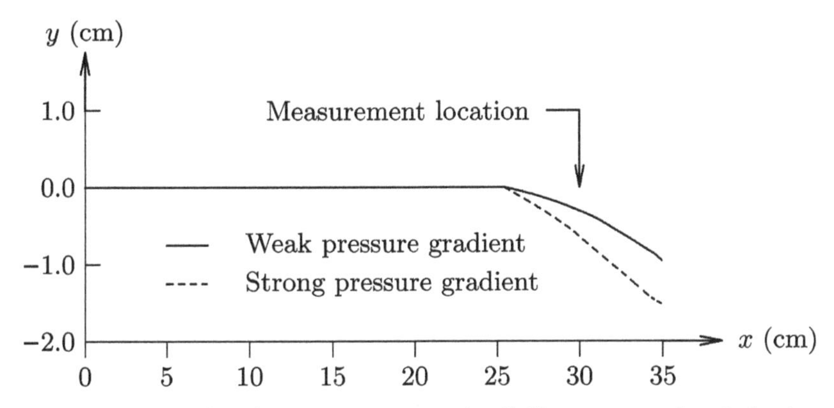


**Figure 4.163.** Location of measurement on tunnel wall. Note exaggerated vertical scale.

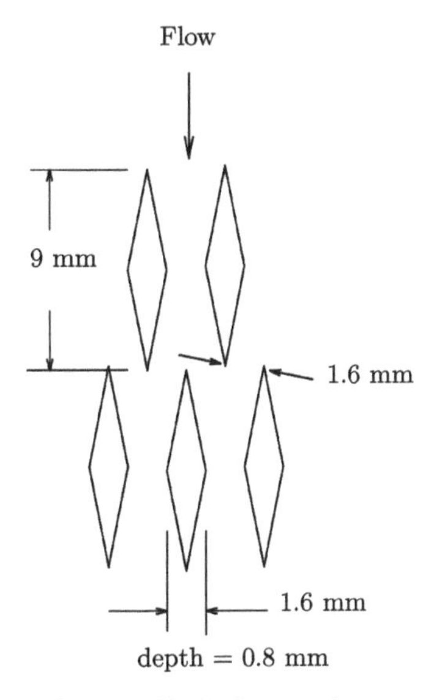


**Figure 4.164.** Surface roughness.

where $\delta^*$ is the incompressible displacement thickness and $x$ is measured along the surface. Experimental diagnostics include schlieren imaging and PIV. Additional discussion regarding the effect of the surface roughness on the structure of the boundary layer is presented in Peltier *et al* (2016).

Figure 4.165 presents the Van Driest transformed velocity $u_{\text{VD}}/u_\tau$ versus $y^+$ for the smooth wall (Tichenor 2010) and rough wall zero pressure gradient boundary layers. The smooth and rough wall profiles show close agreement with the compressible Law of the Wall (3.56) and (3.60), respectively, with $H(ku_\tau/\nu_{\text{w}}) = 8.7$.

Figure 4.166 displays the normalized kinematic streamwise Reynolds stress $\overline{u'u'}/u_e^2$ versus $y/\delta$ for the zero, weak, and strong pressure gradient rough wall cases, together with the zero pressure gradient smooth wall data of Tichenor (2010). The roughness elements increase the streamwise turbulent fluctuations near the wall.

**Table 4.63.** Freestream conditions.

| Quantity | Value | | |
|---|---|---|---|
| | ZPG | WPG | SPG |
| $M_\infty$ | 4.89 | 4.89 | 4.89 |
| $p_{t_\infty}$ (MPa) | 2.24 | 2.24 | 2.24 |
| $T_{t_\infty}$ (K) | 370 | 370 | 370 |
| $T_\infty$ (K) | 63 | 63 | 63 |
| $T_w$ (K) | 300[a] | 300[a] | 300[a] |
| $T_w/T_{aw}$ | 0.89[b] | 0.89[b] | 0.89[b] |
| $H_\infty$ (MJ kg$^{-1}$) | 0.37 | 0.37 | 0.37 |
| $\delta$ (mm) | 9.86 | 12.7 | 15.0 |
| $Re \times 10^{-6}$ (m$^{-1}$)[c] | 50. | 50. | 50. |
| $\beta$ | 0 | −0.4 | -0.6 |

ZPG, zero pressure gradient; SPG, strong pressure gradient; WPG, weak pressure gradient.
[a] Assumed.
[b] Using (3.46) with $Pr_t = 0.89$.
[c] Approximate.

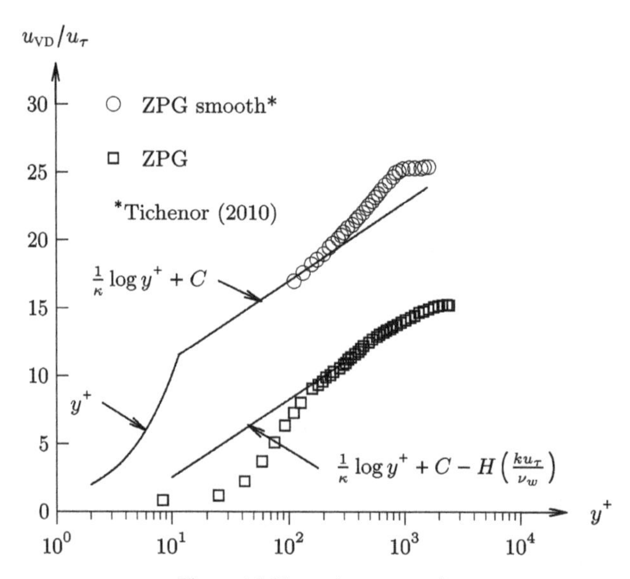


**Figure 4.165.** $u_{VD}/u_\tau$ versus $y^+$.

The favorable pressure gradients decrease the streamwise turbulent fluctuations compared to the zero pressure gradient case except close to the wall.

Figure 4.167 shows the normalized kinematic wall normal Reynolds stress $\overline{v'v'}/u_e^2$ versus $y/\delta$ for the zero, weak, and strong pressure gradient rough wall cases, together with the zero pressure gradient smooth wall data of Tichenor (2010). The zero pressure gradient rough wall shows a very significant increase compared to the

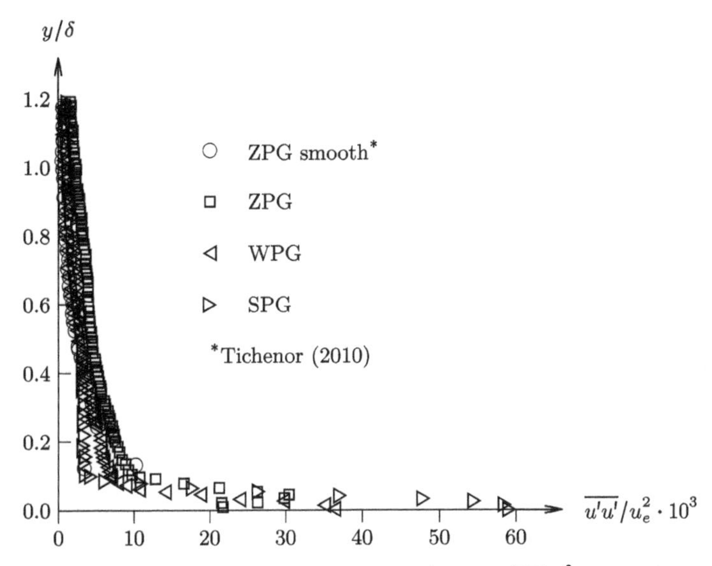


**Figure 4.166.** Kinematic Reynolds streamwise stress $\overline{u'u'}/u_e^2$ versus $y/\delta$.

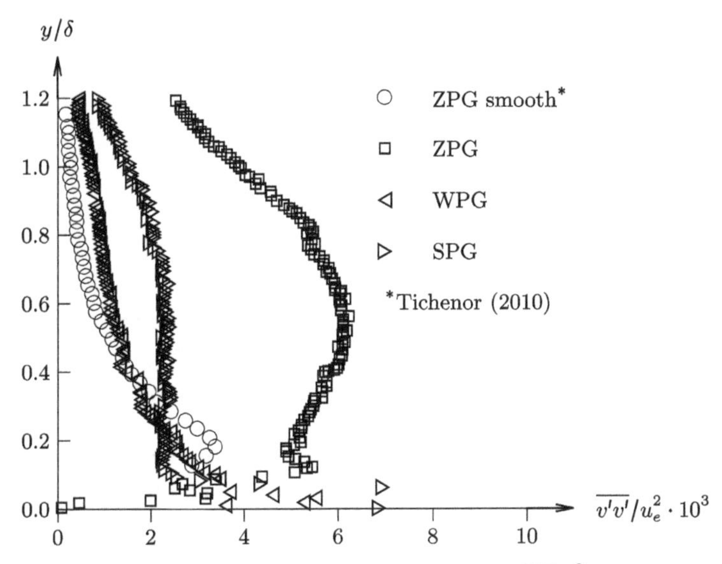


**Figure 4.167.** Kinematic Reynolds wall normal stress $\overline{v'v'}/u_e^2$ versus $y/\delta$.

zero pressure gradient smooth wall across the entire boundary layer, due in part to the compression and expansion waves generated by the diamond roughness.

Figure 4.168 displays the normalized kinematic Reynolds shear stress $\overline{u'v'}/U_e^2$ versus $y/\delta$ for the zero, weak, and strong pressure gradient rough wall cases, together with the zero pressure gradient smooth wall data of Tichenor (2010). The zero pressure gradient rough wall shows a significant increase in peak magnitude of the Reynolds shear stress compared to the smooth wall due to the effect of the

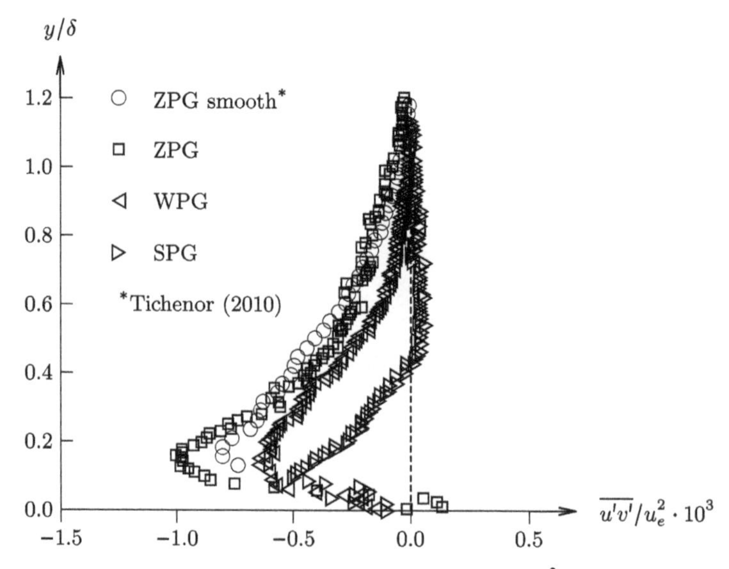


**Figure 4.168.** Kinematic Reynolds shear stress $\overline{u'v'}/u_e^2$ versus $y/\delta$.

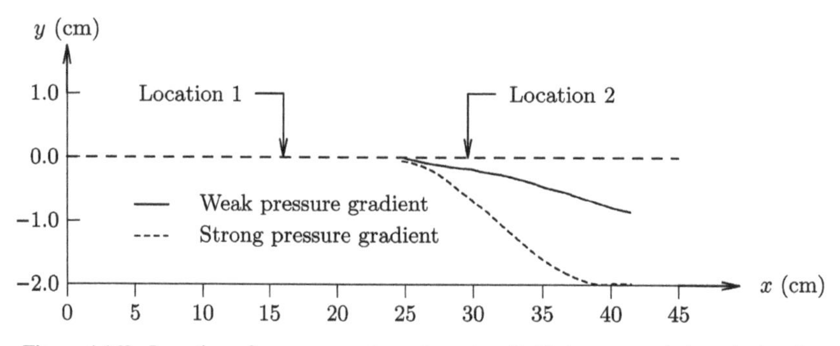


**Figure 4.169.** Location of measurements on tunnel wall. Note exaggerated vertical scale.

diamond roughness elements. The peak magnitude of the Reynolds shear stress decreases with increasingly favorable pressure gradient and moves closer to the wall.

## 4.54 Tichenor *et al* (2013)

Tichenor *et al* (2013) performed an investigation of the effect of streamline curvature–driven favorable pressure gradients on the structure of a hypersonic turbulent boundary layer at Mach 4.9. The experiments were conducted in the high-speed blowdown wind tunnel at the National Aerothermochemistry Laboratory at Texas A&M University (see section A.35). The experiments were obtained on the tunnel wall boundary layer at two locations indicated in figure 4.169. The first station was located upstream of the streamline curvature–driven pressure gradient, and the second station within the favorable pressure gradient region. Two different favorable pressure gradients were considered. The flow conditions[9] are shown in table 4.64. The experimental diagnostic is PIV.

Figure 4.170 displays the mean velocity profile for the three cases using Van Driest transformed variables. The weak pressure gradient modestly reduces the wake contribution, while the strong pressure gradient virtually eliminates the wake contribution. The effect is also observed in terms of the wake defect in figure 4.171.

**Table 4.64.** Freestream conditions.

| Quantity | Value | | |
|---|---|---|---|
| | ZPG | WPG | SPG |
| $M_\infty$ | 4.9 | 4.9 | 4.9 |
| $p_{t_\infty}$ (MPa) | 2.35 | 2.35 | 2.35 |
| $T_{t_\infty}$ (K) | 380 | 380 | 380 |
| $T_{aw}$ (K) | 342 | 320 | 317 |
| $H_\infty$ (MJ kg$^{-1}$) | 0.38 | 0.38 | 0.38 |
| $\delta$ (mm) | 7.8[a] | 10.0[b] | 13.4[b] |
| $Re_\theta \times 10^{-4}$ | 4.3[a] | 5.5[b] | 7.4[b] |
| $Re_{\delta_2} \times 10^{-4}$ | 0.9[a] | 1.1[b] | 1.3[b] |

[a] Measured at location no. 1.
[b] Measured at location no. 2.

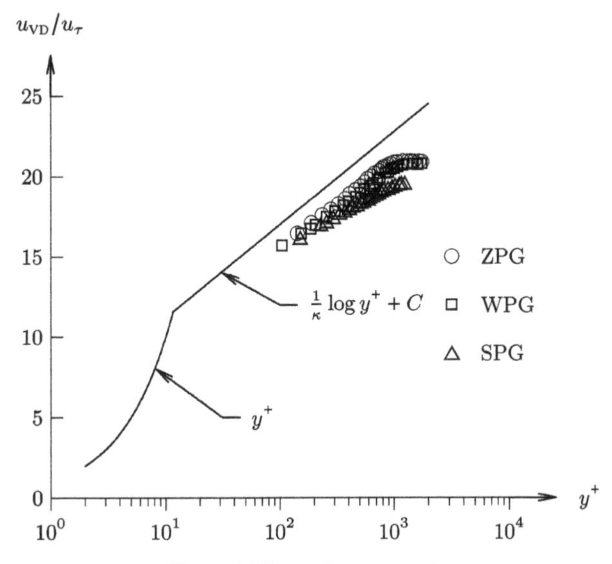


**Figure 4.170.** $u_{VD}/u_\tau$ versus $y^+$.

[9] $Re_\theta = \rho_e U_e \theta/\mu_e$ and $Re_{\delta_2} = \rho_e U_e \theta/\mu_w$ where $\theta$ is the incompressible displacement thickness.

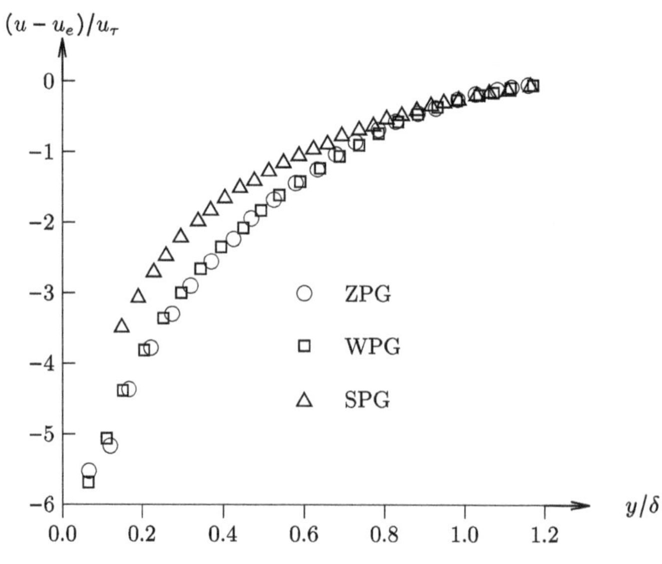


**Figure 4.171.** Mean velocity wake defect.

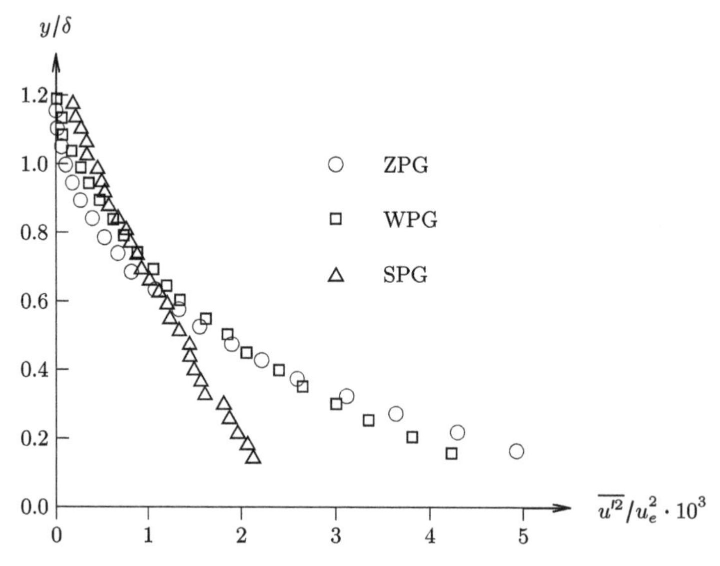


**Figure 4.172.** Axial component of Reynolds stress.

Figures 4.172, 4.173, and 4.174 present the axial, transverse, and shear components of the Reynolds stress for the three cases. The strong pressure gradient significantly reduces the axial Reynolds stress while increasing the transverse Reynolds stress. The Reynolds shear stress is nearly eliminated by the strong

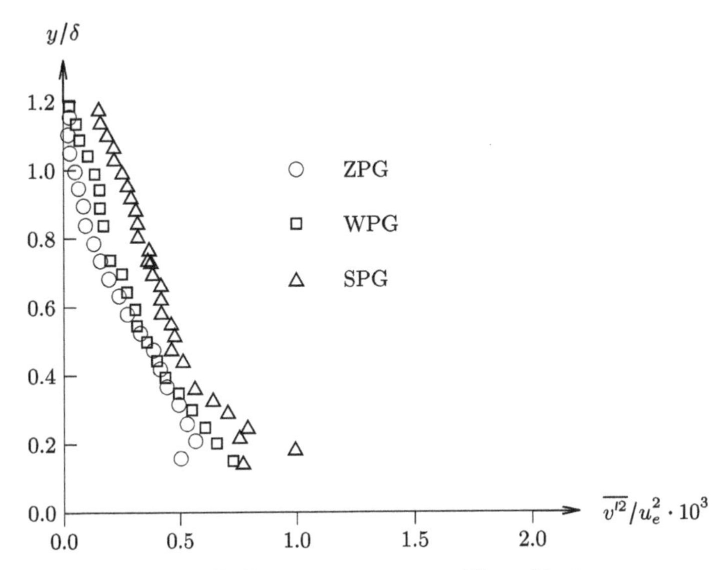


**Figure 4.173.** Transverse component of Reynolds stress.

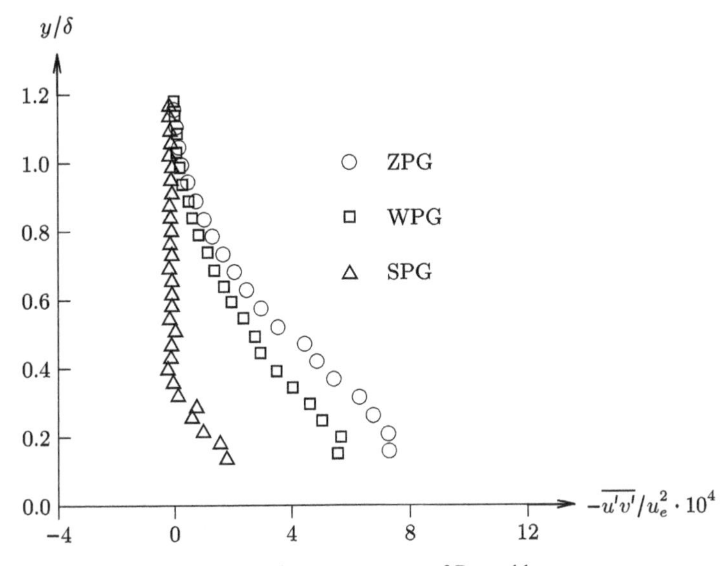


**Figure 4.174.** Shear component of Reynolds stress.

pressure gradient except for $y/\delta < 0.3$. The streamwise autocorrelation[10] is shown in figure 4.175. The strong pressure gradient causes a significant reduction in $R_{u'u'}(\Delta x)$

[10] The streamwise autocorrelation function is defined as

$$R_{u'u'}(\Delta x;\, x,\, y) = \frac{\overline{u'(x,\, y)u'(x + \Delta x,\, y)}}{\overline{u'(x,\, y)u'(x,\, y)}}.$$

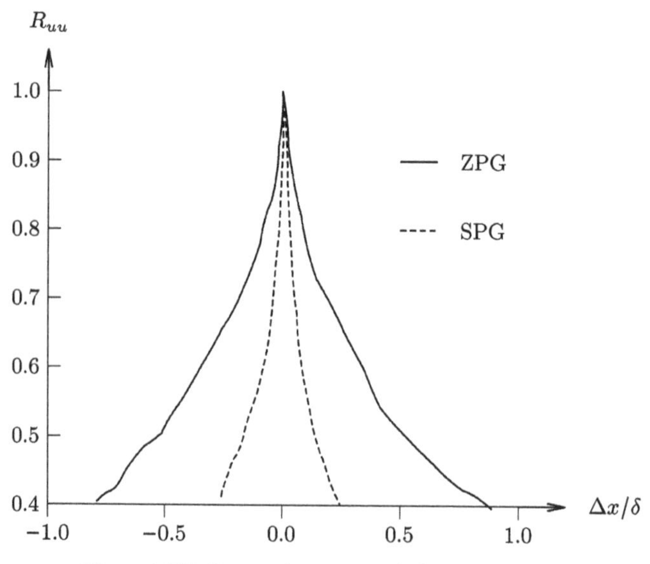


**Figure 4.175.** Streamwise autocorrelation at $y/\delta = 0.5$.

indicating that the large turbulent structures are tilted away from the wall by the favorable pressure gradient.

## 4.55 Neeb *et al* (2015)

Neeb *et al* (2015) investigated the rough wall hypersonic turbulent boundary layer on a 7° half-angle cone at Mach 6.1. The model is illustrated in figure 4.176 and comprised three segments. The first segment incorporated a sharp (0.1 mm radius) metallic nose. The second segment and third segments were fabricated from polyether ether ketone to enable surface temperature and heat transfer measurements. The third segment was roughened using a square bar pattern (figure 4.177) with a wavelength $\lambda$ to width $w$ ratio equal to four, and the top of each square bar aligned with the smooth segments. The bar height $k$ and width $w$ are 0.5 mm. The total model length L = 730 mm with an end diameter $D = 180$ mm. The roughness pattern started at $x/L = 0.44$ and terminated at $x/L = 0.89$. The experiments were performed in the German Aerospace Center hypersonic wind tunnel H2K (section A.11.1). The nominal flow conditions are listed in table 4.65 and indicate both the freestream and boundary layer edge conditions, with the latter obtained from solution of the Taylor–Maccoll equation (Anderson 1990). Experimental diagnostics include surface heat transfer and temperature and PIV.

A total of sixteen runs were completed. Figure 4.178 shows the Van Driest transformed velocity $u_{\text{VD}}/u_\tau$ versus $y^+$ for runs 7 and 8 (smooth wall) and 12 and 14 (rough wall). The smooth wall boundary layer parameters are listed in table 4.66. The profiles satisfy the Law of the Wall and Wake with $\Pi = 0.40$ to 0.41. The rough wall boundary layer parameters are listed in table 4.67 including the offset $H(k^+)$ defined in (3.60) and the equivalent sand grain roughness $k_s$. The profiles indicate a region satisfying the Law of the Wall.

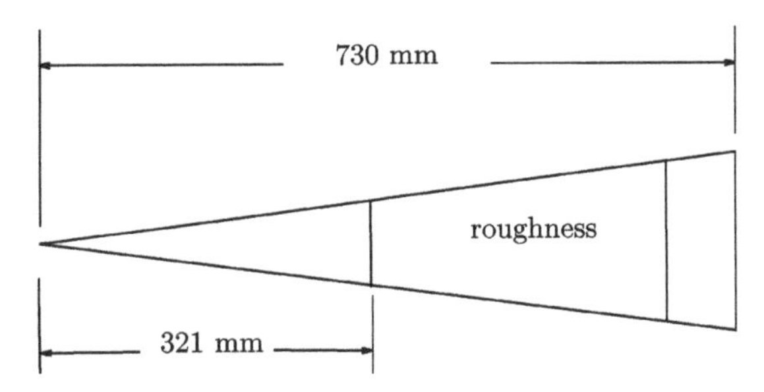


**Figure 4.176.** Model.

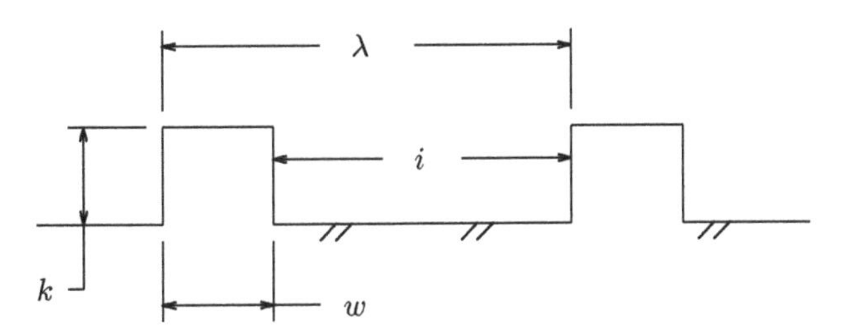


**Figure 4.177.** Roughness pattern.

**Table 4.65.** Freestream conditions.

| Quantity | Freestream | Boundary layer edge |
|---|---|---|
| $M$ | 6.1 | 5.4 |
| $p_{t_\infty}$ (MPa) | 2.03 | 2.00 |
| $T_{t_\infty}$ (K) | 500 | 500 |
| $Re_L \times 10^{-6}$ | 11.5 | 15.1 |

## 4.56 Peltier *et al* (2016)

Peltier *et al* (2016) examined the effect of surface roughness on a flat plate hypersonic turbulent boundary layer in air at Mach 4.9. The experiments were conducted in the high-speed blowdown wind tunnel at the National Aerothermodynamics Laboratory at Texas A&M University (see section A.35). The surface roughness comprised a cross-hatched diamond pattern (figure 4.164). The flow conditions are listed in table 4.68. Experimental diagnostics include PIV and schlieren imaging.

Figure 4.179 displays the mean velocity profile $u/u_e$ versus $y/\delta$ for the rough and smooth walls. The retardation of the velocity profile near the wall due to the surface roughness is evident. Figure 4.180 shows the Van Driest transformed velocity $u_{\text{VD}}/u_\tau$ versus $y^+$ for both cases. The profiles satisfy the Law of the Wall. The vertical offset due to the roughness effect $H(k^+)$ is evident. Small variations that can be seen in the

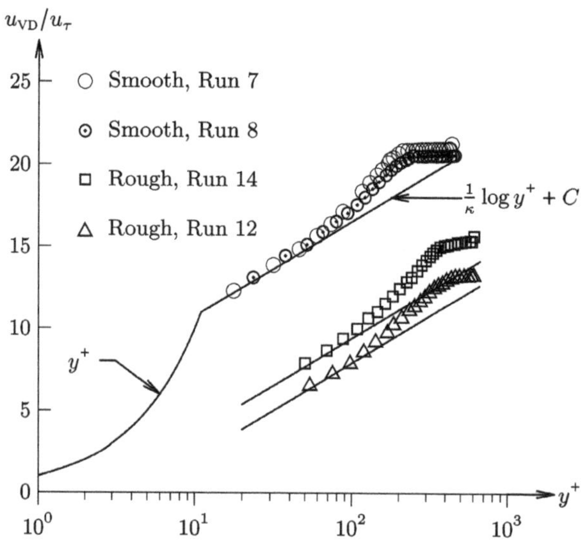


**Figure 4.178.** $u_{VD}/u_\tau$ versus $y^+$.

**Table 4.66.** Smooth wall boundary layer.

| Run | $u_e$ | $u_\tau$ | $c_f \times 10^3$ | $\delta$ | $\Pi$ | $T_w$ |
|---|---|---|---|---|---|---|
| | (m s$^{-1}$) | (m s$^{-1}$) | | (mm) | | (K) |
| 7 | 918 | 52.9 | 1.37 | 3.58 | 0.44 | 340 |
| 8 | 922 | 52.2 | 1.31 | 3.32 | 0.40 | 340 |

**Table 4.67.** Rough wall boundary layer.

| Run | $u_\tau$ | $\Pi$ | $H(k^+)$ | $k^+$ | $k_s^+$ | $k_s$ | $k_s/k$ |
|---|---|---|---|---|---|---|---|
| 12 | 82.1 | 0.41 | 8.6 | 45 | 127 | 1.4 | 2.8 |
| 14 | 72.7 | 0.43 | 7.1 | 40 | 69 | 0.86 | 1.7 |

log region for the rough wall depend on the relative streamwise location with respect to the roughness pattern. Figure 4.181 shows the normalized kinematic Reynolds shear stress $-\overline{u'v'}/u_\infty^2$ versus $x/\delta$ at six fixed elevations $y/\delta$. The variation in $-\overline{u'v'}/u_\infty^2$ with $x$ is largest near the wall and diminishes with distance above the wall.

## 4.57 Tichenor *et al* (2017)

Tichenor *et al* (2017) investigated the effect of streamline curvature–induced adverse pressure gradients on the structure of a Mach 4.9 turbulent boundary layer. The

**Table 4.68.** Freestream conditions.

| Quantity | Rough wall | Smooth wall |
|---|---|---|
| $M_\infty$ | 4.9 | 4.9 |
| $p_{t_\infty}$ (MPa) | 2.34 | 2.34 |
| $T_{t_\infty}$ (K) | 365 | 365 |
| $T_w$ (K) | 310 | 310 |
| $T_w/T_{aw}$ | 0.93[a] | 0.93[a] |
| $H_\infty$ (MJ kg$^{-1}$) | 0.37 | 0.37 |
| $Re \times 10^{-6}$ (m$^{-1}$) | 46 | 46 |
| $\delta$ (mm) | 11.3 | 10.9 |
| $\delta^*$ (mm) | 2.2 | 1.3 |
| $\theta$ (mm) | 1.4 | 1.0 |
| $u_\tau$ (m s$^{-1}$) | 69 | 35 |

[a] Using 3.46.

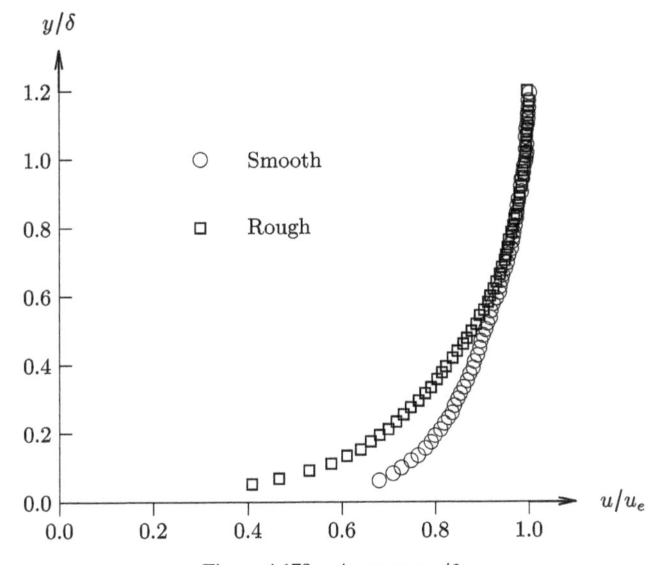


**Figure 4.179.** $u/u_e$ versus $y/\delta$.

experiments were conducted at the National Aerothermodynamics Laboratory blowdown wind tunnel at Texas A&M University (section A.35). The model is shown in figure 4.182. An equilibrium turbulent boundary develops on the flat plate. An adverse pressure gradient is generated by an 8° arc with radius of 50.8 cm beginning at a distance of 27.3 cm from the leading edge of the flat plate. An inverted arc returns the surface to horizontal. The flow conditions are listed in table 4.69. Experimental data were obtained at one station upstream of the adverse pressure gradient and two different stations within the region of streamline curvature. The boundary layer thickness $\delta$, edge streamwise velocity $u_e$, and pressure

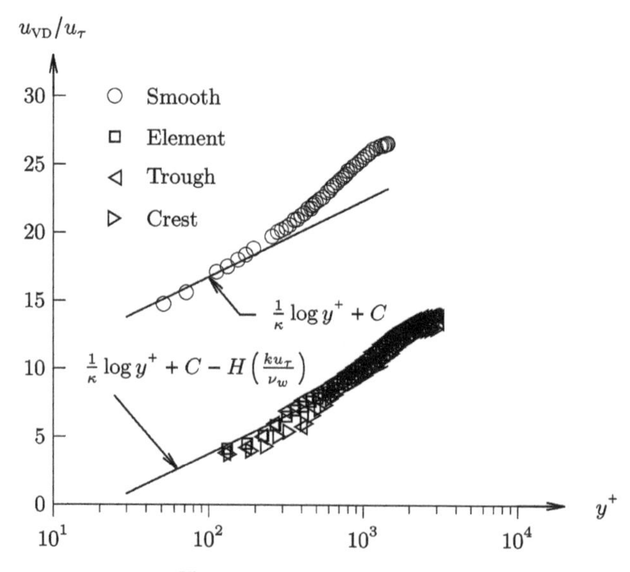


**Figure 4.180.** $u_{VD}/u_\tau$ versus $y^+$.

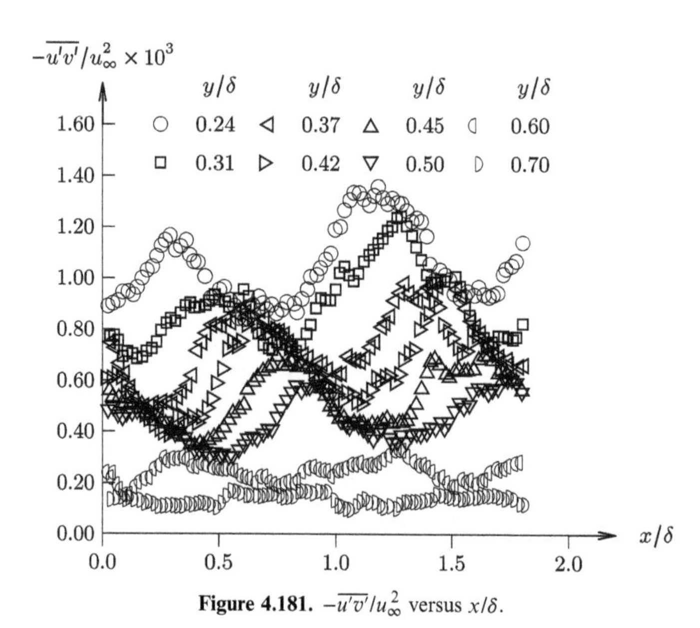


**Figure 4.181.** $-\overline{u'v'}/u_\infty^2$ versus $x/\delta$.



**Figure 4.182.** Model.

**Table 4.69.** Freestream conditions.

| Quantity | Value |
|---|---|
| $M_\infty$ | 4.9 |
| $T_{t_\infty}$ (K) | 364 |
| $p_{t_\infty}$ (MPa) | 2.0 |
| $H_\infty$ (MJ kg$^{-1}$) | 0.37 |
| $Re \times 10^{-6}$ (m$^{-1}$) | 45 |

**Table 4.70.** Boundary layer parameters.

| Quantity | Upstream | Downstream 1 | Downstream 2 |
|---|---|---|---|
| $x$-location (cm) | 15.4 | 29.2 | 30.8 |
| $\delta$ (mm) | 9.7 | 11.4 | 10.7 |
| $u_e$ (m s$^{-1}$) | 780 | 779 | 775 |
| $\beta$ | 0.09 | 0.85 | 1.05 |

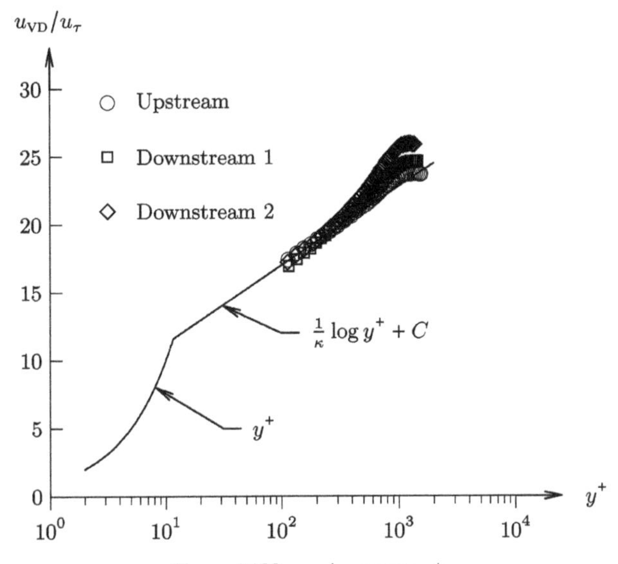


**Figure 4.183.** $u_{VD}/u_\tau$ versus $y^+$.

gradient parameter $\beta$ at each location are shown in table 4.70 where $\beta = (dp/dx)\delta^*/\tau_w$ and $\delta^*$ is the incompressible displacement thickness (Tichenor *et al* 2013). Experimental diagnostics include PIV.

Figures 4.183 and 4.184 display the Van Driest transformed velocity $u_{VD}/u_\tau$ and velocity wake defect, respectively, at the three locations. Agreement with the Law of the Wall is evident.

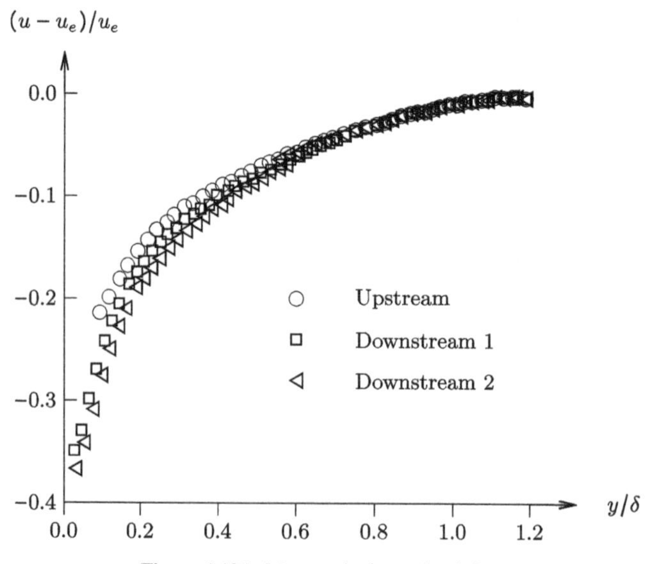


**Figure 4.184.** Mean velocity wake defect.

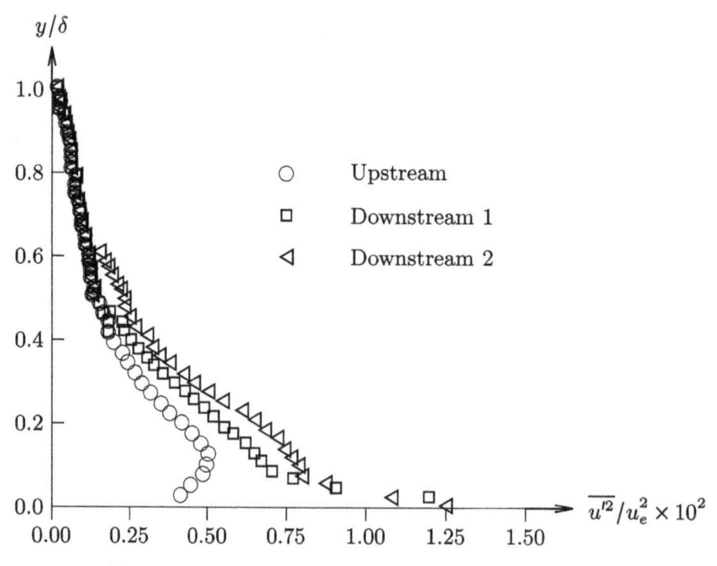


**Figure 4.185.** Streamwise component of Reynolds stress.

Figures 4.185, 4.186, and 4.187 display the Reynolds kinematic streamwise, wall normal, and shear stress at locations upstream and within the adverse pressure gradient. The streamwise and normal turbulent stresses increase with adverse pressure gradient. The Reynolds shear stress shows a 90% increase in peak absolute magnitude.

Figure 4.188 shows the streamwise velocity autocorrelation $R_{uu}$ versus $\Delta x/\delta$ at the upstream and downstream 1 locations for $y/\delta = 0.2$. The autocorrelations are

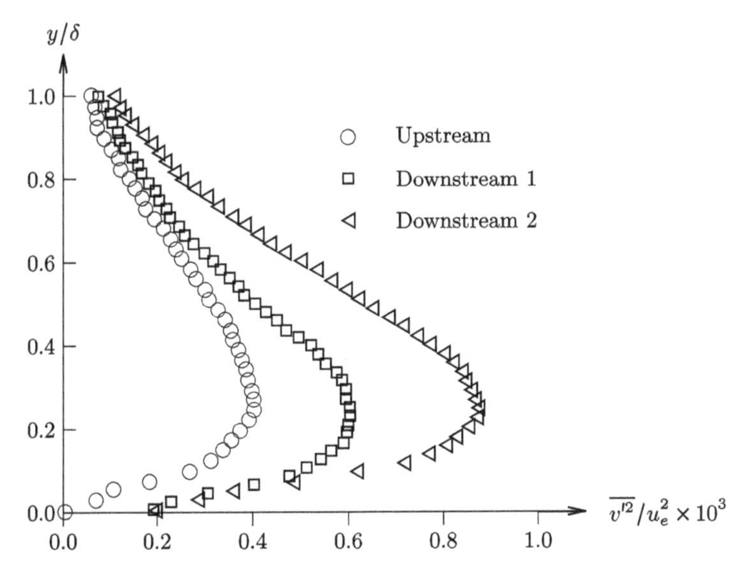


**Figure 4.186.** Transverse component of Reynolds stress.

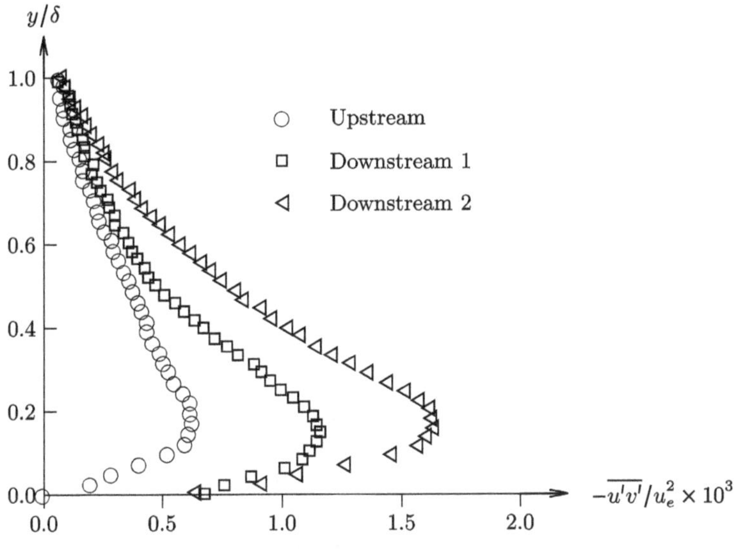


**Figure 4.187.** Shear component of Reynolds stress.

virtually identical, in contrast to the significant differences in $R_{uu}$ between the zero pressure gradient and adverse pressure gradient locations in Tichenor *et al* (2013).

## 4.58 Williams and Smits (2017)

Williams and Smits (2017) conducted an experimental investigation of the effect of the height of cylindrical tripping devices on the structure of a zero pressure gradient hypersonic turbulent boundary layer in air at Mach 7.6. The experiments were

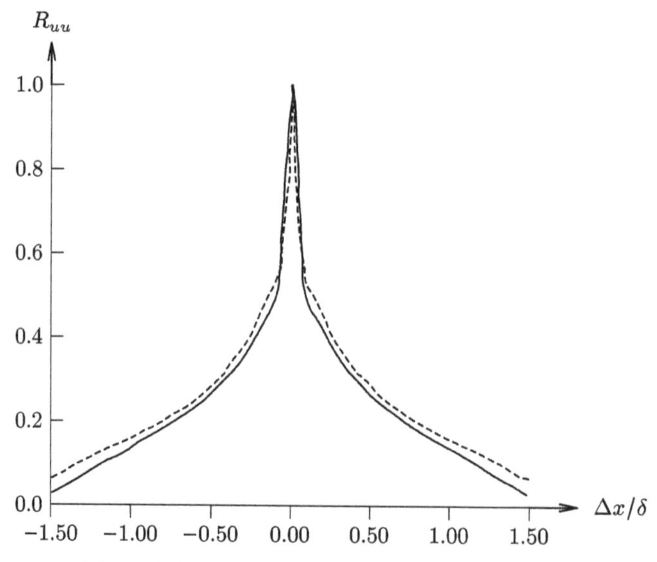


**Figure 4.188.** Streamwise autocorrelation at $y/\delta = 0.2$.

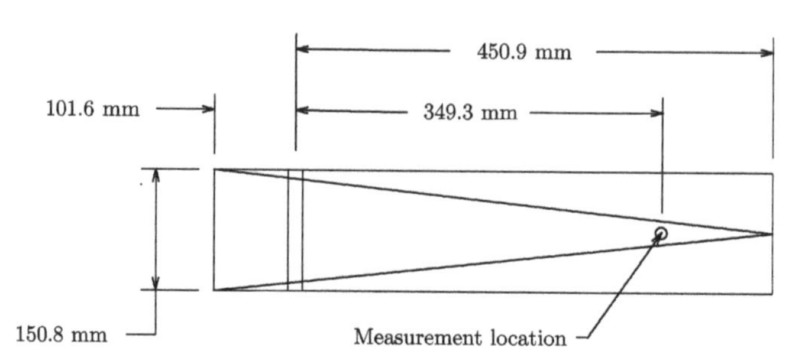


**Figure 4.189.** Model.

performed at the Mach 8 hypersonic boundary layer facility at the Princeton Gas Dynamics Laboratory (section A.28.1). The model is shown in figure 4.189. A spanwise row of cylinders (figure 4.190) was fixed at a distance of 101.6 mm from the leading edge of the flat plate. The cylinder diameter of 3.175 mm and spacing of 9.53 mm were fixed, and the height was varied from 1.5 mm to 4.25 mm. The flow conditions are shown in table 4.71. The tripping devices details are listed in table 4.72 where $k$ is the cylinder height, $k^+ = k\, u_\tau/\nu_w$ where $u_\tau = 41.7$ m s$^{-1}$ at the tripping location, and $\delta^*$ is the displacement thickness. Experimental diagnostic is PIV.

Figure 4.191 displays the Van Driest transformed velocity for cases 1, 3, and 8 at a distance of 349.3 mm from the tripping location. Close agreement with the compressible Law of the Wall 3.55 is evident for all cases. A significant wake component is evident for all cases. The wake strength $\Delta(u_{VD}/u_\tau)$ is shown in figure 4.192 together with the correlation of Coles (1962). The value of $\Delta(u_{VD}/u_\tau)$ for case 1 is anomalously high which suggests that the mean velocity profile may be

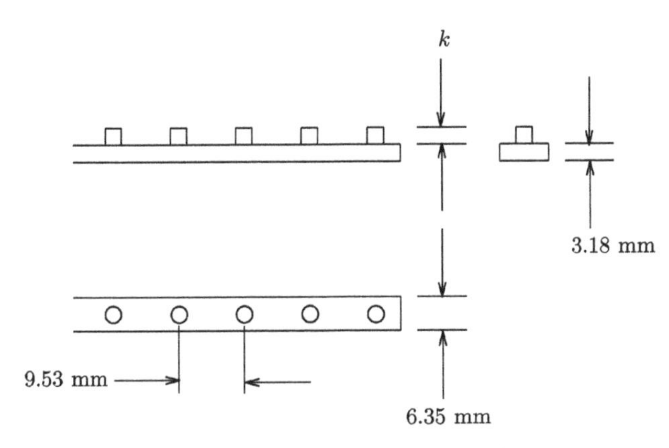


**Figure 4.190.** Cylindrical tripping device.

**Table 4.71.** Freestream conditions.

| Quantity | Value |
|---|---|
| $M_e$ | 7.6 |
| $p_{t_e}$ (MPa) | 9.62 |
| $T_{t_e}$ (K) | 610 |
| $T_w$ (K) | 439 |
| $T_w/T_{aw}$ | 0.80 |

**Table 4.72.** Details of trips.

| | Value | | | | | | | |
|---|---|---|---|---|---|---|---|---|
| Quantity | Case 1 | Case 2 | Case 3 | Case 4 | Case 5 | Case 6 | Case 7 | Case 8 |
| $k^+$ | 27.1 | 42.2 | 52.1 | 51.6 | 50.7 | 60.2 | 67.9 | 79.9 |
| $k/\delta^*$ | 1.22 | 1.87 | 2.30 | 2.39 | 2.67 | 2.69 | 3.06 | 3.53 |
| $Re_\theta \times 10^{-3}$ | 9.76 | 9.11 | 9.34 | 9.28 | 9.06 | 10.2 | 10.7 | 11.4 |

transitional (Williams and Smits 2017). The optimal range of trip heights is $1.9 \leqslant k/\delta^* \leqslant 2.7$ in terms of agreement with the correlation of Coles (1962).

Figure 4.193 presents the Morkovin-scaled root-mean-square streamwise turbulence fluctuations $(\overline{\rho u'u'}/\rho_w u_\tau^2)^{\frac{1}{2}}$ for all cases, together with the direct numerical simulation (DNS) data of Priebe and Martín (2011) and the incompressible experimental data of DeGraaff and Eaton (2000). With the exception of case 1, the experimental data are closely grouped and in close agreement with Priebe and Martín (2011) and DeGraaff and Eaton (2000) for $y/\delta > 0.4$ and 0.2, respectively.

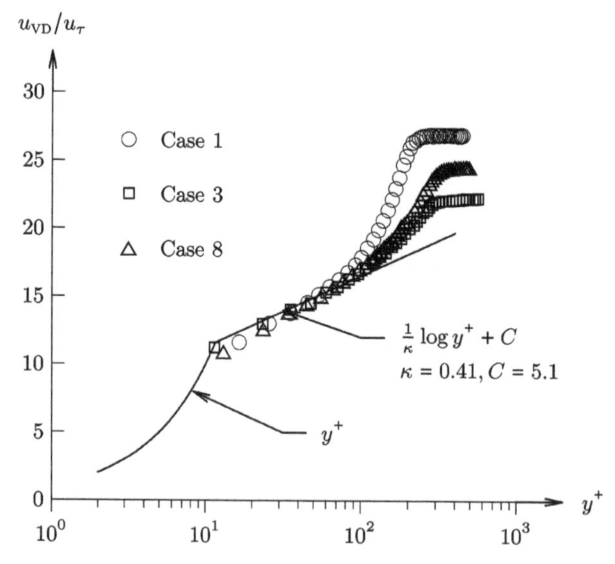


**Figure 4.191.** $u_{VD}/u_\tau$ versus $y^+$ (every other data point shown).

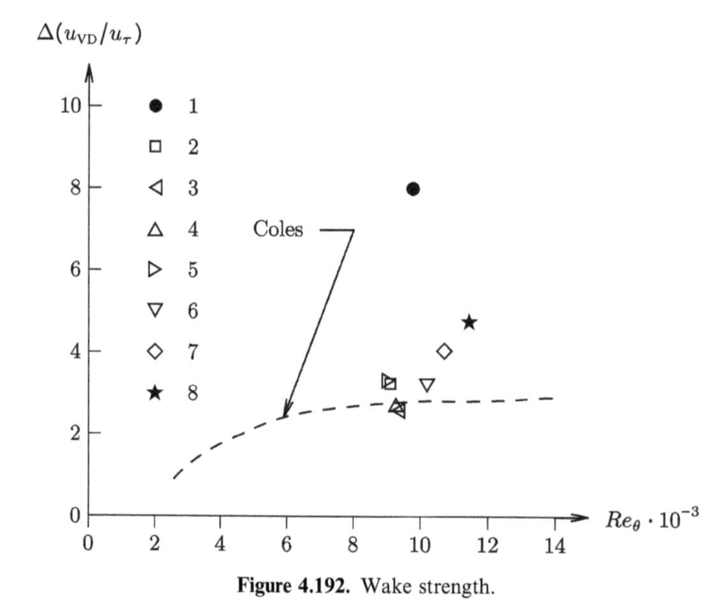


**Figure 4.192.** Wake strength.

## 4.59 Neeb *et al* (2018)

Neeb *et al* (2018) performed an experimental study of a turbulent boundary later on a 7° sharp cone at $M_\infty = 6$ ($M_e = 5.4$) in air at the German Aerospace Center hypersonic wind tunnel H2K (section A.11.1). The test conditions are indicated in table 4.73. Experimental measurements include pitot pressure, wall pressure, and PIV.

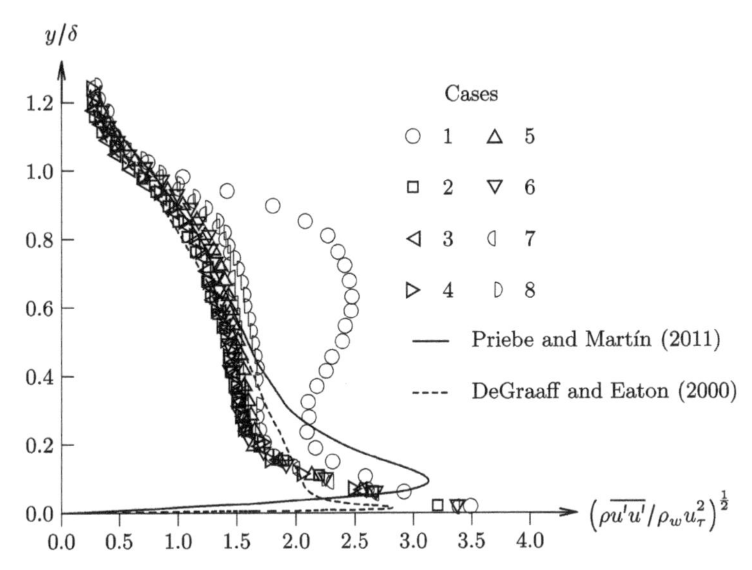


**Figure 4.193.** Morkovin-scaled root-mean-square streamwise turbulence fluctuations (authors showed every other experimental data point shown for clarity).

**Table 4.73.** Freestream conditions.

| | Value | |
|---|---|---|
| Quantity | Case 1[a] | Case 2[b] |
| $M_e$ | 5.4 | 5.4 |
| $p_{t_e}$ (MPa) | 2.01 | 1.89 |
| $T_{t_e}$ (K) | 500 | 475 |
| $T_w/T_{aw}$ | 0.79–0.80 | 0.79 |
| $H_e$ (MJ kg$^{-1}$) | 0.50 | 0.48 |
| $\delta$ (mm) | 3.4–3.7 | 0.151–0.154 |
| $\theta$ (mm) | 3.9 | 0.152 –0.154 |
| $Re \times 10^{-6}$ (m$^{-1}$) | 20.5 | 21.0 |

[a] Denoted as FC1 in paper.
[b] Denoted as FC2 in paper.

Figure 4.194 displays the Van Driest transformed velocity $u_{VD}/u_\tau$ for the PIV measurements. Excellent agreement is observed with the Law of the Wall (3.54). Figure 4.195 shows the outer velocity obtained from (3.61) and similar close agreement is obtained.

Figures 4.196 and 4.197 show the density weighted streamwise and wall normal turbulence intensities $(\overline{\rho u'u'}/\rho_w u_\tau^2)^{\frac{1}{2}}$ and $(\overline{\rho v'v'}/\rho_w u_\tau^2)^{\frac{1}{2}}$ together with experimental data of Ekoto *et al* (2007), Sahoo *et al* (2009), and Williams and Smits (2017). The experimental data do not appear to scale in accordance with Morkovin's hypothesis.

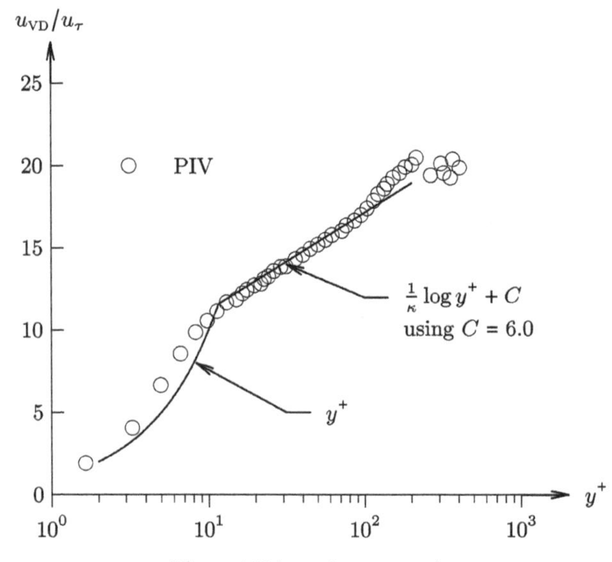


**Figure 4.194.** $u_{VD}/u_\tau$ versus $y^+$.

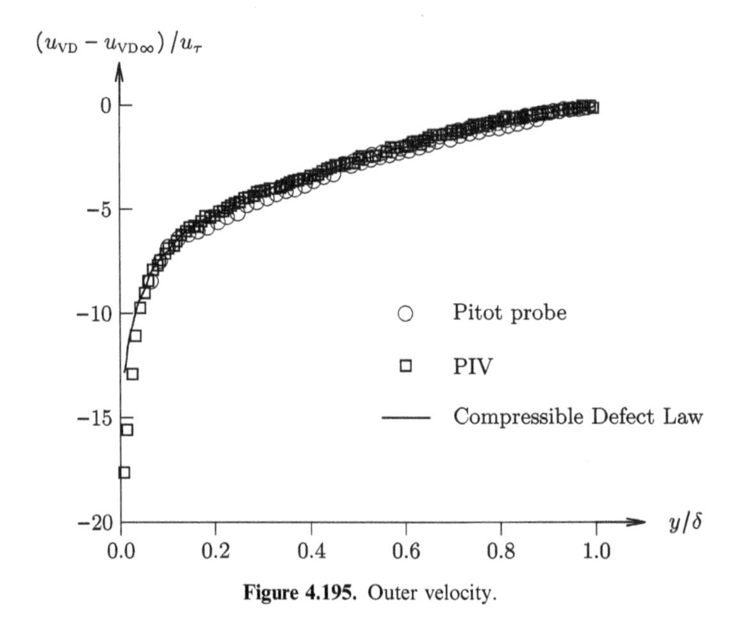


**Figure 4.195.** Outer velocity.

Neeb *et al* (2018) suggest the difference may be attributable to a combination of factors including freestream turbulence and particle lag in the PIV measurements.

## 4.60 Williams *et al* (2018)

Williams *et al* (2018) studied a hypersonic turbulent boundary layer in air on a flat plate at $M_\infty = 7.5$ in the Mach 8 hypersonic boundary layer facility at the Princeton Gas Dynamics Laboratory (section A.28.1). The freestream conditions are shown in

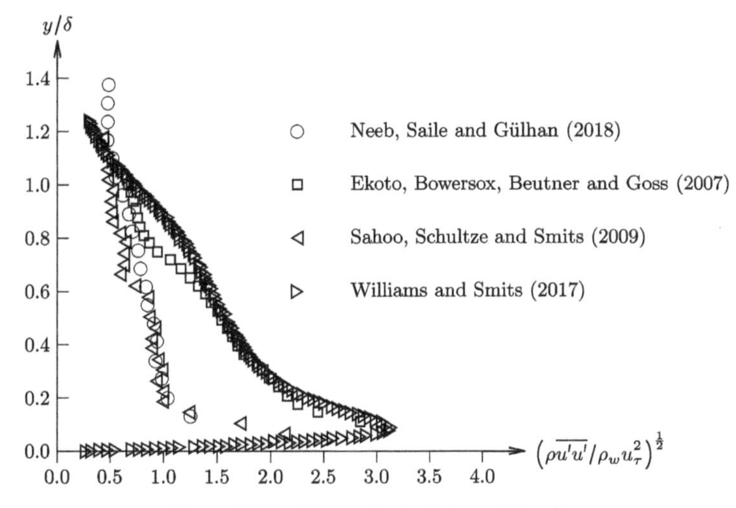


**Figure 4.196.** Streamwise turbulence intensity.

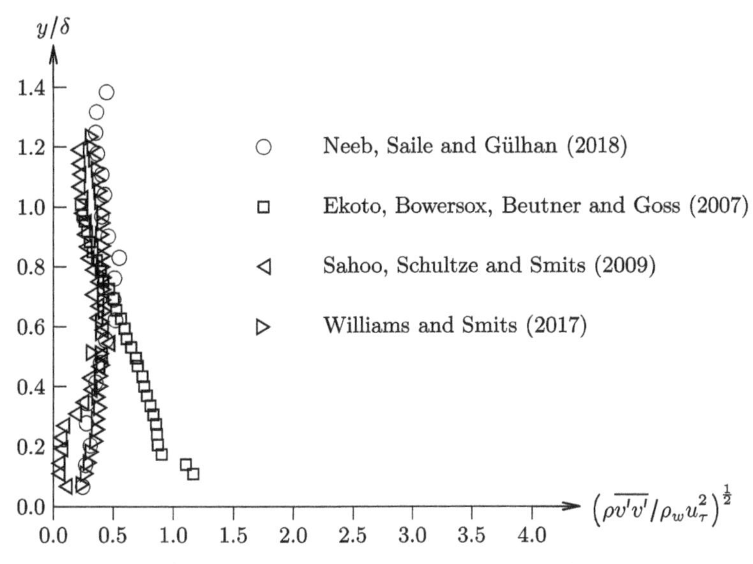


**Figure 4.197.** Wall normal turbulence intensity.

table 4.74. Experimental diagnostics included FRS and PIV. Earlier FRS measurements are presented in Baumgartner (1997).

Figure 4.198 shows the Van Driest transformed velocity $u_{\mathrm{VD}}/u_\tau$ for cases 2 to 4. Excellent agreement is observed between experiment and the Law of the Wall with $\kappa = 0.4$ and $C = 5.1$.

Figure 4.199 displays the density weighted streamwise turbulence intensity $(\overline{\rho u' u'}/\rho_{\mathrm{w}} u_\tau^2)^{\frac{1}{2}}$ in outer layer coordinates $y/\delta$ together with the DNS results of

**Table 4.74.** Freestream conditions.

| Quantity | Value | | | |
|---|---|---|---|---|
| | Case 1 | Case 2 | Case 3 | Case 4 |
| $M_e$ | 7.45 | 7.25 | 7.44 | 7.60 |
| $p_{t_e}$ (MPa) | 7.00 | 7.22 | 7.15 | 9.60 |
| $T_{t_e}$ (K) | 759 | 756 | 740 | 606 |
| $T_w$ (K) | 561 | 562 | 552 | 430 |
| $T_w/T_{aw}$ | 0.82 | 0.83 | 0.82 | 0.79 |
| $\delta$ (mm) | 11.5 | 9.50 | 12.0 | 10.2 |
| $\theta$ (mm) | 0.20 | 0.25 | 0.35 | 0.27 |
| $Re \times 10^{-6}$ (m$^{-1}$) | 17.8 | 19.8 | 18.7 | 34.0 |

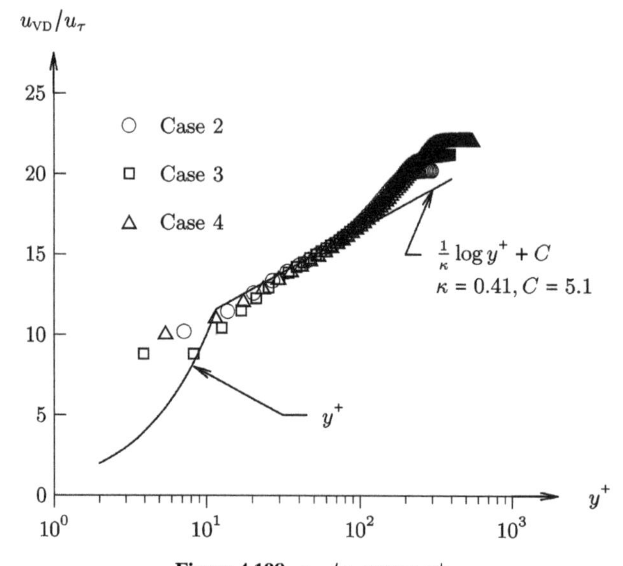


**Figure 4.198.** $u_{VD}/u_\tau$ versus $y^+$.

Priebe and Martín (2011) ($M_\infty = 7.2$, $Re_\theta = 3300$, $T_w/T_{aw} = 0.53$) and the experiments of DeGraaff and Eaton (2000) ($M_\infty = 0$, $Re_\theta = 5200$). All results are in close agreement for $y/\delta \geqslant 0.4$, providing strong support for the validity of Morkovin's density scaling hypothesis in the outer layer. Figure 4.200 shows the same results in inner layer coordinates $y^+ = yu_\tau/\nu_w$. The present experiments and DNS of Priebe and Martín (2011) are in close agreement although the incompressible data show a distinct difference for $y^+ < 50$. Huang *et al* (1995) suggested using a semilocal scaling defined as

$$y^* = y^+ \frac{\mu_w}{\mu}\sqrt{\frac{\rho}{\rho_w}} \tag{4.20}$$

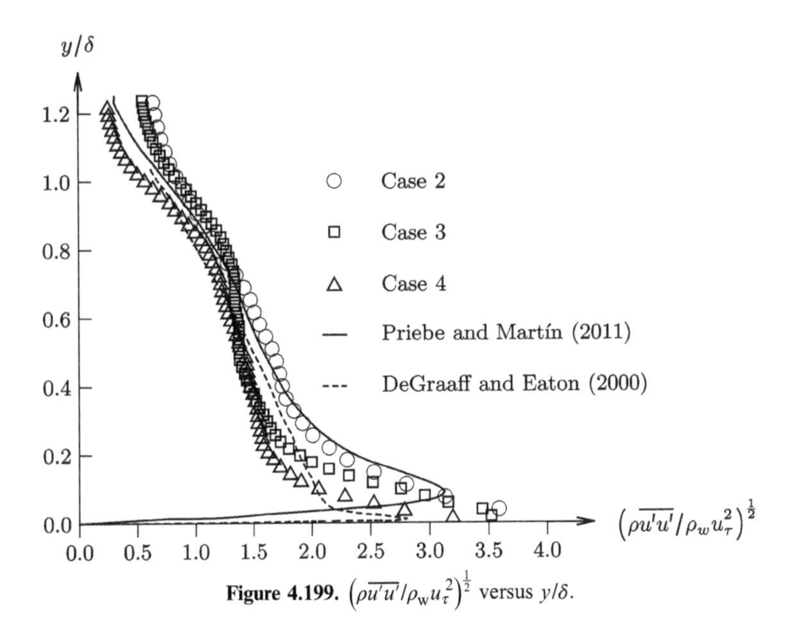


**Figure 4.199.** $\left(\rho\overline{u'u'}/\rho_{w}u_{\tau}^{2}\right)^{\frac{1}{2}}$ versus $y/\delta$.

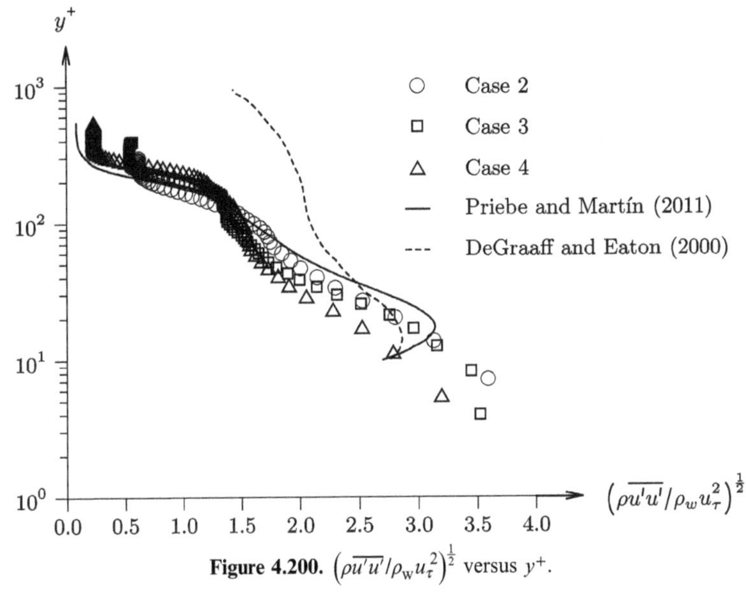


**Figure 4.200.** $\left(\rho\overline{u'u'}/\rho_{w}u_{\tau}^{2}\right)^{\frac{1}{2}}$ versus $y^{+}$.

Assuming $\mu \propto T^{\omega}$ where $\omega$ is a constant ($\approx$0.73 in air) yields

$$y^{*}=y^{+}\left(\frac{T_{w}}{T}\right)^{\omega+\frac{1}{2}} \tag{4.21}$$

Figure 4.201 shows closer agreement of all results for $y^{*} \leqslant 1000$ (note change in scale of ordinate compared to figure 4.200).

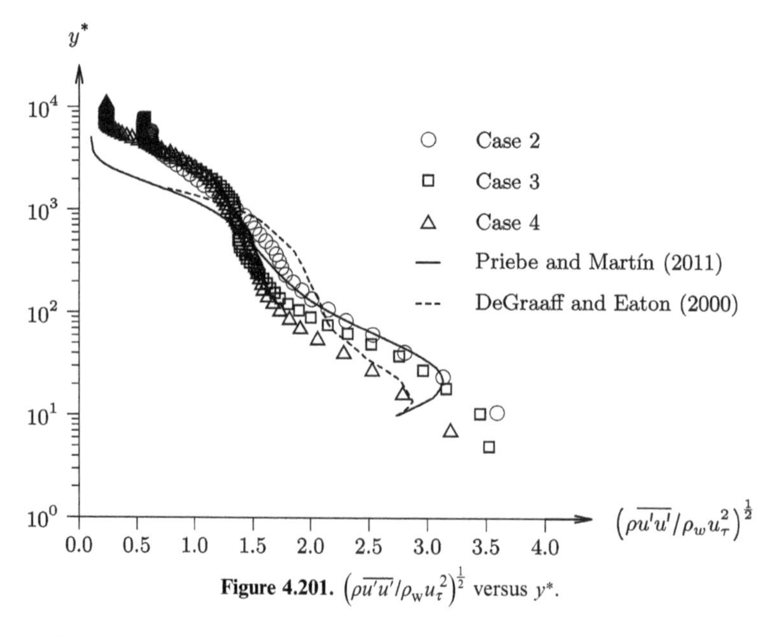


**Figure 4.201.** $\left(\rho\overline{u'u'}/\rho_{\mathrm{w}} u_\tau^2\right)^{\frac{1}{2}}$ versus $y^*$.

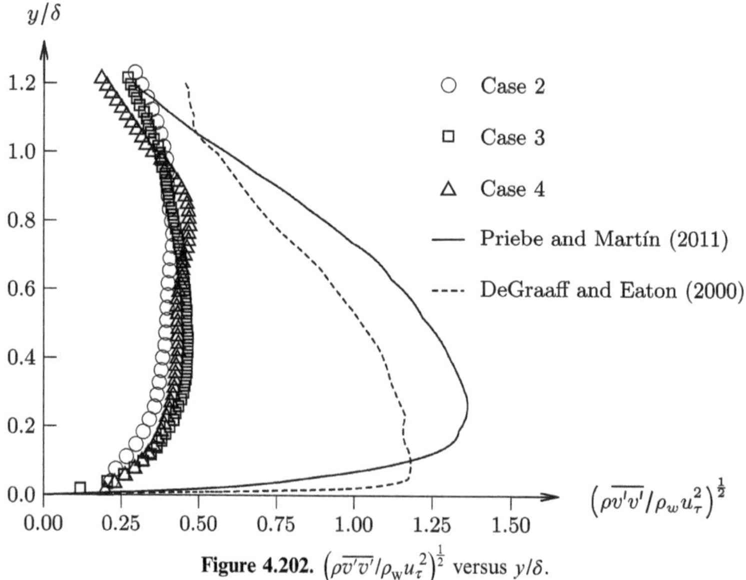


**Figure 4.202.** $\left(\rho\overline{v'v'}/\rho_{\mathrm{w}} u_\tau^2\right)^{\frac{1}{2}}$ versus $y/\delta$.

Figure 4.202 displays the density weighted wall normal turbulence intensity $(\rho\overline{v'v'}/\rho_{\mathrm{w}} u_\tau^2)^{\frac{1}{2}}$ together with the DNS results of Priebe and Martín (2011) and experimental data of DeGraaff and Eaton (2000). The results for cases 2 through 4 are in close agreement but approximately 60% below the DNS data of Priebe and Martín (2011) and experiment of DeGraaff and Eaton (2000). Williams *et al* (2018) note that Tichenor *et al* (2013) and Peltier *et al* (2016) also observed a reduction in the wall normal turbulence intensity of approximately 30% at Mach 5 compared to DNS at the same Mach number. Williams *et al* (2018) indicate that the limitations of

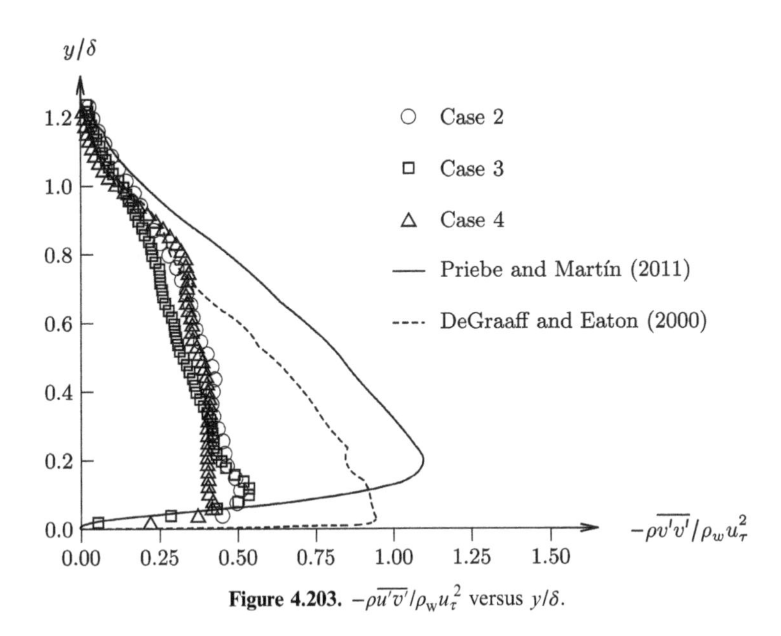


**Figure 4.203.** $-\rho \overline{u'v'}/\rho_{\mathrm{w}} u_\tau^2$ versus $y/\delta$.

the response of the particles used in the PIV measurements to the local flow fluctuations may be more significant for $v'$ than for $u'$. Consequently, Morkovin's hypothesis for the scaling of the wall normal fluctuations cannot be evaluated for the experiment of Williams *et al* (2018). A similar conclusion holds for the density weighted Reynolds shear stress $-\rho \overline{u'v'}/\rho_{\mathrm{w}} u_\tau^2$ shown in figure 4.203.

## 4.61 Ding *et al* (2020)

Ding *et al* (2020) investigated the structure of a zero pressure gradient hypersonic turbulent boundary layer in air at Mach 6. The experiments were conducted in the hypersonic wind tunnel at the National University of Defense Technology, Changsha, People's Republic of China (section A.27.3). The measurements were obtained on the tunnel wall using planar laser scattering with nanoscale $TiO_2$ particles. The freestream conditions[11] are listed in table 4.75.

Figure 4.204 shows the Van Driest transformed velocity $u_{\mathrm{VD}}/u_\tau$ versus $y^+$ together with the Law of the Wall using $C = 5.5$. The experimental velocity profile is slightly above the Law of the Wall.

Figures 4.205 and 4.206 display the normalized root-mean-square streamwise and normal velocity fluctuations $\sqrt{\overline{u'^2}}/u_\tau$ and $\sqrt{\overline{v'^2}}/u_\tau$ versus $y/\delta$. Also included are the DNS results of Duan *et al* (2016) at Mach 5.86 and Priebe and Martín (2011) [12] at

[11] Ding *et al* (2020) indicate $T_{\mathrm{w}}/T_{\mathrm{r}} = 0.89$ where $T_{\mathrm{r}}$ is presumably the recovery temperature (adiabatic wall temperature). However, the results in tables 1 and 2 of Ding *et al* (2020) imply that $T_{\mathrm{r}} = 367.3$ K which is identical to the freestream stagnation temperature $T_{\mathrm{o}}$. The value of $T_{\mathrm{w}}/T_{\mathrm{aw}}$ in table 4.75 is calculated from equation (3.46) assuming $Pr_{\mathrm{t}} = 0.89$.

[12] The data attributed to Priebe and Martín (2011) are taken from Ding *et al* (2020). The data do not appear in Priebe and Martín (2011).

**Table 4.75.** Freestream conditions.

| Quantity | Value |
|---|---|
| $M_e$ | 6.0 |
| $p_{t_e}$ (MPa) | 1.1 |
| $T_{t_e}$ (K) | 367.3 |
| $T_w$ (K) | 326.9 |
| $T_w/T_{aw}$ | 1.0 |
| $H_e$ (MJ kg$^{-1}$) | 0.37 |
| $\delta$ (cm) | 3.21 |
| $Re \times 10^{-6}$ (m$^{-1}$) | 15.5 |

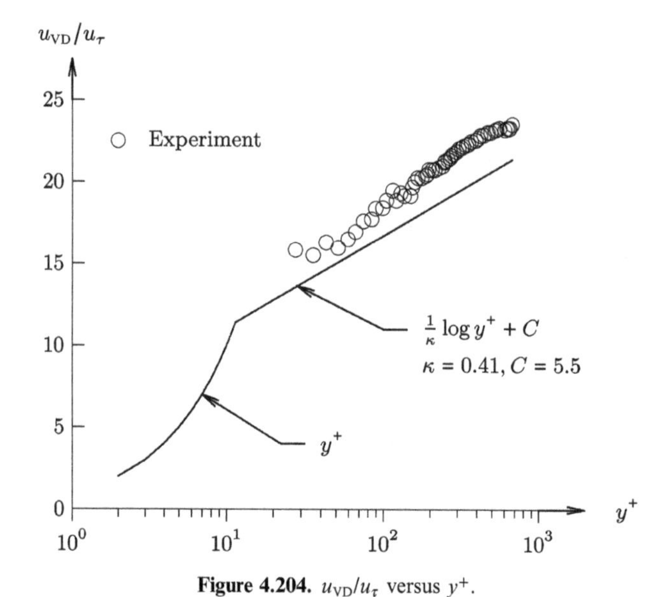


**Figure 4.204.** $u_{VD}/u_\tau$ versus $y^+$.

Mach 7.2. The comparison for $\sqrt{\overline{u'^2}}/u_\tau$ at $y/\delta > 0.4$ among all three results is excellent, and there are modest differences for $y/\delta < 0.4$. The experimental results for $\sqrt{\overline{v'^2}}/u_\tau$ are significantly lower than the DNS of Duan *et al* (2016) across the entire boundary layer.

Figures 4.207 and 4.208 display the Morkovin-scaled experimental streamwise and normal root-mean-square velocity fluctuations $\sqrt{\rho/\rho_w}\sqrt{\overline{u'^2}}/u_\tau$ and $\sqrt{\rho/\rho_w}\sqrt{\overline{v'^2}}/u_\tau$, respectively. Also shown are the similarly scaled DNS results of Duan *et al* (2016) and Priebe and Martín (2011). The experiment and DNS results of Priebe and Martín (2011) are in close agreement, while the DNS results of Duan *et al*

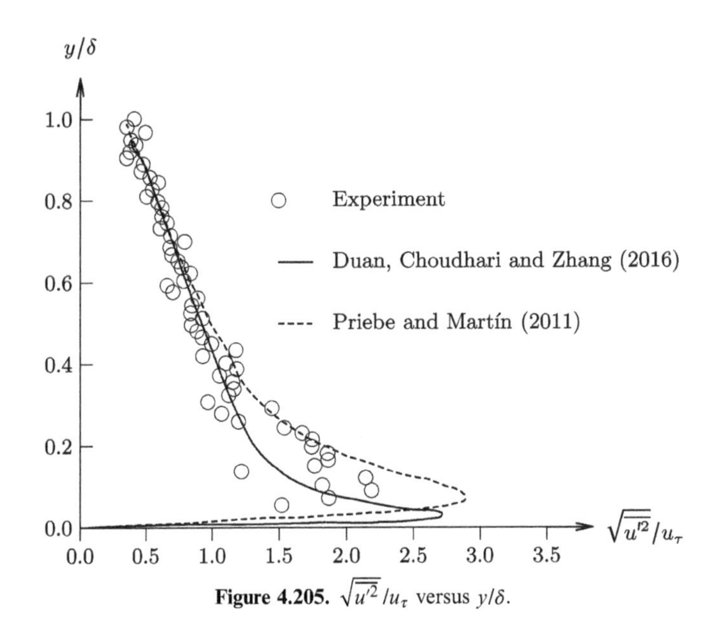


**Figure 4.205.** $\sqrt{\overline{u'^2}}/u_\tau$ versus $y/\delta$.

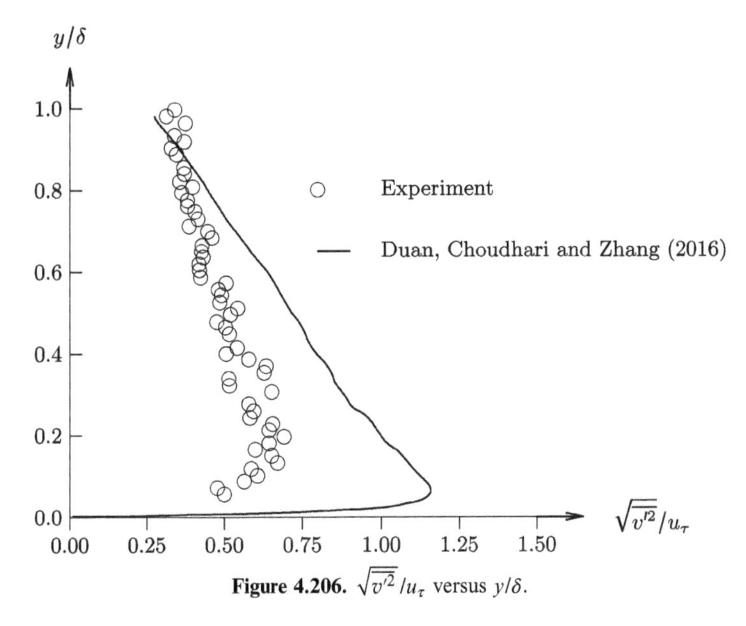


**Figure 4.206.** $\sqrt{\overline{v'^2}}/u_\tau$ versus $y/\delta$.

(2016) for the Morkovin-scaled normal velocity fluctuations are significantly above the experiment.

## 4.62 Williams *et al* (2021)

Williams *et al* (2021) investigated the effect of roughness on a flat plate hypersonic turbulent boundary layer in air at Mach 7.3. Two different types of surface roughness were examined as illustrated in figure 4.209 and covered the entire plate.

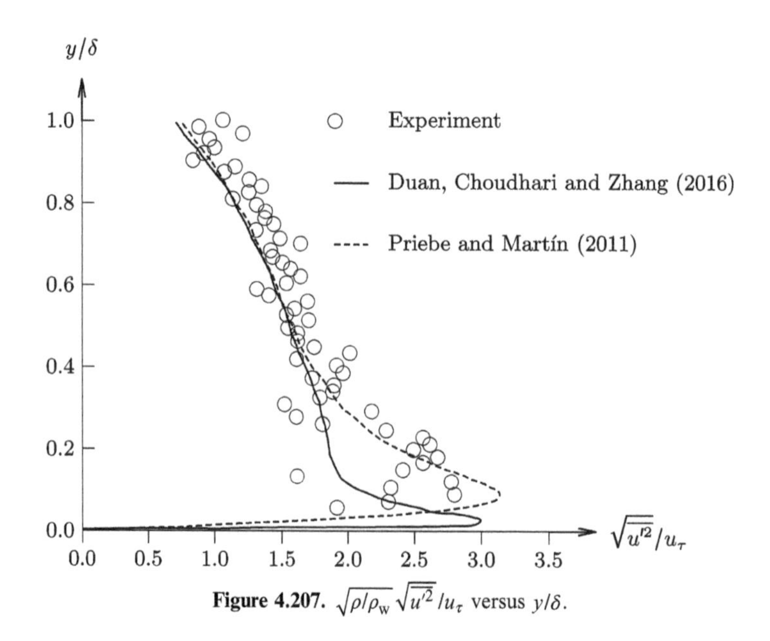


**Figure 4.207.** $\sqrt{\rho/\rho_w}\sqrt{\overline{u'^2}}/u_\tau$ versus $y/\delta$.

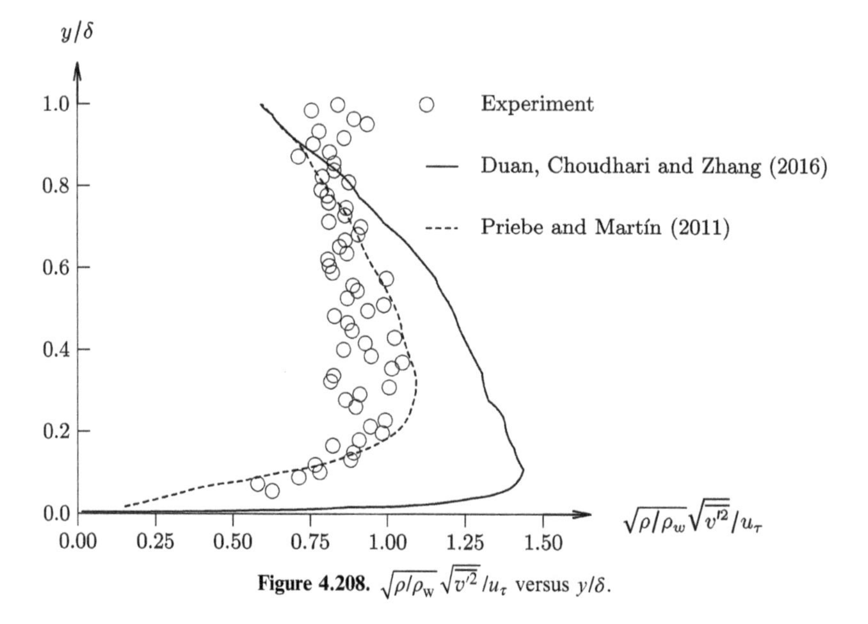


**Figure 4.208.** $\sqrt{\rho/\rho_w}\sqrt{\overline{v'^2}}/u_\tau$ versus $y/\delta$.

The two-dimensional square bar roughness was investigated for three different heights $k$ and streamwise spacing $\lambda$ with solidity $\lambda_s = k/\lambda$ (table 4.76). The diamond mesh was investigated for the same heights $k$ and comparable streamwise spacing $\lambda$, and characterized by the streamwise $\ell_x$ and spanwise $\ell_z$ dimensions of the diamond-shaped roughness with solidity $\lambda_s = k/\ell_x$. The roughness height to boundary layer thickness $k/\delta$ varied from 0.06 to 0.18, and the dimensionless roughness height

**Table 4.76.** Details of roughness.

| *Model* | $k$ (mm) | $\lambda/k$ | $\ell_x$ (mm) | $\ell_z$ (mm) | $\lambda_s$ |
|---|---|---|---|---|---|
| Square bar | 0.75 | 8.33 | — | — | 0.120 |
| | 1.27 | 10.63 | — | — | 0.172 |
| | 1.65 | 5 | — | — | 0.200 |
| Diamond mesh | 0.75 | 7.33 | 5.5 | 9.8 | 0.136 |
| | 1.27 | 10 | 12.7 | 32 | 0.100 |
| | 1.65 | 3.85 | 6.35 | 12.7 | 0.260 |

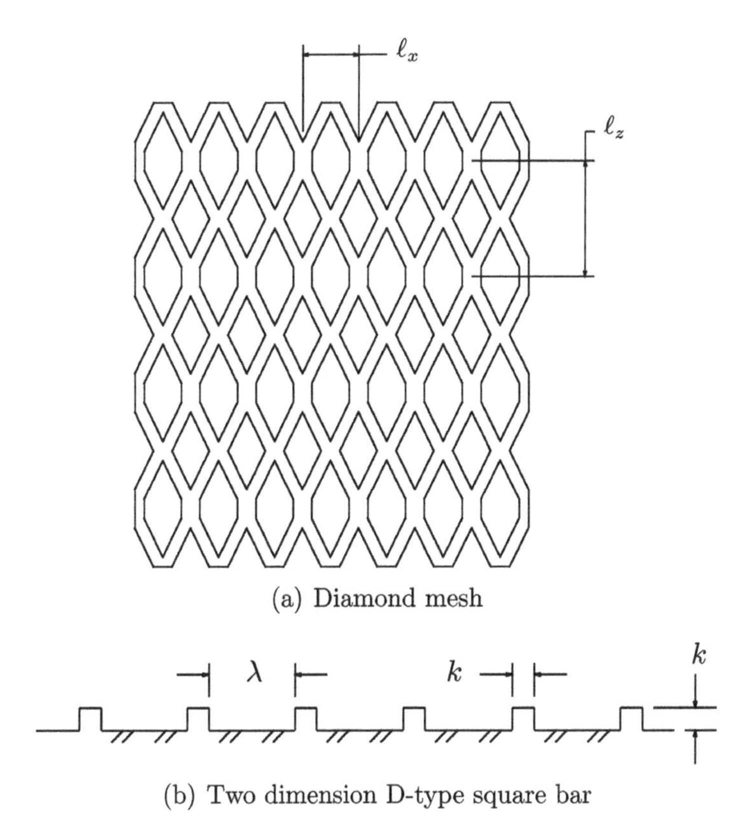


**Figure 4.209.** Roughness models.

$k^+=ku_\tau/\nu_w$ varies from 30 to 112. The freestream conditions are listed in table 4.77. Experimental diagnostics include PIV and schlieren imaging.

Figure 4.210 displays the Van Driest transformed velocity for the smooth (Williams *et al* 2018) and rough surfaces. The effect of surface roughness is a downwards shift of the logarithmic region by $H(k^+)$ as indicated in 3.60. The roughness function $H(k^+)$ displays a monotonic increase with $k^+$ (figure 4.211).

**Table 4.77.** Freestream conditions.

| | Square bar (SB) | | | Diamond mesh (DM) | |
|---|---|---|---|---|---|
| Quantity | Case 1 | Case 2 | Case 3 | Case 1 | Case 2 |
| $M_\infty$ | 7.3 | 7.2 | 7.5 | 7.3 | 7.5 |
| $p_{t_\infty}$ (MPa) | 7.97 | 7.55 | 7.53 | 7.93 | 7.38 |
| $T_{t_\infty}$ (K) | 760 | 790 | 758 | 745 | 709 |
| $T_w/T_{aw}$ | 0.53 | 0.55 | 0.47 | 0.52 | 0.58 |
| $T_w$ (K) | 360 | 393 | 319 | 350 | 367 |
| $H_\infty$ (MJ kg$^{-1}$) | 0.76 | 0.79 | 0.76 | 0.75 | 0.71 |
| $k$ (mm) | 0.75 | 1.27 | 1.65 | 0.75 | 1.65 |
| $k^+$ | 30 | 70 | 112 | 38 | 93 |
| $\delta$ (mm) | 11.1 | 13.0 | 10.0 | 13.1 | 9.2 |
| $Re_\theta$ | 7950 | 8283 | 6950 | 10 023 | 5686 |
| $H(k^+)$ | 4.2 | 9.3 | 11.4 | 5.45 | 11.8 |

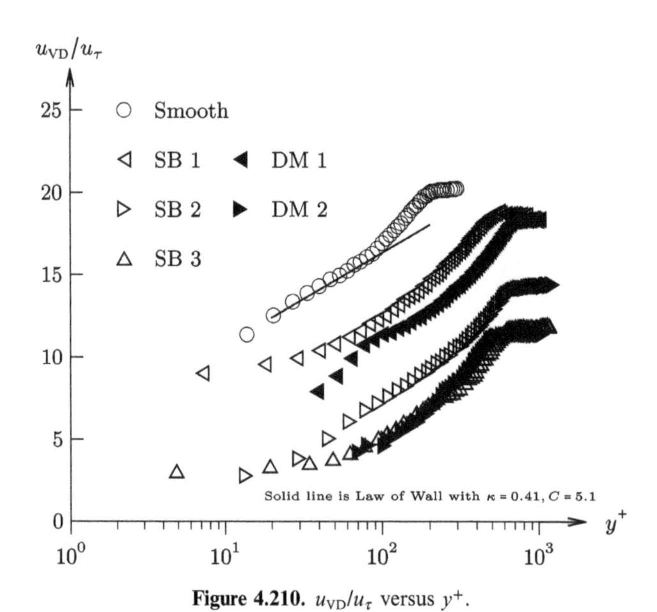


**Figure 4.210.** $u_{VD}/u_\tau$ versus $y^+$.

Figure 4.212 shows the Morkovin-scaled streamwise turbulence intensity $\sqrt{\overline{u'u'}}/u_\tau^*$ versus $y/\delta$ where $u_\tau^* = u_\tau\sqrt{\bar{\rho}/\bar{\rho}_w}$. A significant variation in the profiles is evident suggesting that Morkovin scaling is inapplicable for rough walls. A similar conclusion was reached by Ekoto *et al* (2008) and Ekoto *et al* (2009) at $M_\infty = 2.86$.

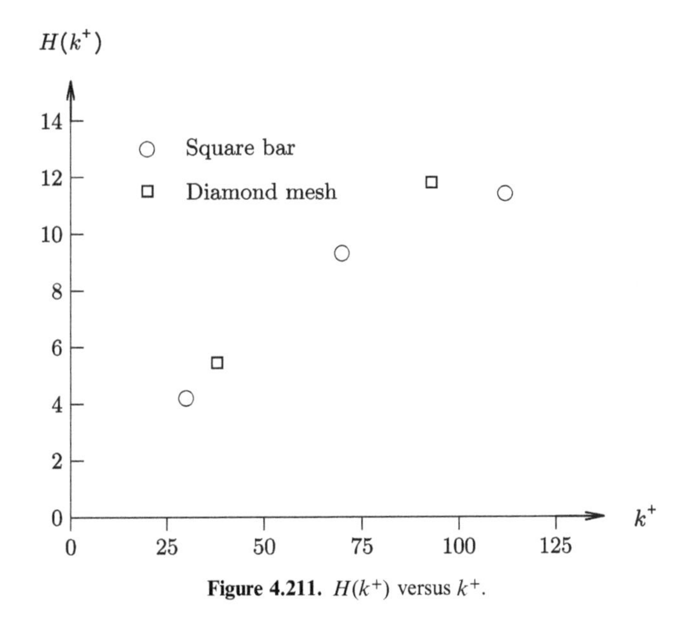


**Figure 4.211.** $H(k^+)$ versus $k^+$.

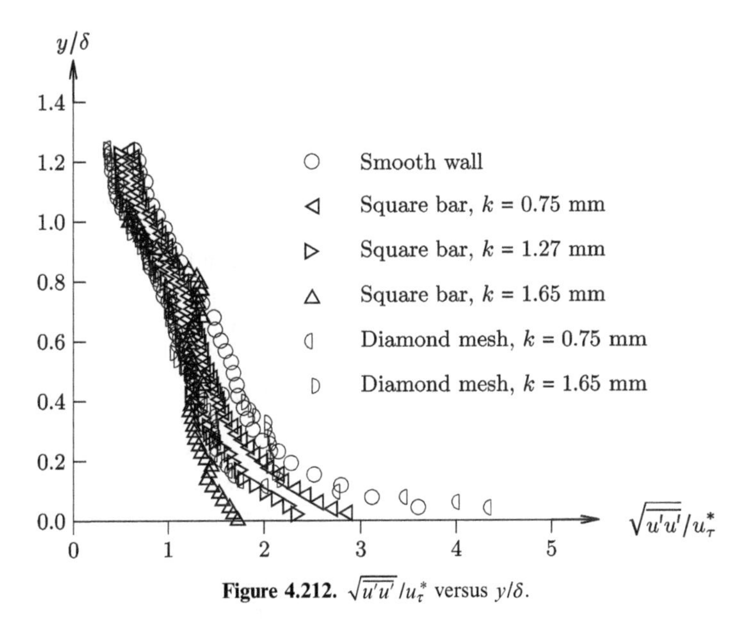


**Figure 4.212.** $\sqrt{\overline{u'u'}}/u_\tau^*$ versus $y/\delta$.

## 4.63 Additional results

Additional results on rough wall turbulent boundary layers include Holden (1982, 1983, 1984), Holden and Moselle (1992), Holden *et al* (2008, 2011), and Holden *et al* (2015).

## References


Adcock J, Peterson J and McRee D 1965 *Experimental Investigation of a Turbulent Boundary Layer at Mach 6, High Reynolds Numbers, and Zero Heat Transfer* Technical Note TN D-2907 (Washington, DC: National Aeronautics and Space Administration)

Anderson J 1990 *Modern Compressible Flow with Historical Perspective* 2nd edn (New York: McGraw-Hill)

Anderson J 2006 *Hypersonic and High-Temperature Gas Dynamics* (Reston, VA: American Institute of Aeronautics and Astronautics)

Auvity B, Etz M and Smits A 2001 Effects of transverse helium injection on hypersonic turbulent boundary layers *Phys. Fluids.* **13** 3025–32

Backx E 1973 Experiments in the Mach 15 turbulent boundary layer on the wall of the longshot conical nozzle *EUROMECH 43 Colloquium on Heat Transfer in Turbulent Boundary Layers with Variable Fluid Properties (Göttingen, Germany)*

Backx E 1974 *Experimental Study of the Turbulent Boundary Layer at Mach 15 and 19.8 in a Conical Nozzle* Technical Note 102 von Kármán Institute for Fluid Dynamics

Backx E 1975 A study of the turbulent boundary layer at Mach 15 and 19.8 on the wall of a conical nozzle *PhD Thesis* von Kármán Institute for Fluid Dynamics

Backx E and Richards B 1976 A high Mach number turbulent boundary-layer study *AIAA J.* **14** 1159–60

Baumgartner M 1997 Turbulence structure in a hypersonic boundary layer *PhD Thesis* Princeton University

Beckwith I, Harvey W and Clark F 1971 *Comparisons of Turbulent-Boundary-Layer Measurements at Mach Number 19.5 with Theory and an Assessment of Probe Errors* NASA Technical Note TN D-6192 (Washington, DC: National Aeronautics and Space Administration)

Berg D 1977 Surface roughness effects on the hypersonic turbulent boundary layer *PhD Thesis* California Institute of Technology

Bertram M and Neal L 1965 Recent experiments in hypersonic turbulent boundary layers *GARD Specialists Meeting on Recent Developments in Boundary Layer Research (Naples, Italy)*

Bloy A 1975 The expansion of a hypersonic turbulent boundary layer at a sharp corner *J. Fluid Mech.* **67** 647–55

Bushnell D, Johnson C, Harvey W and Feller W 1969 *Comparison of Prediction Methods and Studies of Relaxation in Hypersonic Turbulent Nozzle-Wall Boundary Layers* Technical Note D-5433 (Washington, DC: National Aeronautics and Space Administration)

Cary A 1970 *Summary of Available Information on Reynolds Analogy for Zero-Pressure-Gradient, Compressible, Turbulent-Boundary-Layer Flow* Technical Note D-5560 (Washington, DC: National Aeronautics and Space Administration)

Chi S and Spaulding D 1966 Influence of temperature ratio on heat transfer to a flat plate through a turbulent boundary layer in air *Proc. 3rd Int. Heat Transfer Conf. (Chicago, IL)* 41–9

Coles D 1953a Measurements in the boundary layer on a smooth flat plate in supersonic flow *PhD Thesis* California Institute of Technology

Coles D 1953b *Measurements in the Boundary Layer on a Smooth Flat Plate in Supersonic Flow. Part III: Measurements in a Flat Plate Boundary Layer at Jet Propulsion Laboratory* Technical Report 20-71 Jet Propulsion Laboratory, California Institute of Technology

Coles D 1962 The turbulent boundary layer in a compressible fluid *The Turbulent Boundary Layer in a Compressible Fluid* Report R-403-PR (Santa Monica, CA: Rand Corporation)

Coles D 1964 The turbulent boundary layer in a compressible fluid *The Phys. Fluids* **7** 1403–23

Danberg J 1959 *Measurement of the Characteristics of the Compressible Turbulent Boundary Layer with Air Injection* NAVORD Report 6683 (White Oak, Silver Spring, MD: United States Naval Ordnance Laboratory)

Danberg J 1964 *Characteristics of the Turbulent Boundary Layer with Heat and Mass Transfer at $M = 6.7$* Technical Report NOLTR 64-99 (White Oak, Silver Spring, MD:United States Naval Ordnance Laboratory)

Danberg J 1967 *Characteristics of the Turbulent Boundary Layer with Heat and Mass Transfer: Data Tabulation* Technical Report NOLTR 67-6 (White Oak, Silver Spring, MD:United States Naval Ordnance Laboratory)

DeGraaff D and Eaton J 2000 Reynolds-number scaling of the flat-plate turbulent boundary layer *J. Fluid Mech.* **422** 319–46

Demetriades A and Laderman A 1973 Reynolds stress measurements in a hypersonic boundary layer *AIAA J.* **11** 1594–6

Ding H, Yi S, Quyang T and Zhao Y 2020 Research on velocity measurements of the hypersonic turbulent boundary layer based on the nano-tracer-based planar laser scattering technique *Meas. Sci. Technol.* **31** 085302

Duan L, Choudhari M and Zhang C 2016 Pressure fluctuations induced by a hypersonic turbulent boundary layer *J. Fluid Mech.* **804** 578–607

Ekoto I, Bowersox R, Beutner T and Goss I 2008 Supersonic boundary layers with periodic surface roughness *AIAA J.* **46** 486–97

Ekoto I, Bowersox R, Beutner T and Goss I 2009 Response of supersonic turbulent boundary layers to local and global mechanical distortions *J. Fluid Mech.* **630** 225–65

Ekoto I, Bowersox R, Beutner T and Goss L 2007 Response of a supersonic turbulent boundary layer to periodic surface roughness *45th AIAA Aerospace Sciences Meeting and Exhibit* AIAA Paper 2007-1142 (Reston, VA: American Institute of Aeronautics and Astronautics)

Everhart P and Hamilton H 1967 *Experimental Investigation of Boundary Layer* Technical Note D-4188 (Washington, DC: National Aeronautics and Space Administration)

Federov A, Shiplyuk A, Maslov A, Burov E and Malmuth N 2003 Stabilization of a hypersonic boundary layer using an ultrasonic absorptive coating *J. Fluid Mech.* **479** 99–124

Feller W 1973 Effects of upstream wall temperatures on hypersonic tunnel wall boundary layer profile measurements *AIAA J.* **11** 556–8

Fernholz H and Finley P 1977 *A Critical Compilation of Compressible Turbulent Boundary Layer Data* AGARDograph 223 Advisory Group on Aerospace Research and Development

Fernholz H and Finley P 1980 *A Critical Commentary on Mean Flow Data for Two Dimensional Compressible Turbulent Boundary Layers* AGARDograph 253 Advisory Group on Aerospace Research and Development

Fernholz H, Finley P and Mikulla V 1981 *A Further Compilation of Compressible Boundary Layer Data with a Survey of Turbulence Data* AGARDograph 263 Advisory Group on Aerospace Research and Development

Fischer M, Maddalon D, Weinstein L and Wagner R 1970 Boundary-layer surveys on a nozzle wall at a free-stream Mach number 20 including hot-wire fluctuation measurements *3rd Fluid and Plasma Dynamics Conf.* AIAA Paper 70–746 (Reston, VA: American Institute of Aeronautics and Astronautics)

Fischer M, Maddalon D, Weinstein L and Wagner R 1971 Boundary-layer pitot and hot-wire surveys at $M_\infty \approx 20$ *AIAA J.* **9** 826–34

Gibbings J and Mikulla V 1973 *Measurements of Reynolds Stresses in Compressible Flows* Technical Report 34540FM4408 (London:Aeronautical Research Council)

Goyne C, Stalker R and Paull A 1999 Shock tunnel skin friction measurement in a supersonic combustor *J. Propuls. Power* **15** 699–705

Goyne C, Stalker R and Paull A 2003 Skin friction measurements in high enthalpy hypersonic boundary layers *J. Fluid Mech.* **485** 1–32

Hakkinen R 1955 *Measurements of Turbulent Skin Friction on a Flat Plate at Transonic Speeds* Technical Note 3486 (Washington, DC: National Advisory Committee on Aeronautics)

Harvey W and Bushnell D 1969a *Fluctuating Properties of Turbulent Boundary Layers for Mach Numbers up to 9* Technical Note TN D-5496 (Washington, DC: National Aeronautics and Space Administration)

Harvey W and Bushnell D 1969b Velocity fluctuation intensities in a hypersonic turbulent boundary layer *AIAA J.* **7** 760–2

Hill F 1956 Boundary-layer measurements in hypersonic flow *J. Aeronaut. Sci.* **23** 35–42

Hill F 1959a Appendix II: Skin friction and heat transfer measurements at Mach numbers from 8 to 10 in turbulent boundary layers *Bumblebee Aerodynamics Panel Vol. 1* TG-14-37 Johns Hopkins University, Applied Physics Laboratory

Hill F 1959b Turbulent boundary layer measurements at Mach numbers from 8 to 10 *Phys. Fluids* **2** 668–80

Hinze J 1959 *Turbulence* (New York: McGraw-Hill)

Hironimus G 1966 *Hypersonic Shock Tunnel Experiments on the W7 Flat Plate Model—Expansion Side, Turbulent Flow and Leading Edge Transpiration Data* Report AA-1952-Y-2 Cornell Aeronautical Laboratory

Holden M 1972 *An Experimental Investigation of Turbulent Boundary Layers at High Mach Numbers and Reynolds Numbers* Cornell Aeronautical Laboratory Report No. AB-5072-A-1 (Washington, DC: National Aeronautics and Space Administration)

Holden M 1982 Experimental studies of surface roughness, entropy swallowing and boundary layer transition effects on the skin friction and heat transfer distribution in high speed flows *20th Aerospace Sciences Meeting* AIAA Paper 82-0034 (Reston, VA: American Institute of Aeronautics and Astronautics)

Holden M 1983 *Studies of Boundary Layer Transition and Surface Roughness Effects in Hypersonic Flow* Report 6430-A-5 Calspan Advanced Technology Center

Holden M 1984 Experimental studies of surface roughness shape and spacing effects on heat transfer and skin friction in supersonic and hypersonic flows *22nd Aerospace Sciences Meeting* AIAA Paper 84-0016 (Reston, VA: American Institute of Aeronautics and Astronautics)

Holden M and Chadwick K 1995 Studies of laminar, transitional and turbulent hypersonic flows over curved compression surfaces *33rd Aerospace Sciences Meeting and Exhibit* AIAA Paper 1995-93 (Reston, VA: American Institute of Aeronautics and Astronautics)

Holden M and Moselle J 1992 A database of aerothermal measurements in hypersonic flow for CFD validation *17th Aerospace Ground Testing Conf.* AIAA Paper 92-4023 (Reston, VA: American Institute of Aeronautics and Astronautics)

Holden M, Mundy E and MacClean M 2015 A review of experimental studies at CUBRC to examine the effects of surface roughness and blowing on sharp and blunt hypersonic vehicles *20th AIAA International Space Planes and Hypersonic Systems and Technologies Conf.* AIAA Paper 2015-3600 (Reston, VA: American Institute of Aeronautics and Astronautics)

Holden M, Mundy E and MacLean M 2011 Heat transfer measurements to examine surface roughness and blowing effects in hypersonic flows *49th AIAA Aerospace Science Meeting* AIAA Paper 2011-760 (Reston, VA: American Institute of Aeronautics and Astronautics)

Holden M, Mundy E and Wadhams T 2008 A review of experimental studies of surface roughness and blowing on the heat transfer and skin friction to nosetips and slender cones in high Mach number flows *40th Thermophysics Conf.* AIAA Paper 2008-3907 (Reston, VA: American Institute of Aeronautics and Astronautics)

Hopkins E and Inouye M 1971 An evaluation of theories for predicting skin friction and heat transfer on flat plates at supersonic and hypersonic Mach Numbers *AIAA J.* **9** 993–1003

Hopkins E and Keener E 1966 *Study of Surface Pitots for Measuring Turbulent Skin Friction at Supersonic Mach Numbers—Adiabatic Wall* Technical Note TN D-3478 (Washington, DC: National Aeronautics and Space Administration)

Hopkins E and Keener E 1972a Pressure gradient effects on hypersonic turbulent skin friction and boundary layer profiles *Aerospace Sciences Meeting* AIAA Paper 72-215 (Reston, VA: American Institute of Aeronautics and Astronautics)

Hopkins E and Keener E 1972b Pressure gradient effects on hypersonic turbulent skin friction and boundary layer properties *AIAA J.* **10** 1141–2

Hopkins E, Keener E and Louie P 1970 *Direct Measurements of Turbulent Skin Friction on a Nonadiabatic Flat Plate at Mach Number 6.5 and Comparisons with Eight Theories* Technical Note TN D-5675 (Washington, DC: National Aeronautics and Space Administration)

Hopkins E, Rubesin M, Inouye M, Keener E, Mateer G and Polek T 1969 *Summary and Correlation of Skin-Friction and Heat-Transfer Data for a Hypersonic Turbulent Boundary Layer on Simple Shapes* Technical Note TN D-5089 (Washington, DC: National Aeronautics and Space Administration)

Horstman C and Owen F 1972 Turbulent properties of a compressible boundary layer *AIAA J.* **10** 1418–29

Hoydysh W and Zakkay V 1969 An experimental investigation of hypersonic turbulent boundary layers in adverse pressure gradient *AIAA J.* **7** 105–16

Huang P, Coleman G and Bradshaw P 1995 Compressible turbulent channel flows: DNS results and modelling *J. Fluid Mech.* **305** 185–218

Johnson D and Rose W 1975 Measurements of turbulence transport properties in supersonic boundary layer flow using laser velocimeter and hot wire anemometer techniques *AIAA J.* **13** 512–5

Jones R and Feller W 1970 *Preliminary Surveys of the Wall Boundary Layer in a Mach 6 Axisymmetric Tunnel* Technical Report D-5620 (Washington, DC: National Aeronautics and Space Administration)

Keener E and Hopkins E 1972 *Turbulent Boundary-Layer Velocity Profiles on a Nonadiabatic Flat Plate at Mach Number 6.5* Technical Note TN D-6907 (Washington, DC: National Aeronautics and Space Administration)

Keener E and Polek T 1972 Measurements of Reynolds analogy for a hypersonic turbulent boundary layer on a nonadiabatic flat plate *AIAA J.* **10** 845–6

Kemp J and Owen F 1971 Nozzle wall boundary layers at Mach numbers 20 to 47 AIAA Paper 71-161 (Reston, VA: American Institute of Aeronautics and Astronautics)

Kemp J and Owen F 1972a *Experimental Study of Nozzle Wall Boundary Layers at Mach Numbers 20 to 47* Technical Note D-6965 (Washington, DC: National Aeronautics and Space Administration)

Kemp J and Owen F 1972b Nozzle wall boundary layers at Mach numbers 20 to 47 *AIAA J.* **10** 872–9
Kemp J and Sreekanth A 1969 Preliminary results from an experimental investigation of nozzle wall boundary layers at Mach numbers ranging from 27 to 47 *2nd Fluid and Plasma Dynamics Conf.* AIAA Paper 69-686 (Reston, VA: American Institute of Aeronautics and Astronautics)
Kistler A 1959 Fluctuation measurements in a supersonic turbulent boundary layer *Phys. Fluids* **2** 290–6
Kistler A and Chen W 1962 The fluctuating pressure field in a supersonic turbulent boundary layer *J. Fluid Mech.* **16** 41–64
Klebanoff P 1954 *Characteristics of Turbulence in a Boundary Layer with Zero Pressure Gradient* Technical Report 3178 (Washington, DC: National Advisory Committee on Aeronautics)
Klebanoff P 1955 *Characteristics of Turbulence in a Boundary Layer with Zero Pressure Gradient* Technical Report 1247 (Washington, DC: National Advisory Committee on Aeronautics)
Korkegi R 1956 Transition studies and skin-friction measurements on an insulated flat plate at a Mach Number of 5.8 *J. Aeronaut. Sci.* **23** 97–107
Laderman A 1974 Hypersonic viscous flow over a slender cone, part II: turbulence structure of the boundary layer *7th Fluid and Plasma Dynamics Conf.* AIAA Paper 74-534 (Reston, VA: American Institute of Aeronautics and Astronautics)
Laderman A 1976 New measurements of turbulent shear stresses in hypersonic boundary layers *AIAA J.* **14** 1286–91
Laderman A and Demetriades A 1971 *Mean Flow Measurements in a Hypersonic Turbulent Boundary Layer* Technical Report U-4950 (Newport Beach, CA: Philco-Ford Corporation, Aeronutronic Division)
Laderman A and Demetriades A 1973 Hot-wire measurements of hypersonic boundary-layer turbulence *Phys. Fluids* **16** 179–81
Laderman A and Demetriades A 1974 Mean and fluctuating flow measurements in the hypersonic turbulent boundary layer over a cooled wall *J. Fluid Mech.* **63** 121–44
Lee R, Yanta W and Leonas A 1968 Velocity profile, skin friction balance and heat transfer measurements of the turbulent boundary layer at Mach 5 *Proc. 1968 Heat Transfer and Fluid Mechanics Institute* ed A Emery and C Depew (Stanford, CA: Stanford University Press)
Lee R, Yanta W and Leonas A 1969 *Velocity Profile, Skin-Friction Balance and Heat-Transfer Measurements of the Turbulent Boundary Layer at Mach 5 and Zero-Pressure Gradient* Technical Report NOLTR 69-106 (White Oak, Silver Spring, MD: United States Naval Ordnance Laboratory)
Lobb R, Winkler E and Persh J 1955a Experimental investigation of turbulent boundary layers in hypersonic flow *J. Aeronaut. Sci.* **22** 1–9
Lobb R, Winkler E and Persh J 1955b *NOL Hypersonic Tunnel No. 4 Results VII: Experimental Investigation of Turbulent Boundary Layers in Hypersonic Flight* Technical Report NAVORD Report 3880 (White Oak, Silver Spring, MA: United States Naval Ordnance Laboratory)
Maddalon D and Henderson A 1968 Boundary layer transition on sharp cones at hypersonic Mach numbers *AIAA J.* **6** 424–31
Maslov A, Fedorov A, Bountin D, Shiplyuk A, Sidorenko A, Malmuth N and Knauss H 2008 Experimental study of disturbances in transitional and turbulent hypersonic boundary layers *AIAA J.* **46** 1880–3

Materna P 1977 Hot wire anemometry in a hypersonic turbulent boundary layer *10th Fluid and Plasmadynamics Conf.* AIAA Paper 77-702 (Reston, VA: American Institute of Aeronautics and Astronautics)

Matthews R and Trimmer L 1969 *Nozzle Turbulent Boundary-Layer Measurements in the VKF 50-Inch Hypersonic Tunnels* AEDC Technical Report TR-69-118 (Arnold Air Force Station, TN: Arnold Engineering Development Center)

Matting F, Chapman D, Nyholm J and Thomas A 1961 *Turbulent Skin Friction at High Mach Numbers and Reynolds Numbers in Air and Helium* Technical Report TR R-82 (Washington, DC: National Aeronautics and Space Administration)

McGinley C, Spina E and Sheplak M 1994 Turbulence measurements in a Mach 11 helium boundary layer *25th AIAA Fluid Dynamics Conf.* AIAA Paper 94-2364 (Reston, VA: American Institute of Aeronautics and Astronautics)

Mikulla V and Horstman C 1975 Turbulence stress measurements in a nonadiabatic hypersonic boundary layer *AIAA J.* **13** 1607–13

Monaghan R and Cooke J 1951 *Measurement of Heat Transfer and Skin Friction at Supersonic Speeds. Part III. Measurement of Overall Heat Transfer and the Associated Boundary Layers on a Flat Plate at M = 2.43* Technical Note AERO 2129 Royal Aeronautical Establishment

Moore D and Harkness J 1965 Experimental investigations of the compressible turbulent boundary layer at very high Reynolds numbers *AIAA J.* **3** 631–8

Neal L 1966 *A Study of the Pressure, Heat Transfer and Skin Friction on Sharp and Blunt Flat Plates at Mach 6.8* Technical Note D-3312 (Washington, DC: National Aeronautics and Space Administration)

Neeb D, Saile D and Gülhan A 2015 Experimental flow characterization and heat flux augmentation analysis of a hypersonic turbulent boundary layer along a rough surface *Proc. 8th European Symp. on Aerothermodynamics for Space Vehicles*

Neeb D, Saile D and Gülhan A 2018 Experiments on a smooth wall hypersonic boundary layer at Mach 6 *Exp. Fluids* **59** 68

Nikuradse J 1950 *Laws of Flow in Rough Pipes* Technical Memorandum 1292 National Advisory Committee on Aeronautics Translation of 'Strömungsgesetze in rauhen Rohren', VDI Forshungsheft 361, Beilage zu 'Forshung auf dem Gebiete des Ingenieurwesens', Ausgabe B Band 4, July–August, 1933

Owen F and Calarese W 1987 Turbulence measurement in hypersonic flow *Turbulence Measurement in Hypersonic Flow* AGARD Conf. Proc. No. 428 Advisory Group for Aerospace Research and Development, North Atlantic Treaty Organization pp 5-1–17

Owen F and Horstman C 1972 On the structure of hypersonic turbulent boundary layers *J. Fluid Mech.* **53** 611–36

Owen F and Horstman C 1974 Turbulent measurements in an equilibrium hypersonic boundary layer *Aerospace Sciences Meeting* AIAA Paper 74–93 (Reston, VA: American Institute of Aeronautics and Astronautics)

Owen F, Horstman C and Kussoy M 1975 Mean and fluctuating flow measurements of a fully-developed, non-adiabatic, hypersonic boundary layer *J. Fluid Mech.* **70** 393–413

Parrott T, Jones M and Albertson C 1989 *Fluctuating Pressures Measured Beneath a High Temperature, Turbulent Boundary Layer on a Flat Plate at a Mach Number of 5* Technical Paper 2947 (Washington, DC: National Aeronautics and Space Administration)

Peltier S, Humble R and Bowersox R 2011 Response of a hypersonic turbulent boundary layer to local and global mechanical distortions *49th AIAA Aerospace Sciences Meeting* AIAA Paper 2011-680 (Reston, VA: American Institute of Aeronautics and Astronautics)

Peltier S, Humble R and Bowersox R 2016 Crosshatch roughness distortions on a hypersonic turbulent boundary layer *Phys. Fluids* **28** 045105

Perry J and East R 1968a Experimental measurements of cold wall turbulent hypersonic boundary layers *AGARD CP 30: Hypersonic Boundary Layers and Flowfields* pp 2–1–19

Perry J and East R 1968b Experimental measurements of cold wall turbulent hypersonic boundary layers *Experimental Measurements of Cold Wall Turbulent Hypersonic Boundary Layers* Aeronautics and Astronautics Report 275 University of Southampton

Priebe S and Martín M P 2011 Direct numerical simulation of a hypersonic turbulent boundary layer on a large domain *41st AIAA Fluid Dynamics Conf. and Exhibit* AIAA Paper 2011-3432 (Reston, VA: American Institute of Aeronautics and Astronautics)

Raman K 1974 *A Study of Surface Pressure Fluctuations in Hypersonic Turbulent Boundary Layers* Contractor Report CR-2386 (Washington, DC: National Aeronautics and Space Administration)

Roy C and Blottner F 2006 Review and assessment of turbulence models for hypersonic fows *Prog. Aerosp. Sci.* **42** 469–530

Sahoo D, Schultze M and Smits A 2009 Effect of roughness on a turbulent boundary layer in hypersonic flow *39th AIAA Fluid Dynamics Conf.* AIAA Paper 2009-3678 (Reston, VA: American Institute of Aeronautics and Astronautics)

Samuels R, Peterson J and Adcock J 1967 *Experimental Investigation of the Turbulent Boundary Layer at a Mach Number of 6 with Heat Transfer at High Reynolds Numbers* NASA TN D-3858 (Washington, DC: National Aeronautics and Space Administration)

Scaggs N 1966 *Boundary Layer Profile Measurements in Hypersonic Nozzles* ARL 66-0141 Aeronautical Research Laboratories, Office of Aerospace Research, Wright-Patterson AFB, OH

Settles G and Dodson L 1993 *Hypersonic Turbulent Boundary Layer and Free Shear Layer Database* Contractor Report 177610 (Washington, DC: National Aeronautics and Space Administration)

Sheer R and Nagamatsu H 1968 Methods for distinguishing type of hypersonic boundary layer in shock tunnel *6th Aerospace Sciences Meeting* AIAA Paper 1968-50 (Reston, VA: American Institute of Aeronautics and Astronautics)

Shutts W, Hartwig W and Weiler J 1955 *Final Report on Turbulent Boundary-Layer and Skin-Friction Measurements on a Smooth, Thermally Insulated Flat Plate at Supersonic Speeds* Report DRL-364, CM-823 Defense Research Laboratory, University of Texas, Austin

Smith J and Driscoll J 1975 The electron-beam fluorescence technique for measurements in hypersonic turbulent boundary layers *J. Fluid Mech.* **72** 695–719

Softley E and Sullivan R 1968 *Theory and Experiment for the Structure of Some Hypersonic Boundary Layers* AGARD CP 30 Advisory Group for Aerospace Research and Development, North Atlantic Treaty Organization pp 3–1–18

Sommer S and Short B 1955 *Free-Flight Measurements of Turbulent Boundary-Layer Heating at Mach Numbers from 2.8 to 7.0* NACA TN 3391 (Washington, DC: National Aeronautics and Space Administration)

Spaulding D and Chi S 1964 The drag of a compressible turbulent boundary layer on a smooth plate with and without heat transfer *J. Fluid Mech.* **18** 117–43

Speaker W and Ailman C 1966 *Spectra and Space-Time Correlations of the Fluctuating Pressures at a Wall Beneath a Supersonic Turbulent Boundary Layer Perturbed by Steps and Shock Waves* Contractor Report 486 (Washington, DC: National Aeronautics and Space Administration)

Stalmach C 1958 *Experimental Investigation of the Surface Impact Pressure Probe Method of Measuring Local Skin Friction at Supersonic Speeds* Report DRL-410, CF-2675 Defense Research Laboratory, University of Texas, Austin

Stone D and Cary A 1972 *Discrete Sonic Jets Used as Boundary Layer Trips at Mach Numbers of 6 and 8.5* NASA TN D-6802 (Washington, DC: National Aeronautics and Space Administration)

Stroud J and Miller L 1965 *An Experimental and Analytical Investigation of Hypersonic Inlet Boundary Layers. Volume II. Data Reduction Program and Tabulated Experimental Data* AFFDL TR 65-123 Air Force Flight Dynamics Laboratory

Suraweera M, Mee D and Stalker R 2006 Reynolds analogy in high-enthalpy and high-Mach-number turbulent flows *AIAA J.* **44** 917–9

Tichenor N 2010 Characterization of the influence of a favorable pressure gradient on the basic structure of a Mach 5.0 high Reynolds number supersonic turbulent boundary layer *PhD Thesis* Department of Aerospace Engineering, Texas A&M University

Tichenor N, Humble R and Bowersox R 2013 Response of a hypersonic turbulent boundary layer to favorable pressure gradient *J. Fluid Mech.* **722** 187–213

Tichenor N, Neel I, Leidy A and Bowersox R 2017 Influence of streamline adverse pressure gradients on the structure of a Mach 5 turbulent boundary layer *55th AIAA Aerospace Sciences Meeting* AIAA Paper 2017-1697 (Reston, VA: American Institute of Aeronautics and Astronautics)

Townsend A 1956 *The Structure of Turbulent Shear Flow* (Cambridge: Cambridge University Press)

Vaganov A and Stolyarov E 2008 Statistical laws of the pressure fluctuations in a hypersonic turbulent boundary layer *Fluid Dyn.* **43** 265–73

Van Driest E 1956 Problem of aerodynamic heating *Eng. Rev.* **15** 26–41

Voisinet R and Lee R 1972 *Measurements of a Mach 4.9 Zero-Pressure-Gradient Turbulent Boundary Layer with Heat Transfer. Part 1: Data Compilation* Technical Report NOLTR 72-232 (White Oak, Silver Spring, MD: United States Naval Ordnance Laboratory)

Voisinet R, Lee R and Yanta W 1971 *An Experimental Study of the Compressible Turbulent Boundary Layer with an Adverse Pressure Gradient AGARD Conf. Proc.* Advisory Group for Aerospace Research and Development 9-1–10

Wagner R, Maddalon D and Weinstein L 1970 Influence of measured freestream disturbances on hypersonic boundary layer transition *AIAA J.* **8** 1664–70

Wallace J 1967 Hypersonic turbulent boundary layer studies at cold wall temperatures *Proceedings of the Heat Transfer and Fluid Mechanics Institute (La Jolla, CA, June 19–21)*

Wallace J 1968 *Hypersonic Turbulent Boundary Layer Measurements Using an Electron Beam* Technical Report AN-2112-Y1 (Buffalo, NY: Cornell Aeronautical Laboratory)

Wallace J 1969 Hypersonic turbulent boundary-layer measurements using an electron beam *AIAA J.* **7** 757–9

Wallace J and McLaughlin E 1966 *Experimental Investigation of Hypersonic Turbulent Flow and Laminar Leeside Flow on Flat Plates* Technical Report TR AFFDL-TR-66-63 Air Force Flight Dynamics Laboratory

Watson E, Gnos A, Gallo W and Latham E 1966 Boundary layers and hypersonic inlet flowfields *2nd Propulsion Joint Special Conf.* AIAA Paper 66-606 (Reston, VA: American Institute of Aeronautics and Astronautics)

Watson R, Harris J and Anders J 1973 Measurements in a transitional/turbulent Mach 10 boundary layer at high-Reynolds number *11th Aerospace Sciences Meeting* AIAA Paper 73–165 (Reston, VA: American Institute of Aeronautics and Astronautics)

Wegener P, Winkler E and Sibulkin M 1953 A Measurement of turbulent boundary-layer profiles and heat-transfer coefficient at $M = 7$ *J. Aeronaut. Sci.* **20** 221–2

Williams O, Sahoo D, Baumgartner M and Smits A 2018 Experiments on the structure and scaling of hypersonic turbulent boundary layers *J. Fluid Mech.* **834** 237–70

Williams O, Sahoo D, Papageorge M and Smits A 2021 Effects of roughness on a turbulent boundary layer in hypersonic flow *Exp. Fluids* **62** 1–13

Williams O and Smits A 2017 Effect of tripping on hypersonic boundary layer statistics *AIAA J.* **55** 3051–8

Winkler E 1961 Investigation of flat plate hypersonic turbulent boundary layer with heat transfer *Trans. Am. Soc. Mech. Eng.* E **83** 323–9

Winkler E and Cha M 1959 *Investigation of Flat Plate Hypersonic Turbulent Boundary Layer with Heat Transfer at a Mach Number of 5.2* NAVORD Report 6631 (White Oak, Silver Spring, MD: United States Naval Ordnance Laboratory)

Winkler E and Persh J 1954 *NOL Hypersonic Tunnel No. 4. Results VI: Experimental and Theoretical Investigation of the Boundary Layer and Heat Transfer Characteristics of a Cooled Wall Wedge Nozzle at a Mach Number of 5.5* NAVORD Report 3757, Aeroballistic Research Report 245 (White Oak, Silver Spring, MD: United States Naval Ordnance Laboratory)

Yanta W and Lee R 1974 Determination of turbulence transport properties with the laser doppler velocimeter and conventional time averaged mean flow measurements at Mach 3 *7th Fluid and Plasma Dynamics Conf.* AIAA Paper 74-575 (Reston, VA: American Institute of Aeronautics and Astronautics)

Young F 1965 *Experimental Investigation of the Effects of Surface Roughness on Compressible Turbulent Boundary Layer Skin Friction and Heat Transfer* Technical Report DLR-532, CR-21 Defense Research Laboratory, University of Texas, Austin

Zakkay V and Wang C-H 1972 *Turbulent Boundary Layer in an Adverse Pressure Gradient Without Effect of Wall Curvature* NASA CR 112247 (Washington, DC: National Aeronautics and Space Administration)

**IOP** Publishing

Hypersonic Shock Wave Turbulent Boundary Layers
Direct Numerical Simulation, Large Eddy Simulation and Experiment
**Doyle Knight and Nadia Kianvashrad**

# Chapter 5

# Experiments—hypersonic shock wave turbulent boundary layer interactions

The most severe problems of atmospheric flight at high Mach numbers are associated with viscous-inviscid interactions.

Robert Korkegi (1971)

## Abstract

Experimental research in hypersonic shock wave turbulent boundary layer interactions also began in the second half of the twentieth century due to the interest in high speed flight. The present chapter examines a selected set of experimental data on hypersonic shock wave turbulent boundary layer interactions. Similar to the previous chapter, the experiments were chosen both to illustrate the history of research in this field and to provide an understanding of the state of the art. Experiments were selected that demonstrated an equilibrium incoming turbulent boundary layer. There are other experimental studies where the incoming boundary layer is not in equilibrium, e.g., Borovoy *et al* (2011) and Borovoy *et al* (2013), or transitional, e.g.,Leidy *et al* (2020). A selected set of data (typically, the primary measurements) is presented for each experiment.

Experimental research has focused on canonical geometries to elucidate the physics of shock wave turbulent boundary layer interaction without added geometric complexity. Settles and Dolling (1986) and Zheltovodov *et al* (1987) describe families of interactions. These include both dimensionless and dimensional shock wave boundary layer interactions. A family of dimensionless and dimensional canonical geometries is shown in figure 5.1 of Zheltovodov and Knight (2011). An example of a dimensionless canonical geometry is the single unswept fin shown in figure 5.1. The single fin geometry is characterized by the fin angle relative to the

doi:10.1088/978-0-7503-5002-0ch5  

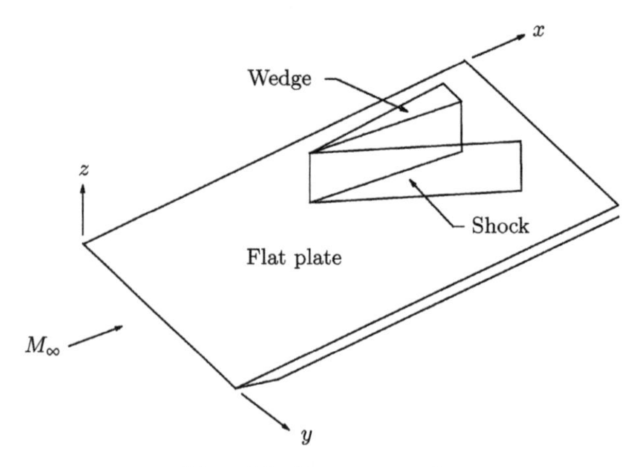


**Figure 5.1.** Sharp unswept fin.

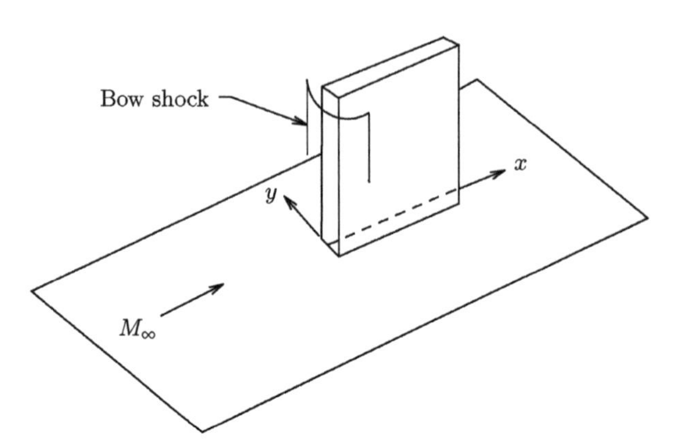


**Figure 5.2.** Blunt fin.

oncoming flow provided the fin height extends far above the incoming turbulent boundary layer, and is therefore dimensionless geometry. An example of a dimensional canonical geometry is the blunt fin shown in figure 5.2. The blunt fin geometry is characterized by a single dimension of the fin width provided that the fin height extends far above the incoming turbulent boundary layer.

Futher detailed information is available in the reviews by Korkegi (1971), Settles and Dodson (1991, 1994), Zheltovodov (1996), Roy and Blottner (2006) and Marvin *et al* (2013). Additional surveys include Zheltovodov and Schülein (1986), Zheltovodov *et al* (1987, 1990), Zheltovodov (1991) and Zheltovodov (2006).

## 5.1 Elfstrom (1972)

Elfstrom (1972) performed an experimental investigation of a hypersonic shock wave turbulent boundary layer interaction on a two dimensional wedge compression corner at Mach 9. The tests were performed in the Imperial College No. 2 Gun Tunnel (section A.13). A schematic of the model is shown in figure 5.3. Wedge angles $\alpha$ of 15° to 38° were examined. The freestream conditions are summarized in table 5.1 where

**Figure 5.3.** Wedge compression corner.

**Table 5.1.** Freestream conditions.

| Quantity | Value |
|---|---|
| Gas | $N_2$ |
| $M_\infty$ | 7.3–9.22 |
| $T_{t_\infty}$ (K) | 1070 |
| $T_w$ (K) | 295–770 |
| $T_w/T_{aw}$ | 0.30–0.68 |
| $H_\infty$ (MJ kg$^{-1}$) | 1.1 |
| $Re_\delta \times 10^{-5}$ | 1–4 |

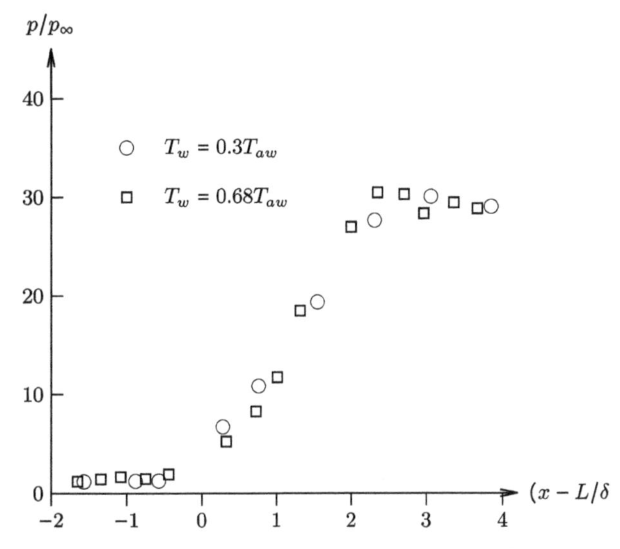


**Figure 5.4.** Effect of wall temperature on pressure distribution for attached flow ($M_\infty = 9.22$, $Re_\delta = 4 \times 10^5$, $\alpha = 26°$).

$Re_\delta$ is the Reynolds number based on the turbulent boundary layer thickness immediately upstream of the interaction. The experimental data comprises surface pressure.

Figures 5.4 and 5.5 display the surface pressure at $M_\infty = 9.22$ and $M_\infty \approx 9$ for attached flow. Figure 5.4 displays results for wall temperature ratios $T_w/T_{aw} = 0.3$ and 0.68 at $\alpha = 26°$ and $Re_\delta = 4 \times 10^5$. The variation in wall temperature has a negligible effect on the surface pressure. Figure 5.5 shows results for Reynolds

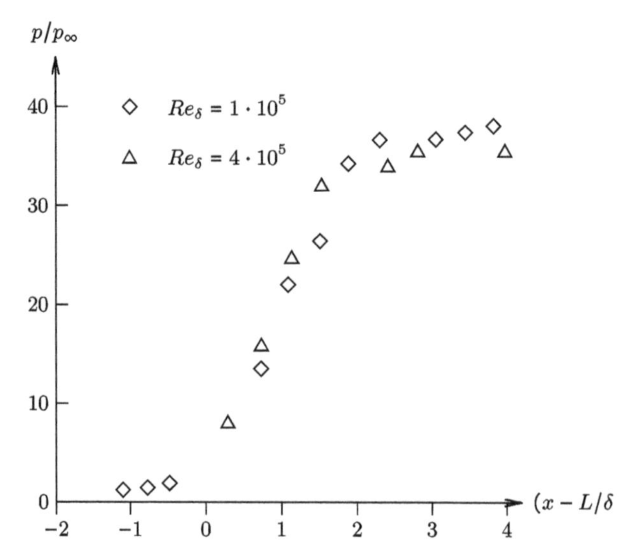


**Figure 5.5.** Effect of Reynolds number on pressure distribution for attached flow ($M_\infty = 9$, $T_w = 0.3T_{aw}$, $\alpha = 30°$).

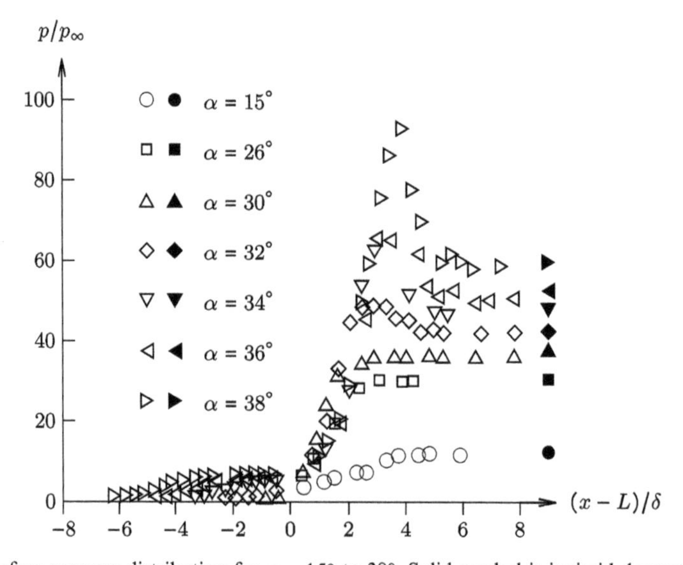


**Figure 5.6.** Surface pressure distribution for $\alpha = 15°$ to 38°. Solid symbol is inviscid downstream pressure.

numbers $Re_\delta = 1 \times 10^5$ and $4 \times 10^5$ for $T_w/T_{aw} = 0.3$ and $\alpha = 30°$. The variation in Reynolds number has a negligible effect on the surface pressure. Note that both results are for attached flows.

Figure 5.6 presents the surface pressure distribution for wedge angles $\alpha = 15°$ to 38° at $M_\infty = 9.22$, $Re_\delta = 4 \times 10^5$ and $T_w/T_{aw} = 0.291$. The separated configurations are characterized by a plateau in the pressure upstream of the corner ($x = L$) and an overshoot in pressure downstream of the corner associated with the interaction of the separation shock and wedge shock.

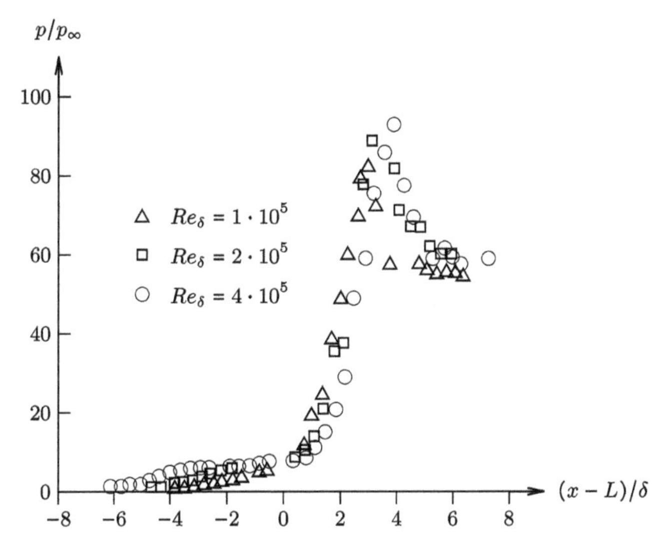


**Figure 5.7.** Effect of $Re_\delta$ on surface pressure for separated flow ($M_\infty = 9$, $T_w = 0.3T_{aw}$, $\alpha = 38°$).

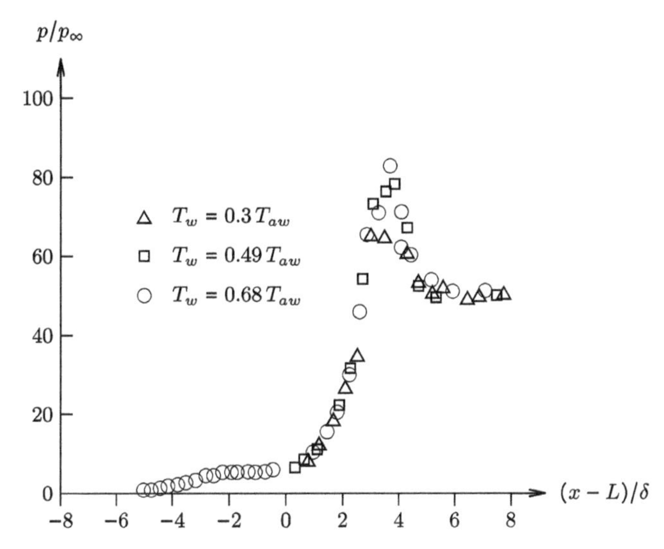


**Figure 5.8.** Effect of wall temperature $T_w/T_{aw}$ on surface pressure for separated flow ($M_\infty = 9.22$, $Re_\delta = 4 \times 10^5$, $\alpha = 36°$).

Figure 5.7 shows the wall pressure for $M_\infty \approx 9$, $T_w/T_{aw} = 0.3$ at $\alpha = 38°$ for $Re_\delta = 1 \times 10^5$ to $4 \times 10^5$. The increase in Reynolds number results in a larger separation region and peak pressure downstream of the corner.

Figure 5.8 displays the wall pressure for $M_\infty = 9.22$, $Re_\delta = 4 \times 10^5$ and $\alpha = 36°$ for $T_w/T_{aw} = 0.3$, 0.49 and 0.68. The only effect of the wall temperature is the peak pressure which increases with $T_w/T_{aw}$.

## 5.2 Coleman and Stollery (1972)

Coleman and Stollery (1972a) conducted an experimental investigation of a hypersonic shock wave turbulent boundary layer interaction on a two dimensional wedge compression corner at Mach 9. The tests were performed in the Imperial College No. 2 Gun Tunnel (section A.13). A schematic of the model is shown in figure 5.3. Wedge angles of 15° to 38° were considered. The freestream conditions are listed in table 5.2. The experiment was selected for the database of Settles and Dodson (1994). The experimental data comprises surface heat transfer measurements and schlieren imaging.

Figures 5.9 and 5.10 display the experimental heat transfer on the plate and wedge for wedge angles from 15° to 38° for Cases 1 and 2, respectively. An enlarged view is presented in figures 5.11 and 5.12. The data is tabulated in Coleman and Stollery (1972b). The principle difference between the two cases is the Reynolds

**Table 5.2.** Freestream conditions.

| | Value | |
|---|---|---|
| Quantity | Case 1 | Case 2 |
| Gas | $N_2$ | $N_2$ |
| $M_\infty$ | 8.96 | 9.22 |
| $T_{t_\infty}$ (K) | 1070 | 1070 |
| $T_w$ (K) | 295 | 295 |
| $T_w/T_{aw}$ | 0.279 | 0.268 |
| $T_\infty$ (K) | 65.5 | 64.5 |
| $H_\infty$ (MJ kg$^{-1}$) | 1.1 | 1.1 |
| $Re \times 10^{-6}$ (m$^{-1}$) | 12.0 | 47.0 |

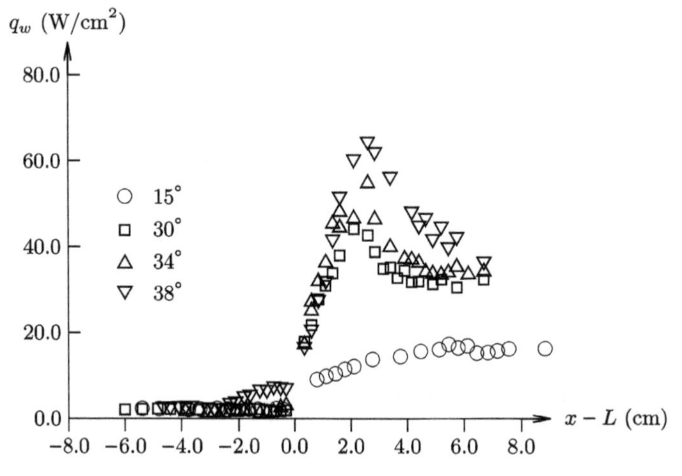


**Figure 5.9.** Heat transfer for Case 1.

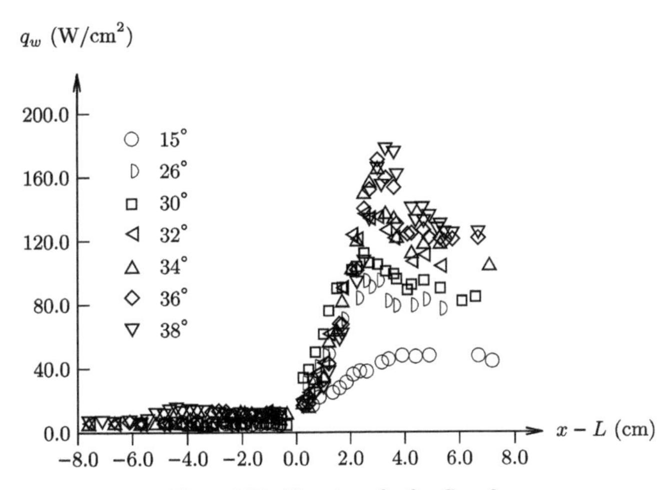


**Figure 5.10.** Heat transfer for Case 2.

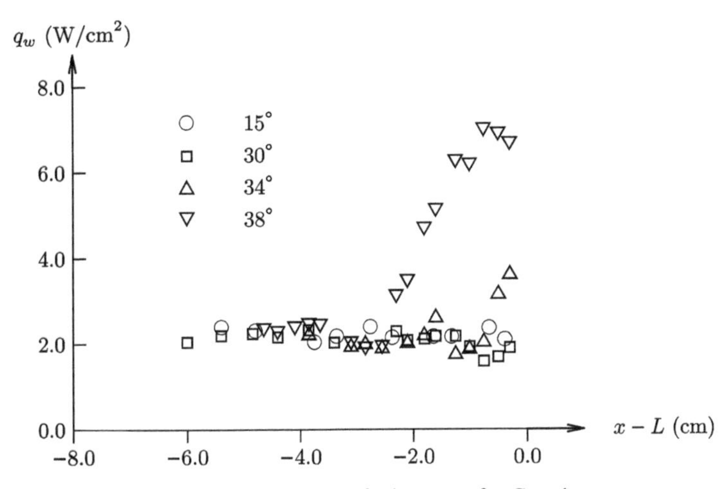


**Figure 5.11.** Heat transfer in corner for Case 1.

number which differs by nearly a factor of four. The higher Reynolds number (Case 2) displays a larger separation region as evidenced by the initial increase in heat transfer upstream of the corner. The angle for incipient separation is approximately the same for both cases at 30° to 32° based upon extrapolation of the separation length measured from schlieren images. Particularly noteworthy is the *increase* in heat transfer in the separation region which is contrary to experimental results for two dimensional laminar separation. Coleman and Stollery (1972a) cite several studies of hypersonic separated flows with similar behavior of the surface heat transfer in the separated region including Holloway *et al* (1965) and Holden (1972). Also, the peak heat transfer is significantly higher for separated flows versus unseparated flows due to the reattachment of the shear layer.

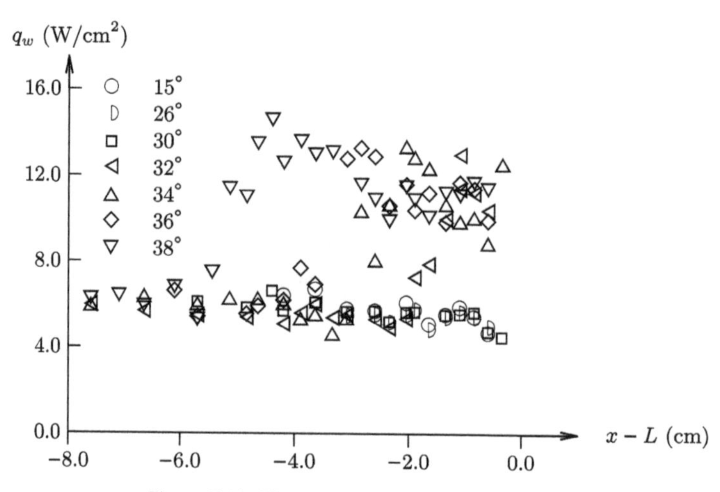


**Figure 5.12.** Heat transfer in corner for Case 2.

## 5.3 Holden (1972)

Holden (1972) performed a series of nominally two dimensional hypersonic shock wave turbulent boundary layer interactions for a compression corner and impinging shock interaction for a Mach number range from 6.25 to 14.9. The experiments were conducted at the Cornell Aeronautical Laboratory (section A.8.1) in the 48 inch and 96 inch shock tunnels. Details of the model geometries are shown in figures 5.13 and 5.14. The model span was 457.2 mm, and the distance from the leading edge to the compression corner and inviscid shock impingement point was varied. The wedge angle $\theta_w$ varied from 27° to 36°, and the shock generator angle $\theta_g$ from 12.5° to 19.8°. The major test conditions are listed in table 5.3. Experimental measurements comprised surface pressure, skin friction and heat transfer, and schlieren imaging.

Figures 5.15 and 5.16 display the surface pressure coefficient $c_p$, skin friction coefficient $c_f$ and heat transfer coefficient $c_h$ versus $x$ for the compression corner at $M_\infty = 8.6$ and $Re_L = 2.25 \times 10^7$ for wedge angles $\theta_w = 27°$, 30°, 33° and 36°. The increase in all three coefficients due to the shock boundary layer interaction is evident, with additional boundary layer separation for the highest wedge angle. Figure 5.17 shows the maximum heat transfer $q_{max}$ normalized by the upstream undisturbed boundary layer heat transfer $q_o$ versus the maximum pressure $p_{max}$ normalized by the corresponding undisturbed pressure. The experimental data is correlated by

$$\frac{q_{max}}{q_o} = \left(\frac{p_{max}}{p_o}\right)^{0.85} \tag{5.1}$$

The corresponding figure in Holden (1972) includes additional experimental data from other sources in agreement with (5.1); however, the data has been excluded in figure 5.17 since references for the other sources is not provided in Holden (1972b).

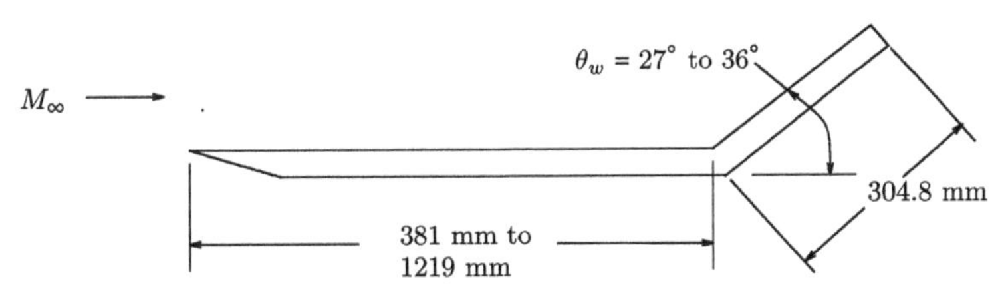


**Figure 5.13.** Compression corner.

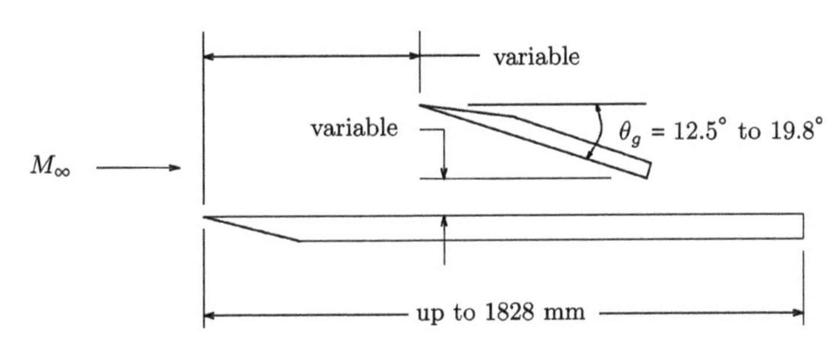


**Figure 5.14.** Impinging shock.

**Table 5.3.** Freestream conditions.

| | Case No. | | | | | | | | | |
|---|---|---|---|---|---|---|---|---|---|---|
| Quantity | 1 | 2 | 3 | 4 | 5 | 6 | 7 | 8 | 9 | 10 |
| $M_\infty$ | 6.25 | 6.50 | 6.60 | 6.60 | 7.90 | 7.90 | 8.60 | 11.40 | 13.0 | 14.9 |
| $T_{t_\infty}$ (K) | 3000 | 778 | 750 | 789 | 867 | 883 | 1011 | 1583 | 1889 | 2333 |
| $T_w/T_{t_\infty}$ | 0.16 | 0.38 | 0.39 | 0.38 | 0.35 | 0.34 | 0.30 | 0.19 | 0.16 | 0.13 |
| $T_\infty$ (K) | 239 | 89 | 83 | 83 | 67 | 67 | 67 | 61 | 61 | 61 |
| $H_\infty$ (MJ kg$^{-1}$) | 3.01 | 0.78 | 0.75 | 0.79 | 0.87 | 0.89 | 1.02 | 1.59 | 1.89 | 2.34 |
| $Re_\infty \times 10^{-6}$ (m$^{-1}$) | 19.4 | 39.4 | 68.9 | 101.7 | 29.5 | 52.5 | 32.2 | 36.1 | 15.4 | 55.8 |

## 5.4 Appels (1973) and Richards and Appels (1973)

Appels (1973) and Richards and Appels (1973) conducted an experimental investigation of a hypersonic shock wave turbulent boundary layer interaction generated by a compression corner at Mach 12. The model, shown in figure 5.18, comprises a flat plate of length $L$ = 38.2 cm and a wedge of length 4.5 cm. The width of the model is 18 cm. The model is oriented at 5° angle relative to the oncoming flow. The wedge angle $\alpha$ was varied from 30° to 42°. The experiments were performed in the von Karman Institute for Fluid Dynamics Longshot Hypersonic Tunnel (section A.48.1). The flow conditions are shown in table 5.4. Experimental diagnostics include surface heat transfer and pressure.

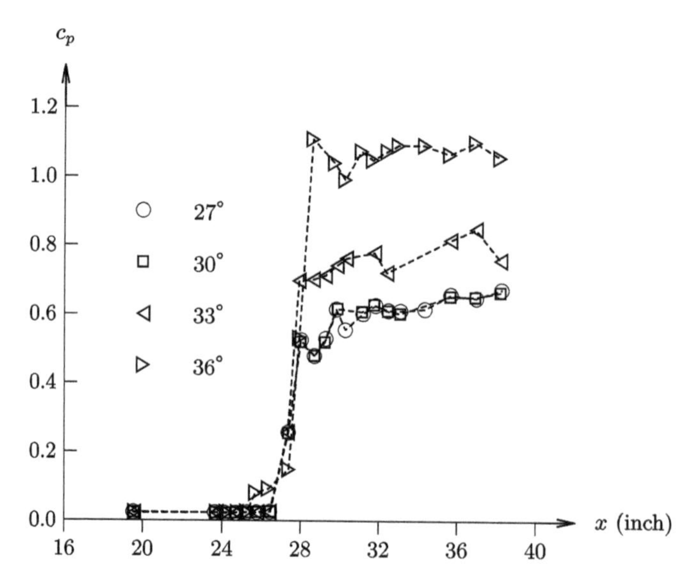


**Figure 5.15.** $c_p$ versus $x$ for compression corner with $\theta = 27°$ to $36°$.

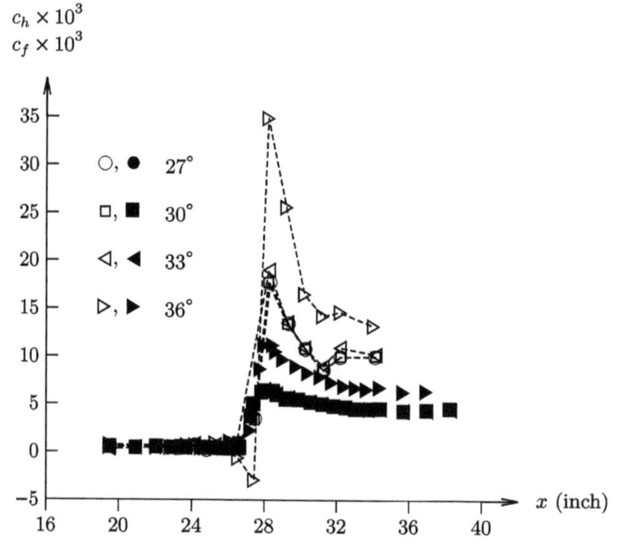


**Figure 5.16.** $c_f$ and $c_h$ versus $x$ for compression corner with $\theta = 27°$ to $36°$. Open symbols ($c_f$) and solid symbols ($c_h$).

Figure 5.19 displays the normalized heat transfer $q_w/q_{w_\infty}$ versus $(x - L)/\delta$ where $q_{w_\infty}$ is the heat transfer on the flat plate immediately upstream of the interaction and $x$ is measured from the leading edge of the flat plate. Peak values of $q_w/q_{w_\infty}$ reach a maximum of 70 at $\alpha = 42°$. Figure 5.20 shows the normalized surface pressure $p/p_\infty$ versus $(x - L)/\delta$. Peak values of $p/p_\infty$ reach a maximum of 100 at $\alpha = 42°$. Both figures correspond to $\delta = 0.65$ cm.

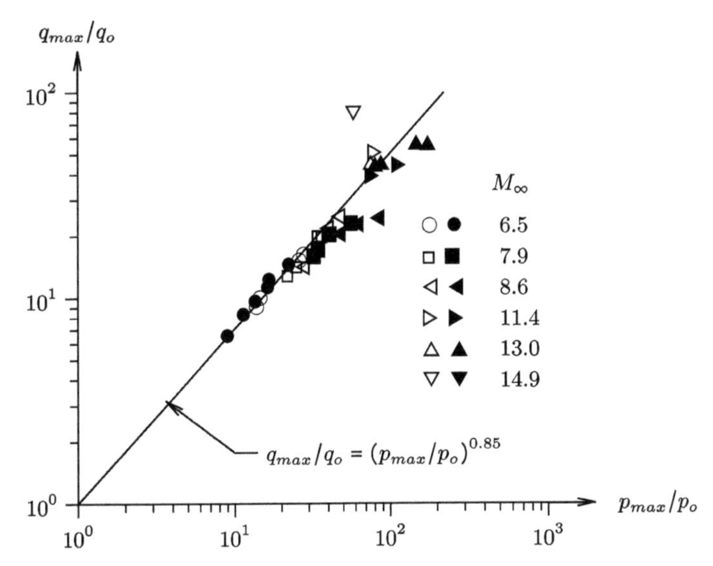


**Figure 5.17.** Correlation of maximum heating for shock wave turbulent boundary layer interactions generated by compression corner (open symbols) or incident shock (solid symbols).

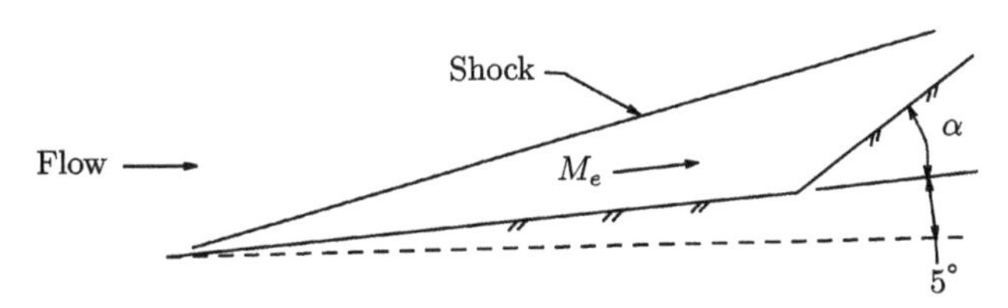


**Figure 5.18.** Model.

**Table 5.4.** Freestream conditions.

| Quantity | Value |
|---|---|
| $M_e$ | 11.7 |
| $p_{t_e}$ (MPa) | 335 |
| $T_{t_e}$ (K) | 2220 |
| $T_w$ (K) | 293 |
| $T_w/T_{aw}$ | 0.15 |
| $Re \times 10^{-6}$ ($m^{-1}$) | 36.6 |
| $\delta$ (cm) | 0.65[a], 0.98[b] |

[a] Natural transition.
[b] Roughness induced.

## 5.5 Coleman (1973)

Coleman (1973a, 1973b) and Coleman and Stollery (1974) conducted an experimental investigation of hypersonic shock wave turbulent boundary layer interaction

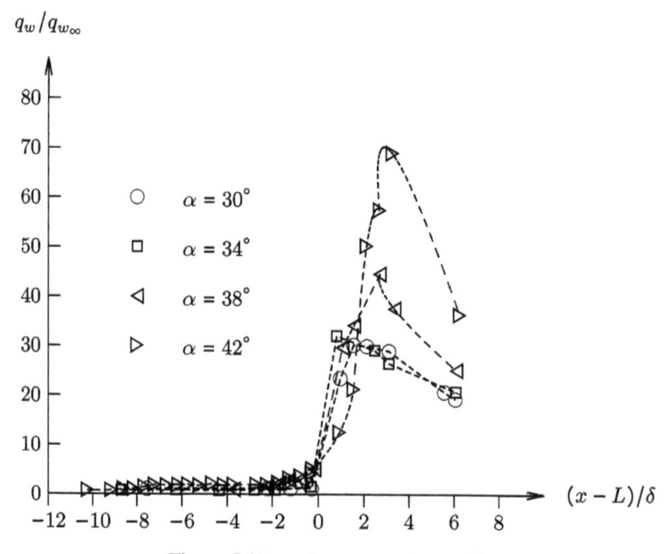


**Figure 5.19.** $q_w/q_{w_\infty}$ versus $(x - L)/\delta$.

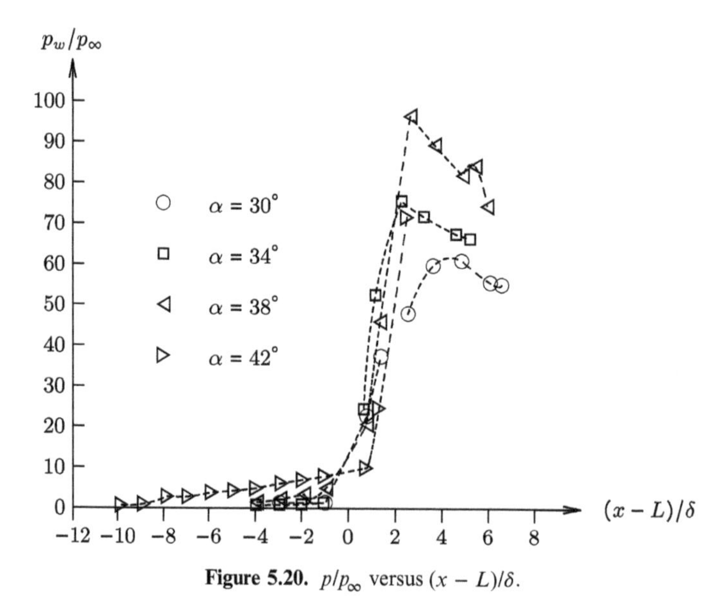


**Figure 5.20.** $p/p_\infty$ versus $(x - L)/\delta$.

for three different axisymmetric geometries at Mach 9.22. The experiments were performed in the Imperial College No. 2 Gun Tunnel (section A.13). The models, shown in figures 5.21(a)–5.21(c), are a hollow cylinder-flare, cone-cylinder-flare and hemisphere cylinder-flare, respectively. The cylinder radius and length are 3.1 cm and 45.7 cm, respectively, for all cases. The cone half-angle is 10°. Flare angles from 15° to 40° were considered. The freestream conditions are listed in table 5.5. Experimental data includes surface pressure and heat transfer, and schlieren visualization.

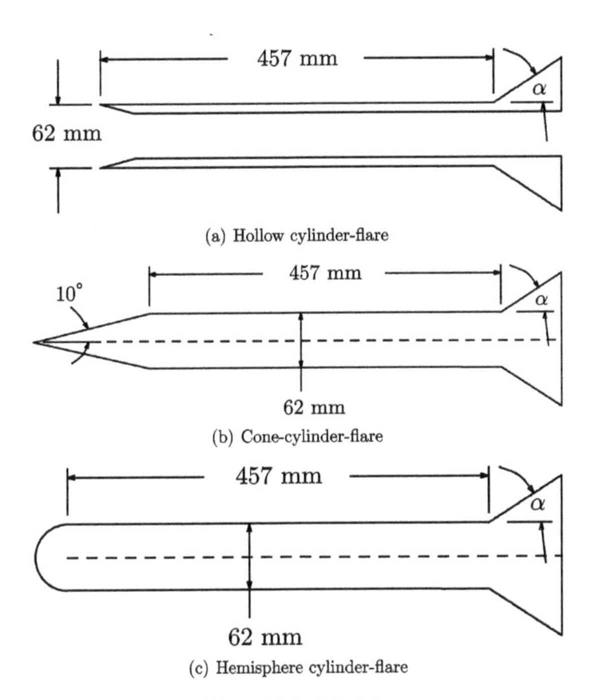


**Figure 5.21.** Models.

**Table 5.5.** Freestream conditions.

| Quantity | Value |
|---|---|
| Gas | $N_2$ |
| $M_\infty$ | 9.22 |
| $T_{t_\infty}$ (K) | 1070 |
| $T_w$ (K) | 295 |
| $T_w/T_{aw}$ | 0.29 |
| $T_\infty$ (K) | 64.5 |
| $H_\infty$ (MJ kg$^{-1}$) | 1.07 |
| $Re_\infty \times 10^{-6}$ (m$^{-1}$) | 47.0 |

Figures 5.22 and 5.23 display the experimental surface pressure $p/p_\infty$ and heat transfer $q_w$ versus $x - L$ for the hollow cylinder-flare for flare angles $\alpha = 15°$ to $40°$ where $x$ is measured from the cylinder leading edge and $L$ is the cylinder length. Also shown in figure 5.22 are the downstream pressure static ratios for the flare at each value of $\alpha$ and the corresponding value for a two dimensional wedge of the same angle. For all angles $\alpha$ except $\alpha = 15°$ the peak pressure ratio exceeds the inviscid value. The $\alpha = 40°$ configuration shows prominent pressure and heat transfer plateaus upstream of the corner corresponding to a separated flow.

Figures 5.24 and 5.25 present the experimental surface pressure $p/p_\infty$ and heat transfer $q_w$ for the cone-cylinder-flare at $\alpha = 15°$, $30°$ and $35°$. The profiles are similar to the hollow cylinder-flare although differing in the peak values for a given

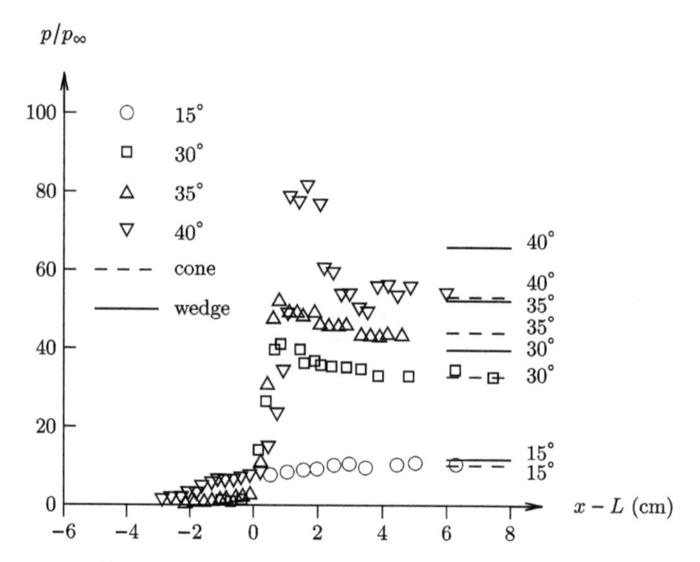


**Figure 5.22.** $p/p_\infty$ versus $x - L$ for hollow-cylinder-flare.

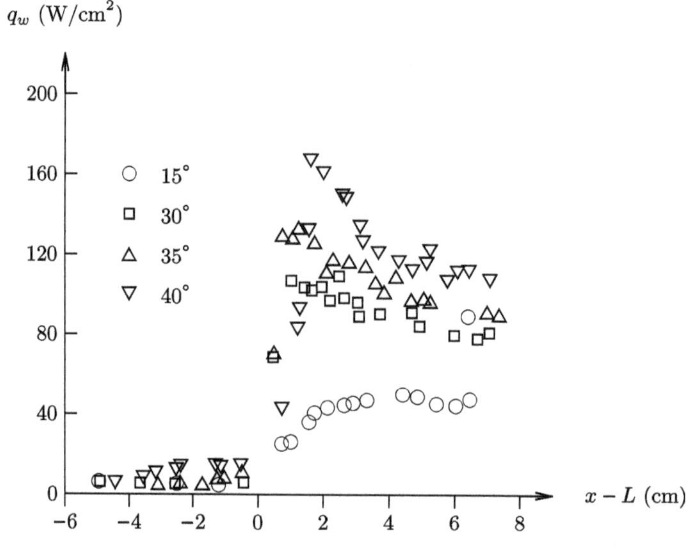


**Figure 5.23.** $q_w$ versus $x - L$ for hollow-cylinder-flare.

flare angle $\alpha$. For example, the peak $p/p_\infty$ for $\alpha = 35°$ are 52.8 and 64.6 for the hollow cylinder-flare and cone-cylinder-flare, respectively, while for $\alpha = 30°$ the corresponding values are 41.0 and 34.3, respectively (i.e., the trend is opposite).

Figures 5.26 and 5.27 display the experimental pressure $p/p_\infty$ and heat transfer $q_w$ for the hemisphere-cylinder-flare. Note that the vertical scales for $p/p_\infty$ and $q_w$ are the same as the cone-cylinder-flare in figures 5.24 and 5.25, respectively. The profiles are significantly different from the hollow cylinder-flare and cone-cylinder-flare. In particular, the peak values are 70% to 75% lower and there is no evidence of a

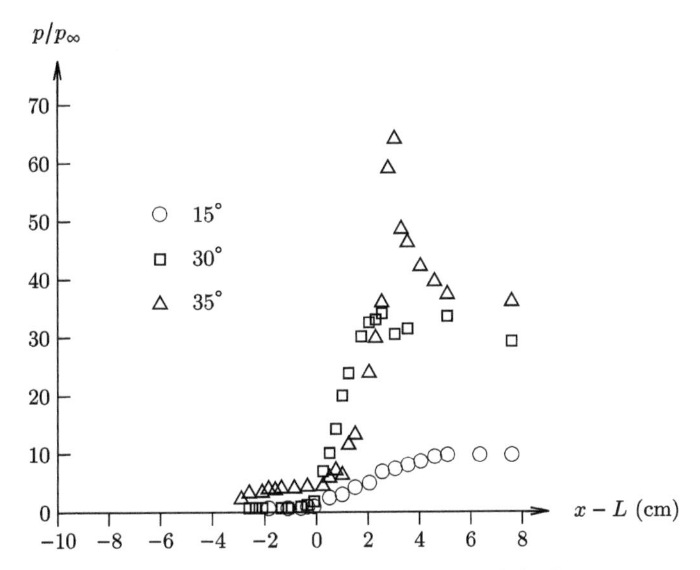


**Figure 5.24.** $p/p_\infty$ versus $x - L$ for cone-cylinder-flare.

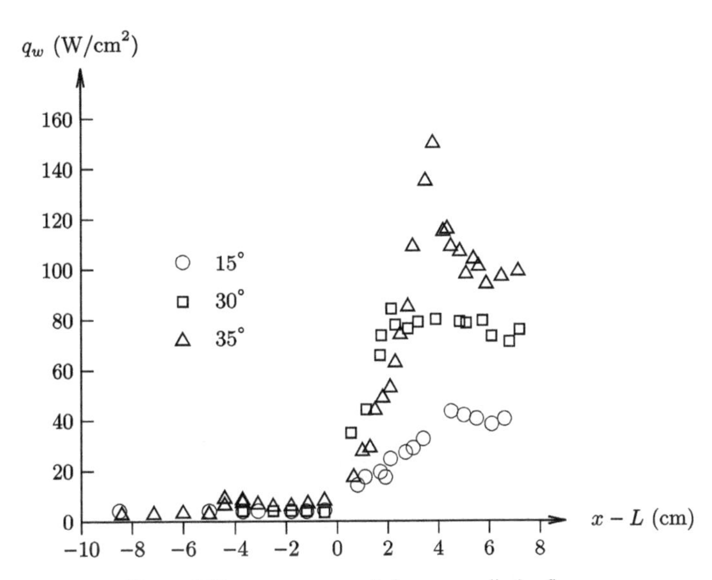


**Figure 5.25.** $q_w$ versus $x - L$ for cone-cylinder-flare.

plateau region in either pressure or heat transfer, implying the boundary layer is unseparated. The effect is a consequence of the detached shock wave generated by the hemisphere and its associated entropy layer.

## 5.6 Kussoy and Horstman (1975)

Kussoy and Horstman (1975) conducted an experimental study of an axisymmetric hypersonic shock wave turbulent boundary layer interaction on a cone-ogive

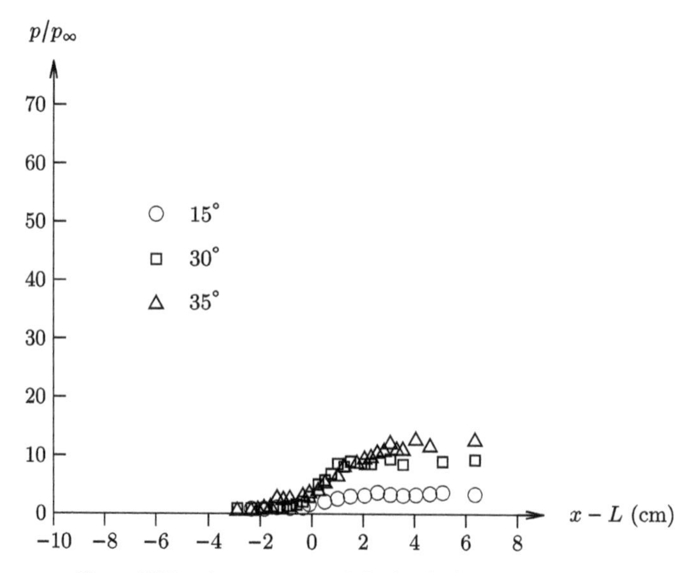


**Figure 5.26.** $p/p_\infty$ versus $x - L$ for hemisphere-cylinder-flare.

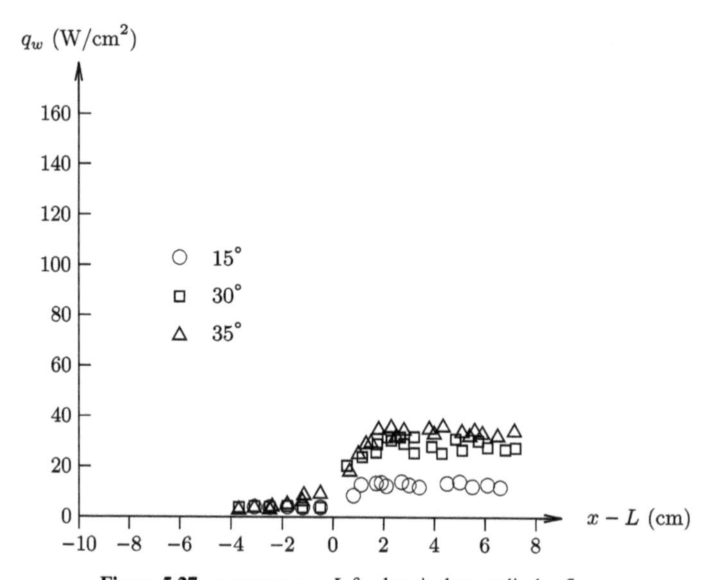


**Figure 5.27.** $p$ versus $x - L$ for hemisphere-cylinder-flare.

cylinder at a nominal freestream Mach number of 7.2. The experiments were performed in the NASA Ames 3.5 ft hypersonic wind tunnel (section A.20.2). The model is shown in figure 5.28. The cone-ogive cylinder length is 3.3 m, and cylinder diameter is 0.203 m. An annular ring with wedge leading edge and outer diameter 0.51 m generates an axisymmetric shock wave turbulent boundary layer interaction on the cylinder. Wedge angles of 7.5° and 15° were considered, corresponding to unseparated and separated shock boundary layer interactions, respectively. The freestream and edge conditions are shown in table 5.6. Experimental measurements

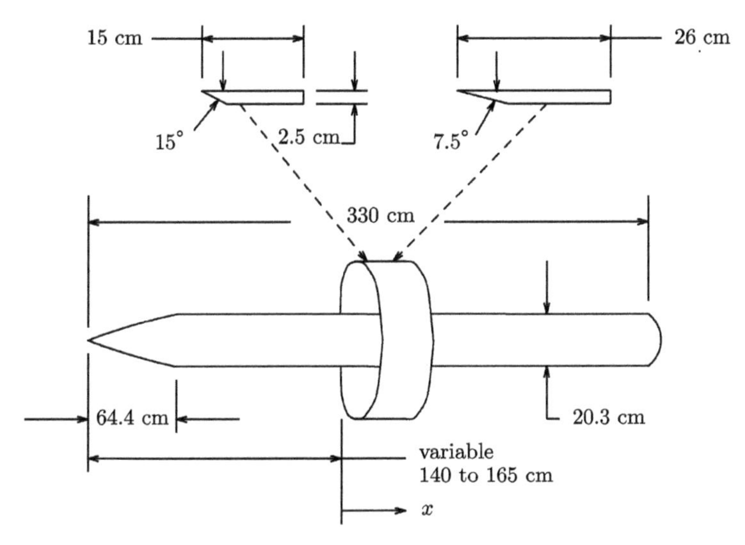


**Figure 5.28.** Model.

**Table 5.6.** Freestream conditions.

| Quantity | Value $\alpha = 7.5°$ | $\alpha = 15°$ |
|---|---|---|
| Gas | Air | Air |
| $M_\infty$ | 7.2 | 7.2 |
| $M_e$ | 6.71 | 6.86 |
| $p_{t_e}$ (MPa) | 1.92 | 2.32 |
| $T_{t_e}$ (K) | 695 | 695 |
| $T_w$ (K) | 300 | 300 |
| $T_w/T_{aw}$ | 0.45 | 0.45 |
| $H_\infty$ (MJ kg$^{-1}$) | 0.70 | 0.70 |
| $\delta$ (cm)[a] | 3.50 | 3.40 |
| $Re_e \times 10^{-6}$ (m$^{-1}$) | 7.1 | 7.7 |

[a] At 85 cm from nosetip.

include surface pressure, heat transfer and skin friction, and boundary layer profiles of pitot pressure, static pressure and total temperature throughout the interaction. The experiment was selected for the database of Settles and Dodson (1994).

The aerothermodynamic loading on the cylinder surface is displayed in figures 5.29–5.31. The data includes surface pressure (Figure 5.29), surface skin friction coefficient $c_f$ (Figure 5.30) and Stanton number $St$ (figure 5.31). Also shown is the Reynolds Analogy Factor $2St/c_f$ (figure 5.32). The expansion fan originating from the trailing edge of the wedge intersects the shock wave boundary layer on the cylinder as seen in the decrease in pressure downstream of the peak values in figure 5.29. The wall shear stress remains positive for $\alpha = 7.5°$ but becomes negative for $\alpha = 15°$ (figure 5.30)

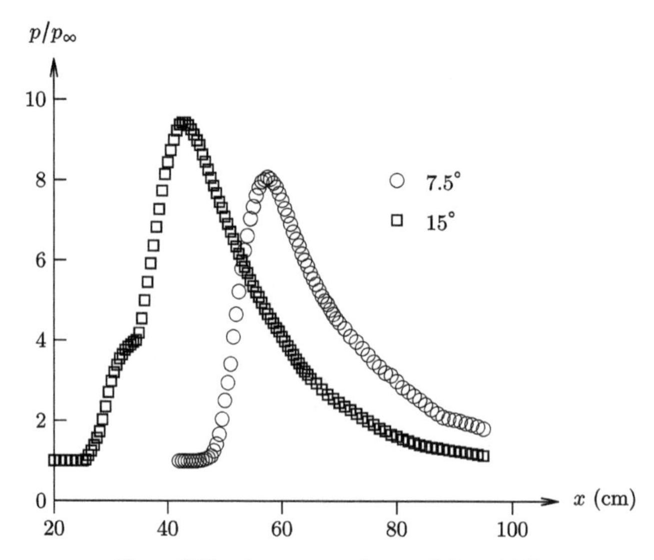


**Figure 5.29.** $p/p_\infty$ versus $x$ for $\alpha = 7.5°$ and 15°.

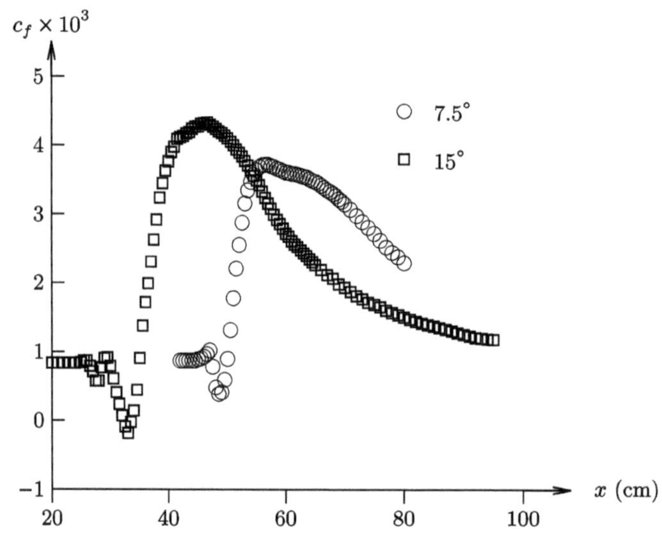


**Figure 5.30.** $c_f$ versus $x$ for $\alpha = 7.5°$ and 15°.

indicating boundary layer separation. The surface heat flux rises throughout the interaction to its peak value for both unseparated and separated flow cases. A significant number of boundary layer profiles of pitot and static pressure, and total temperature throughout the interaction are tabulated in Kussoy and Horstman (1975). The Reynolds Analogy Factor (section 3.4) is shown in figure 5.32 for $\alpha = 7.5°$ and 15°, respectively. The analogy holds for the incoming boundary layer, and in the region downstream of the interaction where the boundary layer has relaxed to near equilibrium. However, the analogy is not satisfied within the interaction region.

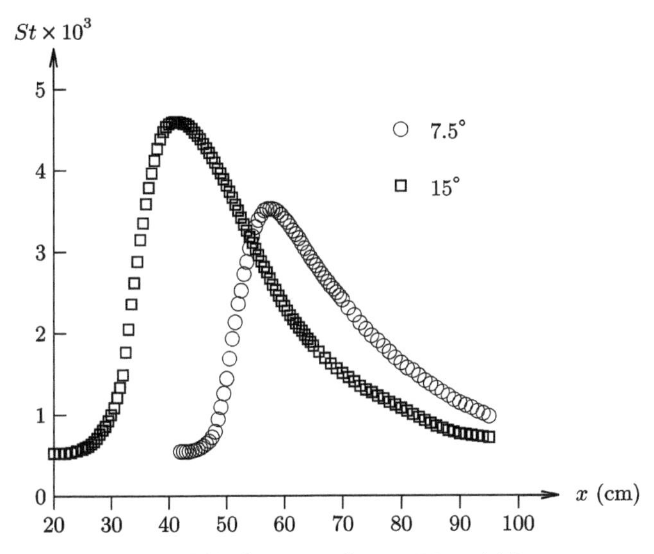


**Figure 5.31.** St versus $x$ for $\alpha$ = 7.5° and 15°.

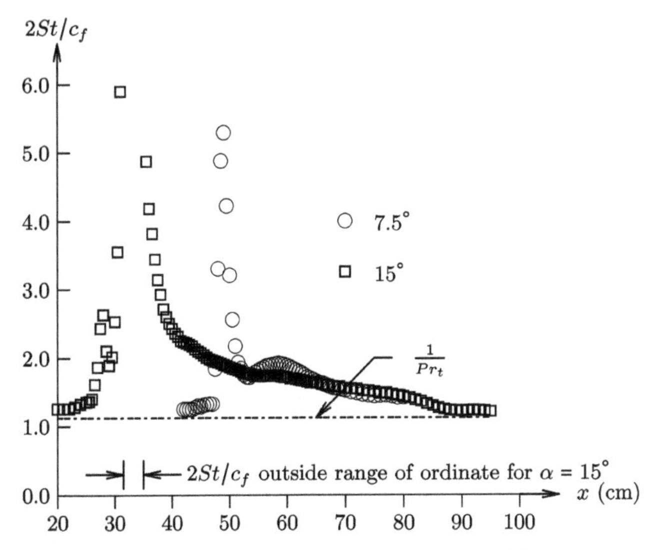


**Figure 5.32.** $2St/c_f$ versus $x$ for $\alpha$ = 7.5° and 15°.

## 5.7 Law (1975)

Law (1975) performed an experimental investigation of a three dimensional hypersonic shock wave turbulent boundary layer interaction at Mach 5.90 generated by a normal sharp fin attached to a flat plate. The experiments were performed in the Mach 6 blowdown wind tunnel at the US Air Force Aerospace Research Laboratories (ARL)[1] (section A.2.3). The model is shown in figure 5.33. The leading

[1] Aerospace Research Laboratories were closed in 1975 and the building was demolished in 2001.

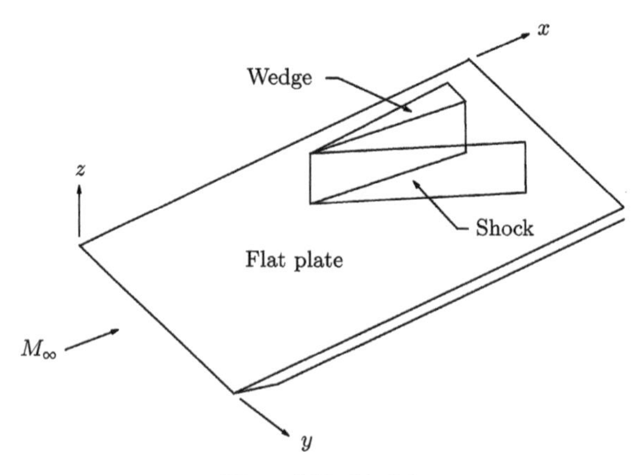


**Figure 5.33.** Model.

**Table 5.7.** Freestream conditions.

| Quantity | Value: Case 1 | Value: Case 2 |
|---|---|---|
| $M_\infty$ | 5.9 | 5.9 |
| $p_{t_\infty}$ (MPa) | 5.61 | 14.6 |
| $T_{t_\infty}$ (K) | 609 | 614 |
| $T_w$ (K) | 297 | 297 |
| $T_w/T_{aw}$ | 0.51 | 0.51 |
| $H_\infty$ (MJ kg$^{-1}$) | 0.61 | 0.62 |
| $\delta$ (cm)[a] | 0.38 | 0.30 |
| $Re_e \times 10^{-6}$ (m$^{-1}$) | 32.8 | 98.4 |

[a] Estimated at leading edge of fin.

edge of the fin is nominally 21.6 cm from the plate leading edge. The flowfield conditions are listed in table 5.7. Fin angles from 4° to 20° were examined corresponding to inviscid static pressure ratios from 1.75 to 9.0. Experimental measurements include surface pressure and heat transfer, and surface oil flow visualization. The experiment was selected for the database of Settles and Dodson (1994).

Figures 5.34 and 5.35 display the normalized surface pressure and heat transfer coefficients for Case Nos. 1 and 2 at $\alpha = 20°$ versus $y/y_s$ where $y$ is the spanwise distance measured from the fin surface and $y_s$ is the corresponding distance to the location of the inviscid shock wave. The separation and primary reattachment lines are shown for Case 1. Peak surface pressure and heat transfer coefficient are located at the line of attachment.

## 5.8 Mikulla and Horstman (1976)

Mikulla and Horstman (1976a) and Mikulla and Horstman (1976b) examined a hypersonic shock wave cold wall turbulent boundary layer interaction at Mach 7.5

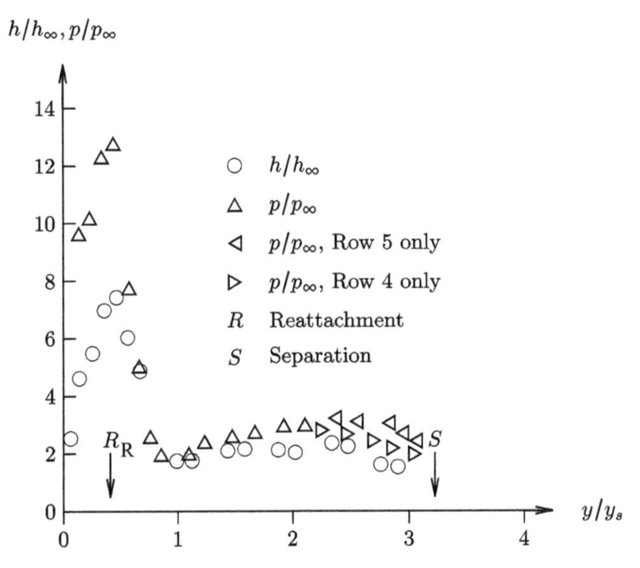


**Figure 5.34.** $p/p_\infty$ and $h/h_\infty$ for Case 1 at $\alpha = 20°$.

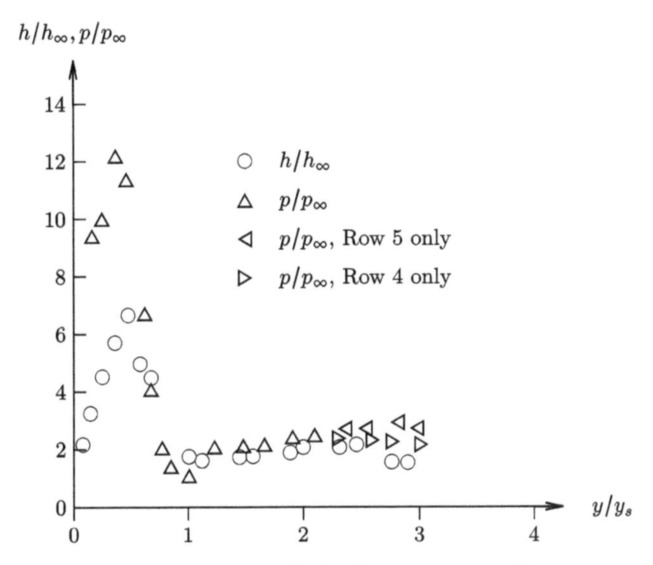


**Figure 5.35.** $p/p_\infty$ and $h/h_\infty$ for Case 2 at $\alpha = 20°$.

for a cone-ogive-cylinder. The experiments were performed in the NASA Ames 3.5 ft hypersonic wind tunnel (section A.20.2). The model is shown in figure 5.36. An annular ring with wedge leading edge generates an axisymmetric shock wave turbulent boundary layer interaction on the cylinder. Wedge angles of 7.5° and 15° were examined, resulting in unseparated and separated shock wave boundary layer interactions, respectively. The freestream and edge conditions are listed in table 5.8. Experimental measurements include mean surface pressure and skin

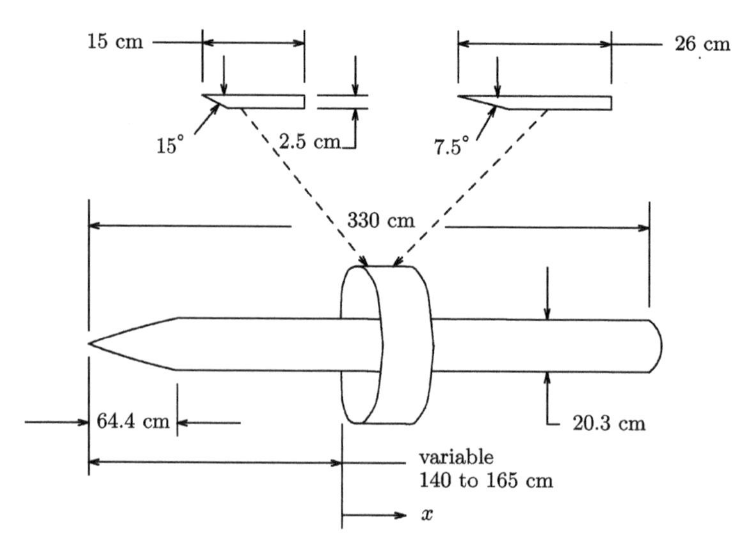


**Figure 5.36.** Model.

**Table 5.8.** Freestream conditions.

| Quantity | Value | |
|---|---|---|
| | $\alpha = 7.5°$ | $\alpha = 15°$ |
| $M_\infty$ | 7.2 | 7.2 |
| $M_e$ | 6.7 | 6.9 |
| $p_{t_e}$ (MPa) | 3.45 | 3.45 |
| $T_{t_e}$ (K) | 695 | 695 |
| $T_w$ (K) | 300 | 300 |
| $T_w/T_{aw}$ | 0.45 | 0.45 |
| $H_\infty$ (MJ kg$^{-1}$) | 0.70 | 0.70 |
| $\delta$ (cm) | 3.2 | 2.7 |
| $Re \times 10^{-6}$ (m$^{-1}$) | 7.2 | 7.2 |

friction, and hot wire (constant temperature) anemometer boundary layer profiles at four streamwise locations as indicated in table 5.9.

Figures 5.37–5.40 show the mean surface pressure and skin friction for the attached and separated cases, respectively, where $x_o$ is the theoretical location of the impingement of the inviscid shock on the cylinder. The locations of the four survey stations is also indicated. The separated case displays the characteristic plateau in the pressure distribution upstream of the peak pressure (Babinsky and Harvey 2011).

Figures 5.41 and 5.42 show the normalized root-mean-square mass flux fluctuations for the attached and separated cases, respectively. A significant increase in the fluctuations occurs at the second survey location corresponding to the impingement of the incident shock wave with peak values exceeding 40% of the local mean mass

**Table 5.9.** Location of boundary layer surveys.

| | Description | |
|---|---|---|
| Location | $\alpha = 7.5°$ | $\alpha = 15°$ |
| 1 | Undisturbed boundary layer | |
| 2 | Minimum $c_f$ | Middle of separation bubble |
| 3 | Peak pressure | |
| 4 | Downstream of interaction | |

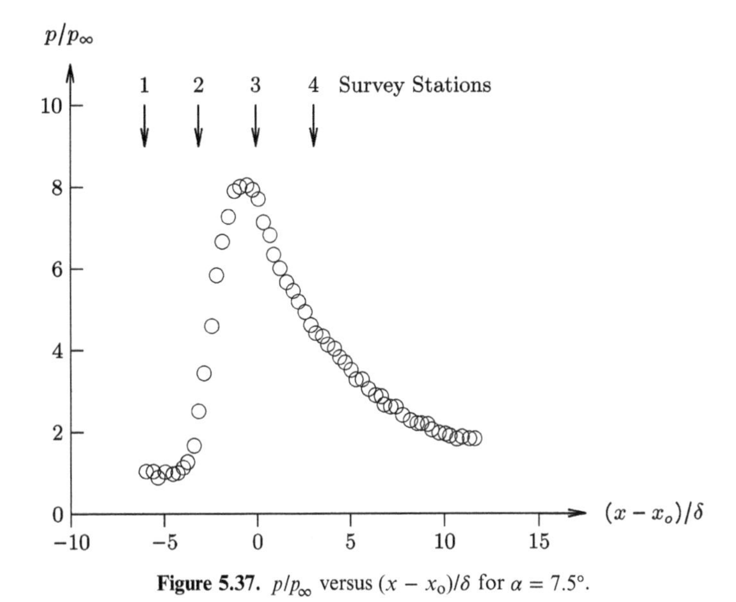


**Figure 5.37.** $p/p_\infty$ versus $(x - x_o)/\delta$ for $\alpha = 7.5°$.

flux. The fluctuations decrease downstream of the interaction as the boundary layer relaxes towards equilibrium. Figures 5.43 and 5.44 display the corresponding profiles for the vertical velocity fluctuations. Peak values reach 30% of the local streamwise velocity at the impingement of the incident shock wave.

Figures 5.45 and 5.46 show the non-dimensional Reynolds stress for the attached and separated cases, respectively. A significant increase is apparent at the second survey station consistent with the results for the mass flux and vertical velocity fluctuations. Most interestingly the relaxation of the Reynolds stress towards equilibrium is far more rapid for the separated case than the attached case. Additional data for power spectra is also presented in Mikulla and Horstman (1976a).

## 5.9 Holden (1977)

Holden (1977) performed an experimental investigation of hypersonic shock wave turbulent boundary layer interaction at high Reynolds numbers. See also Holden

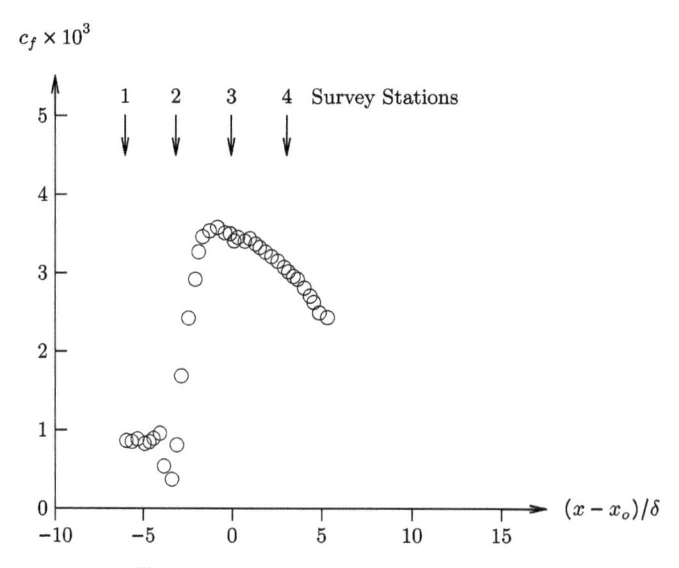


**Figure 5.38.** $c_f$ versus $(x - x_o)/\delta$ for $\alpha = 7.5°$.

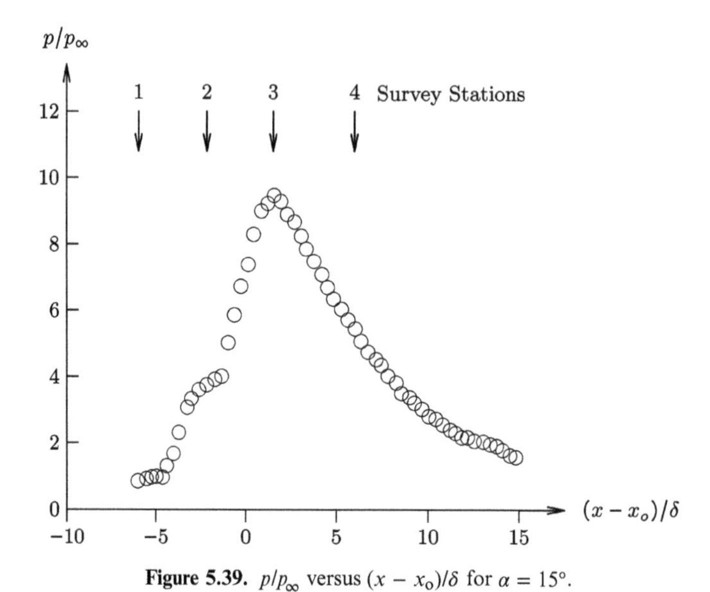


**Figure 5.39.** $p/p_\infty$ versus $(x - x_o)/\delta$ for $\alpha = 15°$.

(1975). The experiments were conducted at the 98-inch Shock Tunnel at Calspan (section A.8.1) with Mach number range from 6.8 to 13 and Reynolds numbers $Re_\delta$ range from $10^5$ to $10^7$. Experimental diagnostics are skin friction, surface pressure, oil flow, and heat transfer measurements. The nominally two dimensional models are a wedge compression corner (Figure 5.47) and a flat plate with an impinging shock (figure 5.48). The freestream conditions are described in table 5.10.

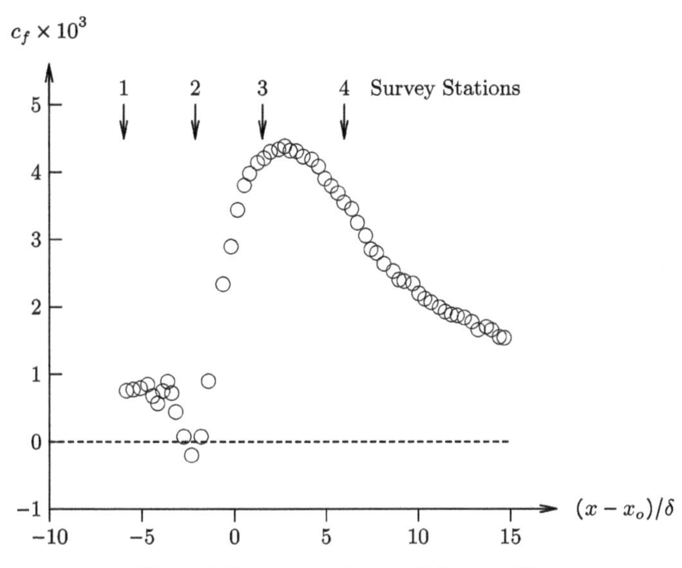


**Figure 5.40.** $c_f$ versus $(x - x_o)/\delta$ for $\alpha = 15°$.

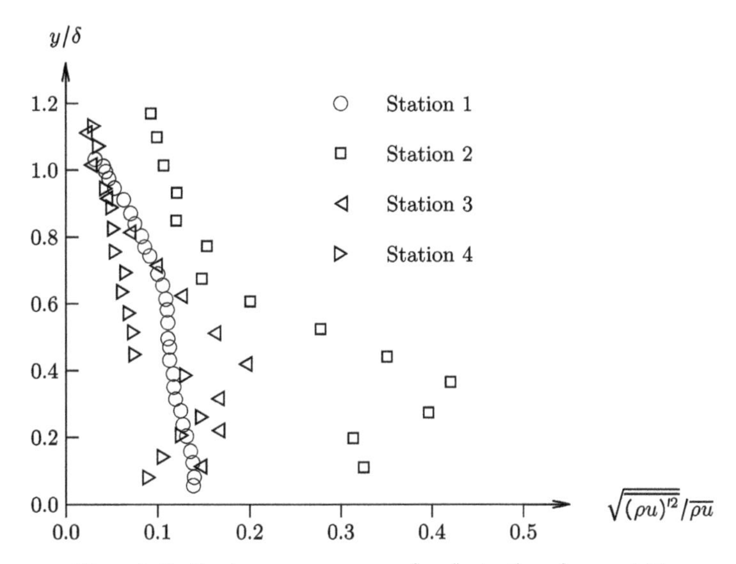


**Figure 5.41.** Root-mean-square mass flux fluctuations for $\alpha = 7.5°$.

Figures 5.49 and 5.50 show the effect of the interaction strength—which is directly related to the wedge angle $\theta_w$—on the surface heat transfer coefficient $c_h$ and pressure coefficient $c_p$ for the compression corner. The location of the corner is indicated. The increase in the wedge angle increases the separation region size. It is reported that the separation bubble is unsteady and thus, large fluctuation exists in the measured surface pressure and heat transfer at the reattachment point.

Figures 5.51 and 5.52 show the effect of the interaction strength—which is directly related to the shock generator angle $\theta_w$—on the surface heat transfer and

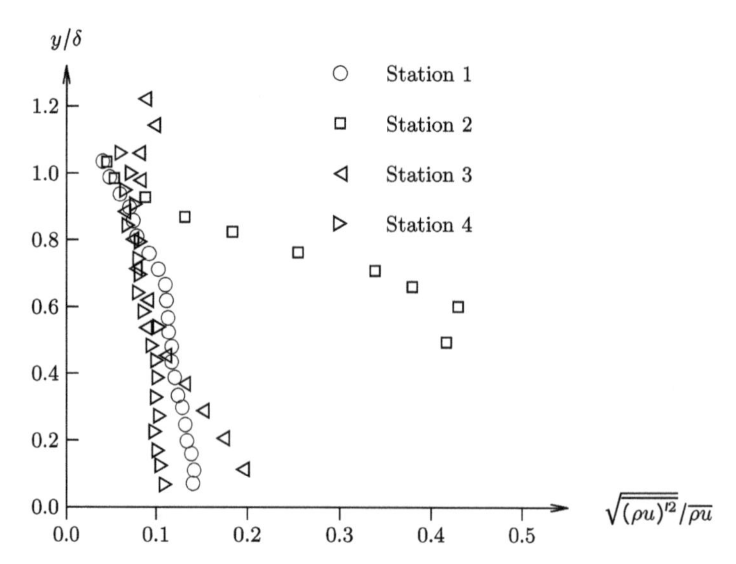


**Figure 5.42.** Root-mean-square mass flux fluctuations for $\alpha = 15°$.

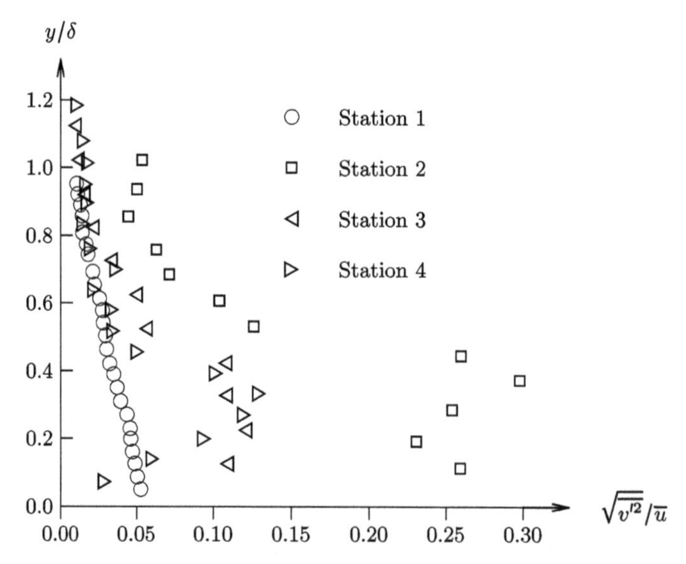


**Figure 5.43.** Root-mean-square vertical velocity fluctuations for $\alpha = 7.5°$.

pressure for an impinging shock wave boundary layer interaction. The location of the inviscid shock impingement is indicated. The separated region for $\theta_w = 17.5°$ is indicated by the plateau region in $c_h$ and $c_p$ for $x = 24$ inch to 25 inch.

## 5.10 Neumann and Hayes (1977)

Neumann and Hayes (1977b) presented an experimental study of the three dimensional shock wave turbulent boundary layer interaction generated by a sharp fin on a

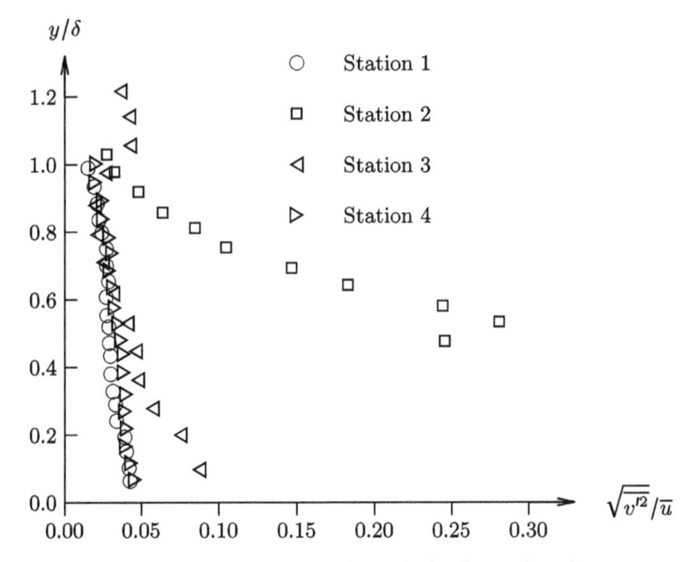


**Figure 5.44.** Root-mean-square vertical velocity fluctuations for $\alpha = 15°$.

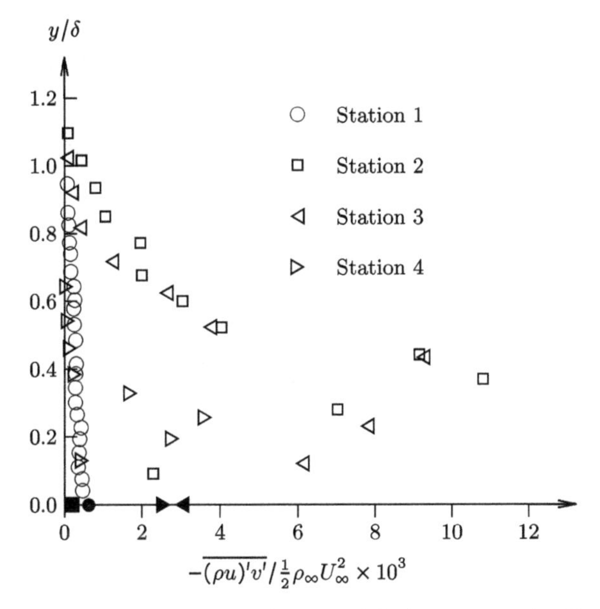


**Figure 5.45.** Reynolds shear stress for $\alpha = 7.5°$ (filled symbols indicate measured wall values of shear stress).

flat plate at Mach 2.95 to 5.85. Experiments at Mach 4.75 and 5.04 were conducted at the Arnold Engineering Development Center VKF Tunnels A and B (section A.5.1), and at Mach 5.85 in the Air Force Flight Dynamics Laboratory Mach 6 High Reynolds Number Facility (section A.2.3). Additional experimental data was obtained at Princeton University at Mach 2.95, and at McDonnell Douglas Corporation at Mach 3.71. See also Hayes (1977) and Neumann and Hayes (1977a).

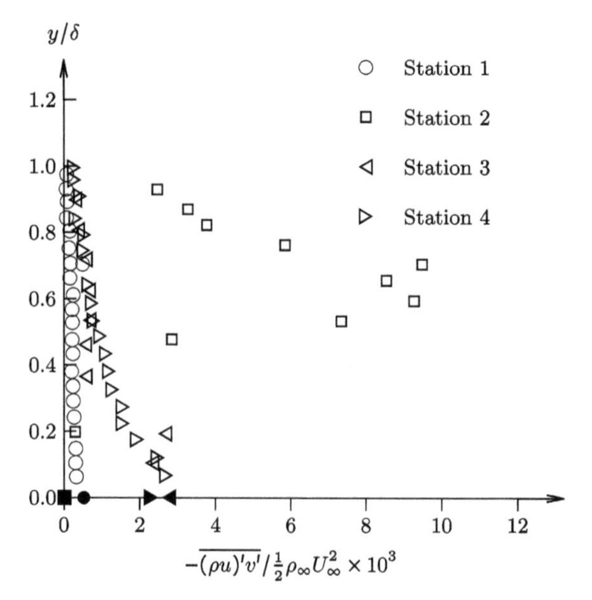


**Figure 5.46.** Reynolds shear stress for $\alpha = 15°$ (filled symbols indicate measured wall values of shear stress).

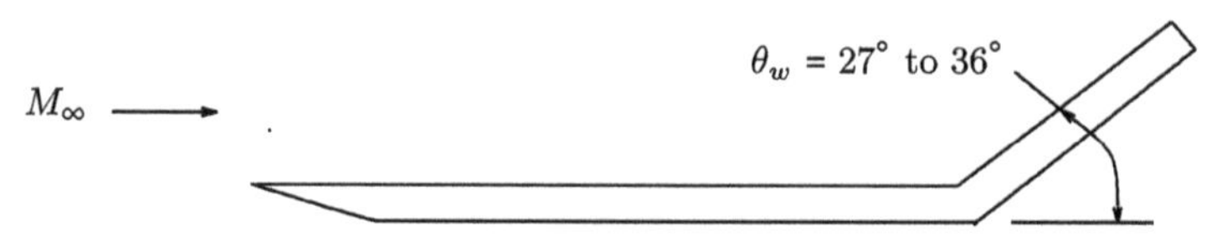


**Figure 5.47.** Compression corner.

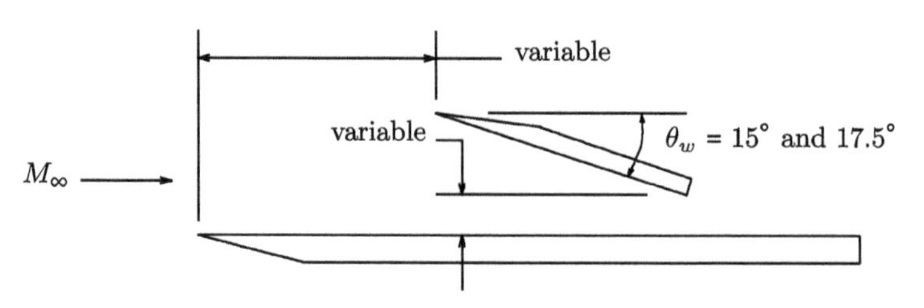


**Figure 5.48.** Impinging shock.

**Table 5.10.** Freestream conditions.

| Quantity | Wedge | Impinging shock |
|---|---|---|
| $M_\infty$ | 8.0 and 8.2 | 8.0 and 8.2 |
| $p_{t_\infty}$ (MPa) | na | na |
| $T_{t_\infty}$ (K) | na | na |
| $T_w$ (K) | 288[a] | 288[a] |
| $Re \times 10^{-6}$ ($m^{-1}$) | 106 and 137 | 137 |

[a] Assumed room temperature.

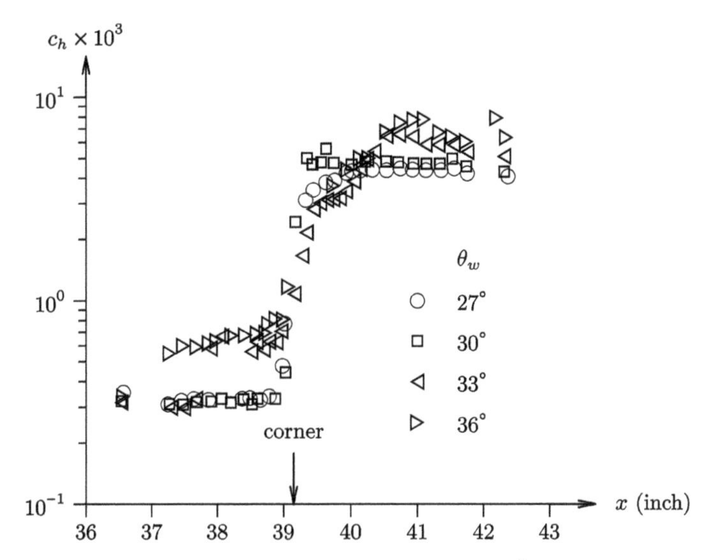


**Figure 5.49.** $c_h$ versus $x$ at $M_\infty = 8.2$ and $Re = 1.37 \times 10^8$ for $\theta_w = 27°$ to 36°.

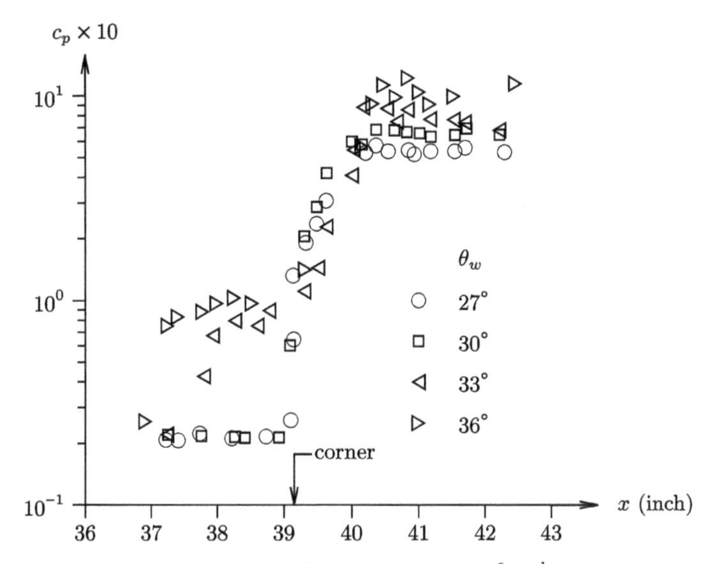


**Figure 5.50.** $c_p$ versus $x$ at $M_\infty = 8.2$ and $Re = 1.37 \times 10^8$ m$^{-1}$ for $\theta_w = 27°$ to 36°.

The model is shown in figure 5.53. The range of freestream conditions is shown in table 5.11.

The experimental data for the peak pressure $p_{max}$ within the interaction on the flat plate for Mach 2.95 to 5.85 was found to be approximated by

$$\frac{p_{max}}{p_\infty} = (M_\infty \sin \theta)^{n_p} \tag{5.2}$$

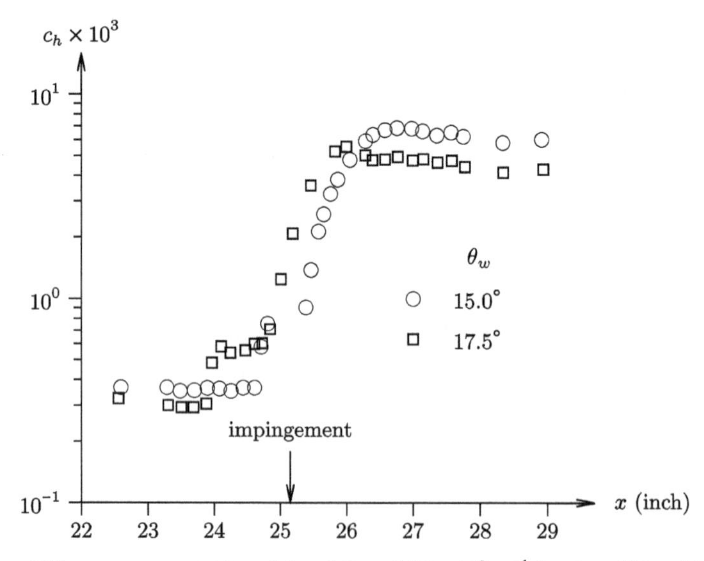


**Figure 5.51.** $c_h$ versus $x$ at $M_\infty = 8.2$ and $Re = 1.37 \times 10^8$ m$^{-1}$ for $\theta_w = 15°$ and 17.5°.

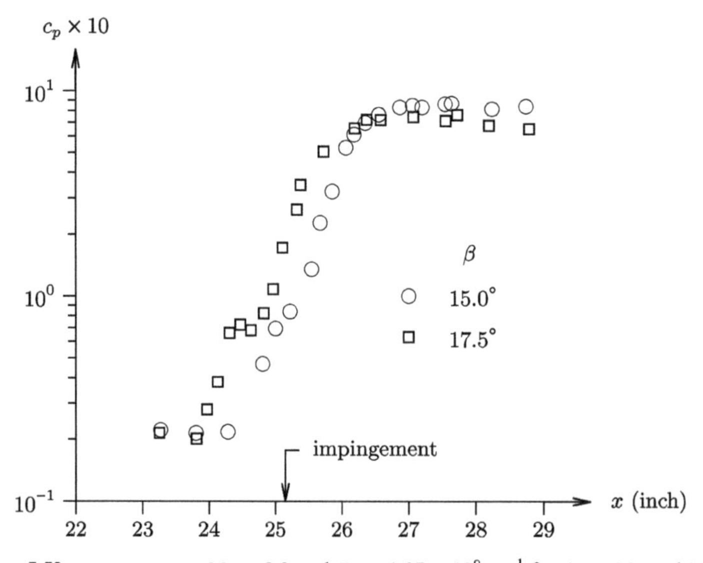


**Figure 5.52.** $c_p$ versus $x$ at $M_\infty = 8.2$ and $Re = 1.37 \times 10^8$ m$^{-1}$ for $\theta_w = 15°$ and 17.5°.

where $\theta$ is the inviscid shock angle and $n_p$ is a function of $x/\delta$ where $x$ is measured from the fin leading edge and $\delta$ is the incoming boundary layer thickness. Figure 5.54 displays the experimental value of $n_p$ versus $x/\delta$ based on the full Mach number range. The location of the peak pressure on the flat plate occurred along a ray from the fin leading edge approximated by Token (1974)

$$\phi = 0.24(\theta - \alpha) + \alpha \tag{5.3}$$

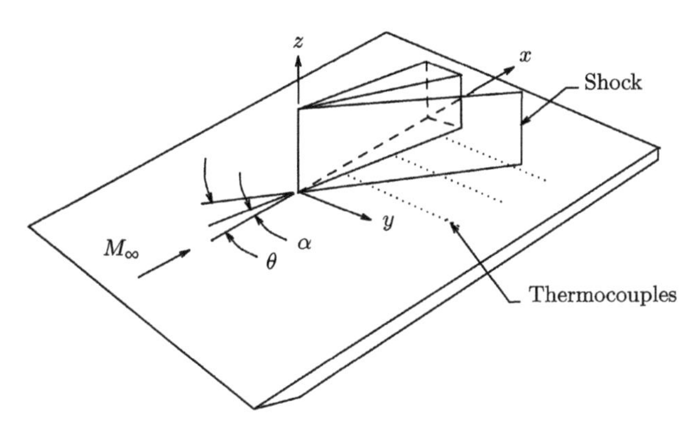


**Figure 5.53.** Model.

**Table 5.11.** Freestream conditions.

| $M_\infty$ | $Re \times 10^{-6}$ m$^{-1}$ | $L^\dagger$ (m) | Reference |
|---|---|---|---|
| 2.95 | 63.0 | n/a | Oskam *et al* (1965) |
| 3.0 | 11.5 | 0.787 | Hayes (1977) |
| 3.01 | 10.6 | 0.267 | Neumann and Token (1974) |
| 3.71 | 4.92 | 0.787 | Token (1974) |
| 3.71 | 11.5 | 0.787 | Token (1974) |
| 3.75 | 10.8 | 0.267 | Neumann and Token (1974) |
| 4.5 | 10.6 | 0.267 | Hayes (1977) |
| 4.75 | 25.5 | 0.483 | Hayes (1977) |
| 5.04 | 24.2 | 0.483 | Hayes (1977) |
| 5.85 | 36.1 | 0.216 | Christophel and Rockwell (1974) |
| 5.85 | 91.9 | 0.216 | Christophel and Rockwell (1974) |
| 5.95 | 16.4 | 1.067 | Neumann and Hayes (1977b) |
| 6.05 | 11.4 | 0.452 | Neumann and Burke (1969) |

† Distance to fin apex.

where $\alpha$ is the fin angle. Figure 5.55 displays the agreement between the above correlation and experiment for the Mach number range from 3.0 to 5.85. A correlation for the peak Stanton number was developed

$$\frac{St_{\text{peak}}}{St_\infty} = n_{st}(M_\infty \sin\theta - 1) + 0.75 \tag{5.4}$$

where $n_{St}$ is a function of $x/\delta$ and shown in figure 5.56.

## 5.11 Zheltovodov *et al* (1979 and later)

Zheltovodov and Pavlov (1979), Zheltovodov (1979a), Zheltovodov *et al* (1983), Zheltovodov and Yakovlev (1986), Zheltovodov *et al* (1987) and Zheltovodov *et al* (1987) conducted an experimental investigation of the interaction of a near

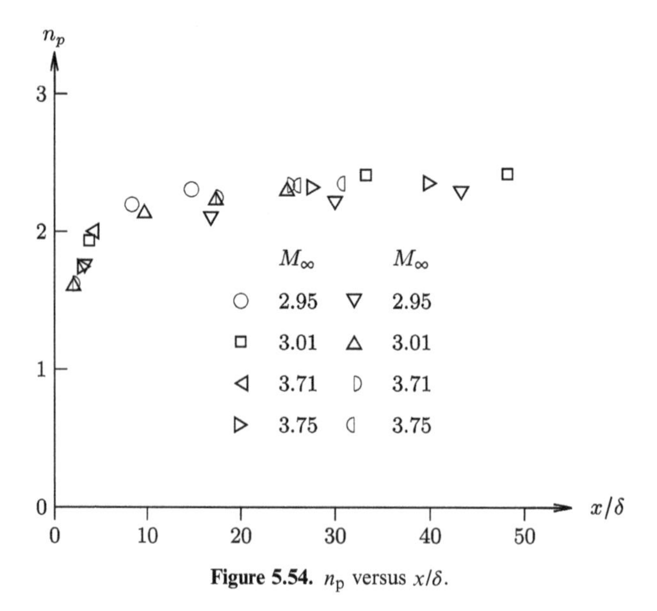


**Figure 5.54.** $n_p$ versus $x/\delta$.

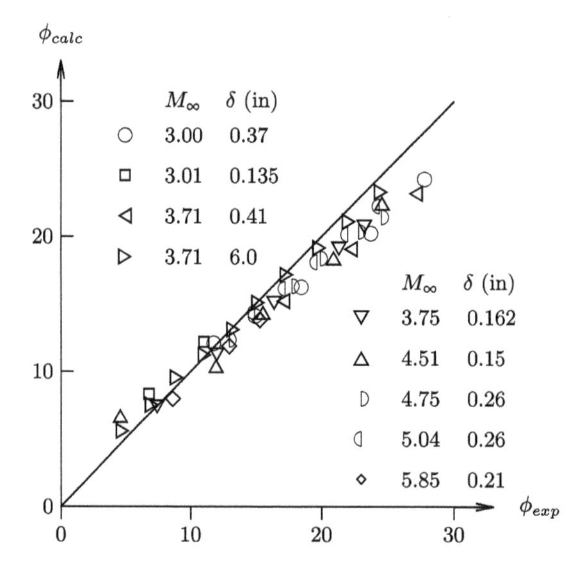


**Figure 5.55.** Correlation of location for peak pressure.

adiabatic turbulent boundary layer in an expansion-compression corner and compression-expansion corner at Mach numbers up to four. The models for the mean flow measurements are shown in figure 5.57 and dimensions in table 5.12. The mean flow, turbulent fluctuation and heat flux measurements were performed in wind tunnels T-313 (section A.18.2), T-325 and T-333 at the Institute of Theoretical and Applied Mechanics, Siberian Division of the Russian Academy of Sciences, Novosibirsk, Soviet Union. The flow conditions for the mean flow measurements are listed in table 5.13. Experimental measurements include surface pressure and heat

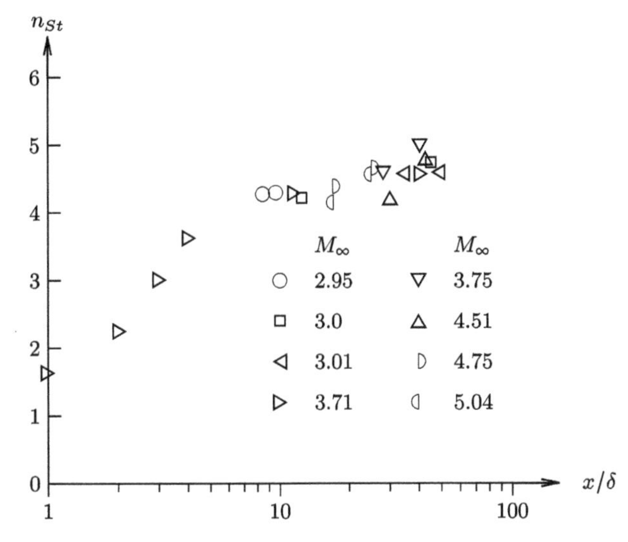


**Figure 5.56.** $n_{St}$ versus $x/\delta$

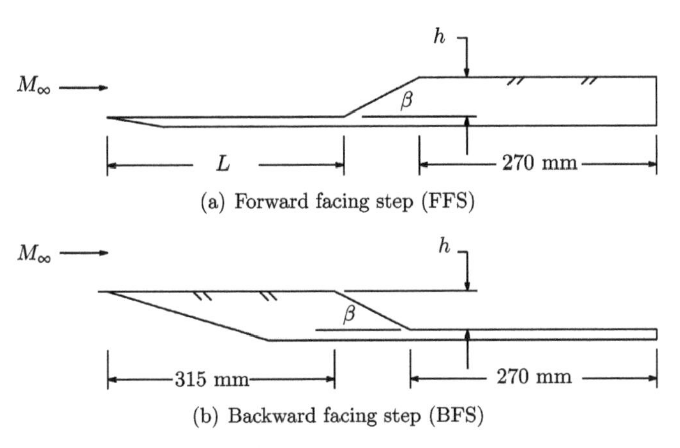


**Figure 5.57.** Models.

transfer, and boundary layer surveys of pitot pressure, static pressure density and turbulence statistics. A compendium of the experimental data is presented in Zheltovodov *et al* (1990). See also Zheltovodov and Schülein (1988), Zheltovodov *et al* (1989, 1990), and Zheltovodov (1991).

Figure 5.58 illustrates the experimental flowfield structure of the expansion-compression interaction for BFS25. The upstream boundary of the expansion fan and the approximate edge of the boundary layer are indicated. The subsequent turning of the flow following the expansion results in a shock wave turbulent boundary layer interaction and formation of a separation bubble indicated by the separtion (S) and reattachment (R) locations. Figure 5.59 displays the surface pressure $p$ normalized by the pressure $p_1$ immediately upstream of the expansion

**Table 5.12.** Model dimensions.

| Case | $\beta$ (deg) | $h$ (mm) | $L$ (mm) |
|---|---|---|---|
| FFS 8° | 8 | 15 | 218 |
| FFS 25° | 25 | 15 | 293 |
| FFS 45° | 45 | 15 | 319 |
| FFS 90° | 90 | 15 | 315 |
| LFFS 90° | 90 | 5 | 315 |
| BFS 8° | 8 | 15 | 315 |
| BFS 25° | 25 | 15 | 315 |
| BFS 45° | 25 | 15 | 315 |

**Table 5.13.** Freestream conditions.

| Quantity | Group 1 | Group 2 |
|---|---|---|
| $M_\infty$ | 3.0 | 4.0 |
| $p_{t_\infty}$ (MPa) | 0.41 to 0.42 | 1.05 |
| $T_{t_\infty}$ (K) | 265 to 294 | 269 to 270 |
| $H_\infty$ (MJ kg$^{-1}$) | 0.27 to 0.30 | 0.27 |
| $Re \times 10^{-6}$ (m$^{-1}$) | 32.4 to 38.3 | 58.2 to 58.3 |

Group 1: FFS 8°, 25°, 45°, 90°; LFFS 90°; BFS 8°, 25° 45°
Group 2: BFS 8°, 25°

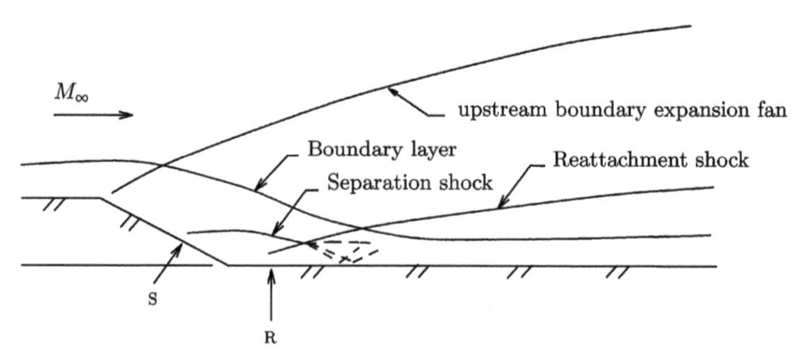


**Figure 5.58.** Flowfield structure for BFS25.

versus distance $x$ measured from the expansion corner. The rapid drop in pressure due to the expansion is evident beginning at $x = 0$. The recompression of the flow results in a shock wave boundary layer interaction and increase in wall pressure to nearly the upstream value. Figure 5.60 shows the wall heat flux $q_w$ normalized by the heat flux $q_{w_1}$ immediately upstream of the interaction versus distance $x$ for three different Reynolds numbers. The rapid expansion reduces the heat flux by 40%, and

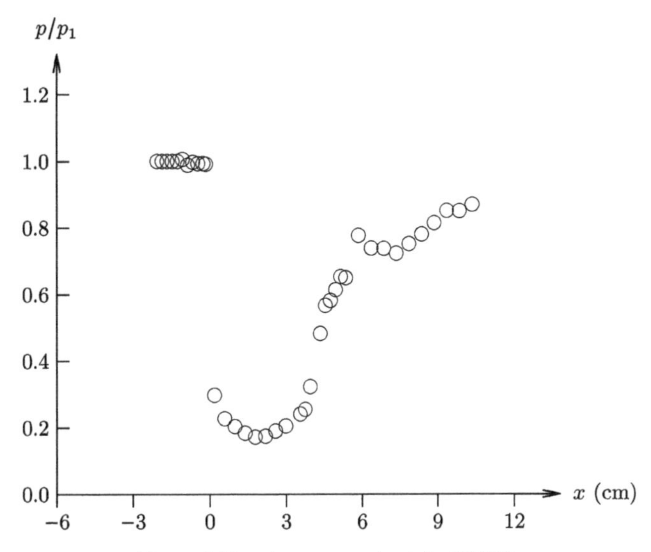


**Figure 5.59.** $p/p_1$ versus $x$ (cm) for BFS25.

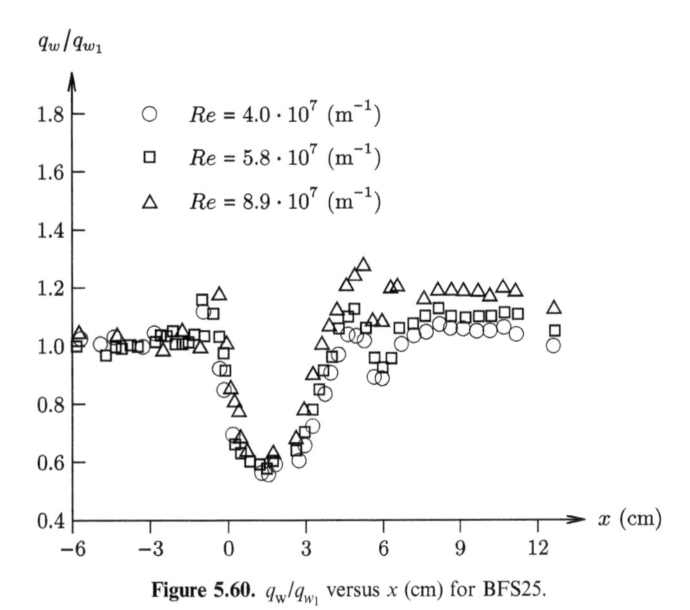


**Figure 5.60.** $q_w/q_{w_1}$ versus $x$ (cm) for BFS25.

the resultant shock boundary layer interaction causes an increase in heat flux to approximately the upstream value.

## 5.12 Zheltovodov (1982)

Zheltovodov (1982), Zheltovodov *et al* (1987) and Zheltovodov *et al* (1990) investigated the interaction of a swept shock wave generated by a sharp fin with a turbulent boundary layer at Mach numbers up to four. The model is shown in figure 5.61. The flow conditions are listed in table 5.14. Experimental diagnostics

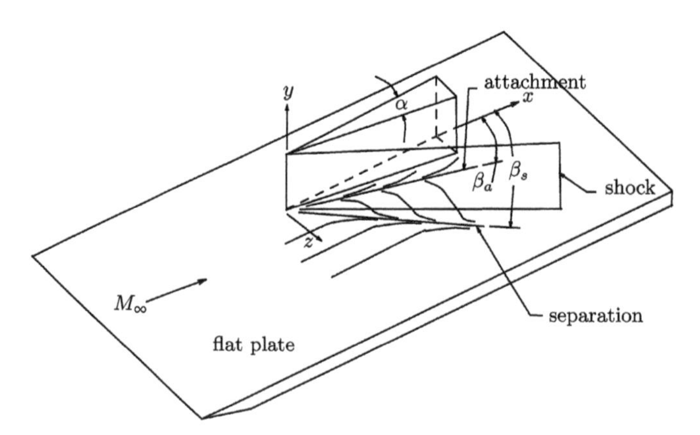


**Figure 5.61.** Model.

**Table 5.14.** Freestream conditions.

| Quantity | Value |
|---|---|
| $M_\infty$ | 2.0 to 4.0 |
| $Re \times 10^{-6}$ ($m^{-1}$) | 30 to 36 |

include surface pressure, surface flow visualization and Laser Doppler Velocimetry (LDF). See also Zheltovodov *et al* (1987), Zheltovodov (1996, 2004) and Zheltovodov (2006).

The interaction of the shock generated by the fin angle $\alpha$ with the turbulent boundary layer on the flat plate results in a separation for sufficiently large pressure rise. Figure 5.62 shows the surface pressure versus $\beta - \alpha$ for $\alpha = 20°$ and 30.6° at Mach 4 where $\beta$ is the cylindrical polar angle measured along the flat plate from the streamwise direction. The former data were obtained at a cylindrical radius $r/\delta_o = 29.6$ from the fin leading edge, and the latter at a streamwise distance $x/\delta_o = 25$ where $\delta_o$ is the boundary layer thickness upstream of the interaction. The plateau region associated with the separated zone is indicated. Figure 5.63 presents the separation line angle $\beta_S$ versus $\alpha$ at Mach 4. The variation with the fin angle is approximately linear.

The flowfield structure shown in figure 5.64 is approximately conical viewed in a spherical polar coordinate system whose origin is located slightly displaced from the intersection of the fin leading edge with the flat plate (Zheltovodov 1979b). In the conical framework the crossflow direction is towards the fin with normal Mach number $M_n$. The shock structure comprises a $\lambda$ shock with a shear layer, expansion and normal shock. Depending on the stregth of the fin generated shock, both a primary $S_1$ and secondary $S_2$ separation occur, with concommitant attachments $A_1$

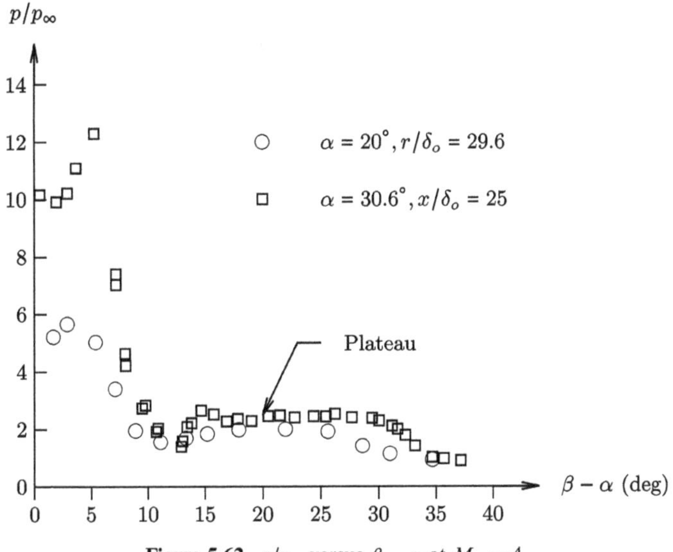


**Figure 5.62.** $p/p_\infty$ versus $\beta - \alpha$ at $M_\infty = 4$.

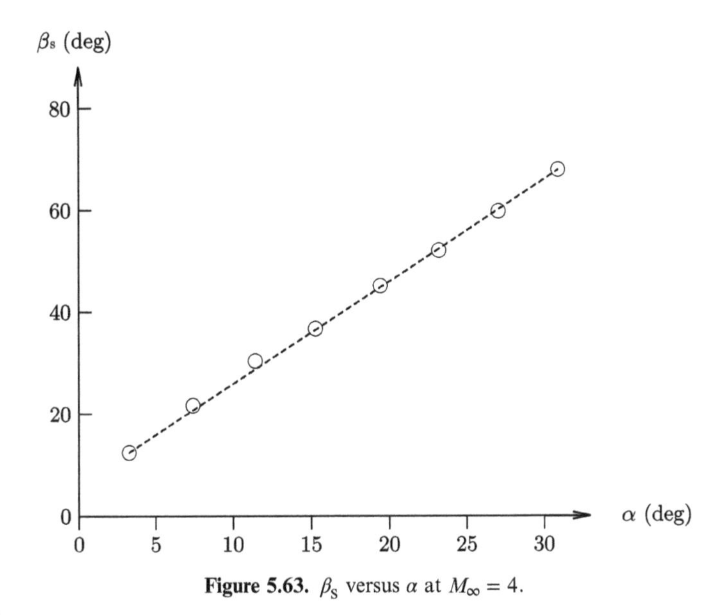


**Figure 5.63.** $\beta_s$ versus $\alpha$ at $M_\infty = 4$.

and $A_2$. A similar model for the flowfield structure is presented in Alvi and Settles (1991) and Alvi and Settles (1992). See also Korkegi (1976).

## 5.13 Holden (1984)

Holden (1984) conducted an experimental study of three dimensional shock wave turbulent boundary layer interactions on a cold wall at Mach 11.3. The experiments were conducted in the Calspan 96 inch shock tunnel (section A.8.1). The models are shown in figures 5.65(a) and 5.65(b). The skewed shock model is defined by the

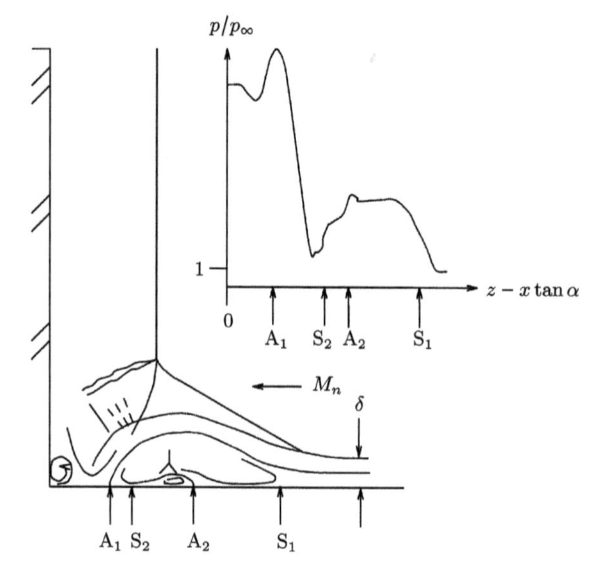


**Figure 5.64.** Flowfield structure.

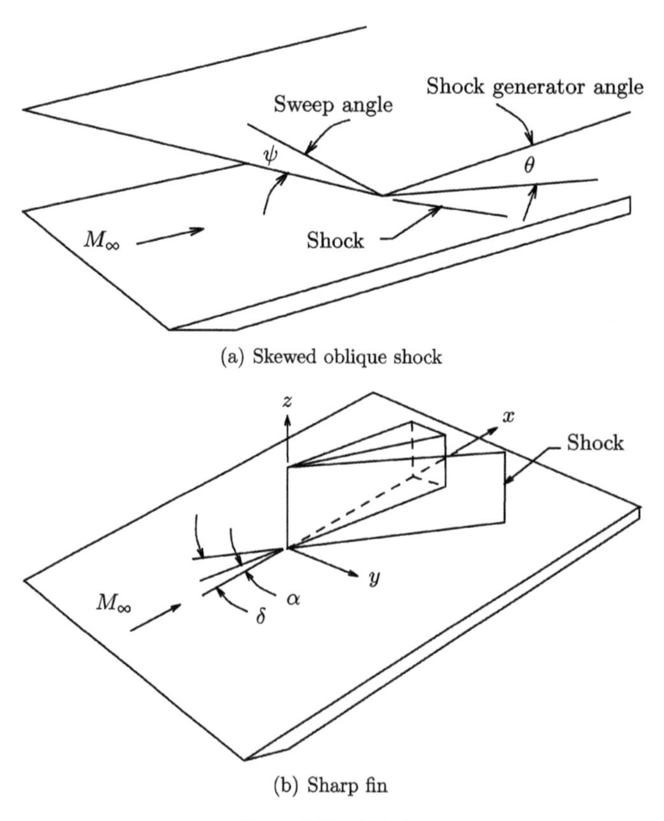


**Figure 5.65.** Models.

sweep angle $\psi$ of the leading edge and the shock generator angle $\theta$ measured in a vertical plane parallel to the freestream. Sweep angles $\psi$ from 0° to 45° were examined at shock generator angles $\theta = 12.5°$ and 15°. Sharp fin angles $\alpha = 5°$ to 12.5° were studied. Nominal freestream conditions are listed in table 5.15. Experimental diagnostics include surface heat transfer and pressure.

Figure 5.66 displays the normalized pressure coefficient $c_p/c_{p_o}$ versus streamwise distance $x - x_s$ for shock generator angle $\theta = 15°$ and sweep angle $\psi = 0°$ to 45° where $x_s$ is the theoretical shock impingement point at the spanwise location corresponding to the experimental data. The streamwise extent of the interaction region is independent of the sweep angle. Figure 5.67 shows the normalized Stanton number for the same configurations. The streamwise extent of the interaction region is similarly independent of the sweep angle.

**Table 5.15.** Freestream conditions.

| Quantity | Value |
|---|---|
| $M_\infty$ | 11.3 |
| $p_{t_\infty}$ (MPa) | 120.5 |
| $T_{t_\infty}$ (K) | 1497 |
| $T_w$ (K) | 300[a] |
| $T_w/T_{aw}$ | 0.20 |
| $T_\infty$ (K) | 62.2 |
| $H_\infty$ (MJ kg$^{-1}$) | 1.66 |
| $Re \times 10^{-6}$ (m$^{-1}$) | 31.6 |

[a] Assumed ambient temperature.

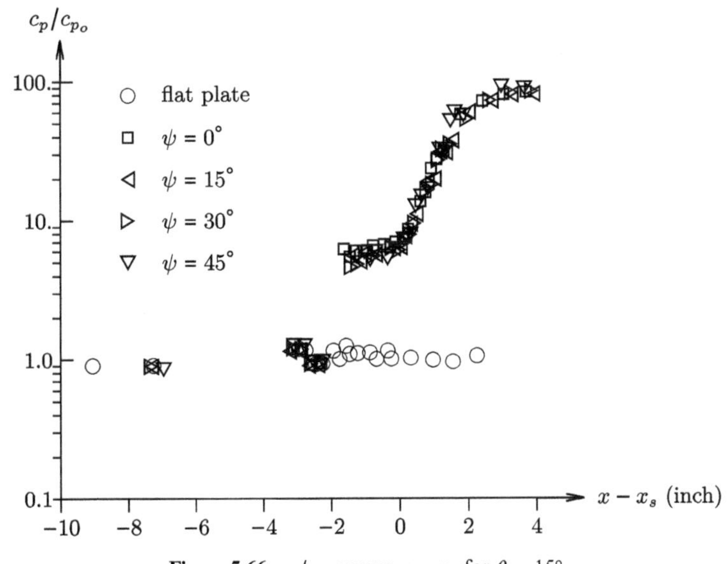


**Figure 5.66.** $c_p/c_{p_o}$ versus $x - x_s$ for $\theta = 15°$.

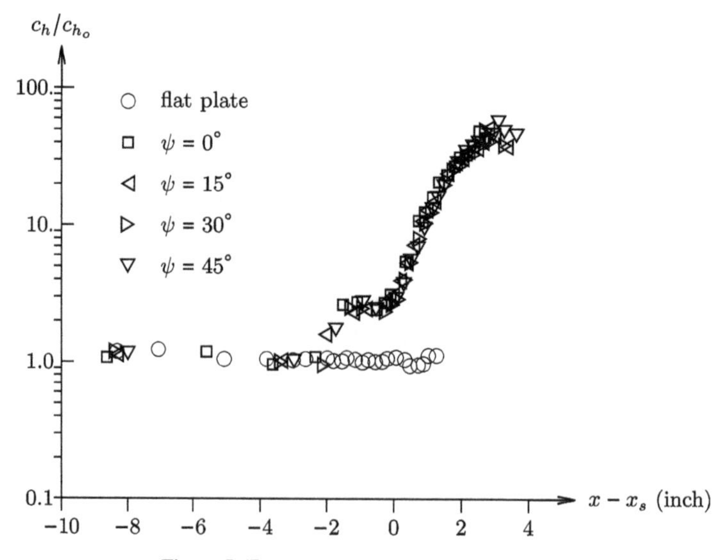


**Figure 5.67.** $c_h/c_{h_o}$ versus $x - x_s$ for $\theta = 15°$.

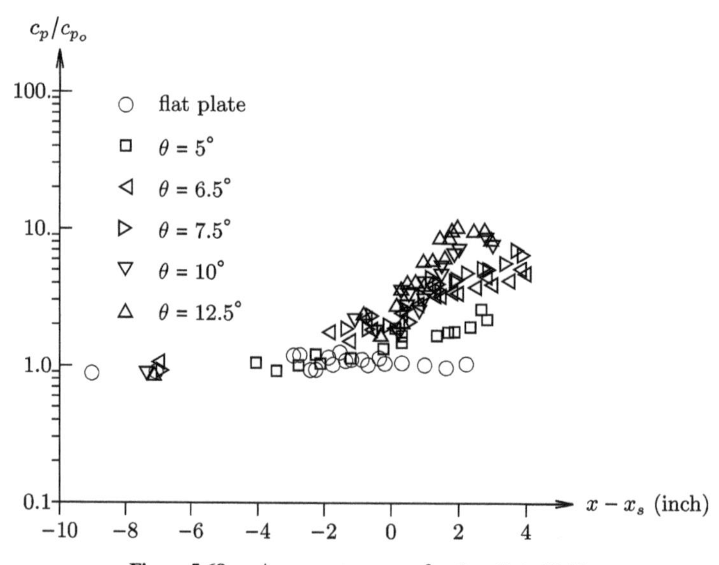


**Figure 5.68.** $c_p/c_{p_o}$ versus $x - x_s$ for $\theta = 5°$ to 12.5°.

Figure 5.68 shows the normalized pressure coefficient $c_p/c_{p_o}$ versus streamwise distance $x - x_s$ for sharp fin angles $\theta = 5°$ to 12°. Figure 5.69 displays the normalized Stanton number for the same configurations. Both heat transfer and pressure distributions display a plateau region for $\theta \geqslant 7.5°$. Figure 5.70 shows that the correlation

$$\frac{q_{\max}}{q_o} = \left(\frac{p_{\max}}{p_o}\right)^{0.85} \tag{5.5}$$

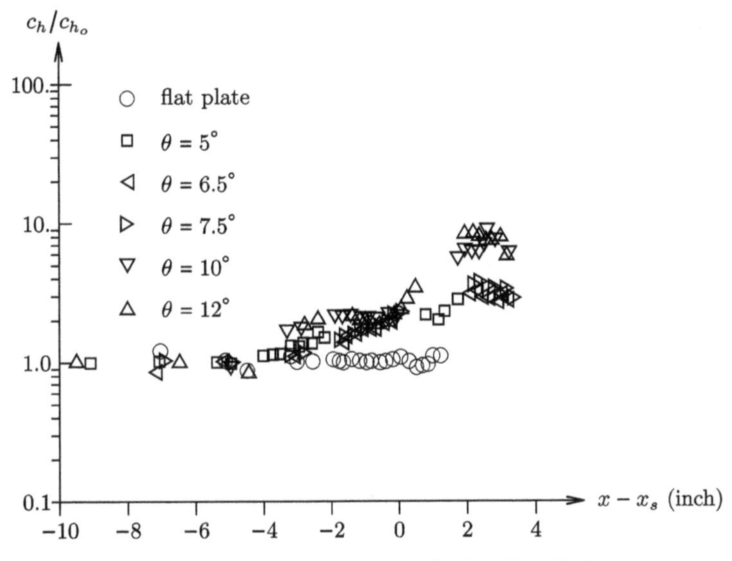


**Figure 5.69.** $c_h/c_{h_o}$ versus $x - x_s$ for $\theta = 5°$ to 12.5°.

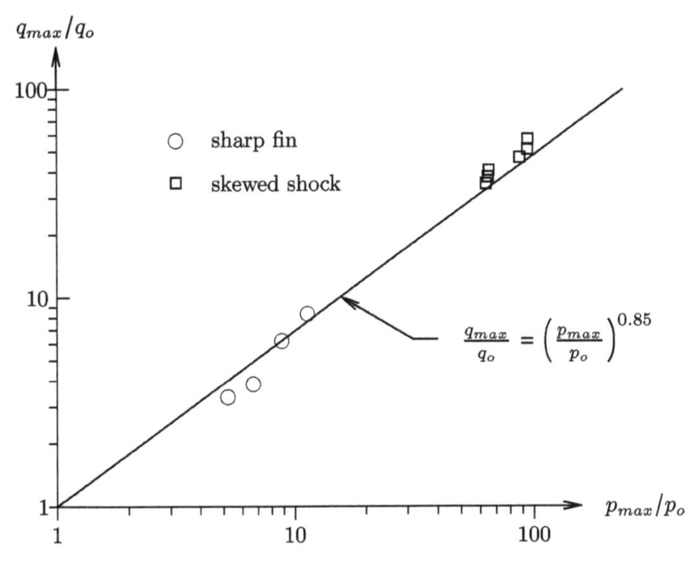


**Figure 5.70.** Correlation of peak heating.

developed earlier by Holden (1972) (figure 5.17) for two dimensional shock wave turbulent boundary layer interactions is also satisfied for the present three dimensional interactions.

## 5.14 Dolling and Rodi (1988)

Dolling and Rodi (1988) conducted an experimental study of fin induced shock wave turbulent boundary layer interactions at Mach 4.9. The experiments were performed in the University of Texas Balcones Research Center Mach 5 wind tunnel. The model is displayed in figure 5.71. A variety of different fin leading edge shapes of constant thickness $t = 0.64$ cm were considered (figure 5.72). The freestream conditions are listed in table 5.16. The fin model is located between 26.7 cm and 27.2 cm from the leading edge of the flat plate. A boundary layer tripping strip comprising 80 grid emery cloth was located 1.27 cm from the leading edge of the flat plate to ensure that the boundary layer entering the shock boundary layer interaction was turbulent. Pitot pressure surveys upstream of the interaction indicated close agreement with the Van Driest transformed Law of the Wall. Experimental diagnostics include surface pressure and surface flow visualization using the kerosene lampblack method (Settles and Teng 1983).

The mean flowfield structure for hemicylinder blunted fins is shown in figure 5.73. The shock wave generated by the fin interacts with the incoming boundary layer causing separation and the formation of a horseshoe vortex around the fin. The mean normalized centerline surface pressure $p/p_\infty$ versus normalized distance upstream $x/t$ is shown in figure 5.74 for different fin leading edge shapes. The inviscid shock detachment half angle for the wedge shaped fins is $\theta_d = 40.9°$, and thus the shock wave is attached for the two smallest values $\theta_L = 15°$ and 30°. At $\theta_L = 43°$ a kink appears in the pressure distribution forming the 'peak-trough' shape associated with the horseshoe vortex.

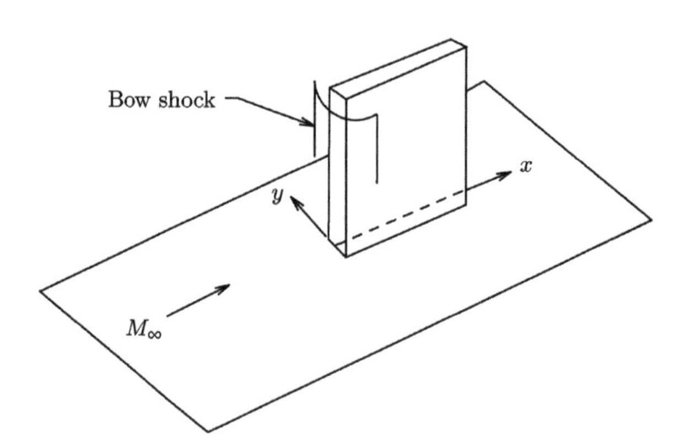


**Figure 5.71.** Model.

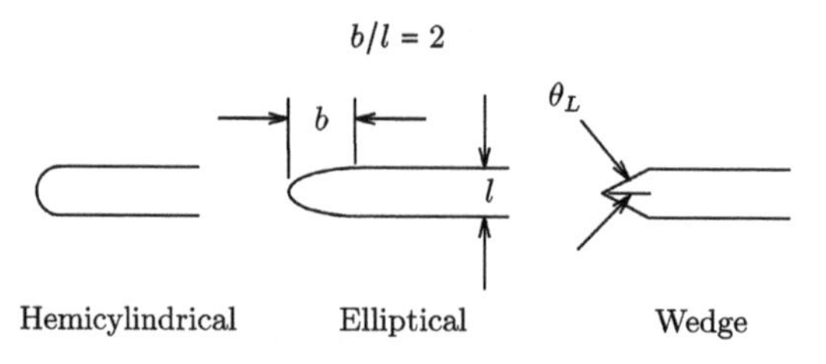


**Figure 5.72.** Fin leading edge shapes ($\theta_L$ = 15°, 30°, 43°, 45°, 53°, 75°, 90°).

**Table 5.16.** Freestream conditions.

| Quantity | Value |
|---|---|
| $M_\infty$ | 4.9 |
| $p_{t_\infty}$ (MPa) | 2.07 |
| $T_{t_\infty}$ (K) | 338 |
| $T_w/T_{aw}$ | $1 \pm 0.05$ |
| $H_\infty$ (MJ kg$^{-1}$) | 0.34 |
| $Re \times 10^{-6}$ (m$^{-1}$) | 52 |

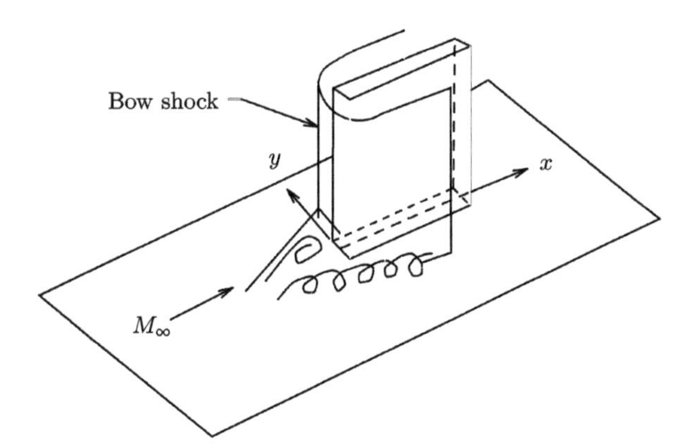


**Figure 5.73.** Flowfield structure.

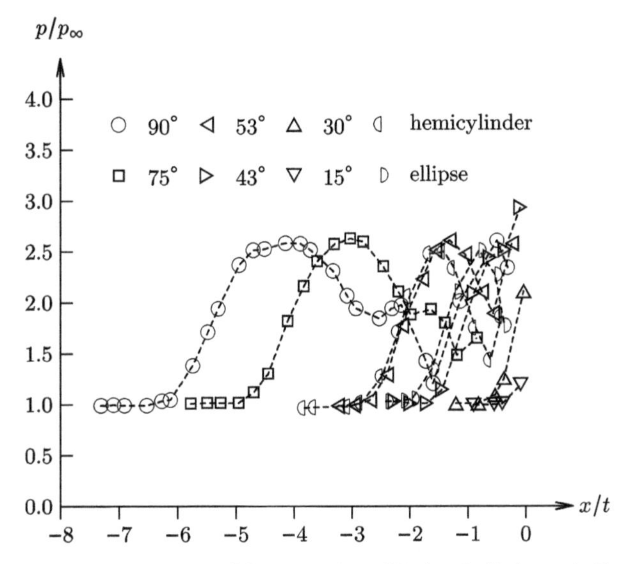


**Figure 5.74.** $p/p_\infty$ versus $x/l$ for $\theta_L = 15°$ to $90°$, hemicylinder and ellipse.

## 5.15 Dolling and Smith (1989)

Dolling and Smith (1989) investigated the unsteady shock dynamics of a cylinder induced shock wave turbulent boundary layer interaction at Mach 5. The experiments were conducted in the University of Texas Balcones Research Center Mach 5 wind tunnel. The model is shown in figure 5.75. Tests were performed with the cylinder attached to a flat plate and the wind tunnel floor. The cylinder diameters $D$ were 1.27 cm and 1.91 cm. The freestream conditions are shown in table 5.17. No trip device was used and the boundary layers developed naturally. The Van Driest transformed mean velocity profile immediately upstream of the interaction showed close agreement with the Law of the Wall for both the flat plate and tunnel wall models. Experimental diagnostics include unsteady surface pressure using two Kulite transducers spaced 0.29 cm along the centerline of the plate and tunnel floor.

The instantaneous surface pressure at a fixed location within the shock wave boundary layer interaction displays an unsteady behavior due to the motion of the

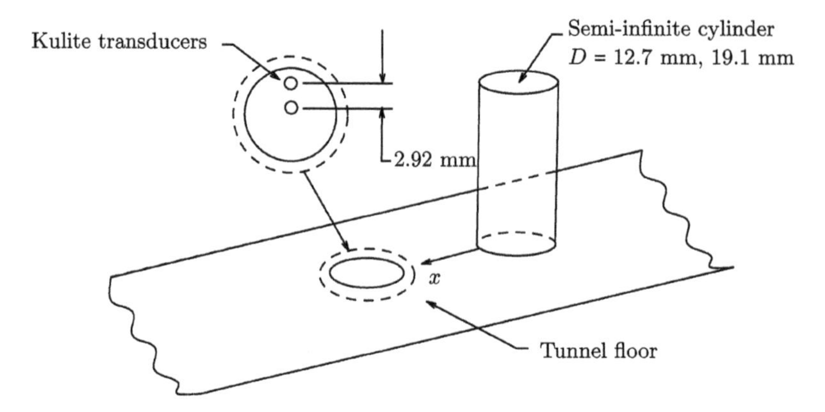


**Figure 5.75.** Model.

**Table 5.17.** Freestream conditions.

| Quantity | Tunnel floor | Flat plate |
|---|---|---|
| $M_\infty$ | 4.90 | 4.96 |
| $p_{t_\infty}$ (MPa) | 2.09 | 2.09 |
| $T_{t_\infty}$ (K) | 330 | 327 |
| $T_w/T_{aw}$ | 1.0 | 1.0 |
| $H_\infty$ (MJ kg$^{-1}$) | 0.33 | 0.33 |
| $Re \times 10^{-6}$ (m$^{-1}$) | 53.3 | 53.1 |
| $\delta$ (mm) | 16.2 | 5.36 |
| $\delta^*$ (mm) | 5.23 | 2.18 |
| $\theta$ (mm) | 0.454 | 0.181 |

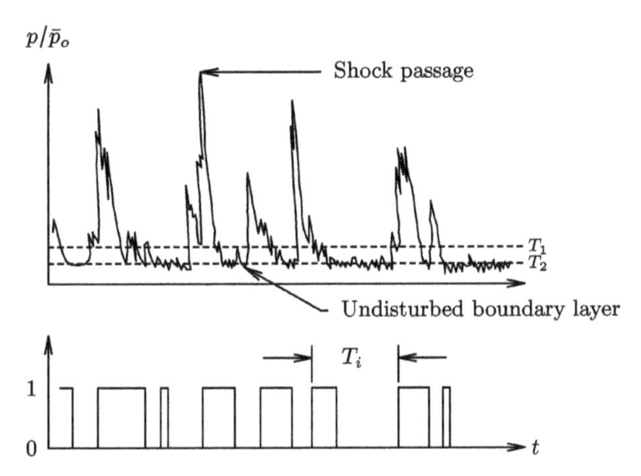


**Figure 5.76.** Unsteady pressure signal and conversion to 'boxcar'.

shock system generated by the cylinder. A schematic is shown in figure 5.76. The pressure signal comprises three components, namely, (1) the high frequency, low amplitude fluctuations of the incoming turbulent boundary layer, (2) the high frequency, low amplitude fluctuations of the boundary layer downstream of the shock, and (3) low frequency, high amplitude fluctuations due to the passage of the shock across the transducer. A conditional sampling method was applied to the pressure signal to isolate the third component and create a 'boxcar' time series. The zero crossing frequency defined by

$$f_c = \frac{1}{T_{\mathrm{m}}} = \left[\frac{1}{N}\sum_{i=1}^{N} T_i\right]^{-1} \tag{5.6}$$

represents the average frequency of the large scale motion of the shock system. The 'boxcar' starts when the instantaneous surface pressure rises above $p_o + 6\sigma_{p_o}$ and is terminated when it drops below $p_o + 3\sigma_{p_o}$ where $p_o$ and $\sigma_{p_o}$ are the mean wall pressure and standard deviation in the incoming boundary layer.

The mean, standard deviation and intermittency of the centerline surface pressure distribution for the initial region of the shock wave boundary layer interaction are shown in figures 5.77–5.79 for both cylinders on the flat plate and tunnel wall. The peak value of the standard deviation in this region reaches nearly 60% of the freestream static pressure.

Figure 5.80 shows the zero crossing frequency normalized by the frequency $u_\infty/\delta$ of the energy containing eddies of the incoming turbulent boundary layer where $\delta$ is the boundary layer thickness. The maximum zero frequency corresponds to the location where the intermittency of the surface pressure signal is approximately 0.6 for both cylinder models on the tunnel wall and flat plate. The maximum value is less than 3% in all cases indicating that the motion of the shock system generated by the fin is slow compared to the frequency of the energy containing eddies passing the transducer location.

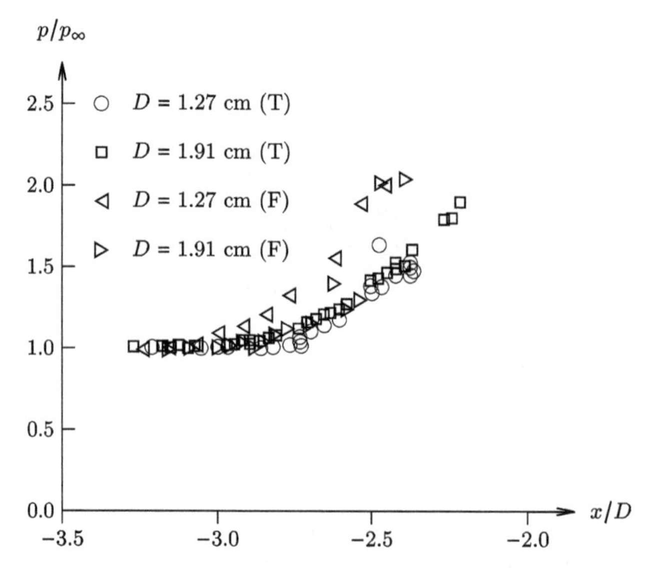


**Figure 5.77.** $p/p_\infty$ versus $x/D$ for tunnel wall (T) and flat plate (F).

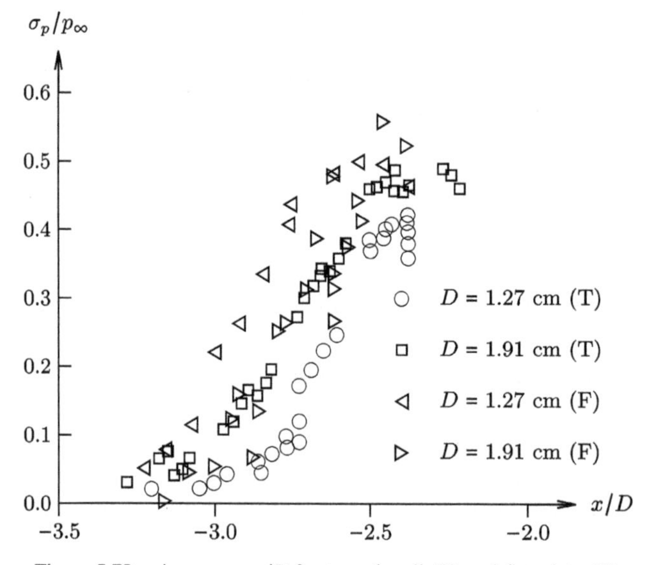


**Figure 5.78.** $\sigma/p_\infty$ versus $x/D$ for tunnel wall (T) and flat plate (F).

## 5.16 Kussoy and Horstman (1989)

Kussoy and Horstman (1989) performed an experimental investgation of an axisymmetric shock wave turbulent boundary layer interaction generated by a flare or fin on an ogive cylinder at Mach 7.2. The experiments were conducted in the NASA Ames 3.5 ft hypersonic wind tunnel (section A.20.2). The models are shown in figures 5.81(a) and 5.81(b). The first model is an ogive-cylinder with a flare at a

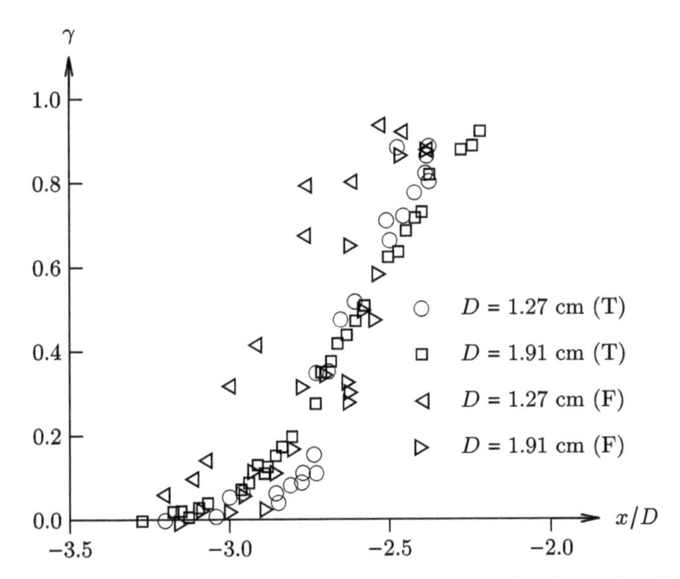


**Figure 5.79.** Intermittency $\gamma$ versus $x/D$ for tunnel wall (T) and flat plate (F).

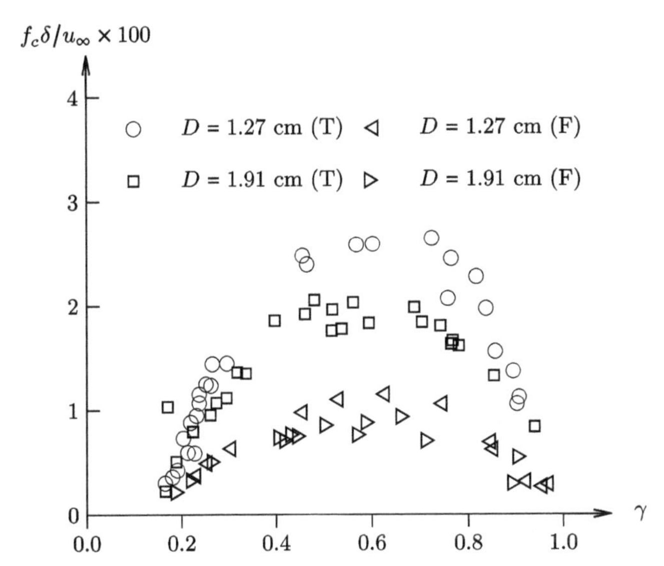


**Figure 5.80.** $f_c\delta/U_\infty$ versus $\gamma$ for tunnel wall (T) and flat plate (F).

distance of 139 cm from the nosetip. Flare angles of 20°, 30°, 32.5° and 35° were considered. The second model is an ogive-cylinder with a fin at a distance of 120 cm from the nosetip. Fin half-angles of 10°, 15° and 20° were considered. The local freestream conditions[2] are listed in table 5.18. Natural transition occurred on the

[2] The values cited are the conditions above the cylinder.

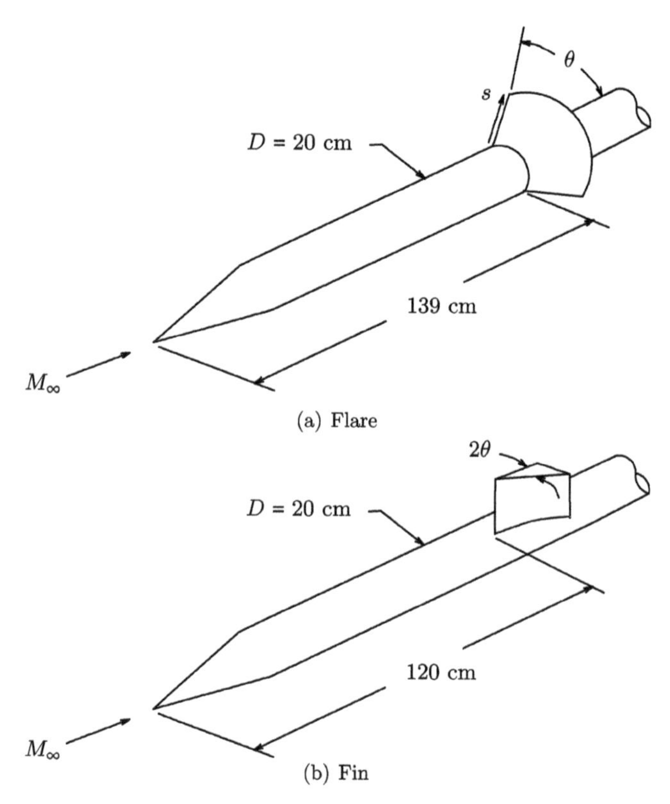


**Figure 5.81.** Models.

**Table 5.18.** Freestream conditions.

| Quantity | Value |
|---|---|
| $M_\infty$ | 7.05 |
| $p_{t_\infty}$ (MPa) | 2.49 |
| $T_{t_\infty}$ (K) | 888 |
| $T_w$ (K) | 311 |
| $T_w/T_{aw}$ | 0.37 |
| $H_\infty$ (MJ kg$^{-1}$) | 0.89 |
| $\delta$ (cm) | 2.5 |
| $\delta^*$ (cm) | 0.74 |
| $\theta$ (cm) | 0.065 |
| $Re \times 10^{-6}$ (m$^{-1}$) | 5.8 |

cylinder between 40 cm and 80 cm from the nosetip. Experimental measurements include surface pressure and heat transfer, and flowfield surveys using pitot pressure, static pressure and total temperature probes. Surface pressure and heat transfer measurements were obtained along a streamwise line for the ogive-cylinder-flare, while for the ogive-cylinder-fin the measurements were obtained in quasi-conical coordinates (figure 5.82) as suggested by the flowfield structure for the three dimensional sharp fin interaction Alvi and Settles (1992). The experiment was selected for the database of Settles and Dodson (1994).

The Van Driest transformed velocity profile upstream of the shock wave turbulent boundary interaction is presented in figure 5.83 together with the Law of the Wall. Close agreement is evident indicating an equilibrium turbulent boundary layer.

Figure 5.84 presents the experimental surface pressure for the ogive-cylinder-flare for flare angles of 20°, 30°, 32.5° and 35°. A plateau region is observed for the 35° case indicating a region of separated flow. Figure 5.85 shows the heat transfer $q_w/q_{w_\infty}$ where $q_{w_\infty} = 0.93$ W/cm$^2$. A similar plateau region is also observed for the heat transfer for the 35° case.

Figure 5.86 displays the surface pressure for the ogive-cylinder fin. In anticipation of a quasi-conical behavior for the surface pressure and heat transfer, data was obtained on two arcs (figure 5.82). The different arcs are distinguished by the open and filled symbols, e.g., ○ and ⊛ for the 10° fin, etc. Quasi-conical behavior is seen for the surface pressure for the 10° and 15° fins; however, the surface pressure for the 20° fin shows significant differences for the two arcs. Similar results hold for the surface heat transfer as shown in figure 5.87.

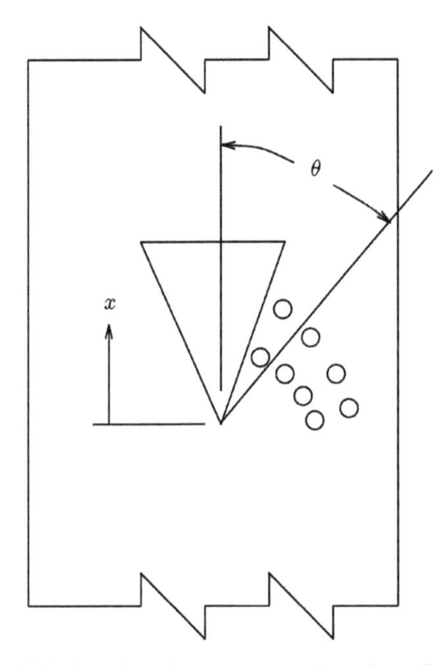


**Figure 5.82.** Location of measurements for ogive-cylinder-fin.

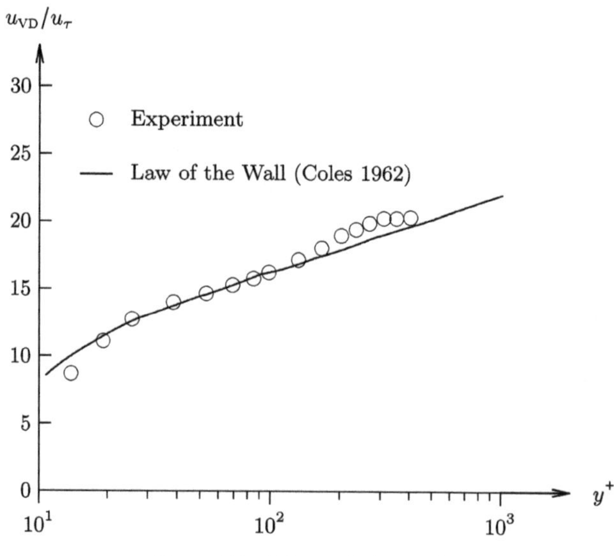


**Figure 5.83.** Inflow velocity profile.

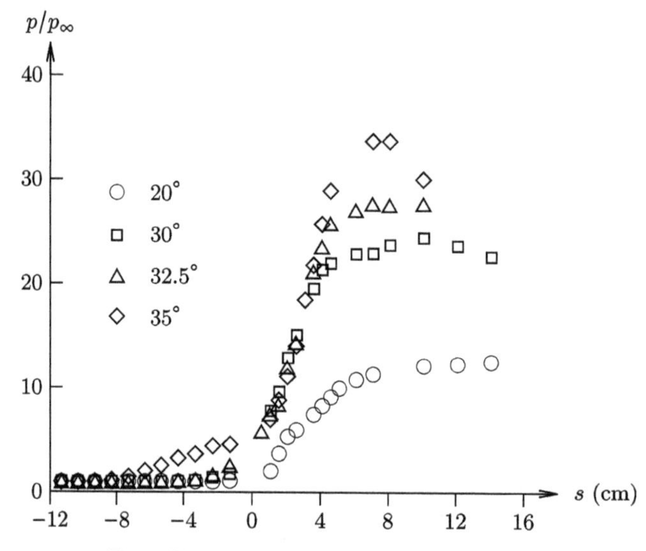


**Figure 5.84.** $p/p_\infty$ versus $s$ for ogive-cylinder-flare.

## 5.17 Alvi and Settles (1990–2)

Alvi and Settles (1990, 1991) and Alvi and Settles (1992) investigated the flowfield structure of the swept shock wave turbulent boundary layer interaction generated by a sharp fin at Mach 3 and 4. The experiments were performed in the Supersonic Wind Tunnel Facility at the Penn State Gas Dynamics Laboratory. The model is shown in figure 5.88. The freestream conditions are listed in table 5.19. Experimental

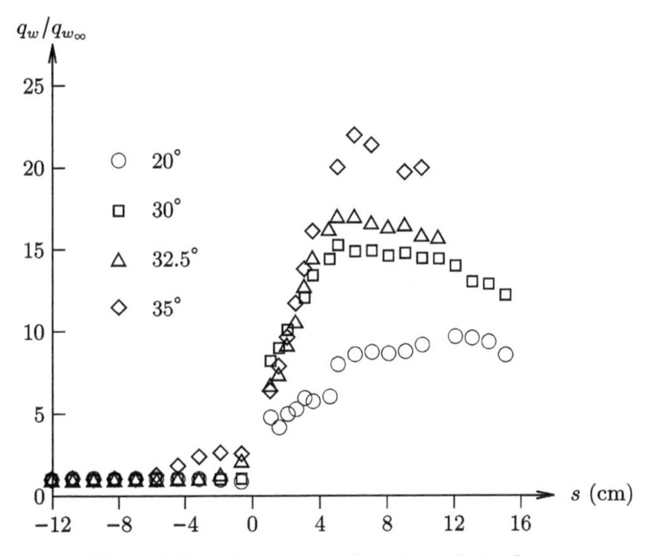


**Figure 5.85.** $q_w/q_{w_\infty}$ versus $s$ for ogive-cylinder-flare.

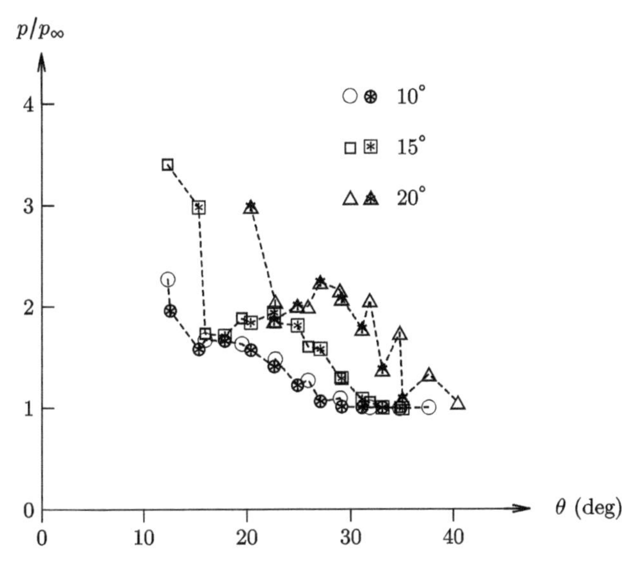


**Figure 5.86.** $p/p_\infty$ versus $\theta$ for ogive-cylinder-fin.

diagnostics include Planar Laser Scattering (PLS), shadowgraphy, and surface flow patterns, skin friction and pressure.

Detailed PLS and shadowgraphy indicate the flowfield arising from the shock wave turbulent boundary layer interaction is conical. Figure 5.89 shows a spherical polar coordinate system whose origin in the experiment is offset from the intersection of the sharp fin and the flat plate. This origin is denoted the Virtual Conical Origin (VCO). The spherical polar coordintes are the spherical radius $r$ and

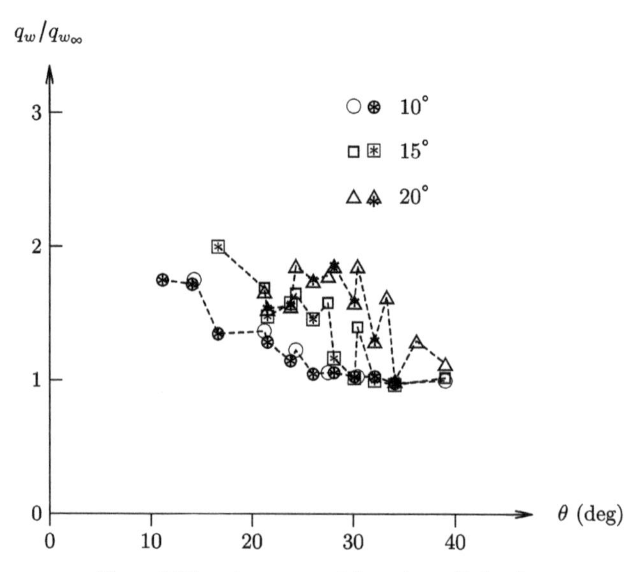


**Figure 5.87.** $q_w/q_{w_\infty}$ versus $\theta$ for ogive-cylinder-fin.

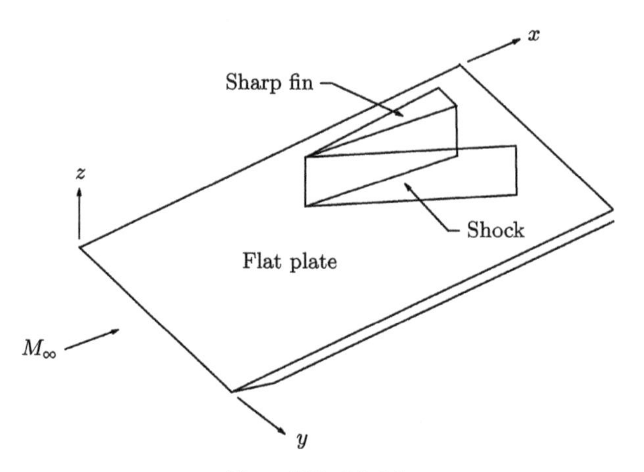


**Figure 5.88.** Model.

**Table 5.19.** Freestream conditions[a].

| Quantity | Case 1 | Case 2 |
|---|---|---|
| $M_\infty$ | 3 | 4 |
| $p_{t_\infty}$ (MPa) | 0.83 | 1.5 |
| $T_{t_\infty}$ (K) | 293 | 293 |
| $T_w/T_{aw}$ | 1.03 | 1.03 |
| $H_\infty$ (MJ kg$^{-1}$) | 0.29 | 0.29 |
| $Re_\infty \times 10^{-6}$ (m$^{-1}$) | 68 | 70 |
| $\delta$ (mm) | 2.7[b] | 2.9[b] |
| $\delta^*$ (mm) | 0.8[b] | 1.0[b] |
| $\theta$ (mm) | 0.17[b] | 0.13[b] |

[a] Same as Lee *et al* (1992), (Alvi 2022).
[b] Upstream of interaction.

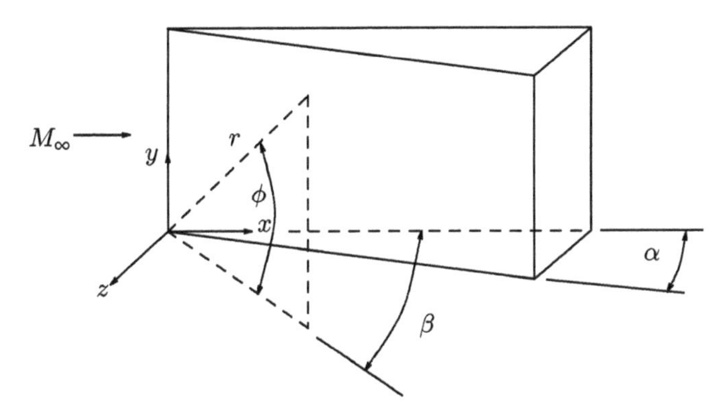


Figure 5.89. Spherical polar coordinate system.

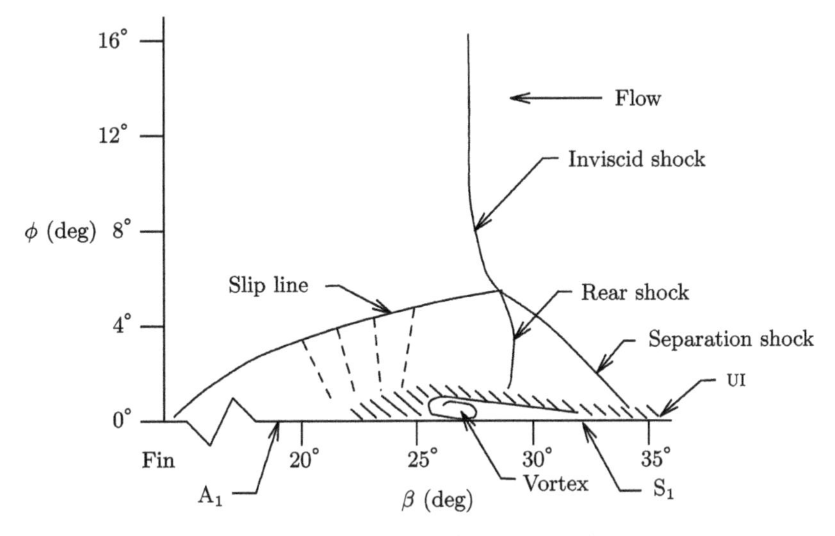


Figure 5.90. Flowfield structure for $M_\infty = 3$ and $\alpha = 10°$.

polar angles $\beta$ measured in the plane of the flat plate, and $\phi$ measured normal to $\beta$. A conical flowfield is described by the two angles $\beta$ and $\phi$ and is independent of the radius (Shapiro 1954). The flowfield structure is strongly affected by the strength $p_2/p_1$ of the fin generated shock where $p_2$ and $p_1$ are the static pressures downstream and upstream of the inviscid shock. Figure 5.90 displays the flowfield structure for $M_\infty = 3$ and $\alpha = 10°$ in the $\beta - \phi$ surface of the spherical polar coordinate system. The interaction of the sharp fin generated shock with the turbulent boundary layer on the flat plate causes an increase in surface pressure beginning at the upstream influence (UI) location. For sufficient shock strength the boundary layer separates ($S_1$) causes a lifting of the boundary layer (shown in crosshatch) and the formation of a vortex. The inviscid flow above the boundary layer is processed through the two oblique shock legs of the $\lambda$-shock and expands downstream of the vortex forming the primary attachment line ($A_1$). A slip line originates from the triple point of the

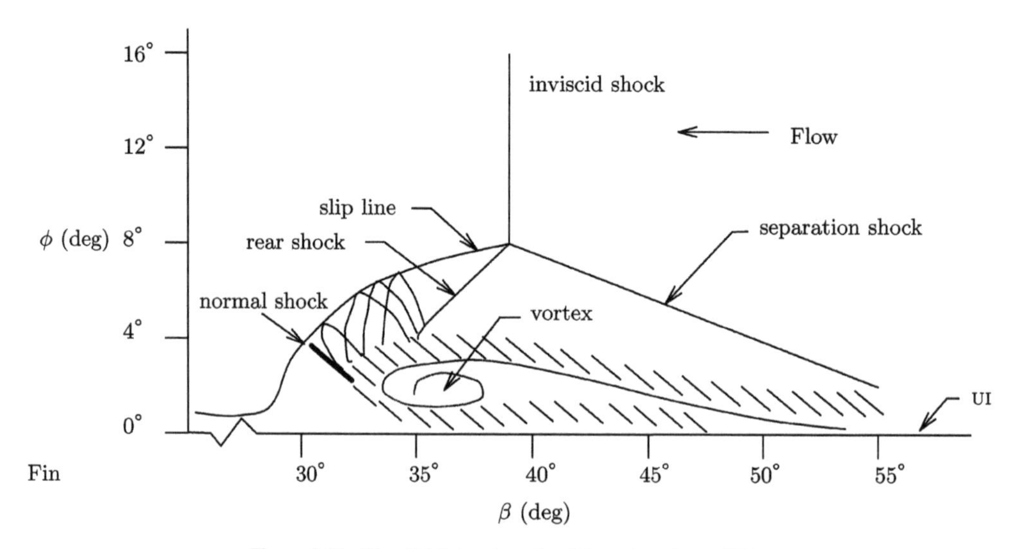


**Figure 5.91.** Flowfield structure for $M_\infty = 4$ and $\alpha = 25°$.

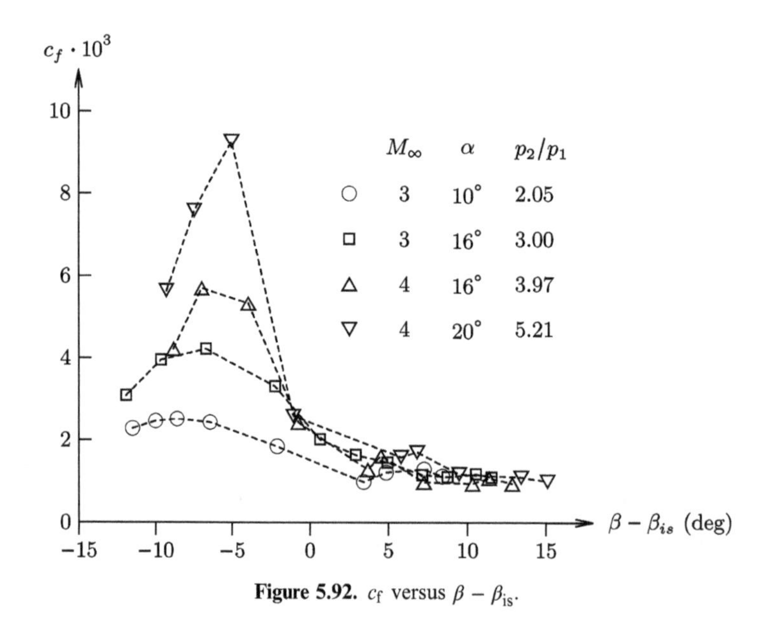


**Figure 5.92.** $c_f$ versus $\beta - \beta_{is}$.

$\lambda$-shock. At higher shock strength the expansion of the flow subsequent to the lambda shock experiences a normal shock. Figure 5.91 displays the flowfield structure for $M_\infty = 4$ and $\alpha = 25°$. A secondary separation and attachment can also occur depending on the shock strength. A detailed description of the flowfield structure is presented in Zheltovodov and Knight (2011). See also Lu *et al* (1990).

Figures 5.92 and 5.93 display the surface skin friction coefficient $c_f$ and non-dimensional surface pressure $p/p_\infty$ versus $\beta - \beta_{is}$ for $M_\infty = 3$ at $\alpha = 10°$ and 16° and

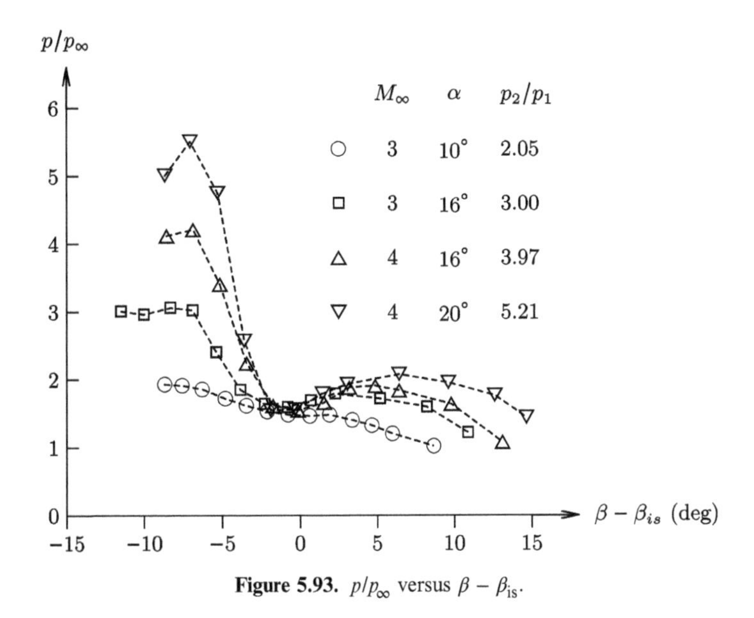


**Figure 5.93.** $p/p_\infty$ versus $\beta - \beta_{\rm is}$.

$M_\infty = 4$ at $\alpha = 16°$ and $20°$ where $\beta_{\rm is}$ is the inviscid shock angle. The peak skin friction coefficient and wall pressure increase with shock strength.

## 5.18 Kim *et al* (1990)

Kim *et al* (1990) presented experimental results for surface skin friction in a swept shock wave turbulent boundary layer interaction generated by a sharp fin at Mach 3 and 4. The experiments were conducted in the Supersonic Wind Tunnel Facility of the Penn State Gas Dynamics Laboratory. The model is shown in figure 5.94. The flow conditions are listed in table 5.20. Experimental diagnostics include Laser Interferometer Skin Friction (LISF) meter measurements of skin friction, shadowgraphy and surface flow visualization. See also Kim and Settles (1988) and Kim and Settles (1990).

The surface skin friction coefficient $c_{\rm f}$ versus angle $\beta - \beta_{\rm is}$ measured on the flat plate is shown in figure 5.95 for fin angles $\alpha = 10°$ and $16°$ at Mach 3 and $\alpha = 16°$ and $20°$ at Mach 4 where $\beta_{\rm is}$ is the angle of the inviscid shock. Qualitatively similar behavior is observed for all interactions, with the peak value of $c_{\rm f}$ increasing with magnitude of the inviscid static pressure ratio $p_2/p_1$ across the shock generated by the sharp fin. The location of the peak $c_{\rm f}$ corresponds approximately to the primary attachment line (figure 5.91).

The topology of the surface streamlines on the flat plate is illustrated in figure 5.96 corresponding to the flowfield structure in figure 5.91. The definition of the angles is shown in table 5.21. The upstream influence defines the location of the initial pressure rise on the flat plate due to the shock wave turbulent boundary layer interaction. The primary separation corresponds to the initial convergence of surface streamlines. A secondary separation occurs between the fin and primary separation line for sufficient shock strength $p_2/p_1$ and is also identified by convergence of

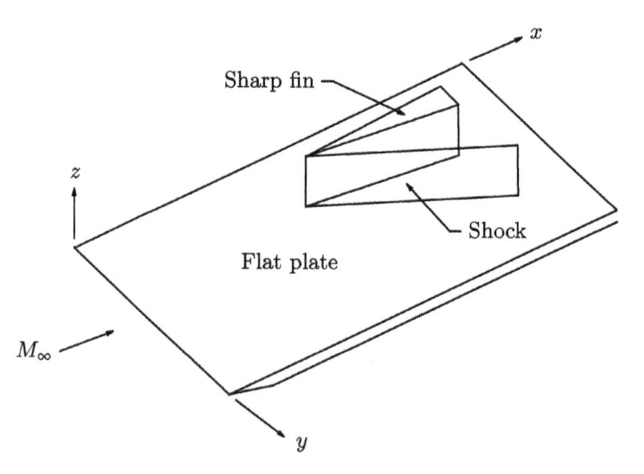


**Figure 5.94.** Model.

**Table 5.20.** Freestream conditions.

| Quantity | Case 1 | Case 2 |
|---|---|---|
| $M_\infty$ | 3.03 ± 0.033 | 3.98 ± 0.071 |
| $p_{t_\infty}$ (MPa) | 0.827 ± 3.0% | 1.524 ± 2.4% |
| $T_{t_\infty}$ (K) | 293.5 ± 2.7% | 293.4 ± 1.6% |
| $T_w/T_{aw}$ | 1.06 ± 2.2% | 1.06 ± 2.5% |
| $H_\infty$ (MJ kg$^{-1}$) | 0.29 | 0.29 |
| $Re_\infty \times 10^{-6}$ (m$^{-1}$) | 61.9 ± 4.0% | 67.9 ± 4.8% |
| $\delta$ (mm) | 4.4[a] | 3.0[a] |
| $\theta$ (mm) | 0.24[a] | 0.16[a] |
| $c_f \times 10^3$ | 1.5[a] | 1.2[a] |

[a] Upstream of interaction.

surface streamlines. The primary attachment line is also evident. Figure 5.97 shows the variation of the difference between the angles of the aforementioned surface features and the inviscid shock angle. A monotonic increase with shock strength $p_2/p_1$ is evindent.

## 5.19 Disimile and Scaggs (1991)

Disimile and Scaggs (1991b) investigated the effect of surface roughness on a Mach 6 shock wave turbulent boundary layer interaction. The experiments were conducted in the Air Force Flight Dynamics Laboratory's High Reynolds Number Mach 6 Facility (section A.2.3). The model is illustrated in figure 5.98. A sharp leading edge flat plate of length $L = 45.72$ cm and width $W = 24.13$ cm was milled with uniform square protuberances of streamwise and spanwise widths $w = 1.016$ mm and depth $c = 0.508$ mm. The streamwise and spanwise gaps between the protuberances was 1.016 mm. A wedge of variable angles 22°, 28° and 34° was affixed at a distance

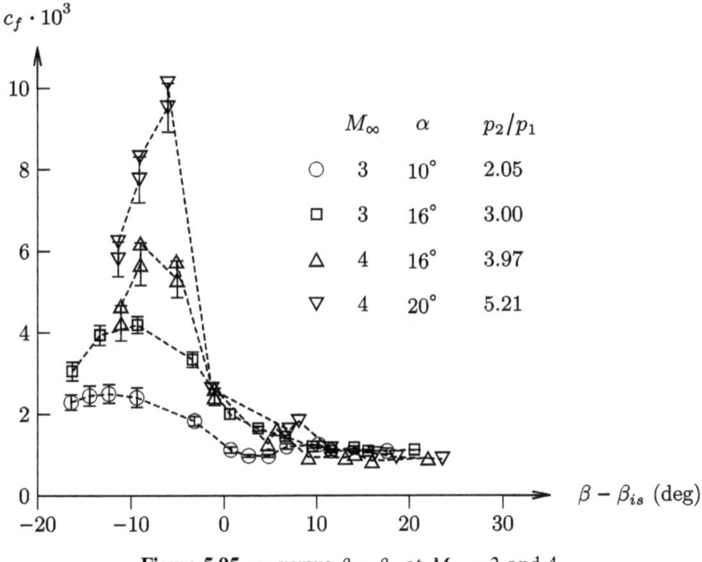


**Figure 5.95.** $c_f$ versus $\beta - \beta_{is}$ at $M_\infty = 3$ and 4.

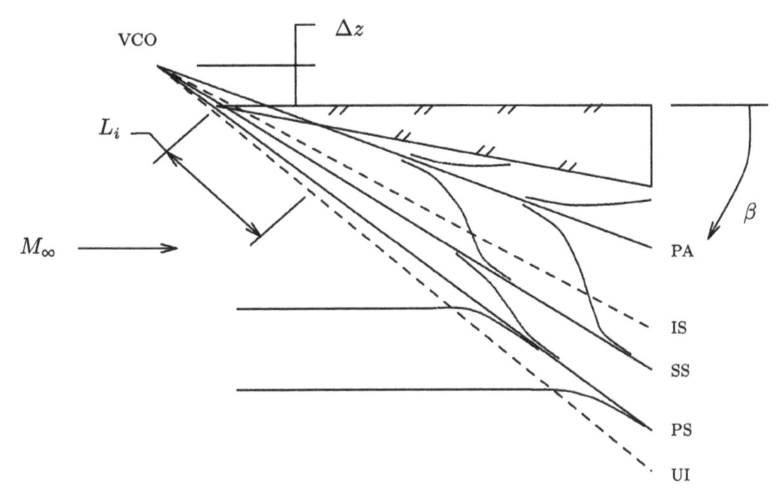


**Figure 5.96.** Surface streamline structure. IS—inviscid shock, PA—primary attachment, PS—primary separation, SS—secondary separation, UI—upstream influence, VCO—virtual conical origin.

40.64 cm from the leading edge of the flat plate. The freestream conditions are listed in table 5.22. Results presented in Disimile and Scaggs (1991b) correspond to a wedge angle of 22°. Additional results are presented in Disimile and Scaggs (1989, 1990, 1991a) and Disimile and Scaggs (1992).

Figure 5.99 shows the dimensionless surface pressure $p/p_\infty$ versus $x$ (cm) for Case Nos. 1 and 3 for both smooth wall and rough wall plates. The smooth wall pressure distribution shows no separation, while the rough wall distribution shows a plateau region characteristic of a separated region.

**Table 5.21.** Surface streamline angles.

| Angle | Definition |
|---|---|
| $\beta_{\text{ui}}$ | Upstream influence |
| $\beta_{\text{ps}}$ | Primary separation |
| $\beta_{\text{ss}}$ | Secondary separation |
| $\beta_{\text{is}}$ | Inviscid shock |
| $\beta_r$ | Primary attachment |

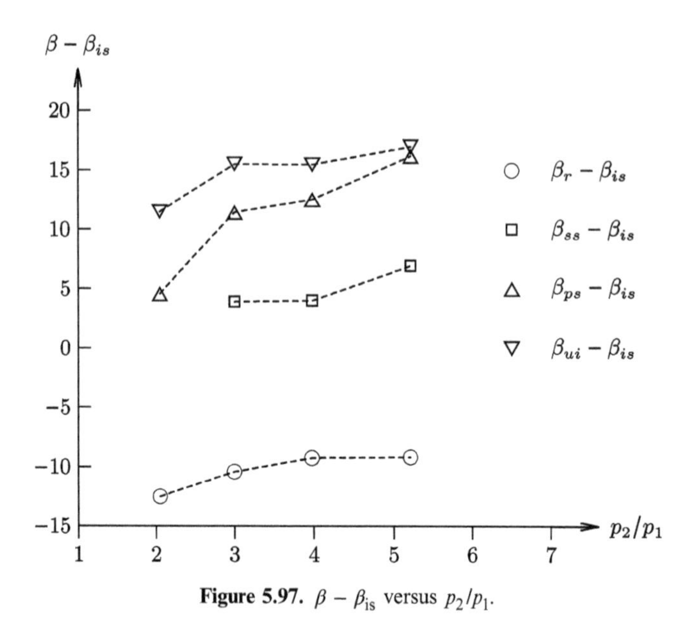


**Figure 5.97.** $\beta - \beta_{\text{is}}$ versus $p_2/p_1$.

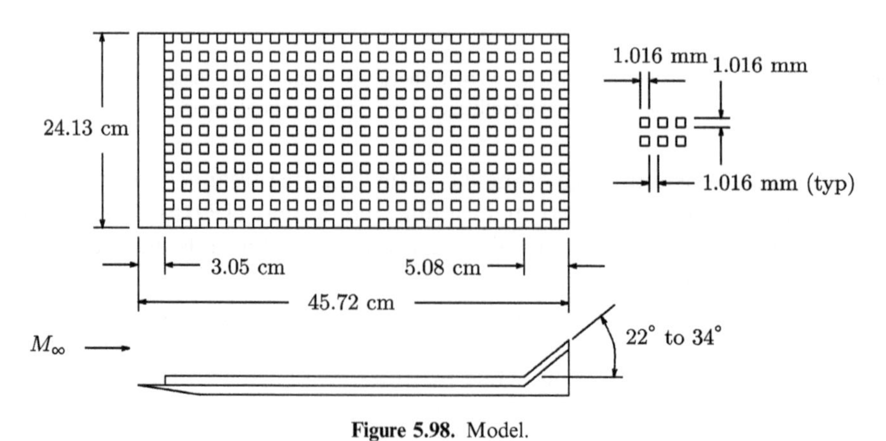


**Figure 5.98.** Model.

**Table 5.22.** Freestream conditions.

| Quantity | Case 1 | Case 2 | Case 3 |
|---|---|---|---|
| $M_\infty$ | 5.9 | 5.9 | 5.9 |
| $p_{t_\infty}$ (MPa) | 4.83 | 9.66 | 13.8 |
| $T_{t_\infty}$ (K) | 517 | 517 | 517 |
| $T_w$ (K) | na | na | na |
| $H_\infty$ (MJ kg$^{-1}$) | 0.52 | 0.52 | 0.52 |
| $Re_\infty \times 10^{-6}$ (m$^{-1}$) | 33 | 66 | 98 |

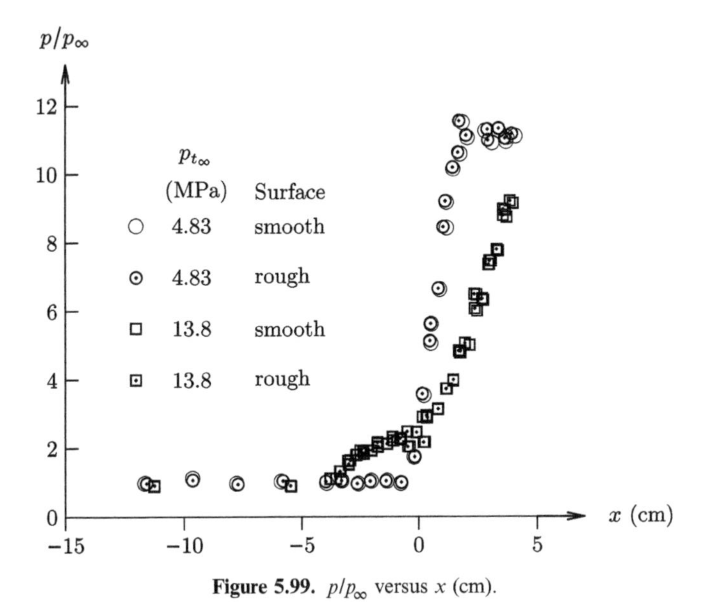


**Figure 5.99.** $p/p_\infty$ versus $x$ (cm).

Figure 5.100 displays the normalized total temperature $T_t/T_{t_\infty}$ versus the distance $z$ measured normal to the surface for Case Nos. 1 to 3 at locations upstream and downstream of the corner located at $x = 0$. A distinct feature is the overshoot in normalized total temperature $T_t/T_{t_\infty}$ downstream of the shock wave turbulent boundary layer interaction.

## 5.20 Holden (1991)

Holden (1991) examined the axisymmetric shock wave turbulent boundary layer interaction generated by a cone flare at a range of Mach numbers from 10.96 to 15.4. The experiments were conducted in the Calspan 96 inch shock tunnel (section A.8.1). The 2.74 m long model is shown in figure 5.101. The cone half angle is 6° and the flare angles are $\alpha = 30°$ and 36°. The freestream conditions ahead of the cone are given in table 5.23. Experimental measurements include surface

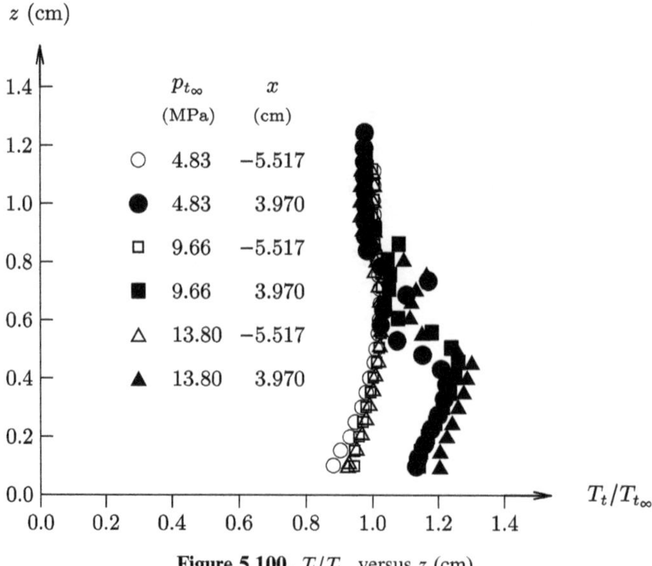


**Figure 5.100.** $T_t/T_{t_\infty}$ versus $z$ (cm).

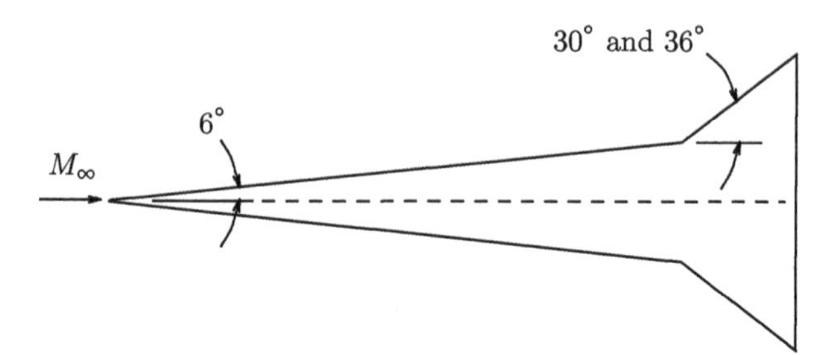


**Figure 5.101.** Model.

pressure and heat transfer, boundary layer profiles of pitot pressure, total temperature and total heat transfer, and schlieren visualization.

Figure 5.102 displays the surface pressure $p/q_\infty$ versus distance $x$ from the nosetip for Case Nos. 1 ($\alpha = 30°$) and 2 ($\alpha = 36°$) at $M_\infty = 11$. A pressure plateau indicative of a separated region is evident for $\alpha = 36°$. The solid and dashed lines indicate the inviscid pressure levels before and after the flare. Figure 5.103 displays the surface pressure for Case Nos. 3 ($\alpha = 30°$), 4 ($\alpha = 36°$) and 5 ($\alpha = 30°$) at $M_\infty = 12.9$, 13.1 and 15.4, respectively. Case No. 4 ($\alpha = 36°$) shows the pressure plateau characteristic of a separated region, while the other two cases do not.

Figures 5.104 and 5.105 display the heat transfer coefficient $c_h$ versus distance $x$ from the nosetip corresponding to figures 5.102 and 5.103, respectively, where the heat transfer coefficient is defined as

$$c_h = \frac{q_w}{\rho_\infty u_\infty (H_\infty - H_w)} \tag{5.7}$$

**Table 5.23.** Freestream conditions.

| | Case | | | | |
|---|---|---|---|---|---|
| Quantity | 1 | 2 | 3 | 4 | 5 |
| Run | 8 | 4 | 7 | 6 | 9 |
| $\alpha$ (deg) | 30 | 36 | 30 | 36 | 30 |
| $M_\infty$ | 11.0 | 11.0 | 13.0 | 13.0 | 15.4 |
| $p_{t_\infty}$ (MPa) | 49.6 | 49.6 | 126.6 | 126.6 | 128.24 |
| $T_{t_\infty}$ (K) | 1509 | 1509 | 1724 | 1724 | 2153 |
| $T_w$ (K) | 288[a] | 288[a] | 288[a] | 288[a] | 288[a] |
| $T_w/T_{aw}$ | 0.21[b] | 0.21[b] | 0.19[b] | 0.19[b] | 0.15[b] |
| $H_\infty$ (MJ kg$^{-1}$) | 1.70 | 1.70 | 1.99 | 1.99 | 2.60 |
| $Re \times 10^{-6}$ (m$^{-1}$) | 12.1 | 12.1 | 14.9 | 14.9 | 4.9 |

[a] Assumed ambient conditions.

[b] Estimated from $T_{aw} = T_{t_\infty}\left[1 + \frac{(\gamma-1)}{2}M_\infty^2\right]^{-1}\left[1 + \frac{(\gamma-1)}{2}Pr_t M_\infty^2\right]$.

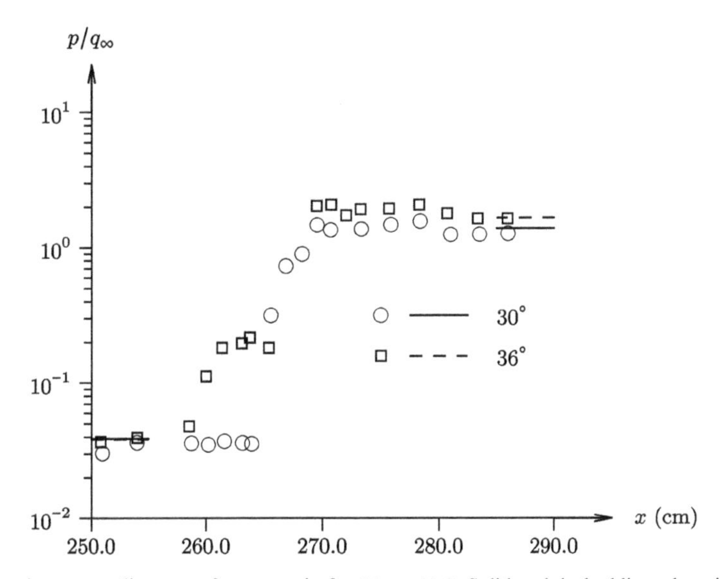


**Figure 5.102.** $p/q_\infty$ versus distance $x$ from nosetip for $M_\infty = 11.0$. Solid and dashed lines show inviscid values upstream and downstream of flare.

A plateau in $c_h$ appears corresponding to the two separated flow cases. The solid and dashed lines indicate the heat transfer coefficient for the flow on the cone immediately upstream of the shock wave turbulent boundary layer interaction based upon the Van Driest II model (Hopkins and Inouye 1971).

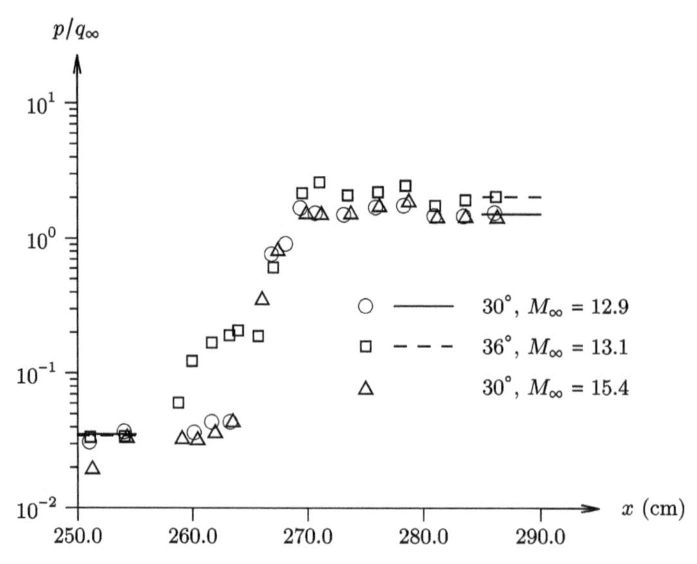


**Figure 5.103.** $p/q_\infty$ versus distance $x$ from nosetip for $M_\infty = 12.9$ to 15.4. Solid and dashed lines show inviscid values upstream and downstream of flare.

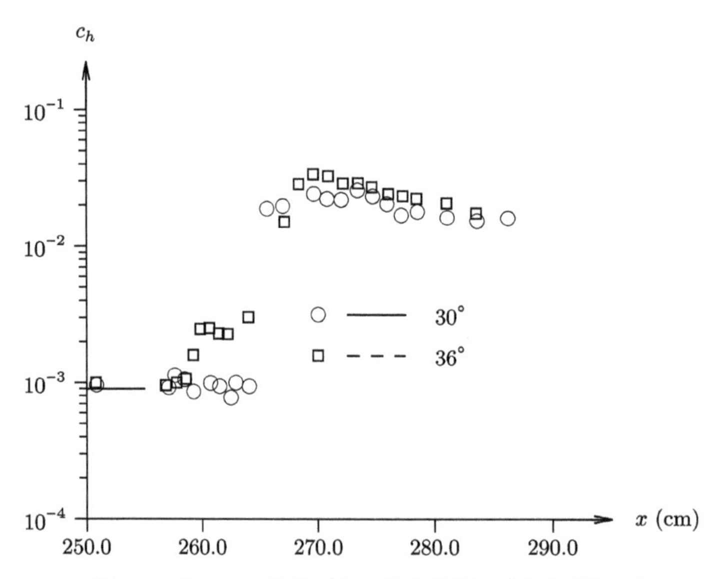


**Figure 5.104.** $c_h$ versus distance $x$ from nosetip for $M_\infty = 11.0$. Solid and dashed lines show upstream $c_h$ based on Van Driest II theory.

## 5.21 Lee, Settles and Horstman (1992)

Lee *et al* (1992) and Lee *et al* (1994) measured the surface heat transfer in a swept shock wave turbulent boundary layer interaction generated by a sharp fin at Mach 3 and 4. The experiments were conducted in the Supersonic Wind Tunnel Facility at the Penn State Gas Dynamics Laboratory. The model is shown in figure 5.106. The

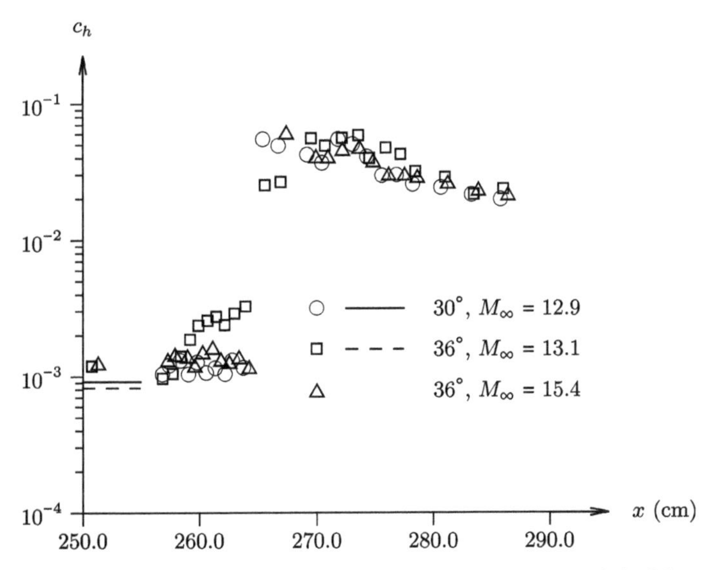


**Figure 5.105.** $c_h$ versus distance $x$ from nosetip for $M_\infty$ = 12.9 to 15.4. Solid and dashed lines show upstream $c_h$ based on Van Driest II theory.

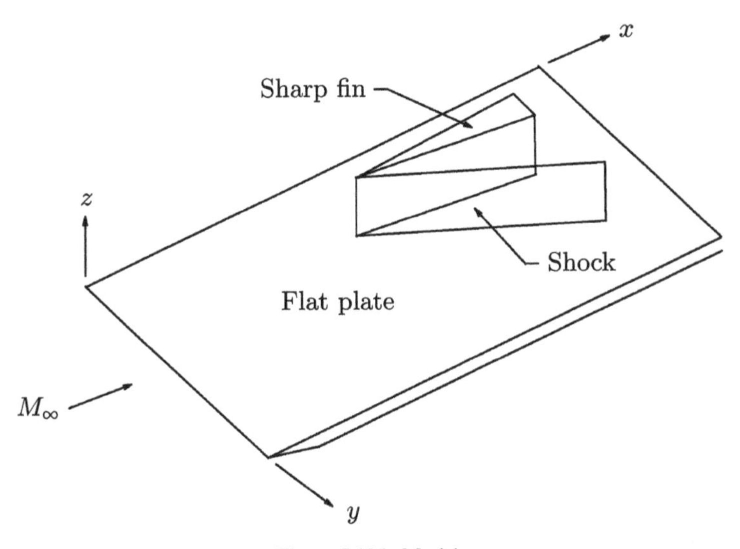


**Figure 5.106.** Model.

freestream conditions are listed in table 5.24. Experimental diagnostics include surface heat transfer and adiabatic wall temperature.

Figure 5.107 shows the surface heat flux coefficient $c_h$ normalized by the heat flux coefficient upstream of the interaction $c_{h_\infty}$ versus $\beta - \beta_{is}$ where $\beta$ is the polar coordinate measured on the flat plate at a fixed radius from the sharp fin leading edge, and $\beta_{is}$ is the inviscid shock angle. Results are shown for Mach 3 and Mach 4 with fin angles $\alpha$ = 10° to 20°. Also shown is the inviscid static pressure ratio $p_2/p_1$

**Table 5.24.** Freestream conditions.

| Quantity | Case 1 | Case 2 |
|---|---|---|
| $M_\infty$ | 3 | 4 |
| $p_{t_\infty}$ (MPa) | $0.83^{a}$ | $1.5^{a}$ |
| $T_{t_\infty}$ (K) | $293^{a}$ | $293^{a}$ |
| $T_w/T_{aw}$ | 1.03 | 1.03 |
| $H_\infty$ (MJ kg$^{-1}$) | 0.29 | 0.29 |
| $Re_\infty \times 10^{-6}$ (m$^{-1}$) | 68 | 70 |
| $\delta$ (mm) | $2.7^{b}$ | $2.9^{b}$ |
| $\delta^*$ (mm) | $0.8^{b}$ | $1.0^{b}$ |
| $\theta$ (mm) | $0.17^{b}$ | $0.13^{b}$ |

[a] Assumed same as Kim *et al* (1990).
[b] Upstream of interaction.

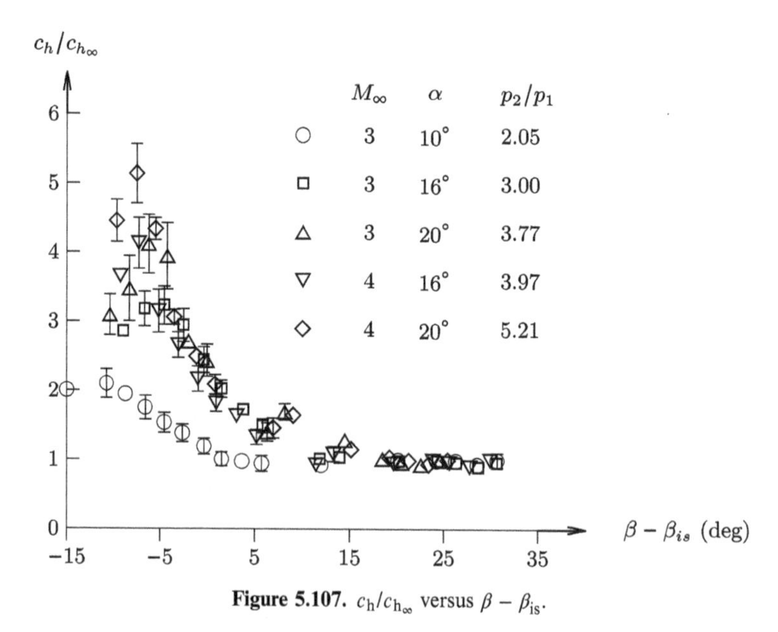


**Figure 5.107.** $c_h/c_{h_\infty}$ versus $\beta - \beta_{is}$.

across the shock wave generated by the sharp fin which is related to the inviscid Mach number normal to the shock $M_n$ according to

$$\frac{p_2}{p_1} = \frac{2\gamma M_n^2 - (\gamma - 1)}{\gamma + 1} \tag{5.8}$$

where $M_n = M_\infty \sin\theta$ and $\theta$ is the shock angle measured relative to the freestream flow. The experimental data indicates that the peak value $c_{h_{peak}}$ increases with $p_2/p_1$. Figure 5.108 shows $c_{h_{peak}}/c_{h_\infty}$ versus $M_n$ with additional experimental data from Oskam *et al* (1965), Rodi and Dolling (1992), Law (1975) and Christophel *et al* (1975). The data is correlated by

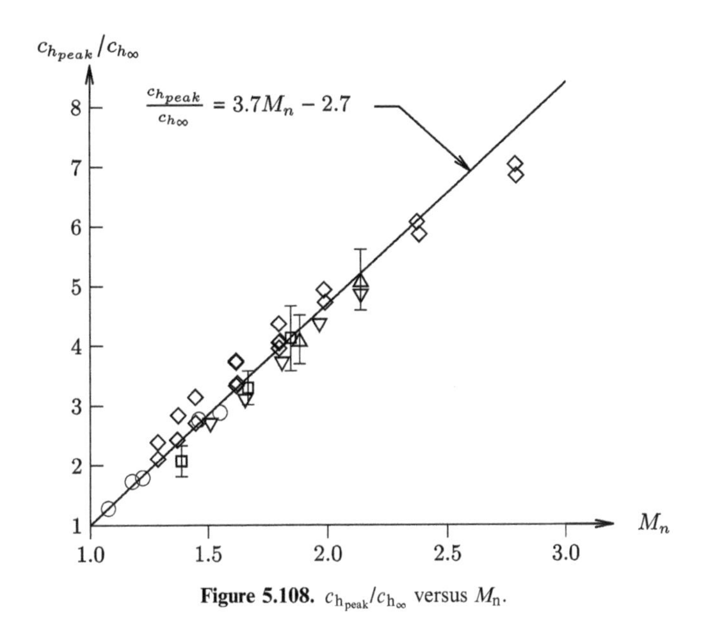


**Figure 5.108.** $c_{h_{peak}}/c_{h_\infty}$ versus $M_n$.

**Table 5.25.** Symbols for figure 5.108.

| Symbol | $M_\infty$ | Reference |
|---|---|---|
| ○ | 3 | Oskam *et al* (1965) |
| □ | 3 | Lee *et al* (1992) |
| △ | 4 | Lee *et al* (1992) |
| ▽ | 5 | Rodi and Dolling (1992) |
| ◇ | 6 | Law (1975), Christophel *et al* (1975) |

$$\frac{c_{h_{peak}}}{c_{h_\infty}} = 3.7M_n - 2.7 \tag{5.9}$$

Note that equation (16) in Lee *et al* (1992) is incorrect; however equation (4) in Lee *et al* (1994) is correct.

## 5.22 Rodi and Dolling (1992, 1995)

Rodi and Dolling (1992) and Rodi and Dolling (1995) performed experiments for a three dimensional shock wave turbulent boundary layer interaction generated by a normal sharp fin on a flat plate at Mach 4.9. Additional details are provided in Rodi *et al* (1991). The experiments were conducted in the Mach 4.9 blowdown tunnel at the University of Texas at Austin (section A.42). The model is shown in figure 5.109. The sharp fin was 15.24 cm long and 8.89 cm high. The fin was mounted 49.35 cm from downstream of the leading edge of the flat plate leading edge. Fin angles from $\alpha = 6°$ to $16°$ were examined. The freestream conditions are listed in table 5.26.

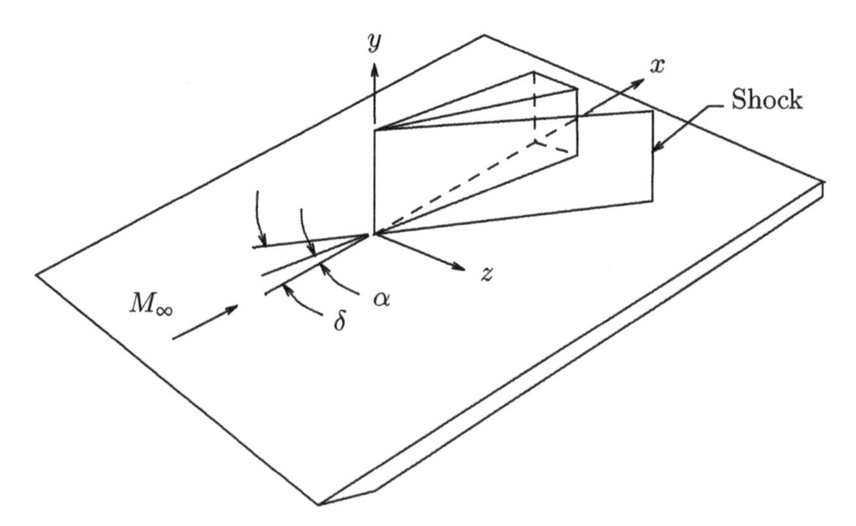


**Figure 5.109.** Model.

**Table 5.26.** Freestream conditions.

| Quantity | Value |
|---|---|
| $M_\infty$ | 4.9 |
| $p_{t_\infty}$ (MPa) | 2.17 |
| $T_{t_\infty}$ (K) | 422 |
| $T_w/T_{aw}$ | 0.80 |
| $H_\infty$ (MJ kg$^{-1}$) | 0.42 |
| $\delta$ (cm) | 0.635 |
| $\delta^*$ (cm) | 0.244 |
| $\theta$ (cm) | 0.023 |
| $Re \times 10^{-6}$ (m$^{-1}$) | 38.1 |

Experimental diagnostics include surface flow visualization, heat transfer and pressure. The experimental data is included in the database of Settles and Dodson (1994).

Figure 5.110 displays the normalized surface pressure $p/p_\infty$ versus the conical angle $\beta$ (figure 5.89). Results are shown for fin angles $\alpha = 8°$ and 16°. Data was obtained along conical rays (fixed $\beta$) and spanwise. The conical behavior of the surface pressure is evident.

Assuming that the mean static temperature field is conical, it is straightforward to show that the Stanton number multiplied by the conical radius $St \cdot r$ is solely function of the conical angle $\beta$. Figure 5.111 presents the scaled Stanton number $St \cdot r/St_\infty \delta$ versus $\beta$ for $\alpha = 8°$ and 16°. Data are taken along two cross planes (i.e., at two different values of the radius $r$ from the leading edge of the sharp fin). The

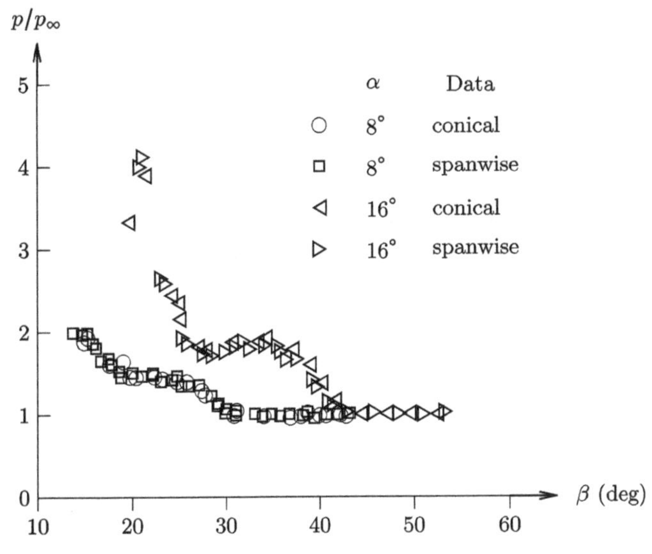


**Figure 5.110.** $p/p_\infty$ versus $\beta$ for $\alpha = 8°$ and 16°.

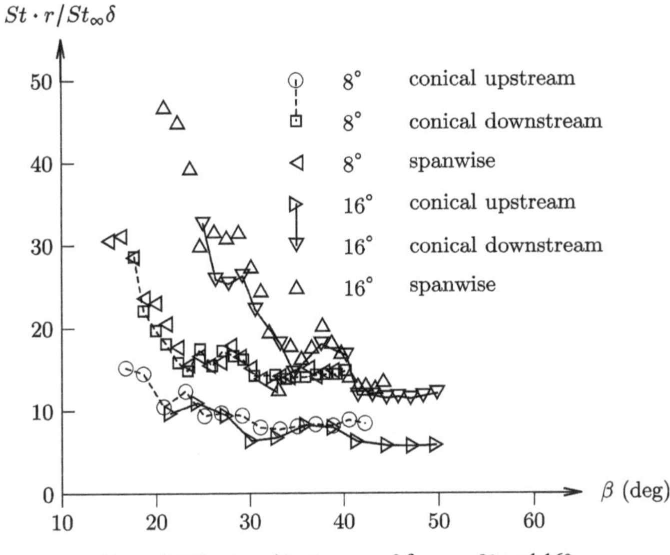


**Figure 5.111.** $St \cdot r/St_\infty \delta$ versus $\beta$ for $\alpha = 8°$ and 16°.

results for both $\alpha = 8°$ and 16° indicate that the Stanton number behavior is not conical, i.e., the mean temperature field is not conical.

## 5.23 Vermeulen and Simeonides (1992)

Vermeulen and Simeonides (1992) performed an experimental study of a nominal two dimensional shock wave turbulent boundary layer interaction generated by a compression corner at Mach 6. The experiments were performed in the von Kármán

Institute for Fluid Dynamics H3 Hypersonic Wind Tunnel (section A.48.2). The model is shown in figure 5.112. Three ramp angles of 10°, 15° and 20° were examined. Two ramp hinge line positions $x_{\mathrm{hl}} = 40$ mm and 90 mm were considered. The plate width is 100 mm. Four different leading edge shapes were tested with thicknesses 26 $\mu$m ± 10 $\mu$m, 40 $\mu$m ± 20 $\mu$m, 98 $\mu$m ± 5 $\mu$m and 316 $\mu$m ± 15 $\mu$m. Additional experiments were performed with 25 mm long strips of grade 1200 sandpaper (0.2 mm thick) distributed along the leading edge of the 98 $\mu$m ± 5 $\mu$m plate with width and spacing of either 5 mm or 10 mm. Three different freestream conditions were examined (table 5.27). Experimental diagnostics included infrared thermography and schlieren imaging.

Figures 5.113 and 5.114 display the heat transfer coefficient $c_{\mathrm{h}}$ defined as

$$c_{\mathrm{h}} = \frac{q_{\mathrm{w}}}{\rho_\infty u_\infty c_{\mathrm{p}}(T_{t_\infty} - T_{\mathrm{w}})} \tag{5.10}$$

on the centerline of the flat plate for compression ramp angles of 10° to 20° at the highest Reynolds number and hinge line locations $x_{\mathrm{hl}} = 40$ mm and 90 mm, respectively[3]. Both figures correspond to the 40 $\mu$m leading edge radius of the flat plate. The magnitude of the heat transfer coefficient upstream of the interaction

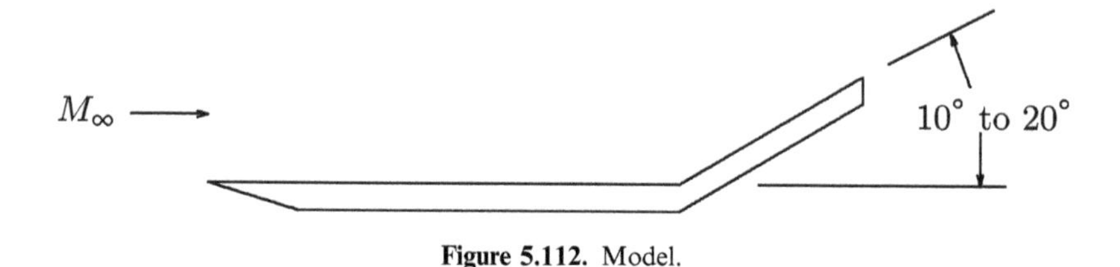


**Figure 5.112.** Model.

**Table 5.27.** Freestream conditions.

| | Value | | |
|---|---|---|---|
| Quantity | Case 1 | Case 2 | Case 3 |
| $M_\infty$ | 6.0 | 6.0 | 6.0 |
| $p_{t_\infty}$ (MPa) | 1.1 | 2.1 | 3.1 |
| $T_{t_\infty}$ (K) | 530 | 530 | 530 |
| $T_{\mathrm{w}}$ (K) | 295 | 295 | 295 |
| $T_{\mathrm{w}}/T_{\mathrm{aw}}$ | 0.90 | 0.90 | 0.90 |
| $H_\infty$ (MJ kg$^{-1}$) | 0.53 | 0.53 | 0.53 |
| $Re \times 10^{-6}$ (m$^{-1}$) | 8.0 | 15.0 | 21.0 |

[3] The experimental data is shown in (Vermeulen and Simeonides 1992) as solid lines. The data has been digitized for presentation.

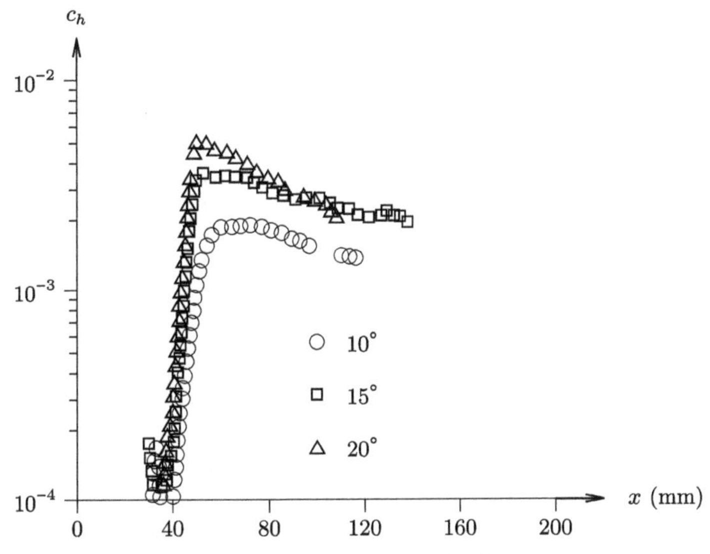


**Figure 5.113.** $c_h$ versus $x$ for $Re = 21 \times 10^6\ m^{-1}$ and $x_{hl} = 40$ mm.

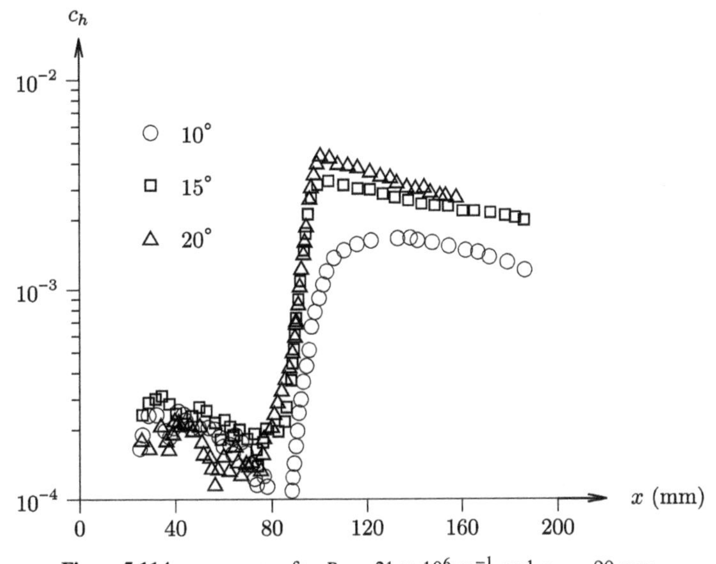


**Figure 5.114.** $c_h$ versus $x$ for $Re = 21 \times 10^6\ m^{-1}$ and $x_{hl} = 90$ mm.

indicates an incoming laminar boundary layer, whereas the magnitude downstream of the hinge line indicates a turbulent boundary layer. The shock wave boundary layer interaction at the compression corner is therefore transitional. Additional detailed measurements of the spanwise heat transfer coefficient indicated the presence of streamwise Görtler vortices and spanwise variations in $c_h$ of typically 20%.

## 5.24 Coët and Chanetz (1993)

Coët and Chanetz (1993) conducted an experimental investigation of two- and three-dimensional hypersonic shock wave boundary layer interactions at Mach 5 and 10. The experiments were performed in the R2Ch and R3Ch blowdown wind tunnels at Chalais-Meudon (section A.25.2). The shock wave turbulent boundary layer model is a two dimensional ramp with deflection angle 35° (figure 5.115). The freestream conditions are listed in table 5.28.

Figure 5.116 displays the Stanton number St versus $x/L$ along the flat plate. The rise in St at $x/L = 0.4$ indicates the separation of the boundary layer. The plateau region from $0.6 \leqslant x/L \leqslant 0.9$ is typical of a separated boundary layer. The effect of wall cooling is a slight decrease in St in the plateau region. Figure 5.117 shows the surface pressure coefficient $c_p$ versus $x/L$.

## 5.25 Garrison *et al* (1993–6)

Garrison *et al* (1993, 1994) and Garrison *et al* (1996) investigated the crossing shock wave turbulent boundary layer interactions at Mach 3 and 4. The experiments were conducted in the Supersonic Wind Tunnel Facility at the Penn State Gas Dynamics Laboratory. The model is shown in figure 5.118. Fin angles $\alpha = 7°$ to $13°$ were investigated at Mach 3, and $\alpha = 7°$ to $15°$ at Mach 4. The freestream conditions are listed in table 5.29. Experimental diagnostics include Planar Laser Scattering (PLS), surface flow visualization and skin friction. See also Garrison and Settles (1992). Additionally, Garrison *et al* (1996) investigated the triple shock wave turbulent boundary layer interaction.

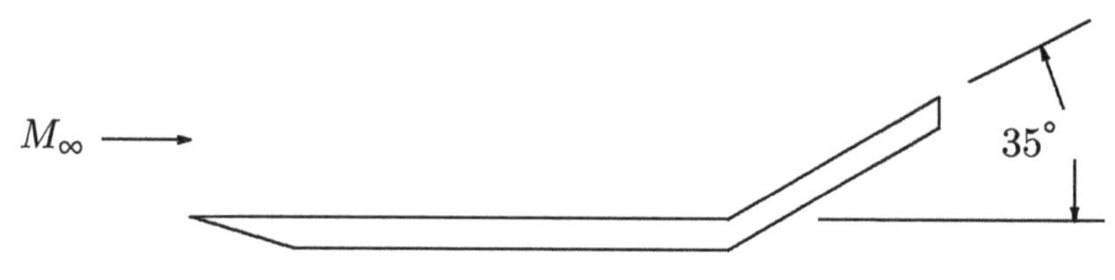


**Figure 5.115.** Model.

**Table 5.28.** Freestream conditions.

| Quantity | Value |
|---|---|
| $M_\infty$ | 5 |
| $p_{t_\infty}$ (MPa) | 3.25 |
| $T_{t_\infty}$ (K) | 500 |
| $T_w/T_{aw}$ | 0.24 and 0.72 |
| $H_\infty$ (MJ kg$^{-1}$) | 0.50 |
| $Re_\infty \times 10^{-6}$ (m$^{-1}$) | 40 |

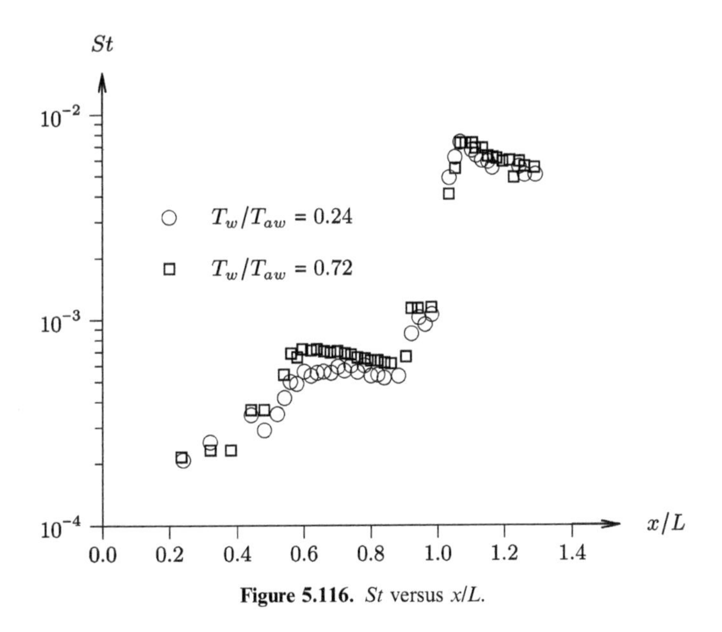


**Figure 5.116.** *St* versus *x*/*L*.

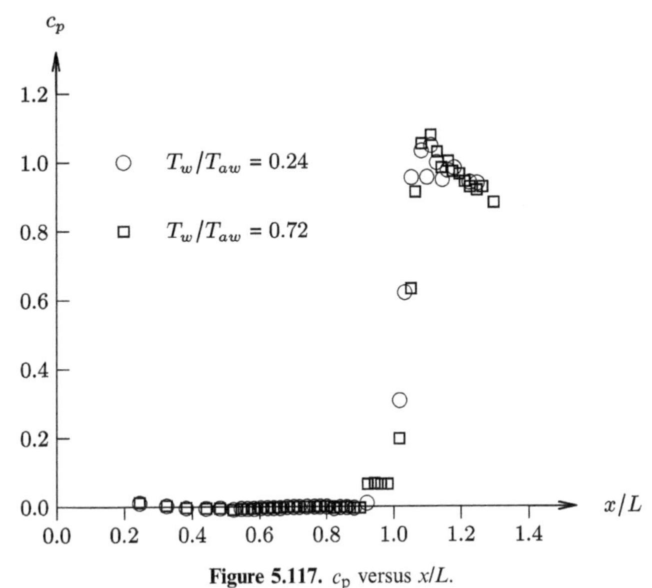


**Figure 5.117.** $c_p$ versus *x*/*L*.

Figure 5.119 shows the surface streamline structure for $M_\infty = 4$ and $\alpha = 15°$. The upstream influence line (UI) defines the beginning of the surface pressure rise due to the shock waves generated by the fins. The primary separation line (PS) is identified by the convergence of surface streamlines (Lighthill 1963). The location of the inviscid shock waves (IS) generated by the fins is shown. The primary attachment (PA), secondary separation (SS) lines are also indicated, saddle point (SPT) and centerline attachment (CA) are also evident.

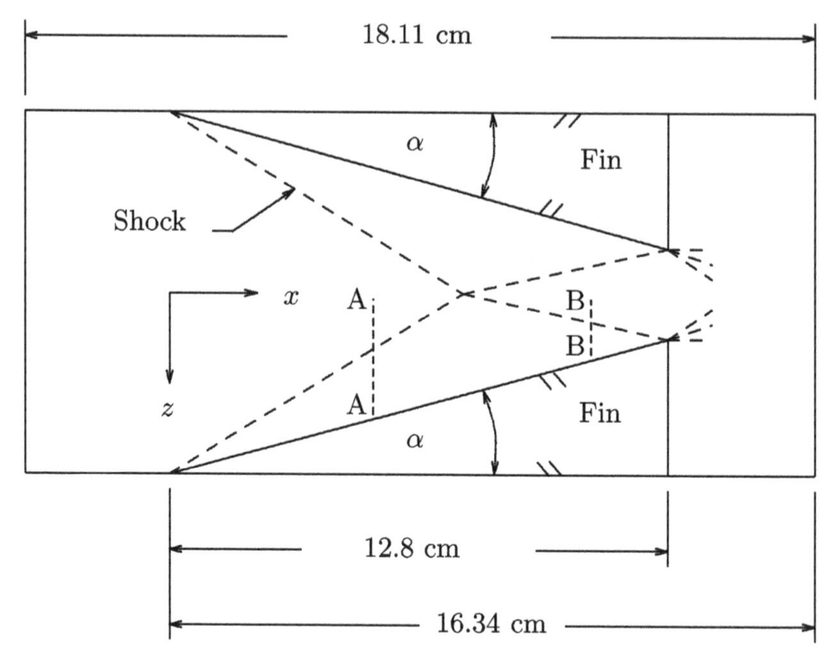


**Figure 5.118.** Model (top view).

**Table 5.29.** Freestream conditions[a].

| Quantity | Case 1 | Case 2 |
|---|---|---|
| $M_\infty$ | 2.9 | 3.85 |
| $p_{t_\infty}$ (MPa) | 0.75 | 1.5 |
| $T_{t_\infty}$ (K) | 295 | 295 |
| $T_w/T_{aw}$ | $\approx 1$ | $\approx 1$ |
| $H_\infty$ (MJ kg$^{-1}$) | 0.30 | 0.30 |
| $Re_\infty \times 10^{-6}$ (m$^{-1}$) | 80 | 80 |
| $\delta$ (mm) | 3.3[b] | 2.9[b] |
| $\delta^*$ (mm) | 1.0[b] | 0.95[b] |
| $\theta$ (mm) | 0.2[b] | 0.13[b] |
| $c_f \times 10^3$ | 1.43 | 1.32 |

[a] Garrison (1994).
[b] Upstream of interaction.

Figure 5.120 displays the flowfield structure for Mach 4 and $\alpha = 15°$ at a location upstream of the intersection of the fin generated shock waves corresponding to the plane A-A in figure 5.118. The details are similar to figure 5.91. Figure 5.121 shows the flowfield structure at a location downstream of the intersection of the incident shock waves corresponding to plane B-B in figure 5.118. A more complex pattern of shock waves and slip lines is evident. Figure 5.122 presents the surface skin friction coefficient $c_f$ versus $x/\delta$ on the centerline of the model. The skin friction decreases

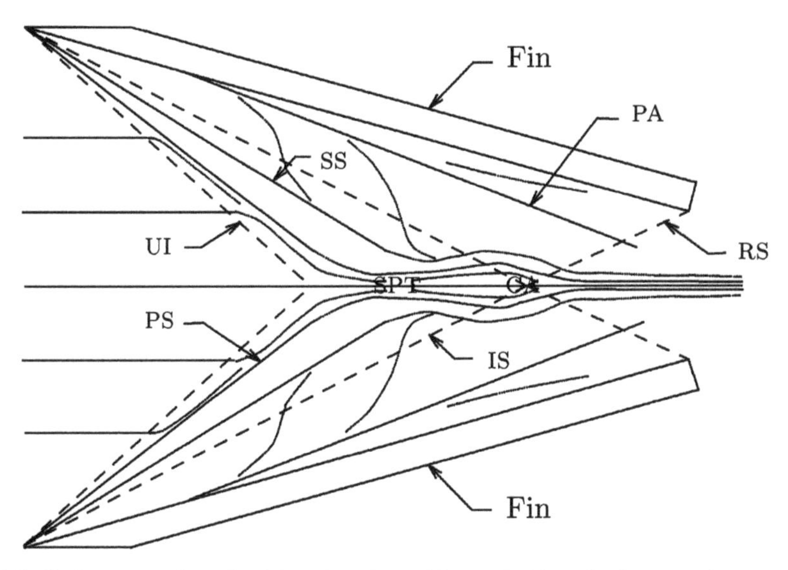


**Figure 5.119.** Surface streamlines for $M_\infty = 4$ and $\alpha = 15°$. IS—incident shock, PA—primary attachment, PS—primary separation, RS—reflected shock, SS— secondary separation, UI— upstream influence.

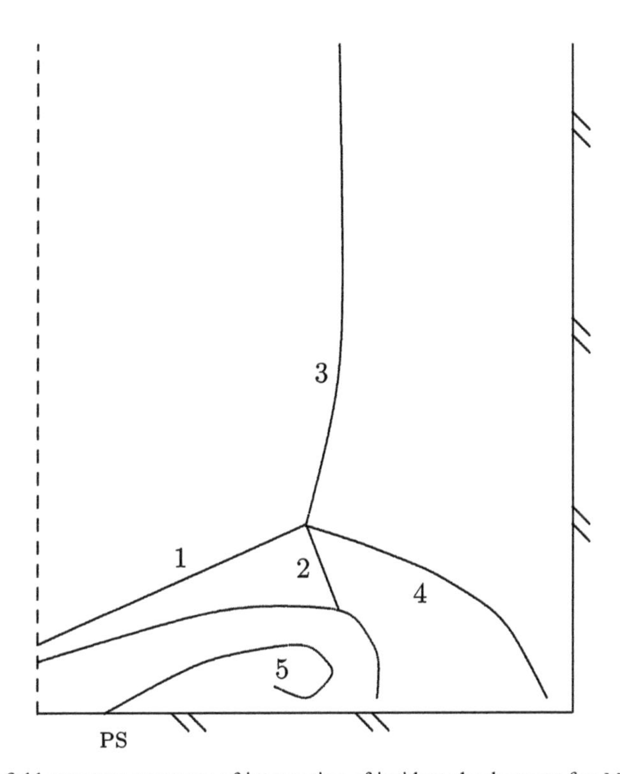


**Figure 5.120.** Flowfield structure upstream of intersection of incident shock waves for $M_\infty = 4$ and $\alpha = 15°$. 1—separation shock, 2—rear shock, 3—incident shock, 4—slip line, 5—vortex, PS—primary separation.

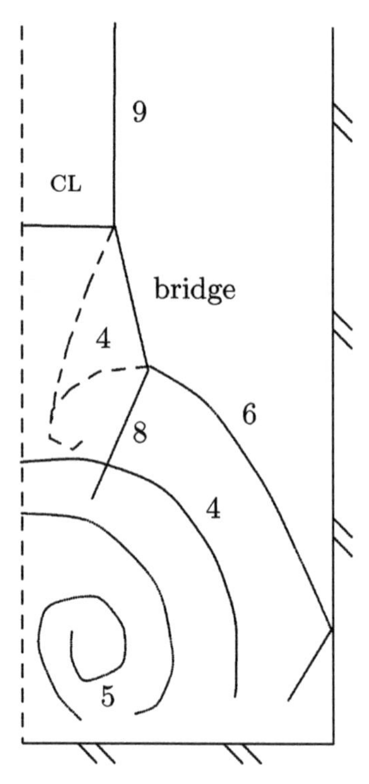


**Figure 5.121.** Flowfield structure downstream of intersection of incident shock waves. 4—slip lines, 5—separation vortex, 6—reflected separation shock, 8—reflected rear shock, 9—reflected incident shock, CL—centerline shock.

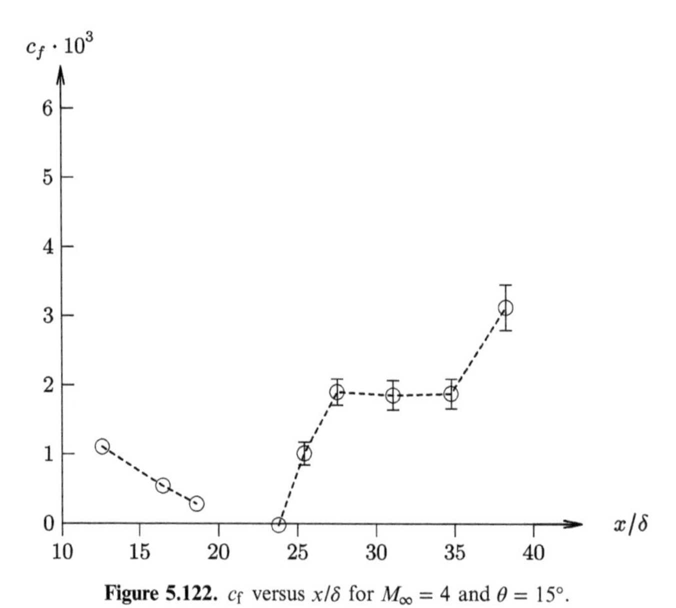


**Figure 5.122.** $c_f$ versus $x/\delta$ for $M_\infty = 4$ and $\theta = 15°$.

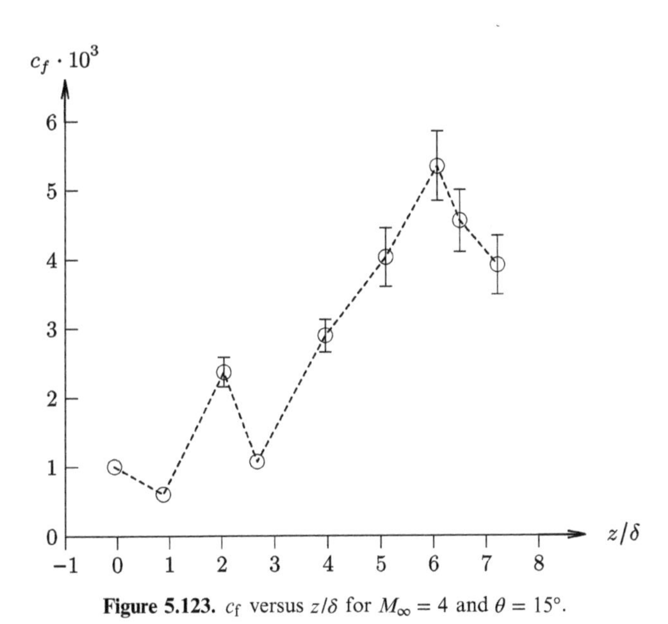


**Figure 5.123.** $c_f$ versus $z/\delta$ for $M_\infty = 4$ and $\theta = 15°$.

towards zero at the saddle point (figure 5.119) and subsequently increases to a plateau downstream of the intersection of the incident oblique shock waves. Figure 5.123 shows the spanwise variation of skin friction coefficient $c_f$ versus $z/\delta$ at $x/\delta = 25.3$ corresponding to a location upstream of the intersection of the incident shock waves. The behavior is similar to figure 5.95.

## 5.26 Kussoy and Horstman (1993)

Kussoy and Horstman (1993) conducted an experimental investigation of a three dimensional shock wave turbulent boundary layer interaction generated by a sharp fin on a flat plate at Mach 8.2. Additional details are provided in Kussoy *et al* (1991), Kussoy and Horstman (1991) and Knight *et al* (1992). The experiments were performed in the NASA Ames 3.5 ft hypersonic wind tunnel (section A.20.2). The model is shown in figure 5.124. The length and width of the plate are 200 cm and 76 cm, respectively. The leading edge of the fin is mounted 176 cm from the leading edge of the flat plate. The freestream conditions are presented in table 5.30. Experimental diagnostics include surface heat transfer, pressure and skin friction.

The surface pressure $p/p_\infty$ versus spanwise distance from the fin $z - x\tan\alpha$ is presented in figure 5.125 at a streamwise distance $x = 18.19$ cm from the leading edge of the fin. The inviscid shock locations for $\alpha = 10°$ and $15°$ are indicated by the dashed lines. The plateau pressure associated with the boundary layer separation is evident. The detailed structure of the single fin shock wave turbulent boundary layer interaction is presented in Alvi and Settles (1992). The surface heat transfer $q_w/q_{w_\infty}$ versus $z - x\tan\alpha$ is shown in figure 5.126 where $q_{w_\infty} = 1.04$ W/cm$^2$. The high heat transfer near the fin ($z - x\tan\alpha = 0$) is associated with the reattachment of the boundary layer. The skin friction coefficient $c_f$ is displayed in figure 5.127 and shows similar behavior to the heat transfer. A substantial amount of additional data for

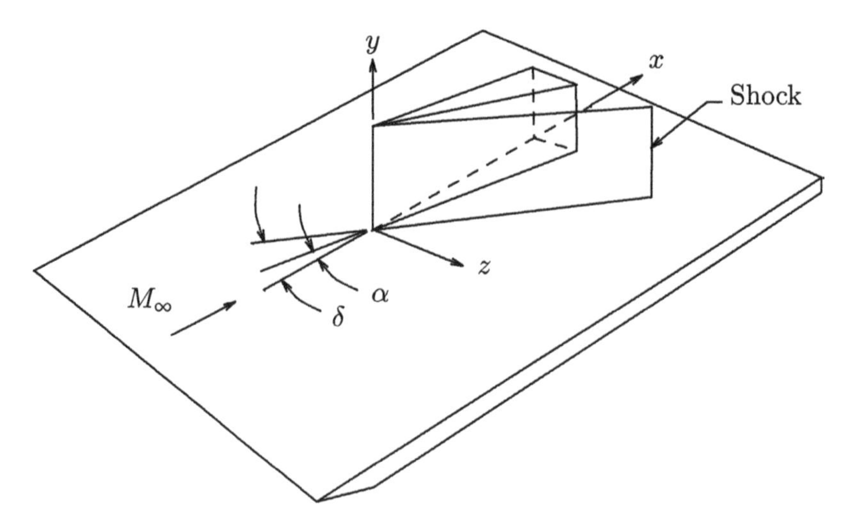


**Figure 5.124.** Model.

**Table 5.30.** Freestream conditions.

| Quantity | Value |
|---|---|
| $M_\infty$ | 8.2 |
| $p_{t_\infty}$ (MPa) | 4.90 |
| $T_{t_\infty}$ (K) | 1111 |
| $T_w$ (K) | 300 |
| $T_w/T_{aw}$ | 0.29 |
| $H_\infty$ (MJ kg$^{-1}$) | 1.1 |
| $\delta$ (cm) | 3.7 |
| $\delta^*$ (cm) | 1.59 |
| $\theta$ (cm) | 0.009 41 |
| $Re \times 10^{-6}$ (m$^{-1}$) | 4.87 |

surface pressure, heat transfer and skin friction, and surveys of pitot pressure are provided in Kussoy and Horstman (1993).

## 5.27 Kussoy *et al* (1993)

Kussoy *et al* (1993) conducted an experimental investigation of a hypersonic crossing shock wave turbulent boundary layer interaction at Mach 8.3. The experiments were performed in the NASA Ames 3.5 ft hypersonic wind tunnel (section A.20.2). The model is shown in figure 5.128. A pair of fins are attached normal to a flat plate. Fin angles of 10° and 15° are considered. Each fin generates an incident oblique shock which interacts with the turbulent boundary layer on the flat plate. The incident shock waves intersect forming a crossing shock wave

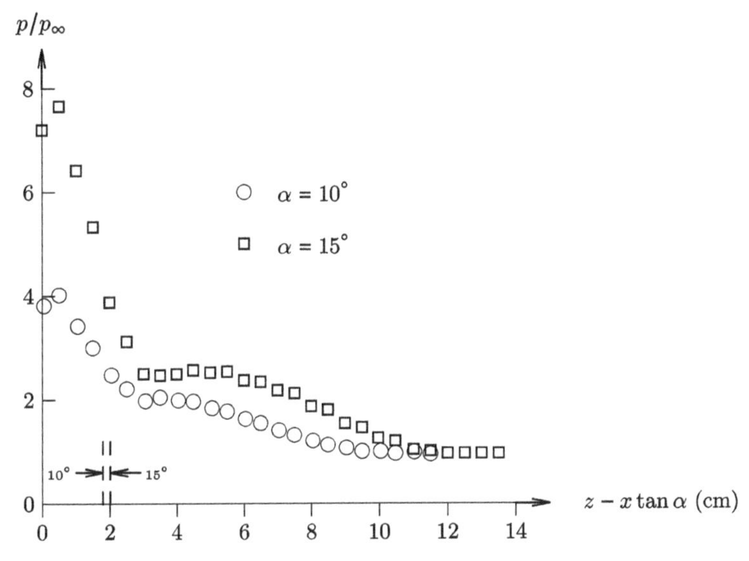


**Figure 5.125.** $p/p_\infty$ versus $z - x \tan\alpha$ at $x = 18.19$ cm.

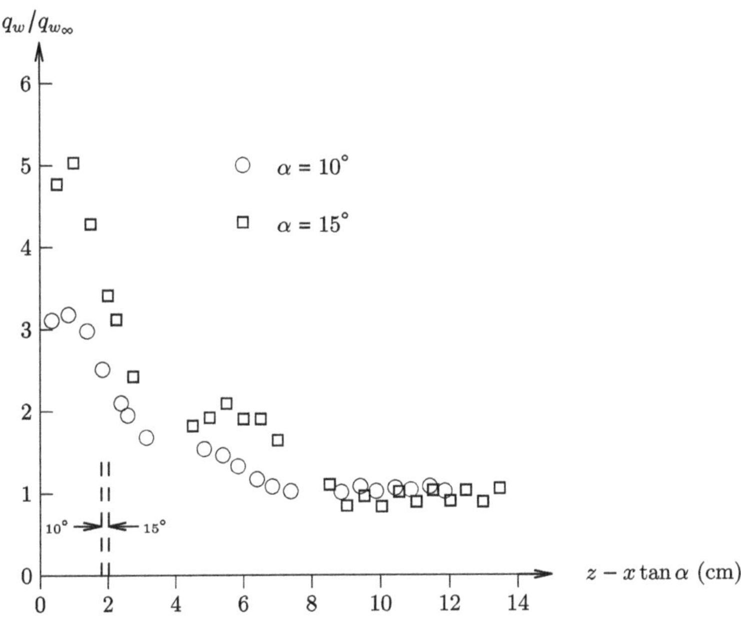


**Figure 5.126.** $q_w/q_{w_\infty}$ versus $z - x\tan\alpha$ at $x = 16.45$ cm.

interaction. The freestream conditions are tabulated in Kussoy and Horstman (1992) and are shown in table 5.31. Experimental diagnostics include surface pressure and heat transfer, oil film visualization on the flat plate surface, and pitot pressure surveys throughout the interaction region. Pitot and static pressure and total temperature surveys were also performed upstream of the interaction.

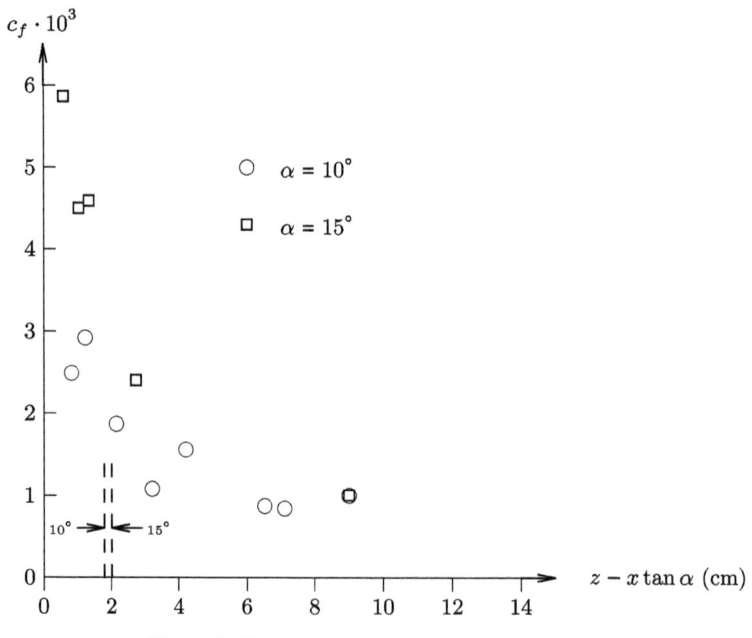


**Figure 5.127.** $c_f$ versus $z - x \tan \alpha$ at $x = 15.5$ cm.

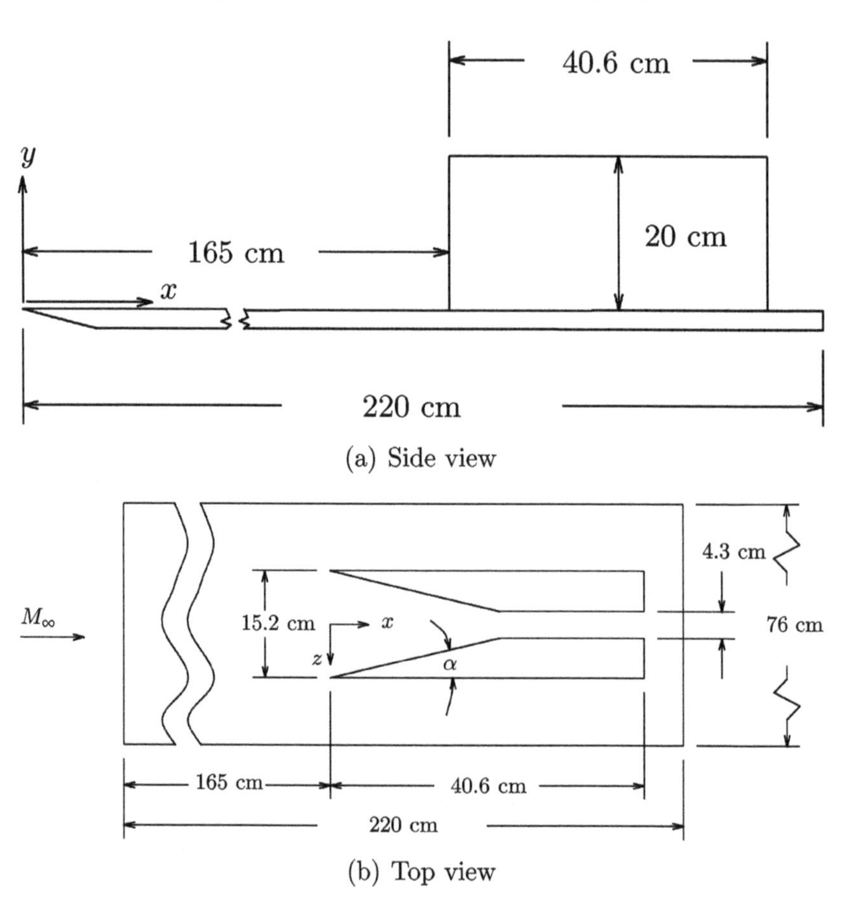


**Figure 5.128.** Model.

**Table 5.31.** Freestream conditions.

| Quantity | Value |
|---|---|
| $M_\infty$ | 8.28 |
| $p_{t_\infty}$ (MPa) | 5.25 |
| $T_{t_\infty}$ (K) | 1177 |
| $T_w$ (K) | 300 |
| $T_w/T_{aw}$ | 0.269 |
| $H_\infty$ (MJ kg$^{-1}$) | 1.2 |
| $\delta$ (cm) | 3.25 |
| $\theta$ (cm) | 0.083 |
| $Re \times 10^{-6}$ (m$^{-1}$) | 5.3 |

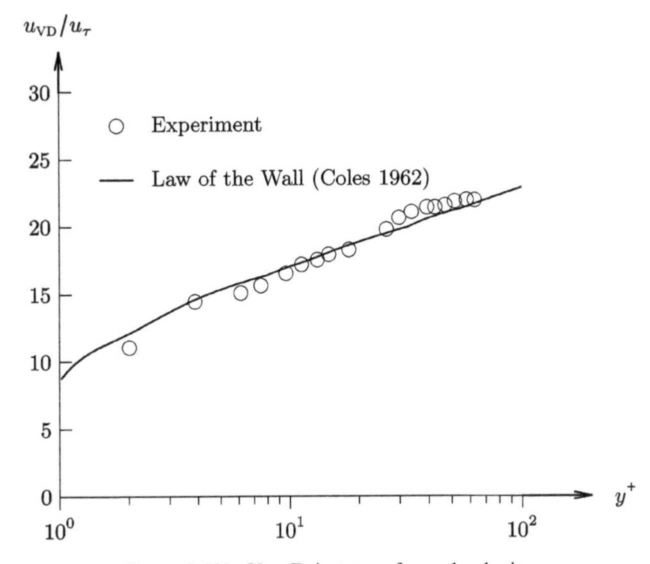


**Figure 5.129.** Van Driest transformed velocity.

Figure 5.129 shows the Van Driest transformed velocity at 162 cm from the leading edge of the flat plate and upstream of the interaction. The experimental data shows close agreement with the Law of the Wall thereby indicating an equilibrium turbulent boundary layer.

The normalized surface pressure $p/p_\infty$ and heat transfer $q_w/q_{w_\infty}$ on the centerline for the 10° and 15° fins are shown in figures 5.130 and 5.131, respectively. The location of the intersection of the inviscid shock waves is indicated by the dashed line. The interaction begins far upstream of the inviscid shock intersection location at $x = 10$ cm and $x = 6$ cm, respectively, where $x$ is measured from the leading edge of the fins. The pressure and heat transfer rise monotonically throughout the interaction to a maximum value downstream of the location of the inviscid shock

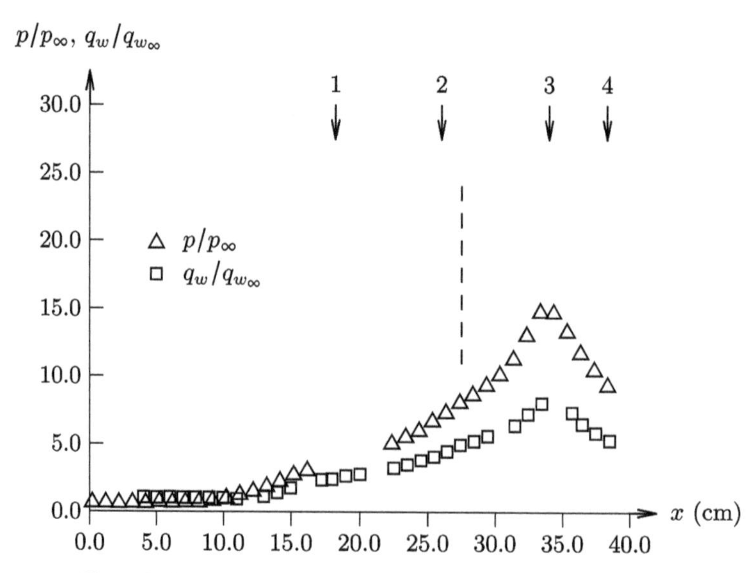


**Figure 5.130.** Pressure and heat transfer on centerline for 10° case.

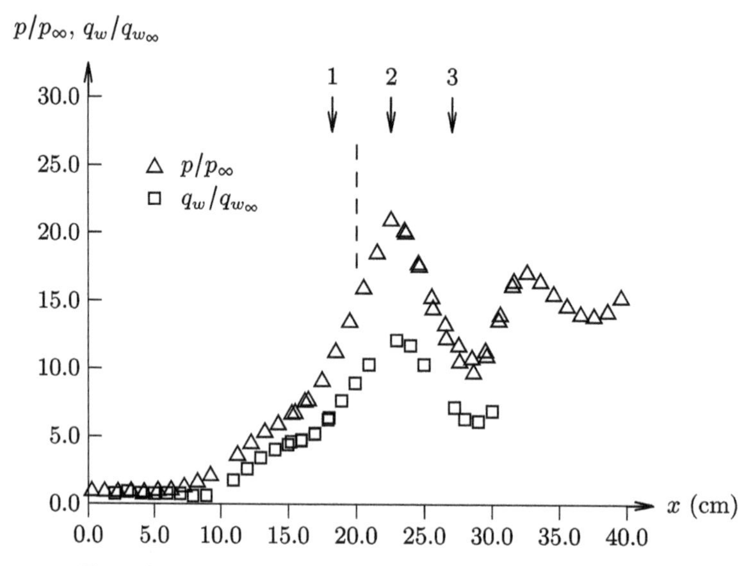


**Figure 5.131.** Pressure and heat transfer on centerline for 15° case.

intersection. Thereafter, the pressure and heat transfer decrease due to the interaction with the expansion fans formed by the wedge corners at $x = 30.91$ cm and 20.34 cm, respectively. Also noted are the locations 1 to 4 and 1 to 3 of the pitot pressure and yaw angle surveys for the 10° and 15° cases.

## 5.28 Simeonides and Haase (1995)

Simeonides and Haase (1995) conducted an experimental investigation of a hypersonic shock wave transitional boundary layer interaction at Mach 14.1. The model comprised a forward flat plate with a nominally sharp leading edge and a rear plate at an angle of 15°. The model is shown in figure 5.132. Two lengths of the forward flat plate were examined (70 mm and 200 mm), and a single length of the rear plate (200 mm). The spanwise width of the model was 200 mm. The experiments were performed in the von Kármán Institute for Fluid Dynamics LongShot Tunnel (section A.48.1). The flow conditions for the longer forward flat plate experiments are listed in table 5.32. Experimental diagnostics include surface heat transfer and surface pressure, and schlieren imaging. See also Simeonides and Wendt (1990) and Simeonides (1996).

Figure 5.133 shows the surface pressure coefficient $c_p$ versus $x$ (mm) for $Re = 6.5 \times 10^6$ m$^{-1}$. The incoming boundary layer is laminar, and transition occurs downstream of reattachment. The plateau pressure in the region 150mm $\leqslant x \leqslant$ 210mm is indicative of a separation region. Figure 5.134 displays the heat transfer coefficient defined as

$$c_h = \frac{q_w}{\rho_\infty c_p u_\infty (T_{t_\infty} - T_w)} \tag{5.11}$$

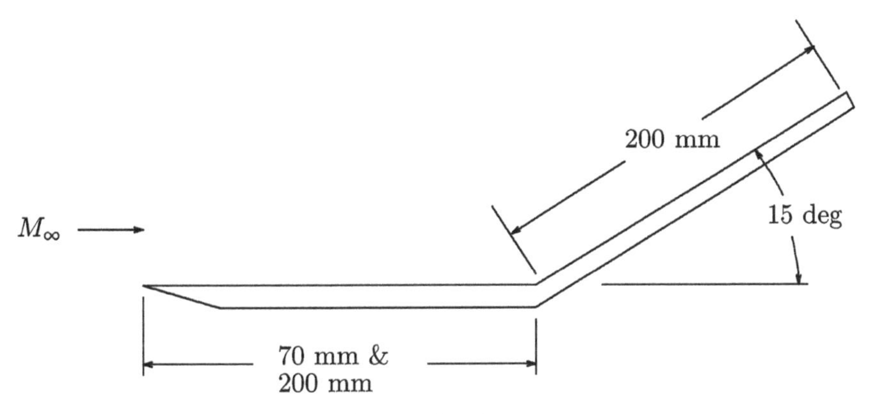


**Figure 5.132.** Model.

**Table 5.32.** Freestream conditions.

| *Quanity* | Value |
|---|---|
| Gas | $N_2$ |
| $M_\infty$ | 14.1 |
| $T_\infty$ (K) | 58.5 |
| $T_w$ (K) | 290 |
| $Re \times 10^{-6}$ (m$^{-1}$) | 6.5 and 13.0 |

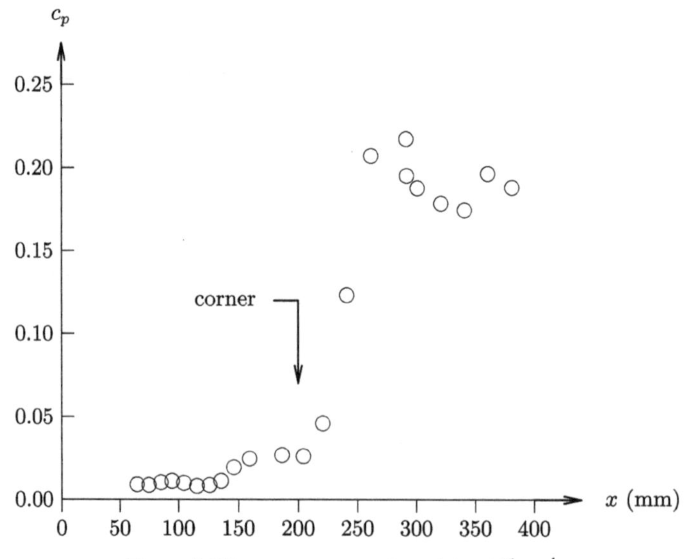


**Figure 5.133.** $c_p$ versus $x$ at $Re = 6.5 \times 10^6$ m$^{-1}$.

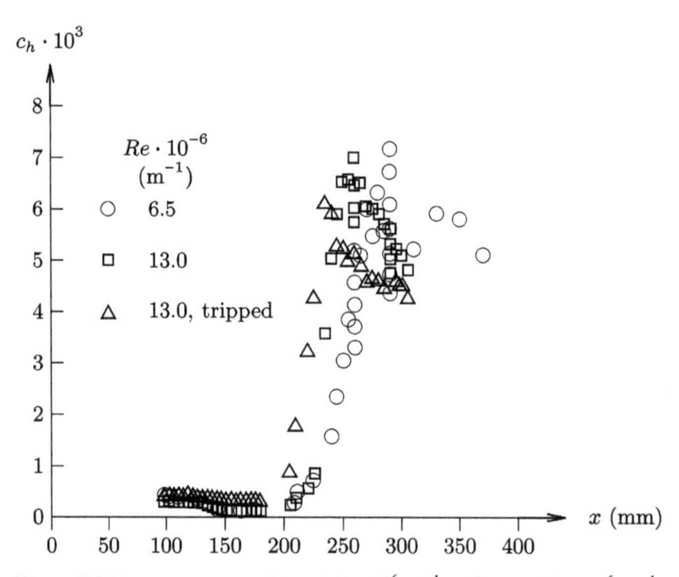


**Figure 5.134.** $c_h$ versus $x$ at $Re = 6.5 \times 10^6$ m$^{-1}$ and $Re = 13 \times 10^6$ m$^{-1}$.

for $Re = 6.5 \times 10^6$ m$^{-1}$ and $Re = 13 \times 10^6$ m$^{-1}$ with incoming laminar (untripped) boundary layers. The decrease in $c_h$ beginning at $x = 140$ mm indicates separation. Also shown are results for a tripped incoming boundary layer; however, Simeonides and Haase (1995) caution that the incoming boundary layer may not be fully turbulent.

## 5.29 Schülein *et al* (1996, 2001, 2006)

Schülein *et al* (1996), Schülein and Zheltovodov (2001) and Schülein (2006) performed a series of experimental investigations of hypersonic shock wave turbulent boundary interactions at Mach 5. The experiments were conducted in the DLR Göttingen Ludwieg Tube Facility (section A.11.3). Three different models were considered as described below. The nominal freestream conditions are listed in table 5.33. Experimental surface diagnostics include heat transfer using the quantitative infrared thermography technique, pressure, and skin friction using the global interferometry technique. Boundary layer surveys comprise pitot and static pressure. Flow visualization includes surface oil imagery and shadowgraphy.

### 5.29.1 Planar incident shock

The model is shown in figure 5.135. A shock generator comprising a flat plate at angle $\beta$ forms an incident shock wave which interacts with the turbulent boundary layer on the lower flat plate. Experiments were performed for shock generator angles $\beta$ from 6° to 14° in increments of 2° to examine both unseparated and separated shock boundary layer interactions. The theoretical inviscid impingement point of the incident shock wave is $x = 350$ mm measured from the leading edge of the flat plate. Natural transition occurs at $x = 120$ mm thereby indicating a fully turbulent inflow boundary layer to the shock wave boundary layer interaction. The boundary layer properties at x = 356 mm in the absence of the shock generator are listed in

**Table 5.33.** Freestream conditions.

| Quantity | Value |
|---|---|
| $M_\infty$ | 5.0 |
| $p_{t_\infty}$ (MPa) | 2.12 |
| $T_{t_\infty}$ (K) | 410 |
| $T_w$ (K) | 300 ± 5 |
| $H_\infty$ (MJ kg$^{-1}$) | 0.41 |

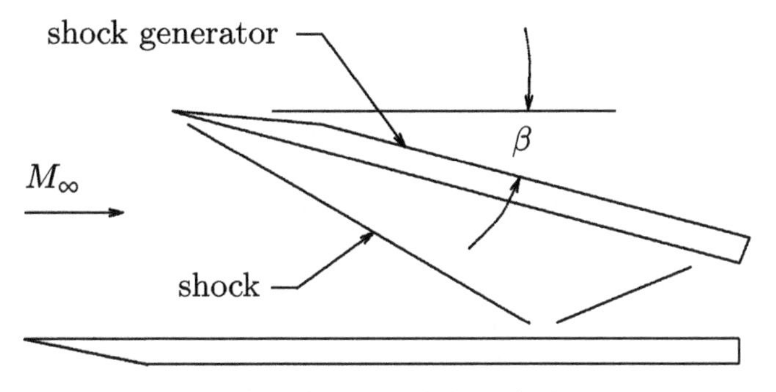


**Figure 5.135.** Planar incident shock.

**Table 5.34.** Inflow conditions.

| Quantity | Value |
|---|---|
| $M_e$ | 4.984 |
| $p_{t_\infty}$ (MPa) | 2.14 |
| $T_{t_\infty}$ (K) | 410.2 |
| $H_\infty$ (MJ kg$^{-1}$) | 0.41 |
| $Re \times 10^{-6}$ (m$^{-1}$) | 39.1 |
| $\delta$ (mm) | 4.813 |
| $\delta^*$ (mm) | 2.000 |
| $\theta$ (mm) | 0.202 |
| $c_f \times 10^3$ | 1.27 ± 0.02 |

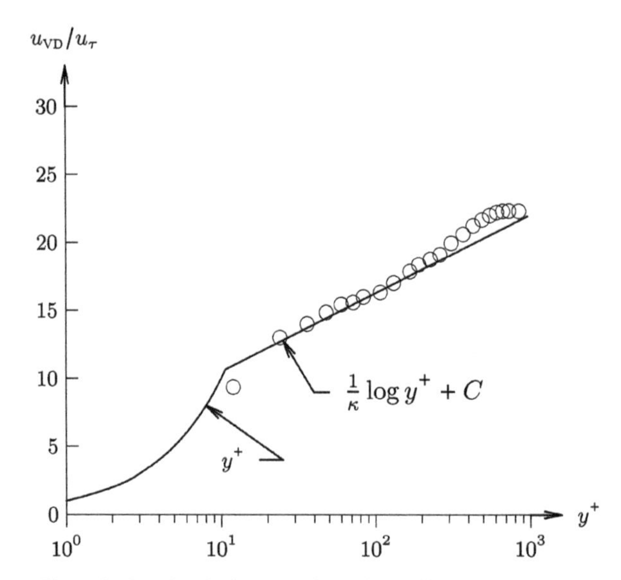


**Figure 5.136.** Van Driest transformed velocity at $x$ = 356 mm.

table 5.34 (Schülein *et al* 1996). The Van Driest transformed velocity at $x$ = 356 mm in the absence of the shock generator is shown in figure 5.136 and is in close agreement with the Law of the Wall.

Figure 5.137 shows the surface pressure versus $x$ for shock generator angles $\beta$ = 6° to 14°. An enlargement of the interaction region is displayed in figure 5.138. No evidence of boundary layer separation is apparent for $\beta$ = 6° and 8°. A small separation region was observed in the oil film visualization at $\beta$ = 10°, and a pressure plateau region typical of a separation zone is evident at $\beta$ = 12° and 14°.

Figure 5.139 displays the skin friction coefficient $c_f$ for shock generator angles $\beta$ = 10° and 14°. The separation zone is defined by the negative values of $c_f$ and increases with shock strength. Figure 5.140 shows the Stanton number $St$ versus $x$ for shock generator angles $\beta$ = 6° to 14°. It is noteworthy that the heat transfer

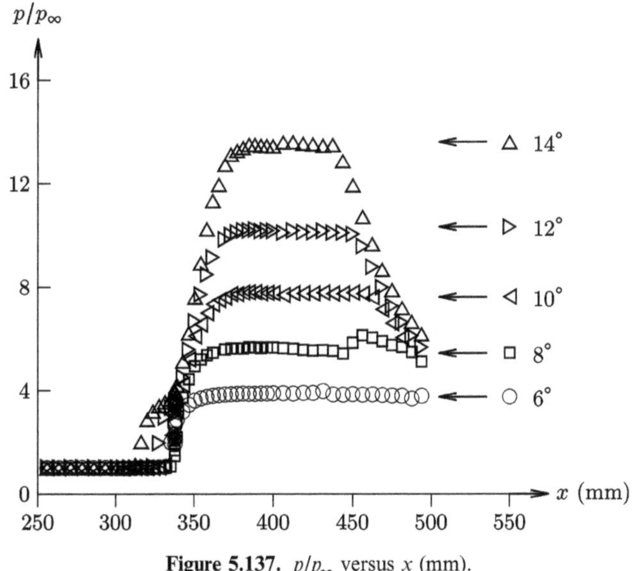


**Figure 5.137.** $p/p_\infty$ versus $x$ (mm).

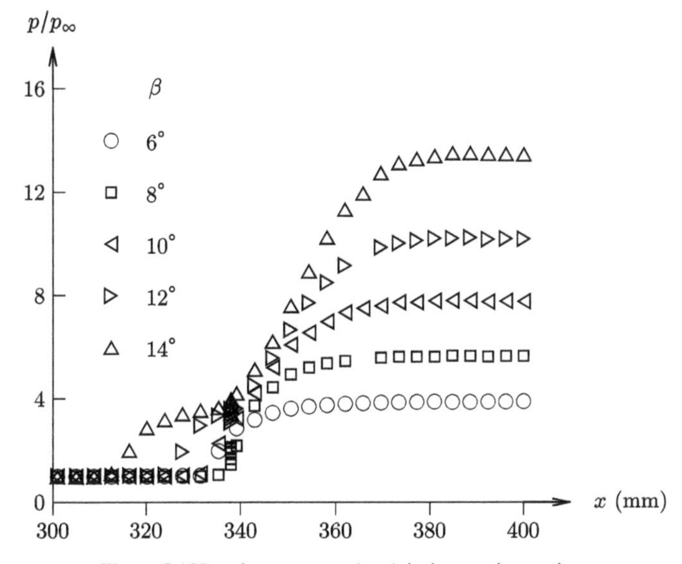


**Figure 5.138.** $p/p_\infty$ versus $x$ (mm) in interaction region.

within the separation region for $\beta = 14°$ is nearly twice the undisturbed upstream value.

### 5.29.2 Sharp fin

The model is shown in figure 5.141. A shock generator comprising a flat plate attached normal to the lower plate forms a swept shock wave which interacts with

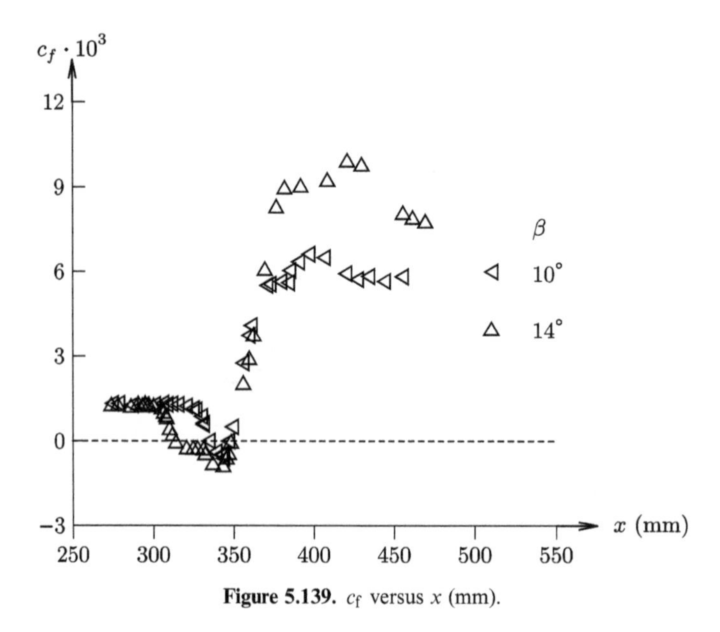


**Figure 5.139.** $c_f$ versus $x$ (mm).

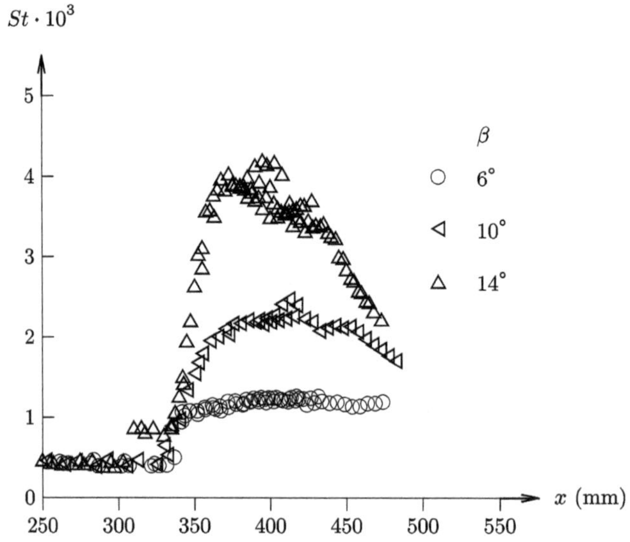

**Figure 5.140.** $St$ versus $x$ (mm).

the turbulent boundary layer on the lower flat plate. Experiments were performed for fin angles $\beta = 2°$ to $27°$. The leading edge of the sharp fin is located from $x = 229.5$ mm to 286 mm from the leading edge of the bottom plate depending on the experiment. Natural transition occurs at approximately $x = 180$ mm from the leading edge of the lower flat plate. The boundary layer properties on the lower flat plate at $x = 266$ mm in the absence of sharp fin are listed in table 5.35.

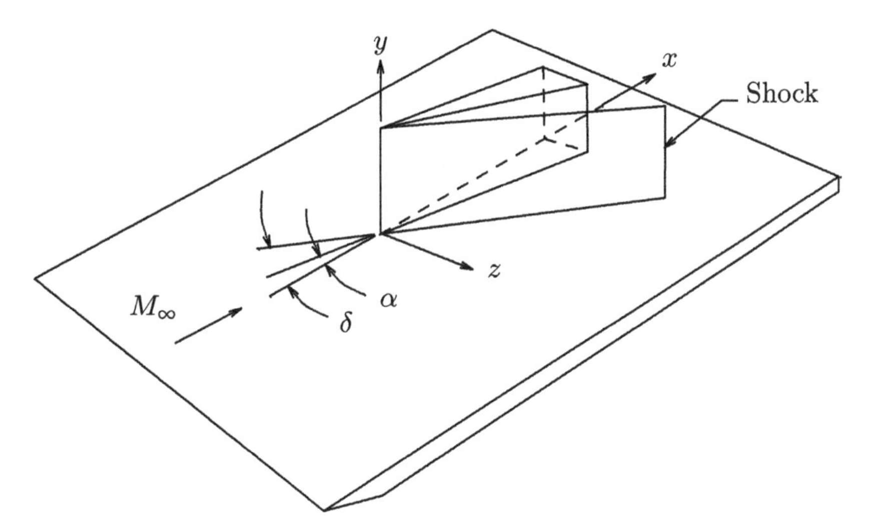


**Figure 5.141.** Sharp fin.

**Table 5.35.** Inflow conditions.

| Quantity | Value |
|---|---|
| $M_e$ | 4.961 |
| $\delta$ (mm) | 3.81 |
| $\delta^*$ (mm) | 1.576 |
| $\theta$ (mm) | 0.157 |
| $c_f \times 10^3$ | 1.382 |

Figure 5.142 shows a schematic of the surface skin friction lines on the bottom plate. The inflow skin friction lines begin to deviate at the Upstream Influence (UI) where the pressure begins to increase. For sufficiently large fin angle $\beta$ the skin friction lines coalesce along a single line PS defining the primary separation line[4]. A primary reattachment line PA forms close to the fin defined by skin friction lines diverging from a single line. For sufficiently large value of $\beta$ secondary separation SS and reattachment lines appear. The separation and reattachment lines display conical behavior since they are defined by constant angles measured in a spherical polar coordinate system (figure 5.143) whose origin (Virtual Conical Origin [VCO]) is observed to be slightly upstream of the fin leading edge.

Figure 5.144 displays the surface pressure $p/p_\infty$ on the bottom flat plate versus $z/x$ for fin angle $\beta = 23°$ where $x$ is measured from the leading edge of the sharp fin and $z$ is the spanwise dimension. Data is shown at $x$-locations from 82.5 mm to 203 mm

[4] Three dimensional steady separation occurs at the confluence of surface skin friction lines and does not necessarily coincide with zero skin friction (Lighthill 1963).

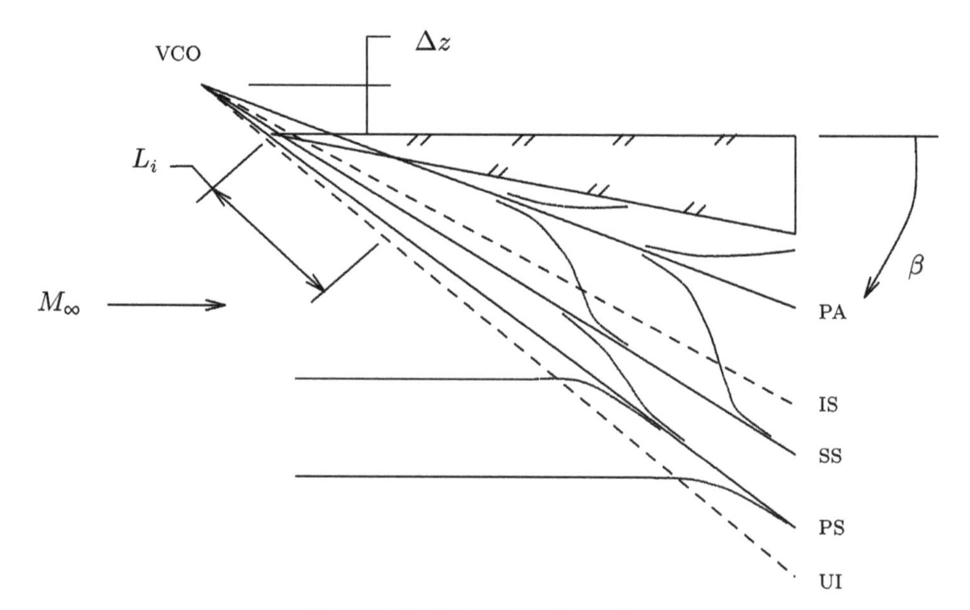


**Figure 5.142.** Surface skin friction lines.

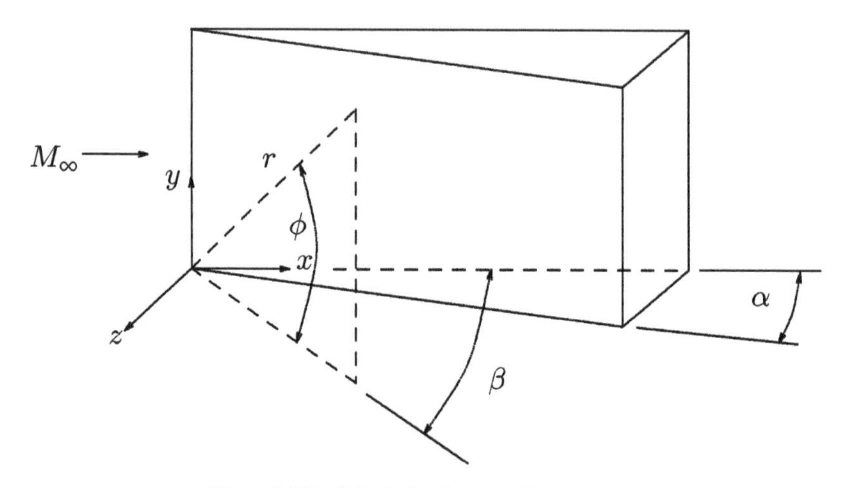


**Figure 5.143.** Spherical polar coordinate system.

measured from the fin leading edge. The pressure distribution displays a conical flow behavior as described in Alvi and Settles (1992). The experimental surface pressure distribution depends solely on the conical angle $\beta = \tan z/x$. One or two separation lines may appear depending upon the shock strength, with corresponding reattachment lines (Zheltovodov and Knight 2011). The location of the first separation line $S1$ corresponds to the initial pressure rise due to the shock. The corresponding reattachment line $R1$ coincides with the peak surface pressure. Within the reversed

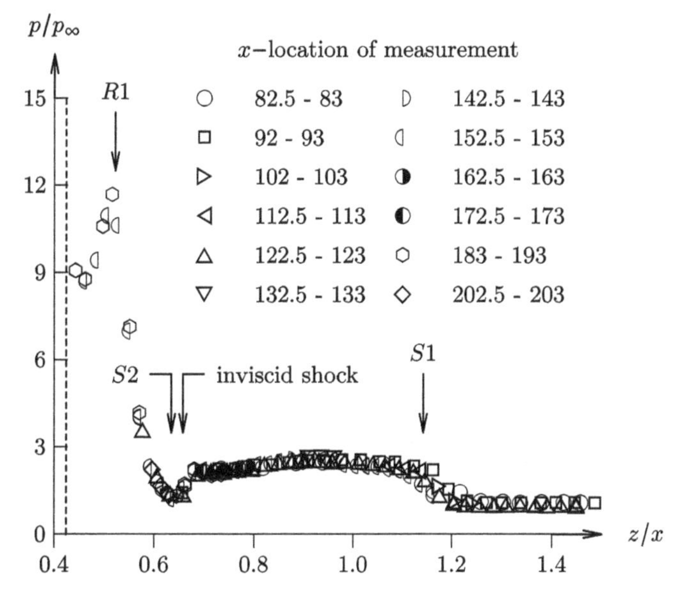


**Figure 5.144.** $p/p_\infty$ versus $z/x$ for $\beta = 23°$ (dashed line indicates fin surface).

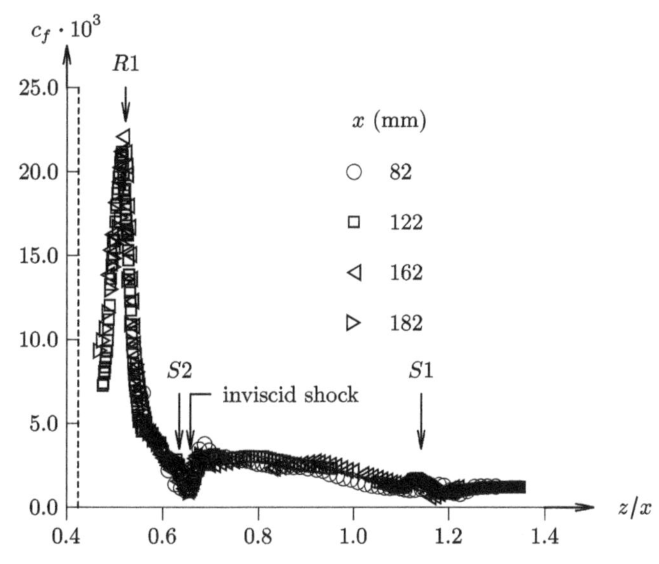


**Figure 5.145.** $c_f$ versus $z/x$ for $\beta = 23°$ (dashed line indicates fin surface).

flow region beween the separation $S1$ and reattachment $R1$ a secondary separation $S2$ occurs with a corresponding reattachment[5].

Figure 5.145 displays the skin friction coefficient $c_f$ on the botton flat plate versus $z/x$ for fin angle $\beta = 23°$. Data is shown for $x$-locations from 82 mm to 182 mm

[5] The reattachment line is located at a slightly larger value of $z/x$ and is not shown.

measured from the fin leading edge. The skin friction coefficient displays approximate conical behavior. The locations of the primary *S*1 and secondary *S*2 separation lines, the primary reattachment line *R*1 and the inviscid shock location are indicated. The primary separation line *S*1 corresponds to an slight increase in the skin friction coefficient. The peak skin friction coefficient occurs at the primary reattachment line *R*1.

Figure 5.146 shows the Stanton number St on the bottom flat plate versus *z*/*x* for fin angle $\beta = 23°$. Data is shown for *x*-locations from 82 mm to 182 mm measured from the fin leading edge. The Stanton number does not display conical behavior; in particular, the peak St deceases with increasing distance *x*. Similarly, the general level of St in the region between the primary separation line *S*1 and inviscid shock also decreases with distance *x*. The peak heat transfer occurs at the reattachment line. Within the interaction region, the Reynolds Analogy Factor $2St/c_f$ differs sharply from the equilibrium value for a flat plate zero pressure gradient boundary layer.

### 5.29.3 Crossing shock

The model is shown in figure 5.147. Two fins are attached normal to a flat plate forming intersecting swept shock waves which interact with the turbulent boundary layer on the lower plate. Experiments were performed for symmetric fin angles $\beta = 8°$ to 23°. The leading edge of the sharp fins is located from 229.5 mm to 286 mm from the leading edge of the lower plate depending on the experiment. Natural transition occurs at approximately 180 mm from the leading edge of the lower plate. The boundary layer properties at $x = 266$ mm in the absence of the two fins are listed in table 5.35.

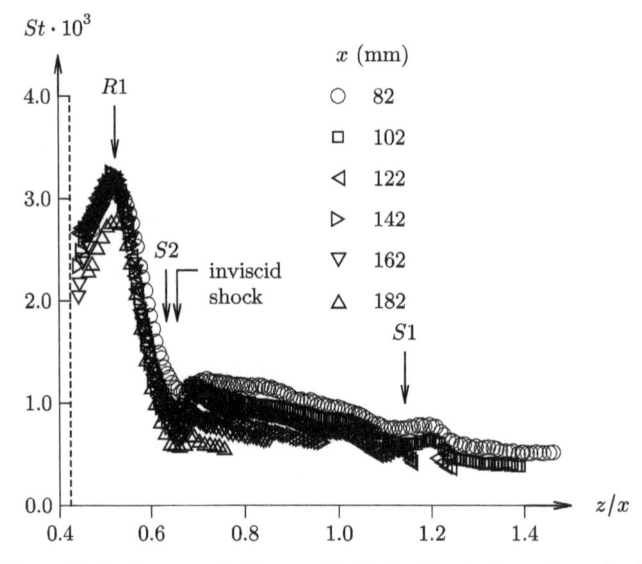


**Figure 5.146.** *St* versus *z*/*x* for $\beta = 23°$ (dashed line indicates fin surface).

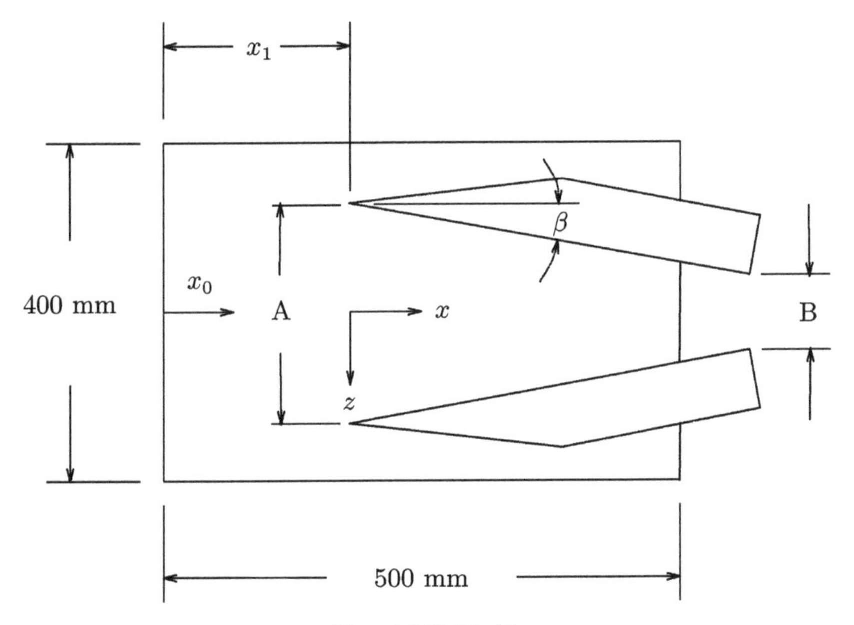


**Figure 5.147.** Model.

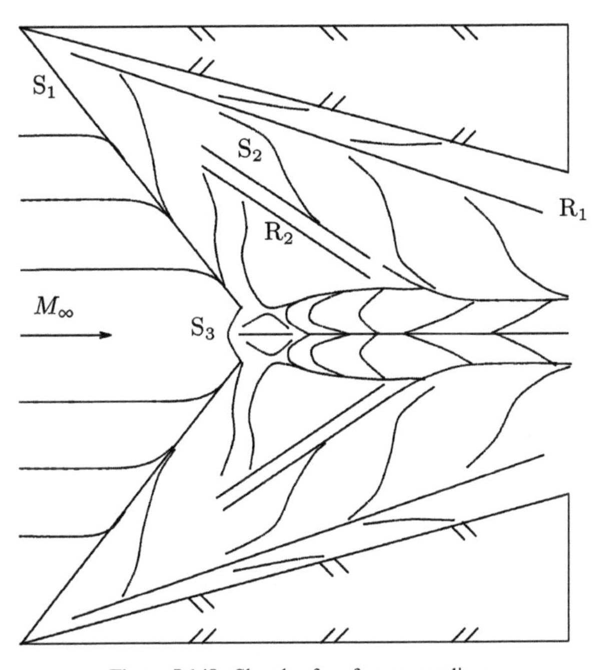


**Figure 5.148.** Sketch of surface streamlines.

Figure 5.148 displays a schematic of the surface streamlines for for the 23° × 23° configuration. The primary separation and reattachment lines *S*1 and *R*1 for the individual single fin interactions upstream of the intersection shocks are evident.

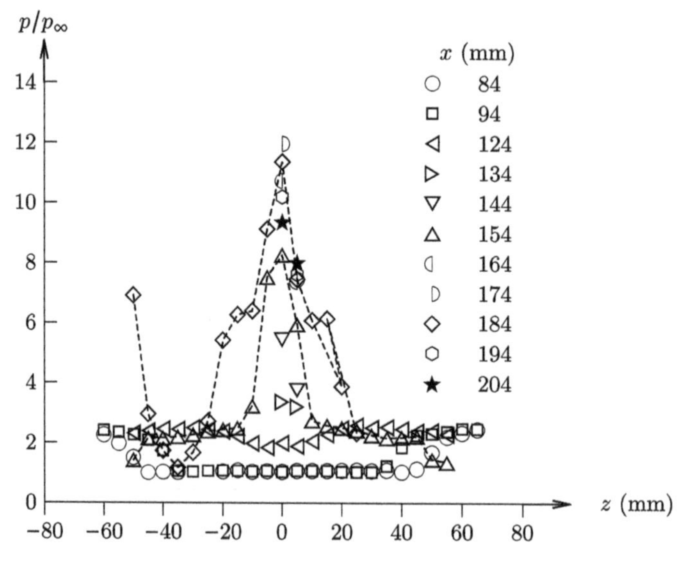


**Figure 5.149.** Surface pressure for 23° × 23° crossing shock.

Also shown is the secondary separation and reattachment lines *S*2 and *R*2 of the individual single fin interactions. A third separation line *S*3 normal to the upstream skin friction lines appears due the interaction of the opposing single fins. A complex structure of skin friction lines appears downstream.

Figure 5.149 presents the surface pressure $p/p_\infty$ versus spanwise coordinate $z$ at streamwise locations from $x$ = 84 mm to 204 mm where $x$ is measured from the fin leading edge and $z$ = 0 corresponds to the middle line between the two fins. The interaction of the individual single fin structures leads to a significant rise in the centerline pressure.

Figure 5.150 shows the Stanton number $St$ versus streamwise distance $x$ on the midline between the symmetric fins where $x = 0$ corresponds to the fin leading edge. The sudden rise in $St$ at $x = 120$ mm corresponds to the beginning of the interaction between the individual fin interaction structures. The theoretical inviscid shock intersection location is indicated by the arrow.

## 5.30 White and Ault (1996)

White and Ault (1996) investigated the interaction of a shock wave with a hypersonic turbulent boundary layer in the vicinity of an expansion corner at Mach 11.5 and unit Reynolds number $5.9 \times 10^7$ m$^{-1}$. See also White and Ault (1995). The experiments were conducted in the Calspan 96 inch shock tunnel (section A.8.1). The impinging shock was generated by a flat plate at angles of 10°, 12°, and 15° relative to the freestream for shock impingement locations upstream, in the vicinity, and downstream of the expansion corner. The experimental model is shown in figure 5.151. Surface heat transfer measurements in the absence of the shock generator indicate that boundary layer transition occurs approximately 76 cm upstream of the corner. The flow conditions are listed in table 5.36 where $\theta_{sg}$ and $\theta_{exp}$ are the shock generator and expansion surface angles, and $x_{sg}$ and $y_{sg}$ are the

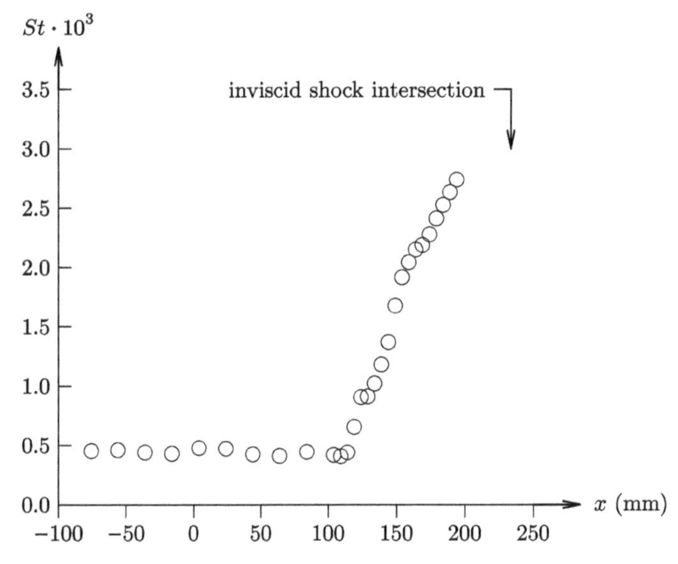


**Figure 5.150.** Stanton number versus $x$ on centerline for 23° × 23° crossing shock.

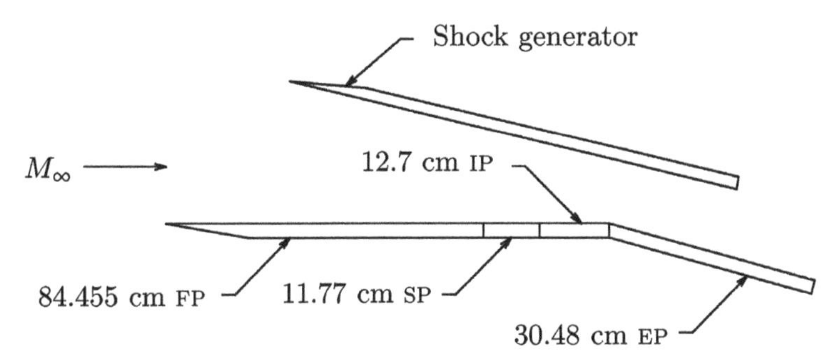


**Figure 5.151.** Model. EP expansion plate, FP flat plate, IP instrumented plate, SP spacer plate.

coordinates of the leading edge of the shock generator relative to the location of the expansion corner. The inviscid shock impingement location on the flat plate is $x_{\text{imp}}$. Run Nos. 2 to 4 were performed without the shock generator. The experimental diagnostics are surface pressure and heat transfer in the interaction region.

Figures 5.152 and 5.153 present the surface heat transfer and pressure for the 15° shock generator angle for different locations $x_{\text{imp}}$ of the impingement of shock wave on the flat plate. The corner is located at $x = 0$. The results indicate that the farther downstream of the impingement location of the shock, the lower the peak values of surface heat transfer and pressure. The same general behavior is observed for all cases.

## 5.31 Babinsky and Edwards (1997)

Babinsky and Edwards (1997) (see also Babinsky (1993) and Babinsky and Edwards (1995)) investigated the effect of surface roughness on a hypersonic shock wave turbulent boundary layer interaction at Mach 5. The model is shown in figure 5.154.

**Table 5.36.** Freestream conditions.

| Run | *Mach No.* | $Re \times 10^{-7}$ (m$^{-1}$) | $\theta_{sg}$ | $\theta_{exp}$ | $x_{sg}$ (cm) | $y_{sg}$ (cm) | $x_{imp}$ (cm) |
|---|---|---|---|---|---|---|---|
| 2 | 11.54 | 5.85 | – | – | – | – | – |
| 3 | 11.55 | 5.44 | – | – | – | – | – |
| 4 | 11.54 | 5.98 | – | – | – | – | – |
| 6 | 11.54 | 5.84 | 12° 14’ | 12° 30’ | −63.75 | 18.42 | −0.35 |
| 7 | 11.55 | 5.47 | 12° 14’ | 12° 30’ | −73.33 | 20.47 | −1.40 |
| 8 | 11.54 | 5.82 | 12° 15’ | 12° 30’ | −55.78 | 16.84 | 0.61 |
| 9 | 11.54 | 5.28 | 12° 30’ | 12° 30’ | −47.78 | 15.11 | 1.08 |
| 10 | 11.54 | 5.63 | 12° 17’ | 12° 33’ | −73.33 | 20.52 | −1.43 |
| 11 | 11.56 | 5.91 | 14° 40’ | 15° 03’ | −71.20 | 23.04 | −1.87 |
| 12 | 11.55 | 5.95 | 14° 40’ | 15° 03’ | −46.28 | 16.66 | 0.69 |
| 13 | 11.54 | 6.05 | 09° 33’ | 10° 00’ | −76.20 | 17.68 | −0.81 |
| 14 | 11.54 | 5.77 | 09° 33’ | 10° 00’ | −63.83 | 15.70 | 0.79 |
| 15 | 11.54 | 5.51 | 09° 33’ | 10° 00’ | −54.66 | 14.22 | 1.97 |
| 16 | 11.55 | 5.83 | 09° 46’ | 10° 00’ | −47.78 | 13.08 | 2.41 |
| 17 | 11.49 | 5.35 | 14° 50’ | 15° 00’ | −58.34 | 19.76 | −0.78 |
| 18 | 11.55 | 5.83 | 14° 40’ | 15° 00’ | −36.83 | 15.82 | 3.47 |
| 20 | 11.55 | 5.92 | 14° 45’ | 15° 00’ | −49.02 | 19.08 | 2.24 |
| 21 | 11.55 | 5.69 | 12° 05’ | 12° 33’ | −50.80 | 16.33 | 2.15 |
| 22 | 11.55 | 5.98 | 12° 02’ | 12° 33’ | −43.03 | 14.73 | 3.12 |
| 23 | 11.55 | 5.63 | 12° 10’ | 12° 33’ | −39.17 | 13.94 | 2.87 |
| 24 | 11.55 | 6.02 | 12° 10’ | 12° 33’ | −50.80 | 16.03 | 1.02 |

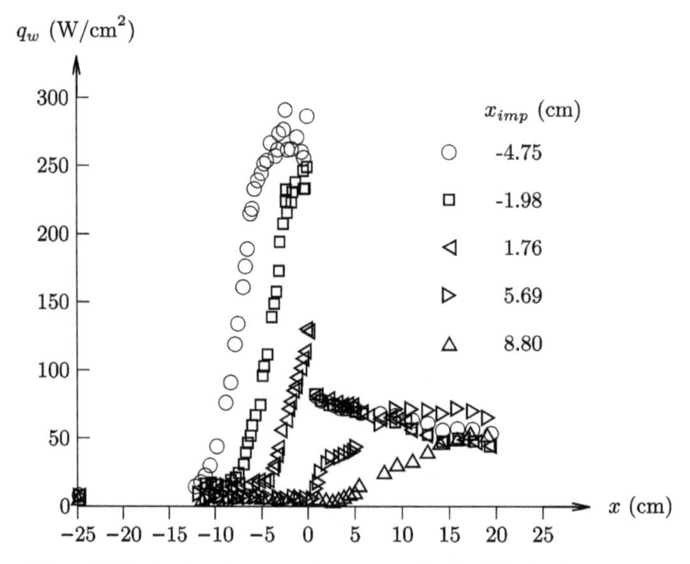


**Figure 5.152.** Surface heat transfer versus $x$ for the 15° shock generator

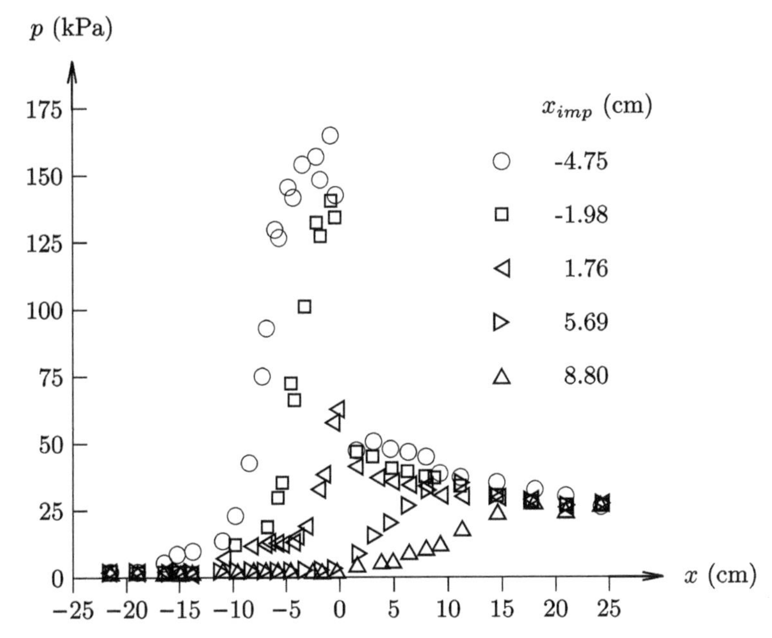


**Figure 5.153.** Surface pressure versus $x$ for the 15° shock generator.

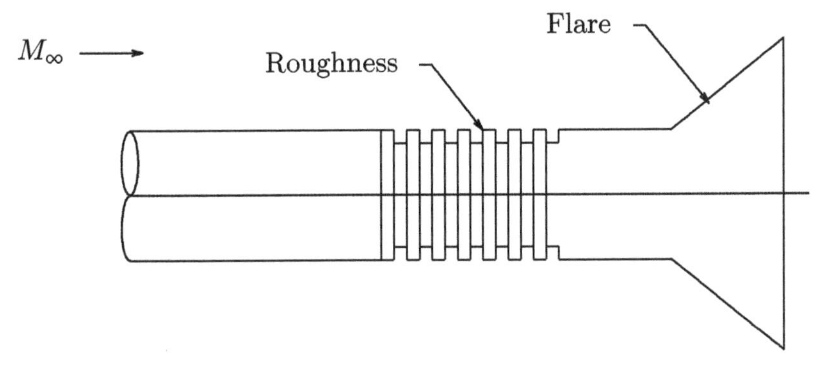


**Figure 5.154.** Model.

A hypersonic turbulent boundary layer develops past a region of roughness on a 50 mm outer diameter cylinder and further downstream interacts with flare of angle $\beta = 15°$ or 20°. Three different roughness elements were examined (figure 5.155). The roughness region terminated 50 mm upstream of the cylinder-flare junction. The experiments were performed in the Defense Research Agency Fort Halstead High Supersonic Tunnel. Experimental diagnostics include surface heat transfer and surface pressure, and pitot pressure profiles within the boundary layer (table 5.37).

Figure 5.156 shows the normalized surface pressure $p/p_\infty$ versus $x$ for $\beta = 20°$ where $x$ is measured from the cylinder corner junction. The location $x = -50$ mm corresponds to the end of the roughness region. Results for smooth and rough walls are virtually indentical. The peak measured pressure is below the theoretical inviscid conical surface pressure $p_{\text{con}}/p_\infty = 10.6$.

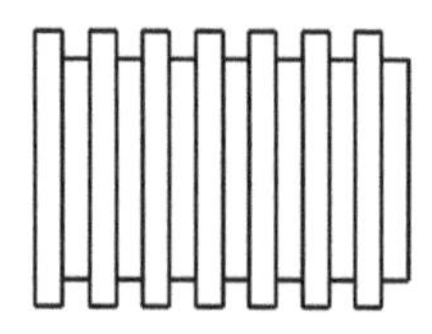
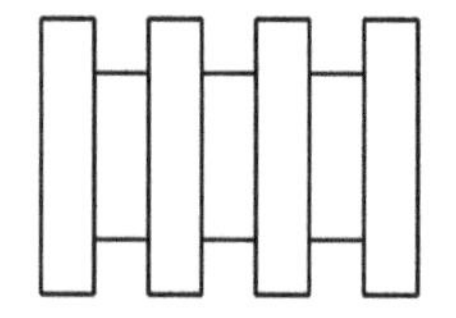
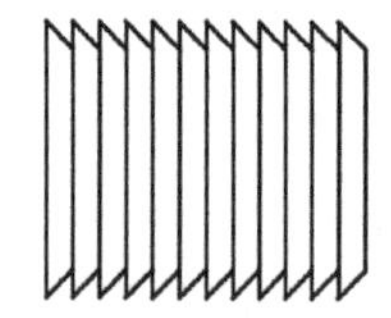

5 mm square cavity | 10 mm square cavity (double length) | 45° sawtooth

**Figure 5.155.** Roughness elements.

**Table 5.37.** Freestream conditions.

| Quantity | Value |
|---|---|
| $M_\infty$ | 5.07 |
| $p_{t_\infty}$ (MPa) | 0.716 |
| $T_{t_\infty}$ (K) | 400 |
| $T_w$ (K) | na |
| $H_\infty$ (MJ kg$^{-1}$) | 0.40 |
| $\delta$ (mm) | 9.8[a] |

[a] At nozzle exit.

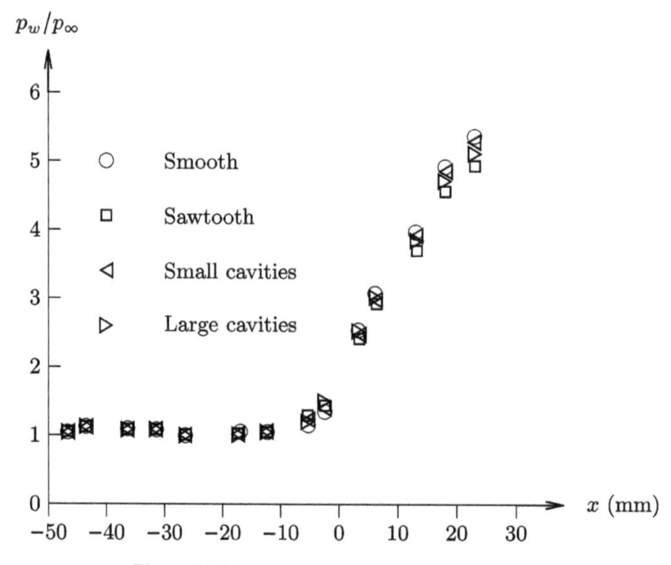


**Figure 5.156.** $p/p_\infty$ versus $x$ (mm) for $\beta = 20°$.

## 5.32 Zheltovodov *et al* (1998)

Zheltovodov *et al* (1998), Zheltovodov and Maksimov (1998, 1999) and Zheltovodov *et al* (2000) examined the interaction of symmetric and asymmetric crossing shock waves with a turbulent boundary at Mach numbers up to four. The model is shown in figure 5.157. Several combinations of fin angles $\beta_1$ and $\beta_2$ were investigated (table 5.38) where $b_1$ and $b_2$ are the spanwise distances between the upper fin tip and throat middle line, and between the lower fin tip and the throat middle line. The throat middle line is the streamwise line bisecting the exit width of 32 mm. The experiments were conducted in the wind tunnel T-333 of the Institute of Theoretical and Applied Mechanics. The freestream conditions are listed in table 5.39. Experimental diagnostics include surface heat transfer and pressure, and surface flow visualization.

Figure 5.158 displays the surface streamlines obtained using a mixture of lampblack and transformer oil for $M_\infty = 3.9$ and $7° \times 15°$. The primary separation lines $S_1$ and $S_2$ due to the incident asymmetric fin shock wave are indicated. The location of the line of upstream influence $U$ indicates the position where the surface pressure begins to increase due to the shock wave turbulent boundary layer interaction. Secondary separation lines $S_3$ and $S_4$ and tertiary separation line $S_5$ are also indicated, together with attachment lines $R_2$ through $R_5$. The location of the throat middle line (midway between the two fins as measured at the position of minimum cross section) and the cross-stream locations I to III are also indicated.

Figures 5.159 and 5.160 present the normalized surface pressure $p/p_\infty$ and heat transfer $c_h/c_{h_\infty}$ versus $x$ on the throat middle line for several fin angle configurations at $M_\infty = 3.9$. The pressure increases due to the interaction of the intersecting shock

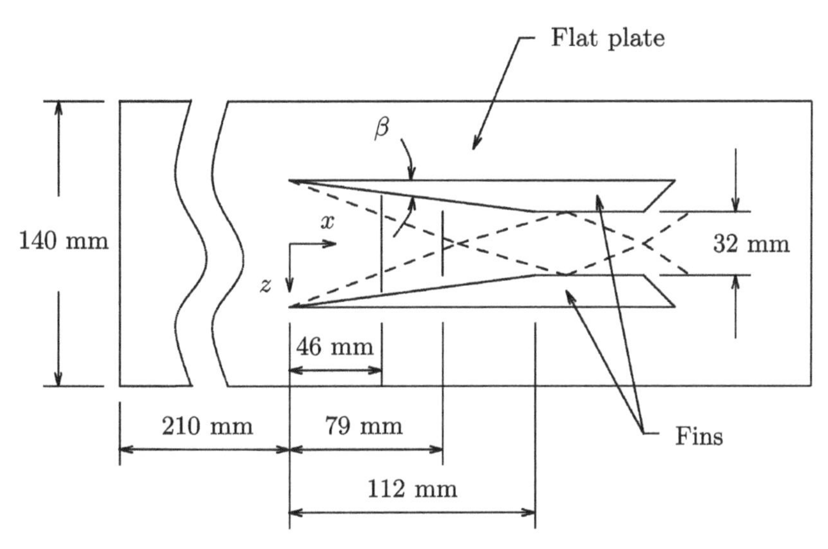


**Figure 5.157.** Model (top view).

**Table 5.38.** Model configurations.

| Length | 7° × 7° | 7° × 11° | 7° × 15° | 11° × 11° | 11° × 15° | 15° × 15° |
|---|---|---|---|---|---|---|
| $b_1$ (mm) | 35.75 | 35.8 | 35.8 | 37.8 | 37.8 | 39.55 |
| $b_2$ (mm) | 35.75 | 37.8 | 39.6 | 37.8 | 39.6 | 39.55 |

**Table 5.39.** Freestream conditions.

| Quantity | Case 1 | Case 2 |
|---|---|---|
| $M_\infty$ | 3.0 | 3.92 ± 0.03 |
| $p_{t_\infty}$ (MPa) | 0.85 to 1.0 | 1.48 ± 0.02 |
| $T_{t_\infty}$ (K) | 260 | 260 ± 2 |
| $T_w$ (K) | na | na |
| $H_\infty$ (MJ kg$^{-1}$) | 0.26 | 0.26 |
| $Re \times 10^{-6}$ (m$^{-1}$) | 80 | 89 ± 4 |
| $\delta$ (mm) | 3.4 | 3.5 |
| $\delta^*$ (mm) | 1.12 | 1.12 |

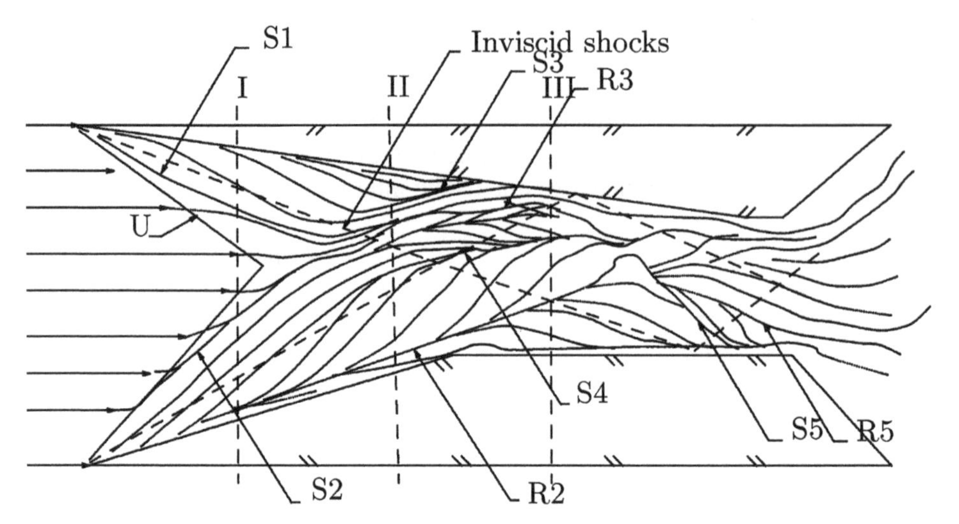


**Figure 5.158.** Surface streamlines for $M_\infty = 3.9$ and 7° × 15°.

waves generated by the fins, with the peak pressure rising with increasing of one or both fin angles, followed by an expansion due to the change in surface geometry of the fin(s). The heat transfer similarly increases due to the interaction of the intersecting shock wave with the turbulent boundary layer on the flat plate.

## 5.33 Bookey *et al* (2005)

Bookey *et al* (2005) conducted an experimental investigation of hypersonic shock wave turbulent boundary layer interactions generated by an 8° compression corner and a 10° sharp fin at Mach 8. The experiments were conducted in the Princeton Gas Dynamics Laboratory Hypersonic Boundary Layer Facility (section A.28.1). The models are shown in figure 5.161. The flat plate segment upstream of the compression corner is 470 mm in length and 152 mm in width. A two dimensional cylinder of diameter 2.4 mm was affixed on the flat plate at a distance of 58.4 mm

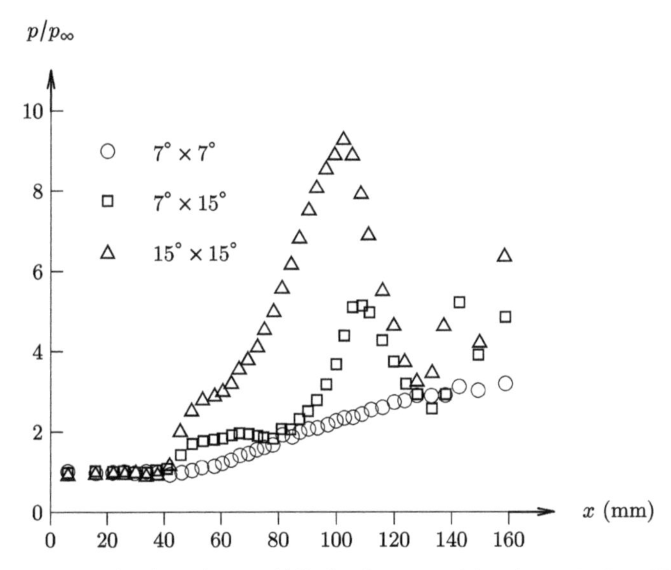


**Figure 5.159.** $p/p_\infty$ versus $x$ (mm) on throat middle line for $M_\infty = 3.9$ and 7° × 7°, 7° × 15° and 15° × 15°.

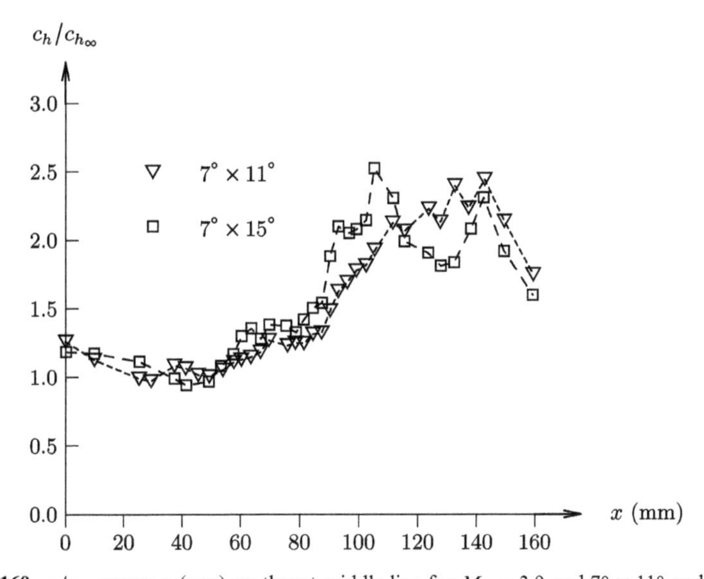


**Figure 5.160.** $c_h/c_{h_\infty}$ versus $x$ (mm) on throat middle line for $M_\infty = 3.9$ and 7° × 11° and 7° × 15°.

from the sharp leading edge to fix the location of transition. The freestream conditions are listed in table 5.40. Experimental diagnostics include Filtered Rayleigh Scattering (FRS) imaging and surface flow visualization.

Figure 5.162 is an instantaneous FRS image of the 8° compression corner. The large scale structures (eddies) of the boundary layer are evident. Figure 5.163 is an instantaneous FRS image of the 10° sharp fin taken in a plane perpendicular to the flat plate and at a distance 1.9$\delta$ from the fin surface. The $\lambda-$ shock structure (Alvi and Settles 1992) is evident, as well as the large eddies in the boundary layer.

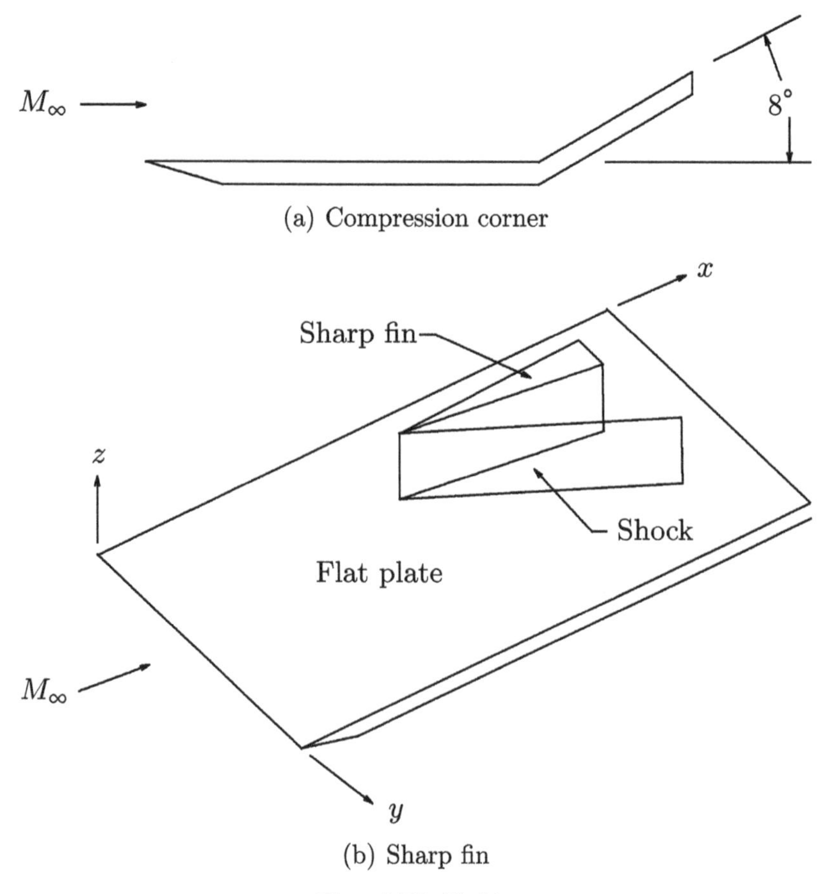


**Figure 5.161.** Models.

**Table 5.40.** Freestream conditions.

| Quantity | Value |
|---|---|
| $M_\infty$ | 8.0 |
| $p_{t_\infty}$ (MPa) | 6.9 |
| $T_{t_\infty}$ (K) | 810 |
| $H_\infty$ (MJ kg$^{-1}$) | 0.81 |
| $Re \times 10^{-6}$ (m$^{-1}$) | 17.5 |
| $\delta$ (mm) | 11.5[a] |
| $\delta^*$ (mm) | 5.9[a] |
| $\theta$ (mm) | 0.2[a] |

[a]At $x$ = 360 mm.

## 5.34 Prince *et al* (2005)

Prince *et al* (2005) conducted an experimental investigation of a hypersonic shock wave turbulent bounday layer interaction at Mach 8.2. The model, illustrated in

**Figure 5.162.** FRS image of Mach 8 compression corner. Image size is $6\delta \times 1.7\delta$ Flow is left to right.

**Figure 5.163.** FRS image of Mach 8 sharp fin at $z = 1.9\delta$ from sharp fin. Image size is $6\delta \times 3\delta$. Flow is left to right.

figure 5.164, comprises a steel flat plate of length 159 mm and a ramp of length 44 mm with an angle adjustable $\beta$ from 0° to 40°. The boundary layer was tripped using vortex generators (Coleman 1973b) affixed at a distance of 20 mm from the sharp leading edge of the flat plate. Surface roughness was provided by sand grains with average height of 0.3 mm on both the flat plate and ramp surface. Ramp angles $\alpha = 25°$, 30°, 35° and 38° were examined. The freestream conditions are listed in table 5.41. Experimental diagnostics include surface heat transfer and pressure, liquid crystal thermography and schlieren imaging. Surface heat transfer data is presented in Prince *et al* (1999).

Figure 5.165 displays the nondimensional surface pressure $p/p_\infty$ versus nondimensional distance from the hinge line $(x - L)/\delta$ for the smooth plate and ramp, rough plate and smooth ramp, and rough plate and rough ramp at $\beta = 38°$. The peak surface pressure (denoted by filled symbols) increases with surface roughness, and moves further downstream consistent with the further downstream of the reattachment location as indicated below.

Figure 5.166 shows the surface heat transfer coefficient $c_h$ versus nondimensional distance from the hinge line $(x - L)/\delta$ for the smooth plate and ramp, rough plate

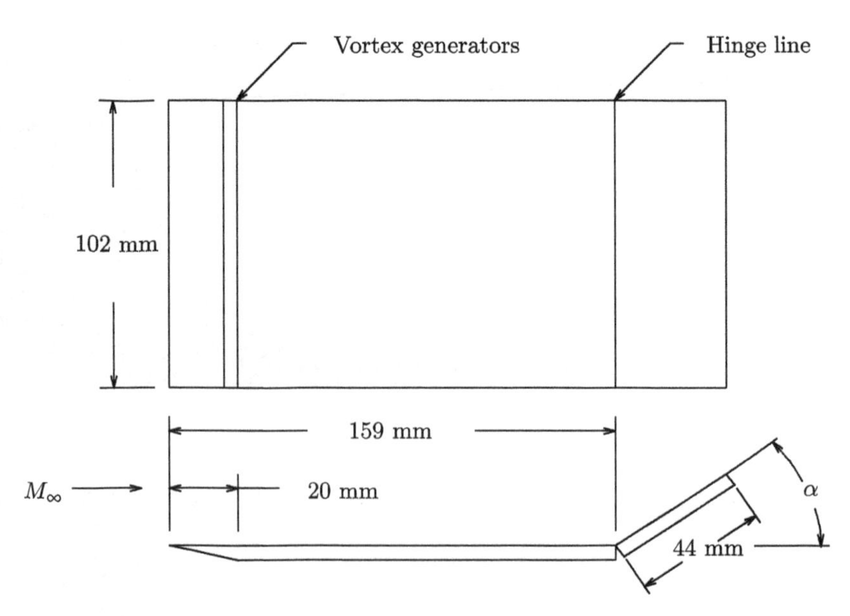


**Figure 5.164.** Model.

**Table 5.41.** Freestream conditions.

| Quantity | Value |
|---|---|
| $M_\infty$ | 8.2 |
| $p_{t_\infty}$ (MPa) | 10.9[a] |
| $T_{t_\infty}$ (K) | 1290[a] |
| $T_w$ (K) | 283[a] |
| $H_\infty$ (MJ kg$^{-1}$) | 1.3 |
| $Re \times 10^{-6}$ | 9.3[a] |
| $\delta$ (mm) | 4.3[a] |

[a] Prince (2021).
[b] Measured at hinge line.

and smooth ramp, and rough plate and rough ramp at $\alpha = 35°$. The peak surface heat transfer (denoted by filled symbols) nominally increases 14% from the smooth to the rough plate and rough ramp cases.

Figure 5.167 shows the magnitude of the separation and reattachment lengths measured from the corner where $x_s$ and $x_r$ are the locations of separation and reattachment, respectively, as measured from the leading edge of the flat plate. Surface roughness on the flat plate alone moves the separation location (□) further upstream for a fixed ramp angle $\alpha$ compared to the smooth plate and ramp (○), and decreases the ramp angle for incipient separation as indicated by the projection of the experimental data to the abscissa. Surface roughness on both the flat plate and ramp move the separation location (△) further upstream for a fixed ramp angle $\alpha$

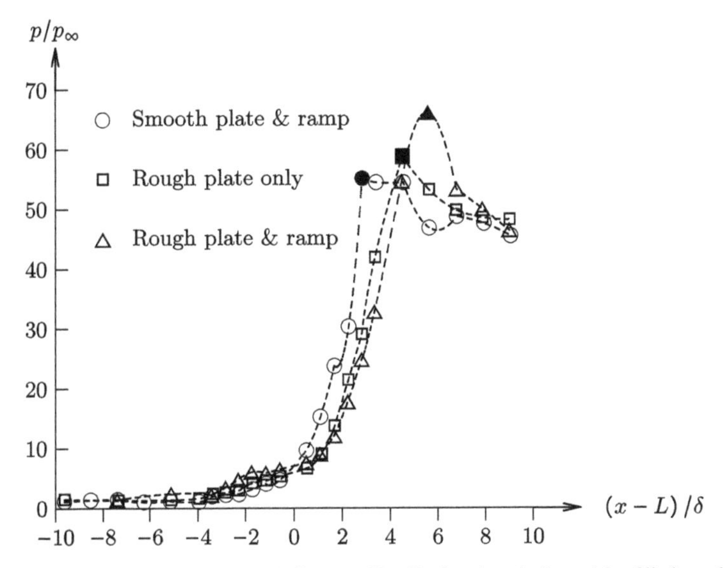


**Figure 5.165.** $p/p_\infty$ versus $(x - L)/\delta$ for $\alpha = 38°$. Peak values indicated by filled symbols.

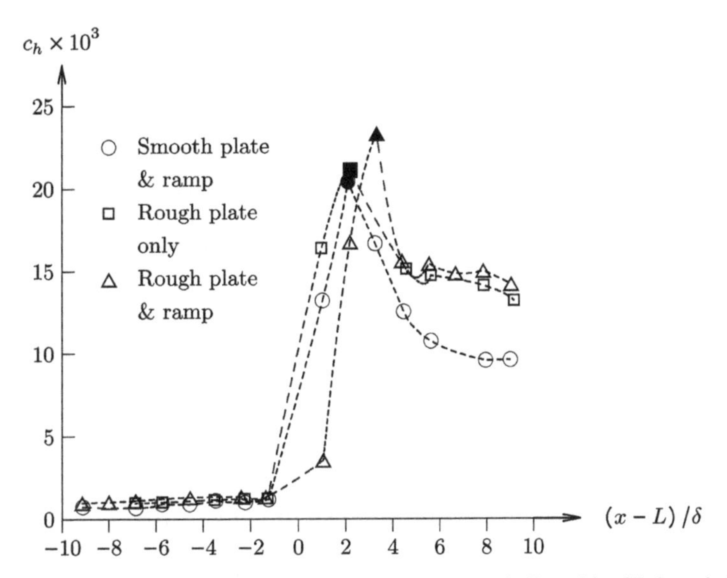


**Figure 5.166.** $c_h$ versus $(x - L)/\delta$ for $\alpha = 35°$. Peak values indicated by filled symbols.

compared to surface roughness on the flat plate alone, and further decrease the ramp angle for incipient separation. Similarly, the location of reattachment at a fixed ramp angle moves downstream with surface roughness on the flat plate alone compared to the smooth plate and ramp, and further downstream with roughness on the ramp. These effects are associated with the decreased momentum in the incoming boundary layer due to the rough surface.

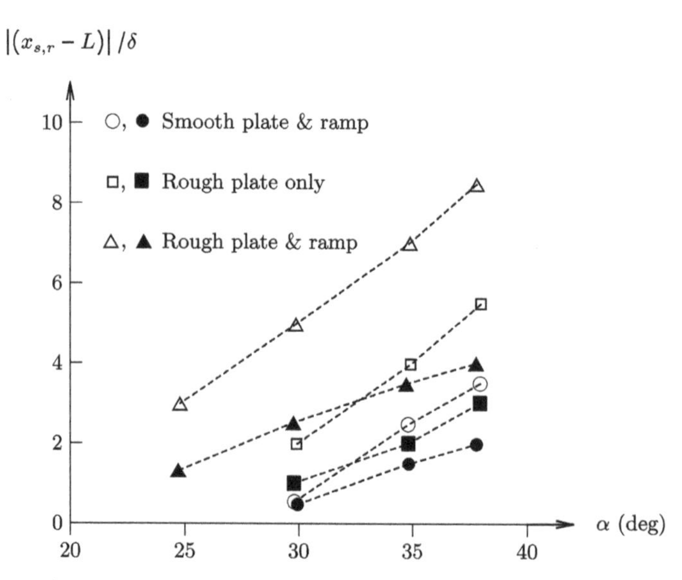


**Figure 5.167.** Separation and reattachment lengths versus $\alpha$. Open symbols are separation length, filled symbols are reattachment length as measured from corner.

## 5.35 Benay *et al* (2006) and Bur and Chanetz (2009)

Benay *et al* (2006) and Bur and Chanetz (2009) (see also Vandomme *et al* (2003)) performed experiments on shock wave transitional boundary layer interaction. The model is an axisymmetric hollow cylinder flare shown in figure 5.168. The overall length of the model is 349 mm comprising a sharp leading edge cylinder of diameter 131 mm and length $L$ = 252 mm, a 15° conical flare of length 101.4 mm, and a 50 mm cylindrical extension to minimize the influence of the base flow on the interaction region. The experiments were conducted in the ONERA R2Ch hypersonic wind tunnel (section A.25.2) at Mach 5. The flow conditions are described in table 5.42. Experimental diagnostics include surface heat transfer and wall pressure, laser doppler velocimetry, pitot boundary layer surveys, schlieren imaging and surface flow visualization. The surface heat transfer and wall pressure were measured using 52 gauges each located on a line parallel to the cylinder axis and opposite sides of the model.

A sketch of the flowfield structure is shown in figure 5.169. A weak leading edge shock is generated by the boundary layer displacement thickness on the cylinder. The interaction of the shock wave generated by the flare results in a separation region and accompanying separation shock wave. The reattachment of the boundary layer on the flare generates a reattachment shock and corresponds to the position of maximum heat transfer and surface pressure.

The wall heat flux density $h$ and nondimensional surface pressure $p/p_\infty$ versus $x/L$ are shown in figures 5.170 and 5.171, respectively, for different values of the Reynolds number $Re_L$ based upon the cylinder length $L$. For $Re_L < 0.83 \times 10^6$ the heat flux decreases slightly at the beginning of the interaction at approximately $x/L = 0.7$ typical of a laminar separation (Chanetz *et al* 1998). For

**Figure 5.168.** Model.

**Table 5.42.** Freestream conditions.

| Quantity | Min | Max |
|---|---|---|
| $M_\infty$ | 5 | 5 |
| $p_{t_\infty}$ (MPa) | 0.09 | 0.70 |
| $T_{t_\infty}$ (deg K) | 500 | 500 |
| $T_w$ (deg K) | 290 | 290 |
| $T_w/T_{aw}$ | na | 0.64[a] |
| $H_\infty$ (MJ kg$^{-1}$) | 0.5 | 0.5 |
| $Re_\infty$ ($10^6$ m$^{-1}$) | 1.1 | 8.8 |

[a] Turbulent boundary layer.

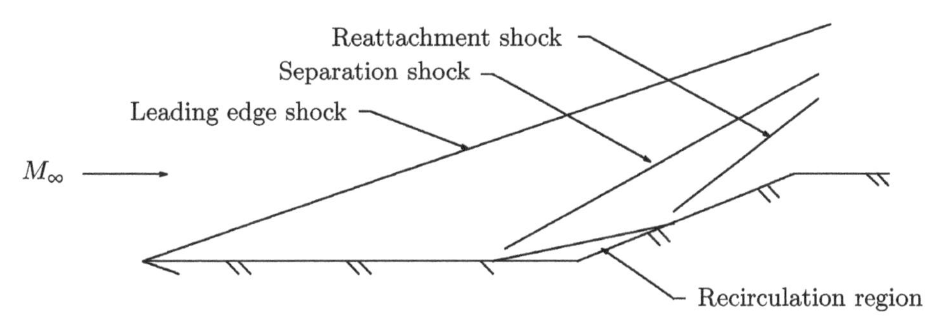


**Figure 5.169.** Flowfield structure.

$Re_L > 0.83 \times 10^6$, the heat flux increases within the separation region indicating boundary layer transition. The initial rise in surface pressure moves further downstream for $Re_L > 0.68 \times 10^6$ indicative of shock wave transitional boundary layer interactions (Coët 1989).

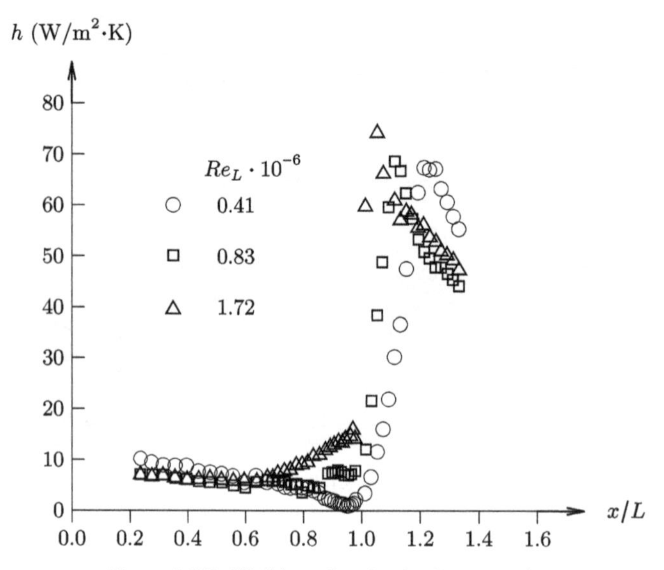


**Figure 5.170.** Wall heat flux density $h$ versus $x/L$.

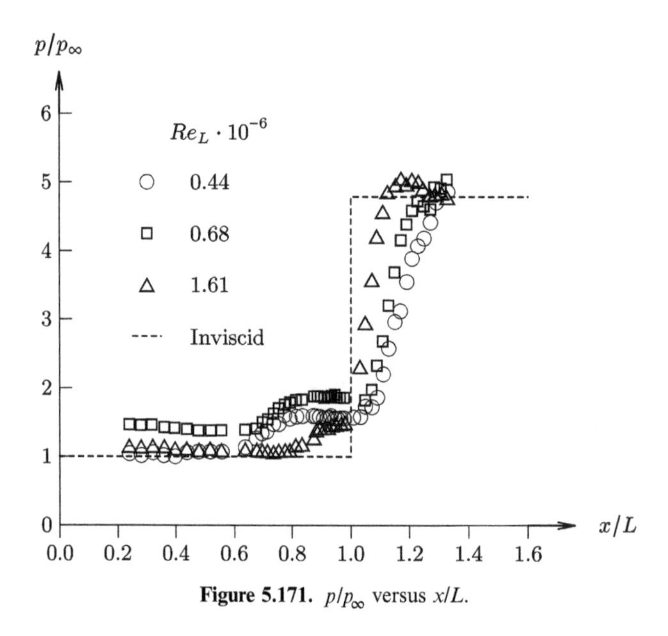


**Figure 5.171.** $p/p_{\infty}$ versus $x/L$.

The transition of the boundary layer from laminar to turbulent during the shock boundary layer interaction results in peak heat transfer significantly above the value that would be achieved for a fully turbulent interaction. Table 5.43 shows the nondimensional maximum heat transfer $q_{w_{max}}/q_{w_{\infty}}$ where $q_{w_{max}}$ is the maximum heat transfer and $q_{w_{\infty}}$ is the heat transfer immediately upstream of the interaction. The correlation of Holden (1972) for a fully turbulent interaction (5.1) based upon the experimental $p_{max}/p_{\infty}$ is also listed. The third column is the ratio of the experimental

**Table 5.43.** Peak heat transfer.

| $Re_L(\times 10^6)$ | $q_{w_{max}}/q_{w_\infty}$ | $\frac{(q_{w_{max}}/q_{w_\infty})}{(p_{max}/p_\infty)^{0.85}}$ |
|---|---|---|
| 0.405 | 12.4 | 3.2 |
| 0.740 | 14.0 | 3.6 |
| 0.827 | 12.6 | 3.2 |
| 1.195 | 13.7 | 3.5 |
| 1.422 | 13.5 | 3.5 |
| 1.462 | 14.5 | 3.7 |
| 1.717 | 14.4 | 3.7 |
| 1.941 | 12.7 | 3.3 |
| 2.031 | 12.8 | 3.3 |
| Holden (1972) | 3.9 | |

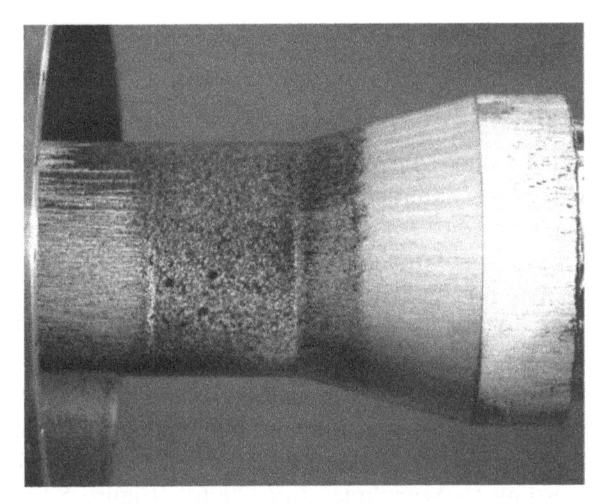

**Figure 5.172.** Surface flow visualization at $Re_L = 0.38 \times 10^6$.

$q_{w_{max}}/q_{w_\infty}$ to the predicted value of $q_{w_{max}}/q_{w_\infty}$ based upon the correlation of Holden (1972) The maximum heat transfer due to the shock wave transitional boundary layer interaction is on average 3.4 times the corresponding maximum heat transfer for a shock wave turbulent boundary layer interaction of the same strength. Similar observation was shown by Longo (2003). The accuracy of Holden's correlation (5.1) for the same experiment was confirmed by artificially tripping the upstream boundary layer to achieve a fully turbulent inflow boundary layer, and the resultant maximum heat transfer agreed closely with (5.1). The higher values of the maximum heat transfer in the experiments are due to the formation of Görtler vortices. Figures 5.172 and 5.173 display surface visualization at $Re_L = 0.38 \times 10^6$ and $1.55 \times 10^6$, respectively, indicating the formation of Görtler vortices at the higher Reynolds number.

**Figure 5.173.** Surface flow visualization $Re_L = 1.55 \times 10^6$.

## 5.36 Murphree *et al* (2006, 2007)

Murphree *et al* (2006) and Murphree *et al* (2007) investigated the shock wave transitional boundary layer interaction due to a circular cylinder fixed normal to a flat plate. The boundary layer transition was 'natural' (i.e., unforced). The experiments were performed at the University of Texas Mach 5 Hypersonic Wind Tunnel. (section A.42) The model is shown in figure 5.174. The cylinder was attached to a 254 mm flat plate at various distances from the leading edge of the flat plate. The cylinder diameters were 9.525 mm and 12.7 mm. The flow conditions are listed in table 5.44. Experimental diagnostics include kerosene-lampblack for surface flow visualization, schlieren imaging, and low and high speed planar laser scattering.

Figure 5.175 shows a sketch of the boundary layer separation line for laminar ('laminar-like' according to Murphree *et al* (2006)) (Figure 5.175(a)), transitional (Figure 5.175(b)), and fully turbulent (figure 5.175(c)) inflow boundary layers, corresponding to distances of the cylinder from the leading edge of 69.85 mm, 101.6 mm and 127 mm. The distance between the primary separation line and cylinder decreases as the incoming boundary layer changes from laminar to transitional to turbulent.

Figure 5.176 shows the nondimensional separation distance $\lambda/d$ versus $x_{\text{sep}}/x_{\text{trans}}$ where $\lambda$ is the distance from the primary separation line to the cylinder leading edge along the flat plate centerline, $d$ is the cylinder diameter, $x_{\text{sep}}$ is the distance from the plate leading edge to the primary separation line and $x_{\text{trans}}$ is the distance from the plate leading edge to the end of transition. The latter value was assumed to be 10.414 mm based on the correlation of Ramesh and Tannehill (2003). Also shown is the data of Kaufman *et al* (1972). The nondimensional separation distance decreases monotonically as the incoming boundary layer changes from laminar to transitional and finally turbulent.

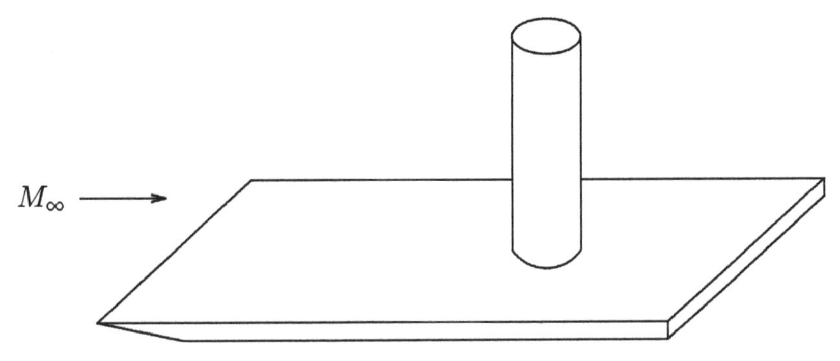


**Figure 5.174.** Model.

**Table 5.44.** Freestream conditions.

| Quantity | 2006 | 2007 |
|---|---|---|
| $M_\infty$ | 5 | 5 |
| $p_{t_\infty}$ (MPa) | 2.50 | 2.48 |
| $T_{t_\infty}$ (deg K) | 363 | 322 |
| $T_w$ (deg K) | 288 | 288 |
| $H_\infty$ (MJ kg$^{-1}$) | 0.36 | 0.32 |
| $Re_\infty$ ($10^6$ m$^{-1}$) | 53.0 | 64.0 |

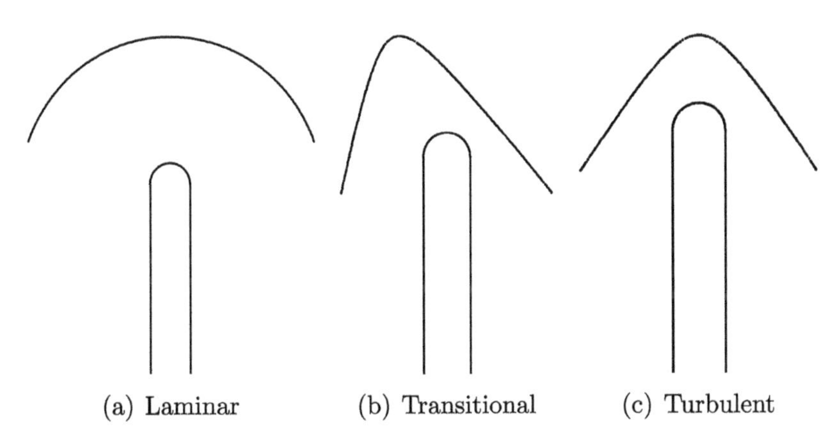


**Figure 5.175.** Surface visualization (a) $Re = 3.7 \times 10^6$, (b) $Re = 5.38 \times 10^6$, (c) $Re = 6.73 \times 10^6$.

## 5.37 Borovoy *et al* (2009, 2012)

Borovoy *et al* (2009) and Borovoy *et al* (2012) conducted experiments for a three dimensional hypersonic shock wave turbulent boundary layer interaction at Mach 5

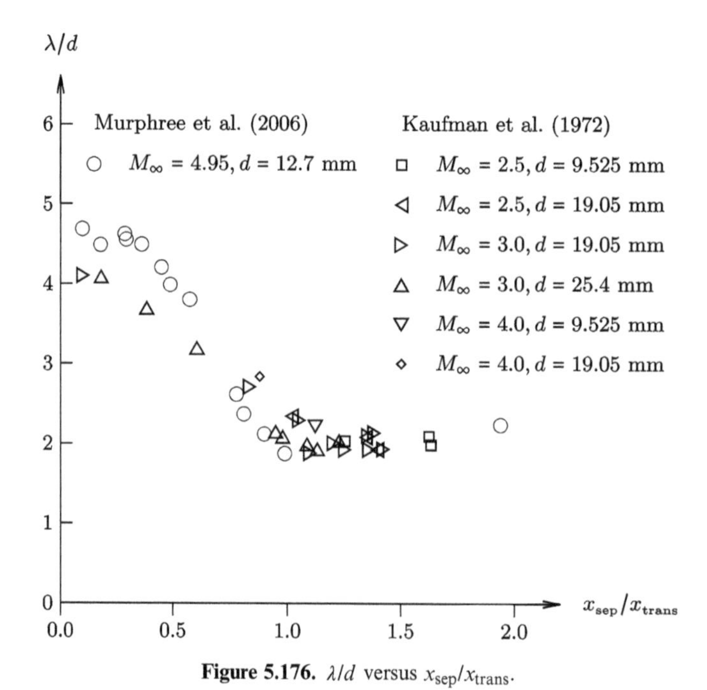


**Figure 5.176.** $\lambda/d$ versus $x_{sep}/x_{trans}$.

and 6. The experiments were performed in the UT-1 shock tunnel at the Central Aerohydrodynamics Institute (TsAGI) (section A.36). The model is shown in figure 5.177. A wedge with sharp leading edge is affixed to a flat plate at a distance $x_o = 129$ mm. The leading edge radius of the flat plate varied from sharp to 4 mm. Wedge angles of 10°, 15° and 20° were evaluated. The flow conditions are listed in table 5.45. Experimental diagnostics include surface heat transfer using fluorescent temperature transmitters (Borovoy *et al* 2009) and thin wall thermocouple sensors (Borovoy *et al* 2005), and surface pressure using fluorscent temperature transmitters (Borovoy *et al* 2009).

Figure 5.178 displays the maximum Stanton number $St_m$ versus dimensionless flat plate radius $r/x_o$ at $x/x_o = 0.64$ where $x$ is the streamwise distance measured from the fin leading edge. Results are shown at wedge angles $\theta = 10°$, 15° and 20° for Mach 5. At each wedge angle the peak Stanton number decreases with increasing radius of the flat plate leading edge due to the effect of the entropy layer formed on the flat plate. Figure 5.179 shows the ratio of the maximum Stanton number $St_m$ normalized by the value $St_{ms}$ for the sharp leading edge flat plate at $x/x_o = 0.64$ for $\theta = 15°$ at Mach 6. Results are shown both with a turbulent boundary layer tripping device upstream of the shock wave boundary layer interaction ('With turbulator') and without the tripping device ('Without turbulator'). A reduction in the maximum Stanton number up to 50% with increasing leading edge radius is observed.

## 5.38 Holden *et al* (2010)

Holden *et al* (2010) summarized a series of experiments on hypersonic shock wave turbulent boundary layer interaction in air at Mach numbers from 7.19 to 11.4. The

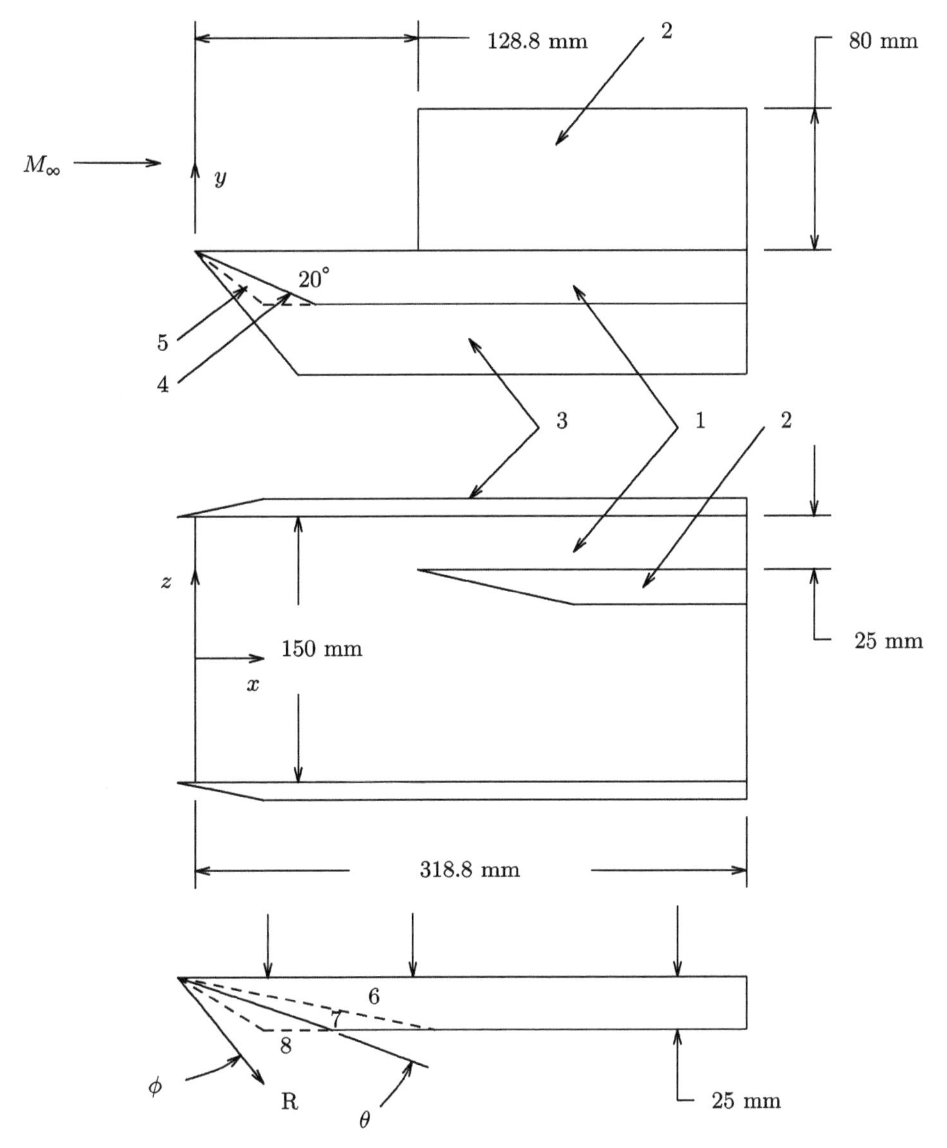


**Figure 5.177.** Model. 1—plate, 2—wedge, 3—edge plates, 4—fixed cover plate, 5—removable cover plate, 6 to 8—wedges with different angles $\theta$.

experiments were performed in the LENS hypersonic tunnels at the Calspan University of Buffalo Research Center (CUBRC) (section A.8.2). Four models were examined, namely, (1) two dimensional compression corner, (2) two dimensional impinging shock, (3) cone-flare and (4) cone-cylinder-flare. The models are shown in figures 5.180, 5.181 5.182 5.183. For Model (1) the length of the flat plate is 102.362 cm, the streamwise length of the ramp is 30.48 cm and the ramp angle is 36°. For Model (2) the length of the flat plate is 127 cm. The leading edge of the shock generator is 45.1867 cm downstream of the leading edge of the flat plate, and

**Table 5.45.** Freestream conditions.

| Quantity | Value | |
|---|---|---|
| | Case 1 | Case 2 |
| $M_\infty$ | 5 | 6 |
| $p_{t_\infty}$ (MPa) | 7.0 | 9.1 |
| $T_{t_\infty}$ (K) | 530 | 570 |
| $T_w$ (K) | 300[a] | 300[a] |
| $T_w/T_{aw}$ | 0.62 | 0.59 |
| $H_\infty$ (MJ kg$^{-1}$) | 0.53 | 0.57 |
| $Re_{x_o} \times 10^{-6}$ | 10.9 | 7.7 |

[a] Presumed value for Ludwieg tunnel.

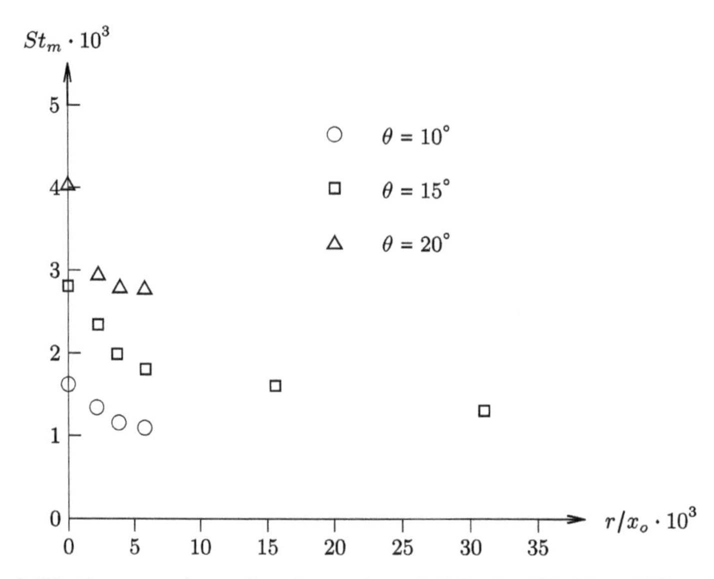


**Figure 5.178.** $St_m$ versus plate radius $r/x_o$ at $x/x_o = 0.64$ for $\theta = 10°$, 15° and 20° at $M = 5$.

25.1943 cm above the flat plate. For Model (3), the cone half-angle is 6°, the flare angle is 42°, the distance from the nose to the flare is 264.67 cm, and the streamwise length of the flare is 25.4 cm. For Model (4), the dimensions are shown in figure 5.183. The freestream conditions are listed in table 5.46. Experimental diagnostics include surface heat transfer, pressure and skin fricion, and schlieren imaging.

Figure 5.184 shows the surface pressure $p/p_\infty$, heat transfer $q_w/q_{w_\infty}$ and shear stress $\tau_w/\tau_{w_\infty}$ versus $x$ for the compression corner (Model 1). The surface pressure and heat transfer increase through the interaction where the corner is denoted by the dotted line. A separated region forms due to the adverse pressure gradient. The size of the

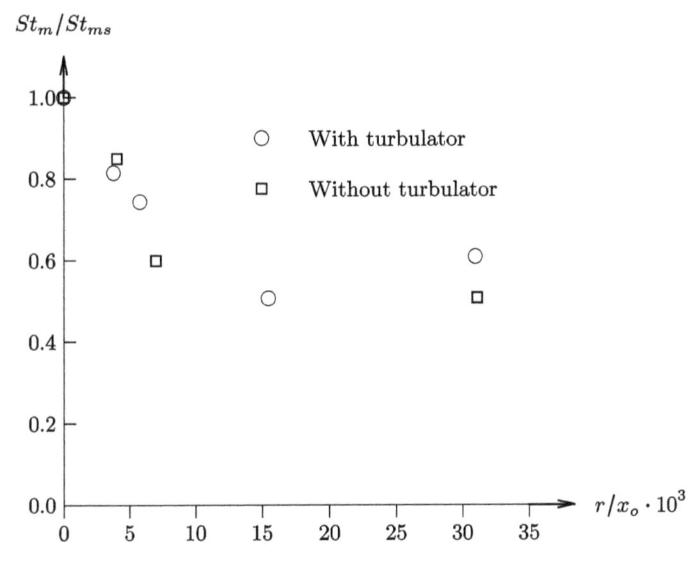


**Figure 5.179.** $St_m/St_{ms}$ versus $r/x_o$ at $x/x_o = 0.64$ for $\theta = 15°$ at $M = 6$.

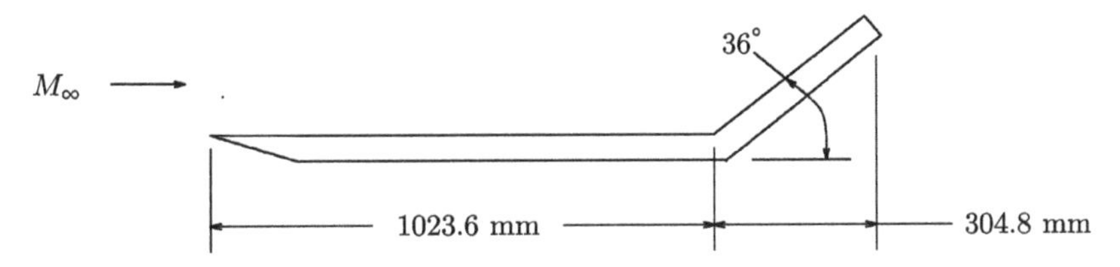


**Figure 5.180.** Compression corner (Model 1).

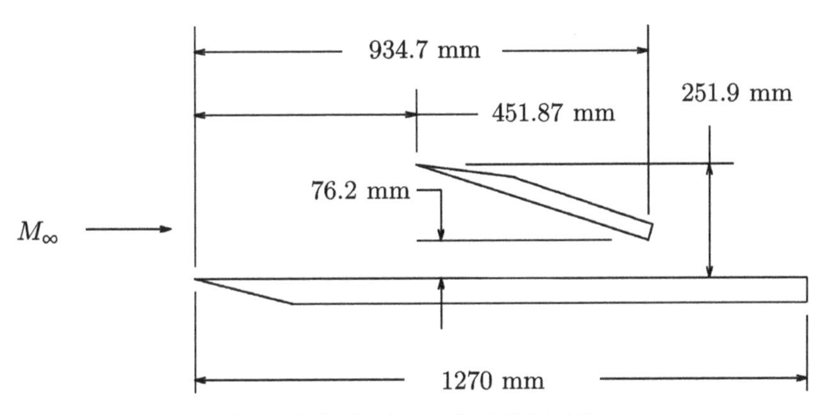


**Figure 5.181.** Impinging shock (Model 2).

separated region is indicated by the horizontal bar. The downstream pressure closely matches the inviscid pressure rise. Additional experimental data for compression ramp angles of 27°, 30°, 33° and 36° at Mach numbers from 8.1 to 8.3 are available in Marvin *et al* (2013).

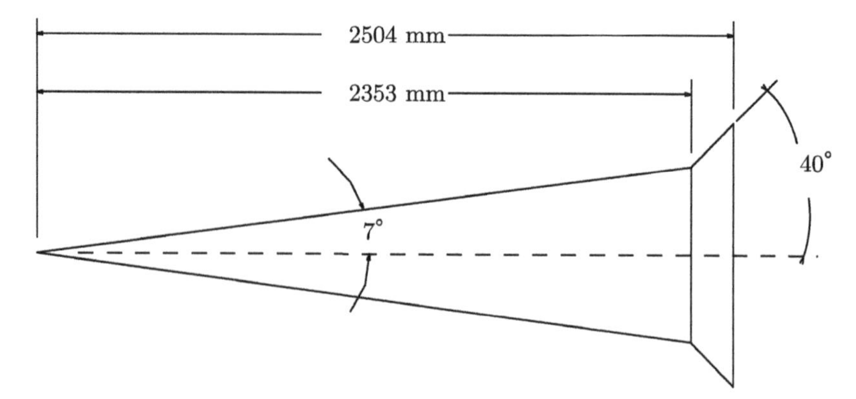


**Figure 5.182.** Cone-flare (Model 3).

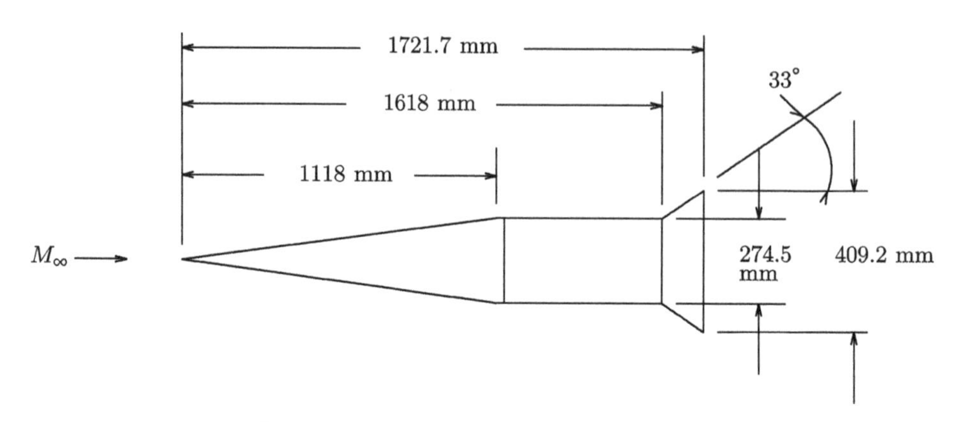


**Figure 5.183.** Cone-cylinder-flare model (Model 4).

**Table 5.46.** Freestream conditions.

| | Value | | | |
|---|---|---|---|---|
| Quantity | Model 1 | Model 2 | Model 3 | Model 4 |
| Run No. | 54 | 49 | 4 | 30 |
| $M_\infty$ | 11.3 | 11.4 | 11.0 | 7.19 |
| $T_\infty$ (K) | 61 | 62.8 | 67.2 | 226.7 |
| $T_w$ (K) | 300[a] | 300[a] | 300[b] | 297[b] |
| $H_\infty$ (MJ kg$^{-1}$) | 1.63 | 1.71 | 1.69 | 2.58 |
| $u_\infty$ (m s$^{-1}$) | 1769 | 1814 | 1804 | 2170 |
| $\rho_\infty$ (kg m$^{-3}$) | 0.082 46 | 0.089 11 | 0.032 70 | 0.067 0 |

[a] Assumed based on similar studies at CUBRC.
[b] See Marvin *et al* (2013).

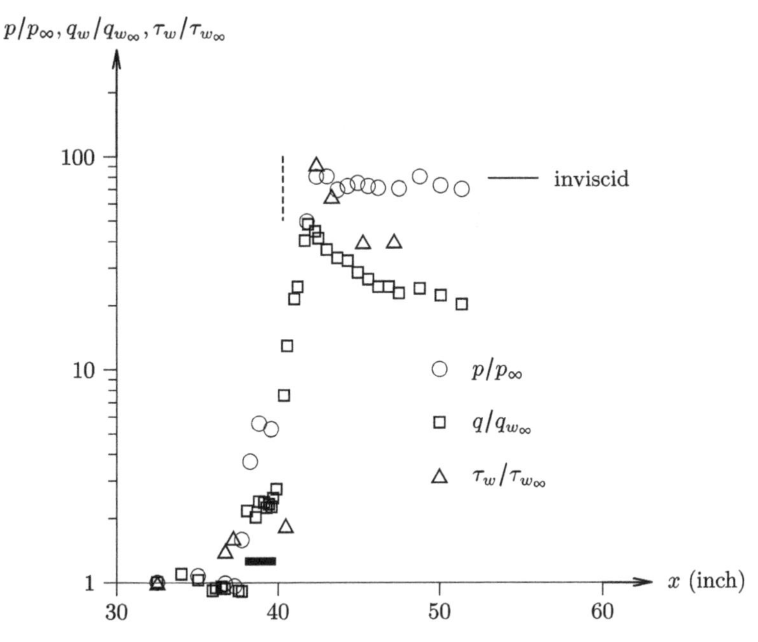


**Figure 5.184.** $p/p_\infty$, $q_w/q_{w_\infty}$ and $\tau_w/\tau_{w_\infty}$ versus $x$.

Figure 5.185 displays the surface pressure $p/p_\infty$, heat transfer $q_w/q_{w_\infty}$ and shear stress $\tau_w/\tau_{w_\infty}$ versus $x$ for the impinging shock (Model 2). The surface pressure and heat transfer increase through the interaction where the inviscid shock impingement location is indicated by the dotted line. A separated region forms due to the adverse pressure gradient. The size of the separated region is indicated by the horizontal bar. The peak pressure closely matches the inviscid pressure rise, and the subsequent decrease in pressure is due to the impingement of the expansion wave emanating from the downstream end of the shock generator.

Figure 5.186 shows the surface pressure $p/q_\infty$ and heat transfer coefficient $c_h$ versus $x$ for the cone-flare (Model 3) where $c_h = q_w/\rho_\infty u_\infty(H_\infty - H_w)$. The surface pressure and heat transfer display a plateau region indicative of a separation region. The location of the cone-flare junction is indicated by the dotted line.

Figure 5.187 presents the surface pressure $p/q_\infty$ and heat transfer coefficient[6] $c_h$ versus $x$ for the cone-cylinder-flare (Model 4). The heat transfer shows boundary layer transition occurring from $x = 15$ inch to $x = 25$ inch. The drop in heat transfer and surface pressure at $x = 44$ inch corresponds to the cone-cylinder juncture. The heat transfer and pressure rise due to the cylinder-flare juncture located at $x = 63.7$ inch are evident. Additional data is available in Marvin *et al* (2013).

[6] The wall temperature is not specified in Holden *et al* (2010) and is assumed to be 300 K.

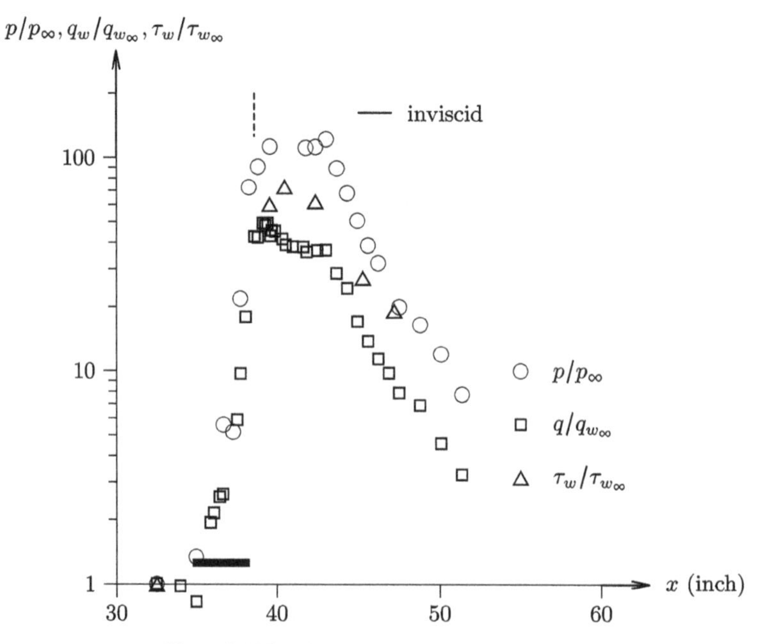


**Figure 5.185.** $p/p_\infty$, $q_w/q_{w_\infty}$ and $\tau_w/\tau_{w_\infty}$ versus $x$.

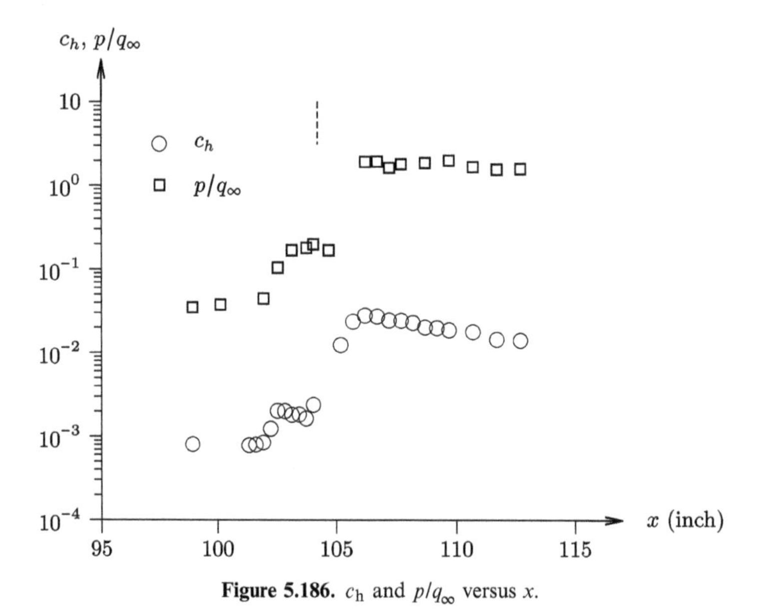


**Figure 5.186.** $c_h$ and $p/q_\infty$ versus $x$.

## 5.39 Borovoy *et al* (2011, 2013)

Borovoy *et al* (2011) and Borovoy *et al* (2013) performed experiments for a nominally two dimensioal hypersonic shock wave turbulent boundary layer interaction at Mach 5 and 6. The experiments were performed in the UT-1 shock tunnel

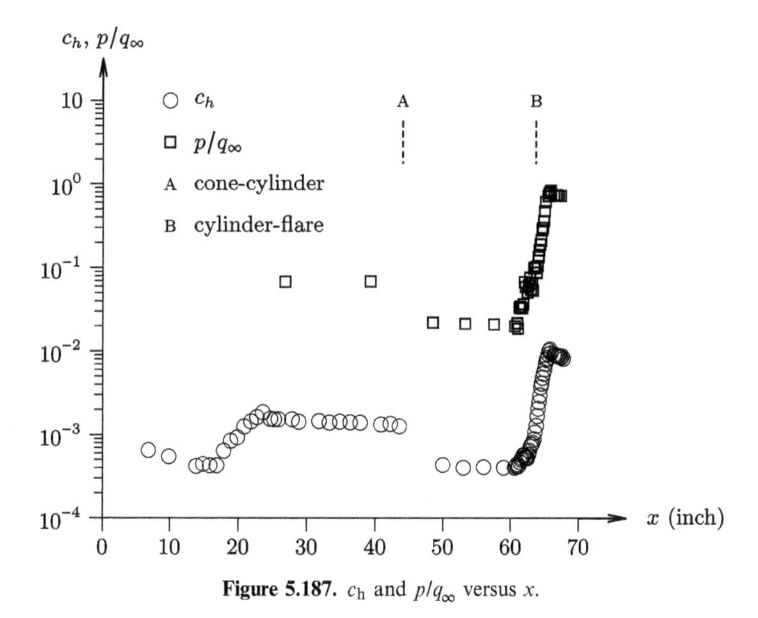


**Figure 5.187.** $c_h$ and $p/q_\infty$ versus $x$.

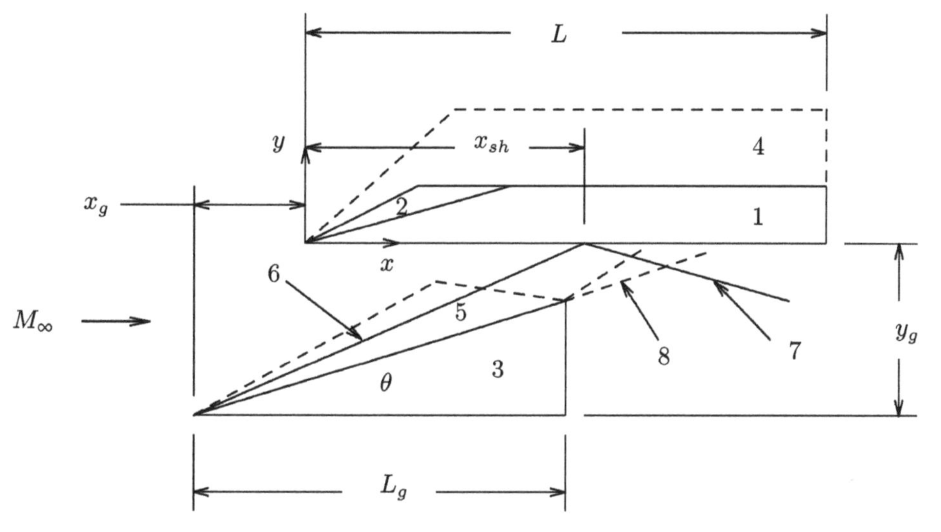


**Figure 5.188.** Model. 1—flat plate, 2—front strip, 3—shock generator, 4—side fences of plate, 5—side fences of shock generator, 6—impinging oblique shock, 7—reflected shock, 8—expansion fan.

at the Central Aerohydrodynamics Institute (TsAGI) (section A.36). The model is shown in figure 5.188. A shock wave generated by a wedge impacts the boundary layer on a flat plate. The flat plate length $L = 319$ mm and width is 150 mm. The leading edge radius $r$ of the flat plate was varied from sharp to 10 mm. The wedge angle $\theta$ was 10°, 15° and 20°. The flow conditions are shown in table 5.47. Experimental diagnostics include surface heat transfer using thin wall sensors and shadowgraph imaging.

**Table 5.47.** Freestream conditions.

| Quantity | Value | | | |
|---|---|---|---|---|
| | Case 1 | Case 2 | Case 3 | Case 4 |
| $M_\infty$ | 5 | 6 | 8 | 10 |
| $p_{t_\infty}$ (MPa) | 7 | 0.8 to 8.5 | 0.8 to 3.5 | 3.6 |
| $T_{t_\infty}$ (K) | 510 | 650 | 750 | 775 |
| $T_w$ (K) | 283 to 291 | 283 to 291 | 283 to 291 | 283 to 291 |
| $T_w/T_{aw}$ | 0.62[a] | 0.49[a] | 0.43[a] | 0.41[a] |
| $H_\infty$ (MJ kg$^{-1}$) | 0.51 | 0.65 | 0.75 | 0.78 |
| $Re \times 10^{-6}$ (m$^{-1}$) | 84 | 3.8 to 44 | 1.8 to 7.4 | 4.3 |

[a]From (3.46) with $Pr_t = 0.89$.

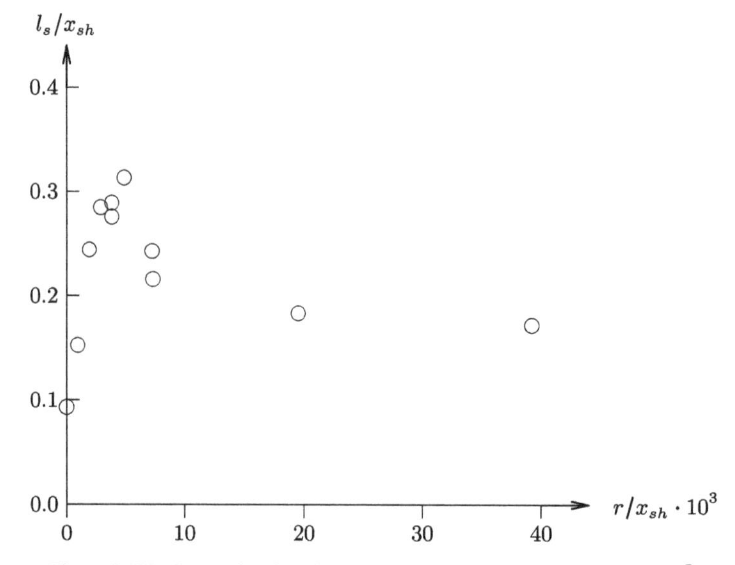


**Figure 5.189.** Separation length versus $r/x_{sh}$ at $M_\infty = 5$, $Re_L = 2.7 \times 10^7$.

Figure 5.189 shows the experimental separation length $l_s/x_{sh}$ normalized by the inviscid shock intersection location $x_{sh}$ measured from the flat plate leading edge versus $r/x_{sh}$. The shock generator angle $\theta = 15°$ (Borovoy 2021). The separation length initially increases with the leading edge radius to a maximum at $r/x_{sh} = 0.005$ due to the entrainment of the high entropy layer formed by the blunt leading edge. Further increase in the leading edge radius results in a plateau for the separation length due to the attentuation of the incident shock strength due to its interaction with the blunt leading edge shock.

Figure 5.190 presents the Stanton number $St$ versus $x$ at $M_\infty = 5$, $Re_L = 2.7 \times 10^7$ at several leading edge radii for the shock wave turbulent boundary layer interaction. The shock generator angle $\theta = 15°$ (Borovoy 2021). Increasing leading edge radii results in a decrease in the peak Stanton number.

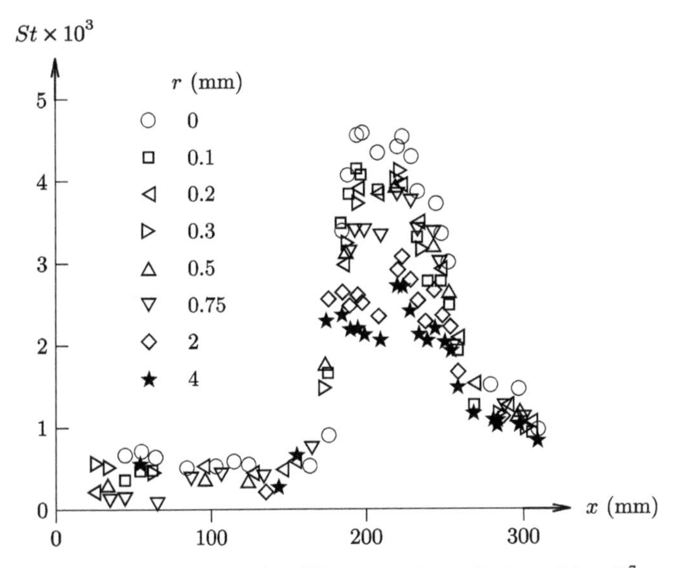


**Figure 5.190.** $St$ versus $x$ for different $r$ at $M_\infty = 5$, $Re_L = 2.7 \times 10^7$.

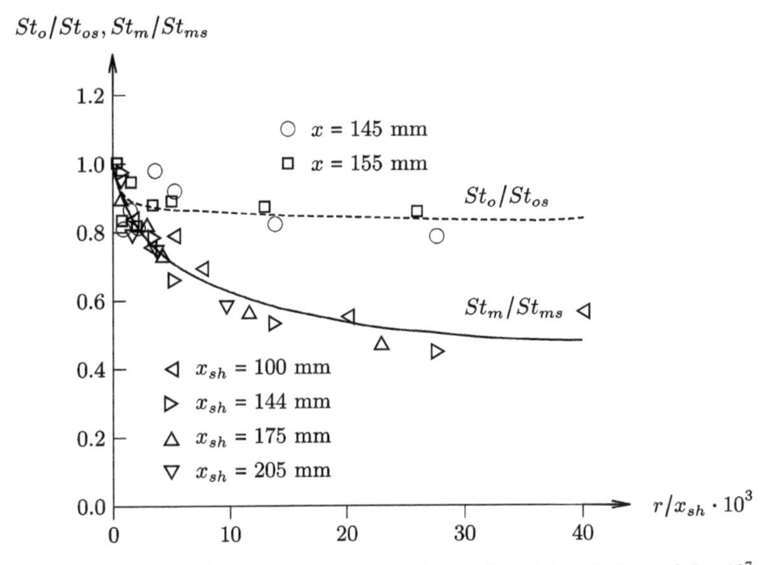


**Figure 5.191.** Effect of plate bluntness on maximum $St$ at $M_\infty = 5$, $Re_L = 2.7 \times 10^7$.

Figure 5.191 displays the maximum Stanton number $St_{\mathrm{m}}$ normalized by the maximum Stanton number $St_{\mathrm{ms}}$ for the sharp leading edge versus $r/x_{\mathrm{sh}}$ for the incident shock wave turbulent boundary layer interaction at $M_\infty = 5$ and $Re_L = 2.7 \times 10^7$. The shock generator angle $\theta = 15°$ (Borovoy 2021). Increasing the leading edge radius $r$ results in a decrease in the maximum Stanton number $St_{\mathrm{m}}$ up to 50% compared to the sharp leading edge. Also shown is the maximum Stanton number $St_{\mathrm{o}}$ normalized by the maximum Stanton number $St_{\mathrm{os}}$ for the sharp leading edge on the flat plate alone. The decrease in $St_0$ due to the nose radius is much less pronounced with a maximum decrease of 18%.

## 5.40 Murray *et al* (2013)

Murray *et al* (2013) conducted an experimental investigation of a hypersonic shock wave turbulent boundary layer interaction at Mach 8.9. The experiments were conducted in the Imperial College Gun Tunnel No. 2 (section A.13). The two models are shown in figure 5.192. Both models used a hollow cylinder to generate an equilibrium turbulent boundary layer on the outer surface of the cylinder. The cylinder length and outer diameter are 1000 mm and 75 mm, respectively. The first model incorporated a cylindrical cowl with two different internal wedge angles of 4.7° and 10° to generate the incident shock wave impinging on the cylinder. The second model included a 36° flare for shock generation. The nominal freestream conditions are listed in table 5.48. Experimental diagnostics include surface pressure and heat transfer, pitot pressure profiles within the incoming boundary layer and schlieren imaging.

Figures 5.193 and 5.194 show the surface pressure and heat transfer on the cylinder for the 4.7° cowl. The initial pressure and heat transfer rise associated with the impingement of the shock wave is followed by a decrease due to the interaction with the expansion wave originating from the cowl training edge. Figures 5.195 and 5.196 display the surface pressure and heat transfer for the 10° cowl. A broad

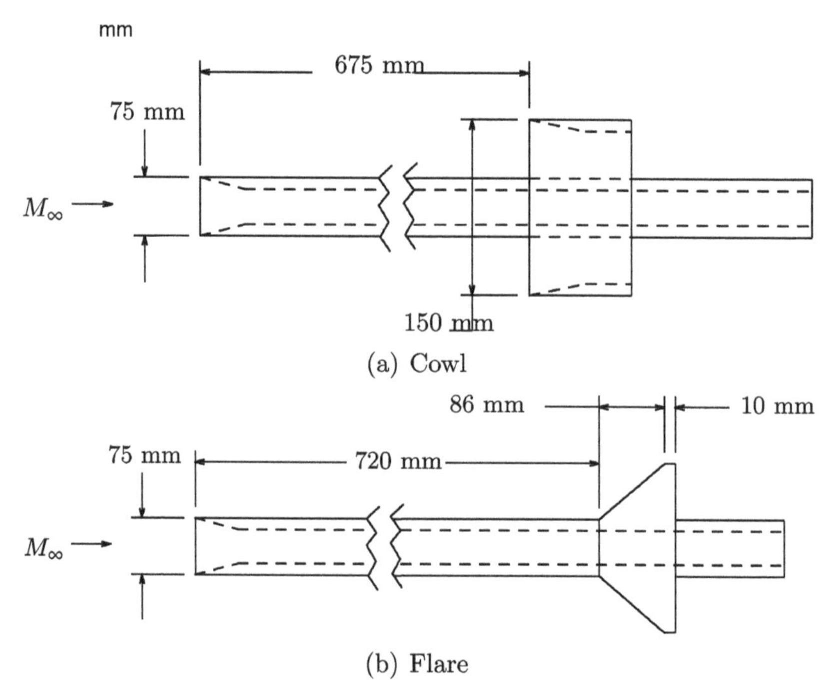


**Figure 5.192.** Models.

**Table 5.48.** Freestream conditions.

| Quantity | Value |
|---|---|
| $M_\infty$ | 8.9 |
| $p_{t_\infty}$ (MPa) | 60.8 |
| $T_{t_\infty}$ (K) | 1150 |
| $T_w$ (K) | 293 |
| $T_w/T_{aw}$ | 0.27 |
| $H_\infty$ (MJ kg$^{-1}$) | 1.2 |
| $\delta$ (cm) | 0.8 |
| $\theta$ (cm) | 0.02 |
| $Re \times 10^{-6}$ (m$^{-1}$) | 48. |

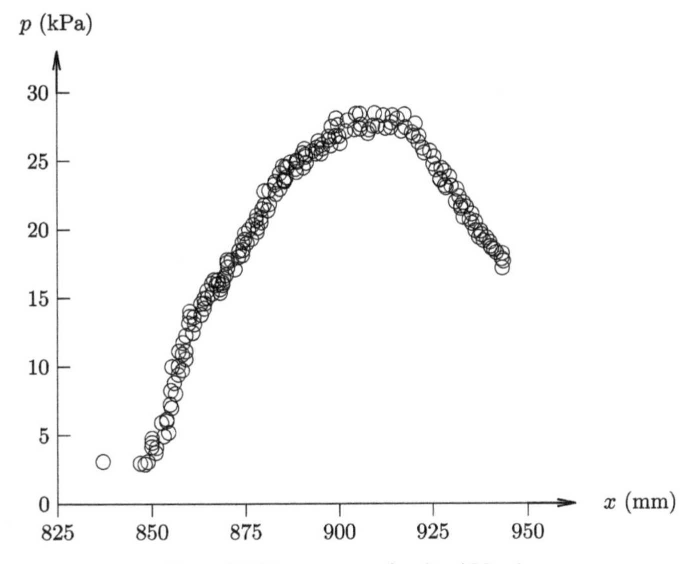


**Figure 5.193.** $p$ versus $x$ for Cowl No. 1.

plateau is evident in the pressure and heat transfer for 760mm $< x <$ 780 mm indicative of a region of separated flow.

Figures 5.197 and 5.198 present the surface pressure and heat transfer for the 36° flare. A broad separated region is evident for 680mm $< x <$ 720 mm. A significantly higher peak heat transfer is observed compared to the two cowl configurations due to the higher pressure rise across the shock.

## 5.41 Willems and Gülhan (2013) and Willems *et al* (2015)

Willems and Gülhan (2013) and Willems *et al* (2015) conducted an experimental investigation of shock wave transitional boundary layer interaction for an incident shock impinging on a flat plate. The experiments were performed at DLR

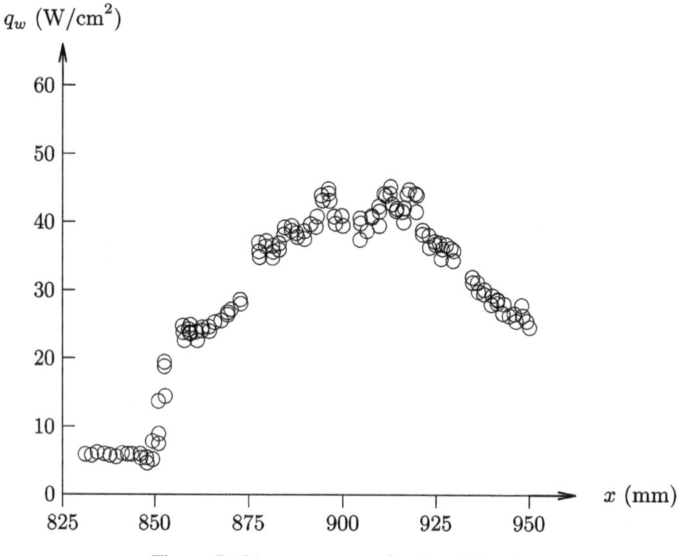


**Figure 5.194.** $q_w$ versus $x$ for Cowl No. 1.

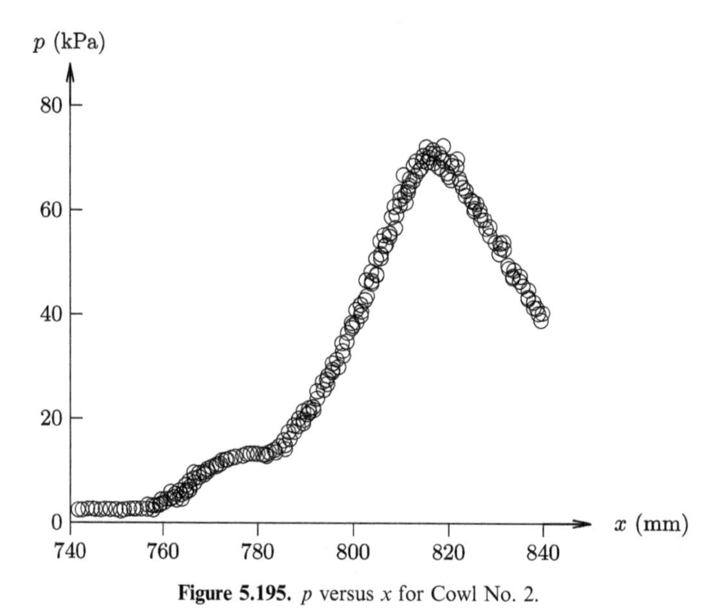


**Figure 5.195.** $p$ versus $x$ for Cowl No. 2.

Hypersonic Wind Tunnel H2K (section A.11.1). The model, shown in figure 5.199, is a flat plate of length 600 mm and width 340 mm. A wedge shock generator of angle 4° and length 190 mm and width 340 mm was located in two positions. The corresponding theoretical inviscid incident shock impingement locations are $x_{\text{shock}}$ = 239 mm and 331 mm. The freestream conditions are shown in table 5.49. Experimental diagnostics include surface heat flux and surface pressure.

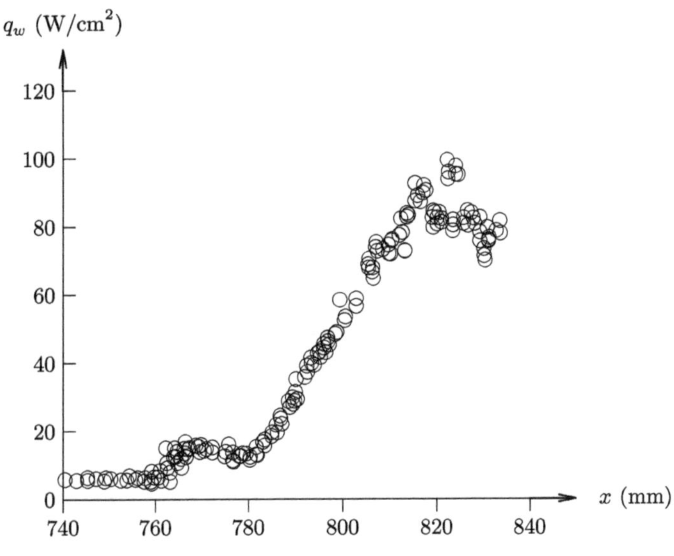


**Figure 5.196.** $q_w$ versus $x$ for Cowl No. 2.

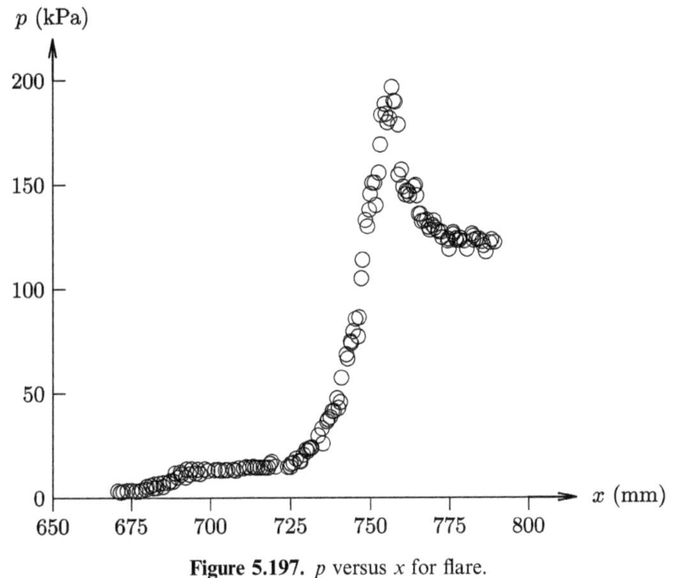


**Figure 5.197.** $p$ versus $x$ for flare.

Table 5.50 displays the ratio $q_{w_{max}}/q_{w_\infty}$ for three experiments performed with the theoretical shock impingement at $x_{shock} = 331$ mm. The data corresponds to Reynolds numbers $Re_{shock} = 1.0 \times 10^6$, $2.0 \times 10^6$ and $4 \times 10^6$.

The value of $q_{w_{max}}/q_{w_\infty}$ according to equation (5.1) is also shown where the maximum pressure ratio $p_{max}/p_\infty = 2.96$ is taken to be the inviscid pressure ratio across the interaction. The ratio $q_{w_{max}}/q_{w_\infty}$ exceeds the value from equation (5.1) by a factor of 2.2 to 3.2 for the two lowest Reynolds numbers indicating that the

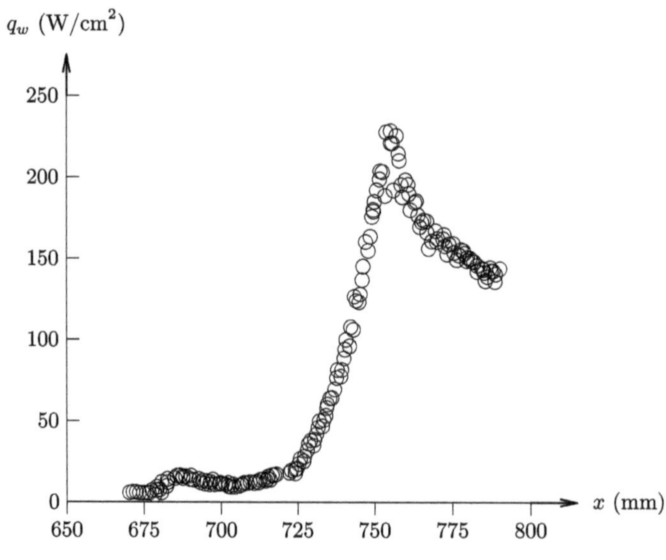


**Figure 5.198.** $q_w$ versus $x$ for flare.

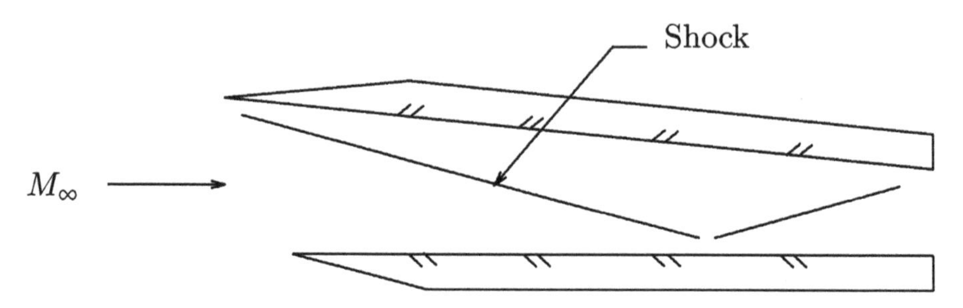


**Figure 5.199.** Model.

**Table 5.49.** Freestream conditions—Willems and Gülhan (2013) and Willems *et al* (2015).

| Quantity | Value | |
|---|---|---|
| | Min | Max |
| $M_\infty$ | 6 | 6 |
| $p_{t_\infty}$ (MPa) | 0.50 | 1.8 |
| $T_{t_\infty}$ (deg K) | 467 | 600 |
| $T_w$ (deg K) | 290 | 290 |
| $H_\infty$ (MJ kg$^{-1}$) | 0.47 | 0.60 |
| $Re_\infty$ ($10^6$ m$^{-1}$) | 3 | 16 |

**Table 5.50.** Peak heat transfer.

| $Re_{shock}(\times 10^6)$ | $q_{w_{max}}/q_{w_\infty}$ | $\frac{(q_{w_{max}}/q_{w_\infty})}{(p_{max}/p_\infty)0.85}$ |
|---|---|---|
| 1 | 8.1 | 3.2 |
| 2 | 5.4 | 2.2 |
| 4 | 2.8 | 1.1 |
| Holden (Holden 1972) | 2.5 | |

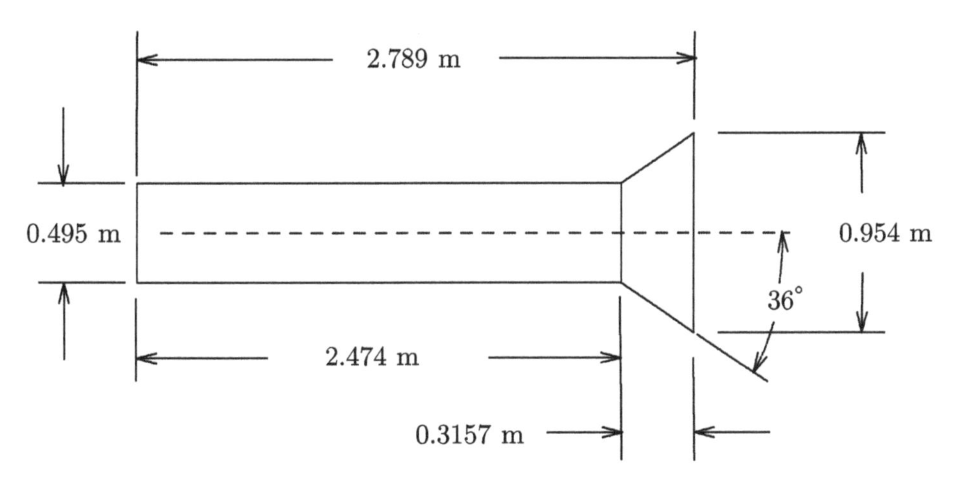


**Figure 5.200.** Hollow cylinder flare.

boundary layer is transitional within the interaction. The $q_{w_{max}}/q_{w_\infty}$ at the highest Reynolds number agres with equation (5.1) implying that the incoming boundary layer is turbulent.

## 5.42 Holden *et al* (2014)—hollow cylinder flare

Holden *et al* (2014) and Holden *et al* (2018) presented an experimental dataset for hypersonic shock wave turbulent boundary layer interaction on a hollow cylinder flare configuration (figure 5.200) for a Mach number range from 4.95 to 7.96. The experiments were performed in the LENS II tunnel at the Calspan University of Buffalo Research Center (CUBRC) (section A.8.2). The test conditions are summarized in table 5.51. Experimental diagnostics are surface pressure and surface heat transfer.

Figure 5.201 displays the surface heat transfer for Case 1 for the entire length of the hollow cylinder flare. Transition is observed to occur close to the leading edge at $x \approx 20$ cm. Details of the heat transfer in the vicinity of the corner are shown in figures 5.202 and 5.203 for Case Nos. 1 to 3 and 4 to 6, respectively. The location of the cylinder-flare junction at $x = 247.37$ cm is indicated by the dashed line. Results

**Table 5.51.** Freestream conditions—hollow cylinder flare.

| | Value | | | | | |
|---|---|---|---|---|---|---|
| Quantity | Case 1 | Case 2 | Case 3 | Case 4 | Case 5 | Case 6 |
| Run No. | 17 | 16 | 11 | 13 | 18 | 21 |
| Gas | Air | Air | Air | Air | Air | Air |
| $M_\infty$ | 4.95 | 4.97 | 5.95 | 6.01 | 6.96 | 7.96 |
| $p_{t_\infty}$ (MPa) | 3.35 | 6.62 | 4.58 | 13.9 | 11.8 | 11.6 |
| $T_{t_\infty}$ (K) | 1262 | 1257 | 1634 | 1585 | 2399 | 2522 |
| $T_w$ (K) | 300 | 300 | 300 | 300 | 300 | 300 |
| $T_w/T_{aw}$ | 0.25 | 0.25 | 0.19 | 0.20 | 0.13 | 0.13 |
| $H_\infty$ (MJ kg$^{-1}$) | 1.27 | 1.27 | 1.65 | 1.60 | 2.42 | 2.54 |
| $Re \times 10^{-6}$ (m$^{-1}$) | 11.3 | 22.3 | 6.67 | 20.5 | 6.57 | 4.06 |

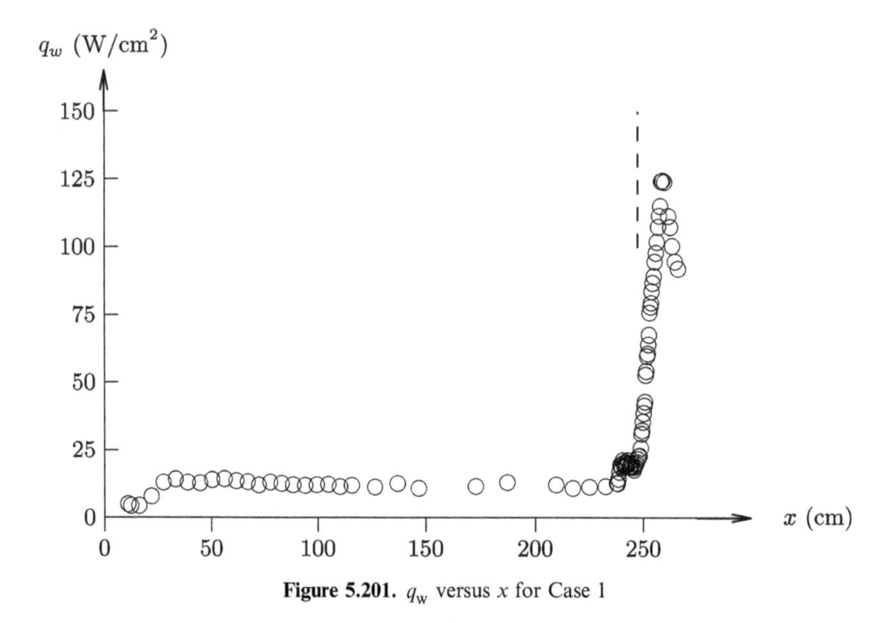


**Figure 5.201.** $q_w$ versus $x$ for Case 1

for the wall pressure are shown in figures 5.204 and 5.205 for Case Nos. 1 to 3 and 4 to 6, respectively. The rapid increase in heat transfer and surface pressure associated with the shock wave turbulent boundary layer interaction is observed.

## 5.43 Holden *et al* (2014)—cone flare

Holden *et al* (2014) and Holden *et al* (2018) presented an experimental dataset for hypersonic shock wave turbulent boundary layer interaction on a cone flare configuration (figure 5.206) at Mach number range from 4.96 to 8.21. The experiments were performed at the Calspan University of Buffalo Research Center

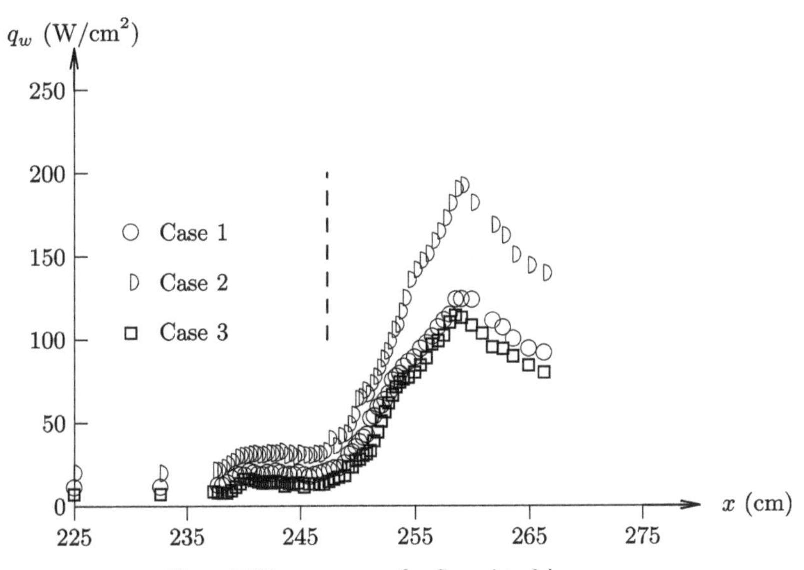


**Figure 5.202.** $q_w$ versus $x$ for Cases 1 to 3 in corner.

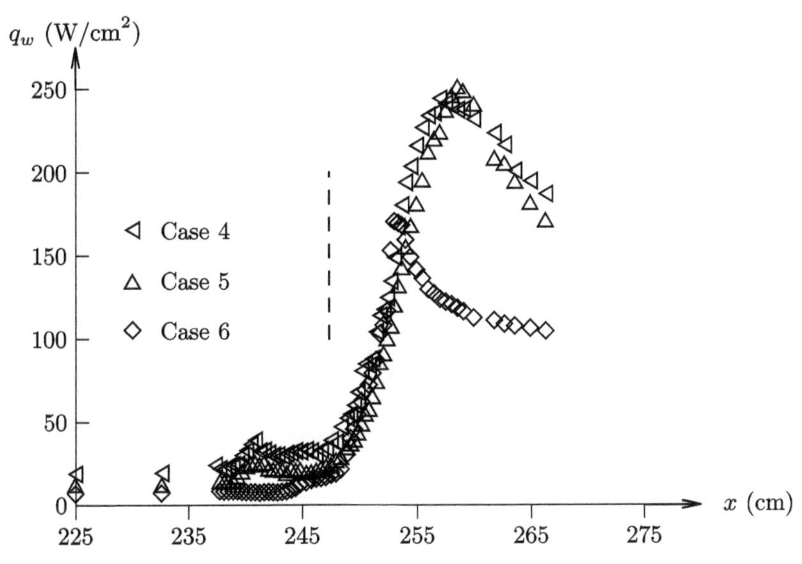


**Figure 5.203.** $q_w$ versus $x$ for Cases 4 to 6 in corner.

(CUBRC) (section A.8.2). The test conditions are summarized in table 5.52. Experimental diagnostics are surface pressure and surface heat transfer.

Figure 5.207 shows the surface heat transfer for Case 1 for the entire length of the cone flare. Transition is seen to occur at $x \approx 30$ cm. Details of the heat transfer in the vicinity of the corner are shown in figures 5.208 and 5.209 for Case Nos. 1 to 5 and 6 to 10, respectively. The location of the cone-flare junction at $x = 235.3$ cm is indicated by the dashed line. Results for the wall pressure are displayed in

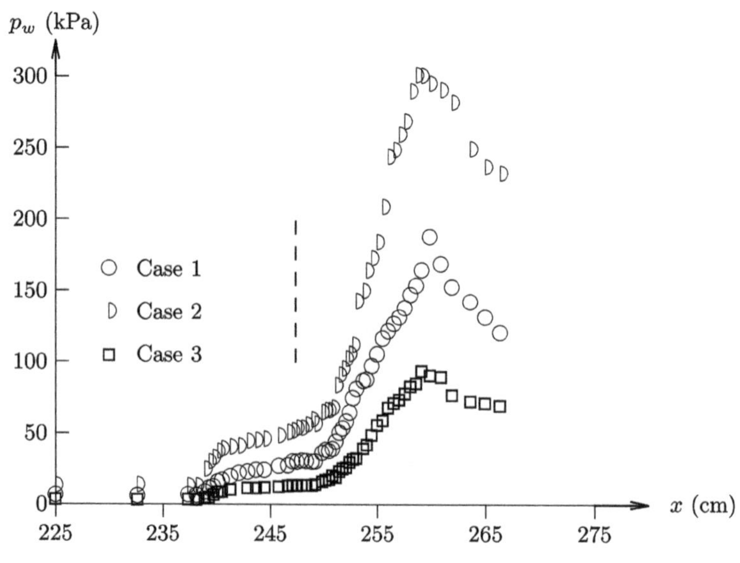


**Figure 5.204.** Wall pressure for Cases 1 to 3.

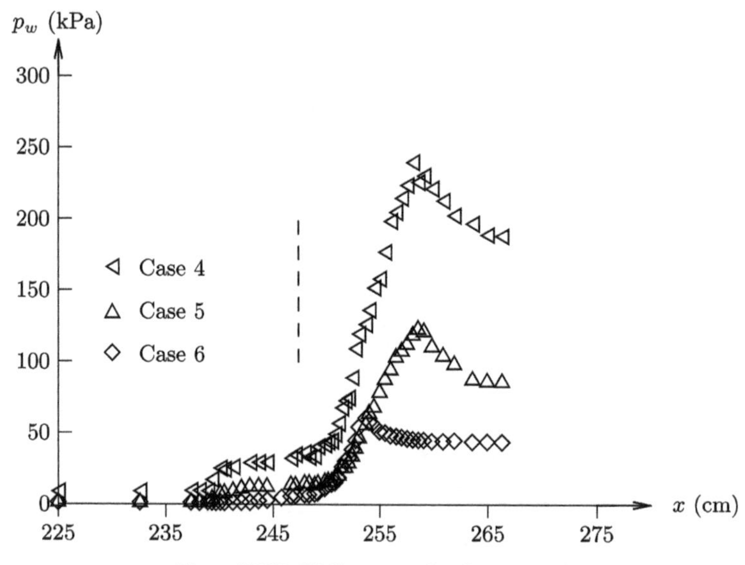


**Figure 5.205.** Wall pressure for Cases 4 to 6.

figures 5.210 and 5.211 for Case Nos. 1 to 5 and 6 to 10, respectively. The rapid rise in heat transfer and surface pressure due to the shock wave boundary layer interaction is evident.

## 5.44 Borovoy *et al* (2016)

Borovoy *et al* (2016) conducted a series of experiments on hypersonic shock wave turbulent boundary layer interactions generated by single and double fins at Mach 5

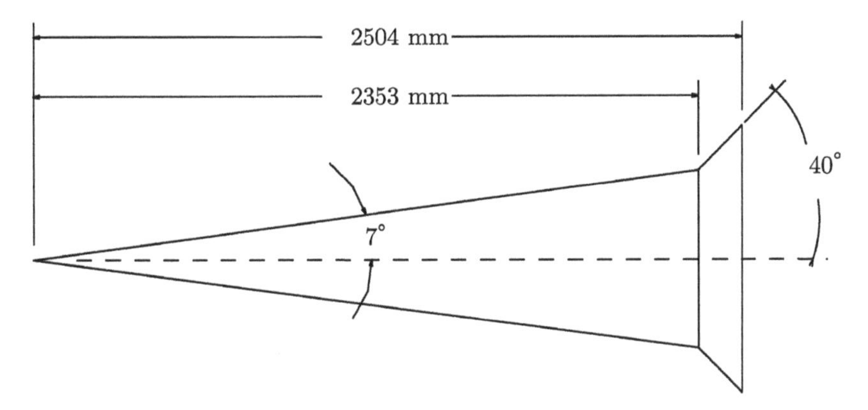


**Figure 5.206.** Cone flare.

**Table 5.52.** Freestream conditions—cone flare.

| Quantity | Value | | | | |
|---|---|---|---|---|---|
| | Case 1 | Case 2 | Case 3 | Case 4 | Case 5 |
| Run No. | 28 | 26 | 45 | 34 | 33 |
| Gas | Air | Air | Air | Air | Air |
| $M_\infty$ | 4.96 | 5.10 | 5.90 | 6.03 | 6.17 |
| $p_{t_\infty}$ (MPa) | 4.52 | 3.67 | 11.12 | 5.65 | 2.24 |
| $T_{t_\infty}$ (K) | 1301 | 470 | 1946 | 1405 | 487 |
| $T_w$ (K) | 300 | 300 | 300 | 300 | 300 |
| $T_w/T_{aw}$ | 0.24 | 0.67 | 0.16 | 0.22 | 0.65 |
| $H_{infty}$ (MJ kg$^{-1}$) | 1.31 | 0.47 | 1.96 | 1.42 | 0.49 |
| $q_{REF}$ (W cm$^{-2}$) | 24.12 | 2.95 | 39.72 | 15.47 | 0.97 |
| $Re \times 10^{-6}$ (m$^{-1}$) | 14.6 | 49.0 | 13.1 | 9.8 | 18.5 |
| | Case 6 | Case 7 | Case 8 | Case 9 | Case 10 |
| Run No. | 43 | 14 | 40 | 41 | 37 |
| Gas | Air | Air | Air | Air | Air |
| $M_\infty$ | 6.92 | 7.18 | 8.03 | 8.1 | 8.21 |
| $p_{t_\infty}$ (MPa) | 10.45 | 5.32 | 8.17 | 11.94 | 8.76 |
| $T_{t_\infty}$ (K) | 2642 | 753 | 1634 | 2358 | 875 |
| $T_w$ (K) | 300 | 300 | 300 | 300 | 300 |
| $T_w/T_{aw}$ | 0.12 | 0.42 | 0.19 | 0.13 | 0.36 |
| $H_\infty$ (MJ kg$^{-1}$) | 2.66 | 0.76 | 1.65 | 2.38 | 0.88 |
| $q_{REF}$ (W/cm$^2$) | 26.53 | 1.65 | 7.29 | 13.88 | 3.45 |
| $Re \times 10^{-6}$ (m$^{-1}$) | 5.2 | 15.1 | 5.2 | 4.4 | 14.0 |

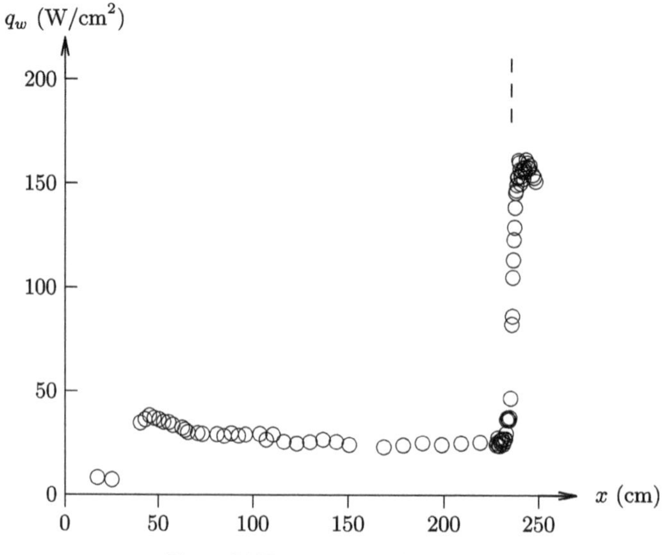


**Figure 5.207.** $q_w$ versus $x$ for Case 1.

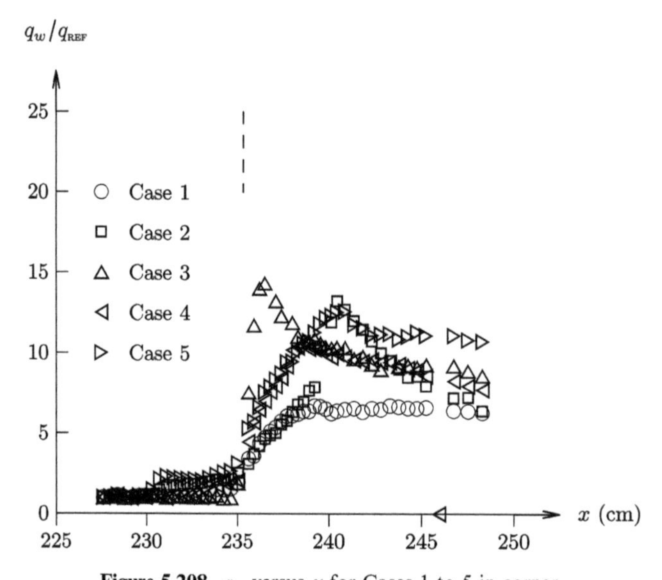


**Figure 5.208.** $q_w$ versus $x$ for Cases 1 to 5 in corner.

to 8. The experiments were performed in the UT-1 shock tunnel at the Central Aerohydrodynamics Institute (TsAGI). The model is shown in figure 5.212. One or two sharp leading edge fins are attached to a flat plate of length $L = 319$ mm at a distance $x_o = 129$ mm from the leading edge. The fin angle $\theta = 10°$, $15°$ and $20°$ for the single fin model, and $\theta = 15°$ for the double fin model. In the latter model, the distance between the fin leading edges $W_o = 100$ mm, and the fin thickness varies to obtain the channel constriction ratios $W_o/W_t = 2$ and 4 where $W_t$ is the minimum width of the channel. An interchangable front pad allows modification of the leading

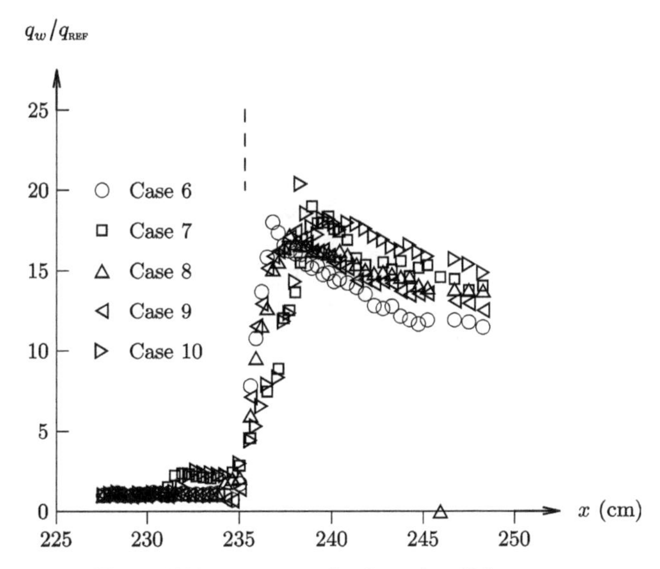


**Figure 5.209.** $q_w$ versus $x$ for Cases 6 to 10 in corner.

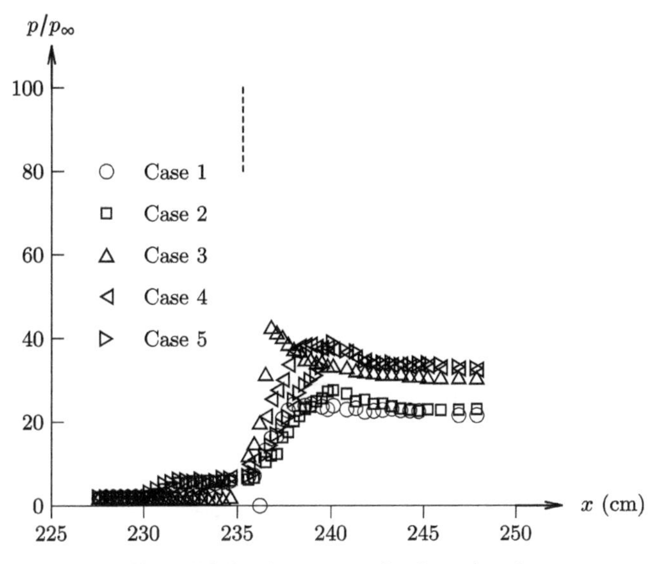


**Figure 5.210.** $p/p_\infty$ versus $x$ for Cases 1 to 5.

edge radius $r$ of the flat plate. The flow conditions are shown in table 5.53. Experimental diagnostics include surface heat transfer using luminescent temperature-sensitive paint, surface pressure and surface flow visualization using oil film seeded with luminescent particles.

Results for the single fin model are presented in Borovoy *et al* (2009) and Borovoy *et al* (2012). The present section describes results for the double fin model at $M_\infty = 5$ wherein boundary layer transition occurs upstream of the leading edge of the fins.

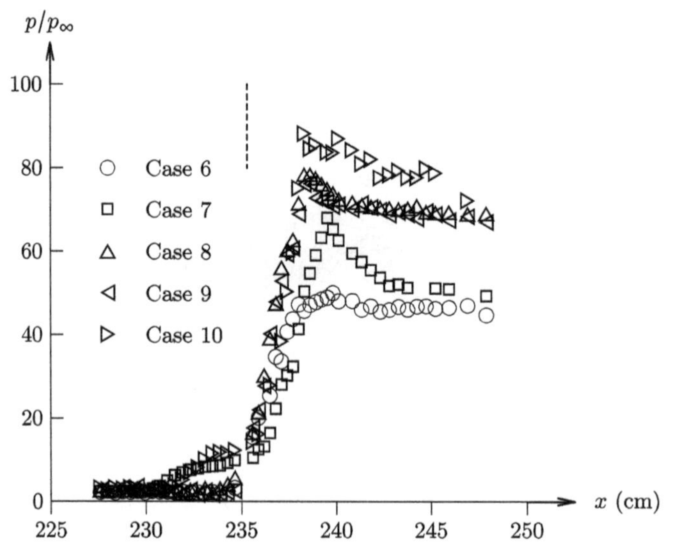


**Figure 5.211.** $p/p_\infty$ versus $x$ for Cases 6 to 10.

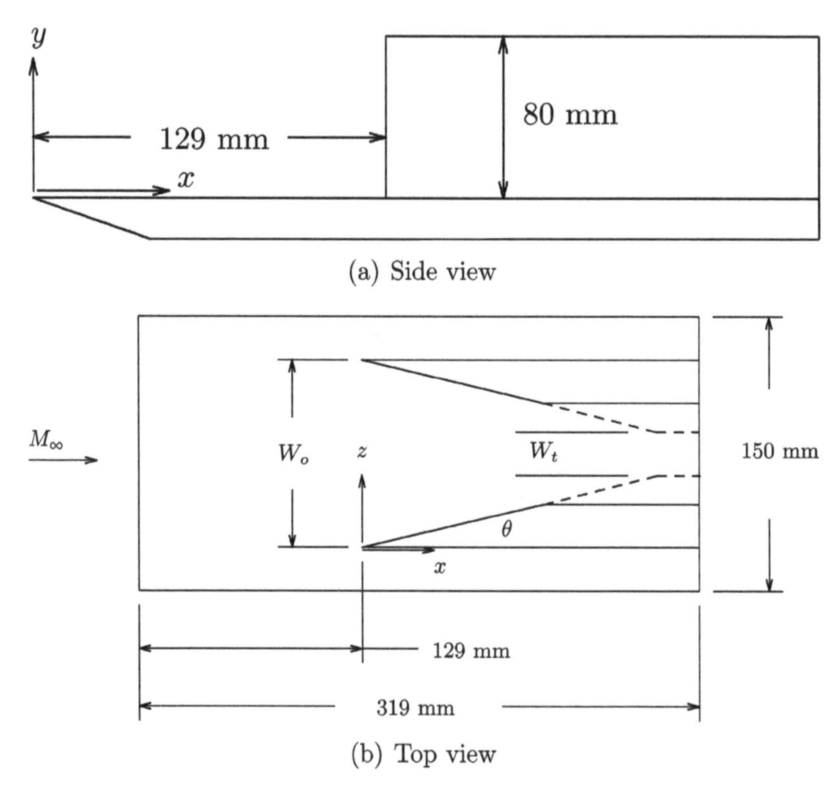


**Figure 5.212.** Model.

**Table 5.53.** Freestream conditions.

| Quantity | Case 1 | Case 2 | Case 3 |
|---|---|---|---|
| $M_\infty$ | 5 | 6 | 8 |
| $T_{t_\infty}$ (K) | 530 | 560 | 690 |
| $H_\infty$ (MJ kg$^{-1}$) | 0.53 | 0.56 | 0.69 |
| $p_{t_\infty}$ (MPa) | 7.0 | 9.1 | 9.1 |
| $Re_L \times 10^{-6}$ | 27 | 19.2 | 7 |

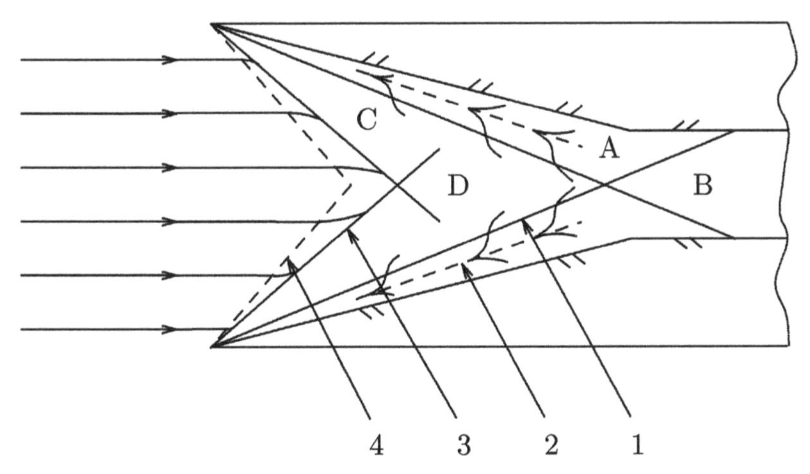


**Figure 5.213.** Regular flow structure. 1—fin shocks, 2—separation lines and separation shocks, 3—reattachment lines, 4—boundary of interference region.

Two different flowfield structures are observed depending upon the radius $r$ of the flat plate. At small leading edge radii, the well known regular structure (figure 5.213) is formed comprising intersecting shock waves downstream of the fin leading edges, primary and secondary separation and reattachment lines (Zheltovodov and Knight 2011). At a critical value of the flat plate leading edge radius, a nonregular structure forms (figure 5.214) comprising a separation line that forms upstream of the fin leading edges similar to a single blunt fin interaction (Dolling and Rodi 1988). The flowfield downstream of the separation shock (5) remains supersonic except close to the wall, and thus the flow configuration does not resemble the classical inlet unstart phenomenon wherein a normal shock is expelled from an inlet and subsonic flow forms downstream (Seddon and Goldsmith 1985).

Figures 5.215 and 5.216 display the pressure coefficient $c_\text{p}$ and Stanton number $St$, respectively, versus $z/x$ at Station 1 located 82 mm downstream of the leading edge of the sharp fins. The streamwise $x$ and spanwise $z$ coordinates are measured from the sharp fin leading edge. The flat plate boundary layer thickness $\delta = 2.9$ mm at $x = 0$. The pressure coefficient shows the pressure rise associated with the shock generated by the sharp fin in the region $0.3 \leqslant z/x \leqslant 0.45$ together with the pressure

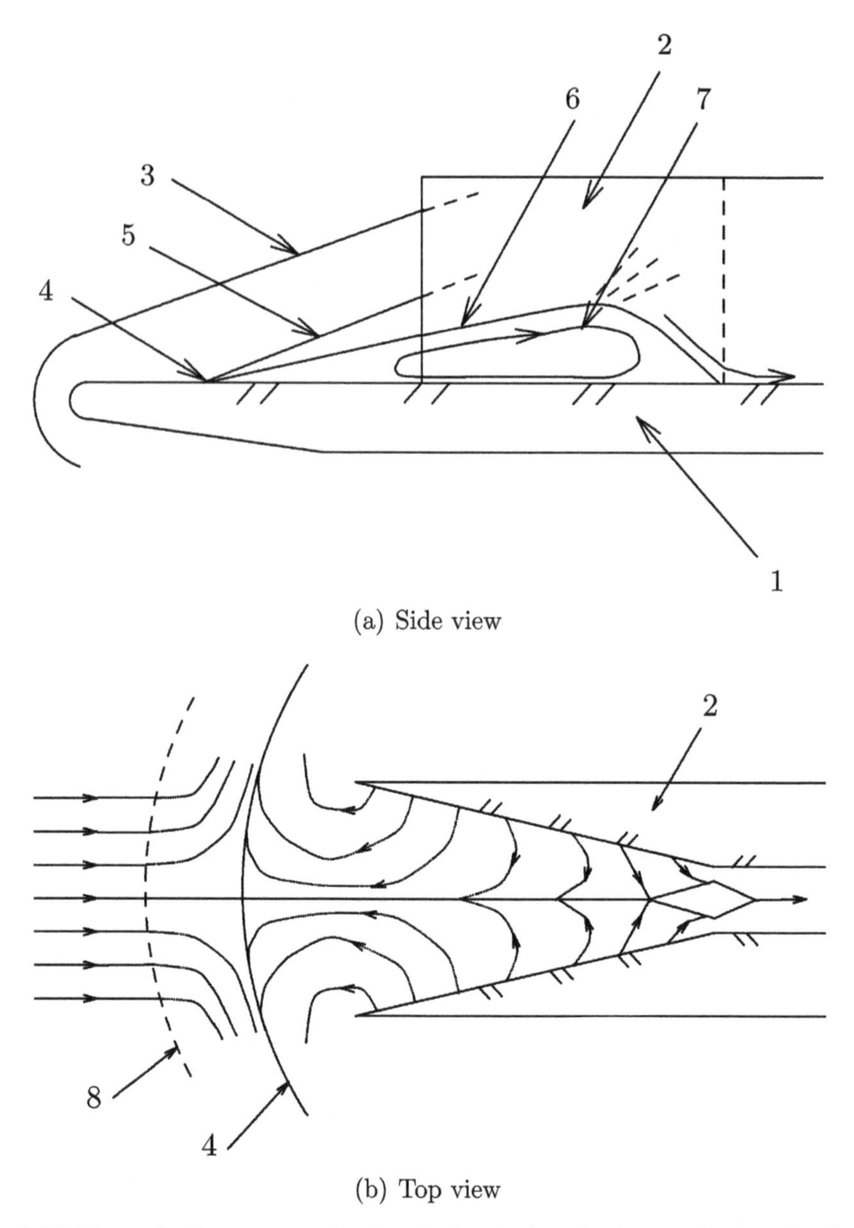


**Figure 5.214.** Nonregular flow structure. 1—plate, 2—fins, 3—bow shock wave, 4—separation line, 5—separation shock, 6—boundary of separation zone, 7—separation zone, 8—boundary of interference zone.

rise associated with the mutual interaction with the opposite fin in the region $0.5 \leqslant z/x \leqslant 0.6$. The regular flow structure (figure 5.213) occurs for $r = 0$ to $r = 0.5$ mm for both contraction ratios, and the nonregular structure occurs at $r = 0.75$ mm for $W_0/W_t = 4$. The peak Stanton number decreases monotonically with increasing flat plate radius.

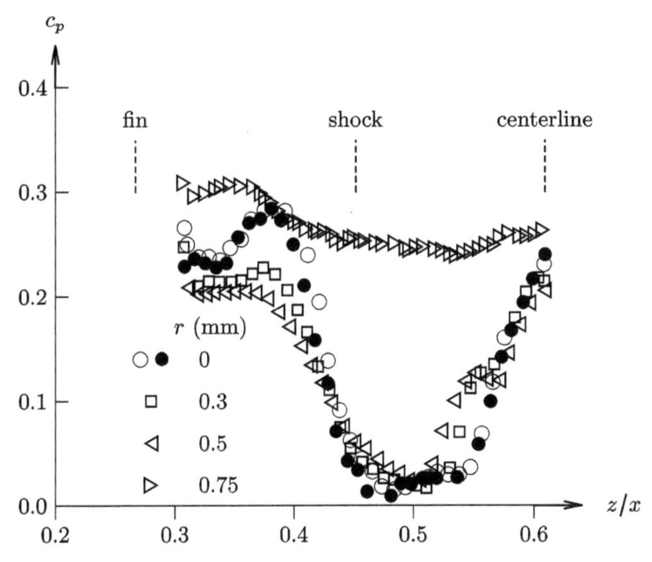


**Figure 5.215.** $c_p$ versus $z/x$ at Station 1 for different leading edge radii and contraction ratios at $M_\infty = 5$ and $Re_L = 2.7 \times 10^7$. Open symbols—$W_o/W_t = 4$, filled symbols—$W_o/W_t = 2$.

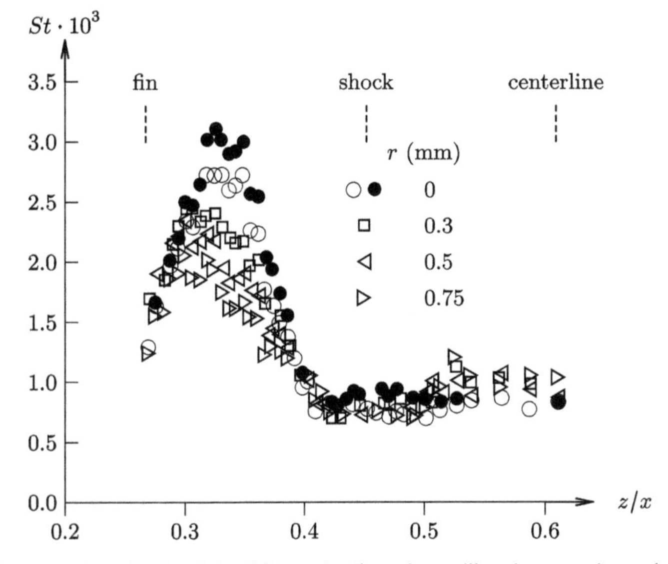


**Figure 5.216.** $St$ versus $z/x$ at Station 1 for different leading edge radii and contraction ratios at $M_\infty = 5$ and $Re_L = 2.7 \times 10^7$. Open symbols—$W_o/W_t = 4$, filled symbols—$W_o/W_t = 2$.

Figures 5.217 and 5.218 show the Stanton number $St$ versus $z/W_t$ at Station 2 located 142 mm downstream of the leading edge of the sharp fins for $W_o/W_t = 2$ and 4, respectively. The spanwise coordinate $z$ is measured from the centerplane of the model. The asymmetrical behavior is attributed to a small (<1°) difference in the effective sharp fin angle $\theta$ due to mounting on the flat plate. The peak Stanton number decreases monotonically with increasing flat plate radius.

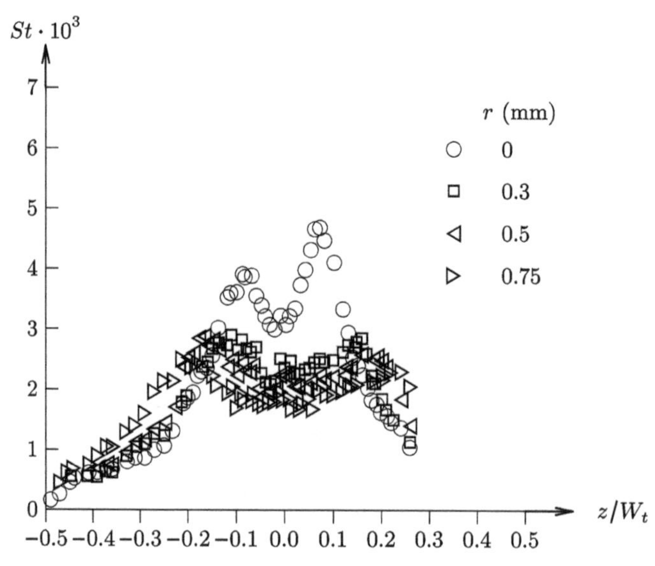


**Figure 5.217.** *St* versus *z*/*x* at Station 2 for different leading edge radii at $W_0/W_t = 2$ at $M_\infty = 5$ and $Re_L = 2.7 \times 10^7$.

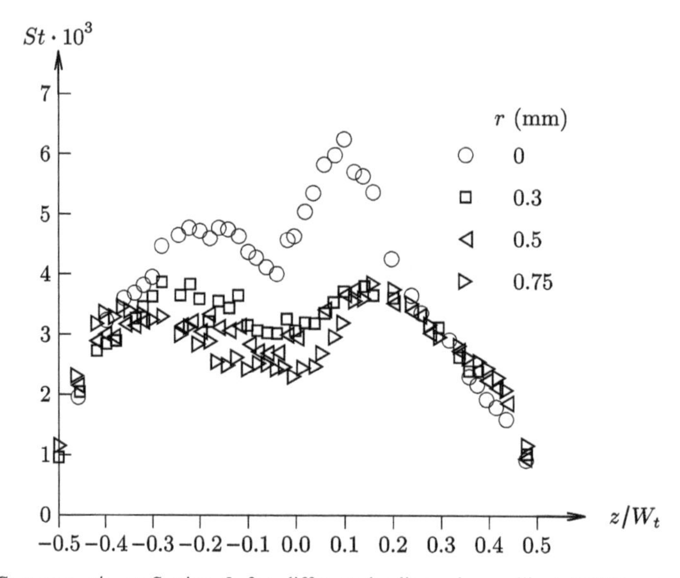


**Figure 5.218.** *St* versus *z*/*x* at Station 2 for different leading edge radii at $W_0/W_t = 4$ at $M_\infty = 5$ and $Re_L = 2.7 \times 10^7$.

Figure 5.219 displays the ratio of the maximum Stanton number $St_m$, normalized by the maximum Stanton number $St_{mo}$ for the flat plate with sharp ($r = 0$) leading edge, versus the normalized flat plate leading edge radius $r/x_o$. Results are shown for both contraction ratios $W_0/W_t$. The effect of leading edge bluntness of the flat plate is reduction in the peak Stanton number by up to 60%.

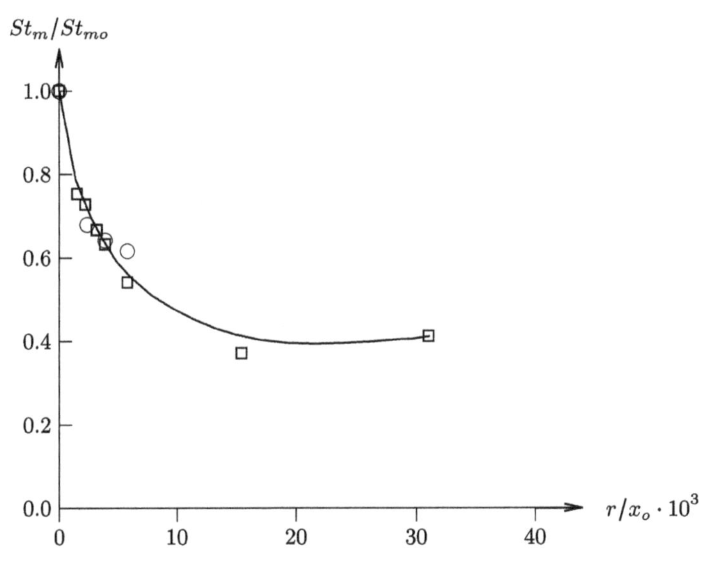


**Figure 5.219.** $St_m/St_{mo}$ versus $r/x_o$ at Station 2 for different leading edge radii at $W_o/W_t = 2$ and 4 at $M_\infty = 5$ and $Re_L = 2.7 \times 10^7$.

## 5.45 Wagner *et al* (2016)

Wagner *et al* (2016) investigated the effect of surface temperature and leading edge bluntness on a hypersonic shock wave boundary layer interaction at Mach 7.4. The model, illustrated in figure 5.220, comprises a heated flat plate of length 502 mm and unheated ramp of length 175 mm. The model width is 360 mm. Ramp angles $\beta = 0°$, 15° and 30° were examined. The experiments were conducted in the DLR High Enthalpy Shock Tunnel Göttingen (HEG) (section A.11.2). The flow conditions are shown in table 5.54. Experimental diagnostics include surface heat transfer, pressure and temperature, and schlieren imaging.

Figure 5.221 displays the Stanton number versus Reynolds number $Re_x$ based upon distance from the leading edge of the sharp leading edge flat plate model for the unheated model at $\beta = 0°$, 15° and 30°. The dashed lines indicate the predicted flat plate Stanton number distribution for laminar and turbulent flow (Van Driest 1956). The results for $\beta = 0°$ indicate that boundary layer transition begins at $x \approx 375$ mm and a fully turbulent boundary layer is achieved at $x \approx 550$ mm. Therefore, the shock wave boundary layer interaction at $\beta = 15°$ and 30° is transitional.

Figure 5.222 presents the Stanton number versus $Re_x$ for the sharp leading edge flat plate model for both unheated ($T_w = 295$ K) and heated ($T_w = 800$ K) configurations at $\beta = 15°$. The dashed lines indicate the Stanton number distribution for laminar and turbulent flow (Van Driest 1956). The effect of wall heating is destabilization of the boundary layer upstream of the shock wave boundary layer interaction.

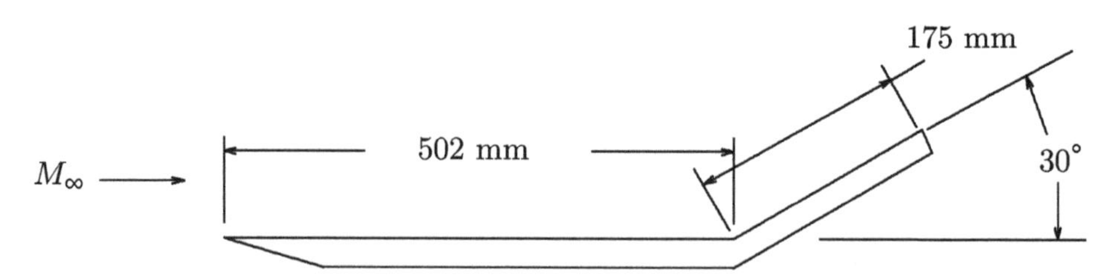


**Figure 5.220.** Model.

**Table 5.54.** Freestream conditions.

| Quantity | Value |
|---|---|
| $M_\infty$ | 7.1 |
| $p_{t_\infty}$ (MPa) | 31.5 |
| $T_{t_\infty}$ (K) | 2687 |
| $T_w$ (K) | 295 |
| $H_{t_\infty}$ (MJ kg$^{-1}$) | 3.18 |
| $Re \times 10^{-6}$ (m$^{-1}$) | 6.65 |

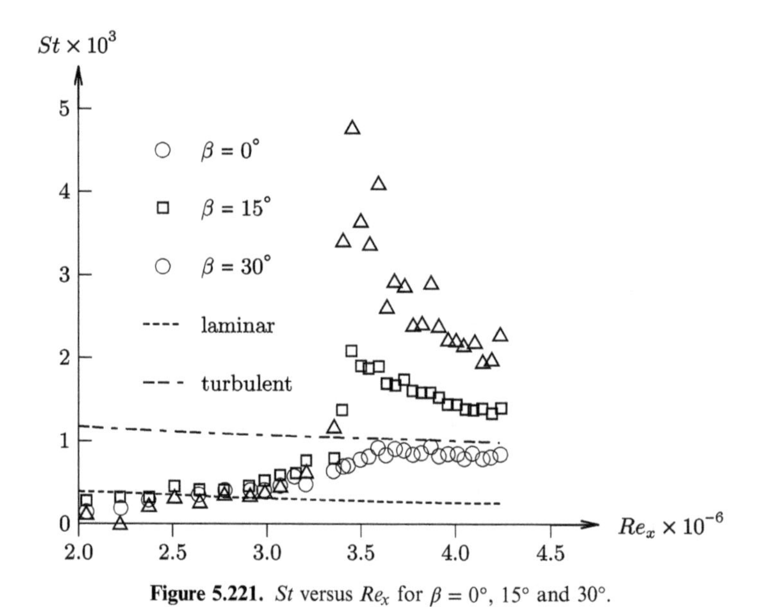


**Figure 5.221.** $St$ versus $Re_x$ for $\beta = 0°$, $15°$ and $30°$.

## 5.46 Schreyer *et al* (2018)

Schreyer *et al* (2018) performed an experimental investigation of a hypersonic shock wave turbulent boundary layer interaction generated by a compression corner at Mach 7.2. The experiments were performed in the Princeton University Mach 8

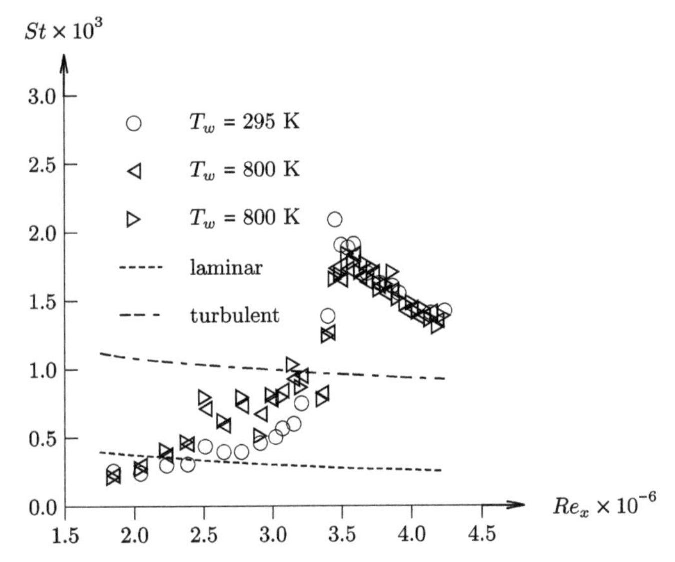


**Figure 5.222.** $St$ versus $Re_x$ for $\beta = 15°$ and $T_w = 295$ K and 800 K.

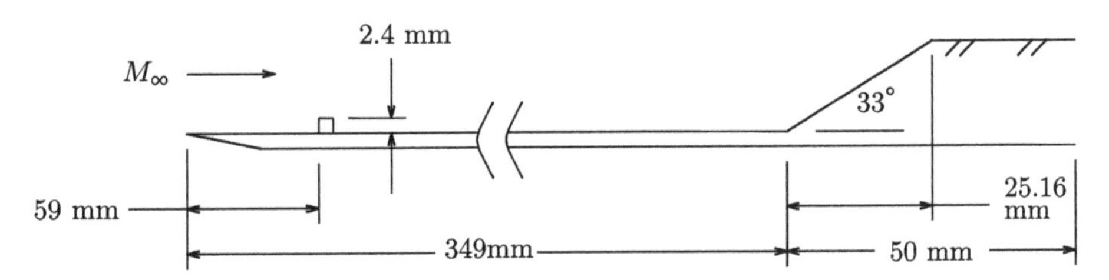


**Figure 5.223.** Model.

Hypersonic Boundary Layer Tunnel (section A.28.1). Compression corner angles of 8° and 33° were examined corresponding to unseparated and separated flow conditions, respectively. The 33° corner angle model is shown in figure 5.223. A 2.4 mm trip wire was affixed to the upstream flat plate at a distance of 59 mm from the sharp leading edge. The 8° and 33° compression ramps were located at 349 mm from the leading edge. The freestream conditions are displayed in table 5.55. Experimental measurements include Particle Image Velocimetry (PIV), Filtered Rayleigh Scattering (FRS) and schlieren imaging. The locations of the boundary layer surveys are listed in table 5.56. See also Schreyer *et al* (2011).

Figure 5.224 displays the upstream Van Driest transformed velocity profile. Close agreement is seen with the Law of the Wall in the region $10 \leqslant y^+ \leqslant 100$ for $\kappa = 0.41$ and $C = 5.1$. Figure 5.225 shows the upstream streamwise turbulence intensity using Morkovin scaling, together with the incompressible experimental data of Klebanoff (1955) ($Re_\theta = 6940$) and DeGraaff and Eaton (2000) ($Re_\theta = 5200$), the hypersonic experimental data of Williams *et al* (2018) ($Re_\theta = 5130$) and the hypersonic DNS results of Priebe and Martín (2011) ($Re_\theta = 3300$). The Morkovin scaled results of Schreyer *et al* (2018) show agreement with the incompressible data for $y/\delta > 0.6$ in

**Table 5.55.** Freestream conditions.

| Quantity | Value |
|---|---|
| Gas | Air |
| $M_\infty$ | 7.2 |
| $p_{t_\infty}$ (MPa) | 7.93 |
| $T_{t_\infty}$ (K) | 760 |
| $T_w$ (K) | 340 |
| $T_w/T_{aw}$ | 0.47 |
| $H_\infty$ (MJ kg$^{-1}$) | 0.76 |
| $u_\tau$ (m s$^{-1}$) | 56 |
| $\delta$ (mm) | 9.8 |
| $\theta$ (mm) | 0.16 |
| $Re \times 10^{-6}$ (m$^{-1}$) | 21.9 |

**Table 5.56.** Location of boundary layer surveys.

| Ramp angle | Location | No. | $x/\delta$ |
|---|---|---|---|
| 8° | Upstream of ramp corner | 1 | −0.54 |
| | Downstream of ramp corner | 2 | 0.80 |
| | | 3 | 2.6 |
| | | 4 | 3.1 |
| | | 5 | 3.6 |
| | | 6 | 4.5 |
| 33° | Upstream of ramp corner | 1 | −2.0 |
| | | 2 | −1.5 |
| | | 3 | −1.0 |
| | | 4 | −0.5 |
| | At corner | 5 | 0 |
| | Downstream of ramp corner | 6 | 0.43 |
| | | 7 | 0.86 |
| | | 8 | 1.28 |
| | | 9 | 1.71 |
| | | 10 | 2.14 |
| | At corner | 11 | 2.57 |
| | Downstream of expansion corner | 12 | 3.06 |
| | | 13 | 3.57 |
| | | 14 | 4.08 |
| | | 15 | 4.59 |
| | | 16 | 5.10 |

$x$ is streamwise distance from ramp corner

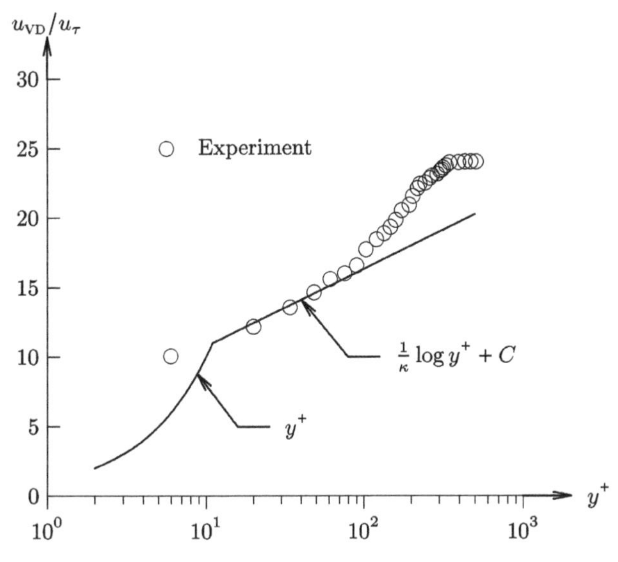


**Figure 5.224.** $u_{VD}/u_\tau$ versus $y^+$.

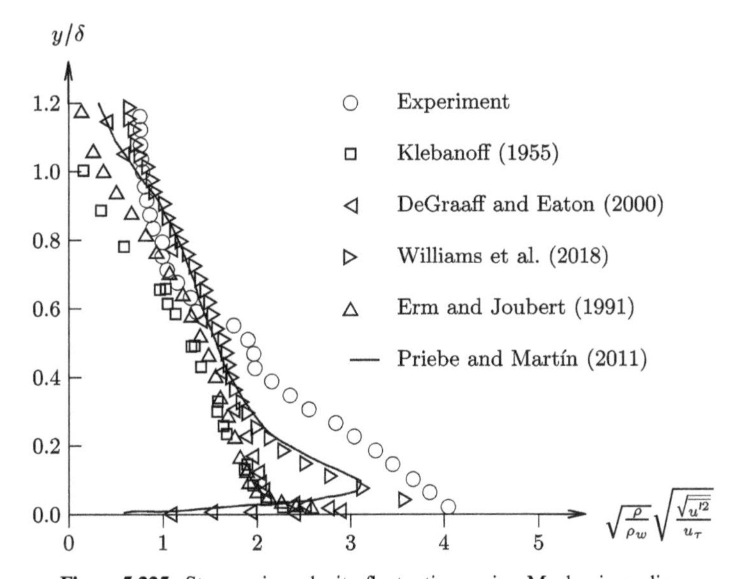


**Figure 5.225.** Streamwise velocity fluctuations using Morkovin scaling.

agreement with Morkovin's hypothesis, but are significantly higher for $y/\delta < 0.4$. Schreyer *et al* (2018) do not discuss possible reasons for such difference between their measurements and the incompressible data, but rather discuss possible instrumentation issues regarding the differences between their results and the measurements of Williams *et al* (2018) taken in the same facility. Figure 5.226 presents the upstream Morkovin scaled wall normal turbulence intensity, together with the additional experimental and DNS results cited above. There is little correlation with either the

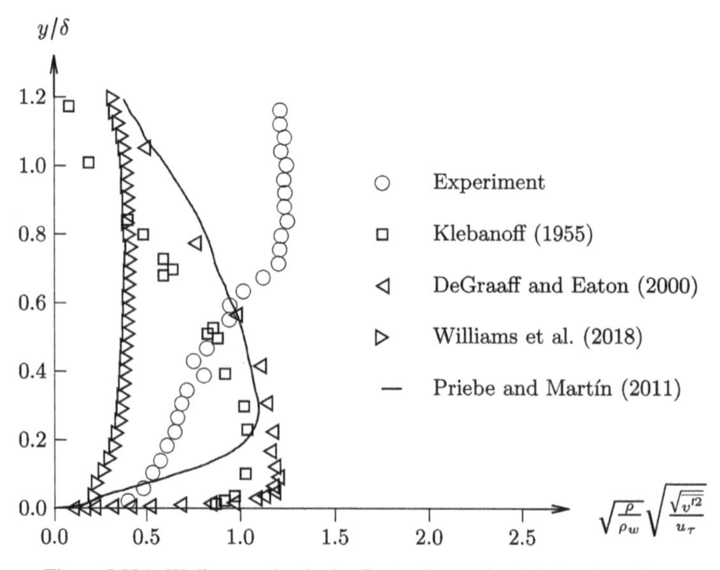


**Figure 5.226.** Wall normal velocity fluctuations using Morkovin scaling.

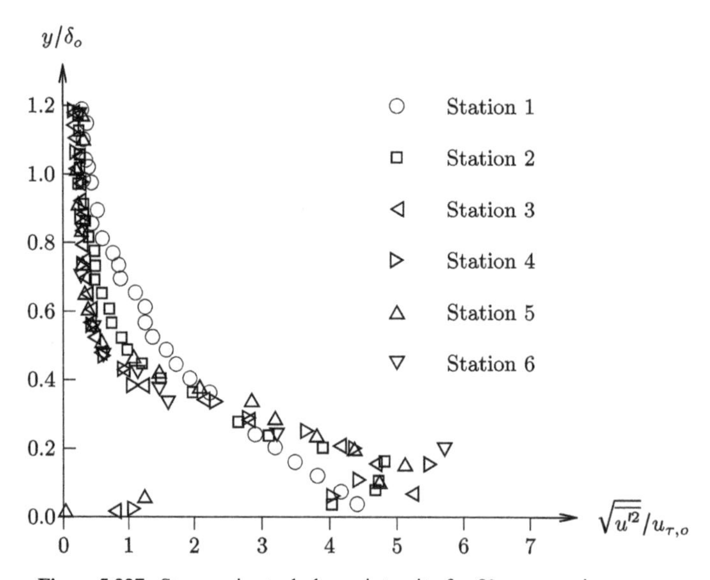


**Figure 5.227.** Streamwise turbulence intensity for 8° compression corner.

incompressible results or the measurements of Williams *et al* (2018) taken in the same facility. Schreyer *et al* (2018) cite limitations of the PIV system.

Figures 5.227 and 5.228 show the experimental normalized streamwise and wall normal turbulence intensity $\sqrt{\overline{u'^2}}/u_{\tau,o}$ and $\sqrt{\overline{v'^2}}/u_{\tau,o}$ versus $y/\delta_o$ for the 8° compression corner where $u_{\tau,o}$ and $\delta_o$ is the upstream wall friction velocity and boundary layer thickness, respectively. Results are presented at the six locations.

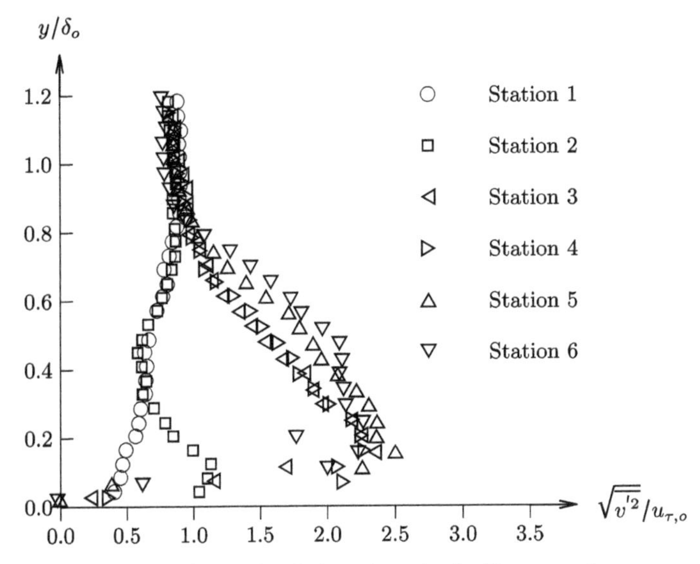


**Figure 5.228.** Wall normal turbulence intensity for 8° compression corner.

The streamwise turbulence intensity is amplified slightly through the interaction, and the experimental and DNS results agree in general except for $y/\delta_o < 0.2$ presumably due to low particle seeding (Schreyer *et al* 2018). The experimental wall normal turbulence intensity shows a dramatic increase across the interaction. Agreement with DNS upstream of the interaction is poor; however, generally good agreement between experiment and DNS is seen downstream (see figures 17(a) and 17b of Schreyer *et al* (2018)). Additional results for the Reynolds stress are presented in Schreyer *et al* (2018) and indicate a significant increase in the near wall region across the interaction.

Figure 5.229 shows the experimental normalized wall parallel turbulence intensity $\sqrt{u'^2}/u_{\tau,o}$ versus $y/\delta_o$ for the 33° compression corner. Results are presented at locations 1 to 5 upstream or at the corner (Figure 5.229(a)) and locations 6 to 10 downstream of the corner (figure 5.229(b)). For all locations the distance $y$ is measured normal to the wall. Downstream of the corner the peak intensity moves closer to the wall. Overall there is negligible change in the peak experimental wall parallel turbulence intensity through the interaction. Figure 5.230 shows the experimental normalized wall normal turbulence intensity $\sqrt{v'^2}/u_{\tau,o}$ versus $y/\delta_o$. Results are shown for locations 1 to 5 (Figure 5.230(a)) and 6 to 10 (figure 5.230(b)). A rapid growth in the peak wall normal turbulence intensity occurs upstream of the corner, while the peak value subsides downstream of the interaction. Additional results for the Reynolds stress are presented in Schreyer *et al* (2018) and indicate a significant increase in the near wall region across the interaction.

## 5.47 Currao *et al* (2020)

Currao *et al* (2020) performed an experimental investigation of a hypersonic shock wave transitional boundary layer interaction at Mach 5.8. The model is shown in

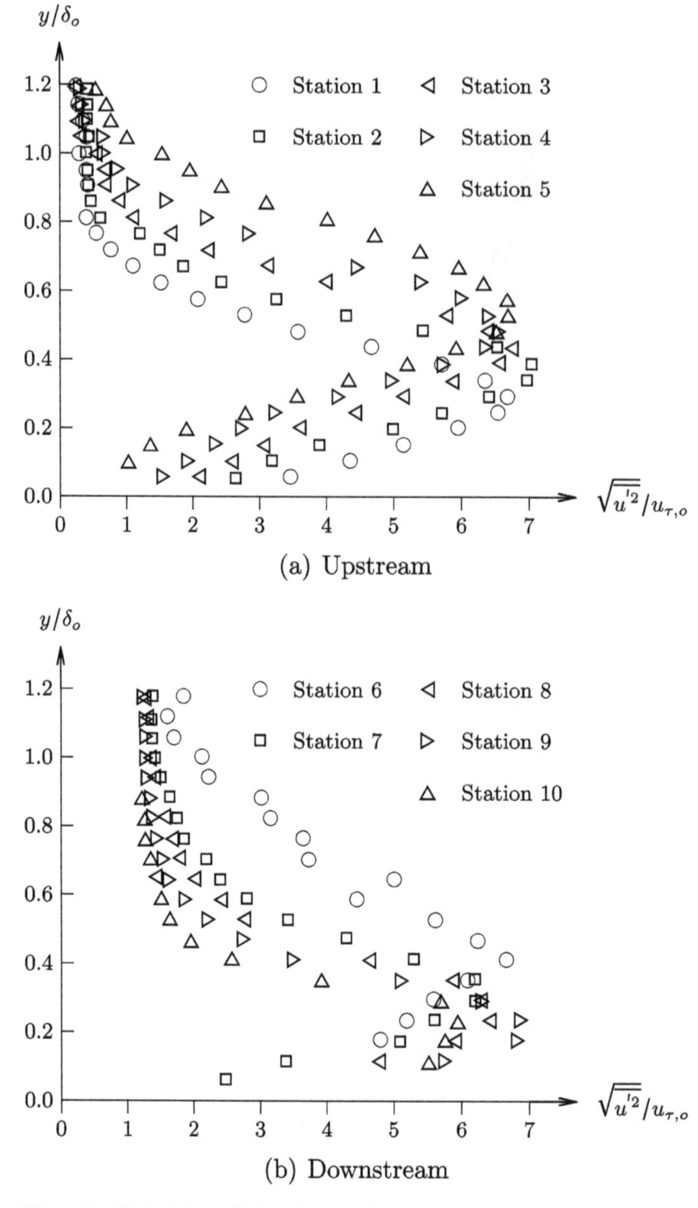


**Figure 5.229.** Wall parallel turbulence intensity for 33° compression corner.

figure 5.231. The experiments were conducted in the free-piston compression-heated Ludwieg tube TUSQ at the University of Southern Queensland (section A.44). An incident shock wave generated by a 10° half angle wedge intersects an incoming laminar boundary layer on a flat plate. The test conditions are listed in table 5.57. Experimental diagnostics include surface pressure measurements using Pressure Sensitive Paint (PSP), surface heat transfer using infrared thermography, and schlieren imaging.

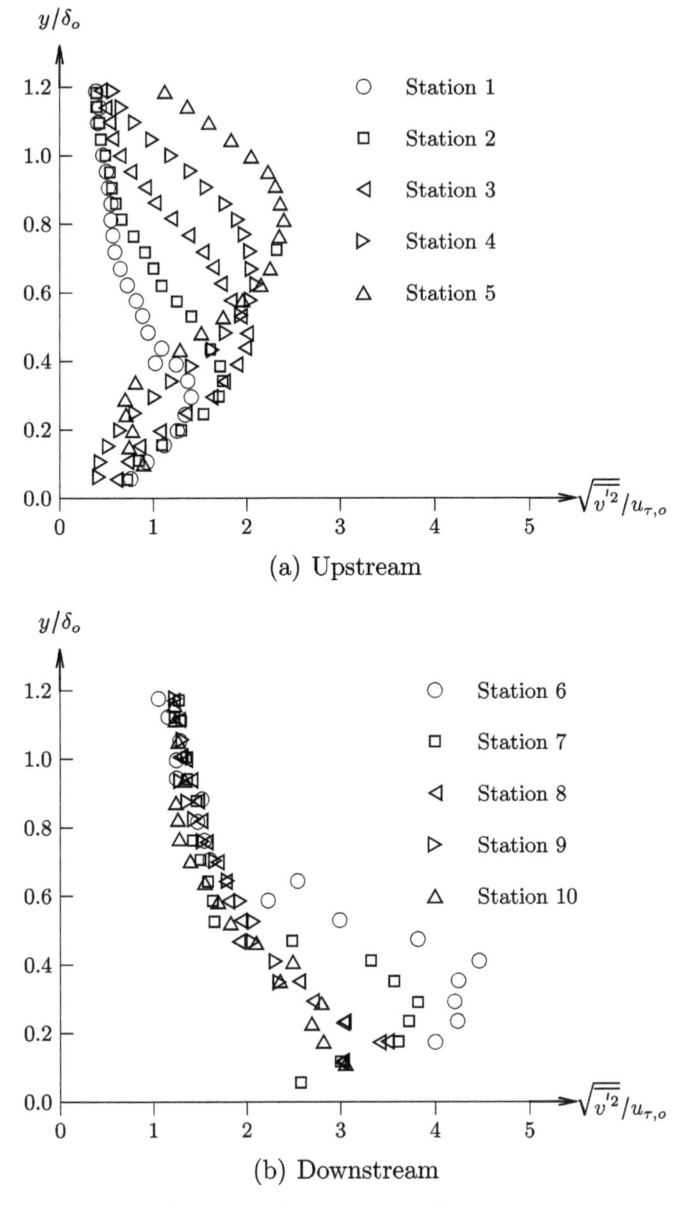


**Figure 5.230.** Wall normal turbulence intensity for 33° compression corner.

The interaction of the shock wave with the incoming laminar boundary layer causes transition due to the formation of Görtler vortices. Figure 5.232 shows the Stanton number versus spanwise distance $z$ at the location of peak heating. A periodic variation in $St$ results from the Görtler vortices with variation about the mean value of approximately ±15%. The spanwise wavelength is $\lambda \approx 3.7\delta$ where $\delta$ is the incoming laminar boundary layer thickness. The resultant peak heat transfer is approximately 2.3 times the calculated peak laminar heat transfer.

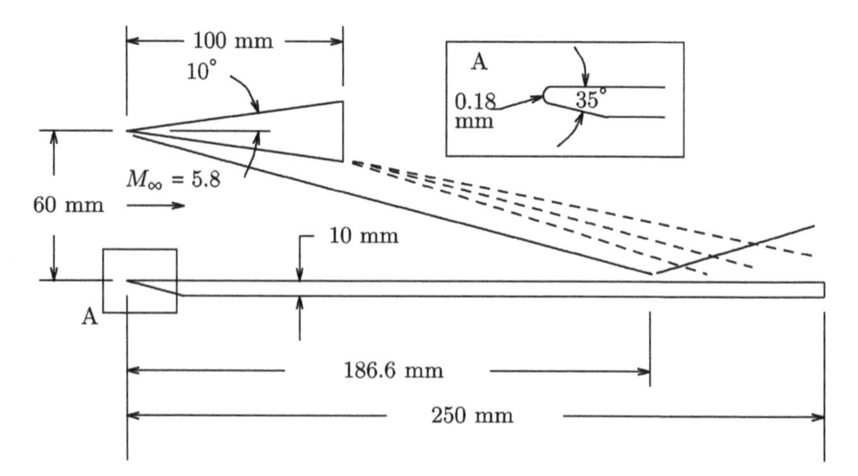


**Figure 5.231.** Model.

**Table 5.57.** Freestream conditions.

| Quantity | Value |
|---|---|
| $M_\infty$ | 5.8 |
| $p_\infty$ (Pa) | 756 |
| $T_\infty$ (K) | 77 |
| $T_w$ (K) | 306 |
| $Re \times 10^{-6}$ (m$^{-1}$) | 6.8 |
| Results shown for Run 577 | |

## 5.48 Chang *et al* (2021)

Chang *et al* (2021) investigated the effect of wall temperature on a hypersonic shock wave boundary layer interaction at Mach 7. The model is shown in figure 5.233 and further details are provided in Chang *et al* (2020). A sharp leading edge flat plate comprised a 70 mm front unheated section and a 250 mm isotropic graphite plate heated section. The width was 200 mm. A shock generator plate of length 215 mm and width 172 mm was mounted above the flat plate. Deflection angles of 10° and 12° were considered. The experiments were conducted in the Stalker T4 Hypersonic Wind Tunnel at the University of Queensland (section A.45.1). The flow conditions are shown in table 5.58. Experimental diagnostics include surface heat transfer and surface pressure, and schlieren visualization.

Figure 5.234 shows the normalized surface pressure $p/p_\infty$ versus $x$ (mm) for Case 1 with the 12° shock generator. The solid line is the theoretical inviscid surface pressure. The incoming boundary layer is laminar with separation at $x$ = 179.5 mm as evidence by the initial plateau. The shock impingement causes transition of the boundary layer (Chang *et al* 2021). The interaction of the reattached boundary layer

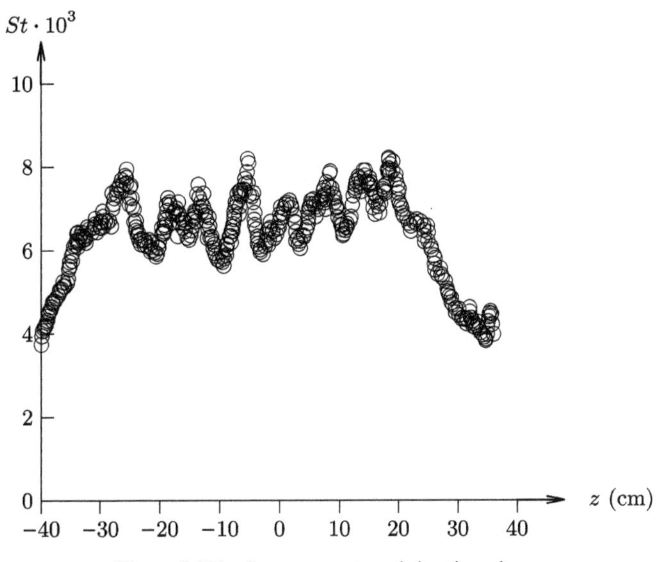


**Figure 5.232.** $St$ versus $z$ at peak heating plane.

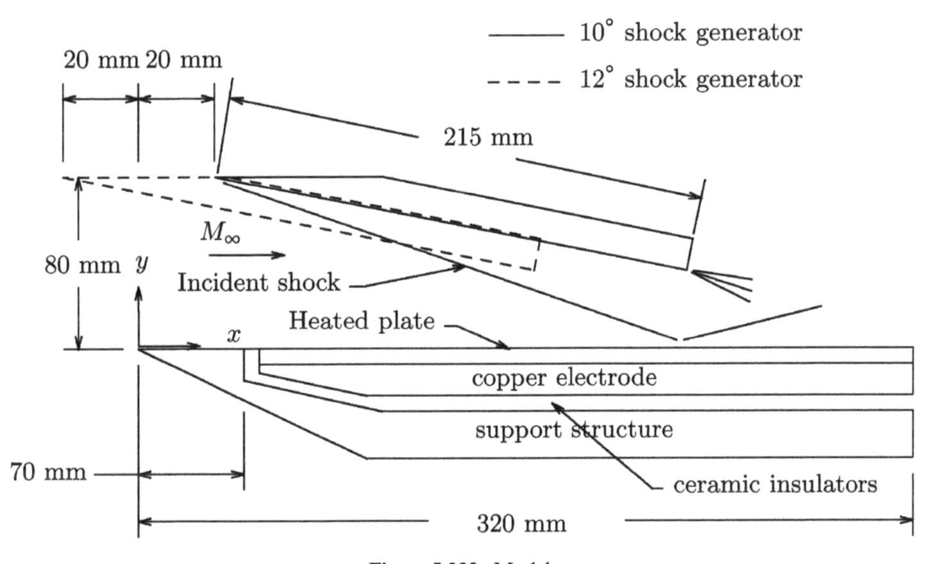


**Figure 5.233.** Model.

with the expansion fan emanating from the shock generator trailing edge results in a decrease in pressure for $x > 300$ mm. Figure 5.235 displays the Stanton number versus $x$ (mm) for the same case. The incoming Stanton number corresponds to a laminar boundary layer (Chang *et al* 2021). The peak Stanton number corresponds to reattachment, and the subsequent decrease in Stanton number is due to the interaction with the expansion fan from the shock generator.

**Table 5.58.** Freestream conditions.

| Quantity | Case 1 | Case 2 | Case 3 | Case 4 |
|---|---|---|---|---|
| $M_\infty$ | 7.00 | 7.07 | 7.29 | 7.36 |
| $p_{t_\infty}$ (MPa) | 15.5 | 4.1 | 5.4 | 5.2 |
| $T_{t_\infty}$ (K) | 2334 | 2147 | 1652 | 1475 |
| $T_w$ (K) | 675[a] | 675[a] | 675[a] | 675[a] |
| $H_\infty$ (MJ kg$^{-1}$) | 2.43 | 2.17 | 1.53 | 1.31 |
| $Re \times 10^{-6}$ (m$^{-1}$) | 4.93 | 1.57 | 3.23 | 3.70 |

[a]Approximate. Other $T_w$ values examined.

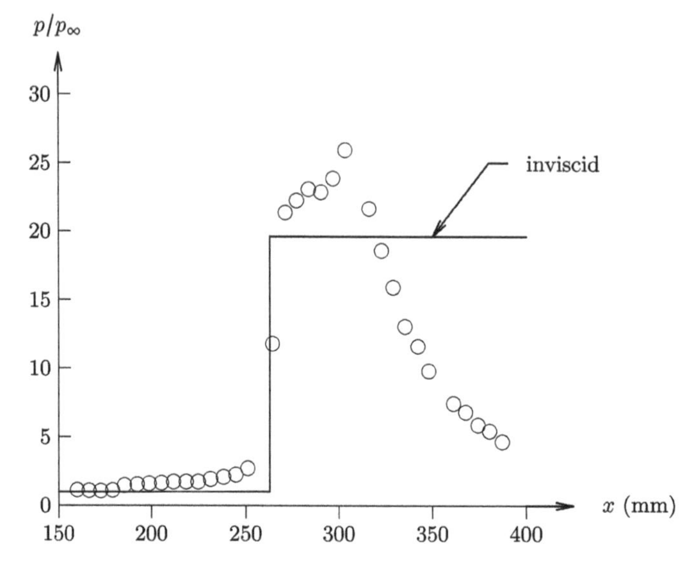


**Figure 5.234.** $p/p_\infty$ versus $x$ (mm) for Case 1 with 12° shock generator angle.

## 5.49 Zhao *et al* (2021)

Zhao *et al* (2021) conducted an experimental investigation of a hypersonic shock wave turbulent boundary layer interaction generated by a blunt fin at Mach 6. The model is shown in figure 5.236. A blunt fin with leading edge diameter $D = 25$ mm and height $H = 200$ mm is attached normal to a flat plate of length $L = 680$ mm and width $W = 380$ mm at a distance of 432 mm from the sharp leading edge of the flat plate. The experiments were performed in the China Academy of Aerospace Aerodynamics (CAAA) FD-20 Hypersonic Gun Tunnel. The flow conditions are listed in table 5.59. Three cases were examined corresponding to fully laminar (Case 1), transitional (Case 2) and fully turbulent (Case 3) interactions. Fluctuating surface pressure measurements were obtained on the flat plate upstream of the blunt fin and aligned with the blunt fin axis using 32 Kulite transducers.

Figure 5.237 shows the Power Spectrum Density (PSD) versus frequency for Case 3 at three locations on the flat plate corresponding to (1) initial laminar boundary

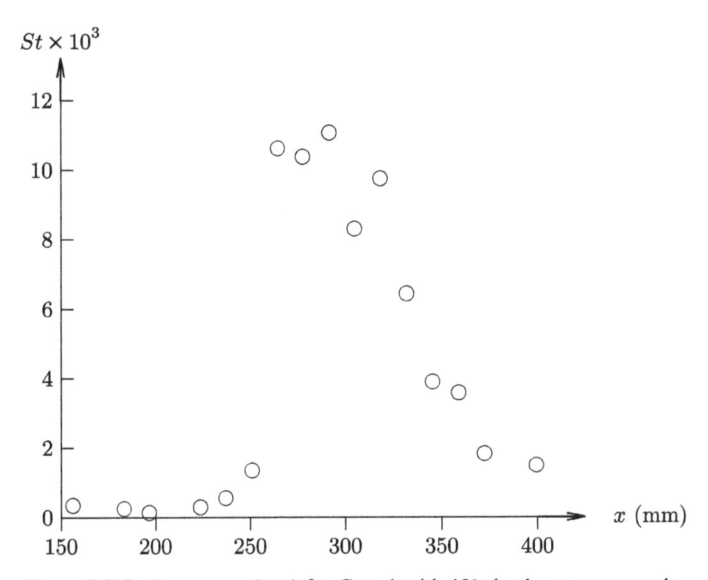


**Figure 5.235.** *St* versus *x* (mm) for Case 1 with 12° shock generator angle.

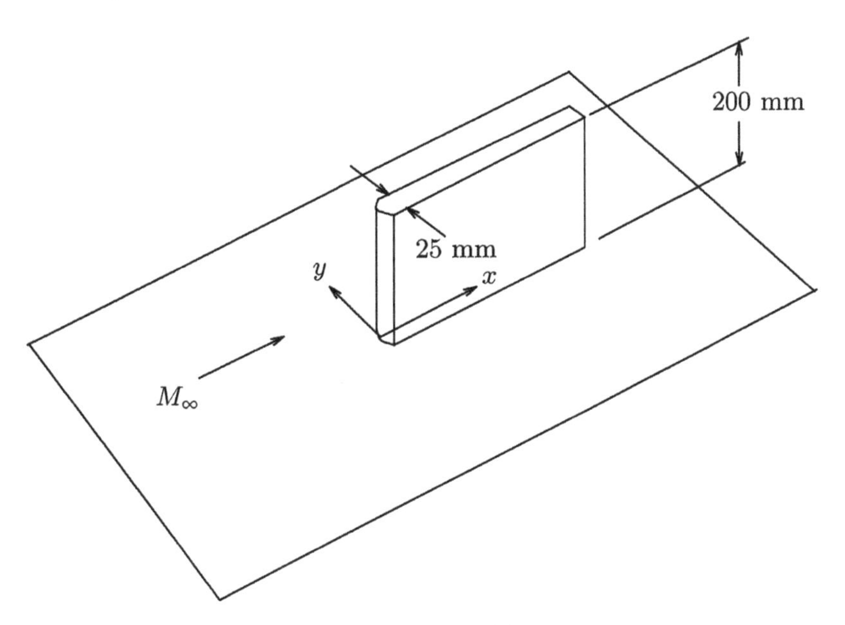


**Figure 5.236.** Model.

layer, (2) turbulent boundary layer upstream of the interaction, and (3) separated region within the interaction. The significant increase in PSD within the separated region compared to the upstream laminar and turbulent boundary layers is evident. A beamforming algorithm analysis (Zhao and Lei 2020, Zhao 2021b) within the three regions indicated the pressure fluctuations within the turbulent region upstream of the interaction were attributable to both acoustic and convective

**Table 5.59.** Freestream conditions.

| | Value | | |
|---|---|---|---|
| Quantity | Case 1 | Case 2 | Case 3 |
| $M_\infty$ | 6 | 6 | 6 |
| $p_{t_\infty}$ (MPa) | 3.5 | 6.0 | 12.0 |
| $T_{t_\infty}$ (K) | 818[a] | 832[a] | 900[a] |
| $q_\infty$ (kPa) | 60.2 | 84.2 | 174 |
| $u_\infty$ (m s$^{-1}$) | 1201.3 | 1211.8 | 1260.0 |
| $H_\infty$ (MJ kg$^{-1}$) | 0.82 | 0.83 | 0.9 |
| $Re \times 10^{-6}$ (m$^{-1}$) | 13 | 15 | 40 |

[a] From $T_{t_\infty} = (U_\infty/M_\infty)^2(\gamma R)^{-1}[1 + \frac{1}{2}(\gamma - 1)M_\infty^2]$.

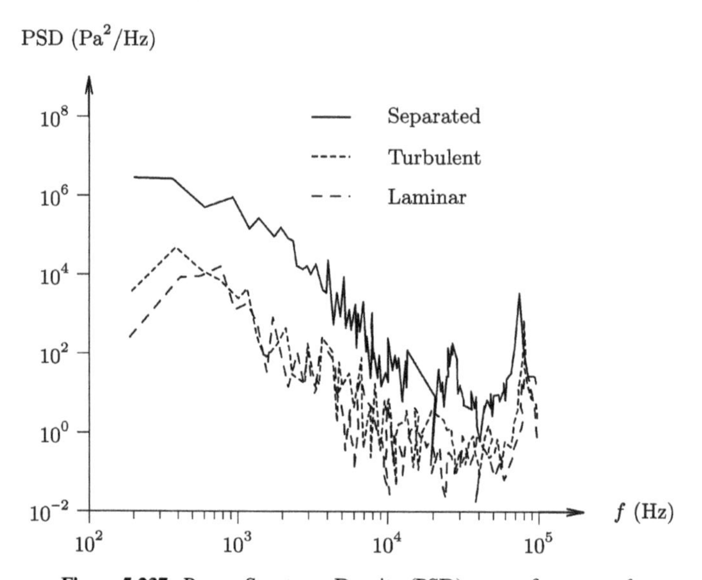


**Figure 5.237.** Power Spectrum Density (PSD) versus frequency $f$.

modes, while the convective mode dominated the pressure fluctuations in the separated region.

## References

Alvi F 2022 private communication

Alvi F and Settles G 1990 Structure of swept shock wave/boundary layer interactions using conical shadowgraphy AIAA Paper 90-1644 (Reston, VA: American Institute of Aeronautics and Astronautics)

Alvi F and Settles G 1991 Physical model of the swept shock/boundary layer interaction flowfield *Physical model of the swept shock/boundary layer interaction flowfield* AIAA Paper 91-1768 (Reston, VA: American Institute of Aeronautics and Astronautics)

Alvi F and Settles G 1992 Physical model of the swept shock/boundary layer interaction flowfield *AIAA J.* **30** 2252–8

Appels C 1973 Turbulent boundary layer separation at Mach 12 *Technical Note 90* von Kármán Institute for Fluid Dynamics

Babinsky H 1993 A study of roughness in turbulent hypersonic boundaryLayers *PhD Thesis* (Cranfield, UK: Cranfield University, College of Aeronautics)

Babinsky H and Edwards J 1995 The influence of large scale roughness on a turbulent hypersonic boundary layer approaching a compression corner AIAA Paper 95-0334 (Reston, VA: American Institute of Aeronautics and Astronautics)

Babinsky H and Edwards J 1997 Large scale roughness influence on turbulent hypersonic boundary layers approaching compression corners *J. Spacecr. Rockets* **34** 70–5

Babinsky H and Harvey J 2011 *Shock Wave-Boundary-Layer Interactions* (New York: Cambridge)

Benay R, Chanetz B, Mangin B, Vandomme L and Perraud J 2006 Shock wave/transitional boundary-layer interactions in hypersonic flow *AIAA J.* **44** 1243–54

Bookey P, Wu M, Smits A and Martín M P 2005 New experimental data of swtbi at dns/les accessible Reynolds numbers AIAA Paper 2005-0309 (Reston, VA: American Institute of Aeronautics and Astronautics)

Borovoy V 2021 private communication

Borovoy V, Egorov I, Mosharov V, Radchenko V, Skuratov A and Struminskaya I 2016 Entropy layer influence on single fin and double fin boundary layer interactions *AIAA J.* **54** 443–57

Borovoy V, Egorov I, Noev A, Radchenko Y, Skuratov A and Struminskaya I 2012 3D shock/ turbulent boundary layer interaction on the plate near a wedge in presence of an entropy layer *TsAGI Sci. J.* **43** 697–718

Borovoy V, Egorov I, Noev A, Skuratov A and Struminskaya I 2011 Two dimensional interaction between an incident shock and a turbulent boundary layer in the presence of an entropy layer *Fluid Dyn.* **46** 917–34

Borovoy V, Egorov I, Skuratov A and Struminskaya I 2005 Interaction between an inclined shock and boundary and high entropy layers on a flat plate *Fluid Dyn.* **40** 911–28

Borovoy V, Egorov I, Skuratov A and Struminskaya I 2013 Two dimensional shock wave boundary layer interaction in the presence of an entropy layer *AIAA J.* **31** 80–93

Borovoy V, Mosharov V, Noev A and Radchenko Y 2009 Laminar-turbulent flow over wedges mounted on sharp and blunt plates *Fluid Dyn.* **44** 382–96

Bur R and Chanetz B 2009 Experimental study on the PRE-X vehicle focusing on the transitional shock-wave/boundary-layer interaction *Aerosp. Sci. Technol.* **13** 393–401

Chanetz B, Benay R, Bousquet J-M, Bur R, Pot T, Grasso F and Moss J 1998 Experimental and numerical study of the laminar separation in hypersonic flow *Aerosp. Sci. Technol.* **2** 205–18

Chang E, Chan W, McIntyre T and Veeraragavan A 2020 Electrically heated flat plate testing in a free-piston driven shock tunnel *Aerosp. Sci. Technol.* **103** 105856

Chang E, Chan W, McIntyre T and Veeraragavan A 2021 Hypersonic shock impingement on a heated flat plate at Mach 7 flight enthalpy *J. Fluid Mech.* **98** R1–1-R1-13

Christophel R and Rockwell W 1974 Tabulated Mach 6 3-D shock wave turbulent boundary layer interaction heat transfer data *Technical Memorandum 74-212-FXG* Air Force Flight Dynamics Laboratory, Wright-Patterson AFB, OH

Christophel R, Rockwell W and Neumann R 1975 Tabulated Mach 6 3-D shock wave turbulent boundary layer interaction heat transfer data (supplement) *Technical Memorandum 74-212-FXG* Air Force Flight Dynamics Laboratory, Wright-Patterson AFB, OH

Coët M C 1989 Etude Expérimentale de l'Interaction Onde de Choc/Couche Limite en Écoulement Hypersonique Bidimensionnel aux Nombres de Mach 5 et 10 *Report 7/4362* Office National d'Études et de Recherches Aéronautiques

Coët M-C and Chanetz B 1993 Experiments on shock wave boundary layer interaction in hypersonic flow *Rech. Aérosp.* **1** 61–704

Coleman G 1973a A study of hypersonic boundary layers over a family of axisymmetric bodies at zero incidence: preliminary report and data tabulation *Imperial College Aero Report 73-06* Imperial College of Science and Technology, London, UK

Coleman G 1973b Hypersonic turbulent boundary layer studies *PhD Thesis* (London, UK: Imperial College of Science and Technology)

Coleman G and Stollery J 1972a Heat transfer from hypersonic turbulent flow at a wedge compression corner *J. Fluid Mech.* **56** 741–52

Coleman G and Stollery J 1972b Heat transfer in hypersonic turbulent separated flows *Aero Report 72-05* Imperial College of Science and Technology, London, UK

Coleman G and Stollery J 1974 Incipient separation of axially symmetric hypersonic turbulent boundary layers *AIAA J.* **12** 119–20

Currao G, Choudhury R, Gai S, Neely A and Buttsworth D 2020 Hypersonic transitional shock wave boundary layer interaction on a flat plate *AIAA J.* **58** 814–29

DeGraaff D and Eaton J 2000 Reynolds-number scaling of the flat-plate turbulent boundary layer *J. Fluid Mech.* **422** 319–46

Disimile P and Scaggs N 1989 An investigation into wedge induced turbulent boundary layer separation on a uniformly roughened surface at Mach 6.0 AIAA Paper 89-2163 (Reston, VA: American Institute of Aeronautics and Astronautics)

Disimile P and Scaggs N 1990 The effect of separation on turbulent boundary layer characteristics over a smooth surface at Mach 6.0 AIAA Paper 90-3028 (Reston, VA: American Institute of Aeronautics and Astronautics)

Disimile P and Scaggs N 1991a Mach 6 turbulent boundary layer characteristics on smooth and rough surfaces AIAA Paper 91-1762 (Reston, VA: American Institute of Aeronautics and Astronautics)

Disimile P and Scaggs N 1991b Wedge induced turbulent boundary layer separation on a roughened surface at Mach 6.0 *J. Spacecr.* **28** 636–45

Disimile P and Scaggs N 1992 Turbulent boundary layer characteristics over a flat plate wedge configuration at Mach 6 *AIAA J.* **30** 270–2

Dolling D and Rodi P 1988 Upstream influence and separation scales in fin-induced shock turbulent boundary layer interaction *J. Spacecr. Rockets* **25** 102–8

Dolling D and Smith D 1989 Separation shock dynamics in Mach 5 turbulent interactions induced by cylinders *AIAA J.* **27** 1698–706

Elfstrom G 1972 Turbulent hypersonic flow of a wedge compression corner *J. Fluid Mech.* **53** 113–27

Garrison T 1994 The interaction between crossing-shock waves and a turbulent boundary layer *PhD Thesis* (University Park, PA: The Pennsylvania State University, Department of Mechanical Energy)

Garrison T and Settles G 1992 Flowfield visualization of crossing shock wave boundary layer interaction AIAA Paper 92-0750 (Reston, VA: American Institute of Aeronautics and Astronautics)

Garrison T, Settles G and Horstman C 1996 Measurements of the triple shock wave/turbulent boundary layer interaction *AIAA J.* **34** 57–64

Garrison T, Settles G, Narayanswami N and Knight D 1993 Structure of crossing shock wave/turbulent boundary layer interactions *AIAA J.* **31** 2204–11

Garrison T, Settles G, Narayanswami N and Knight D 1994 Laser interferometer skin friction measurements of crossing shock wave/turbulent boundary layer interactions *AIAA J.* **32** 1234–41

Garrison T, Settles G, Narayanswami N, Knight D and Horstman C 1996 Flowfield surveys and computations of a crossing shock wave boundary layer interaction *AIAA J.* **34** 50–6

Hayes J 1977 Prediction techniques for the characteristics of fin generated three dimensional shock wave turbulent boundary layer interactions *Technical Report 77-10* Air Force Flight Dynamics Laboratory, Wright-Patterson AFB, OH

Holden M 1972 Shock wave-turbulent boundary layer interaction in hypersonic flow AIAA Paper 72-74 (Reston, VA: American Institute of Aeronautics and Astronautics)

Holden M 1975 Shock wave—turbulent boundary layer interaction in high speed flow *Technical Report 75-0204* Aerospace Research Laboratories, US Air Force Systems Command, Wright-Patterson AFB, OH

Holden M 1977 Shock wave turbulent boundary layer interactions in hypersonic flow AIAA Paper 77-45 (Reston, VA: American Institute of Aeronautics and Astronautics)

Holden M 1984 Experimental studies of quasi-two-dimensional and three-dimensional viscous interaction regions induced by skewed-shock and swept-shock boundary interaction AIAA Paper 84-1677 (Reston, VA: American Institute of Aeronautics and Astronautics)

Holden M 1991 Studies of the mean and unsteady structure of turbulent boundary layer separation in hypersonic flow AIAA Paper 91-1778 (Reston, VA: American Institute of Aeronautics and Astronautics)

Holden M, Carr Z, MacLean M and Wadhams T 2018 Measurements in regions of shock wave/turbulent boundary layer interaction from Mach 5 to 6 at flight duplicated velocities to evaluate and improve the models of turbulence in CFD codes AIAA Paper 2018-3706 (Reston, VA: American Institute of Aeronautics and Astronautics)

Holden M, MacLean M, Wadhams T and Mundy E 2010 Experimental studies of shock wave/turbulent boundary layer interaction in high Reynolds number supersonic and hypersonic flows to evaluate the performance of CFD codes AIAA Paper 2010-4468 (Reston, VA: American Institute of Aeronautics and Astronautics)

Holden M, Wadhams T and MacLean M 2014 Measurements in regions of shock wave/turbulent boundary layer interaction on double cone and hollow cylinder/flare configurations for open and 'blind' code evaluation/validation *Oral Presentation at AVIATION 2014* American Institute of Aeronautics and Astronautics

Holloway P, Sterrett J and Creekmore H 1965 An investigation of heat transfer within regions of separated flow at a Mach Number of 6.0 *Technical Note TN D-3074* (Washington, DC: National Aeronautics and Space Administration)

Hopkins E and Inouye M 1971 An evaluation of theories for predicting skin friction and heat transfer on flat plates at supersonic and hypersonic Mach Numbers *AIAA J.* **9** 993–1003

Kaufman L, Korkegi R and Morton L 1972 Shock impingement caused by boundary layer separation ahead of blunt fins *ARL 72-0118* Aerospace Research Laboratories, Wright-Patterson AFB, OH

Kim K-S, Lee Y, Alvi F and Settles G 1990 Laser skin friction measurements and cfd comparison of weak-to-strong swept shock/boundary layer interactions AIAA Paper 90-0378 (Reston, VA: American Institute of Aeronautics and Astronautics)

Kim K-S and Settles G 1988 Skin friction measurements by laser interferometry in swept shock wave/turbulent boundary layer interactions AIAA Paper 88-497 (Reston, VA: American Institute of Aeronautics and Astronautics)

Kim K-S and Settles G 1990 Skin friction measurements by laser interferometry in swept shock/boundary layer interactions *AIAA J.* **28** 133–9

Klebanoff P 1955 Characteristics of turbulence in a boundary layer with zero pressure gradient *Technical Report 1247* (Washington, DC: National Advisory Committee on Aeronautics)

Knight D, Horstman C and Monson D 1992 The hypersonic shock wave-turbulent boundary layer interaction generated by a sharp fin at Mach 8.2 AIAA Paper 92-0747 (Reston, VA: American Institute of Aeronautics and Astronautics)

Korkegi R 1971 Survey of viscous interactions associated with high Mach number flight *AIAA J.* **9** 771–84

Korkegi R 1976 On the structure of three dimensional shock induced separated flow regions *AIAA J.* **14** 597–600

Kussoy M and Horstman C 1975 An experimental documentation of a hypersonic shock-wave turbulent boundary layer interaction flow–with and without separation *Technical Memorandum X-62412* National Aeronautics and Space Administration, Ames Research Center, Moffett Field, CA

Kussoy M and Horstman C 1989 Documentation of two- and three-dimensional hypersonic shock wave/turbulent boundary layer interaction flows *Technical Memorandum 101075* (Washington, DC: National Aeronautics and Space Administration)

Kussoy M and Horstman K 1991 Documentation of two- and three-dimensional shock-wave/turbulent boundary layer interaction flows at Mach 8.2 *Technical Memorandum 103838* (Washington, DC: National Aeronautics and Space Administration)

Kussoy M and Horstman K 1992 Intersecting shock-wave/turbulent boundary-layer interactions at Mach 8.3 *Technical Memorandum 103909* (Washington, DC: National Aeronautics and Space Administration)

Kussoy M and Horstman K 1993 Three-dimensional hypersonic shock wave/turbulent boundary-layer interaction *AIAA J.* **31** 8–9

Kussoy M, Horstman K and Horstman C 1993 Hypersonic crossing shock-wave/turbulent boundary-layer interaction *AIAA J.* **31** 2197–203

Kussoy M, Kim K and Horstman K 1991 An experimental study of a three-dimensional shock wave/turbulent boundary layer interaction at a hypersonic Mach Number AIAA Paper 91-1761 (Reston, VA: American Institute of Aeronautics and Astronautics)

Law C 1975 3D shock wave-turbulent boundary layer interactions at Mach 6 *Technical Report ARL TR-75-0191* Aerospace Research Laboratories, US Air Force Systems Command, Wright-Patterson AFB, OH

Lee Y, Settles G and Horstman C 1992 Heat transfer measurements and cfd comparison of swept shock wave/boundary layer interactions AIAA Paper 92-3665 (Reston, VA: American Institute of Aeronautics and Astronautics)

Lee Y, Settles G and Horstman C 1994 Heat transfer measurements and computations of swept shock wave/boundary layer interactions *AIAA J.* **32** 726–34

Leidy A, Neel I, Tichenor N and Bowersox R 2020 High-speed schlieren imaging of cylinder-induced hypersonic-shock-wave-boundarylayer interactions *AIAA J.* **58** 3090–9

Lighthill M 1963 *Laminar Boundary Layers* (New York: Oxford) pp 46–113

Longo J 2003 Aerothermodynamics–a critical review at DLR *Aerosp. Sci. Technol.* **7** 429–38

Lu F, Settles G and Horstman C 1990 Mach number effects on conical surface features of swept shock wave boundary layer interactions *AIAA J.* **28** 91–7

Marvin J, Brown J and Gnoffo P 2013 Experimental database with baseline CFD solutions: 2-D and axisymmetric hypersonic shock-wave/turbulent-boundary-layer interactions *Technical Memorandum 2013-216604* (Washington, DC: National Aeronautics and Space Administration)

Mikulla V and Horstman C 1976a Turbulence measurements in hypersonic shock-wave boundary-layer interactions *AIAA J.* **14** 568–75

Mikulla V and Horstman C 1976b Turbulence measurements in hypersonic shock-wave boundary layer interactions AIAA Paper 76-162 (Reston, VA: American Institute of Aeronautics and Astronautics)

Murphree Z, Jagodzinski J, Hood E, Clemens N and Dolling D 2006 Experimental studies of transitional boundary layer shock wave interactions AIAA Paper 2006-0326 (Reston, VA: American Institute of Aeronautics and Astronautics)

Murphree Z, Yüceil K, Clemens N and Dolling D 2007 Experimental studies of transitional boundary layer shock wave interactions AIAA Paper 2007-1139 (Reston, VA: American Institute of Aeronautics and Astronautics)

Murray N, Hillier R and Williams S 2013 Experimental investigation of axisymmetric hypersonic shock-wave/turbulent-boundary-layer interactions *J. Fluid Mech.* **714** 152–89

Neumann R and Burke G 1969 The influence of shock wave boundary layer effects on the design of hypersonic aircraft *Technical Report 68-152* Air Force Flight Dynamics Laboratory, Wright-Patterson AFB, OH

Neumann R and Hayes J 1977a Prediction techniques for the characteristics of the three dimensional shock wave/turbulent boundary layer interaction AIAA Paper 77-46 (Reston, VA: American Institute of Aeronautics and Astronautics)

Neumann R and Hayes J 1977b Prediction techniques for three-dimensional shock-wave/ turbulent boundary layer interactions *AIAA J.* **15** 1469–73

Neumann R and Token K 1974 Prediction of surface phenomena induced by three dimensional interactions on planar turbulent boundary layers *Paper 74-058, Int. Astronautical Federation XXVth Congress*

Oskam B, Bogdonoff S and Vas I 1965 Oblique shock wave/turbulent boundary layer interactions in three dimensions at Mach 3 *Technical Report 76-48* Air Force Flight Dynamics Laboratory, Wright-Patterson AFB, OH

Priebe S and Martín M P 2011 Direct numerical simulation of a hypersonic turbulent boundary layer on a large domain AIAA Paper 2011-3432 (Reston, VA: American Institute of Aeronautics and Astronautics)

Prince S 2021 private communication

Prince S, Vannahme M and Stollery J 1999 Experiments on the hypersonic turbulent shock wave boundary layer interaction and the effects of surface roughness AIAA Paper 99-0147 (Reston, VA: American Institute of Aeronautics and Astronautics)

Prince S, Vannahme M and Stollery J 2005 Experiments on the hypersonic turbulent shock wave boundary layer interaction and the effects of surface roughness *Aeronaut. J.* **109** 177–84

Ramesh M and Tannehill J 2003 Correlations to predict transition in two-dimensional supersonic flows AIAA Paper 2003-3588 (Reston, VA: American Institute of Aeronautics and Astronautics)

Richards B and Appels C 1973 Turbulent heat transfer measurements in a Mach 15 flow *Preprint 1973-2* von Kármán Institute for Fluid Dynamics

Rodi P and Dolling D 1992 An experimental/computational study of sharp fin induced shock wave/turbulent boundary layer interactions at Mach 5: experimental results AIAA Paper 1992-0749 (Reston, VA: American Institute of Aeronautics and Astronautics)

Rodi P and Dolling D 1995 Behavior of pressure and heat transfer in sharp fin-induced turbulent interactions *AIAA J.* **33** 2013–9

Rodi P, Dolling D and Knight D 1991 An experimental/computational study of heat transfer in sharp fin induced turbulent interactions at Mach 5 AIAA Paper 1991-1764 (Reston, VA: American Institute of Aeronautics and Astronautics)

Roy C and Blottner F 2006 Review and assessment of turbulence models for hypersonic fows *Prog. Aerosp. Sci.* **42** 469–530

Schreyer A M, Sahoo D and Smits A 2011 Experimental investigation of a hypersonic shock turbulent boundary layer interaction AIAA Paper 2011-481 (Reston, VA: American Institute of Aeronautics and Astronautics)

Schreyer A-M, Sahoo D, Williams O and Smits A 2018 Experimental investigation of two hypersonic shock/turbulent boundary-layer interactions *AIAA J.* **56** 4830–44

Schülein E 2006 Skin friction and heat flux measurements in shock/boundary layer interaction flows *AIAA J.* **44** 1732–41

Schülein E, Krogmann P and Stanewsky E 1996 Documentation of two dimensional impinging shock/turbulent boundary layer interaction flow *Report IB 223-96 A 49* Deutsches Zentrum für Luft- und Raumfahrt e.V. (DLR), Institut für Strömungsmechanik, Göttingen, Germany

Schülein E and Zheltovodov A 2001 Documentation of experimental data for hypersonic 3-D shock waves/turbulent boundary layer interaction flows *Report IB 223-99 A 26* Deutsches Zentrum für Luft- und Raumfahrt e.V. (DLR), Institut für Strömungsmechanik, Göttingen, Germany

Seddon J and Goldsmith E 1985 *Intake Aerodynamics* (Reston, VA: American Institute of Aeronautics and Astronautics)

Settles G and Dodson L 1991 Hypersonic shock/boundary layer interaction database *Contractor Report 177577* National Aeronautics and Space Administration. PennState Gas Dynamics Laboratory Report PSU-ME-90/91-003, December 1990

Settles G and Dodson L 1994 Supersonic and hypersonic shock/boundary-layer interactions database *AIAA J.* **32** 1377–83

Settles G and Dolling D 1986 Swept shock wave boundary layer interactions *Tactical Missile Aerodynamics* ed M Hemsch and J Nielsen (Reston, VA: American Institute of Aeronautics and Astronautics) pp 297–379

Settles G and Teng H 1983 Flow visualization methods for separated three dimensional shock wave turbulent boundary layer interaction *AIAA J.* **21** 390–7

Shapiro A 1954 *The dynamics and thermodynamics of compressible fluid flow* (New York: Ronald Press)

Simeonides G 1996 Laminar-turbulent transition promotion in regions of shock wave boundary layer interaction Reprint 1996-26, *ESA/MSTP Code Validation Workshop, ESTEC, Noordwijk, The Netherlands*

Simeonides G and Haase W 1995 Experimental and computational investigations of hypersonic flow about compression ramps *J. Fluid Mech.* **283** 17–42

Simeonides G and Wendt J 1990 Compression corner shock wave boundary layer interactions *ICAS 90-3.10.3, Proc. 17th Conf. of the Int. Council of the Aeronautical Sciences, Stockholm, Sweden* pp 1914–26

Token K 1974 Heat transfer due to shock wave turbulent boundary layer interactions on high speed weapons systems *Technical Report 74-77* Air Force Flight Dynamics Laboratory, Wright-Patterson AFB, OH

Vandomme L, Chanetz B, Benay R and Perraud J 2003 Transitional shock wave boundary layer interaction in hypersonic flow at Mach 5 AIAA Paper 2003-6966 (Reston, VA: American Institute of Aeronautics and Astronautics)

Van Driest E 1956 Problem of aerodynamic heating *Aeronaut. Eng. Rev.* **15** 26–41

Vermeulen J and Simeonides G 1992 Parametric studies of shock wave/boundary layer interactions over 2D compression corners at Mach 6 *Technical Note 181* von Kármán Institute for Fluid Dynamics

Wagner A, Schramm J, Hickey J P and Hannemann K 2016 Hypersonic shock wave boundary layer interaction studies on a flat plate at elevated surface temperature *22nd Int. Shock Interaction Symp.* (Glasgow: International Shock Wave Institute)

White M and Ault D 1995 Hypersonic shock wave turbulent boundary layer interactions in the vicinity of an expansion corner AIAA Paper 95-6125 (Reston, VA: American Institute of Aeronautics and Astronautics)

White M and Ault D 1996 Expansion corner effects on hypersonic on shock wave turbulent boundary layer interactions *J. Propuls. Power* **12** 1169–73

Experiments on shock induced laminar-turbulent transition on a flat plate at Mach 6, *Technical Report, European Conf. for Aeronautics and Space Sciences (EUCASS)*

Willems S, Gülhan A and Steelant J 2015 Experiments on the effect of laminar-turbulent transition on the SWBLI in H2K at Mach 6 *Exp. Fluids* **56**

Williams O, Sahoo D, Baumgartner M and Smits A 2018 Experiments on the structure and scaling of hypersonic turbulent boundary layers *J. Fluid Mech.* **834** 237–70

Zhao X J, Hu H and Zhao L 2021 Effect of an adverse pressure gradient on hypersonic wall pressure fluctuations *Phys. Fluids* **33** 106107

Zhao X J and Lei J 2020 Improved technique for evaluation of wall pressure fluctuations from turbulent boundary layer *AIAA J.* **58** 3320–31

Zhao X J and Zhao I 2021 2021 Wall pressure fluctuations beneath hypersonic boundary layer over flat plate *AIAA J.* **59** 1682–91

Zheltovodov A 1979a Analysis of properties of two dimensional separated flows at supersonic speeds—study of near wall viscous gas flows (Анализ своих двухмерных отрывных течений на сверхзвуковых скоростях) Report, Institute of Theoretical and Applied Mechanics, Siberian Academy of Sciences

Zheltovodov A 1979b Physical properties of two- and three-dimensional separated flows at supersonic velocities *Mekh. Zhidk. Gaza* **3** 42–50 in Russian

Zheltovodov A 1982 Regimes and properties of three dimensional separation flows initiated by skewed compression shocks *J. Appl. Mech. Tech. Phys. (Zh. Prikl. Mekh. Tekh. Fiz.)* **3** 116–23 in Russian

Zheltovodov A 1991 Peculiarities of development and modeling possibilities of supersonic turbulent separated flows and jets *Symp. Separated Flows and Jets International Union of Theoretical and Applied Mechanics* ed V Kozlov and A Dovgal (Berlin: Springer) pp 225–36

Zheltovodov A 1996 Shock waves/turbulent boundary layer interactions—fundamental studies and applications *AIAA Paper 96-1977* (Reston, VA: American Institute of Aeronautics and Astronautics)

Zheltovodov A 2004 Advances and problems in modelling of shock wave turbulent boundary layer interactions *Int. Conf. on the Methods of Aerophysical Research* ed V Fomin and A Kharitonov (Novosibirk: Siberian Division) pp 1–12

Zheltovodov A 2006 Some advances in research of shock wave turbulent boundary layer interactions AIAA Paper 2006-0496 (Reston, VA: American Institute of Aeronautics and Astronautics)

Zheltovodov A, Bedarev I, Borisov A, Volkov V, Mazhul I, Maksimov A, Fedorova N and Shpak S 1987 Development and Verification of Calculation Methods in Relation to Supersonic Aerodynamic Problems (Развитие и верификация методов раскета применительно к задачам сверхзвуковой а¸родинамики) Preprint 7-97, Institute of Theoretical and Applied Mechanics, Siberian Academy of Sciences, Novososibirsk, Soviet Union (in Russian)

Zheltovodov A, Dvorzak R and Shafarzhik P 1990 Особенности взаимодействия моделей уплотнения с турбулентным пограничным слоем в условиях транс- и сверхзвуковых скоростей, Известия Сибирского Отделения Академии Наук СССР **6** 31–42 (in Russian)

Zheltovodov A and Knight D 2011 Ideal gas shock wave turbulent boundary layer interactions in supersonic flows and their modeling: three dimensional interactions *Shock Wave Boundary Layer Interactions* ed H Babinsky and J Harvey (Cambridge: Cambridge University Press) pp 202–58

Zheltovodov A and Maksimov A 1998 Symmetric and asymmetric crossing shock waves turbulent boundary layer interactions *Final Report EOARD Contract F61708-97-WE0136* Institute of Theoretical and Applied Mechanics, Novosibirsk, Russian Federation. Accession No. ADA353759

Zheltovodov A and Maksimov A 1999 Hypersonic crossing shock waves/turbulent boundary layer interactions *Final Report EOARD Contract F61775-98-WE091* Institute of Theoretical and Applied Mechanics, Novosibirsk, Russian Federation. Accession No. ADA363672

Zheltovodov A, Maksimov A, Gaitonde D, Visbal M and Shang J 2000 Experimental and numerical study of symmetric interaction of crossing shocks and expansion waves with a turbulent boundary layer *Thermophys. Aeromech.* **7** 155–71

Zheltovodov A, Maksimov A and Schülein E 1987 Development of Turbulent Separated Flows in the Vicinity of Swept Shock Waves *The Interactions of Complex 3-D Flows Institute of Theoretical and Applied Mechanics* ed A Kharitonov (Novosibirsk: Siberian Academy of Sciences) pp 67–91

Zheltovodov A, Maksimov A and Shevchenko A 1998 Topology of three dimensional separation under the conditions of symmetric interaction of crossing shocks and expansion waves with turbulent boundary layer *Thermophys. Aeromech.* **5** 293–312

Zheltovodov A, Mekler P and Shülein E 1987 Peculiarities of development of separated flows in compression corners after the expansion fans (Особенности развития отрывных течений в углах сяты за волнами разрежения) Preprint 10-87, Institute of Theoretical and Applied Mechanics, Siberian Academy of Sciences, Novososibirk, Soviet Union (in Russian)

Zheltovodov A and Pavlov A 1979 Studies of the flow of the supersonic separation zone in front of a step (Исследование техники в сверхзвуковой зоне отрыва перед ступенькой) Preprint 1, Institute of Theoretical and Applied Mechanics, Siberian Academy of Sciences, Novosibirsk, Soviet Union (in Russian)

Zheltovodov A and Schülein E 1986 Three dimensional swept shock waves turbulent boundary layer interaction in the angle configuration (Пространственное бзаимодействие скол~зскакков уплотнений с турбулентным пограничным слоем в угловых конфигурациях) Preprint 34-86, Institute of Theoretical and Applied Mechanics, Siberian Academy of Sciences, Novosibirsk, Soviet Union (in Russian)

Zheltovodov A and Schülein E 1988 Peculiarities of development of turbulent separation in disturbed boundary layers (Особенности развития турбулентного отрыва в возбужденных пограничных замедлениях) *Modeling in Mechanics (Modelirovanie v Mekhanike)* **2** 53–8 (in Russian)

Zheltovodov A, Schülein E and Yakovlev V 1983 Development of a turbulent boundary layer under the conditions of mixed interaction with shock waves and expansion waves (Развитие турбулентного пограничного слоя в условиях смэксанного взаимодействия со скакками уплотнения и волнами разреения) 28–83 Preprint 28-83, Institute of Theoretical and Applied Mechanics, Siberian Academy of Sciences, Novosibirsk, Soviet Union

Zheltovodov A, Trofimov V, Schülein E and Yakovlev V 1990 An experimental documentation of supersonic turbulent flows in the vicinity of sloping forward and back facing steps *Report 2013* Institute of Theoretical and Applied Mechanics, Soviet Academy of Sciences, Siberian Division, Novosibirsk, Soviet Union

Zheltovodov A and Yakovlev V 1986 Stages of Development structure and characteristics of turbulence in compressible separated flows in the vicinity of two dimensional obstacles (Ѐтапы развития, структура и характеристики турбулентности сымаемых отрывных технологий в окрестности двумерных префектур) Preprint 27-86, Institute of Theoretical and Applied Mechanics, Siberian Academy of Sciences, Novosibirsk, Soviet Union (in Russian)

Zheltovodov A, Zaulichniy E and Trofimov V 1990 Development of models for the calculation of heat transfer for supersonic turbulent separated flow conditions (Развитие модели для расчета теплообмена в условиях сверхзвуковых турбулентных отрывных технологий) *J. Appl. Mech. Tech. Phys. (Zh. Prikl. Mekh. Tekh. Fiz.)* **4** 96–104 (in Russian)

Zheltovodov A, Zaulichniy E, Trofimov V, Schülein E and Yakovlev V 1989 Modeling of heat transfer processes for turbulent supersonic separation flow conditions (Моделирование процессов теплообмена в сверхзвуковых турбулентных обрывных техник) *Model. Mech. (Model. Mekh.)* **3** 74–80 (in Russian)

Zheltovodov A, Zaulichniy E, Trofimov V and Yakolev V 1987 Investigation of heat transfer and turbulence in compressible flows (Исследование теплообмена и турбулентности в сжимаемых отрывных течениях) Preprint 22-87, Institute of Theoretical and Applied Mechanics, Siberian Academy of Sciences, Novososibirsk, Soviet Union (in Russian)

**IOP** Publishing

Hypersonic Shock Wave Turbulent Boundary Layers
Direct Numerical Simulation, Large Eddy Simulation and Experiment
**Doyle Knight and Nadia Kianvashrad**

# Chapter 6

# Direct Numerical Simulation and Large Eddy Simulation—boundary layers

## Abstract

Direct Numerical Simulation and Large Eddy Simulation of hypersonic turbulent boundary layers in zero and modest pressure gradients appeared in the early twenty first century. The present chapter examines a selected set of DNS and LES results. Similar to previous chapters, the simulations were selected to illustrate both history of the field and to provide an understanding of the state of the art.

## 6.1 Maeder *et al* (2001)

Maeder *et al* (2001) performed a Direct Numerical Simulation of an adiabatic zero pressure gradient turbulent boundary layer in air at Mach 3, 4.5 and 6. The governing equations are the compressible Navier–Stokes equations for a perfect gas. The flow conditions are listed in table 6.1 where $Re_\theta$ is the Reynolds number based upon the compressible momentum thickness. The simulation is based on the extended temporal direct Navier–Stokes approach (Maeder *et al* 2001).

Figures 6.1 and 6.2 show the computed Morkovin scaled Reynolds streamwise normal stress $\bar{\rho}\ \widetilde{u''u''}/\rho_w u_\tau^2$ and Morkovin scaled Reynolds shear stress $-\bar{\rho}\ \widetilde{u''v''}/\rho_w u_\tau^2$ for Cases 1 to 3 where $u''$ and $v''$ are the Favre fluctuating streamwise and wall normal velocities, respectively. Also shown are the incompressible DNS simulations of Spalart (1988) for an incompressible zero pressure gradient turbulent boundary at $Re_\theta = 670$ and 1410, and the DNS simulations of Moser *et al* (1999) for an incompressible turbulent channel flow at $Re_\tau = 395$ and 590. The simulations of Maeder *et al* (2001) show general agreement with the incompressible DNS results using the Morkovin scaling for $-\bar{\rho}\ \widetilde{u''u''}/\rho_w u_\tau^2$, with particular agreement in the location and magnitude of the peak value. However, no clear trend of agreement is evident between the Morkovin scaled Reynolds shear stress $\bar{\rho}\ \widetilde{u''v''}/\rho_w u_\tau^2$ for Cases 2

doi:10.1088/978-0-7503-5002-0ch6  

**Table 6.1.** Freestream conditions.

| | Value | | |
|---|---|---|---|
| Quantity | Case 1 | Case 2 | Case 3 |
| Gas | Air | Air | Air |
| $M_\infty$ | 3.0 | 4.5 | 6.0 |
| $T_w/T_{aw}$ | 1.0 | 1.0 | 1.0 |
| $Re_\theta \times 10^{-3}$ | 3.028 | 3.305 | 2.945 |

Note: $T_{aw}$ is the laminar value of the adiabatic wall temperature.

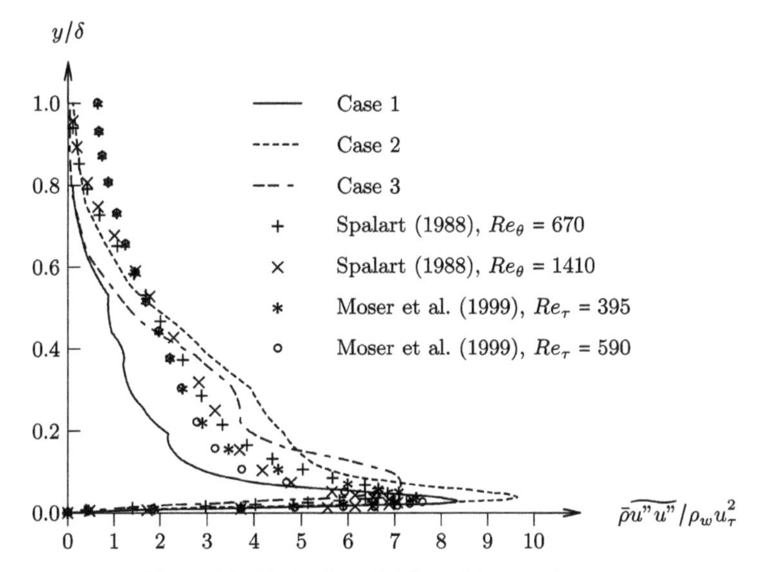


**Figure 6.1.** Morkovin scaled Reynolds normal stress.

(Mach 4.5) and 3 (Mach 6), and in particular the profile for Case 3 (Mach 6) shows twin peaks which are not evident in the incompressible DNS results.

Figure 6.3 displays the computed correlation coefficient $R_{u''T''}$ for Case 1 to 3. It is evident that the Strong Reynolds Analogy No. 3 (3.334) does not hold for any of the cases. Maeder *et al* (2001) note that their simulations show that the total temperature fluctuations $\overline{T_t''^2}$ are comparable to the static temperature fluctuations $\overline{T''^2}$ throughtout the boundary layer, thereby invalidating the fundamental assumption of Strong Reynolds Analogy No. 3.

Figure 6.4 shows the revision of the Strong Reynolds Analogy proposed by Huang *et al* (1995) (3.335)

$$\frac{\left(\sqrt{\overline{T''^2}}/\overline{T}\right)}{(\gamma-1)\overline{M}^2\left(\sqrt{\overline{u''^2}}/\overline{u}\right)} = \frac{1}{Pr_t}\frac{1}{|\, d\overline{T}_t/d\overline{T} - 1\,|} \tag{6.1}$$

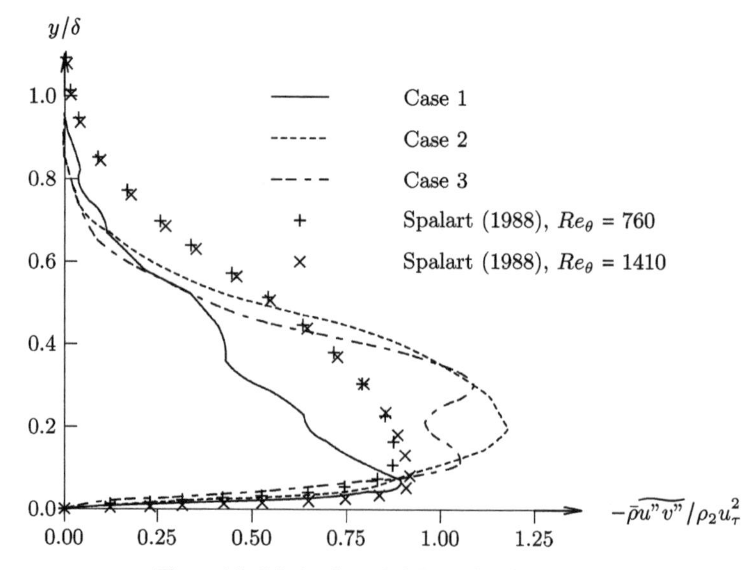


**Figure 6.2.** Morkovin scaled Reynolds shear stress.

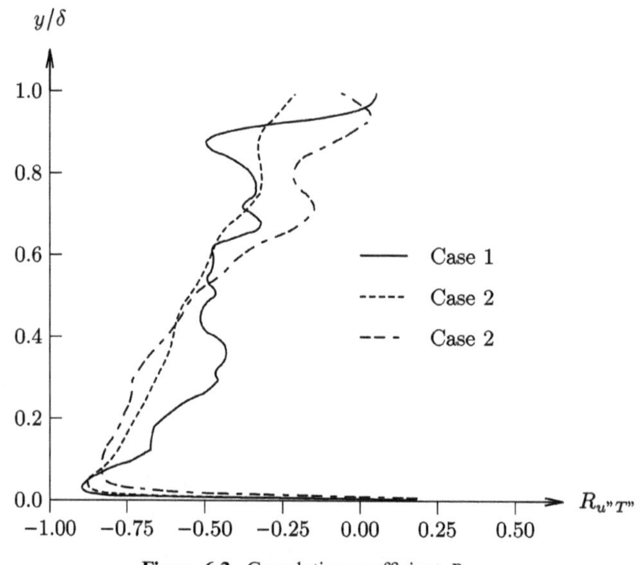


**Figure 6.3.** Correlation coefficient $R_{u''T''}$.

where $Pr_t$ is the local Prandtl number (2.55). The abscissa is the ratio of the left hand side to the right hand side. Generally close agreement with (3.335) is evident.

## 6.2 Li *et al* (2006)

Li *et al* (2006) performed a Direct Numerical Simulation of a near adiabatic zero pressure gradient turbulent boundary layer in air at Mach 6. The governing

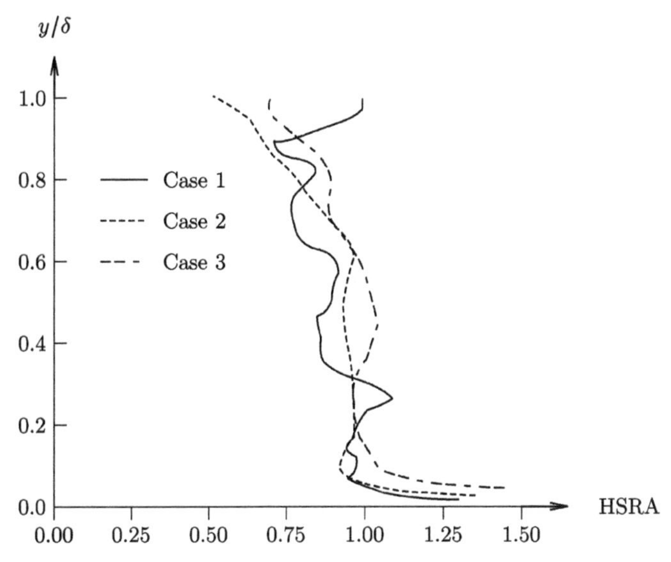


**Figure 6.4.** Huang Strong Reynolds Analogy.

**Table 6.2.** Flow conditions.

| Quantity | Value |
|---|---|
| $M_\infty$ | 6.0 |
| $T_w/T_\infty$ | 6.98 |
| $T_w/T_{aw}$ | 0.94 |
| $Re_\theta \times 10^{-5}$ | 1.095 |
| $Re_\tau$ | 265.0 |

equations are the compressible Navier–Stokes equations for a perfect gas. The flow conditions are shown in table 6.2. The simulation used a two dimensional laminar boundary layer profile as the inflow condition and triggered transition using a blowing/suction disturbance.

Figure 6.5 displays the Van Driest transformed velocity $u_{\text{VD}}/u_\tau$ versus $y^+$. Close agreement with the Law of the Wall is evident. As a point of comparison, the untransformed velocity $u/u_\tau$ is also shown and displays significant disagreement with the Law of the Wall due to the large gradient of density across the boundary layer. The streamwise, wall normal and spanwise normalized turbulence intensity profiles $\sqrt{\overline{u'^2}}/u$, $\sqrt{\overline{v'^2}}/u$ and $\sqrt{\overline{w'^2}}/u$ are shown in figure 6.6 together with experimental data for incompressible flow[1] in the near wall region. The computed intensities agree

[1] No reference is provided for the experimental data.

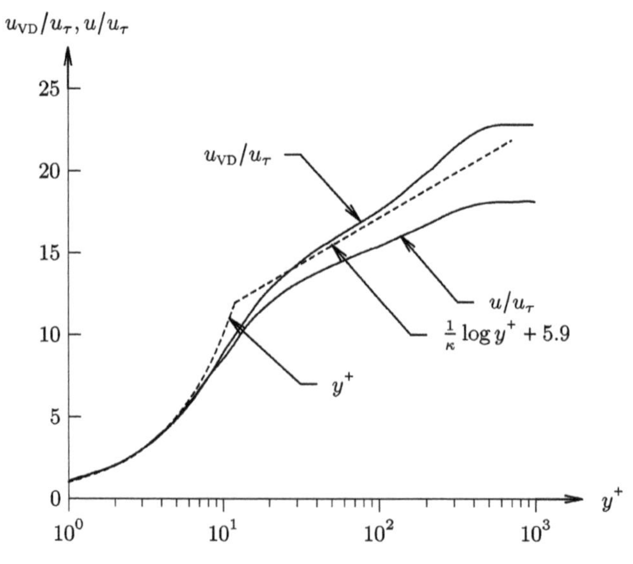


**Figure 6.5.** Van Driest transformed velocity.

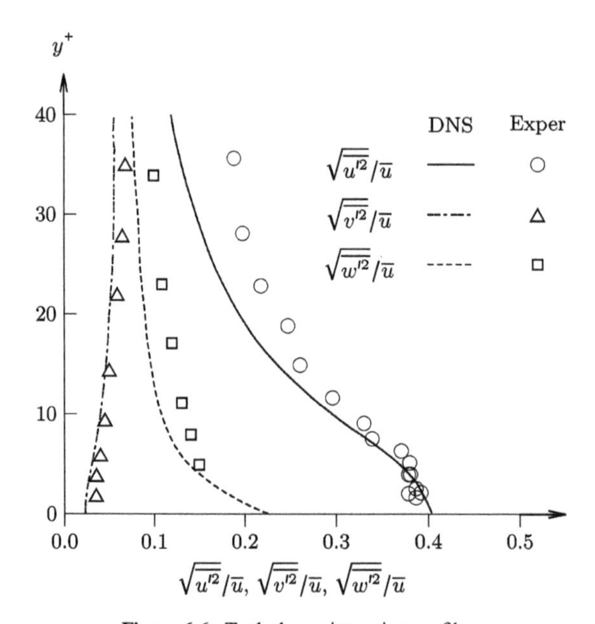


**Figure 6.6.** Turbulence intensity profiles.

closely with experiment for $y^+ < 10$, but diverge for larger $y^+$ due to the density gradient.

## 6.3 Pino Martín (2007)

Martín (2007) performed a series of Direct Numerical Simulations of supersonic and hypersonic zero pressure gradient turbulent boundary layers in air at Mach numbers from 2.98 to 5.95. The governing equations are the compressible Navier–Stokes equations for a perfect gas. The freestream conditions are listed in table 6.3. Simulations were performed using two methods, namely, temporally developing (TDNS) and spatially developing (SDNS).

Figure 6.7 displays the Van Driest transformed velocity for Case 3 ($M_\infty = 4.97$) using SDNS. Close agreement is seen with the Law of the Wall. Similar close agreement was obtained using TDNS. Figure 6.8 shows the root mean square fluctuating Mach number for Case 3 using SDNS. The peak value $\sqrt{\overline{M'^2}} = 0.4$ occurs at $z^+ \approx 10$ corresponding to the edge of the viscous sublayer. Results using TDNS are essentially identical to SDNS. Martín (2007) also compares the computed Morkovin scaled streamwise root mean square velocity fluctuations and Reynolds shear stress with experimental data at $M_\infty = 2.32$ and shows good agreement.

**Table 6.3.** Flow conditions.

| | Value | | | |
|---|---|---|---|---|
| Quantity | Case 1 | Case 2 | Case 3 | Case 4 |
| Run No. | M3 | M4 | M5 | M6 |
| $M_\infty$ | 2.98 | 3.98 | 4.97 | 5.98 |
| $T_w/T_\infty$ | 2.58 | 3.83 | 5.40 | 7.32 |
| $T_w/T_{aw}$ | 0.96 | 0.95 | 0.95 | 0.95 |
| $Re_\theta \times 10^{-3}$ | 2.39 | 3.94 | 6.23 | 8.43 |

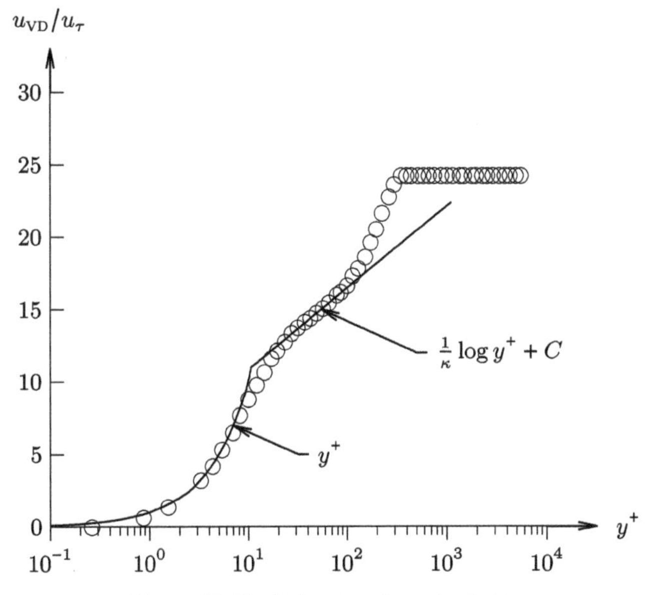


**Figure 6.7.** Van Driest transformed velocity.

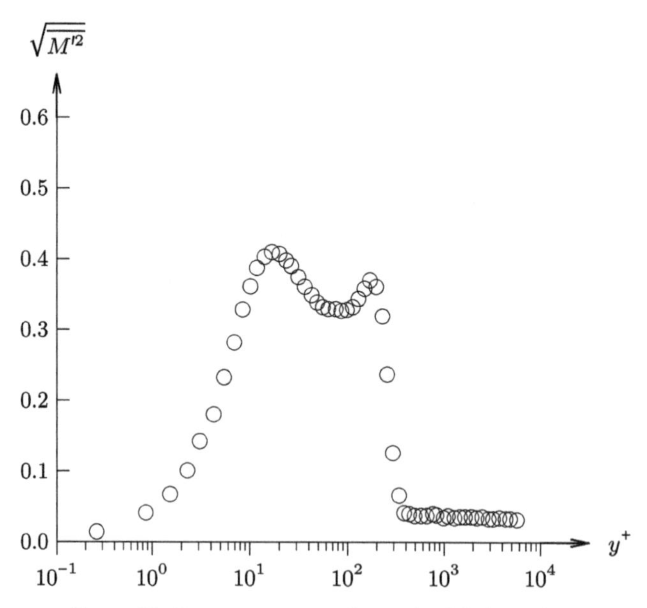


**Figure 6.8.** Root mean square fluctuating Mach number.

**Table 6.4.** Flow conditions.

| Quantity | Value | | | | |
|---|---|---|---|---|---|
| | Case 1 | Case 2 | Case 3 | Case 4 | Case 5 |
| Run No. | M5T1 | M5T2 | M5T3 | M5T4 | M5T5 |
| $M_\infty$ | 5.0 | 5.0 | 5.0 | 5.0 | 5.0 |
| $T_w/T_\infty$ | 1.0 | 1.90 | 2.89 | 3.74 | 5.4 |
| $T_w/T_{aw}$ | 0.18 | 0.35 | 0.53 | 0.68 | 1.0 |
| $Re_\theta \times 10^{-3}$ | 1.28 | 2.30 | 3.01 | 3.82 | 4.84 |
| $c_f \times 10^3$ | 2.46 | 2.05 | 1.81 | 1.50 | 1.28 |
| $c_h \times 10^3$ | 1.39 | 1.14 | 0.98 | 0.84 | na |

na; not applicable.

## 6.4 Duan, Beekman and Pino Martín (2010)

Duan *et al* (2010) performed a series of Direct Numerical Simulations of a hypersonic zero pressure gradient turbulent boundary layer at Mach 5 for a ratio of wall to edge temperatures $T_w/T_\infty$ from 1.0 to 5.4. The governing equations are the compressible Navier–Stokes equations for a perfect gas. The fluid is air. The freestream conditions are listed in table 6.4.

Figure 6.9 displays the Van Driest transformed velocity for Cases 1 to 5. Close agreement is evident for all cases with the Law of the Wall for a range in $z^+$, with the lower limit in $z^+$ of the region of agreement varying with the value of $T_w/T_{aw}$.

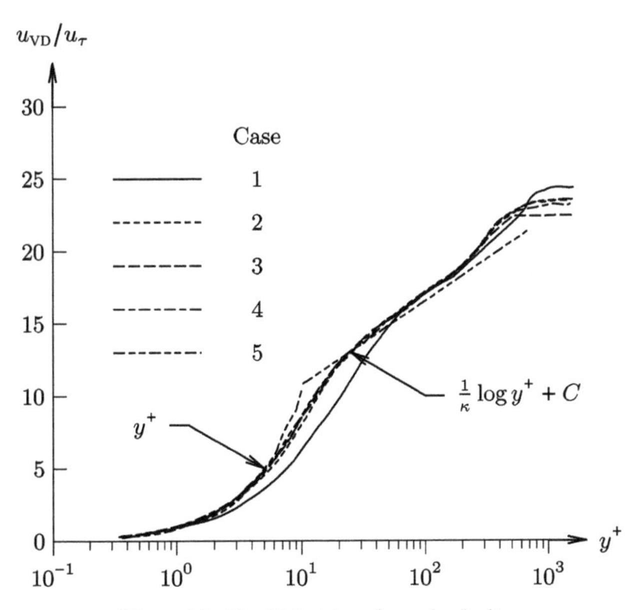


**Figure 6.9.** Van Driest transformed velocity.

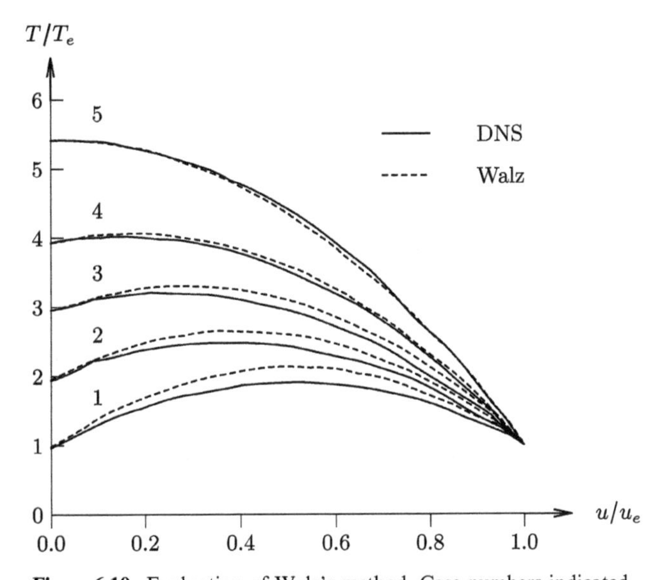


**Figure 6.10.** Evaluation of Walz's method. Case numbers indicated.

Figure 6.10 compares the computed Favre-averaged temperature with Walz's relation (3.171). The agreement is close for all cases.

Figure 6.11 displays the Morkovin scaled streamwise turbulence intensity for all cases, together with the incompressible DNS data of Spalart (1988) at $Re_\theta = 670$

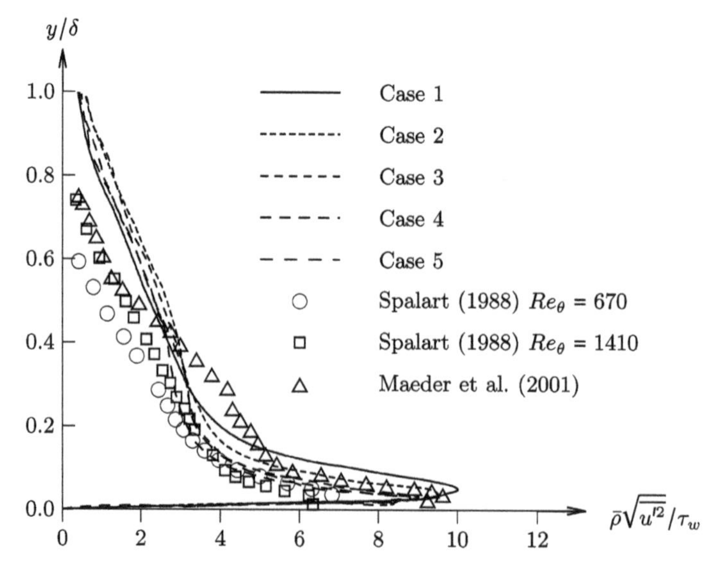


**Figure 6.11.** Morkovin scaled streamwise turbulence intensity.

and 1410, and the compressible DNS data of Maeder *et al* (2001) at $M_\infty = 4.5$. Close agreement is evident throughout the boundary layer. In particular, the location of the peak values agree, although the peak value for Spalart (1988) is below the others.

According to the Strong Reynolds Analogy No. 2 (3.331),

$$\frac{\left(\sqrt{\widetilde{T''^2}}/\tilde{T}\right)}{\left(\gamma-1\right)\overline{M}^2\left(\sqrt{\widetilde{u''^2}}/\tilde{u}\right)} = 1 \tag{6.2}$$

Figure 6.12 displays the computed values using Favre-averaged variables for Cases 1 to 5. It is evident that the Strong Reynolds Analogy doesn't hold for the non-adiabatic cases, and is accurate only for $z/\delta < 0.6$ for the adiabatic case.

Figure 6.13 evaluates the Huang *et al* Strong Reynolds Analogy by taking the ratio of the left side to right side of (3.358). We define

$$\text{HSRA} = \frac{\left(\sqrt{\overline{T'^2}}/\overline{T}\right)}{(\gamma-1)\overline{M}^2\left(\sqrt{\overline{u'^2}}/\overline{u}\right)}[Pr_t \mid d\overline{T}_t/d\overline{T} - 1 \mid]^{-1} \tag{6.3}$$

According to (3.358), HSRA = 1.0. The ratio is within the range of 1.0 to 1.2 over nearly the entire boundary layer for all cases.

Figure 6.14 displays the turbulent Prandtl number (2.56) for all cases using Favre-averaged variables. The value ranges from 0.6 to 1.3 with an average value $Pr_t \approx 1$.

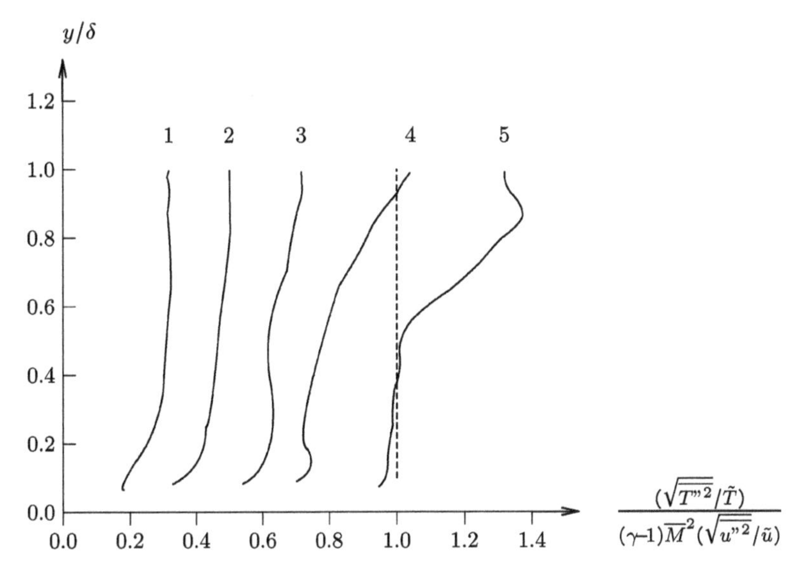


**Figure 6.12.** Assessment of Strong Reynolds Analogy No. 2. Case numbers indicated.

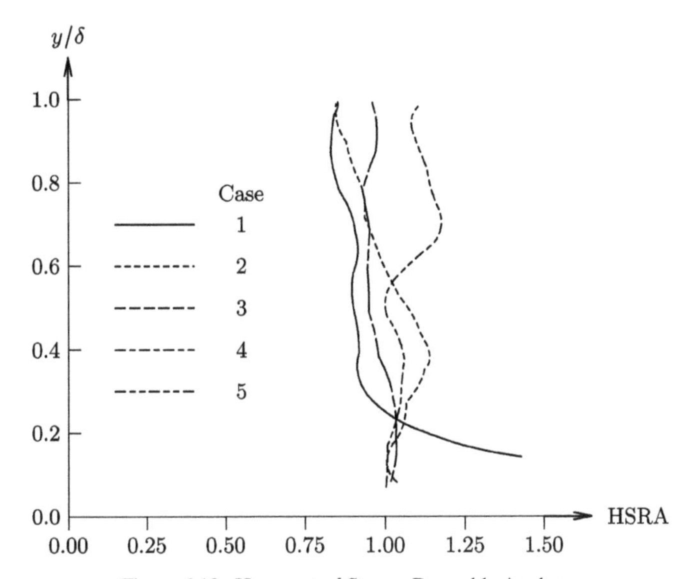


**Figure 6.13.** Huang *et al* Strong Reynolds Analogy.

## 6.5 Duan, Beekman and Pino Martín (2011)

Duan *et al* (2011) performed a series of Direct Numerical Simulations of hypersonic zero pressure gradient turbulent boundary layers in air at Mach numbers from 0.3 to 11.93, stagnation enthalpies from 0.2 MJ kg$^{-1}$ to 6.74 MJ kg$^{-1}$, Reynolds numbers $Re_\theta$ from 1515 to 11 356 at near adiabatic wall temperatures. The freestream conditions are shown in table 6.5. The governing equations are the compressible Navier–Stokes equations for a perfect gas.

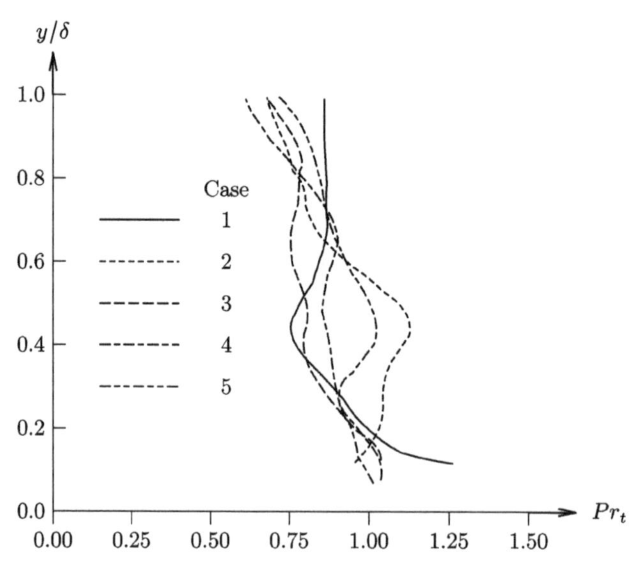


**Figure 6.14.** Turbulent Prandtl number.

**Table 6.5.** Flow conditions.

| Quantity | Value | | | | | | | |
|---|---|---|---|---|---|---|---|---|
| | Case 1 | Case 2 | Case 3 | Case 4 | Case 5 | Case 6 | Case 7 | Case 8 |
| $M_\infty$ | 0.30 | 2.97 | 3.98 | 4.90 | 5.81 | 6.89 | 7.70 | 11.93 |
| $T_w/T_\infty$ | 1.0 | 2.51 | 3.83 | 5.31 | 7.02 | 9.49 | 11.2 | 27.6 |
| $T_w/T_{aw}$ | 1.00 | 0.98 | 1.00 | 1.01 | 1.00 | 1.00 | 0.97 | 1.05 |
| $Re_\theta \times 10^{-3}$ | 1.51 | 3.03 | 4.09 | 4.93 | 5.78 | 7.21 | 7.51 | 11.3 |
| $\delta$ (mm) | 23.0 | 8.85 | 12.0 | 15.1 | 19.7 | 28.1 | 31.8 | 84.7 |
| $\theta$ (mm) | 2.76 | 0.619 | 0.658 | 0.682 | 0.730 | 0.838 | 0.861 | 1.33 |
| Run | M0 | M3 | M4 | M5 | M6 | M7 | M8 | M12 |

The Van Driest transformed velocity for the eight cases is shown in figure 6.15. The legend is presented in figure 6.17. All cases show close agreement with the Law of the Wall for $C = 5.2$. Note that all cases are essentially adiabatic wall, thus supporting the observation (section 3.2.1) that the constant $C$ is a function of the wall temperature boundary condition and is therefore a constant for adiabatic walls independent of freestream Mach number. The mean temperature velocity relation $\overline{T}$ versus $\overline{u}$ shows close agreement with Walz's relation (3.171) (figure 6.16). Density weighting effectively collapses the effect of Mach number for all Reynolds stress components. An example is the density weight Reynolds shear stress $-\sqrt{\overline{\rho}/\overline{\rho}_w}\,(\overline{u'w'}/u_*^2)$ shown in figure 6.17. The correlation $R_{u''T''}$ was found to be approximately –0.6 over most of the boundary layer. The turbulent Prandtl number was observed to be between 0.7 and 1.0 across the boundary layer.

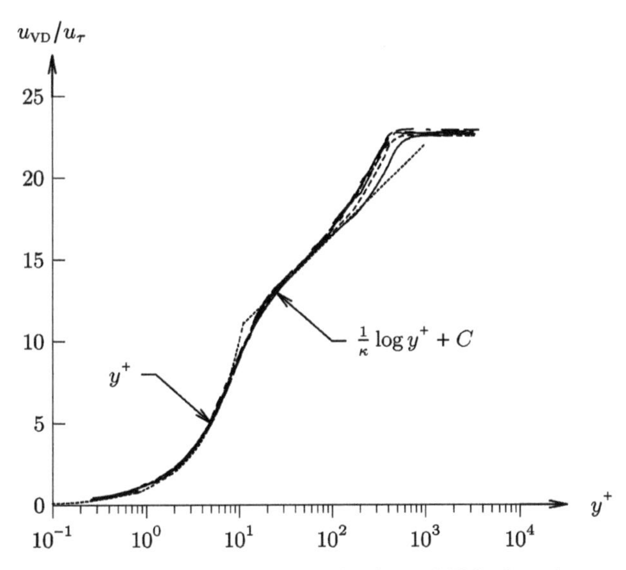


**Figure 6.15.** $u_{VD}/u_\tau$ versus $y^+$. See figure 6.17 for legend.

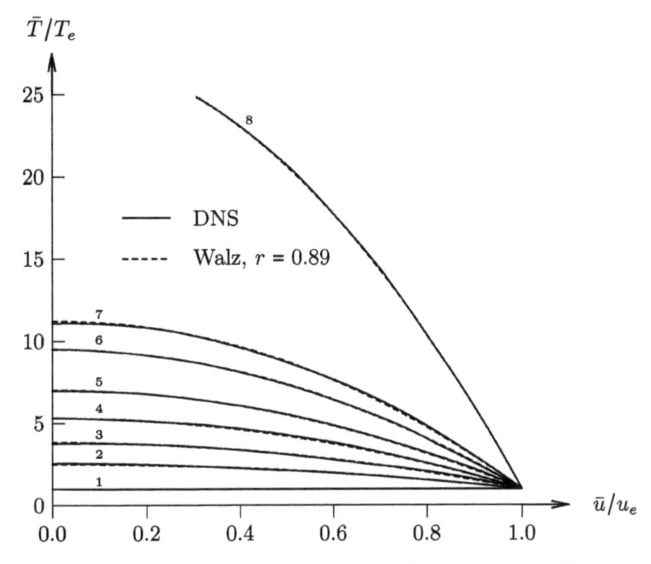


**Figure 6.16.** Temperature versus velocity. Case numbers indicated.

## 6.6 Duan and Pino Martín (2011)

Duan and Martín (2011a) performed a series of Direct Numerical Simulations of hypersonic zero pressure gradient turbulent boundary layers at Mach numbers from 3.4 to 9.4, stagnation enthalpies of 0.78 MJ kg$^{-1}$ and 20 MJ kg$^{-1}$, Reynolds number $Re_\theta$ from 958 to 3058 and wall temperature ratio $T_w/T_{aw} = 0.12$ to 0.17.

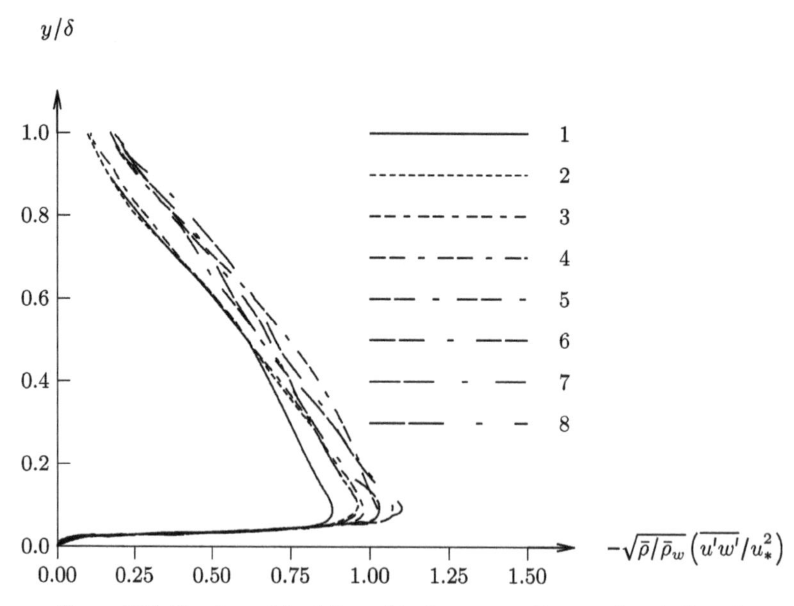


**Figure 6.17.** Density weighted Reynolds shear stress. Case numbers indicated.

**Table 6.6.** Flow conditions.

| | Value | | | | | |
|---|---|---|---|---|---|---|
| Quantity | Case 1 | Case 2 | Case 3 | Case 4 | Case 5 | Case 6 |
| $M_e$ | 3.4 | 3.4 | 9.4 | 9.3 | 3.4 | 9.0 |
| $T_w/T_e$ | 0.54 | 0.53 | 1.86 | 1.94 | 0.53 | 1.80 |
| $T_w/T_{aw}$ | 0.13 | 0.13 | 0.13 | 0.13 | 0.17 | 0.12 |
| $Re_\theta \times 10^{-3}$ | 0.97 | 1.01 | 3.03 | 3.06 | 0.96 | 2.45 |
| $\delta$ (mm) | 1.40 | 1.61 | 9.09 | 8.87 | 8.46 | 1308. |
| $\theta$ (mm) | 0.154 | 0.162 | 0.363 | 0.360 | 0.767 | 42.0 |
| Run | W35SC | W35NC | W8SC | W8NC | LHM3 | LHM10 |

W35SC Wedge35supercatalytic | W35NC Wedge35noncatalytic
W8SC Wedge8supercatalytic | W8NC Wedge8noncatalytic
LHM3 LowH_M3 | LHM10 LowH_M10

The freestream conditions are listed in table 6.6. The governing equations are the compressible Navier–Stokes equations incorporating a five species model for air ($N_2$,$O_2$,NO,N,O) and the thermochemistry model of Park (Park 1990). The thermodynamic database is the NASA Lewis curve fits (Gordon and McBride 1994). The transport properties are calculated using the Gupta-Yos mixing rule (Gupta *et al* 1990, Yos 1963) with Le = 1. No slip isothermal boundary conditions

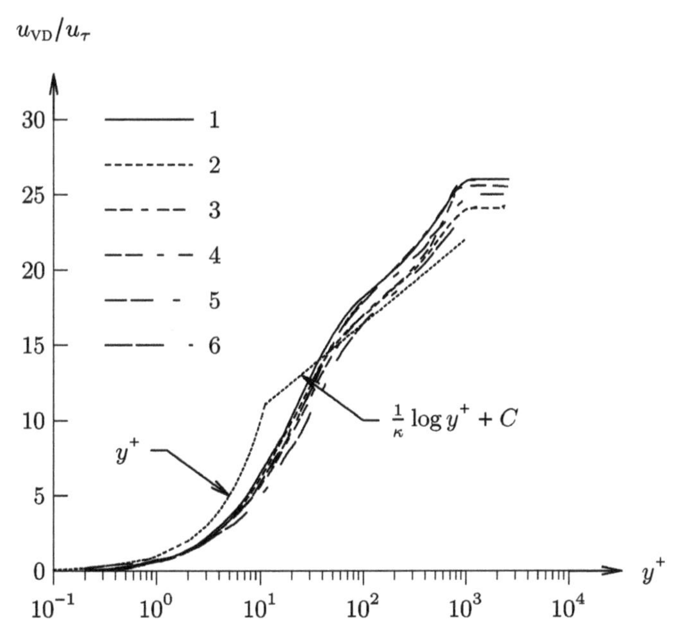


**Figure 6.18.** Van Driest velocity transformed velocity. Case numbers indicated.

are applied at the wall. Both non-catalytic and fully catalytic surface conditions are considered.

The Van Driest transformed velocity shows close agreement with the Law of the Wall (figure 6.18), albeit the constant $C$ is above $C = 5.2$ as may be expected due to the isothermal boundary condition. The mean temperature velocity relation $\overline{T}$ versus $\overline{u}$ differs significantly from Walz's relation (3.171) as indicated in figures 6.19 and 6.20. Duan and Martín proposed a modified enthalpy-velocity relation

$$\tilde{h}=h_{\rm w} + (h_{\rm aw} - h_{\rm w})f\left(\frac{\tilde{u}}{u_e}\right) - \frac{r}{2}u_e^2\left(\frac{\tilde{u}}{u_e}\right)^2 \tag{6.4}$$

where

$$f\left(\frac{\tilde{u}}{u_e}\right) = 0.174\,1\left(\frac{\tilde{u}}{u_e}\right)^2 + 0.825\,9\left(\frac{\tilde{u}}{u_e}\right) \tag{6.5}$$

figure 6.21 displays the dimensionless recovery enthalpy $h_{\rm r}^*$ versus $\tilde{u}/u_e$ where $h_{\rm r}^* = (\tilde{h}-h_{\rm w})/(h_{\rm aw} - h_{\rm w})$. Close agreement with the DNS results is observed, and similarly for Duan *et al* (2010) (not shown). Dunn and Martín also demonstrated that the rms velocity fluctuations are similar to their incompressible counterparts when scaled using the mean density scaling suggested by Morkovin (Morkovin 1962). The original Strong Reynolds Analogy (Morkovin 1962) was not confirmed by the DNS data. The correlation $R_{u''T''}$ is found to be approximately −0.7 over the

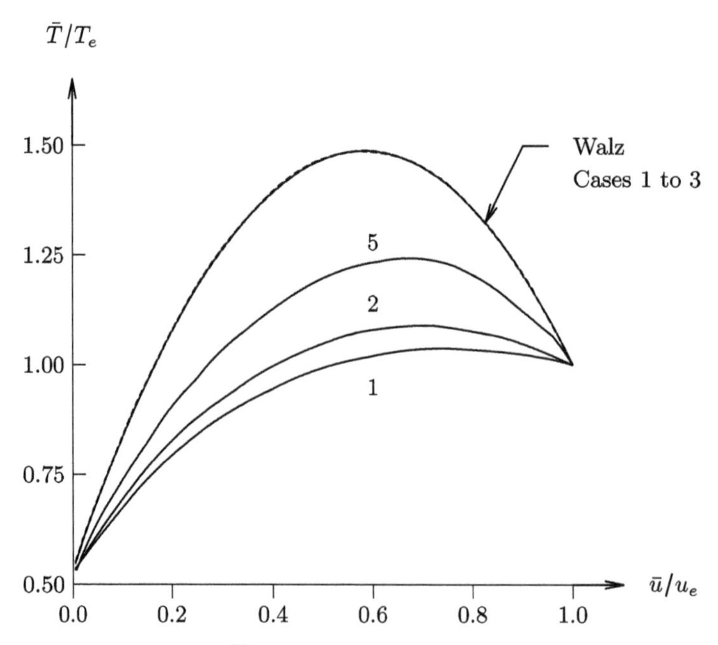


**Figure 6.19.** $\overline{T}/T_e$ versus $\bar{u}/u_e$. Case numbers indicated.

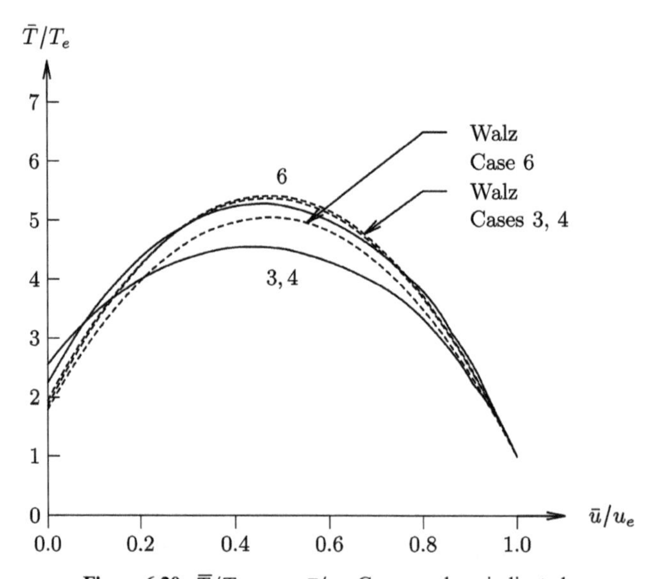


**Figure 6.20.** $\overline{T}/T_e$ versus $\bar{u}/u_e$. Case numbers indicated.

outer portion of the boundary layer. The turbulent Prandtl number was observed to be between 0.7 and 1.0 in the outer 80% of the boundary layer.

Direct Numerical Simulations of the effects of radiation on hypersonic turbulent boundary layers were performed by Duan *et al* (2011) and Duan *et al* (2012).

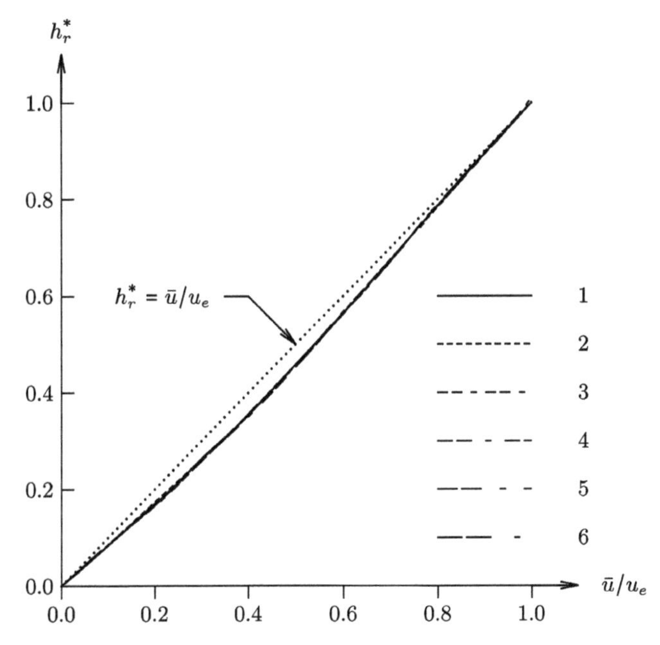


**Figure 6.21.** Dimensionless recovery enthalpy $h_r^*$ versus $\tilde{u}$. Case numbers indicated.

## 6.7 Duan and Pino Martín (2011)

Duan and Martín (2011b) developed an approach for estimating the interaction of turbulence and thermochemical reactions in hypersonic turbulent boundary layers for application to RANS methods. Due to the nonlinear dependence of the species production rate $\dot{\omega}_\alpha^{\text{spe}}(\rho_\alpha, T)$ on the species densities $\rho_\alpha$ and static temperature $T$, the average value of species production rate $\overline{\dot{\omega}_\alpha^{\text{spe}}}$ at a fixed location is not equal to the species production rate based upon the average species density $\overline{\rho_\alpha}$ and average static temperature temperature $\overline{T}$, i.e.,

$$\overline{\dot{\omega}_\alpha^{\text{spe}}} \neq \dot{\omega}_\alpha^{\text{spe}}\left(\overline{T}, \overline{\rho_\alpha}\right) \tag{6.6}$$

The difference between $\overline{\dot{\omega}_\alpha^{\text{spe}}}$ and $\dot{\omega}_\alpha^{\text{spe}}(\overline{T}, \overline{\rho_\alpha})$ is denoted the Turbulence Chemistry Interaction (TCI) (Duan and Martín 2011b).

Duan and Martín (2011b) proposed the following model for the reaction rate for RANS models

$$\overline{\dot{\omega}_\alpha^{\text{spe}}} = \mathcal{M}_\alpha \sum_{j=1}^{J} \left(\nu''_{\alpha,j} - \nu'_{\alpha,j}\right)\left[\overline{k_{f,j}} \prod_{l=1}^{n} \left(\frac{\overline{\rho_l}}{\mathcal{M}_l}\right)^{\nu'_{l,j}} - \overline{k_{b,j}} \prod_{l=1}^{n} \left(\frac{\overline{\rho_l}}{\mathcal{M}_l}\right)^{\nu''_{l,j}}\right] \tag{6.7}$$
$$\text{for } \alpha = 1, \ldots, n$$

where $\overline{\rho_l}$ is the average density of species $l$ computed by the RANS model, and $\overline{k_{f,j}}$ and $\overline{k_{b,j}}$ are determined by

$$\overline{k}_{f,j} = \int_0^{\infty} k_{f,j}(T)\mathcal{P}(T)\, dT \tag{6.8}$$

$$\overline{k}_{b,j} = \int_0^{\infty} k_{f,j}(T)\, k_{\mathrm{eq},j}(T)^{-1}\mathcal{P}(T)\, dT \tag{6.9}$$

where $\mathcal{P}(T)$ is an assumed probability density function (PDF) and $k_{\mathrm{eq},j}(T)$ is the equilibrium constant which depends on temperature. Two different probability density functions were considered. The first is a Gaussian PDF defined by

$$\mathcal{P}(T) = \frac{1}{\sqrt{2\pi T'^2_{\mathrm{rms}}}} \exp\left[-\frac{(T-\overline{T})^2}{2T'^2_{\mathrm{rms}}}\right] \tag{6.10}$$

where $T'_{\mathrm{rms}}$ is the root-mean-square temperature fluctuations. The second is the $\beta$ PDF defined by

$$\mathcal{P}(T) = \frac{r^{\beta_1-1}(1-r)^{\beta_2-1}}{\Gamma(\beta_1)\Gamma(\beta_2)}\Gamma\left(\beta_1+\beta_2\right) \tag{6.11}$$

where

$$\begin{aligned}\beta_1 &= \overline{r}\left[\frac{\overline{r}(1-\overline{r})}{\overline{r'r'}} - 1\right]\\ \beta_2 &= (1-\overline{r})\left[\frac{\overline{r}(1-\overline{r})}{\overline{r'r'}} - 1\right]\end{aligned} \tag{6.12}$$

where

$$r = \frac{T - T_{\min}}{T_{\max} - T_{\min}} \tag{6.13}$$

and thus

$$\overline{r} = \frac{T - T_{\min}}{T_{\max} - T_{\min}} \tag{6.14}$$

and

$$\overline{r'r'} = \frac{T'^2_{\mathrm{rms}}}{(T_{\max} - T_{\min})^2} \tag{6.15}$$

On the basis of a generalization of the Huang Strong Reynolds Analogy and a mixing length hypothesis (section 3.8), Duan and Martín (2011b) deduced a *temperature scaling*

$$T'_{\mathrm{rms}} = \left|\frac{1}{Pr_t}\frac{\partial\overline{T}}{\partial\overline{u}}\right| u'_{\mathrm{rms}} \tag{6.16}$$

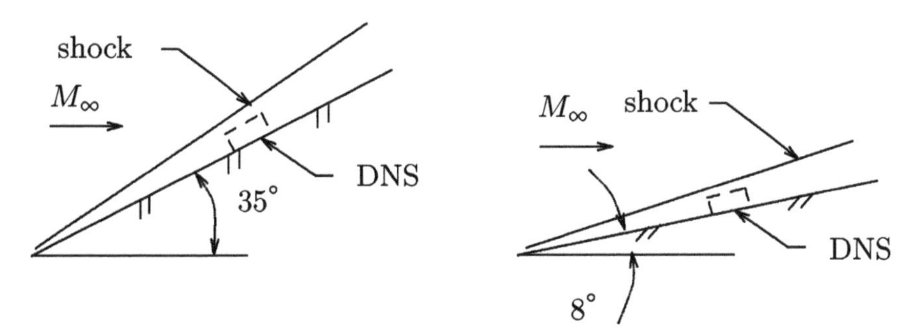


**Figure 6.22.** Flow configurations.

where $u'_{\text{rms}}$ is the root-mean-square streamwise velocity fluctuation and $Pr_t$ is the turbulent Prandtl number. Relating $u'_{\text{rms}}$ to the turbulence kinetic energy $k$ by

$$C_M = \frac{u'_{\text{rms}}}{\sqrt{k}} \tag{6.17}$$

where $C_M$ is approximately constant throughout the boundary layer[2] yields

$$T'_{\text{rms}} = C_M \left| \frac{1}{Pr_t} \frac{\partial \overline{T}}{\partial \overline{u}} \right| \sqrt{k} \tag{6.18}$$

Therefore, a RANS simulation which incorporates a model for $k$ can use (6.7) with $\overline{k}_{f,j}$ and $\overline{k}_{b,j}$ given by (6.8) and (6.9) and the PDF defined by $\overline{T}$ and $T'^2_{\text{rms}}$ from (6.18).

Duan and Martín (2011b) performed four DNS simulations of a hypersonic turbulent boundary layer to examine the accuracy of the model for the average reaction rate (6.7), the root-mean-square temperature fluctuation scaling (6.16) and the structure of $C_M$ in (6.17) within the boundary layer. A five species thermochemistry model ($N_2$, $O_2$, NO, N, O) was used with five reactions

$$\begin{aligned}
&N_2 + M \rightleftharpoons 2N + M && (R_1)\\
&O_2 + M \rightleftharpoons 2O + M && (R_2)\\
&NO + M \rightleftharpoons N + O + M && (R_3)\\
&N_2 + O \rightleftharpoons NO + N && (R_4)\\
&NO + O \rightleftharpoons O_2 + N && (R_5)
\end{aligned} \tag{6.19}$$

The flow configurations are shown in figure 6.22. A boundary layer forms on a wedge oriented at an angle of 8° or 35° with respect to a Mach 21 freestream.

[2] Duan and Martín (2011b) obtained the following empirical expression

$$C_M = \frac{\sqrt{2}}{\sqrt{1 + C\left(1 - e^{-D(z/\delta)^2}\right)^2}}$$

where $C = 1.06$ and $D = 15$.

A RANS simulation (Wright *et al* 2009) is performed to define the edge conditions of the turbulent boundary layer at a fixed distance from the wedge leading edge. The edge conditions are listed in table 6.7.

Figure 6.23 shows the mean forward rate coefficients $\bar{k}_{f,j}$ for reactions R1 to R5 within the boundary layer for Case 1 where $z$ is measured normal to the wall. Results are presented from DNS and from (6.8) using the Gaussian (6.10) and $\beta$ (6.11) probability density functions. Close agreement is seen. Similar close agreement is observed for Case 2 throughout the boundary layer, and in Cases 3 and 4 for $0 < z/\delta < 0.6$ (Duan and Martín 2011b). The Gaussian and $\beta$ probability distribution functions produce virtually identical profiles across the entire boundary layer for all cases.

Figure 6.24 compares the normalized root-mean-square temperature fluctuations $T'_{\text{rms}}/\bar{T}$ for DNS and temperature scaling (6.16) for Case 1 within the boundary

**Table 6.7.** Flowfield edge conditions.

| Quantity | Value | | | |
|---|---|---|---|---|
| | Case 1 | Case 2 | Case 3 | Case 4 |
| $\alpha$ | 35° | 35° | 8° | 8° |
| $M_e$ | 3.4 | 3.4 | 9.4 | 9.3 |
| $T_e$ (K) | 4 474.5 | 4 505.9 | 1 290.9 | 1 234.5 |
| $T_\text{w}$ (K) | 2 400.0 | 2 400.0 | 2 400.0 | 2 400.0 |
| $T_\text{w}/T_\text{aw}$ | 0.13 | 0.13 | 0.13 | 0.13 |
| $Re_\theta$ | 966.2 | 1 011.1 | 3 026.1 | 3 058.1 |
| Wall | SC | NC | SC | NC |

SC supercatalytic
NC noncatalytic

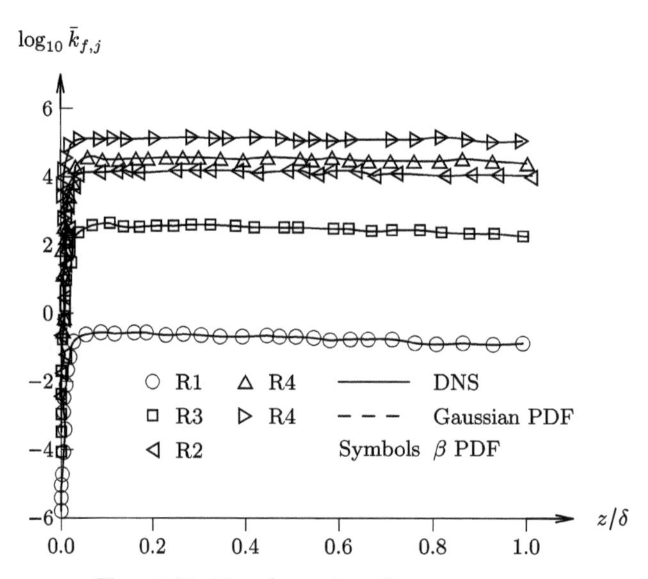


**Figure 6.23.** Mean forward reaction rate constants.

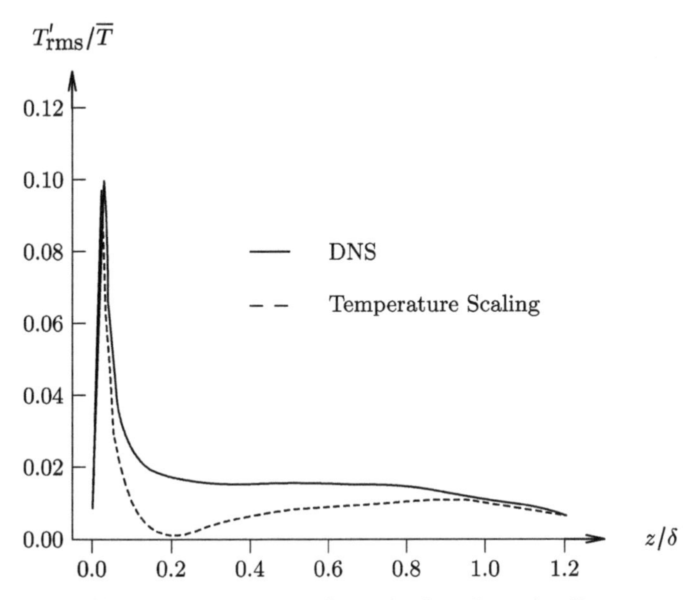


**Figure 6.24.** Temperature fluctuation intensity and scaling.

**Table 6.8.** Flow conditions.

| | Value | |
|---|---|---|
| Quantity | M8T1 | M8T2 |
| $M_\infty$ | 8 | 8 |
| $M_\delta$ | 7.29 | 6.92 |
| $T_\infty$ (K) | 169.44 | 169.44 |
| $T_w/T_\infty$ | 1.9 | 10.03 |
| $T_w/T_{aw}$ | 0.15 | 0.77 |
| $Re_\theta \times 10^{-4}$ | 2.2 | 7.8 |

layer. Close agreement is seen near the wall including the location and magnitude of the peak temperature fluctuations. For $0.2 < z/\delta < 0.8$ the temperature scaling (6.16) underestimates the temperature fluctuations. Similar agreement is observed for Cases 2 to 4 (Duan and Martín 2011b). See also Duan and Martín (2009b, 2009a) and Duan and Martín (2011a).

## 6.8 Liang *et al* (2012, 2013)

Liang *et al* (2012) and Liang and Li (2013) performed a Direct Numerical Simulation of a cold wall hypersonic turbulent boundary layer at Mach 8. The governing equations are the compressible Navier–Stokes equations for a perfect gas. The flow conditions are listed in table 6.8.

Figure 6.25 shows the Van Driest transformed velocity for both cases. The results for M8T1 ($T_w/T_\infty = 1.9$) show close agreement with the Law of the Wall (3.55) for $C = 7.4$, and the results for M8T2 ($T_w/T_\infty = 10.03$) show similarly close agreement with the Law of the Wall for $C = 6.3$. Figure 6.26 displays the normalized streamwise, transverse and spanwise intensity versus $y^+$ for case M8T1. Also shown

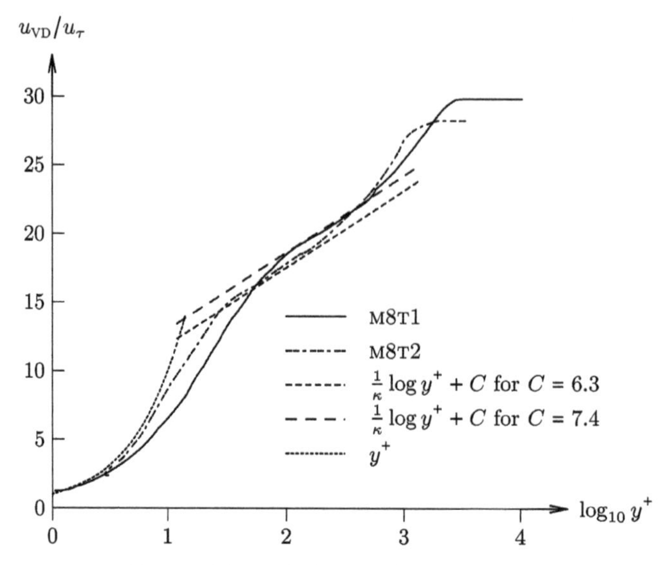


**Figure 6.25.** Van Driest transformed velocity.

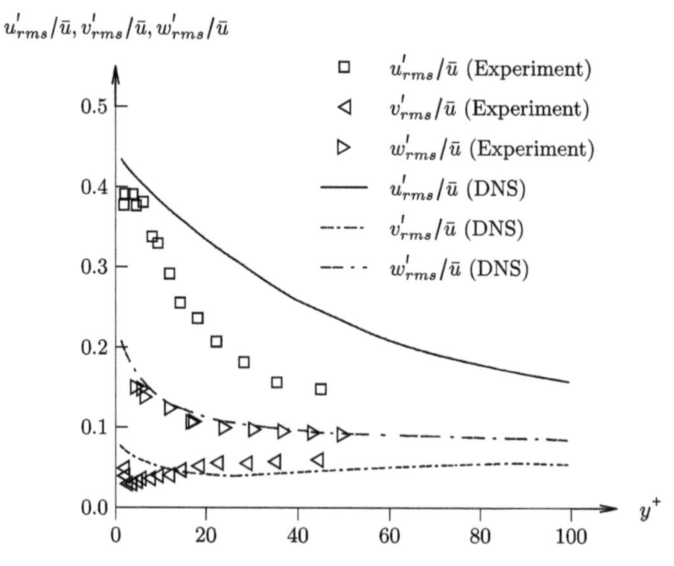


**Figure 6.26.** Turbulence intensity versus $y^+$.

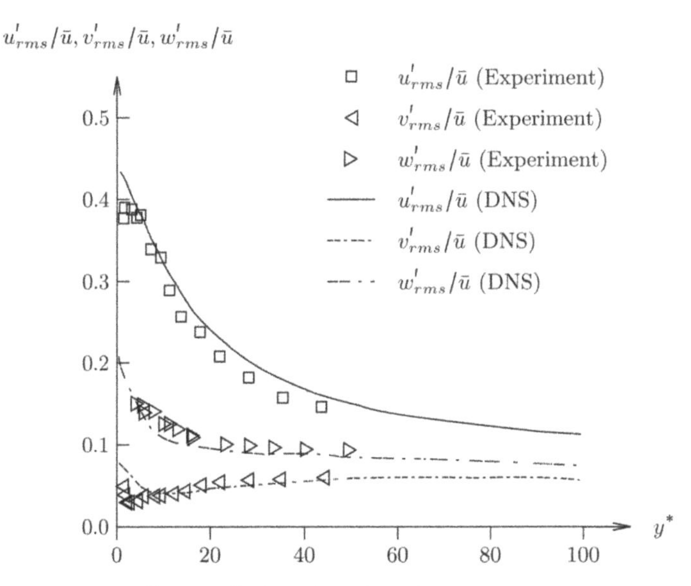


**Figure 6.27.** Turbulence intensity versus $y^*$.

is experimental data for an incompressible turbulent boundary layer[3]. A significant difference is seen between the DNS and experiment for the streamwise and transverse intensities. Rescaling the normalized intensities using the modified wall distance

$$y^* = \frac{y u_\tau^*}{\nu^*} \quad \text{where} \quad u_\tau^* = \sqrt{\frac{\tau_w}{\bar{\rho}(y)}} \quad \text{and} \quad \nu^* = \frac{\bar{\mu}(y)}{\bar{\rho}(y)} \tag{6.20}$$

achieves close agreement between the DNS and incompressible experimental data (figure 6.27).

Figure 6.28 compares the results of DNS with Walz's relation (3.171) for both cases. Close agreement is evident. The Strong Reynolds Analogy Nos. 2 (3.331), 4 (3.334) and 5 (3.336) indicate

$$\frac{\sqrt{\overline{T'^2}}/\bar{T}}{(\gamma - 1)\bar{M}^2\sqrt{\overline{u'^2}}/\bar{u}} = 1 \quad \text{SRA No. 2}$$

$$R_{uT} = -1 \quad \text{SRA No. 4}$$

$$Pr_t = 1 \quad \text{SRA No. 5}$$

Figure 6.29 presents the three functions for case M8T1. It is evident that Strong Reynolds Analogy Nos. 2 and 4 do not hold anywhere within the boundary layer for this highly cooled wall ($T_w/T_{aw} = 0.15$). Reynolds Analogy No. 5 is valid only in the outer portion of the boundary layer.

[3] No reference is provided for the experimental data.

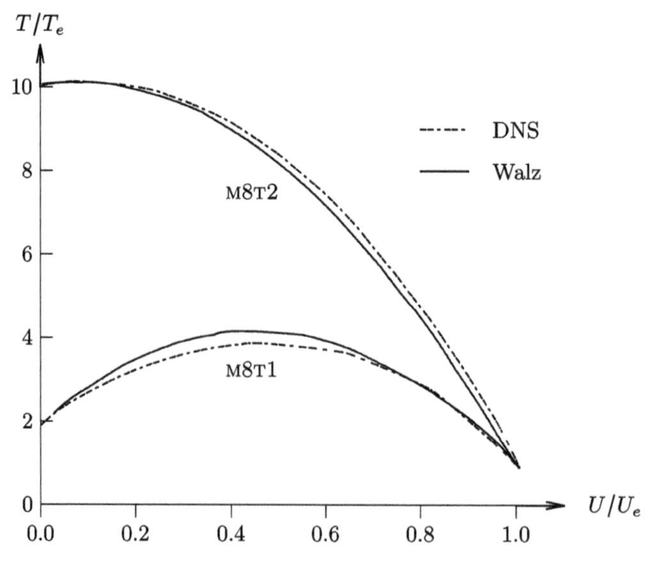


**Figure 6.28.** Evaluation of Walz's equation.

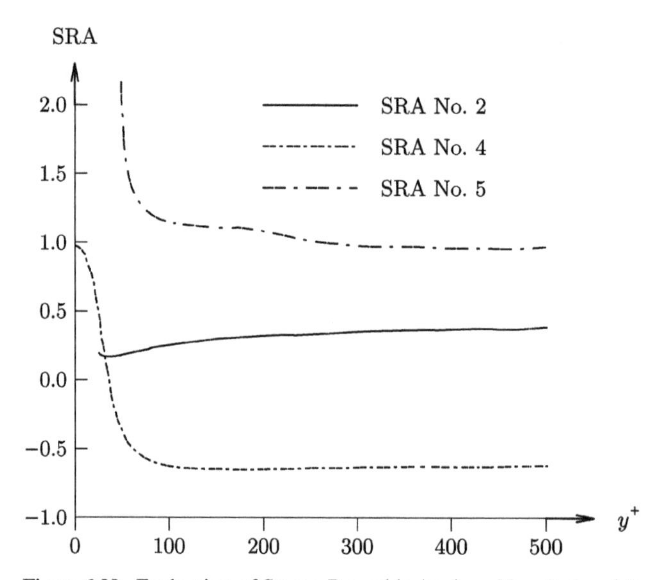


**Figure 6.29.** Evaluation of Strong Reynolds Analogy Nos. 2, 4 and 5.

## 6.9 Chu *et al* (2013)

Chu *et al* (2013) performed a Direct Numerical Simulation of a hypersonic turbulent boundary layer at Mach 4.9 for wall temperature ratios $T_w/T_{aw} = 0.5$, 1.0 and 1.5. The governing equations are the compressible Navier–Stokes equations for a perfect gas. The flow conditions are shown in table 6.9.

**Table 6.9.** Flow conditions.

| Quantity | Value | | |
|---|---|---|---|
| | Case 1 | Case 2 | Case 3 |
| $M_\infty$ | 4.9 | 4.9 | 4.9 |
| $T_w/T_{aw}$ | 0.5 | 1.0 | 1.5 |
| $T_w/T_\infty$ | 2.62 | 5.25 | 7.91 |
| $Re_\delta \times 10^{-5}$ | 0.58 | 1.09 | 1.73 |
| $Re_\theta \times 10^{-3}$ | 3.48 | 5.56 | 7.61 |
| $Re_\tau$ | 532 | 397 | 353 |
| Case ID | M5T2 | M5T5 | M5T8 |

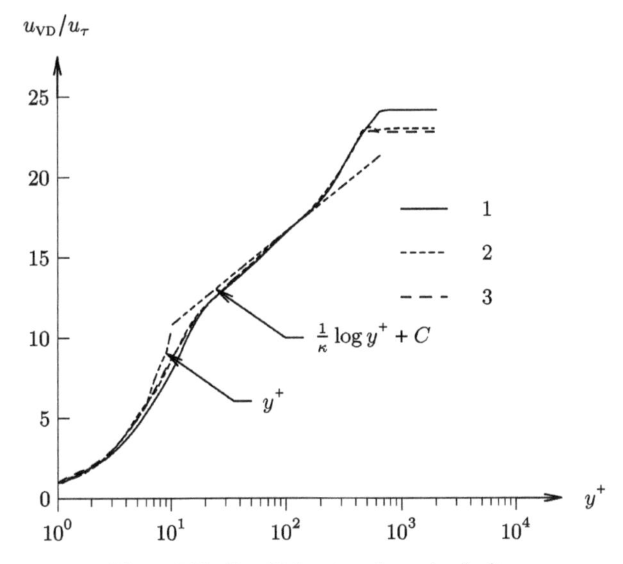


**Figure 6.30.** Van Driest transformed velocity.

Figure 6.30 shows the Van Driest transformed velocity $u_{VD}/u_\tau$ from (3.98) versus $y^+$, together with the Law of the Wall. Close agreement is observed between the simulations and the Law of the Wall for all cases.

Figure 6.31 displays the mean static temperature velocity profiles for all cases together with Walz's relation (3.171). The computed profiles shows close agreement with Walz's relation and also with the relation (6.4) of Duan and Martín (2011a) (not shown), with the latter providing slightly closer agreement with the simulation for the cold wall case.

Figure 6.32 shows the Morkovin scaled streamwise turbulence intensity $(\overline{\rho}/\rho_w)^{\frac{1}{2}}\sqrt{\overline{u'^2}}/u_\tau$ versus $y/\delta$. The profiles for all cases collapse to a single curve. A similar result is obtained for the wall normal and spanwise turbulence intensities.

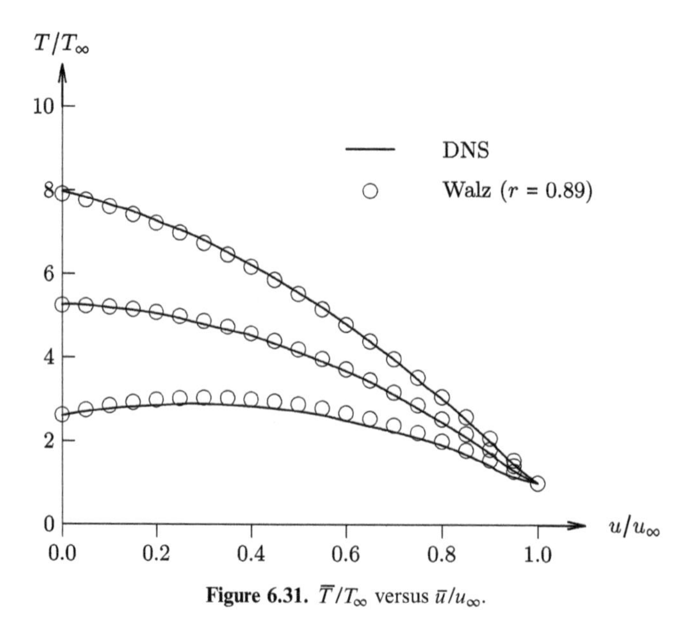


**Figure 6.31.** $\overline{T}/T_\infty$ versus $\overline{u}/u_\infty$.

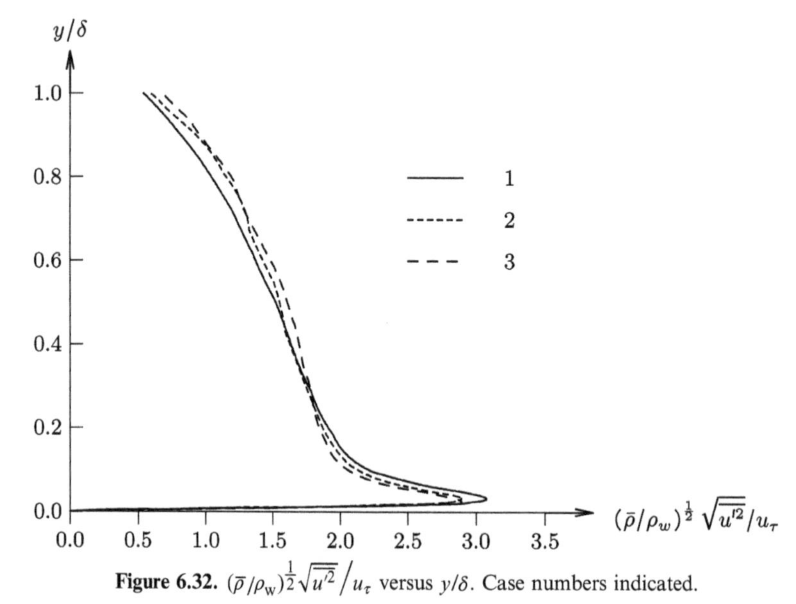


**Figure 6.32.** $(\overline{\rho}/\rho_w)^{\frac{1}{2}}\sqrt{\overline{u'^2}}\big/u_\tau$ versus $y/\delta$. Case numbers indicated.

In the absence of Morkovin scaling, the cold and hot wall profiles differ by typically 50%.

Figure 6.33 displays the non-dimensional root mean square static pressure fluctuations $\sqrt{\overline{p'^2}}\big/\tau_w$ versus $y/\delta$ for the three cases. Normalization by the wall shear stress $\tau_w$ collapses the profiles to essentially a single curve except for $y > 0.6\delta$, while normalization by the mean wall pressure $p_w$ results in a difference between the cold and hot wall as large as 50%.

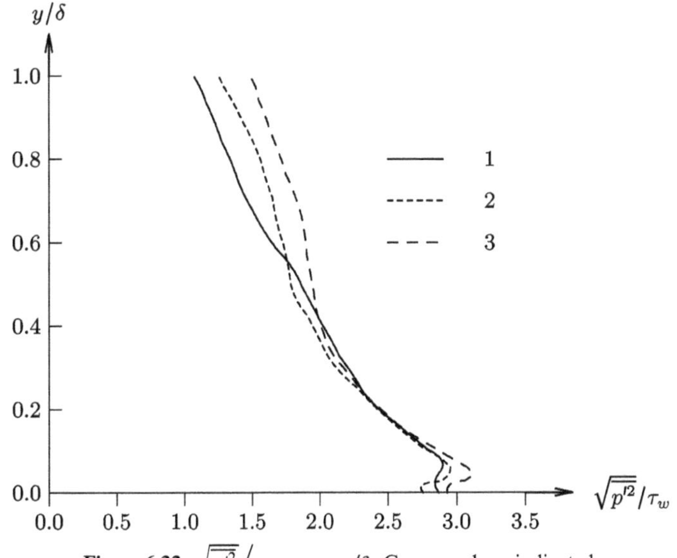


**Figure 6.33.** $\sqrt{\overline{p'^2}}/\tau_w$ versus $y/\delta$. Case numbers indicated.

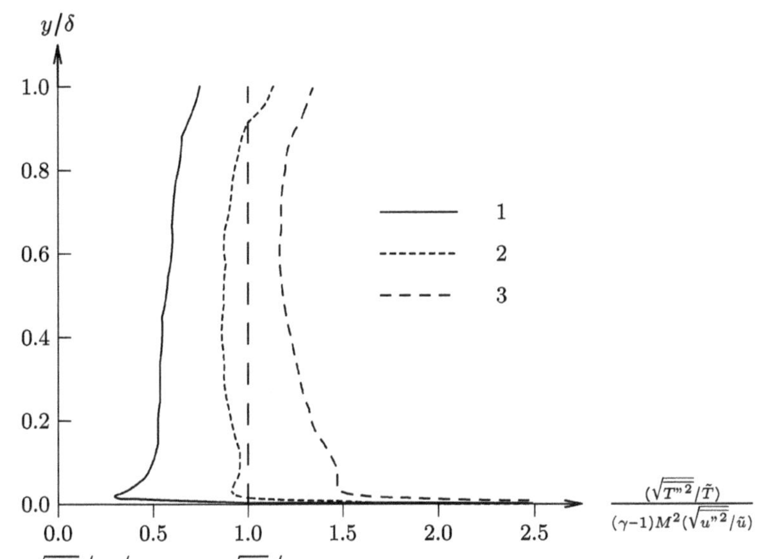


**Figure 6.34.** $(\sqrt{\overline{T''^2}}/\tilde{T})/((\gamma-1)M^2\sqrt{\overline{u''^2}}/\tilde{u})$ versus $y/\delta$. Evaluation of Morkovin's Strong Reynolds Analogy No. 2.

Figure 6.34 examines Morkovin's Strong Reynolds Analogy No. 2 (3.331). The adiabatic wall results (Case 2) indicate (3.331) holds across nearly the entire boundary layer, while the cold and hot wall cases show differences up to 50%. Figure 6.35 examines the modified Strong Reynolds Analogy of Huang *et al* (1995) for all three cases where HSRA is defined in (6.3). Close agreement with (3.358) is observed. Figure 6.36 shows the turbulent Prandtl number $Pr_t$ defined by (2.55)

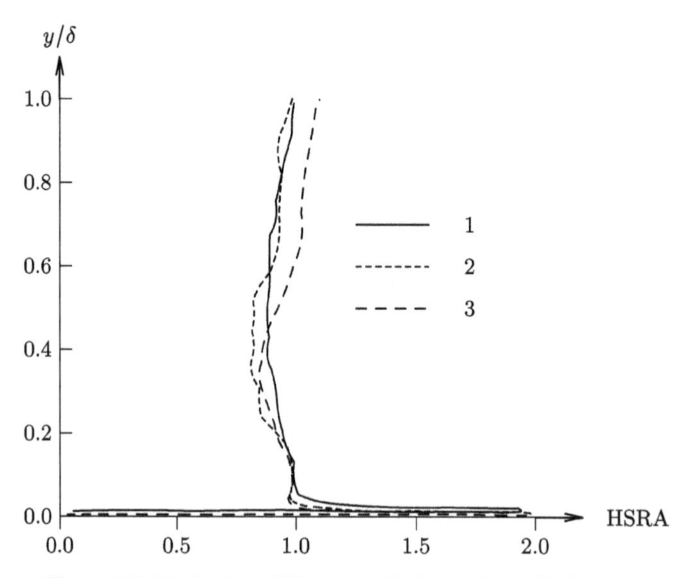


**Figure 6.35.** Evaluation of Huang *et al*'s Strong Reynolds Analogy.

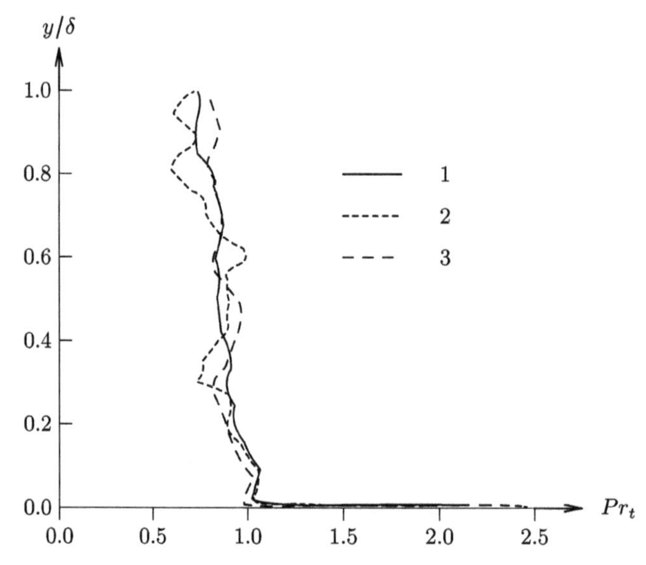


**Figure 6.36.** $Pr_t$ versus $y/\delta$. Case numbers indicated.

versus $y/\delta$. Except close to the wall where the mean static temperature gradient changes sign for the cold wall case, the value $Pr_t \approx 1$.

## 6.10 Duan *et al* (2016)

Duan *et al* (2016) performed a Direct Numerical Simulation of a cold wall hypersonic turbulent boundary layer at Mach 5.86. The governing equations are the compressible Navier–Stokes equations for a perfect gas. The flow conditions are listed in table 6.10.

Figure 6.37 shows the Van Driest transformed velocity $u_{VD}/u_\tau$ (3.98) versus $y^+$. Close agreement in the log region is evident. Figure 6.38 displays the normalized root mean square pressure fluctuations $\sqrt{\overline{p'^2}}/\tau_w$ versus $y/\delta$ for the present simulation, together with the DNS results of Bernardini and Pirozzoli (2011) at $M_\infty = 2$ and 4. The profiles are similar throughout the boundary layer. Figure 6.39 shows the premultiplied pressure spectra $\omega\Phi/\overline{p'^2}$ versus $\omega\delta/u_\infty$ at the wall and $y/\delta = 0.158$ and 0.734. The spectrum peak in the inner region is located at $\omega\delta/u_\infty \approx 8$ ($f\delta/u_\infty \approx 1$) where $f$ is the frequency in Hz. This corresponds to the characteristic frequency of the energy containing eddies of dimension $\delta$. The peak frequency decreases in the outer layer due to intermittency. Further information on propagation speed and structure of the pressure fluctuations in presented in Duan *et al* (2016).

**Table 6.10.** Freestream conditions.

| Quantity | Value |
|---|---|
| $M_\infty$ | 5.86 |
| $T_w/T_{aw}$ | 0.76 |
| $Re_\delta \times 10^{-5}$ | 2.37 |
| $Re_\theta \times 10^{-3}$ | 9.45 |
| $Re_\tau$ | 453.1 |

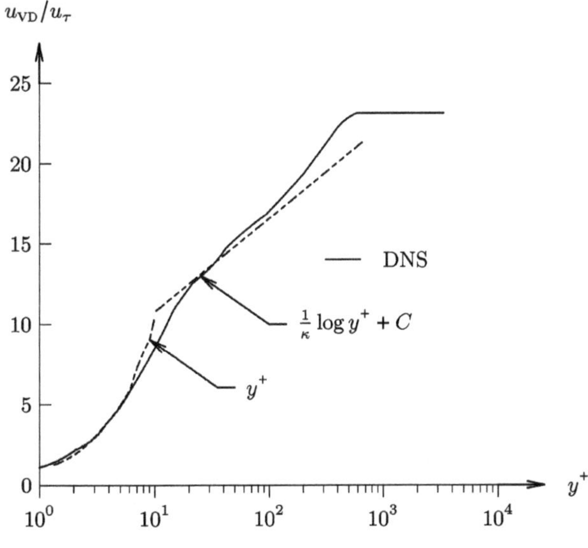


**Figure 6.37.** Van Driest transformed velocity.

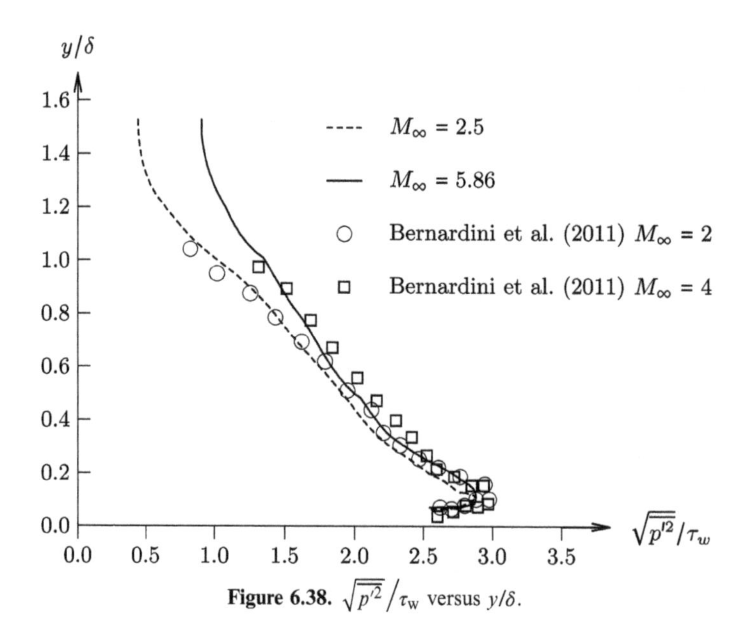


**Figure 6.38.** $\sqrt{p'^2}/\tau_w$ versus $y/\delta$.

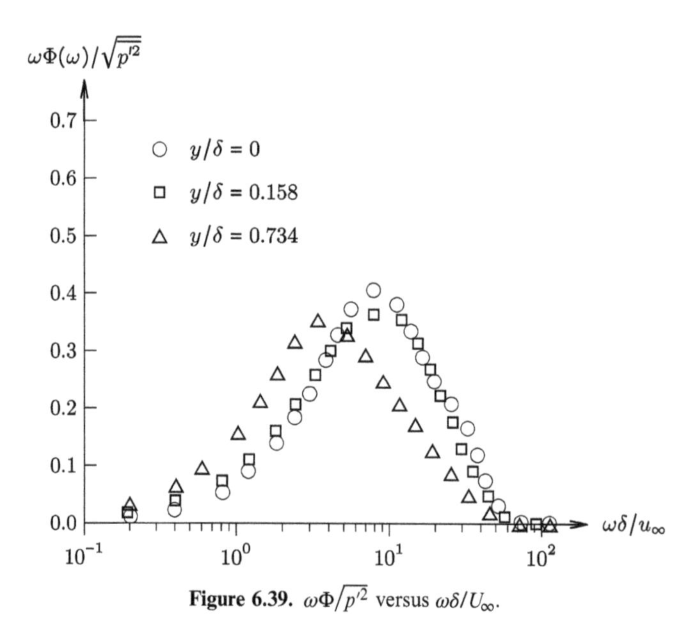


**Figure 6.39.** $\omega\Phi/p'^2$ versus $\omega\delta/U_\infty$.

## 6.11 Zhang *et al* (2017)

Zhang *et al* (2017) performed a Direct Numerical Simulation for a hypersonic cold wall flat plate turbulent boundary layer. The governing equations are the compressible Navier–Stokes equations for a perfect gas. The flow conditions are shown in table 6.11.

**Table 6.11.** Freestream conditions.

| | Value | |
|---|---|---|
| Quantity | Case 1 | Case 2 |
| $M_\infty$ | 5.86 | 5.86 |
| $T_w/T_{aw}$ | 0.25 | 0.76 |
| $Re_\delta \times 10^{-3}$ | 38.4 | 237.4 |
| $Re_\theta$ | 2121 | 9455 |
| $Re_\tau$ | 450 | 453 |
| Case | M6Tw025 | M6Tw076 |

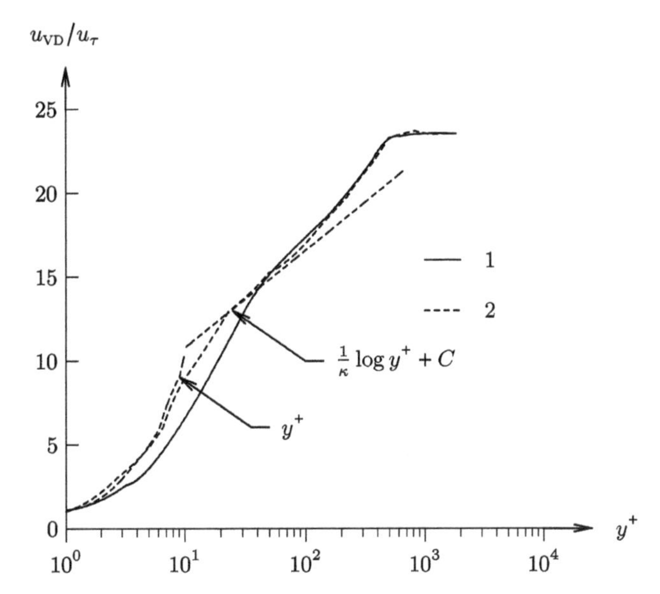


**Figure 6.40.** Van Driest transformed velocity. Case numbers indicated.

Figure 6.40 displays the Van Driest transformed velocity $u_{VD}/u_\tau$ versus $y^+$. Close agreement in the log region is seen between the simulations and Law of the Wall. Figure 6.41 shows the Trettle and Larsson velocity profile $u_{TL}/u_\tau$ (3.113) versus $Y^*$ (3.114). The profiles for both simulations collapse within the viscous sublayer and logarithmic region and agree with the Law of the Wall.

Figure 6.42 shows the mean static temperature—streamwise velocity, together with Walz's relation (3.171) and Zhang *et al*'s relation (3.204). The Zhang *et al* result shows close agreement with both simulations, while the Walz relation agrees closely with the simulation for the higher wall temperature case only.

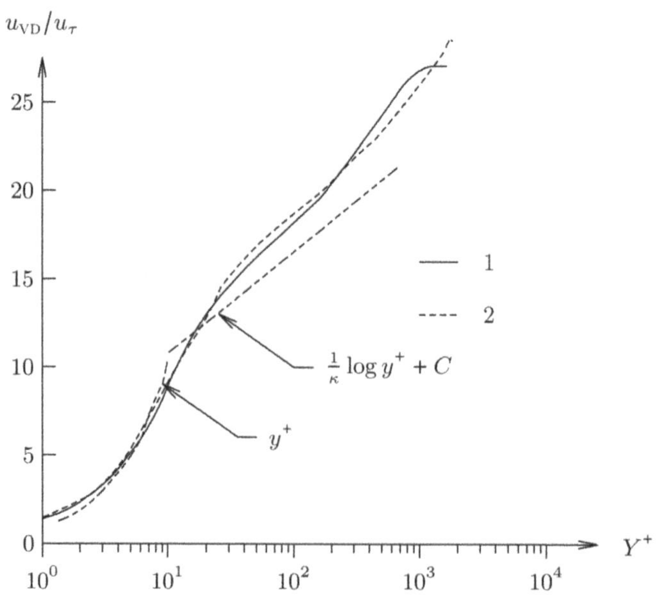


**Figure 6.41.** $u_{\mathrm{TL}}/u_\tau$ versus $Y^+$.

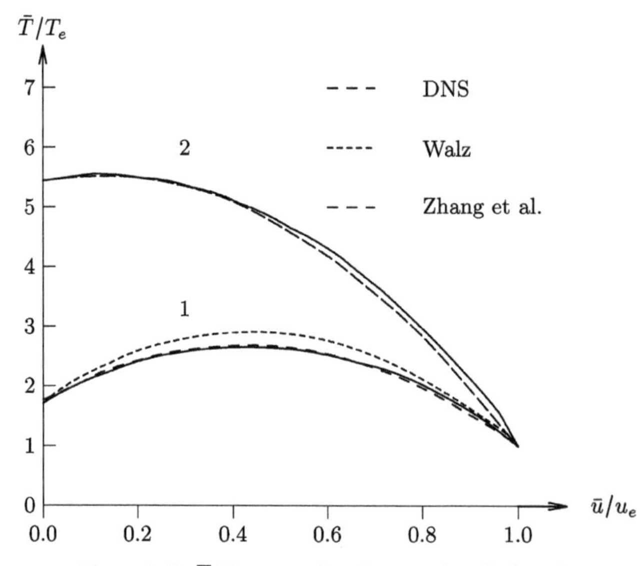


**Figure 6.42.** $\overline{T}/T_e$ versus $\overline{u}/u_e$. Case numbers indicated.

Figure 6.43 displays the modified Strong Reynolds Analogies of Huang *et al* (1995) (3.358) and Zhang *et al* (2014) expressed as

$$\frac{\left(\sqrt{\overline{T'^2}}/\overline{T}\right)\alpha(1 - d\overline{T}_t/d\overline{T})}{(\gamma - 1)\overline{M}^2\left(\sqrt{\overline{u'^2}}/\overline{u}\right)} = 1 \tag{6.21}$$

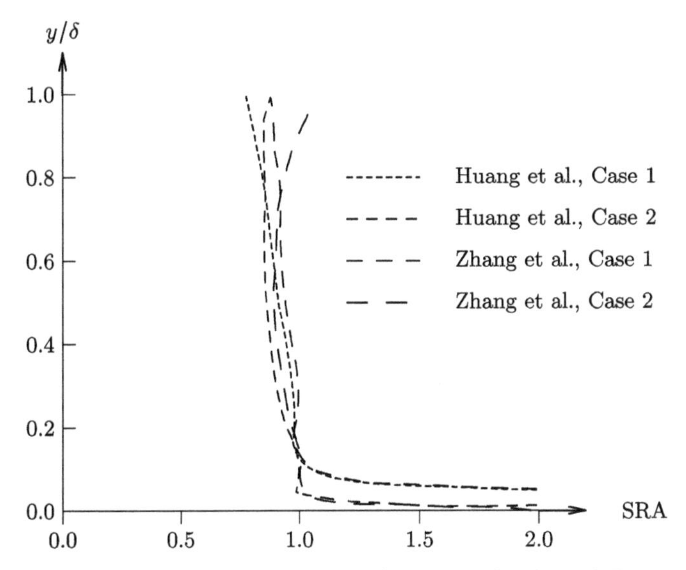


**Figure 6.43.** Modified Strong Reynolds Analogies of Huang *et al* (1995) and Zhang *et al* (2014).

where

$$\alpha = \begin{cases} Pr_t & \text{Huang } et\ al\ (1995) \\ Pr_t(1 + \overline{v}\ \overline{\rho' u'}/\overline{\rho u' v'})/(1 + \overline{v}\ \overline{\rho' T'}/\overline{\rho v' T'}) & \text{Zhang } et\ al\ (2014) \end{cases} \quad (6.22)$$

Both SRA models show close agreement with the simulations.

Figures 6.44 and 6.45 display the Morkovin scaled streamwise turbulence intensity $(\overline{\rho}/\overline{\rho}_w)^{\frac{1}{2}}\sqrt{\overline{u'^2}}\big/u_\tau$ and Reynolds shear stress $(\overline{\rho}/\overline{\rho}_w)^{\frac{1}{2}}\sqrt{\overline{u'v'}}\big/u_\tau$. Also shown are the results of Peltier *et al* (2016) at $M_\infty = 4.9$ and $T_w/T_{aw} = 0.9$; Shadloo *et al* (2015) at $M_\infty = 2$ and $T_w/T_{aw} = 0.5$ and the incompressible DNS data of Schlatter and Örlü (2010). The Morkovin scaling shows a close collapse of the data.

Figure 6.46 displays the root-mean-square pressure fluctuations $\sqrt{\overline{p'^2}}\big/\tau_w$ versus $y/\delta$. The cold wall (Case 1) shows a higher pressure fluctuation near the wall. Figure 6.47 presents the pressure spectrum at the wall $\omega\Phi_p\big/\overline{p'^2}$ versus dimensionless frequency $\omega\delta^*/u_\infty$ where $\Phi_p(\omega)$ is the pressure spectrum and $\delta^*$ is the boundary layer displacement thickness. A strong temperature dependence is evident particularly in the mid-frequency region.

## 6.12 Zhang *et al* (2018)

Zhang *et al* (2018) performed Direct Numerical Simulations of a cold wall hypersonic turbulent boundary layer at Mach 2.5 to 14. The governing equations are the compressible Navier–Stokes equations. The dynamic inflow boundary conditions was obtained using the recycling-rescaling method of Duan *et al* (2016). The flow conditions are shown in table 6.12.

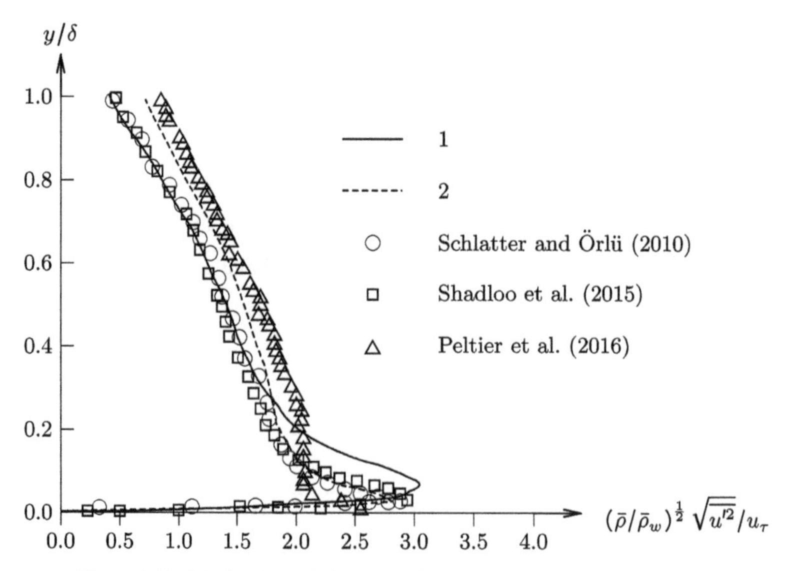


**Figure 6.44.** Morkovin scaled streamwise turbulence intensity versus $y/\delta$.

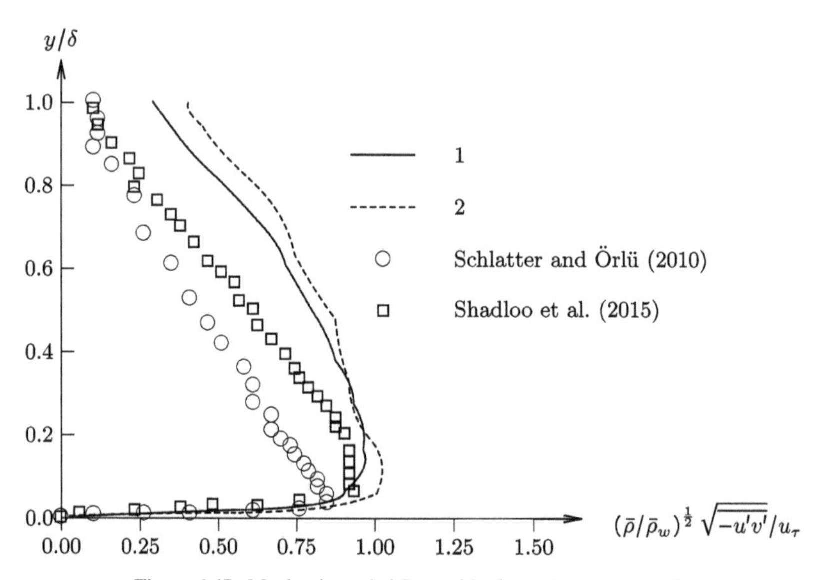


**Figure 6.45.** Morkovin scaled Reynolds shear stress versus $y/\delta$.

The Van Driest mean velocity profiles show close agreement with the Law of the Wall in the logarithmic region. However, figure 6.48 indicates that the coefficient $C$ in (3.54) exhibits a modest dependence on the wall heat flux parameter $B_q$ defined as

$$B_q = \frac{q_w}{\rho_w c_p u_\tau T_w} \tag{6.23}$$

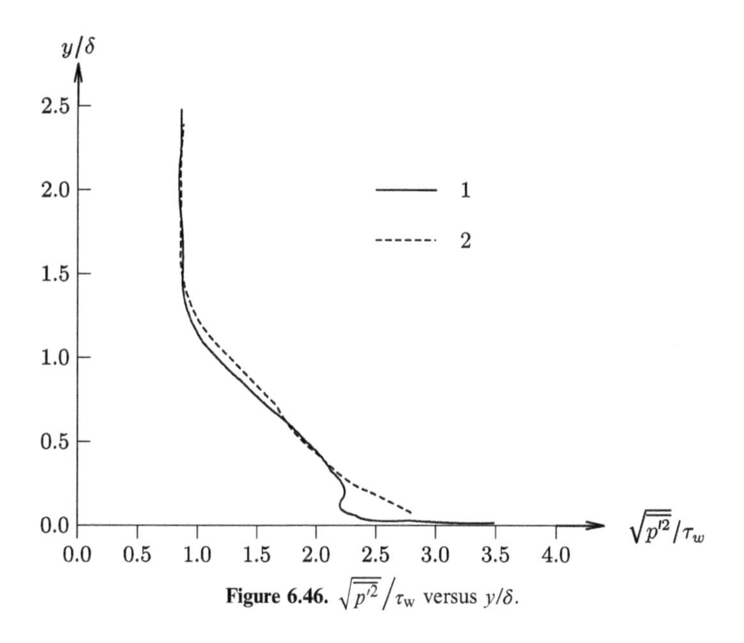


**Figure 6.46.** $\sqrt{\overline{p'^2}}/\tau_{\mathrm{w}}$ versus $y/\delta$.

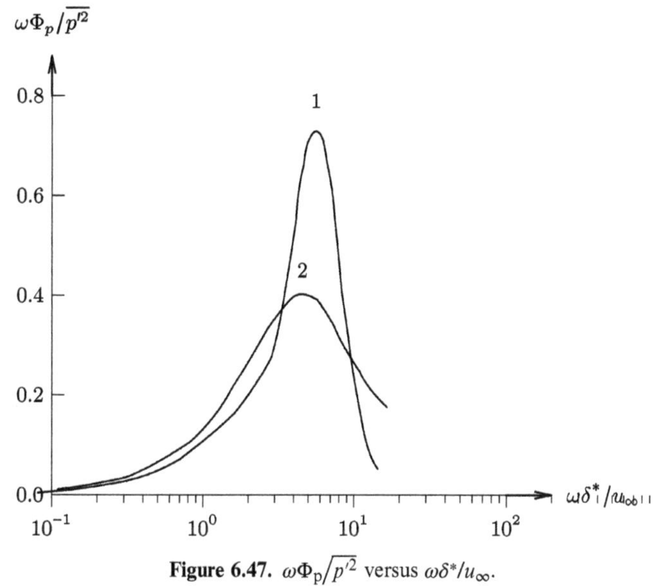


**Figure 6.47.** $\omega\Phi_{\mathrm{p}}/\overline{p'^2}$ versus $\omega\delta^*/u_\infty$.

Also shown are the predictions of Trettel and Larsson (2016) which are in close agreement with the simulation results at low $-B_q$. Patel *et al* (2015) observed that the semi-local Reynolds number $Re_\tau^*$ defined by

$$Re_\tau^* = \sqrt{\frac{\tau_{\mathrm{w}}}{\rho_\infty}}\frac{\delta}{\nu_\infty} \tag{6.24}$$

**Table 6.12.** Freestream conditions.

| | Value | | | | |
|---|---|---|---|---|---|
| Quantity | Case 1 | Case 2 | Case 3 | Case 4 | Case 5 |
| $M_\infty$ | 2.50 | 5.84 | 5.86 | 7.87 | 13.64 |
| $T_{t_\infty}$ (K) | 607.5 | 432 | 433 | 694 | 1811 |
| $p_{t_\infty}$ (MPa) | 0.132 | 0.933 | 0.927 | 3.39 | 79.7 |
| $T_w/T_{aw}$ | 1.0 | 0.25 | 0.76 | 0.48 | 0.18 |
| $Re_\theta \times 10^{-3}$ | 2.84 | 2.12 | 9.46 | 9.71 | 14.4 |
| $Re_\delta \times 10^{-3}$ | 37.6 | 13.8 | 144 | 179 | 200 |
| $-B_q$ | 0. | 0.14 | 0.02 | 0.06 | 0.19 |
| ID | M2p5 | M6Tw025 | M6Tw076 | M8Tw048 | M14Tw018 |

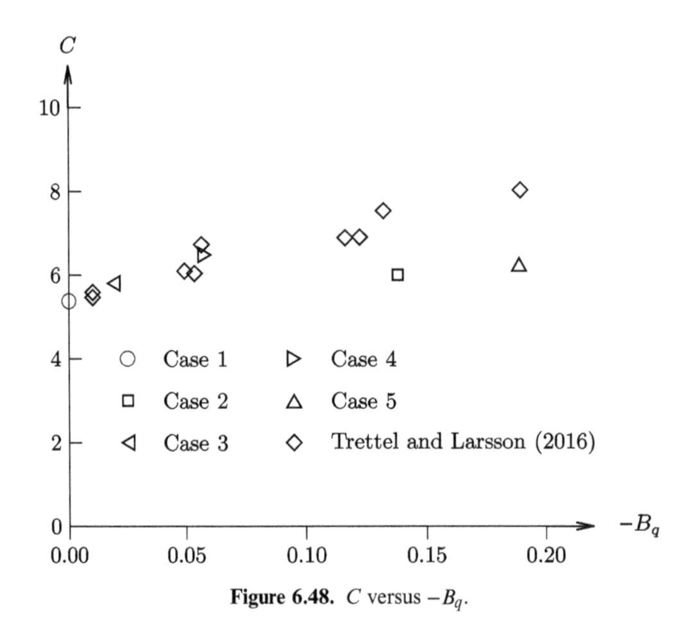


**Figure 6.48.** $C$ versus $-B_q$.

is the governing parameter for wall turbulence statistics with different mean density and viscosity profiles at lower Mach numbers. Figure 6.49 indicates that the present simulations display no significant effect of $Re_\tau^*$ on $C$.

Figure 6.50 compares the mean temperature–velocity relation of Cases 4 and 5 with Walz's relation (3.171) and the modified relation of Zhang *et al* (2014). The comparison with Walz's relation (3.171) for Case 4 ($B_q = -0.06$) is good, while for Case 5 ($B_q = -0.19$) significant differences are evident. The comparison with the modified relation of Zhang *et al* (2014) (3.213) is excellent for both cases.

Figures 6.51 and 6.52 display the Reynolds shear stress normalized using Morkovin scaling versus normalized distance from the wall $y^+$ and $y/\delta$, respectively.

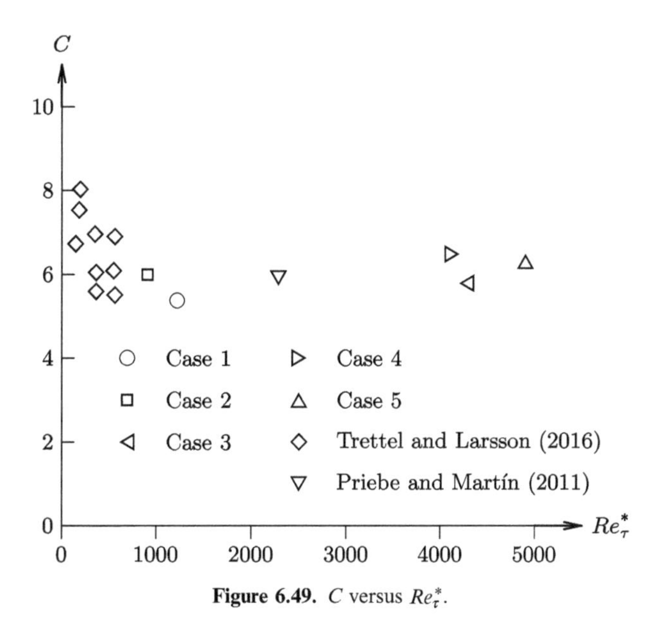


**Figure 6.49.** $C$ versus $Re_\tau^*$.

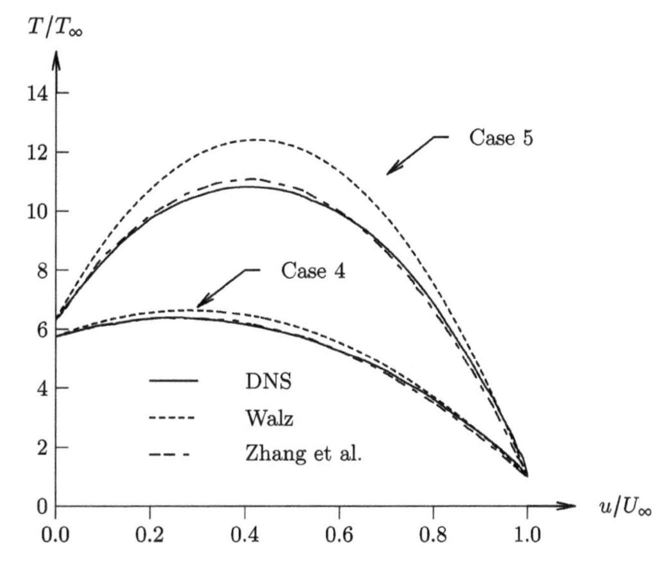


**Figure 6.50.** $T/T_\infty$ versus $UU_\infty$ from DNS, Walz (1969) and Zhang *et al* (2014).

Also shown both figures are the results of Jiménez *et al* (2010) and Silero *et al* (2013) at incompressible flow conditions, with additionally the results of DeGraaff and Eaton (2000), Priebe and Martín (2011) and Williams *et al* (2018) in figure 6.52. A reasonable collapse of the data is observed for both the innner $y^+$ and outer $y/\delta$ scaling with the exception of the data of Williams *et al* (2018). Similar satisfactory collapse of the streamwise, wall normal and spanwise turbulence intensities is observed with the exception of the wall normal intensity of Williams *et al* (2018).

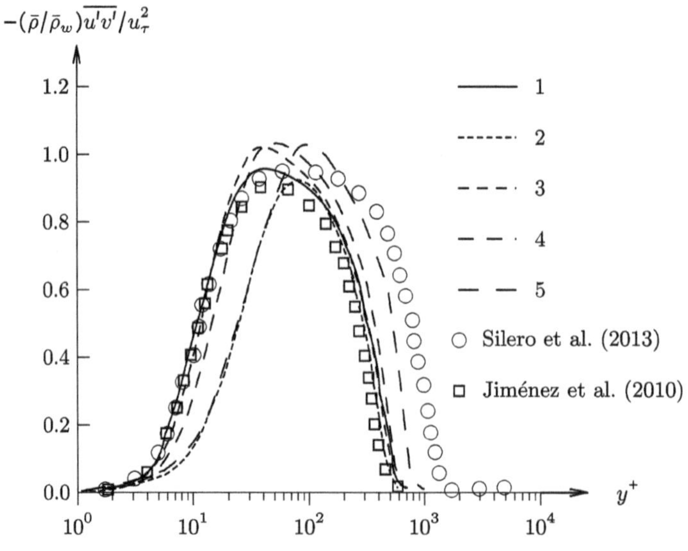


**Figure 6.51.** $-(\overline{\rho}/\overline{\rho}_w)\overline{u'v'}/u_\tau^2$ versus $y^+$. Case numbers indicated.

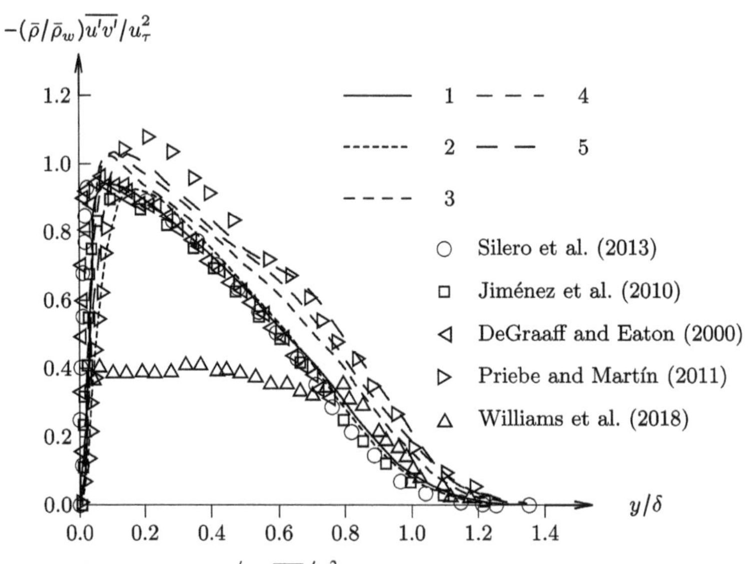


**Figure 6.52.** $-(\overline{\rho}/\overline{\rho}_w)\overline{u'v'}/u_\tau^2$ versus $y/\delta$. Case numbers indicated.

Figure 6.53 displays the turbulent Prandtl number (2.55) versus wall normal distance $y/\delta$ for Cases 1 to 5. Results are omitted for $y/\delta < 0.05$ since the mean temperature profile exhibits an inflection point where the value of $Pr_t$ becomes infinite. The results show an approximate linear decrease in $Pr_t$ from 1.0 at $y/\delta = 0.05$ to 0.8 at $y/\delta = 1$ with a mean value $Pr_t = 0.90$.

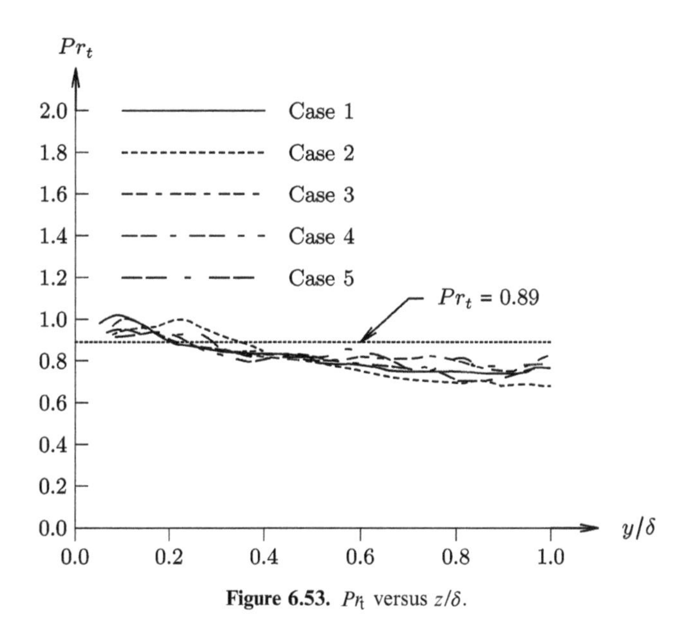


**Figure 6.53.** $Pr_t$ versus $z/\delta$.

**Table 6.13.** Freesteam conditions.

| Quantity | Value | |
|---|---|---|
| | Case 1 | Case 2 |
| $M_\infty$ | 11 | 14 |
| $T_w/T_{aw}$ | 0.2 | 0.2 |
| $Re \times 10^{-7}$ (m$^{-1}$) | 4.1 | 1.1 |

## 6.13 Huang *et al* (2020)

Huang *et al* (2020) performed Direct Numerical Simulations of a cold wall hypersonic turbulent boundary layer at Mach 11 and 14. The governing equations are the compressible Navier–Stokes equations. The simulations were performed using three overlapping computational domains. The dynamic inflow boundary condition for the first computational domain was determined by a recycling-rescaling method (Duan *et al* 2014). The flow conditions are shown in table 6.13. The turbulence statistics were evaluated at three downstream locations for Mach 11 and one downstream location for Mach 14 as indicated in table 6.14.

Figure 6.54 displays the Morkovin scaled turbulence intensities and Reynolds shear stress versus $y^+$. The profiles for Mach 11 and Mach 14 are virtually identical except in the outer region of the boundary layer. Figure 6.55 plots the same variables versus the semi-local scaled distance $y^*$ defined by

**Table 6.14.** Location of turbulence statistics.

| Case | Location | $Re_\theta$ |
|---|---|---|
| 1 | 1 | 9080 |
| | 2 | 14143 |
| | 3 | 18164 |
| 2 | 1 | 14258 |

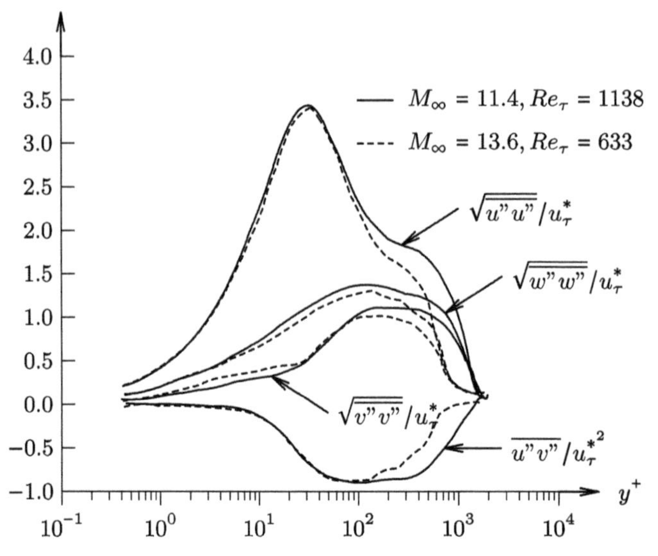


**Figure 6.54.** Normalized turbulence intensities and Reynolds shear stress versus $y^+$.

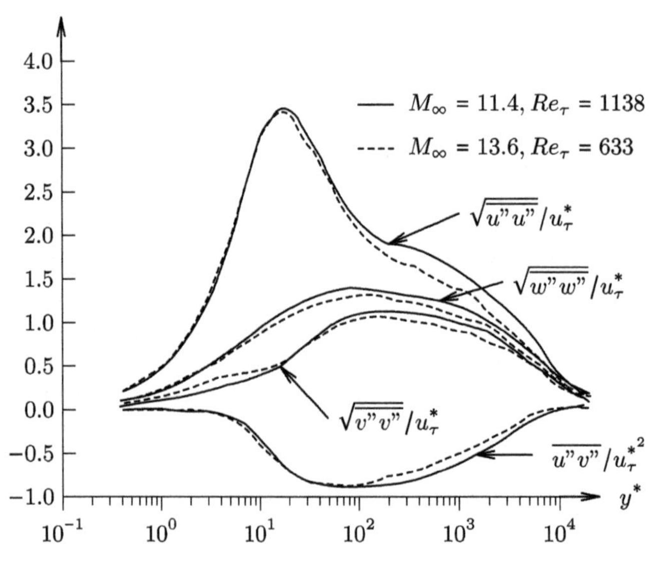


**Figure 6.55.** Normalized turbulence intensities and Reynolds shear stress versus $y^*$.

$$y^* = \frac{\mu}{\mu_w}\sqrt{\frac{\rho}{\rho_w}}\,y^+ \tag{6.25}$$

as suggested by Huang *et al* (1995). The semi-local scaling provides close agreement between the two Mach numbers across the entire boundary layer.

Figure 6.56 shows the turbulent Prandtl number (2.55) versus $y/\delta$. The mean value is $Pr_t = 0.9$ for $y/\delta > 0.2$. The sign of the mean temperature gradient changes near the wall since $T_w < T_{aw}$ and thus $Pr_t$ exhibits a singularity at $y/\delta \approx 0.025$.

## 6.14 Nicholson *et al* (2021a)

Nicholson *et al* (2021a) performed a series of three Direct Numerical Simulations of zero- and favorable pressure gradient hypersonic near adiabatic turbulent boundary layers. The simulations correspond to the experiments of Tichenor *et al* (2013) and are denoted Zero Pressure Gradient (ZPG), Weak Pressure Gradient (WPG) and Strong Pressure Gradient (SPG). The governing equations are the compressible Navier–Stokes equations. The simulations were performed using three overlapping domains. Figure 6.57 shows the computational domains for the SPG case.

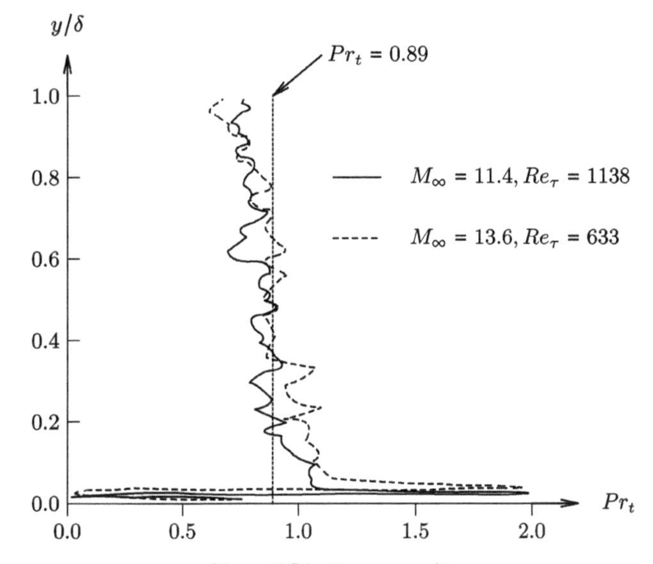


**Figure 6.56.** $Pr_t$ versus $y/\delta$.

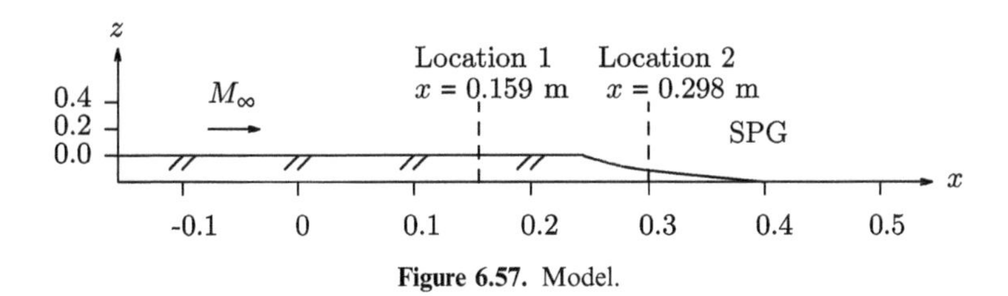


**Figure 6.57.** Model.

The dynamic inflow boundary condition for the first computational domain was determined by a recycling-resclaing method (Duan *et al* 2014). The flow conditions for the simulations are listed in table 6.15.

The ZPG Van Driest transformed velocity profile is in close agreement with the Law of the Wall (Nicholson *et al* 2021a). The ZPG streamwise turbulent intensity $\sqrt{\overline{u''u''}}/u_\tau^*$ agrees closely with the experiment of Tichenor *et al* (2013) where $u_\tau^* = u_\tau\sqrt{\bar{\rho}_w/\bar{\rho}}$ is the Morkovin transformed velocity scale. The ZPG wall normal turbulent intensity $\sqrt{\overline{v''v''}}/u_\tau^*$ and Reynolds shear stress $\overline{u''v''}/u_\tau^{*^2}$ show significant differences with the experiment of Tichenor *et al* (2013) (figures 6.58 and 6.59). Nicholson *et al* (2021a) attributes this discrepancy to particle response limitations of the Particle Image Velocimetry (PIV) method.

Figures 6.60, 6.61 and 6.62 compare the streamwise turbulent intensity, wall normal turbulent intensity and Reynolds shear stress with experiment for the WPG

**Table 6.15.** Freestream conditions.

| Quantity | Value |
|---|---|
| $M_\infty$ | 4.9 |
| $p_{t_\infty}$ (MPa) | 2.43 |
| $T_{t_\infty}$ (K) | 384 |
| $T_w/T_{aw}$ | 0.91 |
| $T_w$ (K) | 317 |
| $Re_\theta$ | 17 406[a] |

[a]At $x = 0.159$ m (figure 6.57).

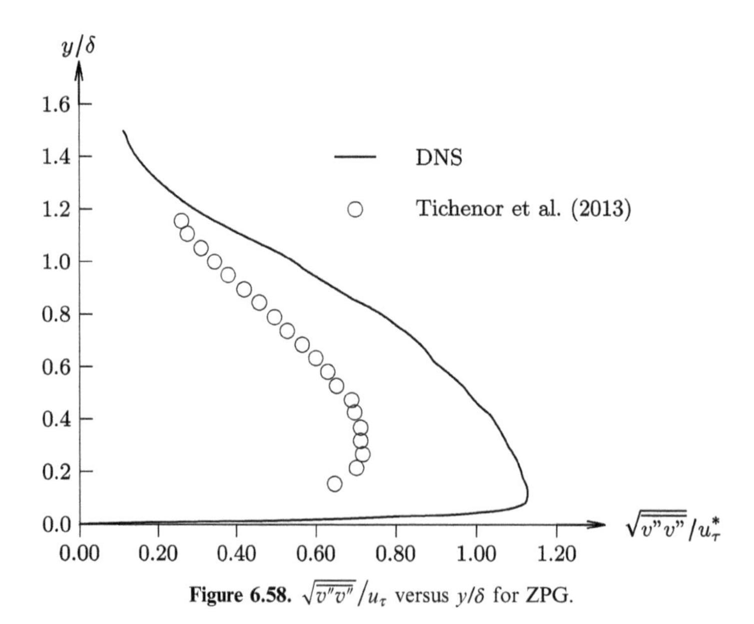


**Figure 6.58.** $\sqrt{\overline{v''v''}}/u_\tau$ versus $y/\delta$ for ZPG.

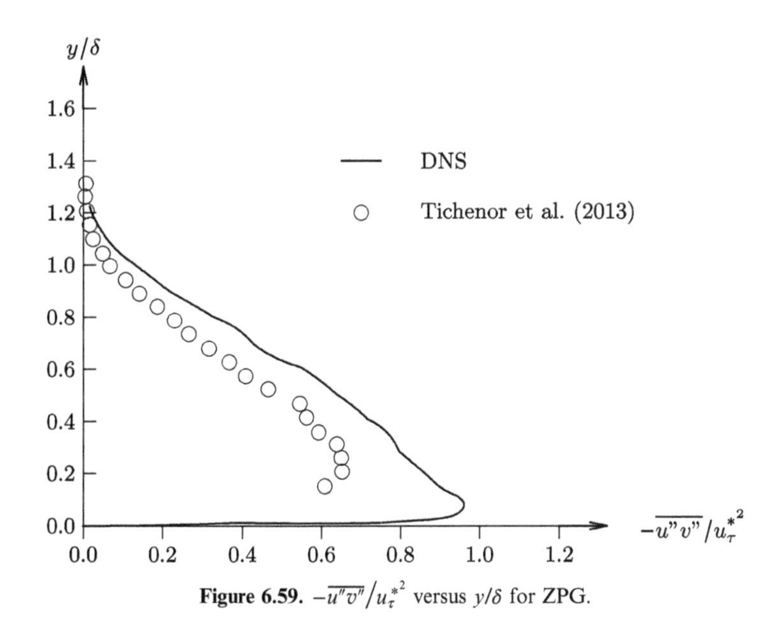


**Figure 6.59.** $-\overline{u''v''}/u_\tau^{*^2}$ versus $y/\delta$ for ZPG.

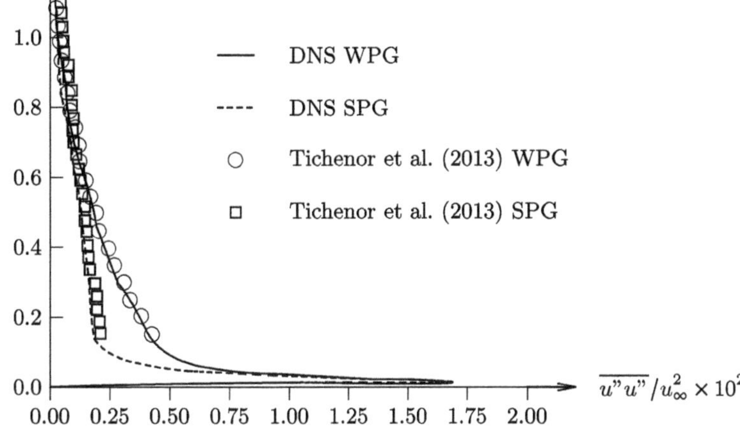


**Figure 6.60.** $\overline{u''u''}/u_\infty^2$ versus $y/\delta$ for WPG and SPG.

and SPG cases at location $x = 29.8$ cm. Close agreement between the simulation and experiment is evident for the streamwise turbulent intensity. However, significant discrepancies are evident between the simulation and experiment for wall normal turbulent intensity and Reynolds shear stress.

Figure 6.63 examines the validity of the Huang *et al* Strong Reynolds Analogy (3.358)

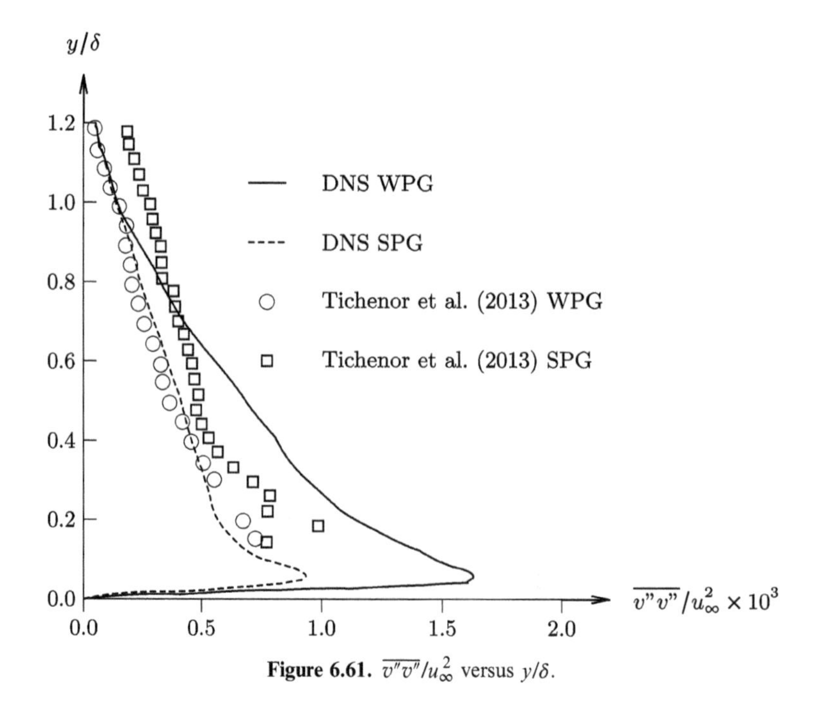


**Figure 6.61.** $\overline{v''v''}/u_\infty^2$ versus $y/\delta$.

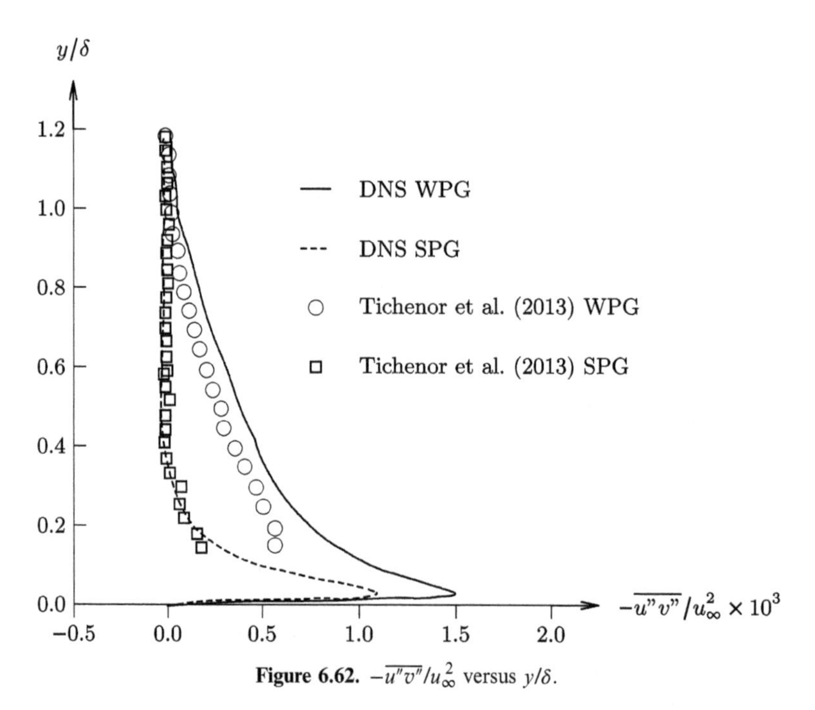


**Figure 6.62.** $-\overline{u''v''}/u_\infty^2$ versus $y/\delta$.

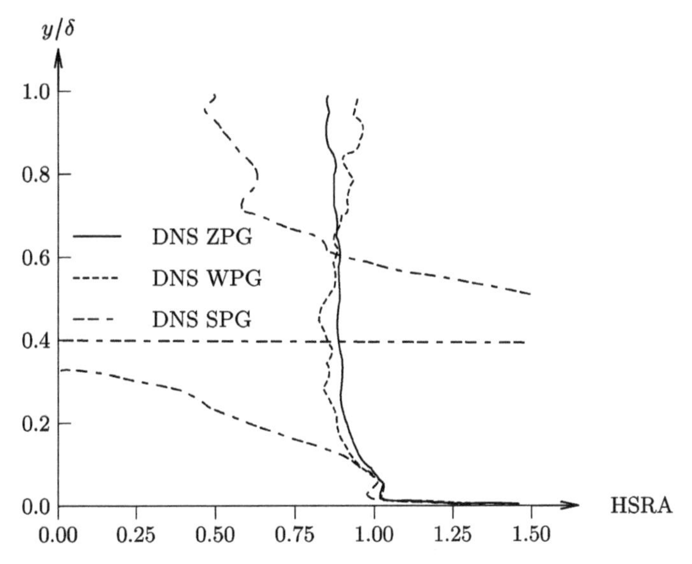


**Figure 6.63.** Evaluation of Huang Strong Reynolds Analogy.

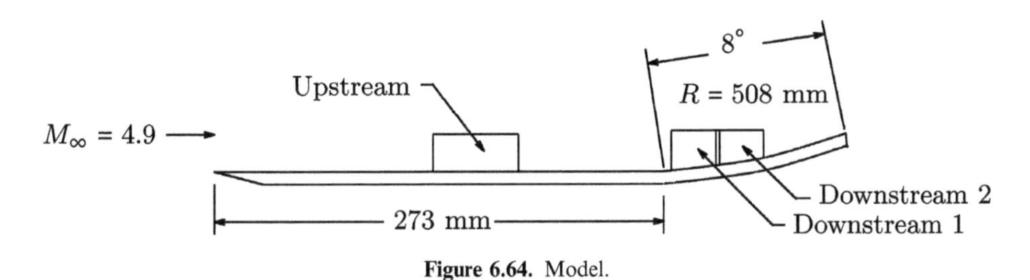


**Figure 6.64.** Model.

$$\frac{\left(\sqrt{\overline{T''^2}}/\overline{T}\right)}{(\gamma - 1)\overline{M}^2\left(\sqrt{\overline{u''^2}}/\overline{u}\right)} = \frac{1}{Pr_t}\frac{1}{\mid d\overline{T}_t/d\overline{T} - 1 \mid} \tag{6.26}$$

where $Pr_t$ is the local Prandtl number (2.55). The ordinate HSRA is the ratio of the left hand side to the right hand side. The WPG and SPG results are at location $x = 29.8$ cm. The Huang *et al* Strong Reynolds Analogy holds for the zero- and weak pressure gradient cases. However, a significant departure is observed for the strong pressure gradient case, due in large measure to the variation of $Pr_t$ across the boundary layer.

## 6.15 Nicholson *et al* (2021b)

Nicholson *et al* (2021b) performed a Direct Numerical Simulation of an adverse pressure gradient hypersonic near adiabatic turbulent boundary layer. The simulation corresponds to the experiment of Tichenor *et al* (2017). The model is shown in figure 6.64. The governing equations are the compressible Navier–Stokes equations. The simulation was performed using three overlapping domains. The flow conditions are listed in table 6.16. The locations for comparison with experiment are indicated in table 6.17 where $\beta = (dp/dx)\delta^*/\tau_w$ is the pressure gradient parameter.

The upstream flat plate boundary layer Van Driest transformed velocity profile is in close agreement with the Law of the Wall (Nicholson *et al* 2021b). The upstream streamwise turbulent intensity $\sqrt{\overline{u''u''}}/u_\tau^*$ is in close agreement with the experiment of Tichenor *et al* (2017). However, the wall normal turbulent intensity $\sqrt{\overline{v''v''}}/u_\tau^*$ and Reynolds shear stress $\overline{u''v''}/u_\tau^{*^2}$ are significantly different from the experiment of Tichenor *et al* (2017) similar to Nicholson *et al* (2021b). Figure 6.65 shows the streamwise turbulent stress $\overline{u'u'}/u_\infty^2$ at locations U, D1 and D2. Close agreement with experiment is evident. Figures 6.66 and 6.67 present the wall normal $\overline{v'v'}/u_\infty^2$ and shear stress $-\overline{u'v'}/u_\infty^2$, respectively. Significant differences between the simulation and experiment are evident similar to Nicholson *et al* (2021b).

Figure 6.68 displays the computed turbulent Prandtl number $Pr_t$ versus $y/\delta$ at the locations listed in table 6.17. The simulations show $Pr_t \approx 0.9$ for $y/\delta \leqslant 0.6$ with slightly lower values for $y/\delta > 0.6$.

Figure 6.69 shows the Huang *et al* Strong Reynolds Analogy (HSRA) (6.3). The abscissa is the ratio of the left side to the right side of (6.1). The Huang *et al* Strong Reynolds Analogy ratio is approximately 0.9 across the boundary layer at all locations.

**Table 6.16.** Freestream conditions.

| Quantity | Value |
|---|---|
| $M_\infty$ | 4.9 |
| $p_{t_\infty}$ (MPa) | 2.43 |
| $T_{t_\infty}$ (K) | 384 |
| $T_w/T_{aw}$ | 0.91 |
| $T_w$ (K) | 317 |
| $Re_\theta$ | 21 674[a] |

[a]At $x = 0.154$ m (figure 6.64).

**Table 6.17.** Location of measurements.

| *Location* | $x$ (m) | $\beta$ |
|---|---|---|
| U | 0.154 | 0.03 |
| D1 | 0.292 | 3.48 |
| D2 | 0.308 | 3.69 |

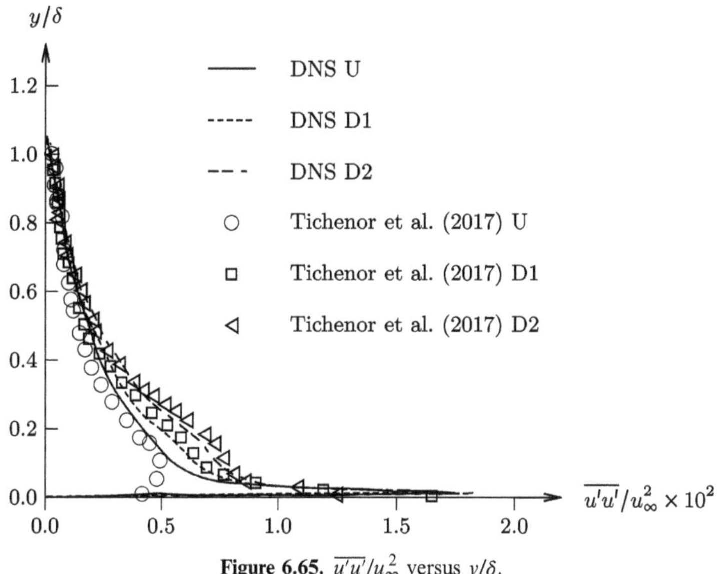


**Figure 6.65.** $\overline{u'u'}/u_\infty^2$ versus $y/\delta$.

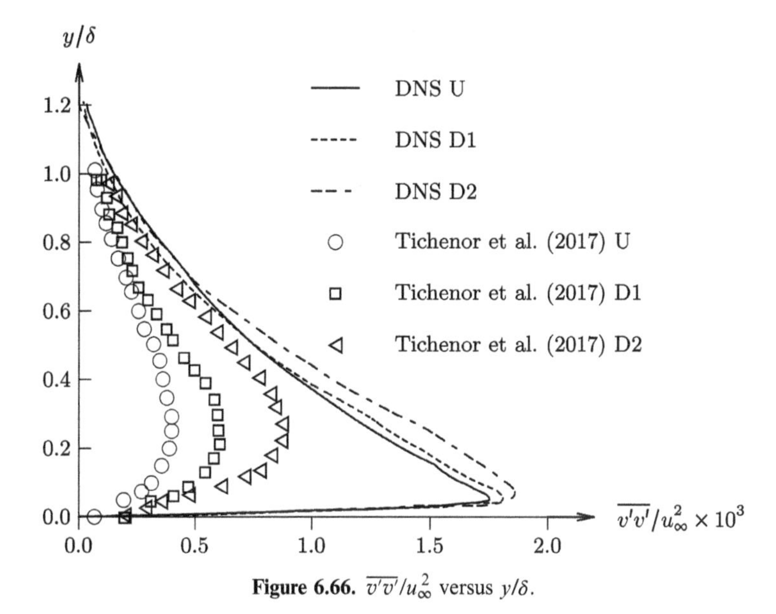


**Figure 6.66.** $\overline{v'v'}/u_\infty^2$ versus $y/\delta$.

## 6.16 Kianvashrad and Knight (2021–2)

Kianvashrad and Knight (2021) and Kianvashrad and Knight (2022) performed Large Eddy Simulations of a hypersonic cold wall flat plate turbulent boundary layer at Mach 6 and 8 and several wall temperature ratios. The governing equations are the compressible Navier–Stokes equations for a perfect gas. The flow conditions

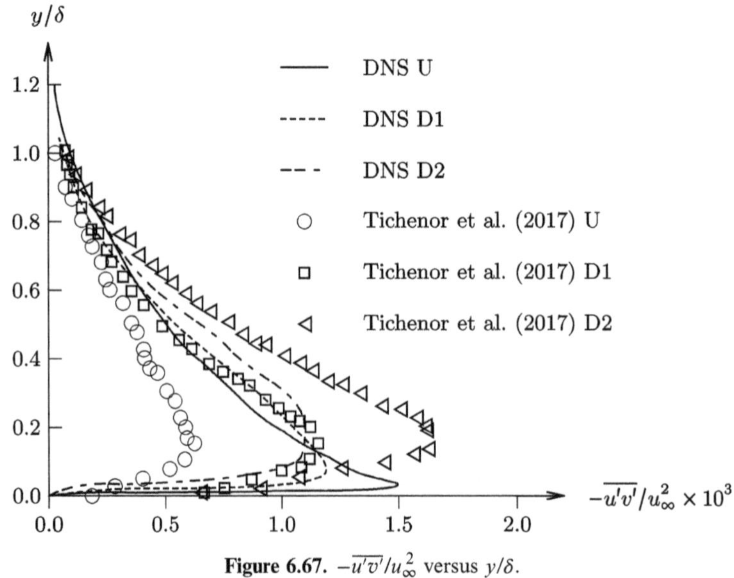


**Figure 6.67.** $-\overline{u'v'}/u_\infty^2$ versus $y/\delta$.

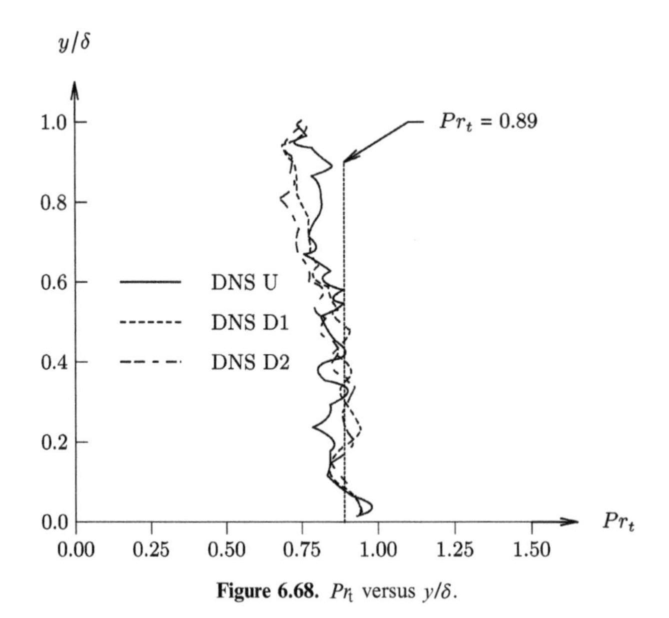


**Figure 6.68.** $Pr_t$ versus $y/\delta$.

are indicated in table 6.18. A novel recycling-rescaling method was developed using the total enthalpy and static pressure along with the velocity at the recycle station to generate the dynamic inflow turbulent boundary layer.

Figure 6.70 displays the Van Driest transformed velocity $u_{VD}/u_\tau$ versus $y^+$ for all cases. The predictions are in close agreement with the Law of the Wall. The constant

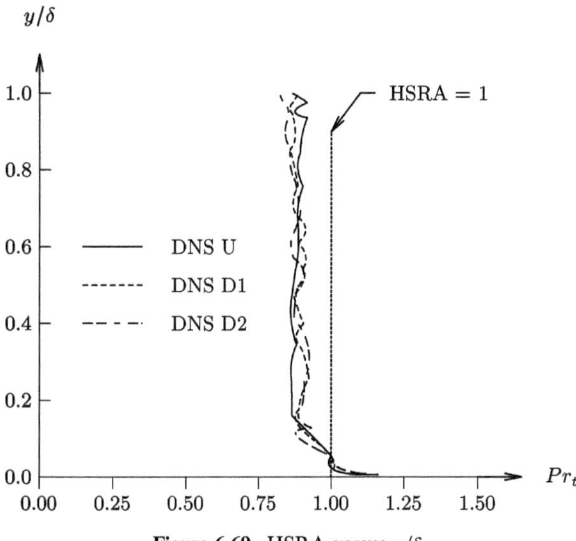


**Figure 6.69.** HSRA versus $y/\delta$.

**Table 6.18.** Freestream conditions.

| Quantity | Case 1 | Case 2 | Case 3 | Case 4 | Case 5 |
|---|---|---|---|---|---|
| $M_\infty$ | 6.0 | 6.0 | 6.0 | 8.0 | 8.0 |
| $T_w/T_{aw}$ | 1.0 | 0.79 | 0.54 | 1.0 | 0.75 |
| $Re_\delta$ | $10^5$ | $10^5$ | $10^5$ | $10^5$ | $10^5$ |

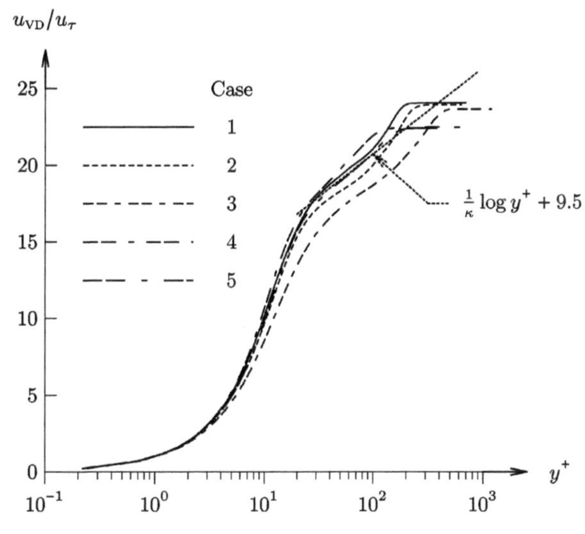


**Figure 6.70.** $u_{VD}/u_\tau$ versus $y^+$.

**Table 6.19.** Constant $C$ and Reynolds Analogy Factor.

| Quantity | Case 1 | Case 2 | Case 3 | Case 4 | Case 5 |
|---|---|---|---|---|---|
| $C$ | 9.5 | 8.4 | 7.2 | 9.5 | 9.1 |
| $2St/c_f$ | — | 0.98 | 1.02 | — | 0.84 |

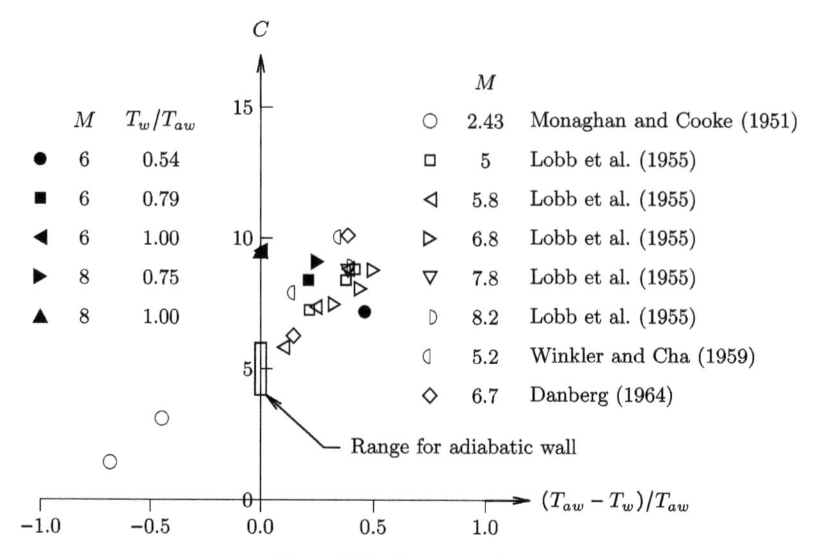


**Figure 6.71.** $C$ versus $T_w/T_{aw}$.

$C$ is listed in table 6.19 for each case. The value of $C$ decreases with decreasing $T_w/T_{aw}$ and is opposite to the trend of Danberg (1964) reproduced in figure 6.71.

The computed Reynolds Analogy Factor $2St/c_f$ for Cases 2 and 3 are listed in table 6.19 and are within experimental undertainty (figure 4.81).

Figure 6.72 shows the Huang *et al* Strong Reynolds Analogy (3.335) for several cases. The ordinate HSRA is the ratio of the left hand side of (3.335) to the right hand side of (3.335). The HSRA is typically between 0.8 and 1.0 except very close to the wall.

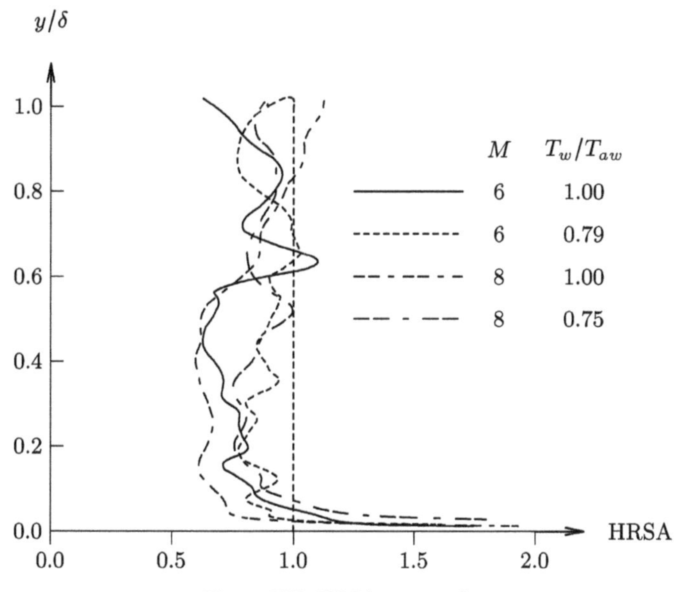


**Figure 6.72.** HRSA versus $y/\delta$.

# References

Bernardini M and Pirozzoli S 2011 Wall pressure fluctuations beneath supersonic turbulent boundary layers *Phys. Fluids* **23** 085102

Chu Y-B, Zhuang Y-Q and Lu X-Y 2013 Effect of wall temperature on hypersonic turbulent boundary layer *J. Turbul.* **14** 37–57

Danberg J 1964 Characteristics of the turbulent boundary layer with heat and mass transfer at $M = 6.7$ *Technical Report NOLTR 64-99* (White Oak, Silver Spring, MD: United States Naval Ordnance Laboratory)

DeGraaff D and Eaton J 2000 Reynolds-number scaling of the flat-plate turbulent boundary layer *J. Fluid Mech.* **422** 319–46

Duan L, Beekman I and Martín M P 2010 Direct numerical simulation of hypersonic turbulent boundary layers. Part 2. Effect of wall temperature *J. Fluid Mech.* **655** 419–45

Duan L, Beekman I and Martín M P 2011 Direct numerical simulation of hypersonic turbulent boundary layers. Part 3. Effect of Mach number *J. Fluid Mech.* **672** 245–67

Duan L, Choudhari M and Wu M 2014 Numerical study of acoustic radiation due to a supersonic turbulent boundary layer *J. Fluid Mech.* **746** 165–92

Duan L, Choudhari M and Zhang C 2016 Pressure fluctuations induced by a hypersonic turbulent boundary layer *J. Fluid Mech.* **804** 578–607

Duan L and Martín M P 2009a Effect of finite-rate chemical reactions on turbulence in hypersonic turbulent boundary layers AIAA Paper 2009-588 (Reston, VA: American Institute of Aeronautics and Astronautics)

Duan L and Martín M P 2009b Procedure to validate direct numerical simulations of wall-bounded turbulence including finite-rate reactions *AIAA J.* **47** 244–51

Duan L and Martín M P 2011a Assessment of turbulence-chemistry interaction in hypersonic turbulent boundary layers *AIAA J.* **49** 172–84

Duan L and Martín M P 2011b Direct numerical simulation of hypersonic turbulent boundary layers. Part 4. Effect of high enthalpy *J. Fluid Mech.* **684** 25–59

Duan L and Martín M P 2011c Effective approach for estimating turbulence-chemistry interaction in hypersonic turbulent boundary layers *AIAA J.* **49** 2239–47

Duan L, Martín M P, Feldick A, Modest M and Levin D 2012 Study of turbulence-radiation interaction in hypersonic turbulent boundary layers *AIAA J.* **50** 447–53

Duan L, Martín M P, Sohn I, Levin D and Modest M 2011 Study of emission turbulence-radiation interaction in hypersonic turbulent boundary layers *AIAA J.* **49** 340–8

Gordon S and McBride B 1994 Computer program for calculation of complex chemical equilibrium compositions and applications *NASA RP 1311* (Washington, DC: National Aeronautics and Space Administration)

Gupta R, Yos J, Thompson R and Lee K 1990 A review of reaction rates and thermodynamic and transport properties for an 11-species air model for chemical and thermal nonequilibrium calculations to 30 000 K *Reference Report 1232* (Washington, DC: National Aeronautics and Space Administration)

Huang J, Nicholson G, Duan L, Choudhari M and Bowersox R 2020 Simulation and modeling of cold wall hypersonic turbulent boundary layers on flat plate AIAA Paper 2020-0571 (Reston, VA: American Institute of Aeronautics and Astronautics)

Huang P, Coleman G and Bradshaw P 1995 Compressible turbulent channel flows: DNS results and modelling *J. Fluid Mech.* **305** 185–218

Jiménez J, Hoyas S, Simens M and Mizuno Y 2010 Turbulent boundary layers at moderate Reynolds numbers *J. Fluid Mech.* **657** 335–60

Kianvashrad N and Knight D 2021 Large eddy simulation of hypersonic cold wall flat plate—part II AIAA Paper 2021-2882 (Reston, VA: American Institute of Aeronautics and Astronautics)

Kianvashrad N and Knight D 2022 Large eddy simulation of hypersonic cold wall flat plate—part II AIAA Paper 2022-0588 (Reston, VA: American Institute of Aeronautics and Astronautics)

Li X-L, Fu D-X and Ma Y-W 2006 Direct numerical simulation of spatially evolving supersonic turbulent boundary layer at Mach 6 *Chin. Phys. Lett.* **23** 1519–22

Liang X and Li X 2013 DNS of a spatially evolving hypersonic turbulent boundary layer at Mach 8 *Sci. China: Phys. Mech. Astron.* **56** 1408–18

Liang X, Li X and Fu D 2012 DNS and analysis of a spatially evolving hypersonic turbulent boundary layer over a flat plate at Mach 8 *Sci. Sin.-Phys. Mech. Astron.* **42** 282–93 (in Chinese)

Lobb R, Winkler E and Persh J 1955 NOL hypersonic tunnel no. 4 results VII: experimental investigation of turbulent boundary layers in hypersonic flight *Technical Report NAVORD Report 3880* (White Oak, Silver Spring, MD: United States Naval Ordnance Laboratory)

Maeder T, Adams N and Kleiser L 2001 Direct numerical simulation of turbulent supersonic boundary layers by an extended temporal approach *J. Fluid Mech.* **429** 187–216

Martín M P 2007 Direct numerical simulation of hypersonic turbulent boundary layers. Part 1. Initialization and comparison with experiments *J. Fluid Mech.* **570** 347–64

Morkovin M 1962 Effects of compressibility on turbulent flows *Mécanique de la Turbulence, Colloques Internationaux du Centre National de la Recherche Scientifique* ed A Favre (Paris: Centre National de la Recherche Scientifique) pp 367–80

Moser R, Kim J and Mansour N 1999 Direct numerical simulation of turbulent channel flow up to $Re_\tau = 5590$ *Phys. Fluids* **11** 943–5

Nicholson G, Huang J, Duan L and Choudhari M 2021a Simulation and modeling of hypersonic turbulent boundary layers subject to adverse pressure gradients due to concave streamline curvature AIAA Paper 2021-2891 (Reston, VA: American Institute of Aeronautics and Astronautics

Nicholson G, Huang J, Duan L, Choudhari M and Bowersox R 2021b Simulation and modeling of hypersonic turbulent boundary layers subject to favorable pressure gradients due to streamline curvature AIAA Paper 2021-1672 (Reston, VA: American Institute of Aeronautics and Astronautics)

Park C 1990 *Nonequilibrium Hypersonic Aerothermodynamics* (New York: Wiley)

Patel A, Peetr J, Boersma B and Pecnik R 2015 Semi-local scaling and turbulence modulation in variable property turbulent channel flows *Phys. Fluids* **27** 095101

Peltier S, Humble R and Bowersox R 2016 Crosshatch roughness distortions on a hypersonic turbulent boundary layer *Phys. Fluids* **28** 045105

Priebe S and Martín M P 2011 Direct numerical simulation of a hypersonic turbulent boundary layer on a large domain AIAA Paper 2011-3432 (Reston, VA: American Institute of Aeronautics and Astronautics)

Schlatter P and Örlü R 2010 Assessment of direct numerical simuation data of turbulent boundary layers *J. Fluid Mech.* **659** 116–26

Shadloo M, Hadjadi A and Hussain F 2015 Statistical behavior of supersonic turbulent boundary layers with heat transfer at $M_\infty = 2$ *Int. J. Heat Fluid Flow* **53** 113–34

Silero J, Jiménez J and Moser R 2013 One point statistics for turbulent boundary layers at Reynolds numbers up to $\delta^+ = 2000$ *Phys. Fluids* **25** 105102

Spalart P 1988 Direct simulation of a turbulent boundary layer up to $Re_\theta$ = 1410 *J. Fluid Mech.* **187** 61–98

Tichenor N, Humble R and Bowersox R 2013 Response of a hypersonic turbulent boundary layer to favorable pressure gradient *J. Fluid Mech.* **722** 187–213

Tichenor N, Neel I, Leidy A and Bowersox R 2017 Influence of streamline adverse pressure gradients on the structure of a Mach 5 turbulent boundary layer AIAA Paper 2017-1697 (Reston, VA: American Institute of Aeronautics and Astronautics)

Trettel A and Larsson J 2016 Mean velocity scaling for compressible wall turbulence with heat transfer *Phys. Fluids* **28**

Walz A 1969 *Boundary Layers of Flow and Temperature* (Cambridge, MA: MIT Press)

Williams O, Sahoo D, Baumgartner M and Smits A 2018 Experiments on the structure and scaling of hypersonic turbulent boundary layers *J. Fluid Mech.* **834** 237–70

Wright M, White T and Mangini N 2009 Data parallel line relaxation (DPLR) user manual: acadia, version 4.01.1 *NASA TM 2009-215388* (Washington, DC: National Aeronautics and Space Administration)

Yos J 1963 Transport properties of nitrogen, hydrogen, oxygen and air to 30 000 K *Technical Report TR AD-TM-63-7* Avco Corporation

Zhang C, Duan L and Choudhari M 2017 Effect of wall cooling on boundary-layer-induced pressure fluctuations at Mach 6 *J. Fluid Mech.* **822** 5–30

Zhang C, Duan L and Choudhari M 2018 Direct numerical simulation database for supersonic and hypersonic turbulent boundary layers *AIAA J.* **56** 4297–311

Zhang Y, Bi W, Hussain F and She Z 2014 A generalized Reynolds analogy for compressible wall-bounded turbulent flows *J. Fluid Mech.* **739** 392–420

# Chapter 7

# Direct Numerical Simulation and Large Eddy Simulation—shock boundary layer interaction

## Abstract

Application of Direct Numerical Simulation and Large Eddy Simulation for shock wave turbulent boundary layer interaction appeared beginning in the early twenty-first century for supersonic flows. Examples of Direct Numerical Simulation include Adams (2000), Pirozzoli *et al* (2004), and Pirozzoli and Grasso (2006). Examples of Large Eddy Simulation include Urbin and Knight (2001), Garnier *et al* (2002), Rizzetta and Visbal (2002), Loginov *et al* (2006), Touber and Sandham (2009a), Touber and Sandham (2009b), Morgan *et al* (2010), Hadjadj (2012), Aubard *et al* (2013), and Jammalamadaka *et al* (2013). Application to hypersonic shock wave turbulent boundary layer interaction began subsequently. Examples are presented in this chapter.

## 7.1 Edwards *et al* (2008)

Edwards *et al* (2008) performed a wall modeled Large Eddy Simulation of a hypersonic shock wave turbulent boundary layer interaction at Mach 5 generated by a 28° compression ramp. The flow configuration is illustrated in figure 7.1. An incoming turbulent boundary layer interacts with the shock wave generated by the ramp resulting in a separated region near the corner. The flow conditions are listed in table 7.1 and correspond to the experiments of Gramann (1989) and Erengil and Dolling (1991). The simulations were performed using a hybrid Reynolds-averaged Navier–Stokes (RANS)/LES model (Baurle *et al* 2003, Xiao *et al* 2003, Fan *et al* 2004, Xiao *et al* 2004).

Figure 7.2 shows the computed and experimental mean surface pressure $p$ versus distance $x/\delta$ measured from the corner where $\delta$ is the boundary layer thickness upstream of the interaction. The variation in computed mean wall pressure with the value of the turbulence model constant $A_{\text{SST}} = 0$ and 0.9 in the RANS shear stress

 

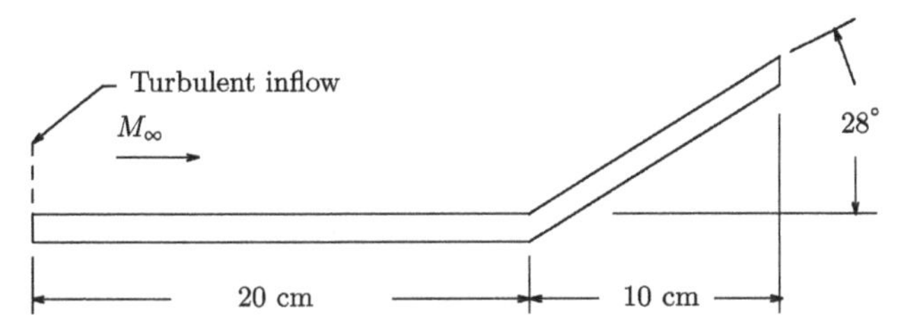


**Figure 7.1.** Model.

**Table 7.1.** Freestream conditions.

| Quantity | Value |
|---|---|
| $M_\infty$ | 5 |
| $T_w/T_{aw}$ | 1.0 |
| $Re_\delta \times 10^{-5}$ | 8.75 |

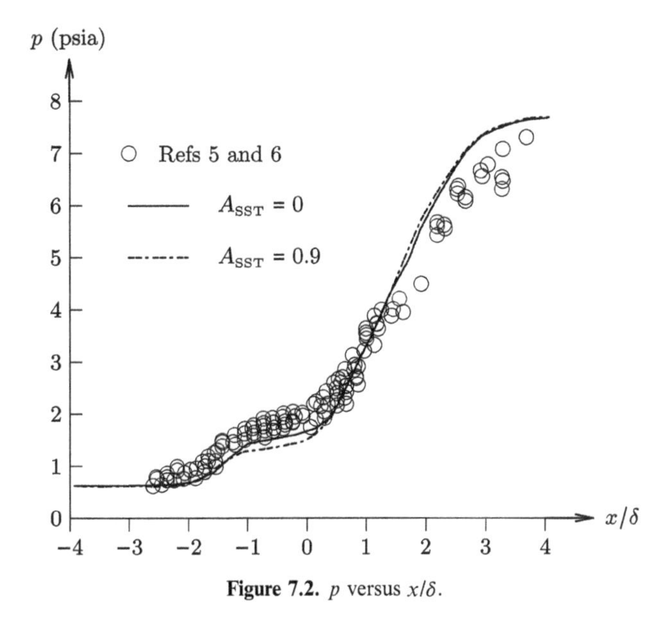


**Figure 7.2.** $p$ versus $x/\delta$.

transport model of Menter (1994) is shown. There is overall reasonable agreement between the experiment and simulation, although the computed plateau pressure within the separated region is underestimated, and the recovery of the pressure downstream of the corner is overestimated. Figure 7.3 displays the computed and experimental root-mean-square wall pressure fluctuations normalized by the local mean pressure $\sqrt{\overline{p'^2}}/p_w$ versus $x/\delta$. The computed results for $A_{SST} = 0$ and 0.9 are shown. The simulations accurately predict the location of the two peaks in pressure

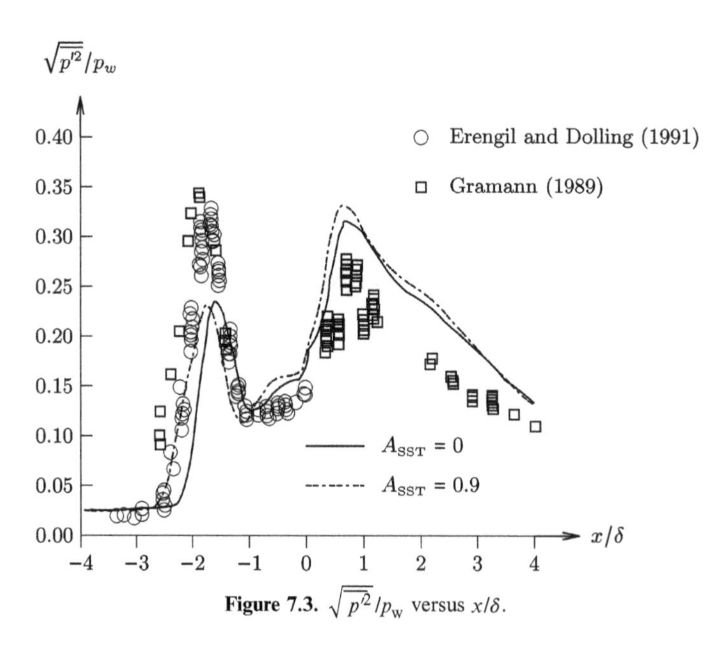


**Figure 7.3.** $\sqrt{p'^2}/p_w$ versus $x/\delta$.

fluctuations at $x/\delta = -1.6$ and 0.8. Additional comparison is provided in Edwards *et al* (2008) for conditionally averaged surface pressure and pitot pressure profiles.

## 7.2 Sandham *et al* (2014)

Sandham *et al* (2014) conducted a combined computational and experimental study of shock wave transitional boundary layer interaction on a flat plate. The experiments were performed in three different facilities: RWG (Rohrwindkanal, Göttingen), H2K (hypersonic wind tunnel, Cologne, section A.11.1) and HEG (high enthalpy shock tunnel, Göttingen, section A.11.2). The model is shown in figure 7.4. The flat plate was 600 mm in length and 340 mm in width, and the 4° shock generator was 210 mm in length and 300 mm in width. The shock generator position was varied with an inviscid shock impingement location from $x = 230$ mm to 337 mm. Experimental diagnostics include surface heat transfer and surface pressure. The freestream conditions are shown in table 7.2. Three Direct Numerical Simulations were performed without the shock impingement, and five with the impinging shock. A random density perturbation was added to the similarity solution of a laminar compressible boundary layer at the inflow boundary to initiate transition to turbulence. The amplitude of the disturbance was adjusted to agree with the experimental transition location identified from the experimental Stanton number distribution for the experiments at RWG and H2K.

Figure 7.5 shows the computed and experimental Stanton numbers for three different shock impingement locations. Also shown are the flat plate laminar and turbulent Stanton numbers in the absence of shock impingement. The Reynolds numbers based upon the inviscid shock impingement location $x_{\text{imp}}$ for the simulations are $1.099 \times 10^6$ (SHU), $1.338 \times 10^6$ (SHC), and $1.577 \times 10^6$ (SHD). There is close agreement between the computation and experiment.

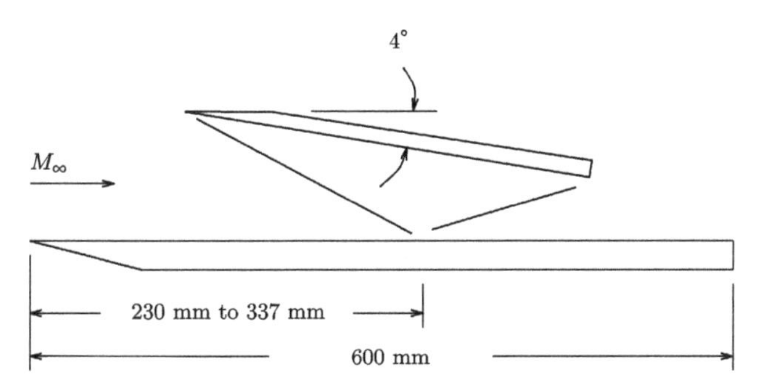


**Figure 7.4.** Model.

**Table 7.2.** Freestream conditions.

| Quantity | RWG | H2K | HEG |
|---|---|---|---|
| Gas | Air | Air | Air |
| $M_\infty$ | 5.98 | 6 | 5.7 |
| $T_{t_\infty}$ (deg K) | 537.8 to 565.8[a] | NS | 2680 |
| $p_{t_\infty}$ (MPa) | 0.405 to 1.66[a] | NS | NS |
| $Re_\infty$ ($10^6$ m$^{-1}$) | 2.8 →[a] 12.1 | 3, 6, and 12 | 13.8 → 14.5 |
| $T_w/T_\infty$ | 4.5 | 4.7 | 0.7 |

[a] From Schülein (2014).
NS, not specified.

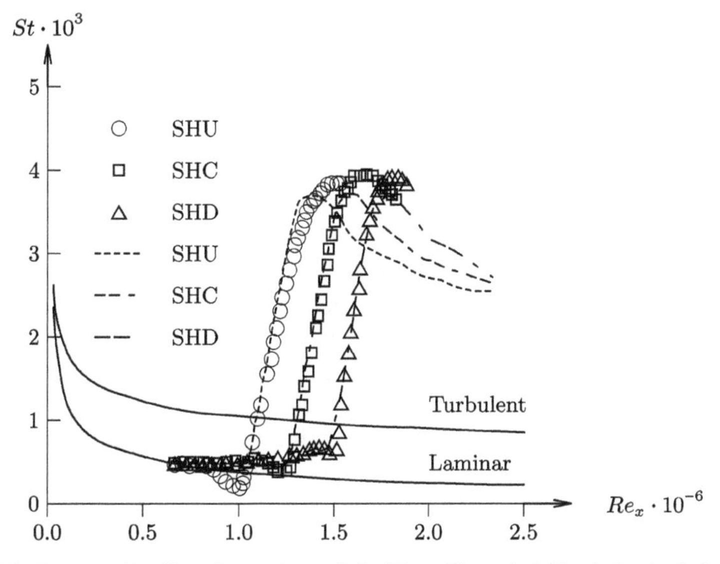


**Figure 7.5.** *St* versus $Re_x$. Experiments in symbols, Direct Numerical Simulation in dashed lines.

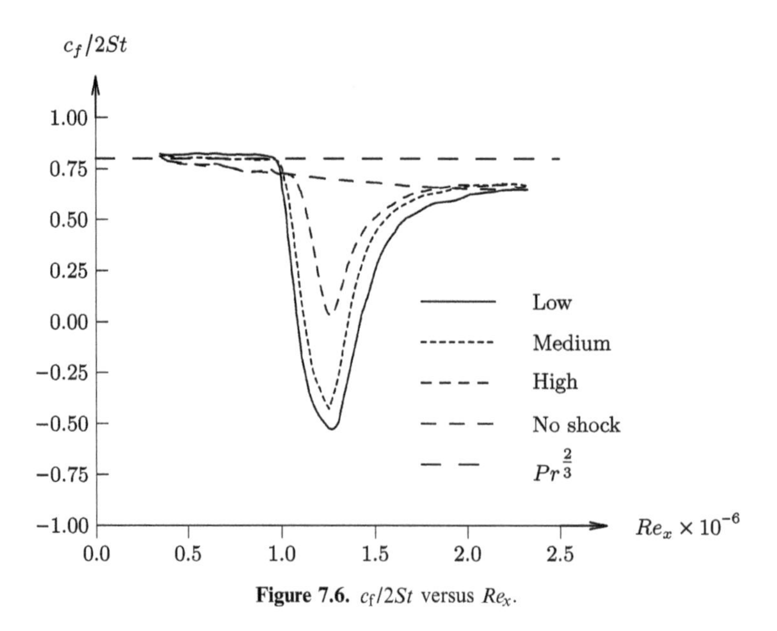


**Figure 7.6.** $c_f/2St$ versus $Re_x$.

Figure 7.6 displays the computed inverted Reynolds Analogy Factor $c_f/2St$ for three simulations corresponding to $Re_{imp} = 1.338 \times 10^6$ with low-, medium-, and high-amplitude initial density perturbations. The laminar flat plate value $c_f/2St = Pr^{\frac{2}{3}}$ is also indicated. A significant effect of the initial density perturbation is evident, with minimum value of $c_f/2St$ increasing with the amplitude of the initial density perturbation.

## 7.3 Fang *et al* (2015)

Fang *et al* (2015, 2017) performed a Large Eddy Simulation of a hypersonic shock wave turbulent boundary layer interaction generated by a single fin. The flow configuration is shown in figure 7.7. An incoming equilibrium Mach 5 turbulent boundary layer interacts with the shock generated by a 23° fin. The flow conditions are listed in table 7.3 and correspond to the experiments of Schülein (2006). The subgrid scale stresses and heat flux were modeled using the dynamic Smagorinsky model (Ghosal *et al* 1995, Moin *et al* 1991). The dynamic inflow boundary layer was generated using a separate flat plate Large Eddy Simulation using a wall blowing and suction technique similar to Rai and Moin (1993). See also Fang *et al* (2014).

Figures 7.8 and 7.9 present the statistics of the inflow equilibrium turbulent boundary layer. Figure 7.8 displays the Van Driest transformed velocity $u_{vd}/u_\tau$ versus $y^+$. Close agreement with the Law of the Wall is observed. Figure 7.9 presents the density scaled streamwise, wall normal, and spanwise root-mean-square turbulent fluctuations. There is close agreement with low-speed experiments (Purtell *et al* 1981, Erm and Joubert 1991) and compressible Direct Numerical Simulation results at $M_\infty = 1.3$ (Pirozzoli *et al* 2010).

The flowfield structure of the single fin interaction displays a conical behavior outside of an initial inception region near the fin leading edge (Alvi and Settles 1992).

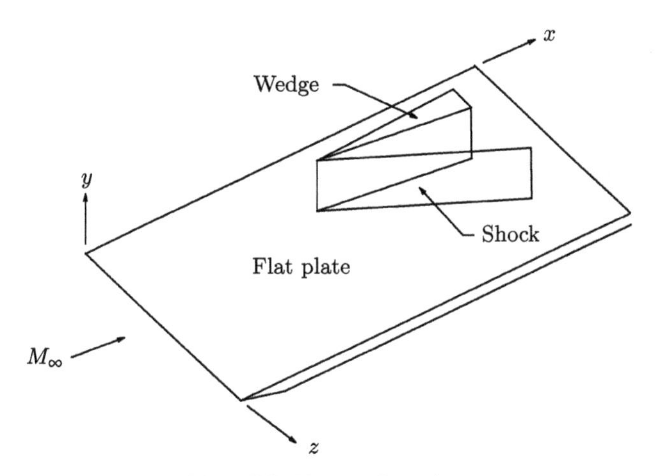


**Figure 7.7.** Flow configuration.

**Table 7.3.** Flow conditions.

| Quantity | Value |
|---|---|
| $M_\infty$ | 5 |
| $p_{t_\infty}$ (MPa) | 2.13 |
| $T_{t_\infty}$ (K) | 410 |
| $T_w/T_{aw}$ (plate) | 0.81 |
| $T_w/T_{aw}$ (fin) | 1.0 |
| $Re_\delta \times 10^{-4}$ | 14.2 |
| $Re_{\delta^*} \times 10^{-4}$ | 5.92 |
| $Re_\theta \times 10^{-3}$ | 6.29 |

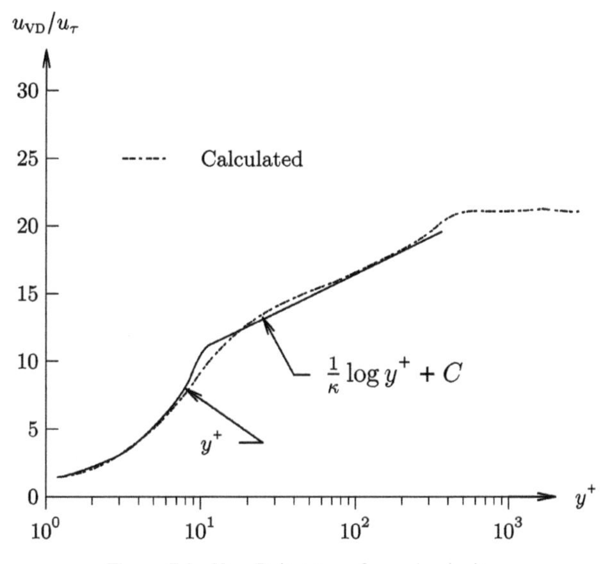


**Figure 7.8.** Van Driest transformed velocity.

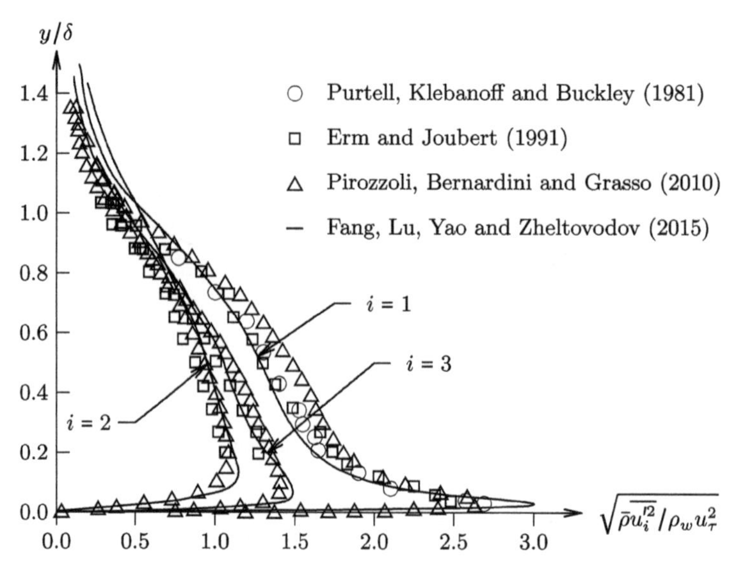


**Figure 7.9.** Density scaled turbulent fluctuations for streamwise ($i$ = 1), spanwise ($i$ = 3), and wall normal ($i$ = 2) versus $y/\delta$.

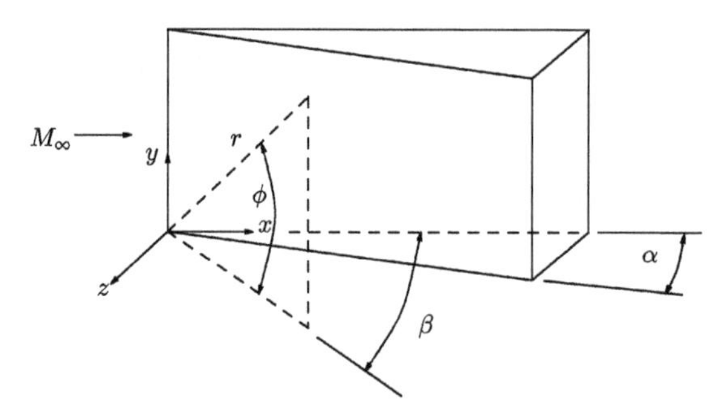


**Figure 7.10.** Spherical polar coordinate system.

The flowfield structure can be viewed in a spherically conical coordinate system whose origin (virtual conical origin) is located upstream of the fin leading edge (figure 7.10).

Figure 7.11 shows the computed and experimental mean surface pressure $p/p_\infty$ versus $z/x$. Outside of an initial inception zone near the fin leading edge, the surface pressure displays conical behavior (Alvi and Settles 1992), and the computed results are in close agreement with experiment. The location of the separation and reattachment locations are indicated. Figure 7.12 displays the computed and experimental mean skin friction coefficient $c_f$ versus $z/x$. Good agreement is observed between computation and experiment for the conical region between the initial separation location $S_1$ and slightly inward of the second separation location $S_2$. However, the peak $c_f$ in the vicinity of the first reattachment location $R_1$ is underpredicted by a factor of four. The computed numerical schlieren image on a spherical arc shows the key features of the model of Alvi and Settles (1992). The computed surface skin

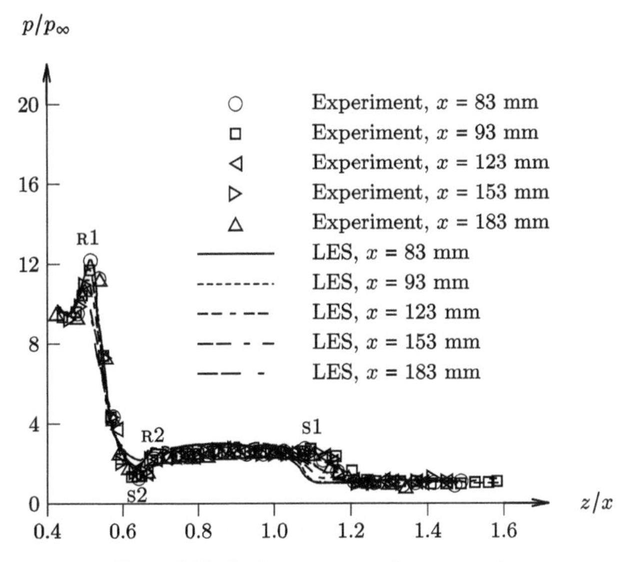


**Figure 7.11.** Surface pressure $p/p_\infty$ versus $z/x$.

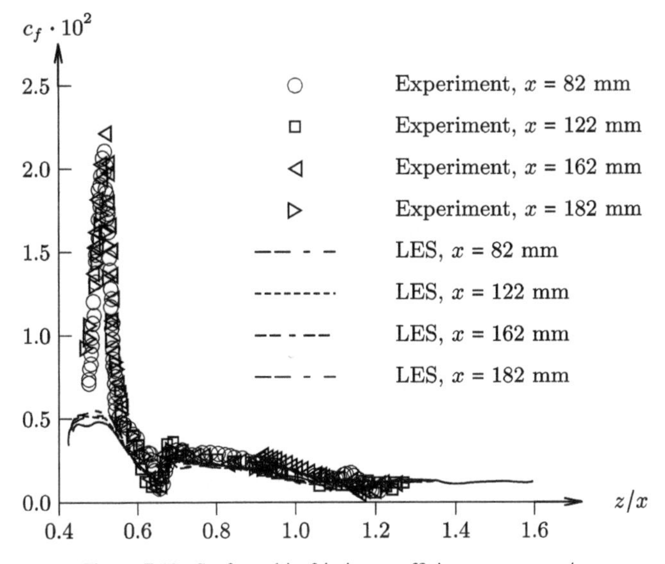


**Figure 7.12.** Surface skin friction coefficient $c_f$ versus $z/x$.

friction lines display primary and secondary separation and attachment lines consistent with the surface skin friction topology described in Zheltovodov and Knight (2011).

## 7.4 Yang *et al* (2018)

Yang *et al* (2018) performed a Wall Modeled Large Eddy Simulation (WMLES) of a shock wave transitional boundary layer interaction at Mach 6 corresponding to the experiments of Schülein (2014) and Willems *et al* (2015). The flow configuration is illustrated in figure 7.13. An incoming laminar boundary layer interacts with an incident shock wave generated by a 4° wedge. The flow conditions are described in table 7.4. The similarity inflow profile is perturbed by broadband acoustic disturbances generated by prescribed density fluctuations to force transition.

Figure 7.14 displays the computed Stanton number versus $Re_x$ for two different WMLES cases with different grid resolution, the Direct Numerical Simulation

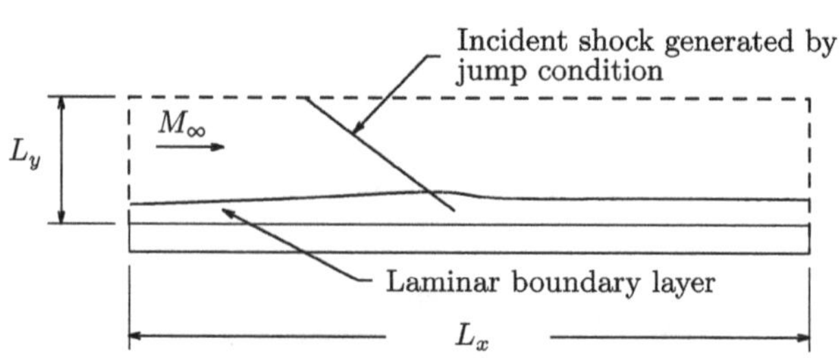


**Figure 7.13.** Flow configuration.

**Table 7.4.** Freestream conditions.

| Quantity | Value |
|---|---|
| $M_\infty$ | 6.0 |
| $T_w/T_\infty$ | 4.5 |
| $T_w/T_{aw}$ | 0.61[a] |
| $Re_{\delta^*}$ | 6830 |

[a]Using equation (3.46) with $Pr_t = 0.89$.

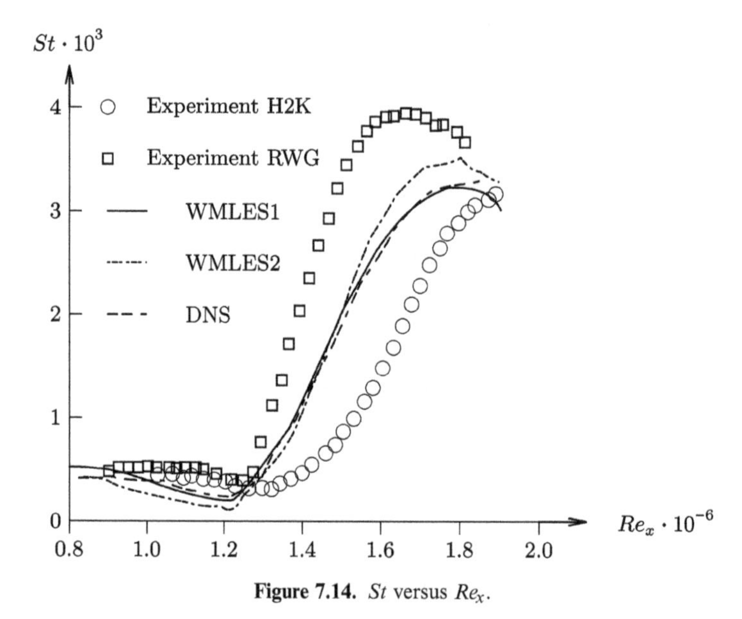


**Figure 7.14.** $St$ versus $Re_x$.

results of Sandham *et al* (2014), and the experiments of Schülein (2014) and Willems *et al* (2015). The WMLES results are in close agreement with the Direct Numerical Simulation computation which used a similar inflow density perturbation. Both the WMLES and Direct Numerical Simulation show differences in comparison to experiment due to the uncertainty in modeling the experimental inflow disturbance level.

## 7.5 Fu *et al* (2020)

Fu *et al* (2020) performed a WMLES of an intersecting shock wave turbulent boundary layer interaction at Mach 8.23 corresponding to the experiments of Kussoy and Horstman (1992) (see also Kussoy *et al* 1993). The flow configuration is shown in figure 7.15. An incoming turbulent boundary layer interacts with the crossing shock waves generated by the pair of sharp fins attached normal to the flat plate. The flow conditions are listed in table 7.5. The inlet boundary conditions upstream of the leading edge of the flat plate are uniform flow with turbulent fluctuations created by a synthetic turbulence method (Wu 2017). The subgrid stress model is a static coefficient Vreman model (Vreman 2004).

Figures 7.16 and 7.17 show the computed and experimental normalized mean surface pressure $p/p_\infty$ and mean heat transfer $q_w/q_{w_\infty}$ versus $x/L_r$ on the flat plate centerline where $L_r = 1$ cm. The streamwise location of the inviscid intersection of the crossing shocks is indicated. Close agreement is evident between the experiment and Large Eddy Simulation. The initial rise in surface pressure and heat transfer is due to the streamwise pressure gradient generated by the intersection of the crossing

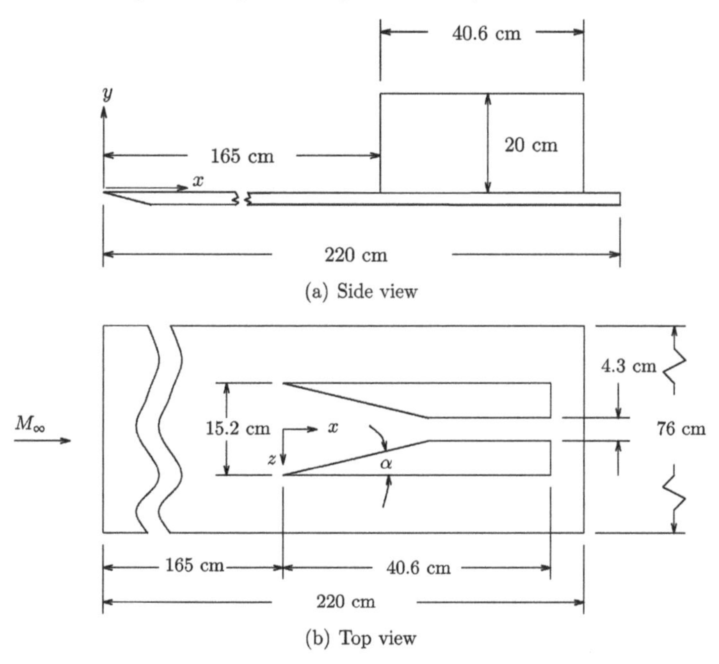


**Figure 7.15.** Model.

**Table 7.5.** Freestream conditions.

| Quantity | Value |
|---|---|
| $M_\infty$ | 8.23 |
| $p_{t_\infty}$ (MPa) | 5.07 |
| $T_{t_\infty}$ (K) | 1177 |
| $T_w$ (K) | 300 |
| $T_w/T_{aw}$ | 0.38[a] |
| $Re_\delta \times 10^{-4}$ | 17.0 |

[a] Using equation (3.46) with $Pr_t = 0.89$.

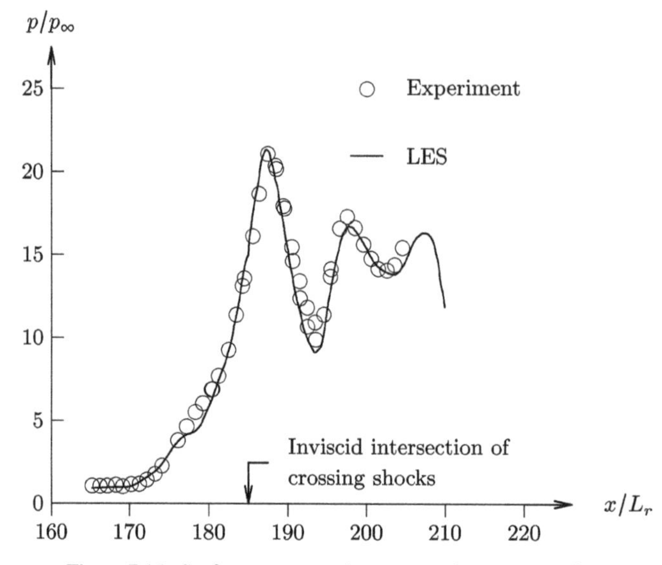


**Figure 7.16.** Surface pressure $p/p_\infty$ versus $x/L_r$ on centerline.

shock wave. The subsequent expansion and decrease in heat flux is due to the expansion generated by the two fins.

Figures 7.18 and 7.19 display the computed and experimental normalized mean surface pressure $p/p_\infty$ and mean heat transfer $q_w/q_{w_\infty}$ versus $z/L_r$ where the spanwise coordinate $z$ is measured from the streamwise centerline of the flat plate. Results are presented for the surface pressure at $x/L_r = 182.2$, 187.5, and 192.0 and the surface heat transfer at $x/L_r = 181.5$, 185.8, and 190.3. The three streamwise locations are upstream, approximately coincident, and downstream of the inviscid intersection of the crossing shocks. The Large Eddy Simulation results are in generally good agreement with experiment.

## 7.6 Vopiani *et al* (2020)

Volpiani *et al* (2020) performed Direct Numerical Simulations of a shock wave turbulent boundary layer interaction on a flat plate at Mach 5 corresponding to the

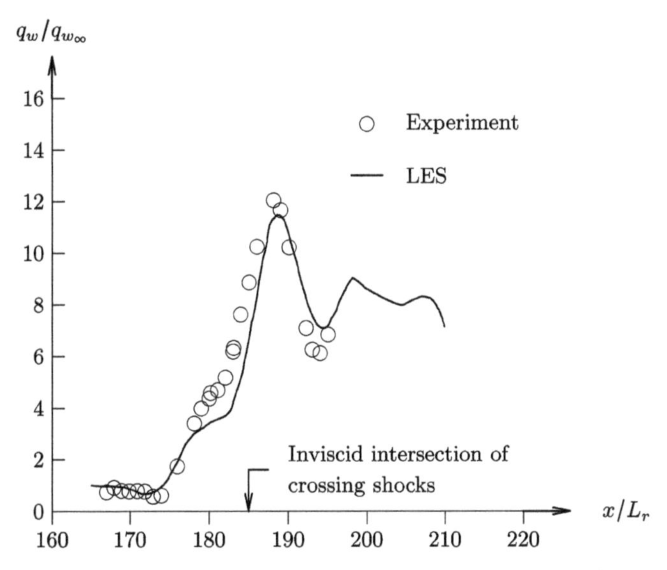


**Figure 7.17.** Surface heat transfer $q_w/q_{w_\infty}$ versus $x/L_r$ on centerline.

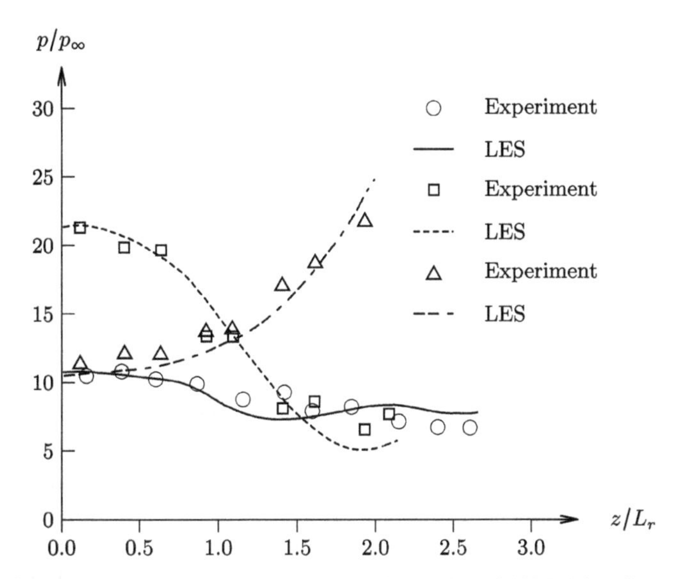


**Figure 7.18.** Surface pressure $p/p_\infty$ at $x/L_r$ = 182.2 (○), 187.5 (□), and 192.0 (Δ) and corresponding Large Eddy Simulation results.

experiments of Schülein (2006) with shock generator angles of $\beta = 6°$, 10°, and 14°. Three simulations were performed for a cold wall with $T_w/T_{aw} = 0.8$ and one simulation for a heated wall at $T_w/T_{aw} = 1.9$. The freestream conditions at the shock impingement location are listed in table 7.6.

Figure 7.20 displays the computed and experimental skin friction coefficient $c_f$ versus the normalized distance $(x - x_i)/\delta_i$ where $x_i$ is the streamwise location of the intersection of the shock wave with the flat plate (assuming inviscid flow) and $\delta_i$ is

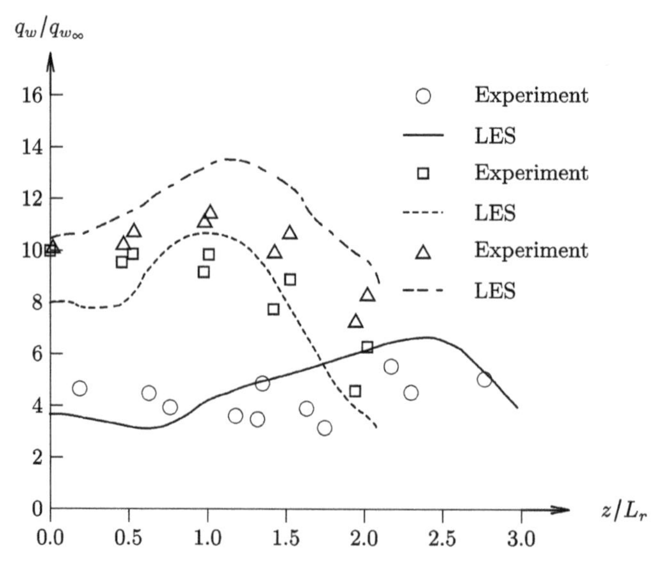


**Figure 7.19.** Surface heat transfer $q_w/q_{w_\infty}$ at $x/L_r$ = 181.5 (○), 185.8 (□), and 190.3 (Δ) and corresponding Large Eddy Simulation.

**Table 7.6.** Freestream conditions.

| Quantity | Case 1 | Case 2 | Case 3 | Case 4 | Experiment |
|---|---|---|---|---|---|
| $M_\infty$ | 5 | 5 | 5 | 5 | 5 |
| $T_w/T_{aw}$ | 0.8 | 0.8 | 0.8 | 1.9 | 0.8 |
| $Re_\delta \times 10^{-5}$ | 0.91 | 1.34 | 1.73 | 1.38 | 1.8 |
| $Re_\theta \times 10^{-3}$ | 3.76 | 5.42 | 6.94 | 3.89 | 7.40 |

the boundary layer thickness at $x_i$ in the absence of the shock wave. Experimental data are shown for $\beta = 10°$ and 14° and computed results for $\beta = 6°$, 10°, and 14°. The computed skin friction shows a significantly smaller separation region for $\beta = 10°$ and 14° compared to experiment. Volpiani *et al* (2020) suggest that three-dimensional effects in the experiment may explain the difference; however, no substantive proof is provided for the statement.

Figure 7.21 shows the computed and experimental normalized surface pressure $p/p_\infty$ versus $(x - x_i)/\delta_i$. The experimental upstream influence point[1] occurs further upstream in the experiment compared to the simulation as expected due to the larger separated region in the experiment. Figure 7.22 displays the computed and experimental Stanton number versus $(x - x_i)/\delta_i$. Significant differences are evident both upstream and within the interaction region.

[1] The upstream influence point is the location of the beginning of the pressure rise.

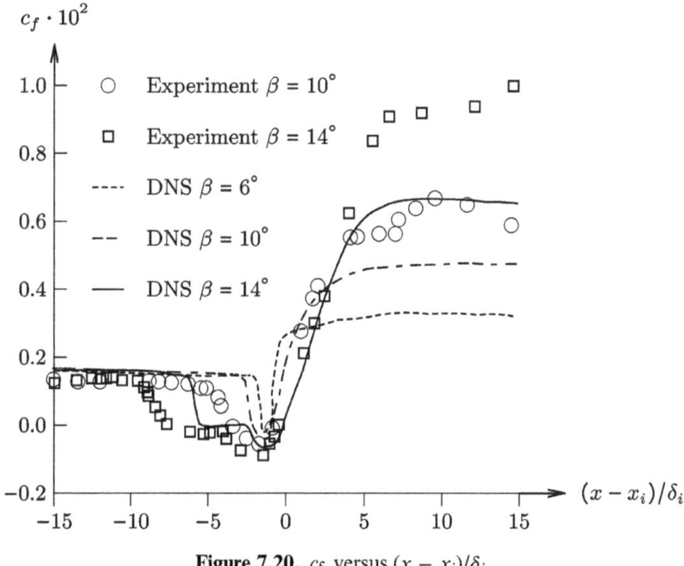


**Figure 7.20.** $c_f$ versus $(x - x_i)/\delta_i$.

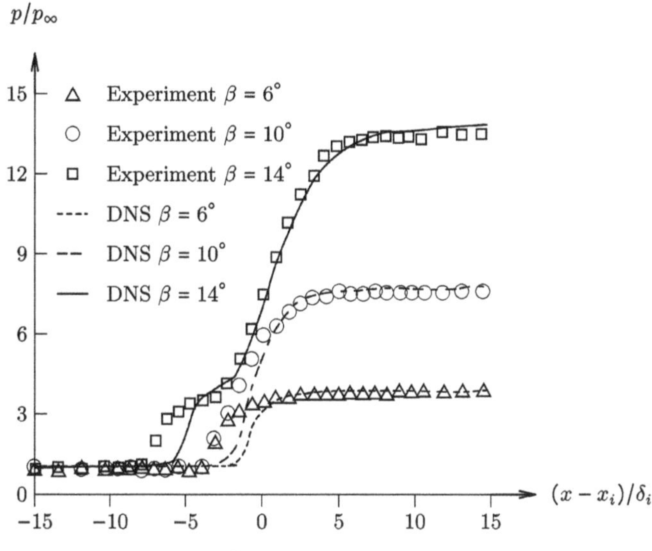


**Figure 7.21.** $p/p_\infty$ versus $(x - x_i)/\delta_i$.

## 7.7 Fu *et al* (2021)

Fu *et al* (2021) conducted a Direct Numerical Simulation and WMLES of an incident shock wave boundary layer interaction at Mach 6 for incident shock angles $\theta = 13.2°$ to $15.7°$ corresponding to wedge angles $\alpha = 5°$ to $8°$. The flow configuration and computational domain are shown in figure 7.23. An incoming laminar boundary layer interacts with the incident shock. The flow conditions are listed in

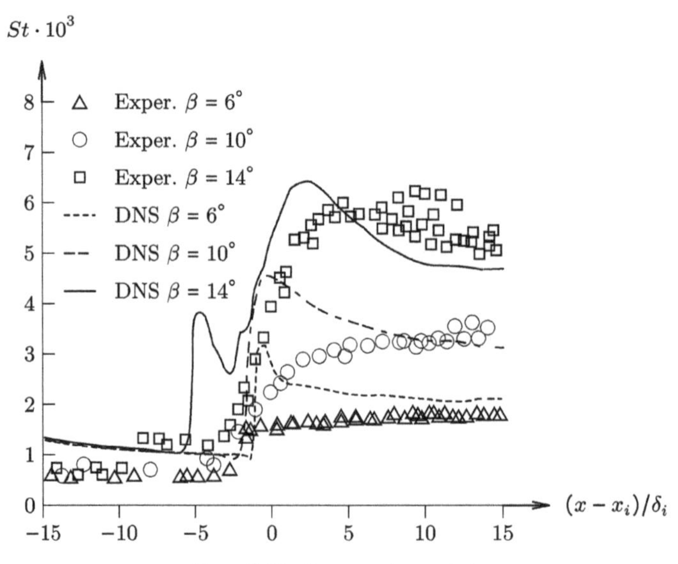


**Figure 7.22.** $St$ versus $(x - x_i)/\delta_i$.

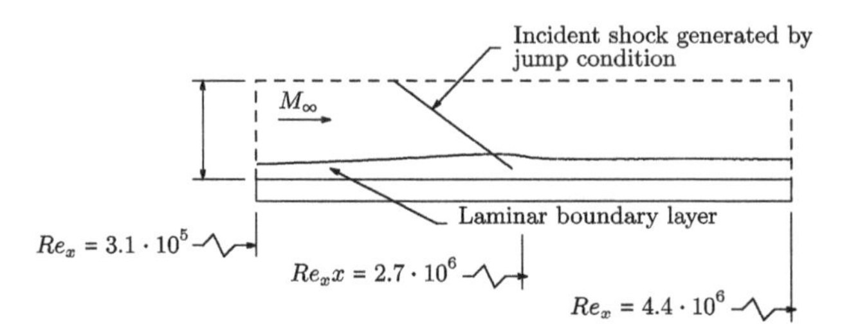


**Figure 7.23.** Model.

**Table 7.7.** Freestream conditions.

| Quantity | Value |
|---|---|
| $M_\infty$ | 6.0 |
| $T_w/T_\infty$ | 4.5 |
| $T_w/T_{aw}$ | 0.61[a] |
| $Re_{\delta^*}$ | 6830[b] |

[a] Using (3.46).
[b] Evaluated at inflow boundary.

table 7.7. The subgrid scale stresses and heat flux were based on the model of Vreman (2004) with an equilibrium near wall model of Yang *et al* (2018).

The boundary layer was observed to transition from laminar to turbulent downstream of the reattachment point of the separated region generated by the incident shock. Figures 7.24 and 7.25 display the computed Direct Numerical Simulation and WMLES mean skin friction coefficient $c_f$ and Stanton number $St$

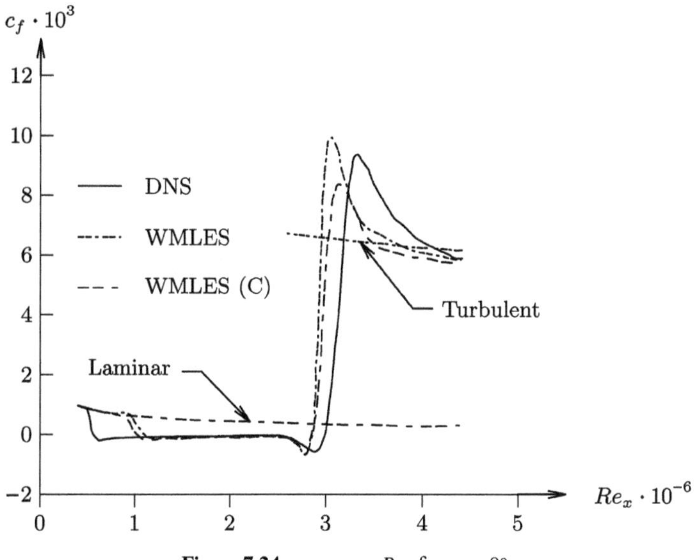


**Figure 7.24.** $c_f$ versus $Re_x$ for $\alpha = 8°$.

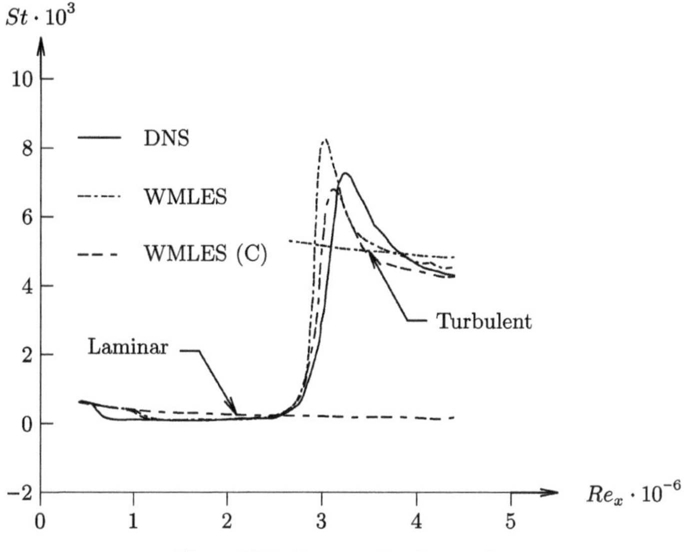


**Figure 7.25.** $St$ versus $Re_x$ for $\alpha = 8°$.

versus $Re_x$, respectively, for $\alpha = 8°$. The laminar profile in the absence of the incident shock is shown indicating an incoming laminar boundary layer. The turbulent profile downstream of the interaction is also shown and is based upon the conditions downstream of the impinging shock wave indicating a relaxation of the boundary layer to an equilibrium condition. The WMLES results show a smaller separation bubble compared to the Direct Numerical Simulation.

## 7.8 Priebe and Martín (2021)

Priebe and Martín (2021) performed a Direct Numerical Simulation of a hypersonic shock wave boundary layer interaction at Mach 7.2 generated by an 8° compression corner. The flow configuration and computational domain are shown in figure 7.26 where $x = 0$ corresponds to the corner. An incoming turbulent boundary layer interacts with the shock wave generated by the compression corner. The flow conditions are shown in table 7.8. The incoming dynamic boundary layer profiles were obtained from an auxiliary Direct Numerical Simulation. The time-averaged surface skin friction indicates fully attached flow, although regions of instantaneous reverse flow occur near the corner.

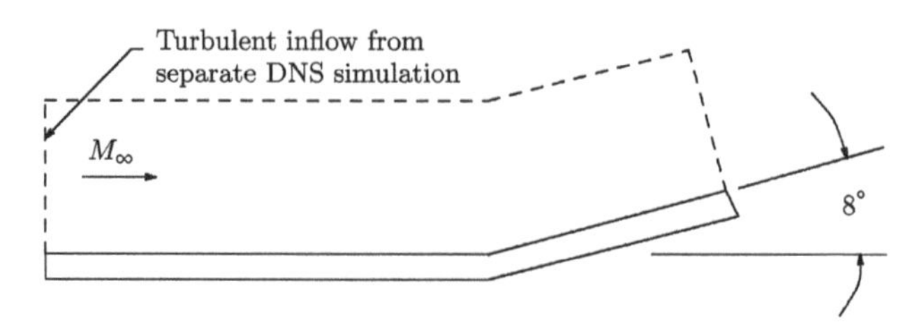


**Figure 7.26.** Flow configuration.

**Table 7.8.** Freestream conditions.

| Quantity | Value |
|---|---|
| $M_\infty$ | 7.21 |
| $T_w/T_{aw}$ | 0.53[a] |
| $Re_\theta$ | 3300 |

[a] Using (3.46) with $Pr_t = 0.89$.

The Reynolds stress anisotropy tensor $b_{ij}$ defined by

$$b_{ij} = \frac{\widetilde{u_i''u_j''}}{\widetilde{u_k''u_k''}} - \frac{1}{3}\delta_{ij} \tag{7.1}$$

provides information on the structure of the turbulent fluctuations in the interaction. The second and third invariants of the anisotropy tensor are

$$\mathrm{II} = -\frac{1}{2}b_{ij}b_{ij} \tag{7.2}$$

$$\mathrm{III} = \frac{1}{3}b_{ij}b_{jk}b_{ki} \tag{7.3}$$

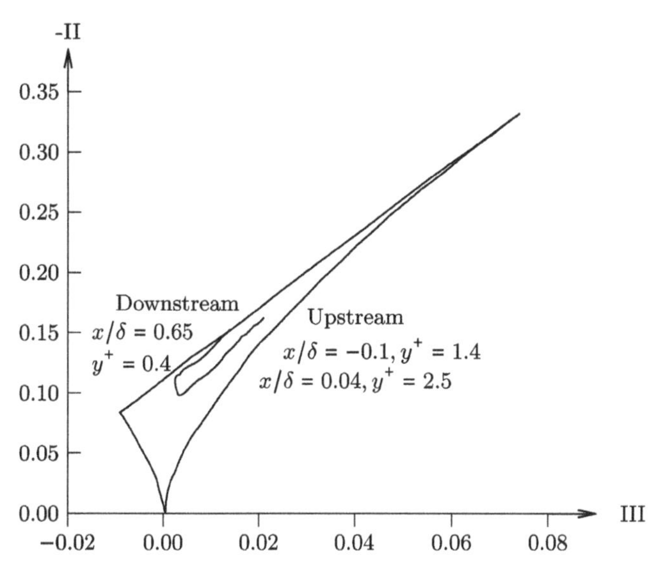


**Figure 7.27.** Traces of the invariant pair along streamline passing through $z^+=1$ at $x/\delta = -4$.

Lumley and Newman (1977) and Choi and Lumley (2001) show that the condition of realizability[2] defines a region of feasible values of II and III known as the Lumley triangle. Figure 7.27 displays traces of the invariant pair (–II, III) along the mean streamline through the interaction originating at $x/\delta = -4$ at $z^+ = 1$. Close to the wall (figure 7.27), the trace follows close to the upper limit of the Lumley triangle corresponding to predominantly two-dimensional turbulence since the wall normal fluctuations are small compared to the streamwise and spanwise fluctuations. Farther from the wall the trace moves follows the bottom right boundary of the Lumley triangle corresponding to 'axisymmetric' turbulence wherein the streamwise velocity fluctuations are large compared to the wall normal and spanwise fluctuations. Farther out from the wall the trace follows close to the bottom corner of the Lumley triangle corresponding to approximately isotropic turbulence.

The Reynolds Analogy Factor $2St/c_f$ is 1.2 upstream of the interaction but peaks to 5.2 at the corner due to the drop in skin friction coefficient (figure 7.28). Thereafter, the Reynolds Analogy Factor relaxes to 1.2 within a distance approximately $10\delta$ from the corner.

Figure 7.29 shows the strong Reynolds analogy no. 2 (3.331) and no. 3 (3.334) and the modified strong Reynolds analogy (3.355) at $x/\delta = 0$. The computed strong

[2] The conditions of realizability applied to the Reynolds stress tensor are (Schumann 1977)

$$\widetilde{u_i''u_i''} \geqslant 0 \quad \text{for} \quad i = 1,\ 2,\ \text{and } 3$$
$$\widetilde{u_i''u_j''}^2 \geqslant \widetilde{u_i''u_i''}\,\widetilde{u_j''u_j''} \quad \text{for} \quad i \neq j$$

Note that the Einstein summation convention is not applied in the above expressions. A stronger requirement is the matrix $\widetilde{u_i''u_j''}$ be positive semidefinite.

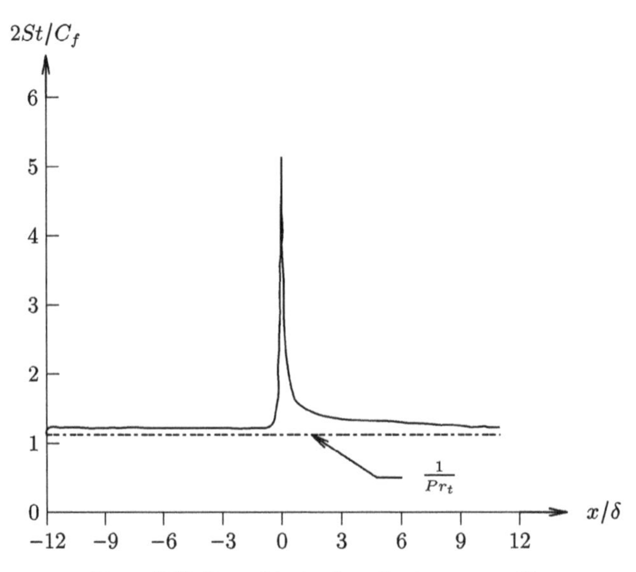


**Figure 7.28.** Reynolds Analogy Factor versus $x/\delta$.

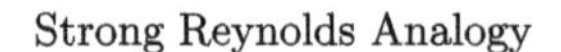


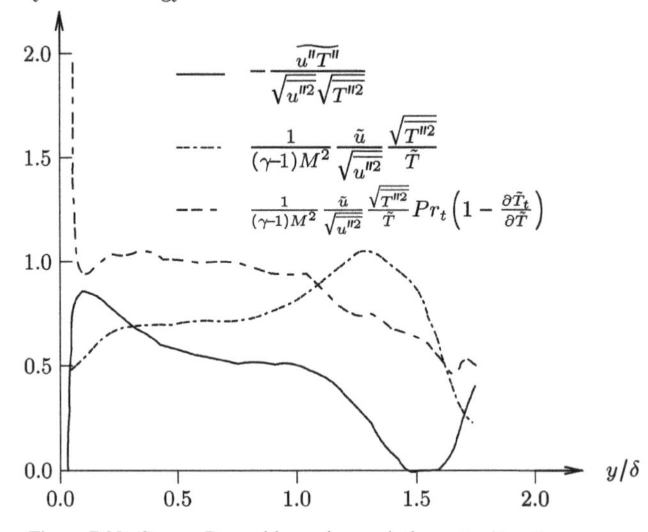


**Figure 7.29.** Strong Reynolds analogy relations at $x/\delta = 0$.

Reynolds analogy nos. 2 and 3 differ significantly from the theoretical values of 1 and −1, respectively, across the entire boundary layer. Similar differences are also observed far upstream and downstream of the corner. The modified strong Reynolds analogy, rewritten as

$$\frac{1}{(\gamma - 1)M^2} \frac{\tilde{u}}{\sqrt{\widetilde{u''^2}}} \frac{\sqrt{\widetilde{T''^2}}}{\tilde{T}} Pr_t \left(1 - \frac{\partial \tilde{T}_t}{\partial \tilde{T}}\right) \tag{7.4}$$

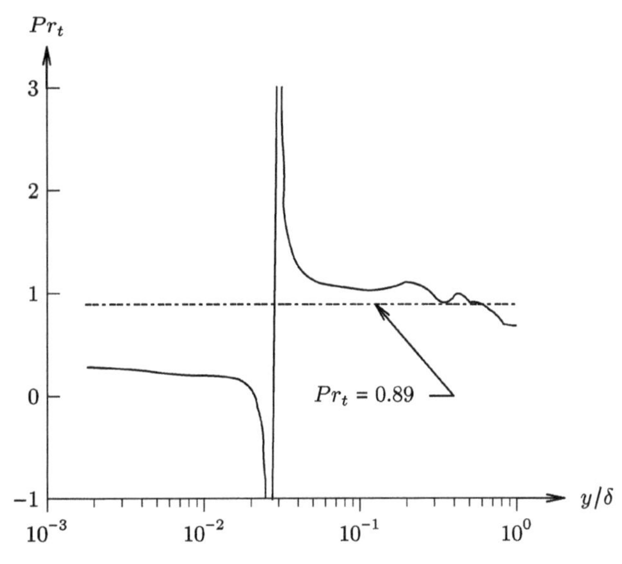


**Figure 7.30.** $Pr_t$ versus $z/\delta$ at $x/\delta = -4$.

is generally satisfied across the boundary layer at the corner, as well as upstream and downstream of the interaction.

Figure 7.30 displays the boundary layer profile of the turbulent Prandtl number (2.55) at $x/\delta = -4$. Whereas $Pr_t \approx 1$ for $y/\delta > 0.05$, there is a singularity close to the wall where $\partial \tilde{T}/\partial y = 0$ since $T_w < T_{aw}$. A similar result is observed downstream of the corner.

## References

Adams N 2000 Direct simulation of the turbulent boundary layer along a compression ramp at M = 3 and $Re_\theta = 1685$ *J. Fluid Mech.* **420** 47–83

Alvi F and Settles G 1992 Physical model of the swept shock/boundary layer interaction flowfield *AIAA J.* **30** 2252–8

Aubard G, Gloerfelt X and Robinet J C 2013 Large-eddy simulation of broadband unsteadiness in a shock/boundary-layer interaction *AIAA J.* **51** 2395–409

Baurle R, Tam J, Edwards J and Hassan H 2003 Hybrid RANS/LES approach to cavity flows: blending, algorithm, and boundary treatment issues *AIAA J.* **41** 1463–80

Choi K and Lumley J 2001 The return to isotropy of homogeneous turbulence *J. Fluid Mech.* **436** 59–84

Edwards J, Choi J-I and Boles J 2008 Large-eddy/Reynolds-averaged Navier-Stokes simulation of a Mach 5 compression-corner interaction *AIAA J.* **46** 977–91

Erengil M and Dolling D 1991 Unsteady wave structure near separation in a Mach 5 compression ramp interaction *AIAA J.* **29** 728–35

Erm L and Joubert P 1991 Low Reynolds number turbulent boundary layers *J. Fluid Mech.* **230** 1–44

Fan C C, Xiao X, Edwards J, Hassan H and Baurle R 2004 Hybrid large-eddy/Reynolds-averaged Navier–Stokes simulations of shock-separated flows *J. Spacecr. Rockets* **41** 897–906

Fang J, Lu L, Yao Y and Zheltovodov A 2015 Large eddy simulation of a three dimensional hypersonic shock wave/turbulent boundary layer interaction of a single fin *53rd AIAA Aerospace Sciences Meeting* AIAA Paper 2015-1062 (Reston, VA: American Institute of Aeronautics and Astronautics)

Fang J, Yao Y, Zheltovodov A and Lu L 2014 Numerical simulations of two-dimensional and three-dimensional shock wave/turbulent boundary layer interactions *17th Int. Conf. of Methods of Aerophysical Research* (Novosibirsk: Khristianovich Institute of Theoretical and Applied Mechanics)

Fang J, Yao Y, Zheltovodov A and Lu L 2017 Investigation of three dimensional shock wave turbulent boundary layer interaction initiated by a single fin *AIAA J.* **55** 509–23

Fu L, Bose S and Moin P 2020 *Wall-modeled LES of Three-Dimensional Intersecting Shock Wave/ Turbulent Boundary-Layer Interactions* Annual Research Briefs 2020 Center for Turbulence Research, Stanford University

Fu L, Karp M, Bose S, Moin P and Urzay J 2021 Shock-induced heating and transition to turbulence in a hypersonic boundary layer *J. Fluid Mech.* **909** A8–1-A8-49

Garnier E, Sagaut P and Deville M 2002 Large eddy simulation of shock/boundary-layer interaction *AIAA J.* **40** 1935–44

Ghosal S, Lund T, Mon P and Akselvoll K 1995 A Dynamic localization model for large eddy simulation o turbulent flows *J. Fluid Mech.* **286** 229–55

Gramann R 1989 Dynamics of separation and reattachment in a Mach 5 unswept compression ramp flow *PhD Thesis* University of Texas at Austin

Hadjadj A 2012 Large-eddy simulation of shock/boundary layer interaction *AIAA J.* **50** 2919–27

Jammalamadaka A, Li Z and Jaberi F 2013 Subgrid-scale models for large-eddy simulations of shock-boundary-layer interactions *AIAA J.* **51** 1174–88

Kussoy M and Horstman K 1992 *Intersecting Shock-Wave/Turbulent Boundary-Layer Interactions at Mach 8.3* Technical Memorandum 103909 (Washington, DC: National Aeronautics and Space Administration)

Kussoy M, Horstman K and Horstman C 1993 Hypersonic crossing shock-wave/turbulent boundary-layer interaction *AIAA J.* **31** 2197–203

Loginov M, Adams N and Zheltovodov A 2006 Large-eddy simulation of shock-wave/turbulent-boundary-layer interaction *J. Fluid Mech.* **565** 135–69

Lumley J and Newman G 1977 The Return to Isotropy of Homogeneous Turbulence *J. Fluid Mech.* **82** 161–78

Menter F 1994 Two equation eddy viscosity turbulence models for engineering applications *AIAA J.* **32** 1598–605

Moin P, Squires K, Cabot W and Lee S 1991 A dynamic subgrid-scale model for compressible turbulence and scalar transport *Phys. Fluids* A **11** 2746–57

Morgan B, Kawai S and Lele S 2010 Large-eddy simulation of an oblique shock impinging on a turbulent boundary layer *40th Fluid Dynamics Conf.* and Exhibit AIAA Paper 2010-4467 (Reston, VA: American Institute of Aeronautics and Astronautics)

Pirozzoli S, Bernardini M and Grasso F 2010 Direct numerical simulation of transonic shock/ boundary layer interaction under conditions of incipient separation *J. Fluid Mech.* **657** 361–93

Pirozzoli S and Grasso F 2006 Direct numerical simulation of impinging shock wave/turbulent boundary layer interaction at $M = 2.25$ *Phys. Fluids* **18** 18065113

Pirozzoli S, Grasso F and Gatski T 2004 Direct numerical simulation and analysis of a spatially evolving supersonic turbulent boundary layer at $M = 2.25$ *Phys. Fluids* **16** 530–45
Priebe S and Martín M P 2021 Turbulence in hypersonic compression ramp flow *Phys. Rev. Fluids* **6** 034601
Purtell L, Klebanoff P and Buckley F 1981 Turbulent boundary layer at low Reynolds number *Phys. Fluids* **24** 802–11
Rai M and Moin P 1993 Direct numerical simulation of transition and turbulence in a spatially evolving boundary layer *J. Comput. Phys.* **109** 169–92
Rizzetta D and Visbal M 2002 Application of large eddy simulation to supersonic compression ramps *AIAA J.* **40** 1574–81
Sandham N, Schülein E, Wagner A, Willems S and Steelant J 2014 Transitional shock-wave/boundary-layer interactions in hypersonic flow *J. Fluid Mech.* **752** 349–82
Schülein E 2006 Skin friction and heat flux measurements in shock/boundary layer interaction flows *AIAA J.* **44** 1732–41
Schülein E 2014 Effects of laminar-turbulent transition on the shock-wave/boundary-layer interaction *44th AIAA Fluid Dynamics Conf.* AIAA Paper 2014-3332 (Reston, VA: American Institute of Aeronautics and Astronautics)
Schumann U 1977 Realizability of Reynolds stress turbulence models *Phys. Fluids* **20** 721–5
Touber E and Sandham N 2009a Comparison of three large-eddy simulations of shock-induced turbulent separation bubbles *Shock Waves* **19** 469–78
Touber E and Sandham N 2009b Large-eddy simulation of low-frequency unsteadiness in a turbulent shock-induced separation bubble *Theor. Comput. Fluid Dyn.* **23** 79–107
Urbin G and Knight D 2001 Large eddy simulation of a supersonic boundary layer using an unstructured grid *AIAA J.* **39** 1288–95
Volpiani P, Bernardini M and Larsson J 2020 Effects of a nonadiabatic wall on hypersonic shock/boundary-layer interactions *Phys. Rev. Fluids* **5** 014602
Vreman A 2004 An eddy viscosity subgrid scale model for turbulent shear flow: algebraic theory and applications *Phys. Fluids* **16** 3670–81
Willems S, Gülhan A and Steelant J 2015 Experiments on the effect of laminar-turbulent transition on the SWBLI in H2K at Mach 6 *Exp. Fluids* **56** 49
Wu X 2017 Inflow turbulence generation methods *Annu. Rev. Fluid Mech.* **49** 23–49
Xiao X, Edwards J and Hassan H 2004 Blending functions in hybrid large-eddy/Reynolds-averaged Navier-Stokes simulations *AIAA J.* **42** 2508–15
Xiao X, Edwards J, Hassan H and Baurle R 2003 Inflow boundary conditions for hybrid large eddy/Reynolds averaged Navier-Stokes simulations *AIAA J.* **41** 1481–9
Yang X, Urzay J, Bose S and Moin P 2018 Aerodynamic heating in wall modeled large eddy simulation of high speed flows *AIAA J.* **56** 731–42
Zheltovodov A and Knight D 2011 Ideal gas shock wave turbulent boundary layer interactions in supersonic flows and their modeling: three dimensional interactions *Shock Wave Boundary Layer Interactions* ed H Babinsky and J Harvey (Cambridge: Cambridge University Press) pp 202–58

**IOP** Publishing



# Chapter 8

# Discussion and future needs

## Abstract

The state of knowledge of hypersonic turbulent boundary layers is summarized. Future needs for both experiment and simulation are discussed.

## 8.1 Equilibrium turbulent boundary layer

By definition an equilibrium flat plate, zero pressure gradient, zero wall mass flux turbulent boundary layer is characterized by its local behavior. The specific dimensionless parameters defining the boundary layer state are listed in table 8.1 where the Reynolds number is based upon a suitable local length scale (e.g., compressible momentum thickness) and the dimensionless roughness scale $k^+$ is normalized by the wall viscous length scale $\nu_w/u_\tau$ assuming uniform roughness. The characterization of the boundary layer mean and statistical properties based upon these local parameters implies that the boundary layer does not retain any memory of its upstream behavior (e.g., laminar to turbulent transition).

In general the Reynolds number dependence is expected to be weak for a fully turbulent boundary layer, and thus the mean and statistical properties are dependent upon the Mach number $M_\infty$ and wall temperature ratio $T_w/T_{aw}$. In the following sections we explore the extent of experiments and simulations using Direct

**Table 8.1.** Dimensionless parameters describing equilibrium boundary layer.

| Symbol | Definition |
|---|---|
| $M_\infty$ | Mach number |
| $Re$ | Reynolds number |
| $T_w/T_{aw}$ | Wall temperature ratio |
| $k^+$ | Wall roughness ratio |

 

Numerical Simulation (DNS) and Large Eddy Simulation (LES) in this parameter space. Tables 8.2 and 8.3 identify the symbols corresponding to the experiments (chapter 4), and table 8.4 identifies the simulations using DNS and LES (chapter 6), respectively, for equilibrium hypersonic turbulent boundary layers.

**Table 8.2.** Experiments of equilibrium boundary layers (1950–80).

| *Symbol* | *Reference* |
|---|---|
| ○ | Wegener, Winkler & Sibulkin (1953) |
| □ | Winkler & Persh (1954) |
| ◁ | Lobb, Winkler & Persh (1955*a*), Lobb, Winkler & Persh (1955) |
| ▷ | Hill (1956) |
| △ | Hill (1959) |
| ▽ | Winkler & Cha (1959), Winkler (1961) |
| ◖ | Danberg (1964) |
| ◗ | Adcock, Peterson & McRee (1965) |
| ⊙ | Young (1965) |
| ⊕ | Scaggs (1966) |
| ⊖ | Samuels, Peterson & Adcock (1967) |
| ⊘ | Wallace (1967) |
| ⊗ | Wallace (1968), Wallace (1969) |
| ⊛ | Bushnell, Johnson, Harvey & Feller (1969) |
| ◯ | Hoydysh & Zakkay (1969) |
| ⬠ | Lee, Yanta & Leonas (1969) |
| ◇ | Matthews & Trimmer (1969) |
| ☆ | Cary (1970) |
| * | Jones & Feller (1970) |
| □ | Beckwith, Harvey & Clark (1971) |
| (small arch symbol) | Fischer, Maddalon, Weinstein & Wagner (1970) |
| | Fischer, Maddalon, Weinstein & Wagner (1971) |
| ▽ | Voisinet, Lee & Yanta (1971) |
| △ | Holden (1972*a*) |
| (square with left arrow) | Hopkins & Keener (1972*a*), Hopkins & Keener (1972*b*) |
| (square with right arrow) | Horstman & Owen (1972), Owen & Horstman (1972) |
| (square with up arrow) | Keener & Hopkins (1972) |
| (square with down arrow) | Keener & Polek (1972) |
| (square with corner flag) | Kemp & Owen (1971), Kemp & Owen (1972*a*), Kemp & Owen (1972*b*) |
| ⊟ | Laderman & Demetriades (1971), Laderman & Demetriades (1973) |
| | Laderman & Demetriades (1974) |
| ⧆ | Stone & Cary (1972) |
| ◫ | Voisinet & Lee (1972) |
| ▣ | Backx (1973), Backx (1974) |
| | Backx (1975), Backx & Richards (1976) |
| ⧅ | Feller (1973) |
| ⧄ | Watson, Harris & Anders (1973) |
| ▣ | Raman (1974) |
| ⊡ | Bloy (1975) |
| ⦶ | Mikulla & Horstman (1975) |
| ⃠ | Owen, Horstman & Kussoy (1975) |
| ◎ | Smith & Driscoll (1975) |
| ⊙ | Laderman (1976) |
| ⊖ | Berg (1977) |
| ⊘ | Materna (1977) |

**Table 8.3.** Experiments of equilibrium boundary layers (1990–present).

| *Symbol* | *Reference* |
|---|---|
| ⦸ | Owen & Calarese (1987) |
| ⦸ | McGinley, Spina & Sheplak (1994) |
| ⌽ | Holden & Chadwick (1995) |
| ⑊ | Auvity, Etz & Smits (2001) |
| // | Goyne, Stalker & Paull (2003) |
| ⑊ | Suraweera, Mee & Stalker (2006) |
| ▯ | Maslov, Fedorov, Bountin, Shiplyuk, Sidorenko, Malmuth & Knauss (2008) |
| ▱ | Vaganov & Stolyarov (2008) |
| ⊲ | Sahoo, Schultze & Smits (2009) |
| ⊳ | Peltier, Humble & Bowersox (2011) |
| ⅄ | Tichenor, Humble & Bowersox (2013) |
| Y | Neeb, Saile & Gülhan (2015) |
| Y | Peltier, Humble & Bowersox (2016) |
| Y | Tichenor, Neel, Leidy & Bowersox (2017) |
| ⅄ | Williams & Smits (2017) |
| ⅄ | Neeb, Saile & Gülhan (2018) |
| ⅄ | Williams, Sahoo, Baumgartner & Smits (2018) |
| ≺ | Ding, Yi, Quyang & Zhao (2020) |
| ≻ | Williams, Sahoo, Papageorge & Smits (2021) |

**Table 8.4.** DNS and LES of equilibrium boundary layers.

| *Symbol* | *Reference* |
|---|---|
| ○ | Maeder, Adams & Kleiser (2001) |
| □ | Li, Fu & Ma (2006) |
| ◁ | Martín (2007) |
| ▷ | Duan, Beekman & Martín (2010) |
| △ | Duan, Beekman & Martín (2011) |
| ▽ | Duan & Martín (2011*a*) |
| ◖ | Duan & Martín (2011*b*) |
| D | Liang, Li & Fu (2012), Liang & Li (2013) |
| ⊙ | Chu, Zhuang & Lu (2013) |
| ⊕ | Duan, Choudhari & Zhang (2016) |
| ⊖ | Zhang, Duan & Choudhari (2017) |
| ⊘ | Zhang, Duan & Choudhari (2018) |
| ⊗ | Huang, Nicholson, Duan, Choudhari & Bowersox (2020) |
| ◯ | Nicholson, Huang, Duan, Choudhari & Bowersox (2021) |
| | Nicholson, Huang, Duan & Choudhari (2021) |
| ⬠ | Kianvashrad & Knight (2021) |

### 8.1.1 Law of the Wall

There is overwhelming evidence that the mean streamwise velocity is accurately described by the Van Driest transformed velocity defined by equation (3.36) over a portion (the 'wall region') of the boundary layer

$$\int_{\tilde{u}_l}^{\tilde{u}} \sqrt{\frac{\bar{\rho}}{\rho_w}}\, d\tilde{u} = \frac{u_\tau}{\kappa} \log \frac{y u_\tau}{\nu_w} + C u_\tau$$

where $\tilde{u}_l$ represents the lower limit of the integration region defined as the location where molecular viscous effects become significant, and $C$ is a constant which depends on the lower limit $\tilde{u}_l$ and hence upon the details of the flow in the viscous sublayer including the wall temperature (e.g., adiabatic or isothermal). As discussed in sections 3.2.3 and 3.2.4, the lower limit $u_l$ is not strictly defined (e.g., is it the location where the molecular shear stress is one percent of the Reynolds shear stress? Or two percent? Or some other fraction?) and never ascertained. Alternately, models have been developed (Trettel and Larsson 2016, Volpiani *et al* 2020) with the lower limit arbitrarily (but nonetheless unambiguously) chosen to be $\tilde{u}_l = 0$ with the addition of a kernel $f(\bar{\rho}/\rho_w)$ in the integrand

$$\int_{\tilde{u}_l}^{\tilde{u}} \sqrt{\frac{\bar{\rho}}{\rho_w}} f(\bar{\rho}/\rho_w)\, d\tilde{u} = \frac{u_\tau}{\kappa} \log \frac{Y u_\tau}{\nu_w} + C u_\tau$$

and an additional rescaling $Y(\bar{\rho}/\rho_w)$ of the distance from the wall with the intention of achieving agreement with experiment for adiabatic and isothermal walls using a single value for the constant $C$. Also, a corresponding expression for hypersonic turbulent boundary layers with wall mass transfer has been developed (section 3.2.2). Figures 8.1 and 8.2 display the experiments and simulations using DNS and LES that examine the Van Driest transformed Law of the Wall. It is evident that the combined experiments and simulations using DNS and LES essentially cover the region $M_\infty < 10$ for $0.2 \leqslant T_w/T_{aw} \leqslant 1$ and that there are few results for $M_\infty > 10$ (figure 8.3).

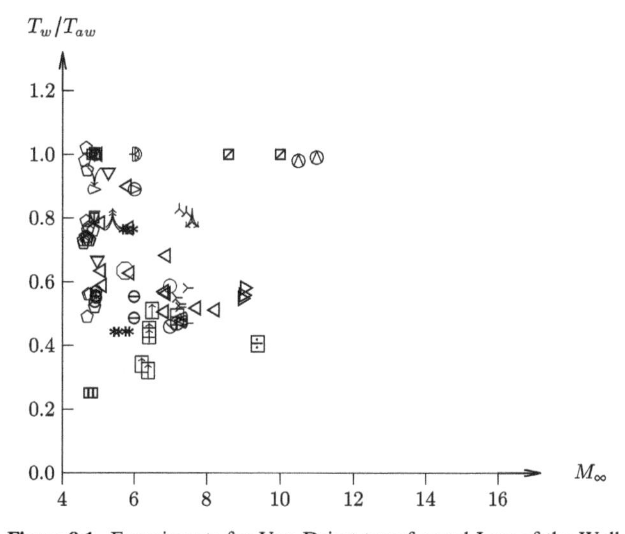


**Figure 8.1.** Experiments for Van Driest transformed Law of the Wall.

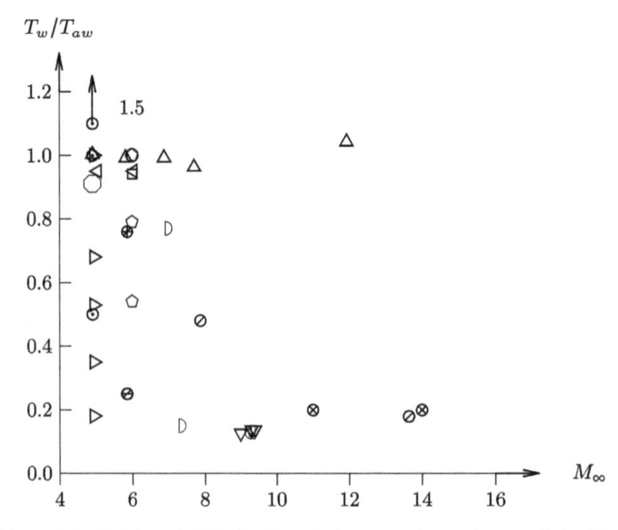


**Figure 8.2.** DNS and LES for Van Driest transformed Law of the Wall.

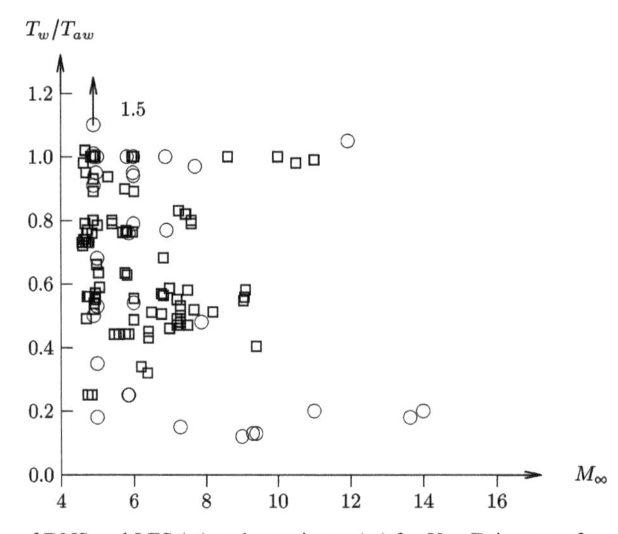


**Figure 8.3.** Overlap of DNS and LES (○) and experiment (□) for Van Driest transformed Law of the Wall.

### 8.1.2 Mean velocity–mean temperature relations

Five models for the mean velocity–mean temperature relation for a flat plate, zero pressure gradient are described in section 3.3. The models are

- Crocco–Busemann relation (3.147)

$$\tilde{T} = T_{\rm w} + (T_{\rm aw} - T_{\rm w})\left(\frac{\tilde{u}}{u_\infty}\right) + (T_\infty - T_r)\left(\frac{\tilde{u}}{u_\infty}\right)^2 \qquad Pr_{\rm t} = Pr = 1$$

or equivalently assuming constant specific heats (3.144)

$$\frac{\tilde{T}_t - T_w}{T_{t_\infty} - T_w} = \frac{\tilde{u}}{u_\infty}$$

- Walz's relation (3.172)

$$\tilde{T} = T_w + (T_r - T_w)\left(\frac{\tilde{u}}{u_\infty}\right) + (T_\infty - T_r)Pr_t\left(\frac{\tilde{u}}{u_\infty}\right)^2$$

- Duan and Martín (2011a) (3.214)

$$\bar{T} = T_w + (T_r - T_w)f\left(\frac{\bar{u}}{u_\infty}\right) + (T_\infty - T_r)\left(\frac{\bar{u}}{u_\infty}\right)^2$$

where

$$f\left(\frac{\bar{u}}{u_\infty}\right) = 0.825\,9\left(\frac{\bar{u}}{u_\infty}\right) + 0.174\,1\left(\frac{\bar{u}}{u_\infty}\right)^2$$

- Zhang *et al* (2014) (3.213)

$$\bar{T} = T_w + (T_r - T_w)\left[(1 - sPr)\left(\frac{\bar{u}}{u_\infty}\right)^2 + sPr\left(\frac{\bar{u}}{u_\infty}\right)\right] + (T_\infty - T_r)\left(\frac{\bar{u}}{u_\infty}\right)^2$$

The above expression is identical to Duan and Martín (2011a) with $sPr = 0.825\,9$ which implies $s = 1.147$ for $Pr = 0.72$.
- Hypersonic velocity–total temperature relation (3.221)

$$\tilde{T} = T_w + (T_\infty - T_w)\left(\frac{\tilde{u}}{u_\infty}\right)^2$$

or equivalently assuming constant specific heats

$$\frac{T_t - T_\infty}{T_{t_\infty} - T_w} = \left(\frac{u}{u_\infty}\right)^2$$

Figures 8.4 and 8.5 indicate the range of experiments and simulations using DNS and LES that examine the velocity–temperature relation. Figure 8.6 shows the overlap experiments and simulations. It is evident that there are negligible data for $T_w/T_{aw} > 0.2$ at $M_\infty > 10$. The experimental and simulation results described in chapter 4 indicate that the models of (Duan and Martín 2011a and Zhang *et al* 2014) are the most accurate and that the Crocco–Busemann and Walz relations become increasingly less accurate as the wall temperature differs from adiabatic.

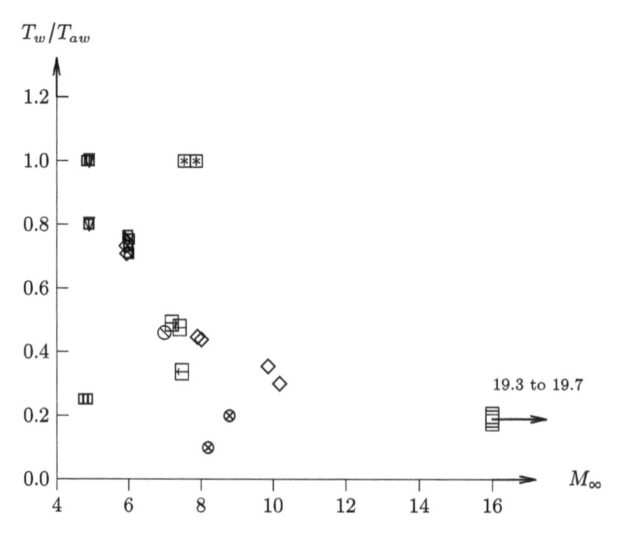


**Figure 8.4.** Experiments for velocity–temperature relation.

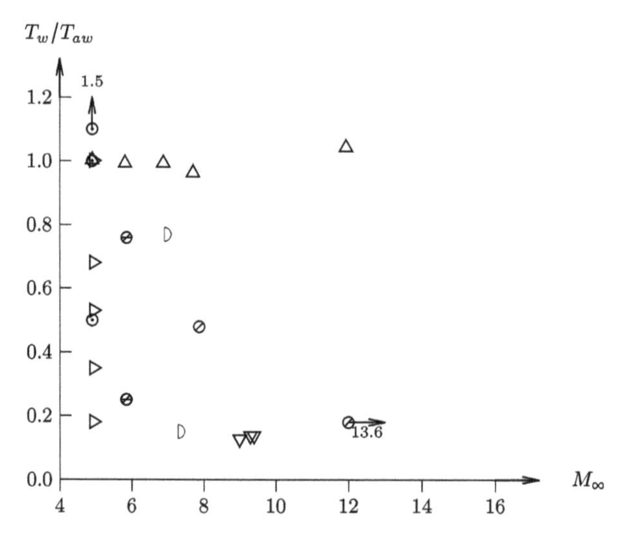


**Figure 8.5.** DNS and LES for velocity–temperature relation.

### 8.1.3 Reynolds Analogy Factor

Two models for the Reynolds Analogy Factor are described in section 3.4. The models are

- Colburn (1933) (3.235)

$$\frac{2St}{c_{\mathrm{f}}} = Pr^{-\frac{2}{3}}$$

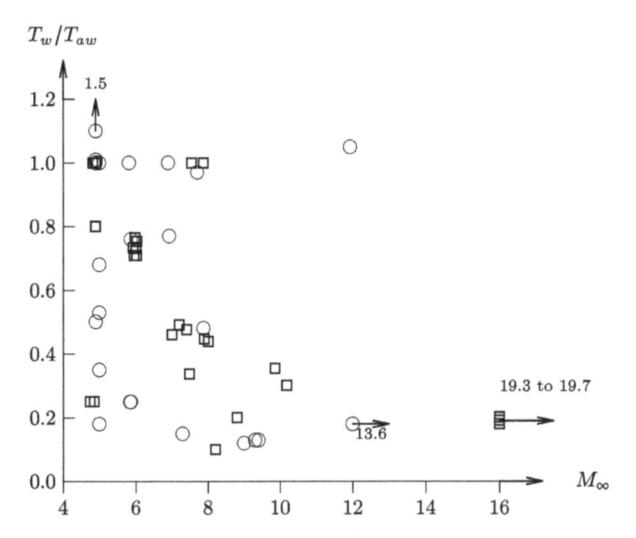


**Figure 8.6.** DNS, LES, and experiments for velocity–temperature relation.

- Van Driest (1951) (3.239)

$$\frac{2St}{c_f} = \frac{1}{Pr_t}$$

The above relationship is also derived from the compressible $k - \epsilon$ turbulence model (section 3.4.4) Assuming a nominal turbulent Prandtl number $Pr_t = 0.89$, the above expression for the Reynolds Analogy Factor yields $2St/c_f = 1.12$.

Experiments examining the Reynolds Analogy Factor are indicated in figure 8.7. The experimental data display a rather wide range of values for $2St/c_f$ ranging from 0.75 to 1.3 (chapter 4). See also the summary of Chi and Spaulding (1966) in section 4.22. There are no experimental data for $T_w/T_{aw} > 0.5$ at $M_\infty > 5$, and no data for $M_\infty > 12$.

Simulations examining the Reynolds Analogy Factor using DNS and LES are shown in figure 8.8. However, the figure is misleading since in principle all DNS and LES cases for non-adiabatic walls provide data for the Reynolds Analogy Factor. Nonetheless, only those papers in chapter 6 including such data are listed.

### 8.1.4 Morkovin's hypothesis

Morkovin (1962) postulated that, at moderate freestream Mach numbers ($M \leqslant 5$), the effects of compressibility on the turbulence stresses in turbulent shear flows can be attributed to the mean variation of density (section 3.5). The generalization of Morkovin's hypothesis is

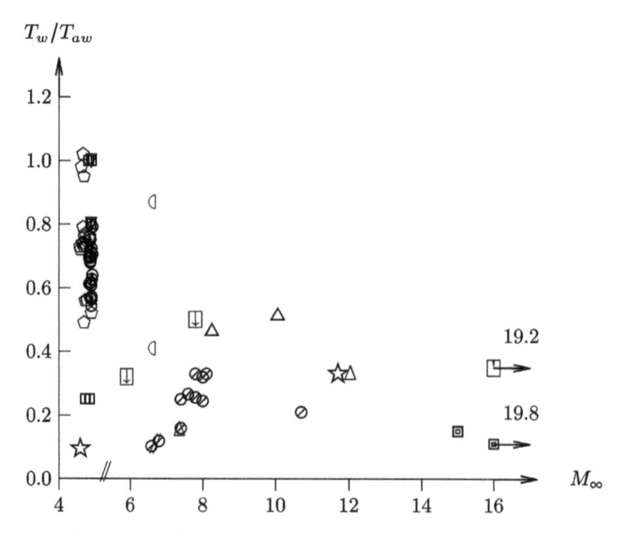


**Figure 8.7.** Experiments for Reynolds Analogy Factor.

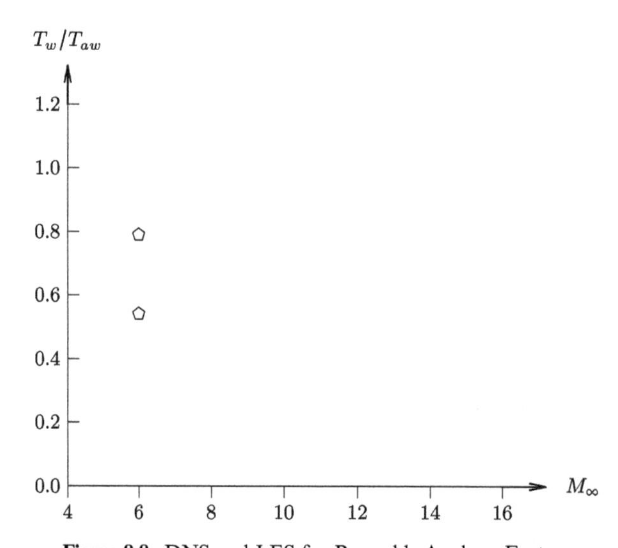


**Figure 8.8.** DNS and LES for Reynolds Analogy Factor.

$$
\begin{aligned}
\frac{\bar{\rho}}{\rho_{\mathrm{w}}} \frac{\overline{u'u'}}{u_\tau^2} &= f\left(\frac{y}{\delta}\right) \\
\frac{\bar{\rho}}{\rho_{\mathrm{w}}} \frac{\overline{u'v'}}{u_\tau^2} &= g\left(\frac{y}{\delta}\right)
\end{aligned}
\tag{8.1}
$$

and so forth where $f(y/\delta)$ and $g(y/\delta)$ are independent of Mach number.

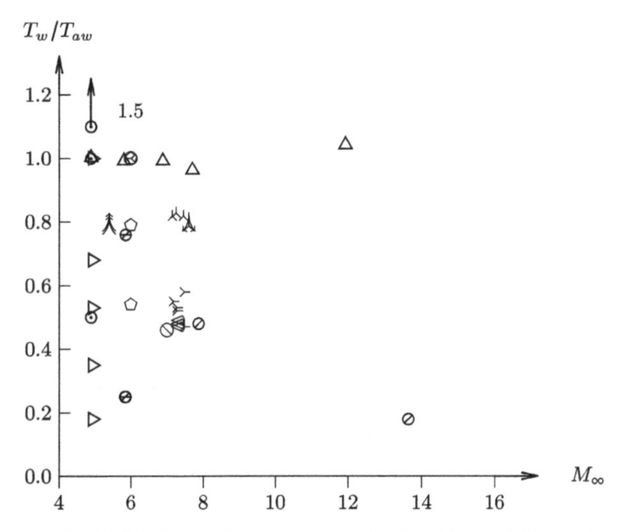


**Figure 8.9.** DNS, LES, and experiment evaluating Morkovin's hypothesis.

Morkovin's hypothesis has been examined for DNS, LES, and experiment, typically in combination of experiment and simulation. Figure 8.9 displays the combined data obtained at Mach number and wall temperature ratio $T_w/T_{aw}$ from chapters 4 and 6. The results in chapters 4 and 6 indicate that Morkovin's hypothesis provides an accurate collapse of the experimental data for the Reynolds stresses examined. Figure 8.9 indicates that there are negligible data for $M_\infty > 8$.

### 8.1.5 Morkovin's strong Reynolds analogy

Morkovin (1962) postulated five strong Reynolds analogies (section 3.6). Several models for the second strong Reynolds analogy have been developed:

- Morkovin (1962) (3.330)

$$\frac{\sqrt{\bar{T}'^2}/\bar{T}}{(\gamma - 1)\bar{M}^2\sqrt{\bar{u}'^2}/\bar{u}} = 1$$

- Cebeci and Smith (1974) (3.342)

$$\frac{\sqrt{\bar{T}'^2}/\bar{T}}{(\gamma - 1)\bar{M}^2\sqrt{\bar{u}'^2}/\bar{u}} = 1 + \frac{(T_w - T_e)}{\bar{T}}\left(\frac{\bar{u}}{u_e}\right)\frac{1}{(\gamma - 1)\bar{M}^2}$$

- Huang *et al* (1995) (3.355)

$$\frac{\left(\sqrt{\bar{T}'^2}/\bar{T}\right)}{(\gamma - 1)\bar{M}^2\left(\sqrt{\bar{u}'^2}/\bar{u}\right)} = \frac{1}{Pr_t}\frac{1}{|\, d\bar{T}_t/d\bar{T} - 1\,|}$$

- Gaviglio (1987) (3.356)

$$\frac{\left(\sqrt{\bar{T'^2}}/\bar{T}\right)}{(\gamma - 1)\bar{M}^2\left(\sqrt{\bar{u'^2}}/\bar{u}\right)} = \frac{1}{(1 - d\bar{T}_t/d\bar{T})}$$

- Gaviglio (1987) and Brun *et al* (2008) (3.358)

$$\frac{\left(\sqrt{\bar{T'^2}}/\bar{T}\right)}{(\gamma - 1)\bar{M}^2\left(\sqrt{\bar{u'^2}}/\bar{u}\right)} = \frac{1}{Pr_m}\frac{1}{\mid d\bar{T}_t/d\bar{T} - 1 \mid}$$

where

$$Pr_m = \frac{(\mu + \mu_t)c_p}{k + k_t}$$

- Duan *et al* (2011) (3.359)

$$\sqrt{\bar{T'^2}} = -\frac{1}{Pr_t}\frac{\partial\bar{T}}{\partial\bar{u}}\sqrt{\bar{u'^2}}$$

Figures 8.10 and 8.11 show the experimental and simulation data using DNS and LES, respectively. The results are limited to $M_\infty \leqslant 10$. The data indicate that the model of Huang *et al* (1995) provides the best representation of the available data.

### 8.1.6 Turbulence structure

Early experimental data for hypersonic turbulent boundary layers focused on mean measurements of pitot pressure, total temperature and surface heat flux, skin friction, and temperature. Beginning in the 1970s, experimental data for turbulence structure (i.e., Reynolds stress, turbulent heat flux, etc) appeared. Examples include Fischer *et al* (1970) and Fischer *et al* (1971) (section 4.21), Horstman and Owen (1972) and Owen and Horstman (1972) (section 4.26), Laderman and Demetriades (1971, 1973) and Laderman and Demetriades (1974) (section 4.30), Mikulla and Horstman (1975) (section 4.38), Owen *et al* (1975) (section 4.39), Smith and Driscoll (1975) (section 4.40), Laderman (1976) (section 4.41), Berg (1977) (section 4.42), Materna (1977) (section 4.3). Additional measurements of turbulence structure were performed in the 1990s through the present as detailed in chapter 4.

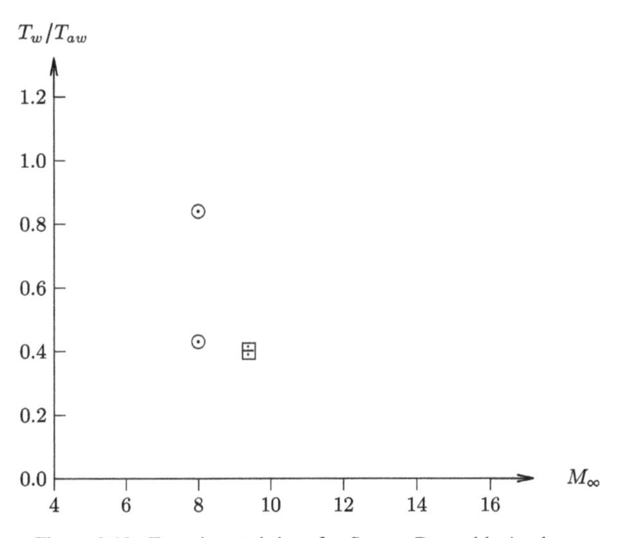


**Figure 8.10.** Experimental data for Strong Reynolds Analogy.

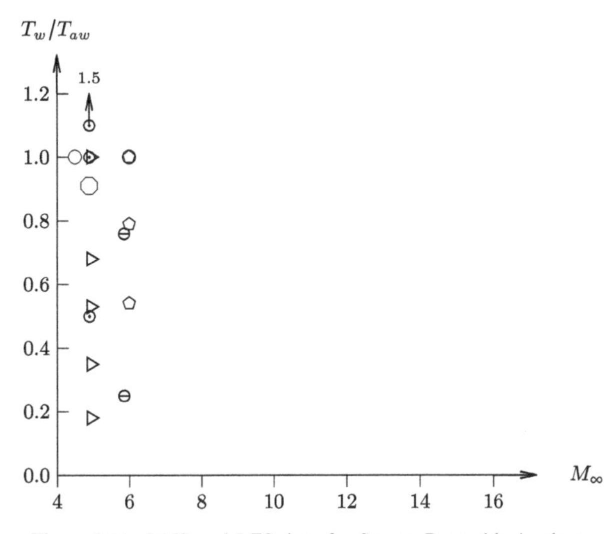


**Figure 8.11.** DNS and LES data for Strong Reynolds Analogy.

The structure of an equilibrium flat plate, zero pressure gradient hypersonic turbulent boundary layer[1] depends upon a specific set of dimensionless parameters, namely, the Mach number $M_\infty$, wall temperature to adiabatic wall temperature ratio $T_w/T_{aw}$, Reynolds number based upon some suitable local length scale (e.g., $Re_\theta$), and surface condition (i.e., hydraulically smooth or rough surface, with the latter defined by the roughness geometry and dimensionless roughness height). The dependence on Reynolds number is weak, and thus the primary dimensionless

[1] Assuming thermochemical non-equilibrium effects are unimportant, e.g., $H < 2.5$ MJ kg$^{-1}$ for air.

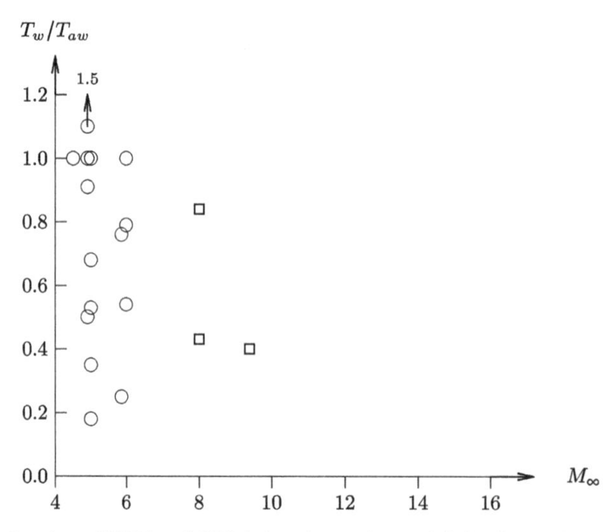


**Figure 8.12.** Overlap of DNS and LES (○) and experiment (□) for Strong Reynolds Analogy.

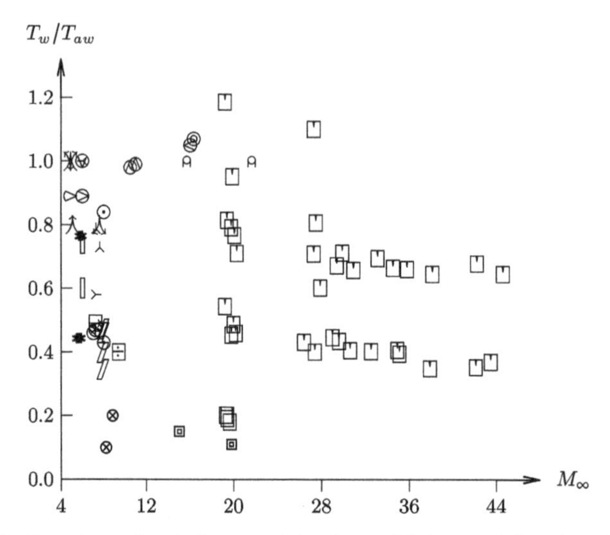


**Figure 8.13.** Experimental turbulence statistics for equilibrium turbulent boundary layers.

parameters are $M_\infty$ and $T_w/T_{aw}$ in addition to the surface condition. Figure 8.13 identifies the experimental data for turbulence statistics from Mach 4 to Mach 22 with the symbols defined in table 8.5. An enlargement for the region from Mach 4 to Mach 12 is shown in figure 8.14. Each symbol represents an experiment with one or more boundary layer profiles of a turbulence statistic (e.g., $\sqrt{\overline{u'^2}}$). Overall, there is a significant lack of detailed experimental turbulence statistics over the entire $M_\infty$ versus $T_w/T_{aw}$ range, with particularly acute absence of data in the Mach 4 to 8

**Table 8.5.** Experimental turbulence statistics (figures 8.13 and 8.14).

| *Symbol* | *Reference* |
|---|---|
| ⊗ | Wallace (1968), Wallace (1969) |
| ∗ | Jones & Feller (1970) |
| ⍝ | Fischer, Maddalon, Weinstein & Wagner (1970) |
| | Fischer, Maddalon, Weinstein & Wagner (1971) |
| □ | Beckwith, Harvey & Clark (1971) |
| ⊟ | Horstman & Owen (1972)[a] |
| | Owen & Horstman (1972) |
| ⊞ | Laderman & Demetriades (1971) |
| | Laderman & Demetriades (1973) |
| | Laderman & Demetriades (1974) |
| ◰ | Kemp & Owen (1972*a*) |
| ⊡ | Raman (1974) |
| ▣ | Backx (1975) |
| ⦶ | Mikulla & Horstman (1975) |
| ⦸ | Owen, Horstman & Kussoy (1975) |
| ◎ | Smith & Driscoll (1975) |
| ⊙ | Laderman (1976) |
| ⊖ | Berg (1977) RW |
| ⊖ | Materna (1977) |
| ⊘ | Owen & Calarese (1987) |
| ⊘ | McGinley, Spina & Sheplak (1994) |
| ▯ | Maslov, Fedorov, Bountin, Shiplyuk, Sidorenko, Malmuth & Knauss (2008) |
| ▱ | Vaganov & Stolyarov (2008) |
| ◁ | Sahoo, Schultze & Smits (2009) RW |
| ▷ | Peltier, Humble & Bowersox (2011) RW |
| ⅄ | Tichenor, Humble & Bowersox (2013) |
| Y | Peltier, Humble & Bowersox (2016) RW |
| Y | Tichenor, Neel, Leidy & Bowersox (2017) |
| ⅄ | Williams & Smits (2017) |
| ⅄ | Neeb, Saile & Gülhan (2018) |
| ⅄ | Williams, Sahoo, Baumgartner & Smits (2018) |
| ≺ | Ding, Yi, Quyang & Zhao (2020) |
| ≻ | Williams, Sahoo, Papageorge & Smits (2021) RW |

RW Rough wall

region for $T_w/T_{aw} < 0.5$. Moreover, experimental data on auto- and cross-correlation are particularly sparse.

Figure 8.15 identifies the DNS and LES data for turbulence statistics[2] for Mach 4 to Mach 12 with the symbols defined in table 8.4. Overall, there is a significant lack of data for the region Mach 9 to Mach 12 except for $T_w/T_{aw} \approx 0.2$. Figure 8.16 overlaps the DNS, LES, and experimental data for turbulence statistics. It is again observed that there is a substantial absence of data in the Mach 9 to Mach 12 region for $T_w/T_{aw} > 0.2$. Furthermore, there are no experimental results for turbulence statistics in the region Mach 4 to Mach 7 for $T_w/T_{aw} < 0.4$ for validation of the DNS results. Also, there is only a single LES simulation in the $M_\infty - T_w/T_{aw}$ domain indicated in figure 8.15.

[2] It is assumed that all of the DNS and LES cases include, in principle, the complete Reynolds stress tensor and turbulent heat flux vector.

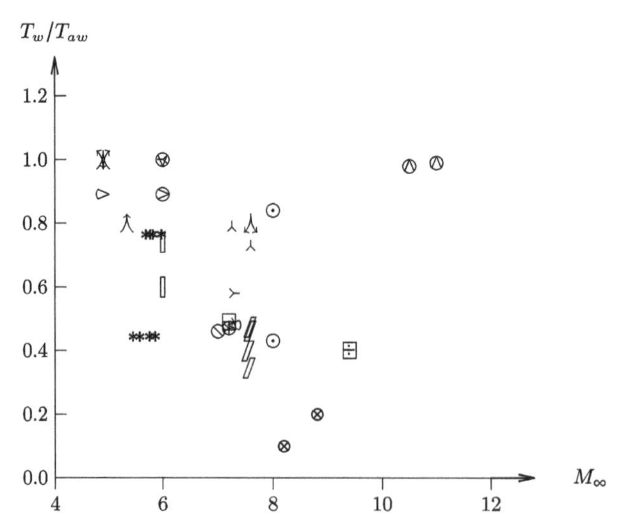


**Figure 8.14.** Experimental turbulence statistics for equilibrium turbulent boundary layers (enlarged).

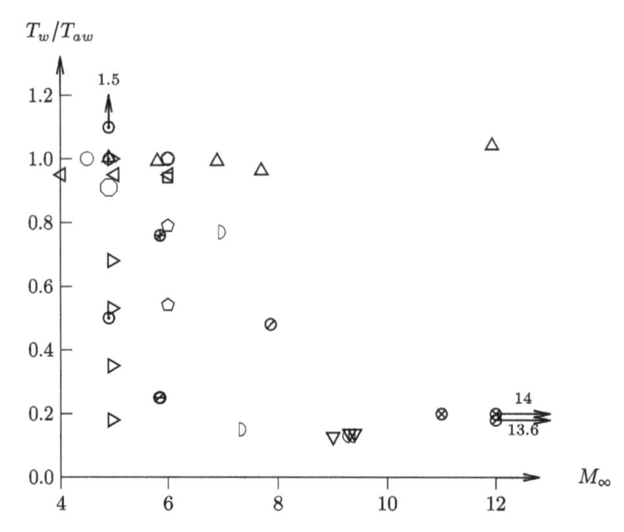


**Figure 8.15.** DNS and LES turbulence statistics for equilibrium turbulent boundary layers.

## 8.2 Shock wave boundary layer interaction

Figure 8.17 shows the experimental data for shock wave turbulent boundary layer interactions. The data are identified in table 8.6. There is a significant lack of experimental data for Mach 8 to 12 in the range $T_w/T_{aw} > 0.2$. It should also be noted that several of the experiments in chapter 5 do not provide wall temperature for the incoming equilibrium turbulent boundary layer and are excluded from the figure.

Figure 8.18 shows the DNS and LES data for hypersonic shock wave turbulent boundary layer interactions. The data are identified in table 8.7. There is a

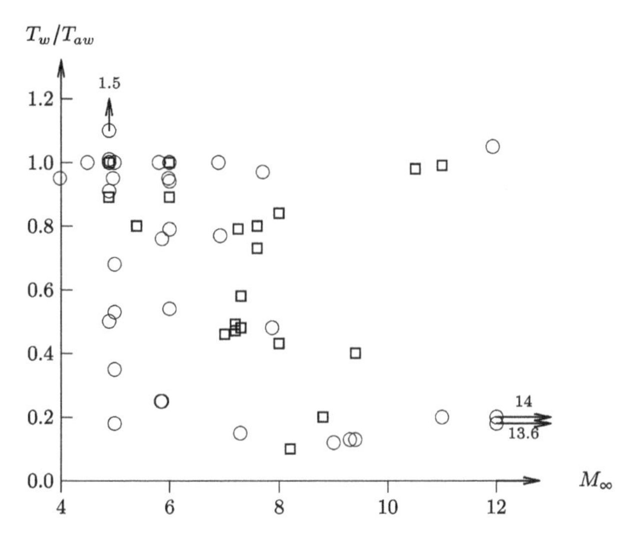


**Figure 8.16.** Turbulence statistics for DNS and LES (○) and experiment (□) for equilibrium turbulent boundary layers.

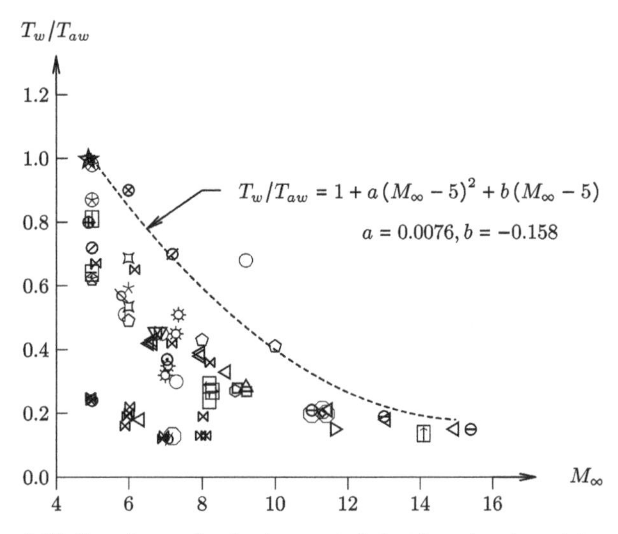


**Figure 8.17.** Experiments for shock wave turbulent boundary layer interaction.

significant lack of DNS and LES results for Mach 6 to 8 in the range $T_w/T_{aw} > 0.6$, and no data whatsoever above Mach 8. Figure 8.19 shows an overlap of the DNS, LES, and experiment data.

## 8.3 Summary

The aforementioned compilation of DNS, LES, and experiment results for description of equilibrium hypersonic turbulent boundary layers (i.e., Law of the Wall,

**Table 8.6.** Experiments for hypersonic shock wave turbulent boundary layer interaction (figure 8.17).

| *Symbol* | *Reference* |
|---|---|
| ○ | Elfstrom (1972) |
| □ | Coleman & Stollery (1972) |
| ◁ | Holden (1972*b*) |
| ▷ | Appels (1973), Richards & Appels (1973) |
| △ | Coleman (1973*a*), Coleman (1973*b*) |
| | Coleman & Stollery (1974) |
| ▽ | Kussoy & Horstman (1975) |
| ◖ | Law (1975) |
| ◗ | Mikulla & Horstman (1976*a*), Mikulla & Horstman (1976*b*) |
| ◇ | Holden (1984) |
| ☆ | Dolling & Rodi (1988) |
| * | Dolling & Smith (1989) |
| ⊙ | Kussoy & Horstman (1989) |
| ⊖ | Holden (1991) |
| ⊕ | Rodi & Dolling (1992), Rodi & Dolling (1995) |
| ⊗ | Vermeulen & Simeonides (1992) |
| ⊘ | Coët & Chanetz (1993) |
| ⊟ | Kussoy & Horstman (1993) |
| ⊟ | Kussoy, Horstman & Horstman (1993) |
| ⍐ | Simeonides & Haase (1995) |
| ⍗ | Schülein, Krogmann & Stanewsky (1996) |
| | Schülein & Zheltovodov (2001), Schülein (2006) |
| ⌜ | Prince, Vannahme & Stollery (2005) |
| ⊞ | Benay, Chanetz, Mangin, Vandomme & Perraud (2006) |
| | Bur & Chanetz (2009) |
| ⊛ | Murphree, Jagodzinski, Hood, Clemens & Dolling (2006) |
| | Murphree, Yüceil, Clemens & Dolling (2007) |
| ⋆ | Borovoy, Mosharov, Noev & Radchenko (2009) |
| | Borovoy, Egorov, Noev, Radchenko, Skuratov & Struminskaya (2012) |
| ◯ | Holden, MacLean, Wadhams & Mundy (2010) |
| ⬠ | Borovoy, Egorov, Noev, Skuratov & Struminskaya (2011) |
| | Borovoy, Egorov, Skuratov & Struminskaya (2013) |
| ⬡ | Murray, Hillier & Williams (2013) |
| ⌑ | Willems & Gülhan (2013), Willems, Gülhan & Steelant (2015) |
| ⋈ | Holden, Wadhams & MacLean (2014) |
| | Holden, Carr, MacLean & Wadhams (2018) |
| ∅ | Wagner, Schramm, Hickey & Hannemann (2016) |
| ⌀ | Schreyer, Sahoo, Williams & Smits (2018) |
| ⍉ | Currao, Choudhury, Gai, Neely & Buttsworth (2020) |
| ☼ | Chang, Chan, McIntyre & Veeraragavan (2021) |

mean velocity–temperature relation, Reynolds Analogy Factor, Morkovin's hypothesis, Morkovin's Strong Reynolds Analogy, and turbulence structure) indicates typically a substantial lack of data for high hypersonic Mach numbers (e.g., $M_\infty \gtrsim 8$) and cold walls ($T_w/T_{aw} \gtrsim 0.2$). Further DNS, LES, and experimental results for this region would significantly expand knowledge of equilibrium hypersonic turbulent boundary layer structure.

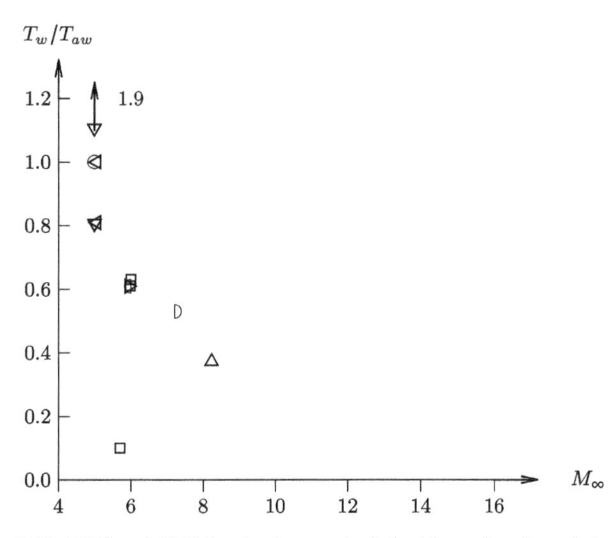


**Figure 8.18.** DNS and LES for shock wave turbulent boundary layer interaction.

**Table 8.7.** DNS and LES for hypersonic shock wave turbulent boundary layer interaction (figure 8.18).

| *Symbol* | *Reference* | *Method* |
|---|---|---|
| ○ | Edwards, Choi & Boles (2008) | WMLES |
| □ | Sandham, Schülein, Wagner, Willems & Steelant (2014) | DNS |
| ◁ | Fang, Lu, Yao & Zheltovodov (2015) | LES |
| | Fang, Yao, Zheltovodov & Lu (2017) | |
| ▷ | Yang, Urzay, Bose & Moin (2018) | WMLES |
| △ | Fu, Bose & Moin (2020) | WMLES |
| ▽ | Volpiani, Bernardini & Larsson (2020) | DNS |
| ( | Fu, Karp, Bose, Moin & Urzay (2021) | DNS, WMLES |
| D | Priebe & Martín (2021) | DNS |

DNS Direct Numerical Simulation LES Large Eddy Simulation
WMLES Wall Modeled Large Eddy Simulation

Experimental data for hypersonic shock wave turbulent boundary layer interactions is essentially limited to the region

$$\frac{T_w}{T_{aw}} < 1 + 0.007\,6(M_\infty - 5)^2 - 0.158(M_\infty - 5)$$

as indicated in figure 8.17. DNS and LES results for hypersonic shock wave turbulent boundary layer interaction are few in number and, with two exceptions, limited to $M_\infty \leqslant 6$. Clearly, DNS and LES results for hypersonic shock wave turbulent boundary layer interactions at higher Mach numbers are critically needed.

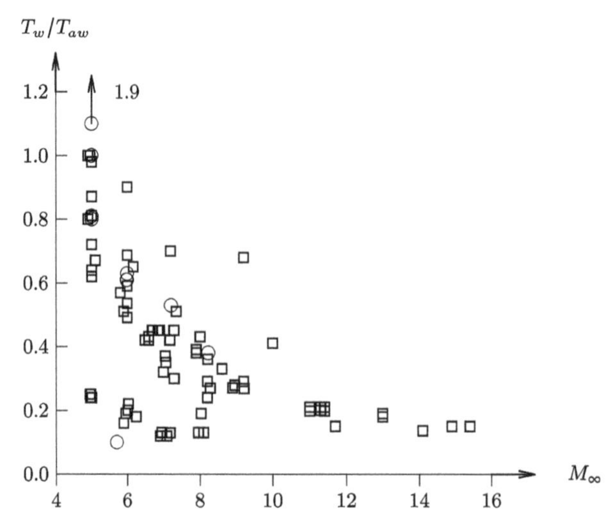


**Figure 8.19.** Shock wave turbulent boundary layer interactions for DNS and LES (○) and experiment (□).

# References

Adcock J, Peterson J and McRee D 1965 *Experimental Investigation of a Turbulent Boundary Layer at Mach 6, High Reynolds Numbers, and Zero Heat Transfer* Technical Note TN D-2907 (Washington, DC: National Aeronautics and Space Administration)

Appels C 1973 *Turbulent Boundary Layer Separation at Mach 12* Technical Note 90 von Kármán Institute for Fluid Dynamics

Auvity B, Etz M and Smits A 2001 Effects of transverse helium injection on hypersonic turbulent boundary layers *Phys. Fluids* **13** 3025–32

Backx E 1973 Experiments in the Mach 15 turbulent boundary layer on the wall of the longshot conical nozzle *EUROMECH 43 Coll. on Heat Transfer in Turbulent Boundary Layers with Variable Fluid Properties* (Göttingen: DFVLR-AVA)

Backx E 1974 *Experimental Study of the Turbulent Boundary Layer at Mach 15 and 19.8 in a Conical Nozzle* Technical Note 102 von Kármán Institute for Fluid Dynamics

Backx E 1975 A study of the turbulent boundary layer at Mach 15 and 19.8 on the wall of a conical nozzle *PhD Thesis* von Kármán Institute for Fluid Dynamics

Backx E and Richards B 1976 A high Mach number turbulent boundary-layer study *AIAA J.* **14** 1159–60

Beckwith I, Harvey W and Clark F 1971 *Comparisons of Turbulent Boundary-Layer Measurements at Mach Number 19 with Theory and an Assessment of Probe Errors* NASA Technical Note TN D-6192 (Washington, DC: National Aeronautics and Space Administration)

Benay R, Chanetz B, Mangin B, Vandomme L and Perraud J 2006 Shock wave/transitional boundary-layer interactions in hypersonic flow *AIAA J.* **44** 1243–54

Berg D 1977 Surface roughness effects on the hypersonic turbulent boundary layer *PhD Thesis* California Institute of Technology

Bloy A 1975 The expansion of a hypersonic turbulent boundary layer at a sharp corner *J. Fluid Mech.* **67** 647–55

Borovoy V, Egorov I, Noev A, Radchenko Y, Skuratov A and Struminskaya I 2012 3D shock/turbulent boundary layer interaction on the plate near a wedge in presence of an entropy layer *TsAGI Sci. J.* **43** 697–718

Borovoy V, Egorov I, Noev A, Skuratov A and Struminskaya I 2011 Two dimensional interaction between an incident shock and a turbulent boundary layer in the presence of an entropy layer *Fluid Dyn.* **46** 917–34

Borovoy V, Egorov I, Skuratov A and Struminskaya I 2013 Two Dimensional shock wave boundary layer interaction in the presence of an entropy layer *AIAA J.* **31** 80–93

Borovoy V, Mosharov V, Noev A and Radchenko Y 2009 Laminar-turbulent flow over wedges mounted on sharp and blunt plates *Fluid Dyn.* **44** 382–96

Brun C, Boiarciuc M, Haberkorn M and Comte P 2008 Large eddy simulation of compressible channel flow *Theor. Comput. Fluid Dyn.* **22** 189–212

Bur R and Chanetz B 2009 Experimental study on the PRE-X vehicle focusing on the transitional shock-wave/boundary-layer interaction *Aerosp. Sci. Technol.* **13** 393–401

Bushnell D, Johnson C, Harvey W and Feller W 1969 *Comparison of Prediction Methods and Studies of Relaxation in Hypersonic Turbulent Nozzle-Wall Boundary Layers* Technical Note D-5433 (Washington, DC: National Aeronautics and Space Administration)

Cary A 1970 Summary of available information on Reynolds analogy for zero-pressure-gradient, compressible, turbulent-boundary-layer flow *Technical Note D-5560* (Washington, DC: National Aeronautics and Space Administration)

Cebeci T and Smith A 1974 *Analysis of Turbulent Boundary Layers* (New York: Academic)

Chang E, Chan W, McIntyre T and Veeraragavan A 2021 Hypersonic shock impingement on a heated flat plate at Mach 7 flight enthalpy *J. Fluid Mech.* **98** R1-1–13

Chi S and Spaulding D 1966 Influence of temperature ratio on heat transfer to a flat plate through a turbulent boundary layer in air *Proc. 3rd Int. Heat Transfer Conf. (Chicago, IL)* 41–9

Chu Y-B, Zhuang Y-Q and Lu X-Y 2013 Effect of wall temperature on hypersonic turbulent boundary layer *J. Turbul.* **14** 37–57

Coët M-C and Chanetz B 1993 Experiments on shock wave boundary layer interaction in hypersonic flow *Rech. Aérosp.* **1** 61–704

Colburn A 1933 A method of correlating forced convection heat transfer data and a comparison with fluid friction *Trans. Am. Inst. Chem. Eng.* **29** 174–211

Coleman G 1973a *A Study of Hypersonic Boundary Layers Over a Family of Axisymmetric Bodies at Zero Incidence: Preliminary Report and Data Tabulation* Imperial College Aero Report 73-06 (London: Imperial College of Science and Technology)

Coleman G 1973b Hypersonic turbulent boundary layer studies *PhD Thesis* Imperial College of Science and Technology

Coleman G and Stollery J 1972 Heat transfer from hypersonic turbulent flow at a wedge compression corner *J. Fluid Mech.* **56** 741–52

Coleman G and Stollery J 1974 Incipient separation of axially symmetric hypersonic turbulent boundary layers *AIAA J.* **12** 119–20

Currao G, Choudhury R, Gai S, Neely A and Buttsworth D 2020 Hypersonic transitional shock wave boundary layer interaction on a flat plate *AIAA J.* **58** 814–29

Danberg J 1964 *Characteristics of the Turbulent Boundary Layer with Heat and Mass Transfer at M = 6.7* Technical Report NOLTR 64-99 (White Oak, Silver Spring, MD: United States Naval Ordnance Laboratory)

Ding H, Yi S, Quyang T and Zhao Y 2020 Research on velocity measurements of the hypersonic turbulent boundary layer based on the nano-tracer-based planar laser scattering technique *Meas. Sci. Technol.* **31** 085302

Dolling D and Rodi P 1988 Upstream influence and separation scales in fin-induced shock turbulent boundary layer interaction *J. Spacecr. Rockets* **25** 102–8

Dolling D and Smith D 1989 Separation shock dynamics in Mach 5 turbulent interactions induced by cylinders *AIAA J.* **27** 1698–706

Duan L, Beekman I and Martín M P 2010 Direct numerical simulation of hypersonic turbulent boundary layers. Part 2. Effect of wall temperature *J. Fluid Mech.* **655** 419–45

Duan L, Beekman I and Martín M P 2011 Direct numerical simulation of hypersonic turbulent boundary layers. Part 3. Effect of Mach number *J. Fluid Mech.* **672** 245–67

Duan L, Choudhari M and Zhang C 2016 Pressure fluctuations induced by a hypersonic turbulent boundary layer *J. Fluid Mech.* **804** 578–607

Duan L and Martín M P 2011a Direct numerical simulation of hypersonic turbulent boundary layers. Part 4. Effect of high enthalpy *J. Fluid Mech.* **684** 25–59

Duan L and Martín M P 2011b Effective approach for estimating turbulence-chemistry interaction in hypersonic turbulent boundary layers *AIAA J.* **49** 2239–47

Edwards J, Choi J-I and Boles J 2008 Large-eddy/Reynolds-averaged Navier-Stokes simulation of a Mach 5 compression-corner interaction *AIAA J.* **46** 977–91

Elfstrom G 1972 Turbulent hypersonic flow of a wedge compression corner *J. Fluid Mech.* **53** 113–27

Fang J, Lu L, Yao Y and Zheltovodov A 2015 Large eddy simulation of a three dimensional hypersonic shock wave/turbulent boundary layer interaction of a single fin *53rd AIAA Aerospace Sciences Meeting* AIAA Paper 2015-1062 (Reston, VA: American Institute of Aeronautics and Astronautics)

Fang J, Yao Y, Zheltovodov A and Lu L 2017 Investigation of three dimensional shock wave turbulent boundary layer interaction initiated by a single fin *AIAA J.* **55** 509–23

Feller W 1973 Effects of upstream wall temperatures on hypersonic tunnel wall boundary layer profile measurements *AIAA J.* **11** 556–8

Fischer M, Maddalon D, Weinstein L and Wagner R 1970 Boundary-layer surveys on a nozzle wall at a free-stream Mach number of about 20 including hot-wire fluctuation measurements *3rd Fluid and Plasma Dynamics Conference* AIAA Paper 70-746 (Reston, VA: American Institute of Aeronautics and Astronautics)

Fischer M, Maddalon D, Weinstein L and Wagner R 1971 Boundary-layer pitot and hot-wire surveys at $M_\infty \approx 20$ *AIAA J.* **9** 826–34

Fu L, Bose S and Moin P 2020 *Wall-Modeled LES of Three-Dimensional Intersecting Shock Wave/Turbulent Boundary Later Interactions* Annual Research Briefs 2020 Center for Turbulence Research, Stanford University

Fu L, Karp M, Bose S, Moin P and Urzay J 2021 Shock-induced heating and transition to turbulence in a hypersonic boundary layer *J. Fluid Mech.* **909** A8–49

Gaviglio J 1987 Reynolds analogies and experimental study of heat transfer in the supersonic boundary layer *Int. J. Heat Mass Transf.* **30** 911–26

Goyne C, Stalker R and Paull A 2003 Skin friction measurements in high enthalpy hypersonic boundary layers *J. Fluid Mech.* **485** 1–32

Hill F 1956 Boundary-layer measurements in hypersonic flow *J. Aeronaut. Sci.* **23** 35–42

Hill F 1959 Turbulent boundary layer measurements at Mach numbers from 8 to 10 *Phys. Fluids* **2** 668–80

Holden M 1972a *An Experimental Investigation of Turbulent Boundary Layers at High Mach Numbers and Reynolds Numbers* Contractor Report 112147 (Washington, DC: National Aeronautics and Space Administration)

Holden M 1972b Shock wave-turbulent boundary layer interaction in hypersonic flow *10th Aerospace Sciences Meeting* AIAA Paper 72-74 (Reston, VA: American Institute of Aeronautics and Astronautics)

Holden M 1984 Experimental studies of quasi-two-dimensional and three-dimensional viscous interaction regions induced by skewed-shock and swept-shock boundary interaction *17th Fluid Dynamics, Plasma Dynamics, and Lasers Conf.* AIAA Paper 84-1677 (Reston, VA: American Institute of Aeronautics and Astronautics)

Holden M 1991 Studies of the mean and unsteady structure of turbulent boundary layer separation in hypersonic flow *22nd Fluid Dynamics, Plasma Dynamics and Lasers Conf.* AIAA Paper 84-0016 (Reston, VA: American Institute of Aeronautics and Astronautics)

Holden M, Carr Z, MacLean M and Wadhams T 2018 Measurements in regions of shock wave/turbulent boundary layer interaction from Mach 5 to 6 at flight duplicated velocities to evaluate and improve the models of turbulence in CFD codes *2018 Fluid Dynamics Conf.* AIAA Paper 2018-3706 (Reston, VA: American Institute of Aeronautics and Astronautics)

Holden M and Chadwick K 1995 Studies of laminar, transitional and turbulent hypersonic flows over curved compression surfaces *33rd Aerospace Sciences Meeting and Exhibit* AIAA Paper 1995-93 (Reston, VA: American Institute of Aeronautics and Astronautics)

Holden M, MacLean M, Wadhams T and Mundy E 2010 Experimental studies of shock wave/turbulent boundary layer interaction in high Reynolds number supersonic and hypersonic flows to evaluate the performance of CFD codes *40th Fluid Dynamics Conference and Exhibit* AIAA Paper 2010-4468 (Reston, VA: American Institute of Aeronautics and Astronautics)

Holden M, Wadhams T and MacLean M 2014 Measurements in regions of shock wave/turbulent boundary layer interaction on double cone and hollow cylinder/flare configurations for open and 'blind' code evaluation/validation *AVIATION 2014*

Hopkins E and Keener E 1972a Pressure gradient effects on hypersonic turbulent skin friction and boundary layer profiles *Aerospace Sciences Meeting* AIAA Paper 72-215 (Reston, VA: American Institute of Aeronautics and Astronautics)

Hopkins E and Keener E 1972b Pressure gradient effects on hypersonic turbulent skin friction and boundary layer properties *AIAA J.* **10** 1141–2

Horstman C and Owen F 1972 Turbulent properties of a compressible boundary layer *AIAA J.* **10** 1418–29

Hoydysh W and Zakkay V 1969 An experimental investigation of hypersonic turbulent boundary layers in adverse pressure gradient *AIAA J.* **7** 105–16

Huang J, Nicholson G, Duan L, Choudhari M and Bowersox R 2020 Simulation and modeling of cold wall hypersonic turbulent boundary layers on flat plate *AIAA Scitech 2020 Forum* AIAA Paper 2020-0571 (Reston, VA: American Institute of Aeronautics and Astronautics)

Huang P, Coleman G and Bradshaw P 1995 Compressible turbulent channel flows: DNS results and modelling *J. Fluid Mech.* **305** 185–218

Jones R and Feller W 1970 Preliminary surveys of the wall boundary layer in a Mach 6 axisymmetric tunnel *Preliminary Surveys of the Wall Boundary Layer in a Mach 6*

*Axisymmetric Tunnel* Technical Report D-5620 (Washington, DC: National Aeronautics and Space Administration)

Keener E and Hopkins E 1972 *Turbulent Boundary-Layer Velocity Profiles on a Nonadiabatic Flat Plate at Mach Number 6.5* Technical Note TN D-6907 (Washington, DC: National Aeronautics and Space Administration)

Keener E and Polek T 1972 Measurements of Reynolds analogy for a hypersonic turbulent boundary layer on a nonadiabatic flat plate *AIAA J.* **10** 845–6

Kemp J and Owen F 1971 Nozzle wall boundary layers at Mach numbers 20 to 47 *AIAA J.* **10** 872–9

Kemp J and Owen F 1972a Experimental study of nozzle wall boundary layers at Mach Numbers 20 to 47 *Experimental Study of Nozzle Wall Boundary Layers at Mach Numbers 20 to 47* Technical Note D-6965 (Washington, DC: National Aeronautics and Space Administration)

Kemp J and Owen F 1972b Nozzle wall boundary layers at Mach Numbers 20 to 47 *AIAA J.* **10** 872–9

Kianvashrad N and Knight D 2021 Large eddy simulation of hypersonic cold wall flat plate—Part II *AIAA Aviation 2021 Forum* AIAA Paper 2021-2882 (Reston, VA: American Institute of Aeronautics and Astronautics)

Kussoy M and Horstman C 1975 *An Experimental Documentation of a Hypersonic Shock-Wave Turbulent Boundary Later Interaction Flow—With and Without Separation* Technical Memorandum X-62412 National Aeronautics and Space Administration, Ames Research Center

Kussoy M and Horstman C 1989 *Documentation of Two- and Three-Dimensional Hypersonic Shock Wave/Turbulent Boundary Layer Interaction Flows* Technical Memorandum 101075101075 (Washington, DC: National Aeronautics and Space Administration)

Kussoy M and Horstman K 1993 Three-dimensional hypersonic shock wave/turbulent boundary-layer interaction *AIAA J.* **31** 8–9

Kussoy M, Horstman K and Horstman C 1993 Hypersonic crossing shock-wave/turbulent boundary-layer interaction *AIAA J.* **31** 2197–203

Laderman A 1976 New measurements of turbulent shear stresses in hypersonic boundary layers *AIAA J.* **14** 1286–91

Laderman A and Demetriades A 1971 *Mean Flow Measurements in a Hypersonic Turbulent Boundary Layer* Technical Report U-4950 (Newport Beach, CA: Philco-Ford Corporation, Aeronutronic Division)

Laderman A and Demetriades A 1973 Hot-wire measurements of hypersonic boundary-layer turbulence *Phys. Fluids* **16** 179–81

Laderman A and Demetriades A 1974 Mean and fluctuating flow measurements in the hypersonic turbulent boundary layer over a cooled wall *J. Fluid Mech.* **63** 121–44

Law C 1975 *3D Shock Wave-Turbulent Boundary Layer Interactions at Mach 6* Technical Report ARL TR-75-0191 Aerospace Research Laboratories, United States Air Force Systems Command, Wright-Patterson AFB, OH

Lee R, Yanta W and Leonas A 1969 *Velocity Profile, Skin-Friction Balance and Heat-Transfer Measurements of the Turbulent Boundary Layer at Mach 5 and Zero-Pressure Gradient* Technical Report NOLTR 69-106 (White Oak, Silver Spring, MD: United States Naval Ordnance Laboratory)

Li X-L, Fu D-X and Ma Y-W 2006 Direct numerical simulation of spatially evolving supersonic turbulent boundary layer at Mach 6 *Chin. Phys. Lett.* **23** 1519–22

Liang X and Li X 2013 DNS of a spatially evolving hypersonic turbulent boundary layer at Mach 8 *Sci. China: Phys. Mech. Astron.* **56** 1408–18

Liang X, Li X and Fu D 2012 DNS and analysis of a spatially evolving hypersonic turbulent boundary layer over a flat plate at Mach 8 *Sci. Sin.-Phys. Mech. Astron.* **42** 282–93 in Chinese

Lobb R, Winkler E and Persh J 1955a Experimental investigation of turbulent boundary layers in hypersonic flow *J. Aeronaut. Sci.* **22** 1–9

Lobb R, Winkler E and Persh J 1955b *NOL Hypersonic Tunnel No. 4 Results VII: Experimental Investigation of Turbulent Boundary Layers in Hypersonic Flight* Technical Report NAVORD Report 3880 (White Oak, Silver Spring, MD: United States Naval Ordnance Laboratory)

Maeder T, Adams N and Kleiser L 2001 Direct numerical simulation of turbulent supersonic boundary layers by an extended temporal approach *J. Fluid Mech.* **429** 187–216

Martín M P 2007 Direct numerical simulation of hypersonic turbulent boundary layers. Part 1. Initialization and comparison with experiments *J. Fluid Mech.* **570** 347–64

Maslov A, Fedorov A, Bountin D, Shiplyuk A, Sidorenko A, Malmuth N and Knauss H 2008 Experimental study of disturbances in transitional and turbulent hypersonic boundary layers *AIAA J.* **46** 1880–3

Materna P 1977 Hot wire anemometry in a hypersonic turbulent boundary layer *10th Fluid and Plasmadynamics Conf.* AIAA Paper 77-702 (Reston, VA: American Institute of Aeronautics and Astronautics)

Matthews R and Trimmer L 1969 *Nozzle Turbulent Boundary-Layer Measurements in the VKF 50-Inch Hypersonic Tunnels* AEDC Technical Report TR-69-118 Arnold Engineering Development Center, Arnold Air Force Station

McGinley C, Spina E and Sheplak M 1994 Turbulence measurements in a Mach 11 helium boundary layer *25th AIAA Fluid Dynamics Conf.* AIAA Paper 70-746 (Reston, VA: American Institute of Aeronautics and Astronautics)

Mikulla V and Horstman C 1975 Turbulence stress measurements in a nonadiabatic hypersonic boundary layer *AIAA J.* **13** 1607–13

Mikulla V and Horstman C 1976a Turbulence measurements in hypersonic shock-wave boundary-layer interactions *AIAA J.* **14** 568–75

Mikulla V and Horstman C 1976b Turbulence measurements in hypersonic shock-wave boundary layer interactions *AIAA J.* **14** 568-575

Morkovin M 1962 Effects of compressibility on turbulent flows *Mécanique de la Turbulence* (Paris: Centre National de la Recherche Scientifique) pp 367–80

Murphree Z, Jagodzinski J, Hood E, Clemens N and Dolling D 2006 Experimental studies of transitional boundary layer shock wave interactions *44th AIAA Aerospace Sciences Meeting and Exhibit* AIAA Paper 2006-0326 (Reston, VA: American Institute of Aeronautics and Astronautics)

Murphree Z, Yüceil K, Clemens N and Dolling D 2007 Experimental studies of transitional boundary layer shock wave interactions *45th AIAA Aerospace Sciences Meeting and Exhibit* AIAA Paper 2007-1139 (Reston, VA: American Institute of Aeronautics and Astronautics)

Murray N, Hillier R and Williams S 2013 Experimental investigation of axisymmetric hypersonic shock-wave/turbulent-boundary-layer interactions *J. Fluid Mech.* **714** 152–89

Neeb D, Saile D and Gülhan A 2015 Experimental flow characterization and heat flux augmentation analysis of a hypersonic turbulent boundary layer along a rough surface *Proc. 8th European Symp. on Aerothermodynamics for Space Vehicles*

Neeb D, Saile D and Gülhan A 2018 Experiments on a smooth wall hypersonic boundary layer at Mach 6 *Exp. Fluids* **59** 68

Nicholson G, Huang J, Duan L and Choudhari M 2021 Simulation and modeling of hypersonic turbulent boundary layers subject to adverse pressure gradients due to concave streamline curvature *AIAA Aviation 2021 Forum* AIAA Paper 2021-2891 (Reston, VA: American Institute of Aeronautics and Astronautics)

Nicholson G, Huang J, Duan L, Choudhari M and Bowersox R 2021 Simulation and modeling of hypersonic turbulent boundary layers subject to favorable pressure gradients due to streamline curvature *AIAA Scitech 2021 Forum* AIAA Paper 2021-1672 (Reston, VA: American Institute of Aeronautics and Astronautics)

Owen F and Calarese W 1987 *Turbulence Measurement in Hypersonic Flow* AGARD Conf. Proc. No. 428 Advisory Group for Aerospace Research and Development, North Atlantic Treaty Organization 5-1–17

Owen F and Horstman C 1972 On the structure of hypersonic turbulent boundary layers *J. Fluid Mech.* **53** 611–36

Owen F, Horstman C and Kussoy M 1975 Mean and fluctuating flow measurements of a fully-developed, non-adiabatic, hypersonic boundary layer *J. Fluid Mech.* **70** 393–413

Peltier S, Humble R and Bowersox R 2011 Response of a hypersonic turbulent boundary layer to local and global mechanical distortions *49th AIAA Aerospace Sciences Meeting* AIAA Paper 2011-680 (Reston, VA: American Institute of Aeronautics and Astronautics)

Peltier S, Humble R and Bowersox R 2016 Crosshatch roughness distortions on a hypersonic turbulent boundary layer *Phys. Fluids* **28** 045105

Priebe S and Martín M P 2021 Turbulence in hypersonic compression ramp flow *Phys. Rev. Fluids* **6** 034601

Prince S, Vannahme M and Stollery J 2005 Experiments on the hypersonic turbulent shock wave boundary layer interaction and the effects of surface roughness *Aeronaut. J.* **109** 177–84

Raman K 1974 *A Study of Surface Pressure Fluctuations in Hypersonic Turbulent Boundary Layers* Contractor Report CR-2386 (Washington, DC: National Aeronautics and Space Administration)

Richards B and Appels C 1973 *Turbulent Heat Transfer Measurements in a Mach 15 Flow* von Kármán Institute for Fluid Dynamics

Rodi P and Dolling D 1992 An experimental/computational study of sharp fin induced shock wave/turbulent boundary layer interactions at Mach 5: experimental results *30th Aerospace Sciences Meeting and Exhibit* AIAA Paper 1992-749 (Reston, VA: American Institute of Aeronautics and Astronautics)

Rodi P and Dolling D 1995 Behavior of pressure and heat transfer in sharp fin-induced turbulent interactions *AIAA J.* **33** 2013–9

Sahoo D, Schultze M and Smits A 2009 Effect of roughness on a turbulent boundary layer in hypersonic flow *39th AIAA Fluid Dynamics Conference* AIAA Paper 2009-3678 (Reston, VA: American Institute of Aeronautics and Astronautics)

Samuels R, Peterson J and Adcock J 1967 *Experimental Investigation of the Turbulent Boundary Layer at a Mach Number of 6 with Heat Transfer at High Reynolds Numbers* NASA TN D-3858 (Washington, DC: National Aeronautics and Space Administration)

Sandham N, Schülein E, Wagner A, Willems S and Steelant J 2014 Transitional shock-wave/boundary-layer interactions in hypersonic flow *J. Fluid Mech.* **752** 349–82

Scaggs N 1966 *Boundary Layer Profile Measurements in Hypersonic Nozzles* ARL 66-0141 Aeronautical Research Laboratories, Office of Aerospace Research, United States Air Force

Schreyer A-M, Sahoo D, Williams O and Smits A 2018 Experimental investigation of two hypersonic shock/turbulent boundary-layer interactions *AIAA J.* **56** 4830–44

Schülein E 2006 Skin friction and heat flux measurements in shock/boundary layer interaction flows *AIAA J.* **44** 1732–41

Schülein E, Krogmann P and Stanewsky E 1996 *Documentation of Two Dimensional Impinging Shock/Turbulent Boundary Layer Interaction Flow* Report IB 223-96 A 49 Deutsches Zentrum für Luft- und Raumfahrt e.V. (DLR), Institut für Strömungsmechanik

Schülein E and Zheltovodov A 2001 *Documentation of Experimental Data for Hypersonic 3-D Shock Waves/Turbulent Boundary Layer Interaction Flows* Report IB 223-99 A 26 Deutsches Zentrum für Luft- und Raumfahrt e.V. (DLR), Institut für Strömungsmechanik

Simeonides G and Haase W 1995 Experimental and computational investigations of hypersonic flow about compression ramps *J. Fluid Mech.* **283** 17–42

Smith J and Driscoll J 1975 The electron-beam fluorescence technique for measurements in hypersonic turbulent boundary layers *J. Fluid Mech.* **72** 695–719

Stone D and Cary A 1972 *Discrete Sonic Jets Used as Boundary Layer Trips at Mach Numbers of 6 and 8.5* NASA TN D-6802 (Washington, DC: National Aeronautics and Space Administration)

Suraweera M, Mee D and Stalker R 2006 Reynolds analogy in high enthalpy and high Mach number turbulent flows *AIAA J.* **44** 917–9

Tichenor N, Humble R and Bowersox R 2013 Response of a hypersonic turbulent boundary layer to favorable pressure gradient *J. Fluid Mech.* **722** 187–213

Tichenor N, Neel I, Leidy A and Bowersox R 2017 Influence of streamline adverse pressure gradients on the structure of a Mach 5 turbulent boundary layer *55th AIAA Aerospace Sciences Meeting* AIAA Paper 2017-1697 (Reston, VA: American Institute of Aeronautics and Astronautics)

Trettel A and Larsson J 2016 Mean velocity scaling for compressible wall turbulence with heat transfer *Phys. Fluids* **28** 026102

Vaganov A and Stolyarov E 2008 Statistical laws of the pressure fluctuations in a hypersonic turbulent boundary layer *Fluid Dyn.* **43** 265–73

Van Driest E 1951 Turbulent boundary layer in compressible fluids *J. Aeronaut. Sci.* **18** 145-160

Vermeulen J and Simeonides G 1992 *Parametric Studies of Shock Wave/Boundary Layer Interactions Over 2D Compression Corners at Mach 6* Technical Note 181 von Kármán Institute for Fluid Dynamics

Voisinet R and Lee R 1972 Measurements of a Mach 4.9 zero-pressure-gradient turbulent boundary layer with heat transfer. Part 1—data compilation *Technical Report NOLTR 72-232* (White Oak, Silver Spring, MD: United States Naval Ordnance Laboratory)

Voisinet R, Lee R and Yanta W 1971 *An Experimental Study of the Compressible Turbulent Boundary Layer with an Adverse Pressure Gradient* Advisory Group for Aerospace Research and Development 9-1–10

Volpiani P, Bernardini M and Larsson J 2020 Effects of a nonadiabatic wall on hypersonic shock/boundary-layer interactions *Phys. Rev. Fluids* **5** 014602-1–20

Wagner A, Schramm J, Hickey J P and Hannemann K 2016 Hypersonic shock wave boundary layer interaction studies on a flat plate at elevated surface temperature *22nd Int. Shock Interaction Symp. (Glasgow)*

Wallace J 1967 Hypersonic turbulent boundary layer studies at cold wall temperatures *1967 Heat Transfer and Fluid Mechanics Institute (La Jolla, CA)* ed P Libby, B Olfe and C V Atta (Stanford, California: Stanford University Press)

Wallace J 1968 *Hypersonic Turbulent Boundary Layer Measurements Using an Electronic Beam* Technical Report AN-2112-Y1 Cornell Aeronautical Laboratory, Inc.

Wallace J 1969 Hypersonic turbulent boundary-layer measurements using an electron beam *AIAA J.* **7** 757–9

Watson R, Harris J and Anders J 1973 Measurements in a transitional/turbulent Mach 10 boundary layer at high-Reynolds Number *11th Aerospace Sciences Meeting* AIAA Paper 73-165 (Reston, VA: American Institute of Aeronautics and Astronautics)

Wegener P, Winkler E and Sibulkin M 1953 A Measurement of turbulent boundary-layer profiles and heat-transfer coefficient at $M = 7$ *J. Aeronaut. Sci.* **20** 221–2

Willems S and Gülhan A 2013 Experiments on shock induced laminar-turbulent transition on a flat plate at Mach 6 *European Conf. for Aeronautics and Space Sciences (EUCASS)*

Willems S, Gülhan A and Steelant J 2015 Experiments on the effect of laminar-turbulent transition on the SWBLI in H2K at Mach 6 *Exp. Fluids* **56** 49

Williams O, Sahoo D, Baumgartner M and Smits A 2018 Experiments on the structure and scaling of hypersonic turbulent boundary layers *J. Fluid Mech.* **834** 237–70

Williams O, Sahoo D, Papageorge M and Smits A 2021 Effects of roughness on a turbulent boundary layer in hypersonic flow *Exp. Fluids* **62** 1–13

Williams O and Smits A 2017 Effect of tripping on hypersonic boundary layer statistics *AIAA J.* **55** 3051–8

Winkler E 1961 Investigation of flat plate hypersonic turbulent boundary layer with heat transfer *Trans. Am. Soc. Mech. Eng. E* **83** 323–9

Winkler E and Cha M 1959 *Investigation of Flat Plate Hypersonic Turbulent Boundary Layer with Heat Transfer at a Mach Number of 5.2* NAVORD Report 6631 (White Oak, Silver Spring, MD: United States Naval Ordnance Laboratory)

Winkler E and Persh J 1954 *NOL Hypersonic Tunnel No. 4 Results VI: Experimental and Theoretical Investigation of the Boundary Layer and Heat Transfer Characteristics of a Cooled Wall Wedge Nozzle at a Mach Number of 5.5* NAVORD Report 3757 (White Oak, Silver Spring, MD: United States Naval Ordnance Laboratory)

Yang X, Urzay J, Bose S and Moin P 2018 Aerodynamic heating in wall modeled large eddy simulation of high speed flows *AIAA J.* **56** 731–42

Young F 1965 *Experimental Investigation of the Effects of Surface Roughness on Compressible Turbulent Boundary Layer Skin Friction and Heat Transfer* Technical Report DLR-532, CR-21 Defense Research Laboratory, University of Texas, Austin

Zhang C, Duan L and Choudhari M 2017 Effect of wall cooling on boundary-layer-induced pressure fluctuations at Mach 6 *J. Fluid Mech.* **822** 5–30

Zhang C, Duan L and Choudhari M 2018 Direct numerical simulation database for supersonic and hypersonic turbulent boundary layers *AIAA J.* **56** 4297–311

Zhang Y, Bi W, Hussain F and She Z 2014 A generalized Reynolds analogy for compressible wall-bounded turbulent flows *J. Fluid Mech.* **739** 392–420

**IOP** Publishing

Hypersonic Shock Wave Turbulent Boundary Layers
Direct Numerical Simulation, Large Eddy Simulation and Experiment
**Doyle Knight and Nadia Kianvashrad**

# Appendix A

# Hypersonic test facilities

## A.1 Introduction

### A.1.1 Scope of survey

The first design of a hypersonic wind tunnel was performed at the German WVA Supersonic Laboratory in Kochel, Bavaria, at the close of World War II (Lukasiewicz 1973). The design incorporated a 1 m × 1 m test section operating at atmospheric stagnation pressure with a maximum Mach number of 10 (Smelt 1946). With the defeat of Germany in 1945, the project was never completed.

This appendix provides an historical survey of hypersonic wind tunnels, both currently operating and decommissioned, in governmental research organizations and universities. Additional hypersonic facilities in industry, current and decommissioned, are omitted due to limitations on length of the appendix. The tunnel operating conditions listed herein have typically been obtained from the original references describing the installation and calibration of the facility. Significant modifications to currently operating tunnels may have been performed, and the reader is directed to the individual facilities for further information. Notation 'na' indicates information not available to the authors.

The facilities cited herein do not constitute a complete inventory of of hypersonic wind tunnels. Indeed Wittliff (1987) cites a total of 82 hypersonic wind tunnels as of 1963 ! The interested reader may consult earlier surveys by Vincente and Foy (1963), Pirrello *et al* (1971), Penaranda and Freda (1985), Wendt (1987) and Wittliff (1987) for additional information.

### A.1.2 Notation

The specification 'Run time' or 'Test time' is provided wherever available for each facility. The term 'Run time' may include the actual duration of the test including startup and shutdown. The term 'Test time' represents the useful period of time for measurements (i.e., the period when the freestream conditions in the test section are

 

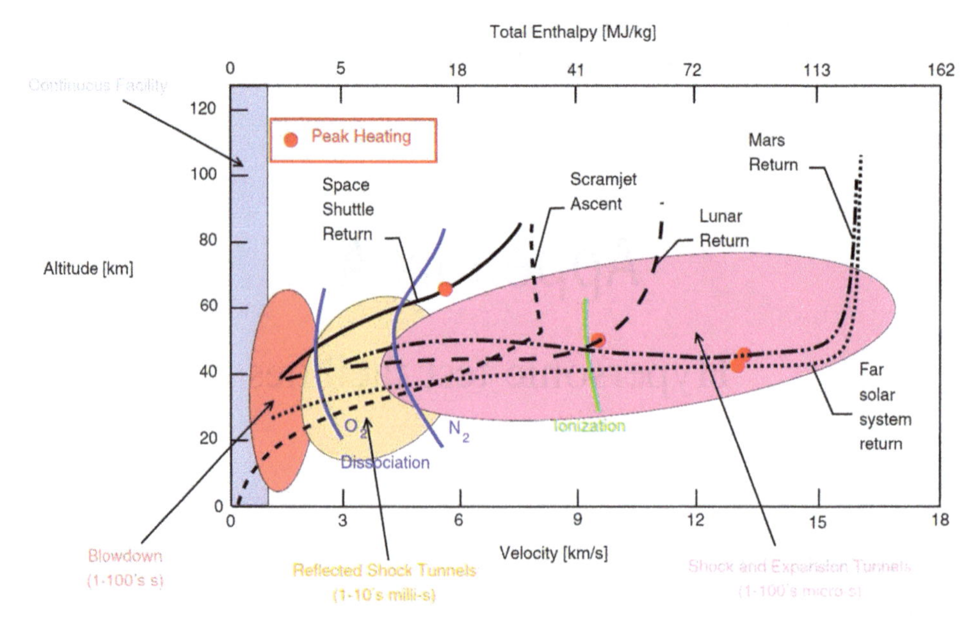


**Figure A.1.** Examples of mission trajectories CC.

essentially constant) as specifically provided by the facility. The notation $M_\infty$, $p_{t_\infty}$, $T_{t_\infty}$ and $H_\infty$ refer to the Mach number, total pressure, total temperature and stagnation enthalpy per unit mass in the test section.

### A.1.3 Flight envelope

The flight envelope of various space missions is illustrated in figure A.1 from Collen *et al* (2021) adapted from Fletcher (2005). Total (stagnation) enthalpies as large as 100 MJ $kg^{-1}$ and higher can be experienced. Hypersonic wind tunnels are designed to replicate some portion of the total envelope of flight conditions. Cox (1962) categorizes hypersonic wind tunnels into two categories. The first group is *hypersonic tunnels* wherein the Mach number is the principal parameter simulated, while the gas is heated only enough to avoid liquefaction and hence the total enthalpy corresponding to flight conditions is not met. The second group is *hypervelocity tunnels* which attempt to reproduce the flight total enthalpy as well as Mach number.

### A.1.4 Blowdown tunnel

The typical conventional blowdown wind tunnel comprises a high pressure gas source, pressure regulator and settling (stagnation) chamber, convergent-divergent nozzle, test section, second throat and (optional) vacuum chamber. One or more ejectors are used in some blowdown tunnels. The gas in the high pressure storage tank is typically at ambient temperature, and consequently the gas in the settling chamber may be heated to avoid condensation in the test section. The concept is illustrated in figure A.2. Maximum stagnation temperatures are limited by the

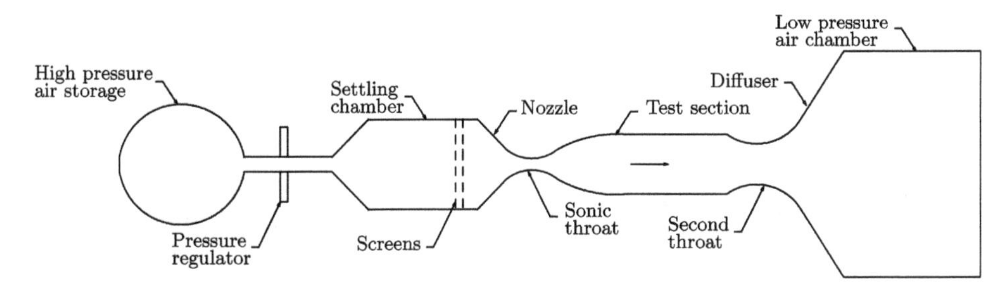


**Figure A.2.** Blowdown wind tunnel.

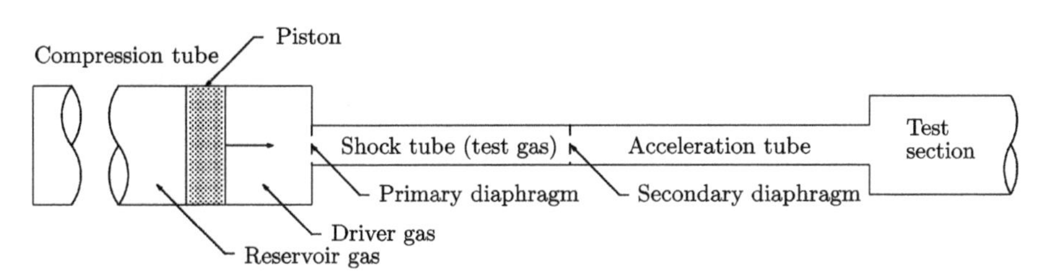


**Figure A.3.** Free piston expansion tube.

settling chamber material (e.g., the melting temperature of stainless steel is 1600 K to 1700 K). Helium has been used as the test gas since it can be expanded from ambient stagnation temperature (e.g., 300 K) to Mach numbers up to 35 without condensation (Erickson 1963), and doesn't require sensitive intrusive measurement instruments to be cooled. However, the non-equilibrium effects associated with hypersonic flight in air (e.g., thermochemical reaction, vibrational-translational energy transfer, etc) cannot be examined.

### A.1.5 Expansion tunnel

The expansion tunnel is a shock tunnel with three or more separate sections that achieves high stagnation enthalpy without the need for a reflected shock at the nozzle. The concept was initially proposed by Resler and Bloxsom (1952) and analyzed by Trimpi (1962), Trimpi and Callis (1965) and Trimpi (1966). The components are illustrated in figure A.3. The driver tube, driven tube and acceleration tube are initially pressurized to different conditions with the highest pressure in the driver tube. Similar to the reflected shock tunnel, the primary diaphragm separating the driver and driven tube bursts and the primary shock wave propagates into the driven tube. The primary interface separates the driver and driven (test) gases. The primary shock intersects the secondary diaphragm causing a secondary shock to propagate into the acceleration tube followed by the secondary interface separating the driven (test) gas and acceleration tube gas, and an unsteady expansion wave. The test gas is accelerated by the expansion wave and a region of high velocity, low temperature gas propagates into the test section. Further expansion of

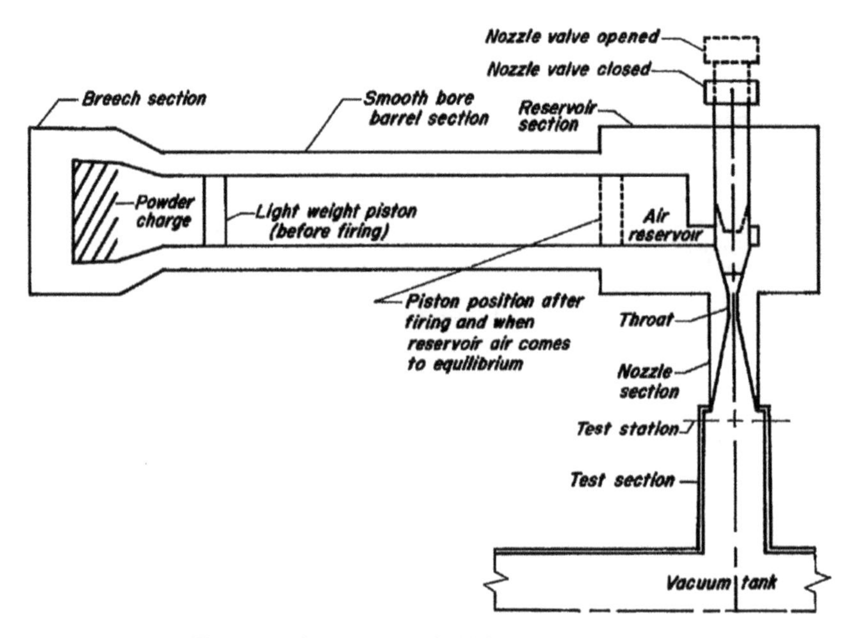


**Figure A.4.** Ames aeronautical laboratory gun tunnel.

the gas can be achieved by a diverging nozzle. The test gas is not brought to rest, and consequently the parameters of the driver, test and acceleration gases can be chosen to insure that the static temperature of the test gas remains sufficiently low to minimize (or prevent) non-equilibrium effects.

### A.1.6 Gun tunnel

Eggers *et al* (1955) describe the design of the original gun tunnel developed at Ames Aeronautical Laboratory. A schematic is shown in figure A.4. A powder charge is ignited accelerating a lightweight piston down an initially closed barrel, creating a shock wave that compresses the gas to a volume of high pressure, high temperature stagnant gas at the end of the barrel. A valve is opened allowing the gas to expand through a converging diverging nozzle into the test section. The gas empties into a vacuum tank. Modern gun tunnels use a high pressure driver gas to burst the primary diaphragm initially separating the driver gas and piston, thereby accelerating the piston down the barrel. A secondary diaphragm initially separates the test gas from the test section. The secondary diaphragm bursts when sufficient pressure is reached in the test gas, allowing the gas to expand through the converging diverging nozzle into the test section and vacuum chamber (figure A.5).

### A.1.7 Hotshot tunnel

The fundamental components of a hotshot tunnel are illustrated in figure A.6. Energy stored in a capacitor bank or flywheel is deposited into the test gas by electrical discharge into an arc chamber. A diaphragm, initially separating the arc

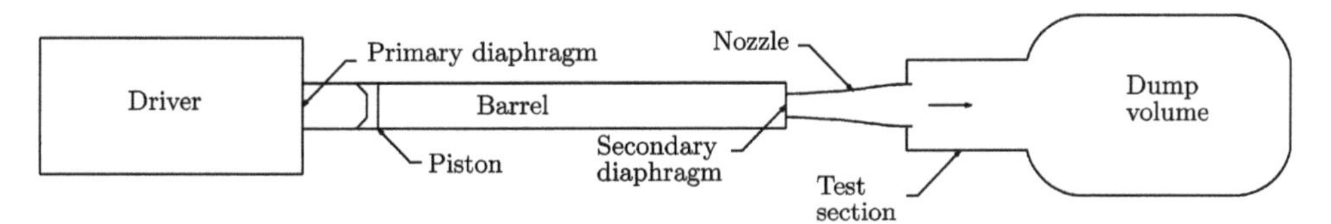


**Figure A.5.** Modern gun tunnel.

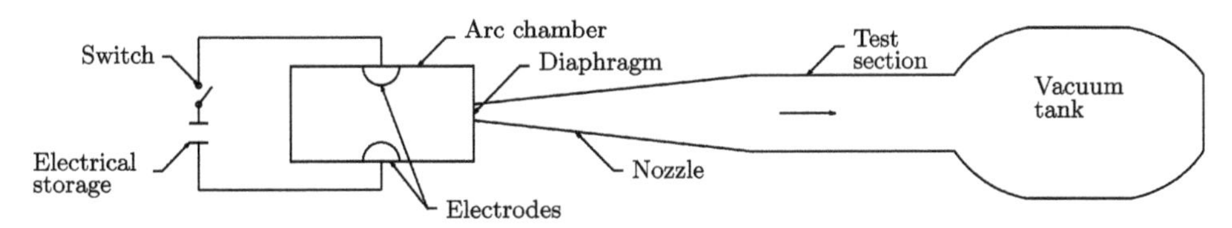


**Figure A.6.** Hot shot hypersonic wind tunnel.

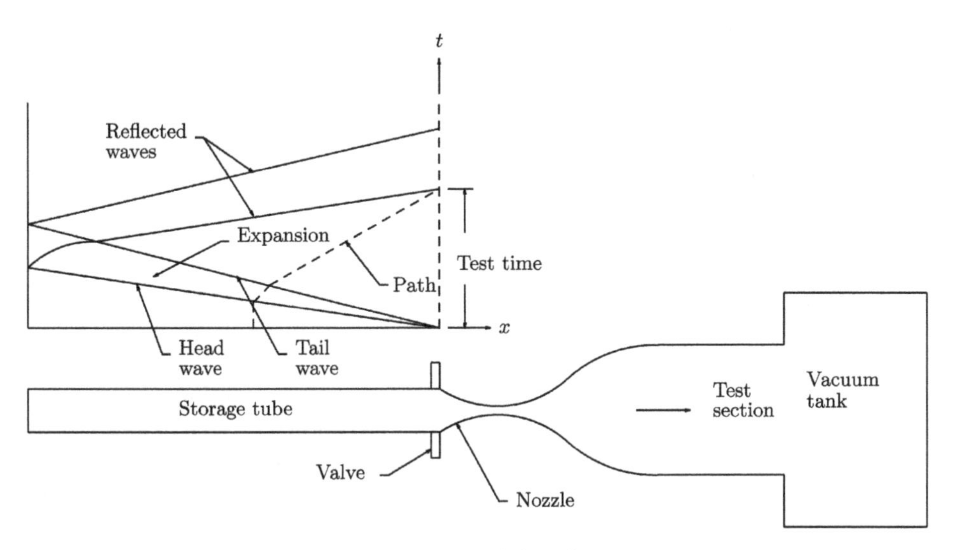


**Figure A.7.** Ludwieg tube.

chamber and stagnation (plenum) chamber, is vaporized and the heated gas expands through the plenum chamber into the test section and recovered in a vacuum tank. The stagnation temperature and pressure decrease during the test.

### A.1.8 Ludwieg tube

The Ludwieg tube was invented by Hubert Ludwieg as a low cost hypersonic wind tunnel (Ludwieg 1955, 1957). The concept is illustrated in figure A.7 and a simplified mathematical model is presented in Schrijer and Bannink (2008). The components are a long pressurized driver tube, converging-diverging nozzle, test section and dump tank. A fast acting valve or diaphragm is located either upstream of the nozzle or downstream of the test section. The driver tube is initially pressurized and the dump tank evacuated. Upon opening of the valve (or rupture of the diaphragm), an expansion wave moves upstream into the driver tube and the driver gas accelerates through the converging-diverging nozzle to the desired Mach number in the test section. The driver gas can be heated to permit higher test section Mach numbers and avoid condensation. Typically test times are hundreds of milliseconds.

### A.1.9 Reflected shock tunnel

The operating principles of the conventional reflected shock tunnel are illustrated in figure A.8. The tunnel comprises a driver section, driven section, diverging nozzle extending into an enclosed test section and a vacuum chamber (dump tank). The driver section contains the driver gas (typically helium) and the driven section contains the test gas (e.g., air, $CO_2$, $N_2$). Prior to operation the driver, driven and test section are separated by diaphragms. Upon reaching the desired driver gas pressure, the primary diaphragm separating the driver and driven sections ruptures, and a shock wave propagates into the driven section towards the nozzle and accelerates the test gas. The contact surface separating the driver and test gas also moves towards the nozzle at a velocity less than the shock wave. Simultaneously, an expansion wave propagates to the left into the driver gas. The details of this initial process are described in Liepmann and Roshko (1957). Upon reaching the nozzle the shock wave bursts the diaphragm and essentially reflects from the nozzle entrance since the cross-sectional area of the nozzle throat is small compared to the cross-sectional area of the driven section. This momentarily brings the test gas to rest upstream of the nozzle throat, thereby achieving a stagnated flow with high total pressure and total temperature. A wave system *a* to *d* in figure A.8 passes through the nozzle prior to the establishment of steady flow in the test section (Smith 1966). The reflected shock intersects the contact surface, and depending on the initial conditions in the driver and driven sections, either reflects as a shock or expansion from the contact surface. A special case known as 'tailored interface' condition can be achieved depending on the initial conditions in the driver and driven sections whereby there is no reflection

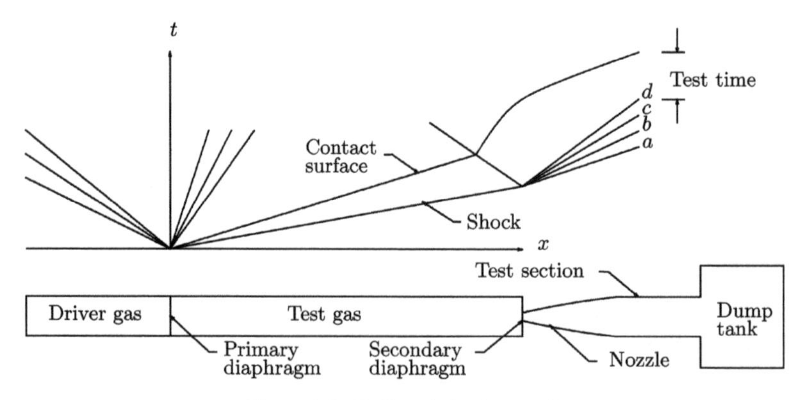


**Figure A.8.** Reflected shock tunnel.

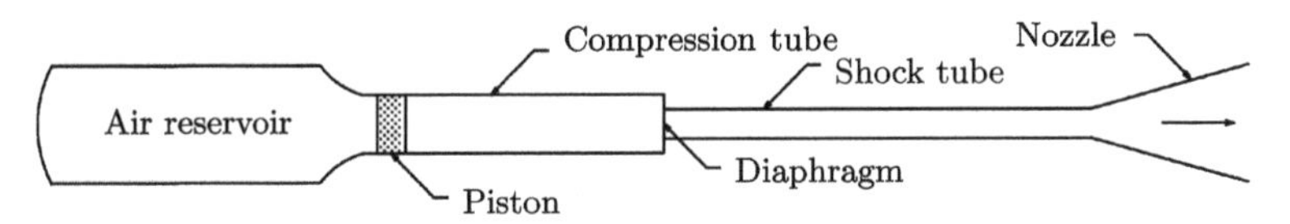


**Figure A.9.** Free piston driven shock tunnel.

from the interface back towards the nozzle. The tailored interface concept was first suggested by Hertzberg (Chan *et al* 1965). Typical test times are several milliseconds.

A variation of the reflected shock tunnel is the free piston driven reflected shock tunnel which uses a moving piston to accelerate the driver gas. The concept, originally proposed by Stalker (1967) (see also Stalker (1972)), is shown in figure A.9. A large piston is initially held in place at the exit of a high pressure air reservoir. Upon release the piston accelerates into the compression tube containing the driver gas. The diaphragm initially separating the driver and test (driven) gas is designed to rupture prior to the arrival of the piston at the junction of the driver and driven sections, thereby allowing the piston to continue to compress the driver gas. A shock wave propagates into the test gas in the same manner as the conventional reflected shock tunnel, and reflects from the driven tube-nozzle interface thereby creating the high stagnation pressure and high stagnation temperature test gas which accelerates into the test section. Typical test times are milliseconds.

Another variation of the reflected shock tunnel is the detonation driven shock tunnel (Yu *et al* 1992). The driver section initially contains a combustible mixture (e.g., hydrogen and ozygen). The mixture is ignited at the diaphragm initially separating the driver and driven sections and creates a Chapman-Jouguet detonation wave (Williams 1985) which propagates into the driven section. A damping section is attached to the end of the driver section to mitigate the reflected detonation wave.

## A.2 Aerospace Research Laboratories

The Aerospace Research Laboratories (ARL) were established in 1948 at Wright-Patterson AFB, OH by the United States Air Force at the recommendation of Theodore von Kármán. The laboratories were located in Building 450 in Area B just inside the fenceline running alongside National Road. The Aerospace Research Laboratories were disestablished in 1975 and the resources were assigned to the new Air Force Wright Aeronautical Laboratories (AFWAL).

### A.2.1 ARL 20 inch hypersonic wind tunnel

Gregorek and Lee (1962) and Gregorek (1962) detail the design performance, operational characteristics and initial calibration of the Aerospace Research Laboratory 20 inch Hypersonic Wind Tunnel. The configuration is shown in figure A.10. The wind tunnel operates in the blowdown mode. A Chicago-Pneumatic air compressor with dryer provides high pressure air at up to 20 MPa to a 39.6 $m^3$ storage system. A 1.5 MW electrical resistance air heater increases the stagnation temperture up to 1555 K. The air exhausts through an axisymmetric

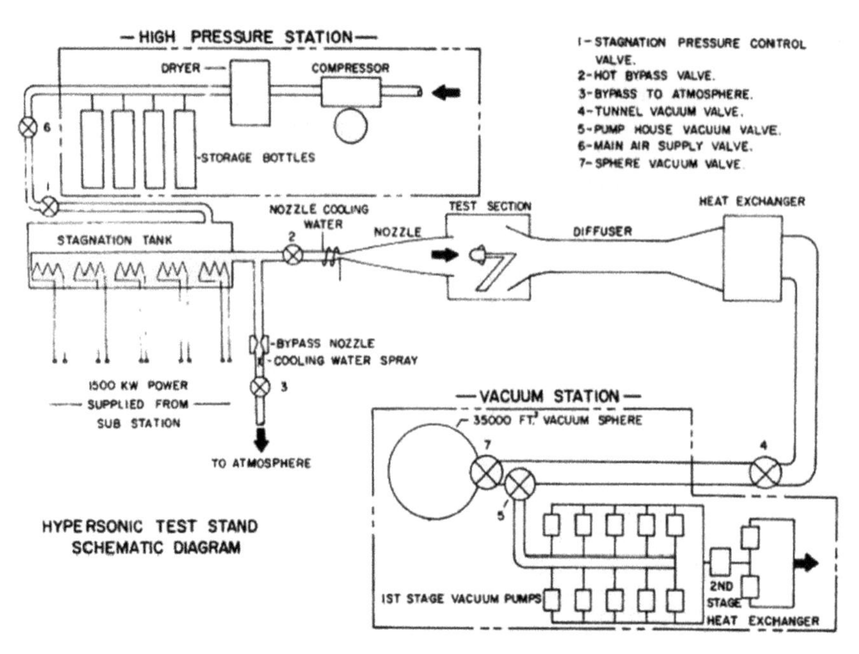


**Figure A.10.** ARL 20 inch Hypersonic Wind Tunnel.

**Table A.1.** Operating conditions.

| Quantity | Value |
|---|---|
| $M_\infty$ | 8–14 |
| $T_{t_\infty}$ (K) | 1555[a] |
| $p_{t_\infty}$ (MPa) | 20[b] |
| Test time (s) | 25–60 |

[a] Heater maximum.
[b] Storage tank maximum.

converging-diverging nozzle to an open jet with design Mach numbers from 8 to 14 depending on the nozzle. The flow exhausts to a 991 m$^3$ vacuum sphere initially evacuated to a pressure less than 266 Pa by twelve Allis Chalmers 27-D vacuum pumps. The operating conditions are listed in table A.1. Test times are estimated assuming $p_t$ = 11 MPa, $T_t$ = 1555 K and total pressure ratio of 0.60 across the heat exchanger.

### A.2.2 ARL 30 inch hypersonic wind tunnel

Scaggs *et al* (1963) describe the design performance, operational characteristics and initial calibration of the Aerospace Research Laboratory 30 inch Hypersonic Wind Tunnel. The configuration is shown in figure A.11. The wind tunnel operates in the blow down mode. An air compressor with dryer provides high pressure air at up to

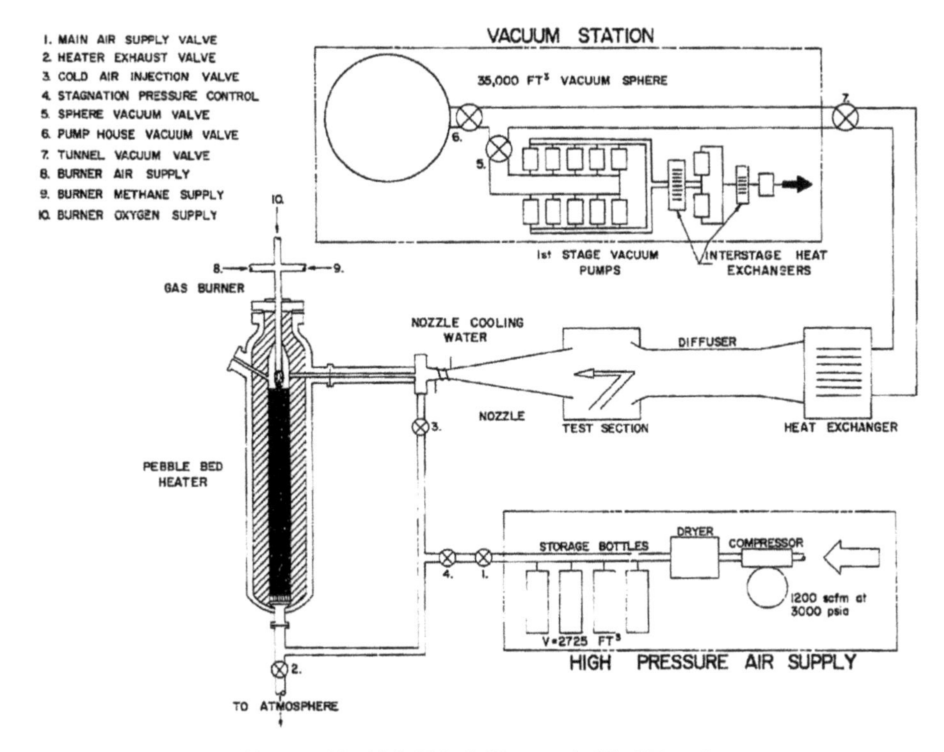


**Figure A.11.** ARL 30 inch Hypersonic Wind Tunnel.

**Table A.2.** Operating conditions.

| Quantity | Value |
|---|---|
| $M_\infty$ | 16–22 |
| $T_{t_\infty}$ (K) | 2444[a] |
| $p_{t_\infty}$ (MPa) | 20[b] |
| Test time (s) | 40–50 |

[a] Heater maximum.
[b] Storage tank maximum.

20 MPa to a 39.6 m$^3$ storage system. A gas fired, pebble bed, regenerative storage heater increases the stagnation temperature up to 2444 K. The air exhausts through an axisymmetric converging-diverging nozzle to an open jet with Mach numbers from 16 to 22. The air exhausts to a 991 m$^3$ vacuum sphere. The operating conditions are listed in table A.2. Test times are estimated assuming a total pressure ratio of 0.6 across the heat exchanger.

### A.2.3 ARL Mach 6 High Reynolds Number Facility

Fiore and Law (1975), Stetson (1980), Parobek *et al* (1985) describe the operational characteristics of the Air Force Wright Aeronautical Laboratories (AFWAL) Mach

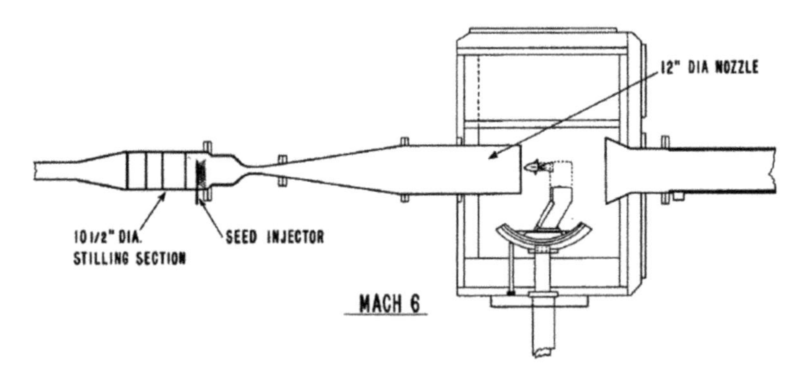


**Figure A.12.** AFWAL Mach 6 High Reynolds Number Tunnel.

**Table A.3.** Operating conditions.

| Quantity | Value |
|---|---|
| $M_\infty$ | 6 |
| $T_{t_\infty}$ (K) | 639 |
| $p_{t_\infty}$ (MPa) | 4.8 to 14.5[a] |
| $Re_\infty \times 10^{-6}$ (m$^{-1}$) | 98[b] |

[a] Fiore and Law (1975).
[b] At maximum $p_{t_\infty}$.

6 High Reynolds Number Tunnel. The stagnation chamber, nozzle and test section layout is shown in figure A.12. The open jet test flow exit nozzle diameter is 31.2 cm with an effective test core diameter of 25.4 cm (Stetson 1980). Stagnation temperatures up to 639 K are obtained using an electric pebble bed heater containing 50 000 stainless steel balls. The wind tunnel exhausts to atmosphere. The operating conditions are listed in table A.3. The wind tunnel was later designated as the Air Force Wright Aeronautical Laboratories Mach 6 High Reynolds Number Tunnel upon the disestablishment of Aerospace Research Laboratory in 1975.

## A.3 AFRL Mach 6 Ludwieg Tube

Kimmel *et al* (2017), Running *et al* (2019) and Hill *et al* (2022) describe the design and operation of the United States Air Force Research Laboratory (AFRL) Mach 6 Ludwieg Tube located at Wright-Patterson Air Force Base, Ohio. A perspective of the facility is shown in figure A.13. The driver tube comprises two 6.1 m sections and two 4.57 m sections assembled in two legs joined by two 90° elbows forming a 180° bend. The configuration is denoted a reflexed driver tube (Friehmelt *et al* 1993, Wolf *et al* 2007, Schrijer and Bannink 2008). The driver tube diameter is 24.77 cm and the nozzle exit diameter is 76 cm. The facility typically operates using dry air. The operating conditions are listed in table A.4.

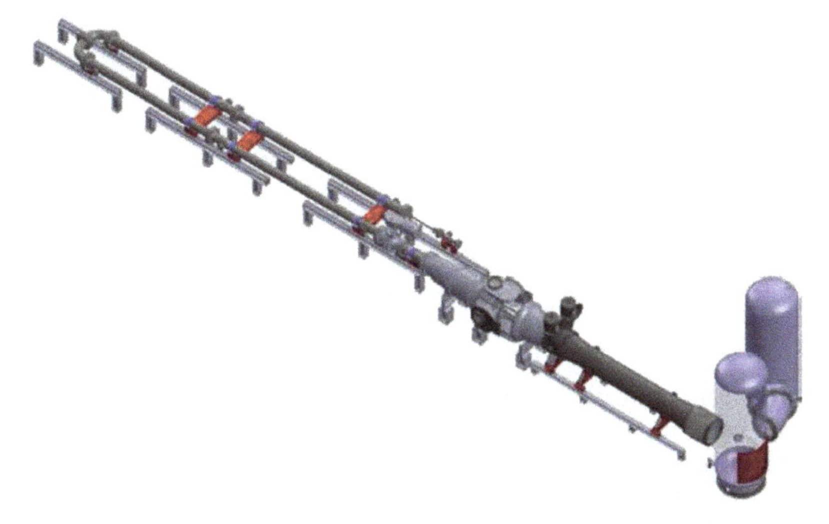

**Figure A.13.** AFRL Mach 6 Ludwieg Tube.

**Table A.4.** Operating conditions.

| Quantity | Min | Max |
|---|---|---|
| $M_\infty$ | 6 | 6 |
| $p_{t_\infty}$ (MPa) | 0.345 | 4.0 |
| $T_{t_\infty}$ (K) | 288 | 505 |
| $Re_\infty \times 10^{-6}$ (m$^{-1}$) | 2.3 | 30.1 |
| Test time (ms) | 100[a] | |

[a]Two periods each at 100 ms.

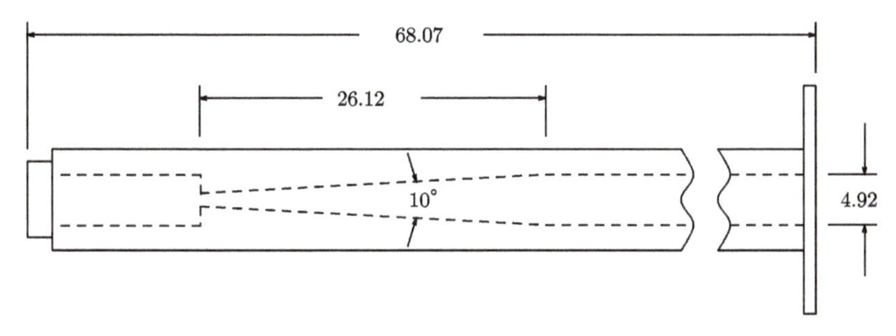


**Figure A.14.** Test section. Dimensions in cm.

## A.4 Applied Physics Laboratory JHU

Faro *et al* (1951) describes the hypersonic tunnel developed at the Applied Physics Laboratory Johns Hopkins University (JHU). The test section is shown in figure A.14. A 10° conical nozzle connects to a cylindrical test section. The conical nozzle length is 26.12 cm and the overal length is 68.07 cm. The cylindrical test section diameter is

**Table A.5.** Operating conditions.

| Quantity | Value |
| --- | --- |
| $M_\infty$ (air) | 10 |
| $M_\infty$ (He) | 16.2 |
| $p_{t_{max}}$ (MPa, air) | 20.7 |
| $p_{t_{max}}$ (MPa, other) | 13.8 |
| $T_{t_\infty}$ (K) | 722 to 777[a] |

[a] Hill (1956).

4.92 cm. The conical nozzle is faired to a circular contour at the throat. The tunnel could be operated in both blowdown and continuous mode, and operated with air, helium or nitrogen (Hill 1959). The operating conditions are shown in table A.5.

## A.5 Arnold Engineering Development Center

The Arnold Engineering Development Center (AEDC) von Kármán Gas Dynamics Facility of the United States Air Force comprises nine wind tunnels with operating ranges from subsonic to hypersonic. The three hypersonic wind tunnels are denoted Tunnel B, Tunnel C and Tunnel 9. Tunnels B and C are located at Arnold Air Force Base in Tullahoma, Tennessee, and Tunnel 9 at White Oak near Silver Spring, Maryland. Each wind tunnel provides unique capabilities in terms of Mach number and Reynolds number. The overall capabilities are shown in figure A.15.

### A.5.1 Tunnels B and C

The AEDC Hypersonic Tunnels B and C are closed circuit, variable density, continuous flow facilities. Details are presented in Wittliff (1987) and Mills (2015). The tunnels were designed in the 1950s and became operational between 1958 and 1961. The tunnels are the largest continuous flow hypersonic tunnels in the United States. A schematic of Tunnel B is shown in figure A.16. Tunnel C configuration is similar. The main compressor system comprises six axial and seven centrifugal compressors arrranged in nine stages, with an additional tenth stage for pressure control, drying and jet simulation. The test section is 1.27 m in diameter. Operating characteristics are listed in table A.6.

### A.5.2 Tunnel 9

Boyd and Ragsdale (1991) and Marren and Lafferty (2002) describe the AEDC Tunnel 9. The facility began operation in 1978 as part of the United States Naval Surface Weapons Center at White Oak, Maryland. It was transferred to the United States Air Force Arnold Engineering Development Center in 1997. The overall configuration is shown in figure A.17. The facility is blowdown type and operates using nitrogen gas. The test section diameter is 152.4 cm. Operating conditions for the several nozzle configurations are listed in table A.7.

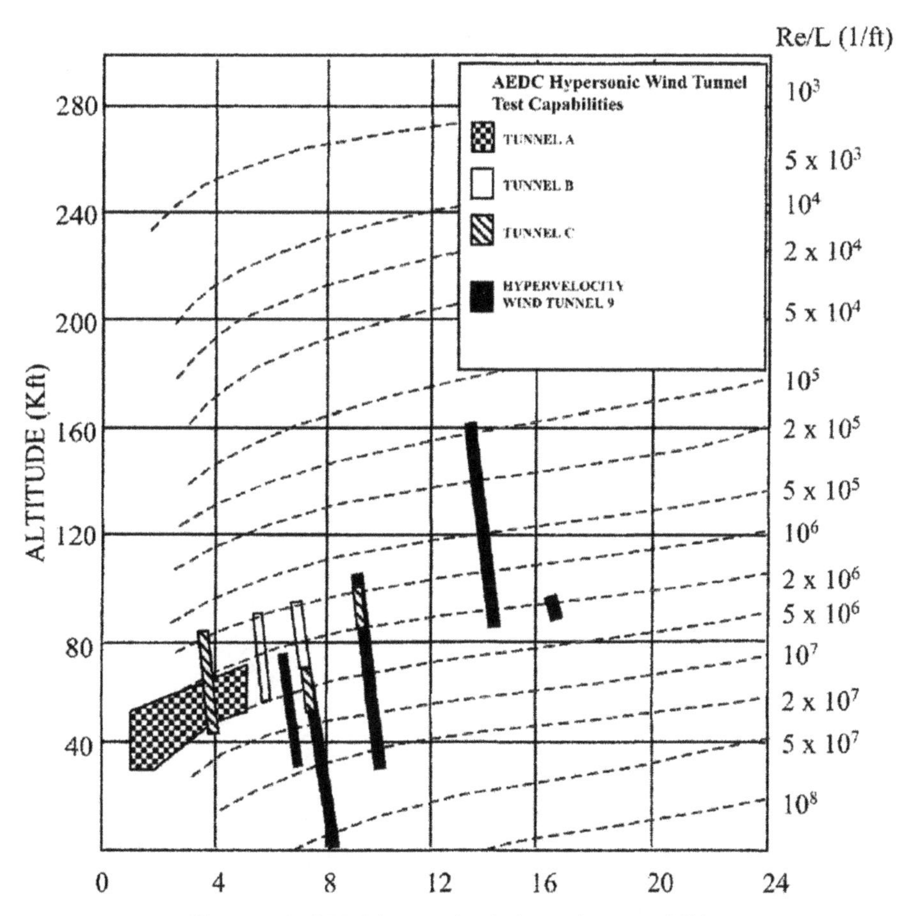


**Figure A.15.** AEDC hypersonic wind tunnel test capabilities.

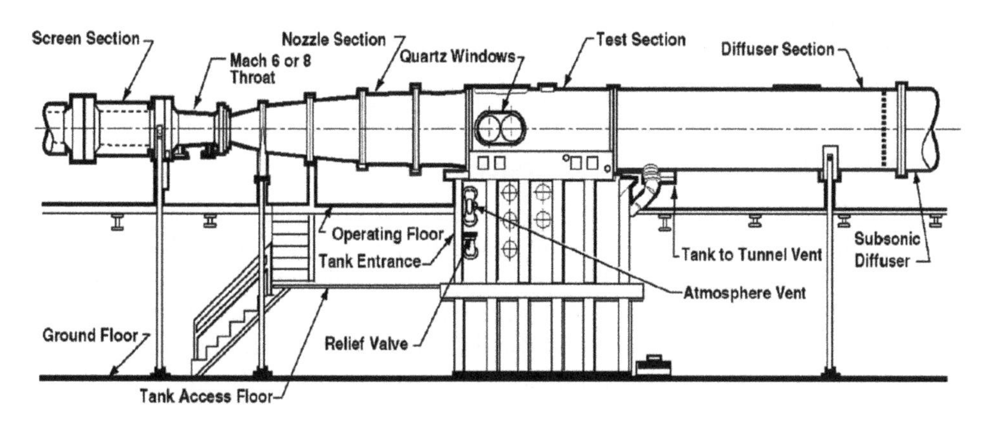


**Figure A.16.** AEDC Tunnel B.

**Table A.6.** Operating conditions.

| Quantity | Tunnel B | Tunnel C |
|---|---|---|
| $M_\infty$ | 6 and 8 | 10 |
| $p_{t_\infty}$ (MPa) | 0.28 to 2.1[a] | 1.38 to 13.1 |
| | 0.69 to 6.2[b] | |
| $T_{t_\infty}$ (K) | 750[c] | 1056[c] |
| $Re_\infty \times 10^{-6}$ ($m^{-1}$) | 1.0 to 15.4[d] | 1.0 to 15.4[d] |

[a] At Mach 6.
[b] At Mach 8.
[c] Maximum.
[d] Wittliff (1987).

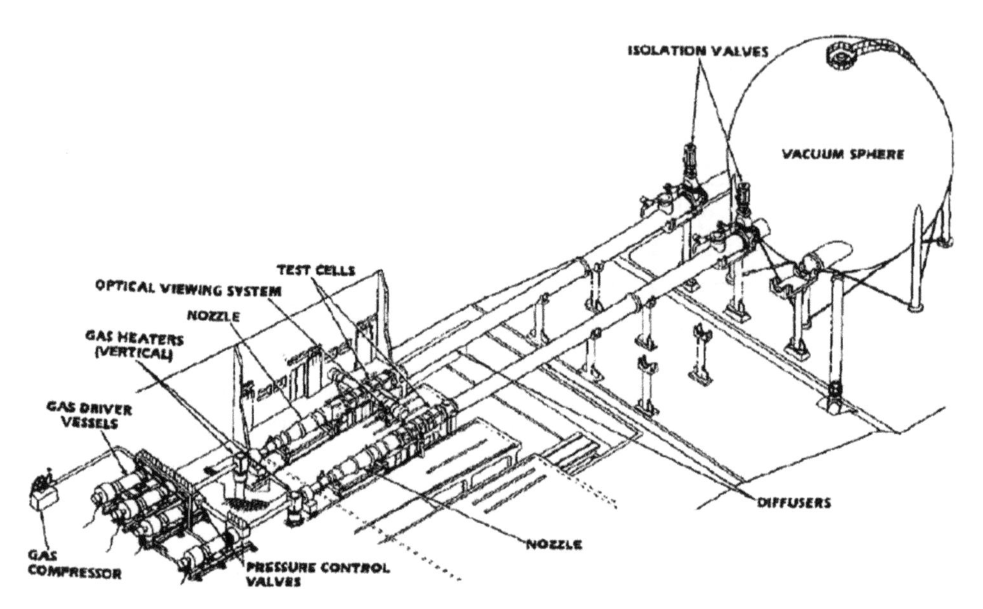


**Figure A.17.** AEDC Tunnel 9.

## A.6 Australian National University

Stalker (1972) describe the design performance, operational characteristics and initial calibration of the free piston shock tunnels at the Australian National University. The concept of a free piston driven shock tunnel was proposed by Stalker (1960). Figure A.18 presents schematics of the three tunnels, namely T1, T2, and T3. T3 began operation in 1968 (Stalker 2006). The operating conditions are listed in table A.8. T3 was decommissioned in the early 2000s. It was purchased by Don Fry, owner and chairman of NQEA Australia and Adjunct Professor of the School of Engineering at the University of Queensland, and stored at the University of Queensland veterinary farm. In 2014 the University of Oxford purchased the compression and driver tube assembly (see section A.26.2).

**Table A.7.** Operating conditions.

| | $p_{t_\infty}$ (MPa) | | $T_{t_\infty}$ (K) | $Re_\infty \times 10^{-6}$ ($m^{-1}$) | | Duration (s) | |
|---|---|---|---|---|---|---|---|
| Mach | Min | Max | | Min | Max | Min | Max |
| 7 | 17.9 | 81.4 | 1922 | 12.1 | 51.8 | 1 | 6 |
| 8 | 6.9 | 82.7 | 866 | 11.5 | 164 | 0.33 | 5 |
| 10 | 3.4 | 96.5 | 1005 | 2.8 | 65.6 | 0.23 | 8 |
| 14 | 0.7 | 137.9 | 1783 | 0.24 | 12.5 | 0.7 | 15 |
| 16.5 | 16.5 | 144.8 | 1855 | 2.8 | 11.5 | 3.5 | 6 |

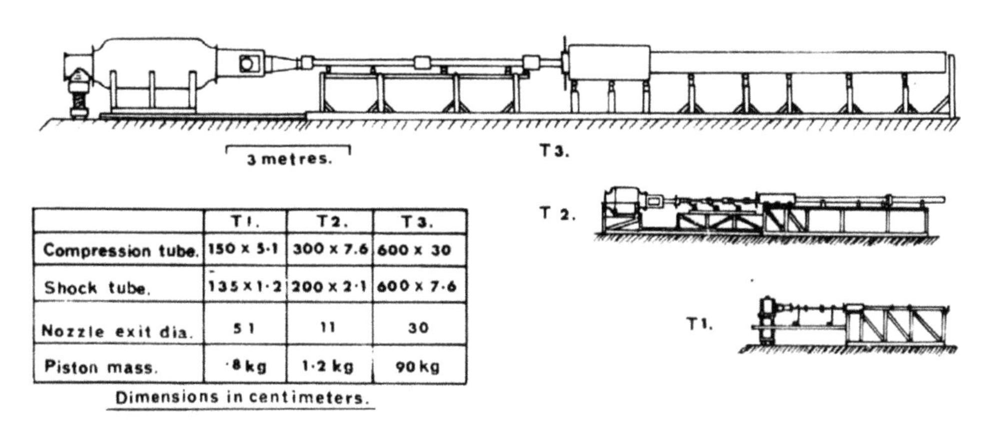


| | T1. | T2. | T3. |
|---|---|---|---|
| Compression tube. | 150 x 5·1 | 300 x 7·6 | 600 x 30 |
| Shock tube. | 135 x 1·2 | 200 x 2·1 | 600 x 7·6 |
| Nozzle exit dia. | 5 1 | 11 | 30 |
| Piston mass. | ·8 kg | 1·2 kg | 90 kg |

**Figure A.18.** Free piston shock tunnels at Australian National University. Reproduced with permission (Stalker 1972).

**Table A.8.** Operating conditions.

| Quantity | Value |
|---|---|
| $M_\infty$ | 4–11 |
| $T_{t_\infty}$ (K) | 1000[a] |
| $p_{t_\infty}$ (MPa) | 4[b] |

[a] Heater maximum.
[b] Supply maximum.

## A.7 California Institute of Technology

### A.7.1 Hypersonic wind tunnels

Baloga and Nagamatsu (1955) describe two hypersonic wind tunnels at the Graduate Aerospace Laboratories of the California Institute of Technology (GALCIT). The schematic of the facility is shown in figure A.19. The closed return, continuously operating tunnel comprised two legs with test section cross sections 12.7 cm by 12.7 cm. The tunnel was powered by sixteen compressors in a serial/

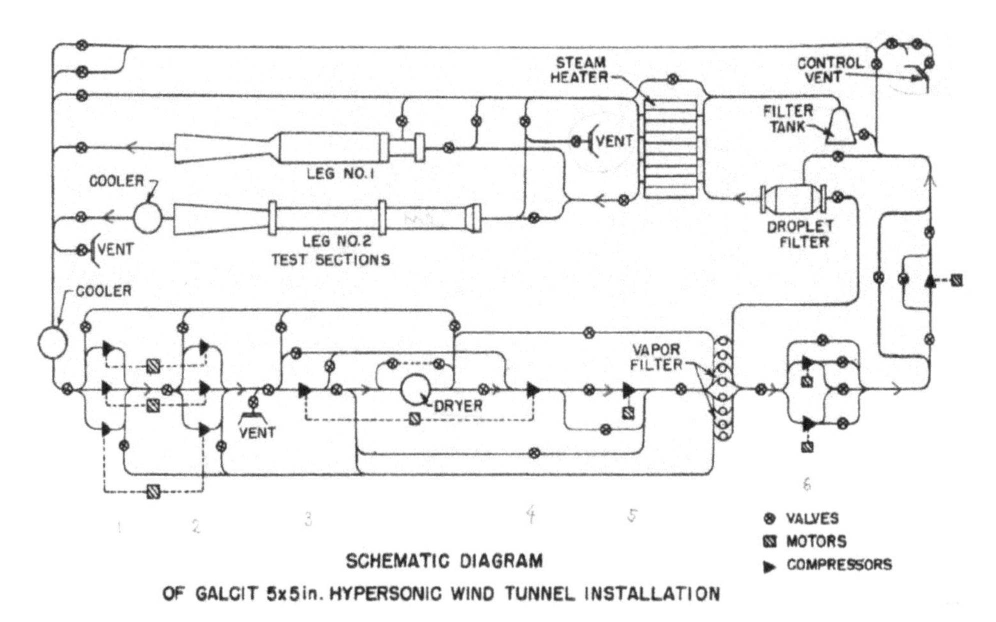


**Figure A.19.** GALCIT hypersonic wind tunnels.

**Table A.9.** Operating conditions.

| Quantity | Leg 1 | Leg 2 |
|---|---|---|
| $M_\infty$ | 2 to 7 | 7 to 11 |
| $p_{t_\infty}$ (MPa) | 7.3[a] | 7.3[a] |
| $T_{t_\infty}$ (K) | 478[a] | 866 |

[a] Maximum.

parallel arrangement. Air was heated by a steam heater in Leg 1 and an electric heater in Leg 2. The operating conditions are listed in table A.9.

### A.7.2 T5 hypervelocity tunnel

The GALCIT T5 hypervelocity tunnel is a free piston driven, reflected shock tunnel. A schematic of the facility is shown in figure A.20. The tunnel comprises a secondary reservoir, compression tube, shock tube, nozzle, test section and vacuum tank. The compression tube is 30 m in length with a diameter of 30 cm. The shock tube is 12 m in length with a diameter of 9 cm. The driver gas is a mixture of helium and argon. Air, nitrogen, carbon dioxide, hydrogen and gas mixtures are used as test gases. The contoured nozzle exit diameter is 314 mm. In a typical run, the secondary reservoir initially contains air at 13 MPa, the compression tube contains helium at 150 kPa, and the shock tube contains air at 90 kPa. The diaphragm separating the compression tube and shock tube is designed to burst at up to 110 MPa.

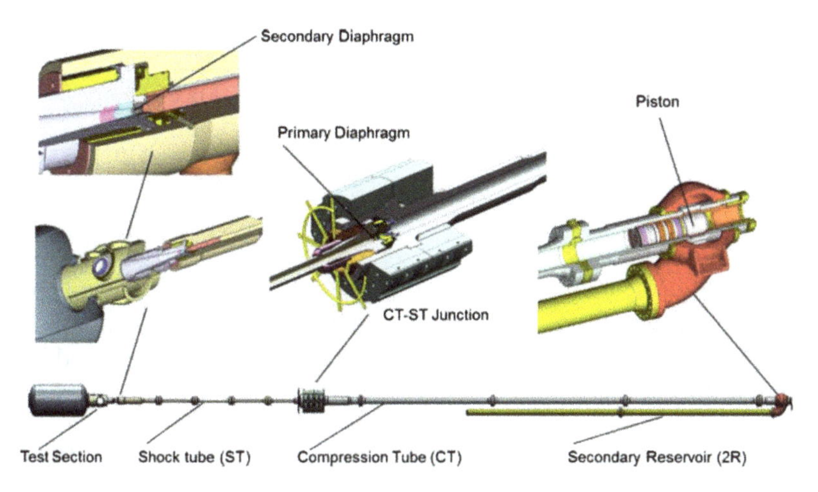


**Figure A.20.** GALCIT T5 tunnel.

**Table A.10.** Operating conditions.

| Quantity | Value |
|---|---|
| $M_\infty$ | 5.2 |
| $H_{t_\infty}$ (MJ kg$^{-1}$) | 25[a] |
| $p_{t_\infty}$ (MPa) | 100[a] |
| $T_{t_\infty}$ (K) | 10 000[b] |
| Test time (ms) | 1–2 |

[a] Maximum (Marineau and Hornung 2009).
[b] Maximum (Hannemann *et al* 2016).

Compressed air accelerates the piston to approximately 300 m s$^{-1}$ which heats the helium to approximately 4000 K at the moment of rupture of the diaphragm separating the compression and shock tubes. The piston continues to move and while decelerating to approximately 20 m s$^{-1}$ maintains an approximate constant driver gas pressure. The shock speed in the shock tube is typically 4 km s$^{-1}$. The facility began operation on December 17, 1990. Operating conditions are listed in table A.10. Details are presented in Hornung (1992), Hornung *et al* (1995) and Hannemann *et al* (2016).

### A.7.3 Hypervelocity expansion tube

The Hypervelocity Expansion Tube (HET) at GALCIT is an expansion tunnel. The overall length is 9.14 m comprising the driver (1.22 m), driven (3.96 m) and acceleration (3.96 m) sections, plus the test section and dump tank. The internal diameter of the three sections is 152 mm and the wall thickness is 0.5 mm. The primary diaphragm separating the driver and driven sections is 1.52 mm thick aluminum with a burst pressure of 4.3 MPa. The secondary diaphragm separating

**Table A.11.** Operating conditions.

| Quantity | Min | Max |
|---|---|---|
| $M_\infty$ | 3.5 | 7.5 |
| $H_\infty$ (MJ kg$^{-1}$) | 2.0 | 8.8 |

the driven and acceleration sections is 12.7 $\mu$m Mylar. The operating conditions are listed in table A.11. Test gases are air and nitrogen. Details are presented in Dufrene (2006) and Dufrene *et al* (2007).

## A.8 Calspan University of Buffalo Research Center (CUBRC)

The Calspan University of Buffalo Research Center (CUBRC) is the descendent of the Cornell Aeronautical Laboratory established on January 1, 1946 in Buffalo, New York. The laboratory was originally founded by the Curtiss-Wright Aircraft Company in 1942, and donated to Cornell University coincident with the founding of the Graduate School of Aeronautical Engineering at Cornell on November 1, 1945 (*Curtiss-Wright Corp. Gives Lab to Cornel; To Aid Air Research* 1945). In 1972, the laboratory was reorganized as a for-profit corporation and renamed Calspan. In 1983, the aerospace laboratories were reorganized as an independent, non-profit research and development company in collaboration with the University of Buffalo, and renamed Calspan University of Buffalo Research Center (CUBRC).

### A.8.1 Calspan 48 inch and 96 inch shock tunnels

Holden (1975) describes the operation of the 48 inch and 96 inch reflected shock tunnels at Calspan. The 48 inch shock tunnel began operation on 1 July 1959 and the 96 inch shock tunnel in January 1964 when the laboratory was known as the Cornell Aeronautical Laboratory (Holden 1990). The upgraded 96 inch shock tunnel began operation in September 1992 (Albrechcinski *et al* 1995). The 96 inch tunnel used a 19.37 cm diameter driver tube of length 4.88 m, and a variable length driven tube comprised of segments of 1.52 m, 2.74 m and 4.27 m with a typical total length of 14.94 m. Selection of specific properties of the driver and driven gas achieves a tailored interface condition (Wittliff *et al* 1959) whereby the reflected shock is transmitted through the contact surface without creation of a secondary reflected shock. The tailored interface concept was first suggest by Hertzberg (Chan *et al* 1965) and introduced at Calspan in 1957 (Holden 1990). The operating conditions of the 48 inch and 96 inch tunnels are listed in table A.12.

### A.8.2 LENS I, II

Holden (1990), Holden and Parker (2002) and Holden (2010) describe the design and operational characteristics of the Large Energy National Shock Tunnel (LENS) I and II. Both facilities operate as reflected shock tunnels. The incident shock is

**Table A.12.** Operating conditions.

| | 48 inch | | 96 inch | |
|---|---|---|---|---|
| Quantity | Min | Max | Min | Max |
| $M_\infty$ | 6 | 18 | 8 | 16 |
| $H_\infty$ (MJ kg$^{-1}$) | 0.29[a] | 4.65[a] | 3.3 | 12.8[b] |
| $p_{t_\infty}$ (MPa) | | | 33 | 97[b] |
| $Re \times 10^{-6}$ (m$^{-1}$) | 0.33[a] | 164.[a] | | |
| Runtime (ms) | | | 1 | 10[b] |

[a] Holden (2010).
[b] Albrechcinski *et al* (1995).

**Table A.13.** Operating conditions.

| | LENS I | | LENS II | |
|---|---|---|---|---|
| Quantity | Min | Max | Min | Max |
| $M_\infty$ | 8 | 18 | 3 | 10 |
| $H_\infty$ (MJ kg$^{-1}$) | 0.42 | 10.5 | 0.29 | 3.8 |
| $Re_\infty \times 10^{-6}$ (m$^{-1}$)[a] | 0.03 | 328 | 0.032 | 3280 |
| Test time (ms) | 5[b] | 18[c] | 30[d] | 80[d] |

[a] Holden *et al* (2006).
[b] Mach 10 to 18 (Holden and Parker 2002).
[c] Mach 7 to 10 (Holden and Parker 2002).
[d] Holden and Parker (2002).

generated by the sudden release of the driver gas by rupture of a double diaphragm assembly. Additionally, LENS II can be operated as a Ludwieg tube (Mach 2 to 4) or an expansion tunnel Holden (2010). Both tunnels use contoured nozzles and operate at the tailored interface condition. The operating conditions are summarized in table A.13. The length and diameter of the driver and driven tubes for LENS I and II are also listed (table A.14).

### A.8.3 LENS XX

Dufrene *et al* (2010) describe the LENS XX expansion tunnel comprising a driver section, a driven section, and an expansion tube, nozzle and test section. The internal diameter of the tubes is 60.96 cm with a 2.44 m diameter test section. The large diameter minimizes the viscous/boundary layer effect and creates a large clean core flow. The total length of the LENS XX tunnel is 73.15 m. This tunnel can operate over a large range of conditions with little or no dissociation in the test gas. The shock wave first accelerates the gas and heat the gas at the same time. The unsteady expansion wave increases the gas velocity to the desired speed. The test time in this tunnel is about 4 ms with Mach numbers over 30, Reynolds number per

**Table A.14.** Facility dimensions.

| Quantity | LENS I | LENS II |
|---|---|---|
| Driver length (m) | 7.62 | 18.29 |
| Driver diameter (cm) | 27.94 | 60.96 |
| Driven length (m) | 18.29 | 30.48 |
| Driven diameter (cm) | 20.32 | 60.96 |

**Table A.15.** Operating conditions[a].

| Quantity | Min | Max |
|---|---|---|
| $M_\infty$ | 8 | 25 |
| $H_\infty$ (MJ $kg^{-1}$) | 3.0 | 74.3 |
| $Re_\infty \times 10^{-6}$ ($m^{-1}$) | $3.3 \times 10^{-3}$ | 3280 |

[a] Holden (2010).

meter of $10^7$, and gas velocities greater than 13 km $s^{-1}$. The four-chamber configuration increases the possible total enthalpy to more than 120 MJ $kg^{-1}$. The tunnel is designed for hydrogen as the driver gas which is electrically heated. However, other driver gas such as helium and argon can be used to achieve specific flight conditions. All tubes are honed stainless steel to minimize wall imperfection and corrosion. The first diaphragm is a double diaphragm system and the second diaphragm is movable to achieve a longer test time for each test condition. There are two test sections available. LENS XX tunnel is capable of simulating Earth, Jovian, Martian, and Titan atmospheres. The operating conditions are listed in table A.15.

## A.9 Central Aerohydrodynamic Institute (TsAGI)

The Central Aerohydrodynamic Institute (Центральный первоклассный ню и аэрогидродинамический институт ЦАГИ) is located in Zhukovsky, Russian Federation.

### A.9.1 T-116

The T-116 Wind Tunnel is a supersonic and hypersonic blowdown test facility (*TsAGI T-116 Wind Tunnel* 2022). Air is heated for hypersonic tests by resistance heaters and expanded through a converging-diverging nozzle to a test section with cross section 1 m × 1 m and length 2.35 m. A separate independent air supply system is provided for supersonic conditions (Mach 1.8 to 4.0) and two separate air supply systems for hypersonic flow Mach numbers 5 to 7 and 7 to 10. A three level suction ejector discharges the air to atmosphere. Models up to 1 m in length with wingspan up to 0.5 m can be accommodated. Operating conditions are shown in table A.16.

**Table A.16.** Operating conditions.

| Quantity | Min | Max |
|---|---|---|
| $M_\infty$ | 1.8 | 10 |
| $p_{t_\infty}$ (MPa) | 0.11 | 8 |
| $T_{t_\infty}$ (K) | 290 | 1075 |
| $Re_\infty \times 10^{-6}$ (m$^{-1}$) | 2 | 42 |
| Runtime (s) | $\leqslant$300 | |

**Table A.17.** Operating conditions.

| Quantity | Min | Max |
|---|---|---|
| $M_\infty$ | 7.5 | 18.6 |
| $p_{t_\infty}$ (MPa) | 0.8 | 20.0 |
| $T_{t_\infty}$ (K) | 600 | 3400 |
| $Re_\infty \times 10^{-6}$ (m$^{-1}$) | 0.15 | 8.5 |
| Runtime (s) | 30 | 180 |

### A.9.2 T-117

The T-117 is a hypersonic blowdown test facility (*TsAGI T-117 Wind Tunnel* 2022). Air is heated by resistance heaters and expanded through a contoured converging-diverging nozzle to an Eiffel-type test section. A variety of nozzles with exit diameter of 1 m are used to achieve a Mach number range from 7.5 to 18.6. The test section has a cross section 2.4 m × 1.9 m and length 2.5 m. Models up to 1 m in length with wingspan up to 0.4 m can be accommodated. Operating conditions are listed in table A.17.

### A.9.3 T-131B

The T-131B Hypersonic Wind Tunnel is part of the T-131 Complex which also includes The T-131V Aerodynamic Test Bench (*TsAGI T-132 Wind Tunnel Complex* 2022). Air is heated by a gas flame heaterand expanded through a converging-diverging nozzle to an axisymmetric test section with diameter 1.2 m and length 2.3 m. The nozzle exit diameter is 0.4 m. Models up to 2 m in length can be accommodated. The operating conditions are listed in table A.19.

### A.9.4 UT-1M

The UT-1M wind tunnel at TsAGI (Central Aerohydrodynamics Institute) in the Russian Federation is a shock tunnel operating as a Ludwieg tube. The operating conditions are listed in table A.18 (Vaganov *et al* 2014, Borovoy *et al* 2013).

**Table A.18.** Operating conditions.

| Quantity | Value |
|---|---|
| Gas | Air |
| $M_\infty$ | 5, 6, 8 and 10 |
| $T_{t_\infty}$ (deg K) max | 800 |
| $p_{t_\infty}$ (MPa) max | 20 |
| $Re_\infty \times 10^{-6}$ (m$^{-1}$) | $\leqslant$70 |
| Duration (ms) | 40 |

**Table A.19.** Operating conditions.

| Quantity | Value |
|---|---|
| $M_\infty$ | 5 to 10 |
| $p_{t_\infty}$ (MPa) | $\leqslant$11 |
| $T_{t_\infty}$ (K) | $\leqslant$2350 |
| $Re_\infty \times 10^{-6}$ (m$^{-1}$) | $\leqslant$10 |
| Runtime (s) | 180 |

## A.10 Cranfield Hypersonic Gun Tunnel

Kumar (1995) and Buttsworth (2016) describe the operation of the Cranfield Hypersonic Gun Tunnel. A schematic of the facility is shown in figure A.21. The facility is a free piston shock tunnel also known as a gun tunnel. High pressure gas from a reservoir causes a double diaphragm to burst and accelerate a free piston (0.1 kg) along the compression tube. The piston motion generates a shock wave that reflects from the end of the compression tube. The heated working gas expands through a contoured nozzle to the design Mach number in the test section. Currently there are two nozzles for test section Mach numbers of 8.2 and 12.2. A third nozzle for a nominal test section Mach number of 10 is under development (Prince 2022). Operating conditions are shown in table A.20 (Buttsworth 2016).

## A.11 Deutsches Zentrum für Luft- und Raumfahrt (German Aerospace Center)

### A.11.1 Hypersonic Wind Tunnel H2K

Willems *et al* (2015) and Anonymous (1990) describe the design and operational characteristics of the DLR Hypersonic Wind Tunnel H2K. The facility was constructed in 1968 (Anonymous 1990). The tunnel is a blow down open jet type and is equipped with five nozzles providing test section Mach numbers from 4.8 to 11.2 (Anonymous 1990). The test gas is preheated using electric heaters with a maximum power of 2.5 MW (Anonymous 1990). The test gas is air and exhausts to a

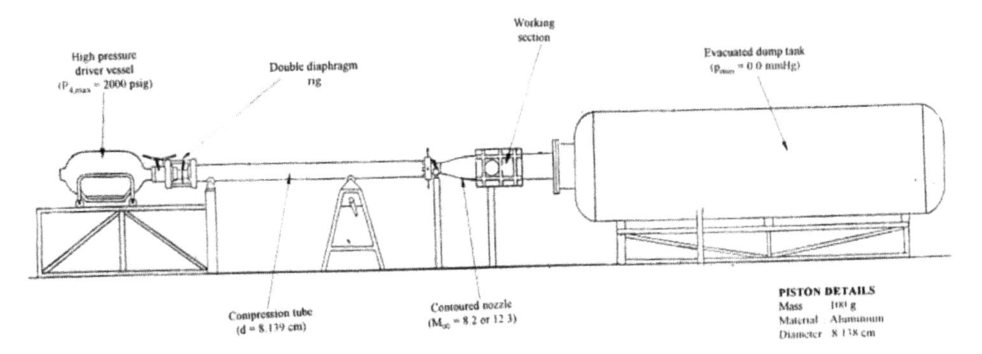


**Figure A.21.** Cranfield Hypersonic Gun Tunnel. Copyright Professor Simon Prince.

**Table A.20.** Operating conditions.

| Quantity | $M_\infty = 8.2$ | $M_\infty = 12.2$ |
|---|---|---|
| $p_{t_\infty}$ (MPa) | 11[a] | 11[a] |
| $T_{t_\infty}$ (K) | 1290[a] | 1290[a] |
| $Re_\infty \times 10^{-6}$ (m$^{-1}$) | 9.44 | 3.45 |
| Test duration (ms) | 80 | 50 |

[a] Maximum (Prince 2022).

**Table A.21.** Operating conditions.

| | Value | |
|---|---|---|
| Quantity | Min | Max |
| $M_\infty$ | 4.8 | 11.2 |
| $T_{t_\infty}$ (K) | 300 | 1300 |
| $p_{t_\infty}$ (MPa) | 0.50 | 5.0 |
| $Re_\infty \times 10^{-6}$ (m$^{-1}$) | 0.24 | 55.0 |
| Test time (s) | 30 | |

vacuum sphere. The operating conditions are listed in table A.21. The nozzle exit diameter is 60 cm with a useful test flow region diameter of 40 cm (Wendt 1987).

### A.11.2 High enthalpy shock tunnel HEG

Eitelberg (1994), Beck *et al* (1995), Hannemann and Beck (2002), Hannemann *et al* (2016) and Hannemann *et al* (2018) discuss the design and performance of the free piston DLR High Enthalpy Shock Tunnel HEG. The overall structure of the reflected shock tunnel is shown in figure A.22. The facility was constructed to simulate high enthalpy and non-equilibrium effects in hypersonic flows. At the time of its commissioning in 1991 it was the largest facility of its type in the world. The

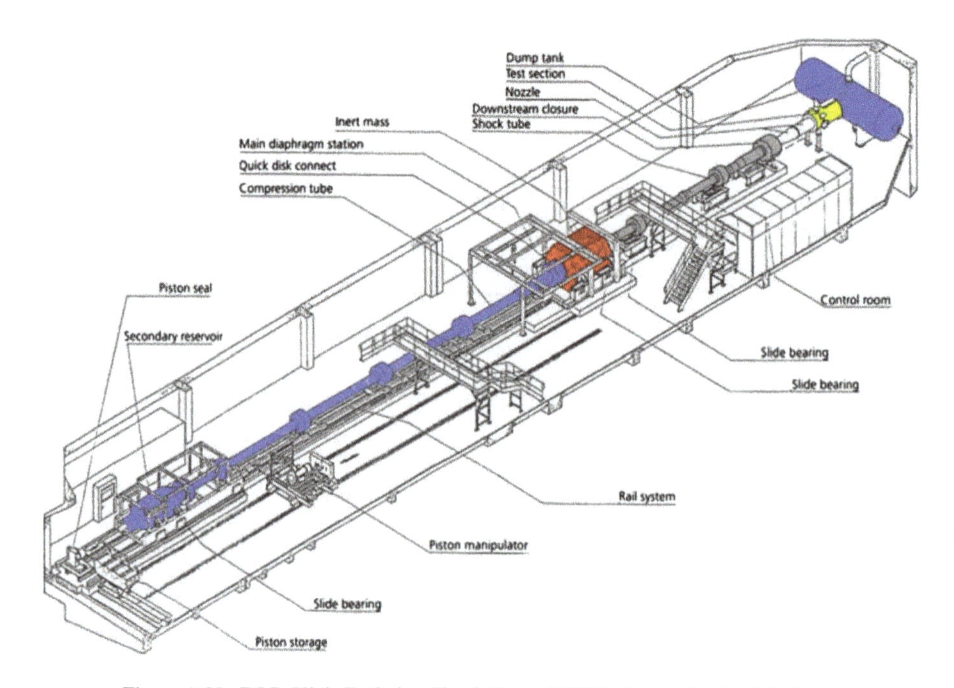


**Figure A.22.** DLR High Enthalpy Shock Tunnel HEG. Copyright Jan Martinez.

**Table A.22.** Operating conditions.

| Quantity | Min | Max |
|---|---|---|
| $M_\infty$ | 5.7 | 10.3 |
| $H_\infty$ (MJ kg$^{-1}$) | 1.5 | 23 |
| $p_{t_\infty}$ (MPa) | 6.8 | 188 |
| $T_{t_\infty}$ (K) | 1640 | 9900 |
| $U_\infty$ (km s$^{-1}$) | 1744 | 6200 |
| $p_\infty$ (Pa) | 660 | 97 800 |
| $T_\infty$ (K) | 208 | 1450 |
| $\rho_\infty$ (gm m$^{-3}$) | 1.7 | 1160 |

overall length is 62 m and the mass is 280 tons. The main sections are the compression tube (33 m), shock tube (17 m) and convergent-divergent nozzle (3.75 m). Four pistons ranging from 275 kg to 848 kg are used. The facility has a range of operating conditions depending upon the nozzle (table A.22). The minimum or maximum are overall values and do not correspond to an individual test configuration. The test time is approximately 1 ms.

### A.11.3 DLR Ludwieg tube facility

Schülein and Zheltovodov (1998), Schülein (2003) and Schülein (2022) describe the design and operation of the DLR Ludwieg Tube Tunnel. The facility is located in

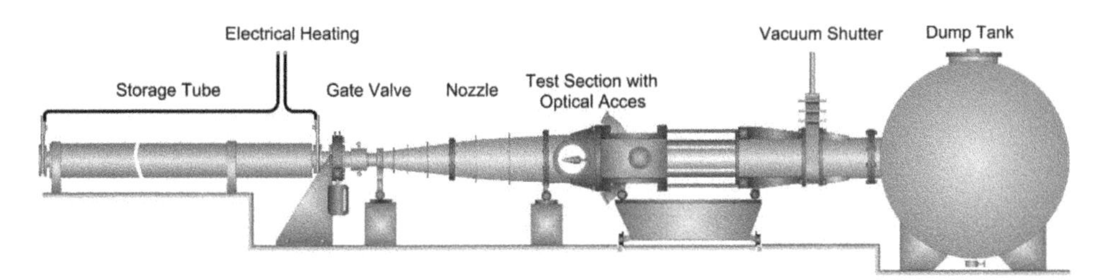


**Figure A.23.** DLR Ludwieg Tube. Copyright Erich Schülein.

**Table A.23.** Operating conditions.

| Quantity | Tube A | Tube A | Tube A |
|---|---|---|---|
| $M_\infty$ | 2 | 3 | 4 |
| $p_{t_\infty}$ (MPa) | 0.05 to 0.7 | 0.06 to 0.6 | 0.08 to 1.0 |
| $T_{t_\infty}$ (K) | 236 to 262 | 241 to 267 | 258 to 287 |
| $Re_\infty \times 10^{-6}$ (m$^{-1}$) | 10 to 110 | 6 to 70 | 4 to 60 |
| Test section (m$^2$) | 0.34 × 0.35 | 0.5 × 0.5 | 0.5 × 0.5 |
| Quantity | Tube B | Tube B | Tube B |
| $M_\infty$ | 5 | 6 | 6.85 |
| $p_{t_\infty}$ (MPa) | 0.4 to 2.9 | 0.4 to 3.4 | 0.4 to 3.6 |
| $T_{t_\infty}$ (K) | 340 to 610 | 410 to 640 | 440 to 655 |
| $Re_\infty \times 10^{-6}$ (m$^{-1}$) | 5 to 55 | 3 to 28 | 2 to 17 |
| Test section (m) | ⊘0.5 | ⊘0.5 | ⊘0.5 |
| Test time (ms) | 300 to 350 | | |

Göttingen, Germany. A schematic of the facility is shown in figure A.23. Two different storage (driver) tube configurations A (unheated) and B (heated) provide a test section Mach number range from 2 to 7. The driver tube length is 80 m providing a test time from 300 ms to 350 ms. Reentry vehicle models up to 300 mm to 350 mm in length can be tested in the facility depending on angle of attack. Maximum length for slender bodies or flat plate models is approximately 700 mm. The operational characteristics are listed in table A.23.

## A.12 HYPULSE

Erdos *et al* (1994, 1997) and Chue *et al* (2002) describe the unique dual mode hypersonic tunnel at General Applied Sciences Laboratory (GASL) located in Ronkonkoma, NY. The facility is capable of operation as either a reflected shock tunnel or an expansion tunnel. The tunnel can also operate with a conventional light gas driver or a detonation driver. The facility is the former NASA Langley Research Center expansion tube originally designed and constructed in the 1960s for studies of radiative gas dynamics and real gas effects in hypersonic flows (Miller and Jones 1983). It was decommissioned at NASA Langley Research Center in 1983 and relocated to GASL in 1989. The operating mode of the tunnel depends on the desired

**Table A.24.** Operating conditions.

| Quantity | Min | Max |
|---|---|---|
| $M_\infty$ | 7 | 21 |
| $H_\infty$ (MJ kg$^{-1}$) | 7 | 21 |

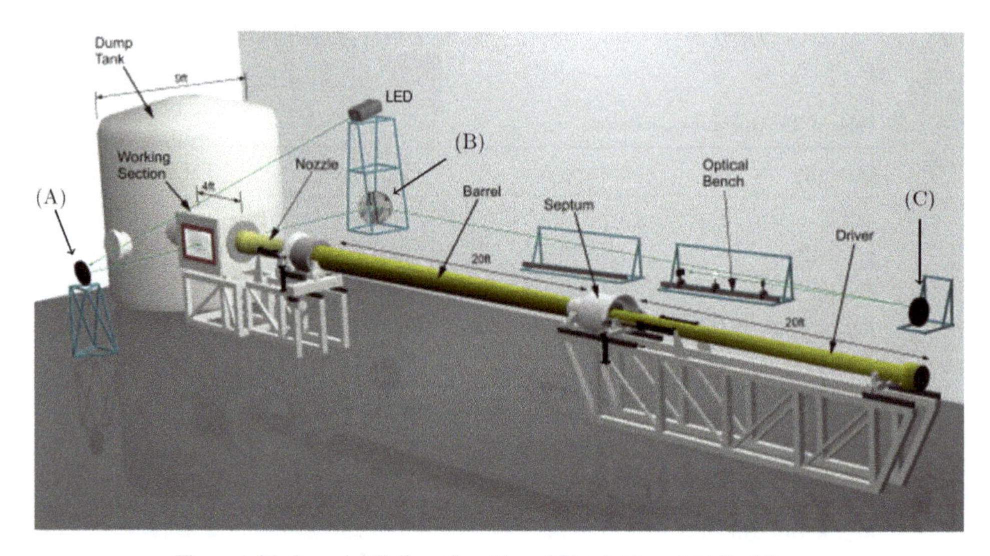


**Figure A.24.** Imperial College Gun Tunnel No. 2. Copyright Paul Bruce.

test Mach number. For Mach numbers less than seven, the tunnel is operated in the reflected shock mode using helium as the driver gas. For Mach numbers between seven and twelve, the tunnel is operated in the detonation driven, reflected shock mode. For Mach numbers from thirteen to twenty five, the expansion tube (or tunnel) with the detonation driver is used. The operating conditions are listed in table A.24.

## A.13 Imperial College Gun Tunnel No. 2

Stollery *et al* (1960), Buttsworth (2016) and Taylor (2019) summarize the design and operational characteristics of the Imperial College Gun Tunnel No. 2. A schematic of the tunnel is shown in figure A.24. The Imperial College Gun Tunnel No. 1 was constructed in the 1950s and subsequently moved to Cranfield University. The test section diameter is 0.4 m and the length is 1.0 m. The working gas is nitrogen. Indicative operating conditions are shown in table A.25.

## A.14 Instituto Superior Técnico (IST) ESTHER

Lino da Silva *et al* (2016), Felgueiras Luis (2018) and Lino da Silva *et al* (2020) describe the design and operational characteristics of the European Shock Tube for High Enthalpy Research (ESTHER) located at the Instituto Superior Técnico, Lisbon, Portugal. A schematic of the facility is shown in figure A.25. The shock tube

**Table A.25.** Operating conditions.

| Quantity | Value |
|---|---|
| Gas | $N_2$ |
| $M_\infty$ | 9 |
| $p_{t_\infty}$ (MPa) | 60[a] |
| $T_{t_\infty}$ (K) | 1150[a] |
| $Re_\infty \times 10^{-6}$ ($m^{-1}$) | 7 to 47 |
| Run time (ms) | 20 |
| Test time (ms) | 6 |

[a] Maximum.

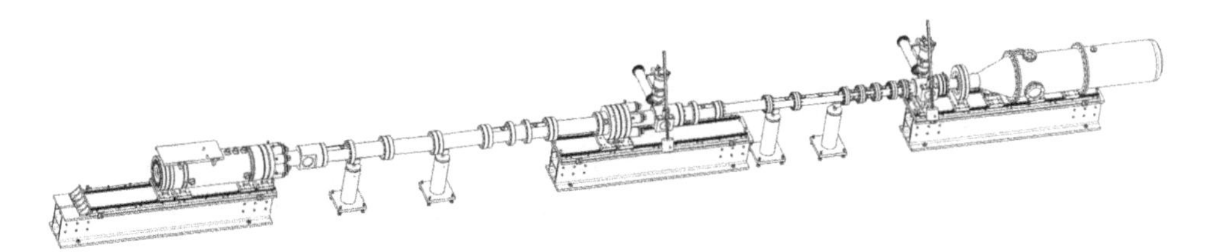

**Figure A.25.** ESTHER. Copyright Mario Lina da Silva.

uses a combustion driver ($He/H_2/O_2$) with an initial pressure up to 10 MPa and final pressure of 60 MPa. The combustion is nominally a deflagration (subsonic combustion), although detonation (supersonic combustion) may occasionally occur (Lino da Silva *et al* 2016). The primary diaphragm initially separates the combustion tube from the intermediate compression initially containing helium at pressures from 1 kPa to 100 kPa. The secondary diaphragm initially separates the compression tube from the shock tube initially filled with the test gas at typical pressures between 10 Pa and 100 Pa. The combustion chamber is 1.6 m in length with an internal diameter of 200 mm. The compression tube is 5 m in length with an internal diameter of 130 mm. The shock tube is 5.4 m in length with an internal diameter of 80 mm. The test section exits to a dump tank with volume $10^3$ L. Further information is provided in Oliveira (2021).

## A.15 Institut de Saint-Louis shock tunnels

Gnemmi *et al* (2016) describe the two high energy shock tunnels STA and STB at the Institut de Saint-Louis. A schematic of the shock tunnel is shown in figure A.26, and a photograph of shock tunnel STA in figure A.27. The shock tube inner diameter is 800 m, and the overall length of each tunnel is 30 m. The driver section is 3.6 m (STA) and 4.0 m (STB). The driven section is 18.4 m for both tunnels. The dump tanks are 10 $m^3$ (STA) and 20 $m^3$ (STB). The operating conditions are listed in table A.26.

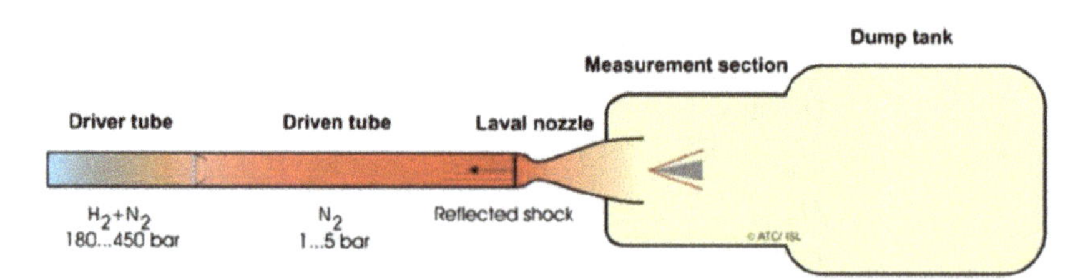


**Figure A.26.** Diagram of ISL shock tunnels. Reproduced from Gnemmi *et al* (2016) with permission of Springer.

**Figure A.27.** ISL shock tunnel STA. Reproduced from Gnemmi (2016) with permission of Springer.

**Table A.26.** Operating conditions.

| Quantity | Value |
|---|---|
| $M_\infty$ | 3 to 14 |
| $p_{t_\infty}$ (MPa) | 400[a] |
| $H_\infty$ (MJ kg$^{-1}$) | 8[b] |
| $T_{t_\infty}$ (K) | 7000[a] |
| $Re_\infty \times 10^{-6}$ (m$^{-1}$) | 1[c] to 100[c] |
| Test time (ms) | <1 |

[a] Maximum (Wendt 1987).
[b] Maximum (Gnemmi *et al* 2016).
[c] Wendt (1987).

## A.16 Japan HIEST

Itoh *et al* (2002) and Hannemann *et al* (2016) describe the chacteristics and operating conditions of the free piston High Enthalpy Shock Tunnel (HIEST) at the Japan Aerospace Exploration Agency (JAXA) Kakuda Space Center. The facility was designed to conduct aerothermodynamic tests at velocities from

**Figure A.28.** HIEST. Reproduced from Hannemann *et al* (2016) with permission of Springer.

**Table A.27.** Operating conditions.

| Quantity | Value |
|---|---|
| $U_\infty$ (km s$^{-1}$) | 3 to 7 |
| $H_\infty$ (MJ kg$^{-1}$) | 25[a] |
| $p_{t_\infty}$ (MPa) | 150[a] |
| Test time (ms) | 2[b] |

[a] Maximum (Hannemann *et al* 2016).
[b] Runtime is 2 ms at highest stagnation enthalpy.

3 km s$^{-1}$ to 7 km s$^{-1}$ corresponding to the flight trajectory of the HOPE (H-II Orbiting Plane) and scramjet tests in the Mach number range from Mach 8 to 16 and dynamic pressures from 50 kPa to 100 kPa. The facility is shown in figure A.28. The compression tube length is 42 m with a diameter of 0.6 m, and the shock tube length is 17 m with a diameter of 180 mm (Hannemann *et al* 2016). A range of piston masses from 220 kg to 780 kg are used. The tunnel can operate with either a conical or contoured nozzle with exit diameters of 1.2 m and 0.8 m, respectively. The operating conditions are described in table A.27. See also Itoh *et al* (1997, 1998, 2002) and Itoh (2002).

## A.17 Jet Propulsion Laboratory 21 inch Hypersonic Wind Tunnel

Laumann and Herrera (1967) details the design and operational capabilities of the Jet Propulsion Laboratory 21 inch Hypersonic Wind Tunnel. The configuration is shown in figure A.29. The wind tunnel operates in a continuous mode. A sequence of

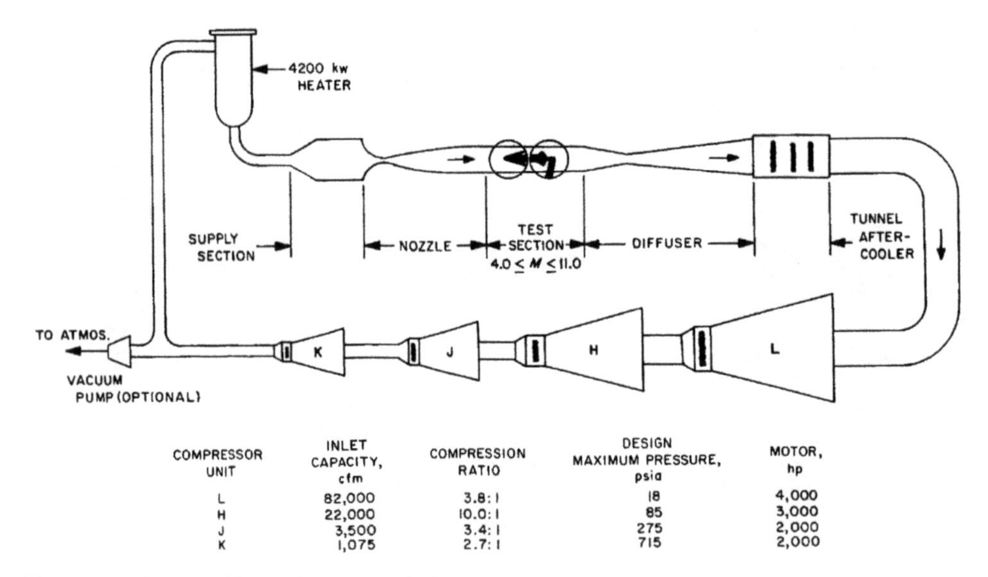


| COMPRESSOR UNIT | INLET CAPACITY, cfm | COMPRESSION RATIO | DESIGN MAXIMUM PRESSURE, psia | MOTOR, hp |
|---|---|---|---|---|
| L | 82,000 | 3.8:1 | 18 | 4,000 |
| H | 22,000 | 10.0:1 | 85 | 3,000 |
| J | 3,500 | 3.4:1 | 275 | 2,000 |
| K | 1,075 | 2.7:1 | 715 | 2,000 |

**Figure A.29.** Jet Propulsion Laboratory 21 inch Hypersonic Wind Tunnel. Reproduced from Laumann and Herrera (1967). Copyright 1967 Jet Propulsion Laboratory. Courtesy of NASA.

**Table A.28.** Operating conditions.

| Quantity | Value |
|---|---|
| $M_\infty$ | 4–11 |
| $T_{t_\infty}$ (K) | 1000[a] |
| $p_{t_\infty}$ (MPa) | 4[b] |

[a] Heater maximum.
[b] Supply maximum.

air compressors provides a supply pressure up to 4 MPa. A 4.2 MW heater supplies air at stagnation temperatures up to 1000 K. A flexible two dimensional nozzle provides test section Mach numbers from 4 to 11. The operating conditions are listed in table A.28.

## A.18 Khristianovich Institute of Theoretical and Applied Mechanics

The Khristianovich Institute of Theoretical and Applied Mechanics (Институт теоретической и прикладной механики имени С.А.Христиановича СО РАН) is located in Academgorodok, Novosibirsk, Russian Federation.

### A.18.1 AT-303

Kharitonov *et al* (2006) describe the design and operation of the Khristianovich Institute of Theoretical and Applied Mechanics AT-303 Hypersonic Wind Tunnel. The unique test gas source with adiabatic compression was developed by the

Lavrentyev Institute of Hydrodynamics and Technology Institute of High Rate Hydrodynamics of the Siberian Branch of the Russian Academy of Sciences. The blow down facility incorporates a converging-diverging 8° half-angle conical nozzles with exit diameters of 300 mm for Mach 8 to 15, and 600 mm for Mach 15 to 20. The free jet test section is $1.75\ \mathrm{m} \times 2\ \mathrm{m} \times 2.5\ \mathrm{m}$, and the vacuum tank capacity is $110\ \mathrm{m}^3$. The operating conditions are listed in table A.29. See also Adamov *et al* (2006).

### A.18.2 T-313

The Khristianovich Institute of Theoretical and Applied Mechanics T-313 Hypersonic Wind Tunnel is a supersonic to hypersonic wind tunnel (*T-313 Hypersonic Wind Tunnel* 2022). The facility is equipped with an ohmic heater with power up to 6 MW. Profiled two dimensional nozzles for Mach numbers from 1.75 to 7.0 are used. The test section is $60\ \mathrm{cm} \times 60\ \mathrm{cm} \times 200\ \mathrm{cm}$. Operating conditions are listed in table A.30.

### A.18.3 T-326

Maslov (2003) describe the design and operation of the Khristianovich Institute of Theoretical and Applied Mechanics T-326 hypersonic wind tunnel. The facility is a blow down hypersonic wind tunnel. Air is heated using an ohmic heater. The nozzle exit diameter is 200 mm, and the open jet test section length is 400 mm. The operating conditions are indicated in table A.31.

**Table A.29.** Operating conditions.

| Quantity | Min | Max |
|---|---|---|
| $M_\infty$ | 8 | 20 |
| $p_{t_\infty}$ (MPa) | 100 | 300 |
| $T_{t_\infty}$ (K) | 1300 | 2500 |
| Runtime (ms) | 20 | 200 |

**Table A.30.** Operating conditions.

| Quantity | Value |
|---|---|
| $M_\infty$ | 1.75 to 7.0 |
| $p_{t_\infty}$ (MPa) | 1.6[a] |
| $T_{t_\infty}$ (K) | 873[a] |
| $Re_\infty \times 10^{-6}$ (m$^{-1}$) | 5 to 70 |

[a] Maximum.

**Table A.31.** Operating conditions.

| Quantity | Value |
|---|---|
| $M_\infty$ | 6 to 14 |
| $p_{t_\infty}$ (MPa) | 20[a] |
| $T_{t_\infty}$ (K) | 800[a] |
| $Re_\infty \times 10^{-6}$ ($m^{-1}$) | 5 to 70 |

[a] Maximum.

**Table A.32.** Operating conditions.

| Quantity | Value |
|---|---|
| $M_\infty$ | 4 to 20 |
| $p_{t_\infty}$ (MPa) | 101[a] |
| $T_{t_\infty}$ (K) | 4000[a] |
| Run time (s) | 0.05 to 0.5 |

[a] Maximum.

### A.18.4 IT-302M

Maslov *et al* (2013) and Goldfeld *et al* (2016) describe the design and operation of the (*IT-302M hot shot hypersonic wind tunnel IT-302M Hypersonic Hot-Shot Wind Tunnel* 2022). The energy is supplied by a 1.75 MJ capacitor bank. The facility has a variety of nozzles including axisymmetric (400 mm diameter) for Mach 6, 7.5 and 8; axisymmetric (300 mm diameter) for Mach 4, 5, 6, 7 and 8; and conical (330 mm diameter) for Mach 9 to 20. The Eiffel chamber test section dimensions are 1 m × 1 m × 1.5 m. The operating conditions are listed in table A.32.

### A.18.5 Transit-M

Gromyko *et al* (2013) discusses the design and operation of the Khristianovich Institute of Theoretical and Applied Mechanics Transit-M Hypersonic Wind Tunnel. Pressurized air is heated using resistance heaters and expanded through a converging-diverging nozzle to the open jet test section. The air is exhausted through a diffuser to a vacuum chamber with volume 6.5 $m^3$. Typical operating conditions are shown in table A.33.

## A.19 Laboratoire de Recherche Balistiqueet Aerodynamiques C2

Grönig (1992) summarizes the design and operational characteristics of the light piston (100 g) reflected shock tunnel C2 at the Laboratoire de Recherche Balistique et Aerodynamiques (LRBA) in Vernon, France. The facility operates with two different contoured nozzles with providing test section Mach numbers of 16 and 20.1. The tunnel was constructed in 1961 and operated at the Laboratoire de

**Table A.33.** Operating conditions[a].

| Quantity | Min | Max |
|---|---|---|
| $M_\infty$ | 4 | 8 |
| $p_{t_\infty}$ (MPa) | 0.2 | 2.0 |
| $T_{t_\infty}$ (K) | 370 | 410 |
| $Re_\infty \times 10^{-6}$ ($m^{-1}$) | 4 | 24 |
| Runtime (ms) | 110 | 200 |

[a] Gromyko *et al* (2013).

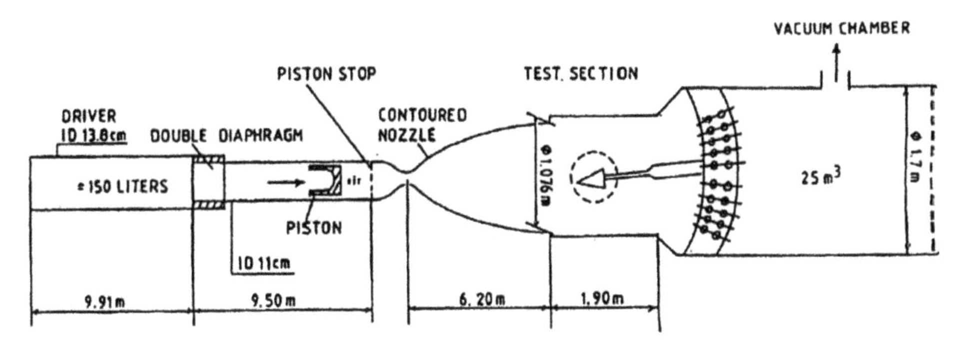


**Figure A.30.** LRBA C2. Reproduced from Grönig (1992) with permission of Springer.

**Table A.34.** Operating conditions.

| Quantity | $M = 16$ | $M = 20.1$ |
|---|---|---|
| $p_{t_\infty}$ (MPa) | 3.5 | 4 to 60 |
| $T_{t_\infty}$ (K) | 2400[a] | 2400[a] |
| Test time[b] (ms) | 10 | 20 |

[a] Maximum (Grönig 1992).
[b] Anonymous (1990).

Recherche Balistique et Aerodynamiques beginning in 1974 (Anonymous 1990). A schematic of the facility is shown in figure A.30 and the operating conditions in table A.34.

## A.20 NASA Ames Research Center

### A.20.1 NASA Ames Electric Arc Shock Tube (EAST)

The NASA Ames Research Center Electric Arc Shock Tube (EAST) facility comprises a shock tube and a shock tunnel. The shock tunnel diameter is 76.2 cm, and the shock tube diameter is 10.16 cm (Chandel *et al* 2018). A 40 kV high voltage DC power supply and a 1.25 MJ capacitor bank operates both tunnels. The EAST facility was commissioned in 1968. The tunnel operates at test section velocities from 1.3 km $s^{-1}$ to 46 km $s^{-1}$ with static pressures between 13.3 Pa and 530 kPa with a variety of test gases simulating planetary atmospheres.

### A.20.2 NASA Ames 3.5 foot Hypersonic Wind Tunnel

The NASA Ames 3.5 foot Hypersonic Wind Tunnel is a closed circuit blowdown facility comprising a heater, converging-diverging nozzle, test section, diffuser and vacuum spheres. A schematic of the facility is shown in figure A.31. High pressure air is heated by passing through an alumina pebble bed chamber. A series of interchangable axisymmetric contoured nozzles with exit diameter of 1.06 m provide test section Mach numbers from 5.2 to 10.4 (Cummings *et al* 1965). A small mass flow rate of helium is injected tangentially at the throat to reduce the frictional losses in the nozzle (Raman 1974). The open jet test section is 3.66 m in diameter and 12.2 m in length, and the vacuum sphere system has a total capacity of 25 485 $m^3$ (Cummings *et al* 1965). After each test the air in the vacuum spheres is purified and recovered for the next test, and the helium is extracted by a cyrogenic separation process for reuse. Typical operating conditions are listed in table A.35.

### A.20.3 NASA Ames Mach 50 Helium Tunnel

Kemp (1970) describes the design and operation of the NASA Ames M-50 Helium Tunnel. A schematic of the facility is shown in figure A.32. The facility is a blowdown tunnel using helium. The gas is heated using an electrical resistance heater and then expands through a converging diverging nozzle with exit diameter of 77.2 cm into an open jet test section of length 3.048 m. The gas exhausts through a diffuser and steam ejector, and the helium is subsequently purified and stored for reuse. The operating conditions are listed in table A.36.

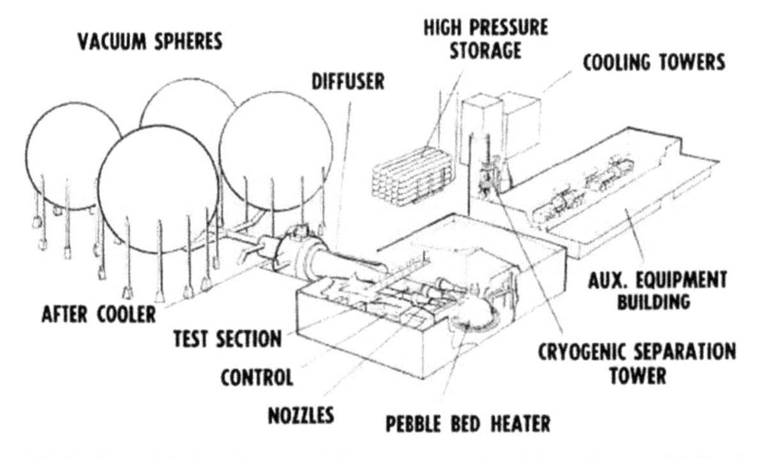


**Figure A.31.** NASA Ames 3.5 foot Hypersonic Tunnel. Reproduced from Raman (1974). Copyright US Government. Image stated to be in public domain.

**Table A.35.** Operating conditions[a].

| Quantity | Min | Max |
|---|---|---|
| $M_\infty$ | 5.2 | 10.4 |
| $p_{t_\infty}$ (MPa) | 0.64 | 11 |
| $T_{t_\infty}$ (K) | 700 | 1200 |
| Runtime (s) | 30 | 60 |

[a]Raman (1974).

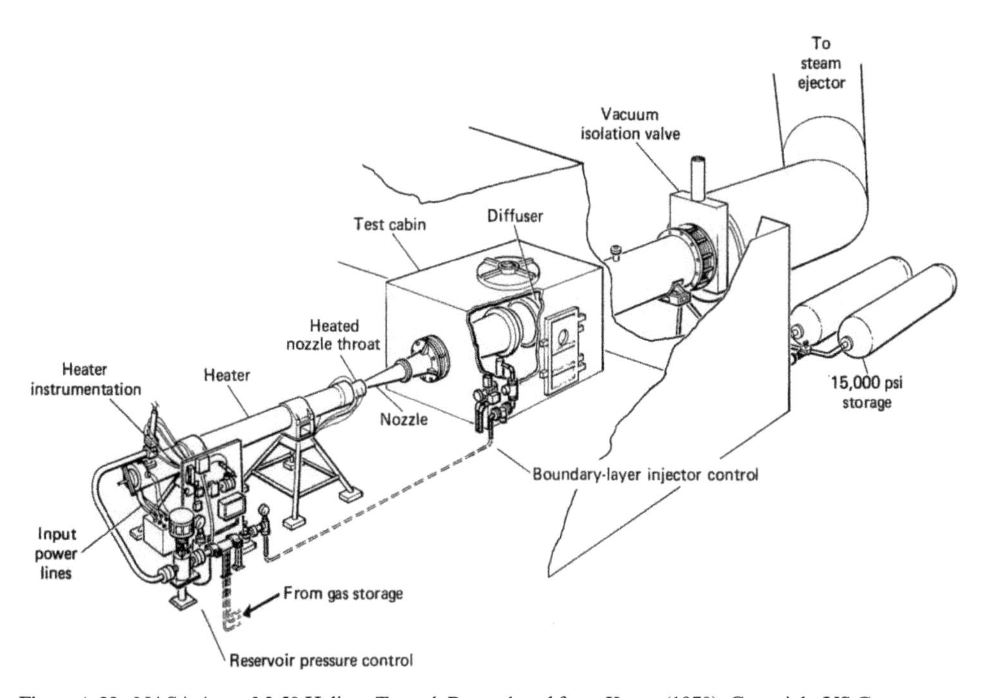


**Figure A.32.** NASA Ames M-50 Helium Tunnel. Reproduced from Kemp (1970). Copyright US Government. Image stated to be in public domain.

**Table A.36.** Operating conditions.

| Quantity | Min | Max |
|---|---|---|
| Gas | He | He |
| $M_\infty$ | 40 | 43 |
| $p_{t_\infty}$ (MPa) | 10.3 | 13.8 |
| $T_{t_\infty}$ (K) | 555 | 1056 |
| $Re_\infty \times 10^{-6}$ ($m^{-1}$) | 2.0 | 3.2 |
| Runtime (min) | 3 | 20 |

## A.21 NASA Glenn Research Center

The NASA Glenn Research Center Hypersonic Tunnel Facility (HTF) is described by Woike and Willis (2002) and Thomas *et al* (2010). The facility began operation in 1966 as the Hydrogen Heat Transfer Facility (HHTF) intended for development of nuclear reactor engines. From 1969 to 1971 the facility was converted into the Hypersonic Tunnel Facility including conversion of the original pebble bed heater to a graphite core heater. A schematic of the facility, located at the NASA Glenn Research Center Plum Brook Station in Sandusky, Ohio, is shown in figure A.33. The tunnel is a blow down, free jet wind tunnel simulating realistic flight enthalpy flight conditions from Mach 5 to 7. Nitrogen gas is heated by a 3 MW graphite core. The heater comprises fifteen cylindrical graphite blocks stacked in series. Each block is 1.8 M in diameter and 0.6 m in height with 1945 holes ranging in size from 19 mm to 29 mm drilled axially. The total mass is 27 200 kg. The heated nitrogen gas is mixed with oxygen to generate simulated air with additional nitrogen as necessary to achieve the desired total tempeature. The heated air exhausts through axisymmetric nozzles into the test chamber comprising a domed cylinder 6.1 m in height and 7.6 m in diameter. The nozzle exit diameter are 1.07 m. The operating conditions are listed in table A.37.

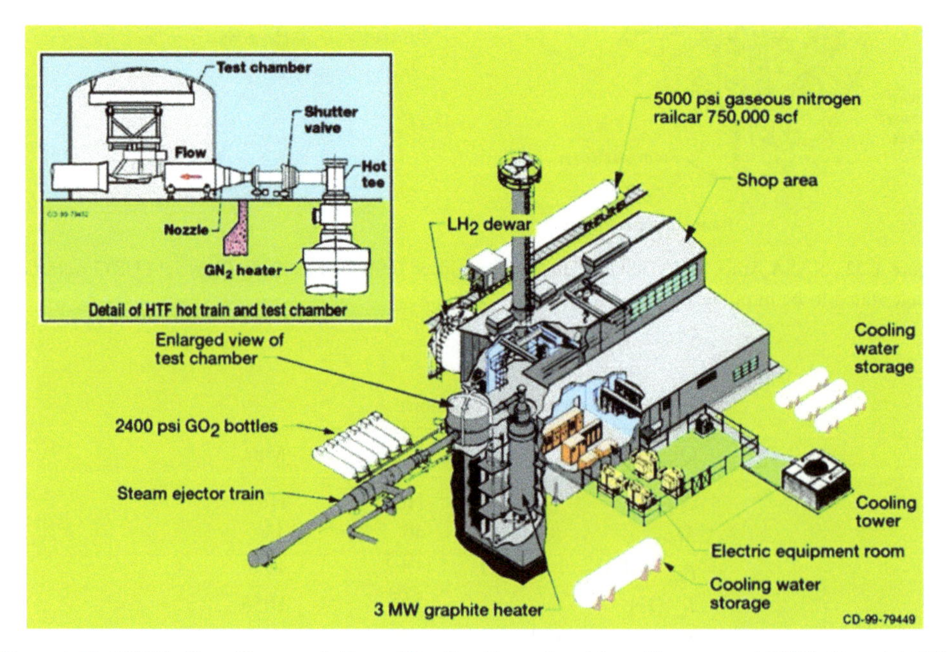


**Figure A.33.** NASA Glenn Hypersonic Tunnel Facility. Reproduced from Thomas *et al* (2010). Copyright US Government. Image stated to be in public domain.

**Table A.37.** Operating conditions.

| | $M = 5$ | | $M = 6$ | | $M = 7$ | |
|---|---|---|---|---|---|---|
| Quantity | Min | Max | Min | Max | Min | Max |
| $p_{t_\infty}$ (MPa) | 0.48 | 2.83 | 1.00 | 8.27 | 2.96 | 8.27 |
| $T_{t_\infty}$ (K) | 1222 | 1344 | 1647 | 1839 | 2128 | 2167 |

Maximum $p_{t_\infty}$ corresponds to minimum $T_{t_\infty}$ condition

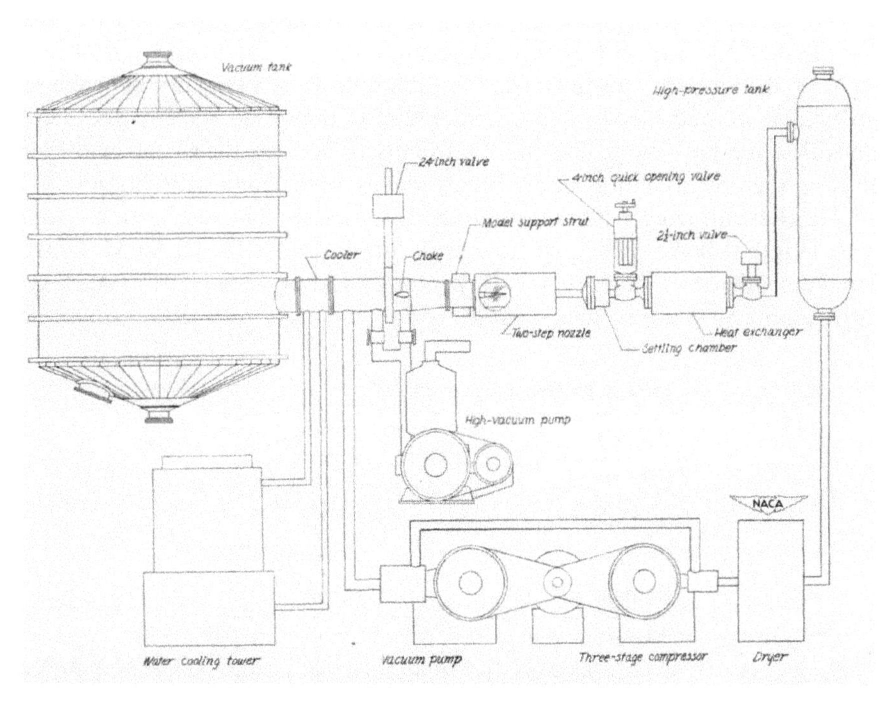


**Figure A.34.** Langley Aeronautical Laboratory Hypersonic Tunnel. Reproduced from McLellan *et al* (1950). Copyright US Government. Image stated to be in public domain.

# A.22 NASA Langley Research Center

## A.22.1 Langley Aeronautical Laboratory Hypersonic Tunnel

McLellan *et al* (1950) describe the Langley Aeronautical Laboratory 11 inch Hypersonic Tunnel. The facility began operation on November 26, 1947. A schematic is shown in figure A.34. The facility is a blow down tunnel. High pressure air at an initial stagnation pressure up to 5.1 MPa is stored in a 11.3 $m^3$ tank. The air exhausts through a heat exchanger which increases the temperature up to 728 K. The air expands through a two dimensional nozzle into the test section whose cross section is 25.4 cm × 25.4 cm. The air exhausts to a 340 m$^3$ vacuum tank and is pumped back to the high pressure storage tank. The facility was tested with a unique double expansion nozzle with test section Mach number 6.98 which proved unsatisfactory. A separate single expansion nozzle for Mach 7 proved successful. The operating conditions are shown in table A.38.

### A.22.2 Mach 6 Quiet Tunnel

Beckwith *et al* (1990) and Wilkinson *et al* (1992) describe the design and operation of the NASA Langley Mach 6 Quiet Tunnel, also denoted the NASA Langley Advanced Mach 6 Axisymmetric Pilot Nozzle located in the NASA Langley Nozzle Test Chamber (M6NTC). A schematic of the facility is shown in figure A.35. The axisymmetric nozzle (exit diameter 17.78 cm) is a slow expansion design with a unique bleed system in the throat region to remove the incoming nozzle wall boundary layer and maintain laminar flow on the nozzle wall into the expansion, thereby avoiding eddy Mach waves associated with boundary layer turbulence that can trigger transition on the test models. The test section is an open jet with length 1.0 m. The quiet nozzle is now at the Texas A&M University National Aerothermochemistry and Hypersonics Laboratory (section A.35.1). The operating conditions are shown in table A.39.

**Table A.38.** Operating conditions.

| Quantity | Value |
|---|---|
| $M_\infty$ | 6 to 10 |
| $p_{t_\infty}$ (MPa) | 5.1[a] |
| $T_{t_\infty}$ (K) | 616 to 728 |
| Running time (s) | ≈30 |

[a] Maximum.

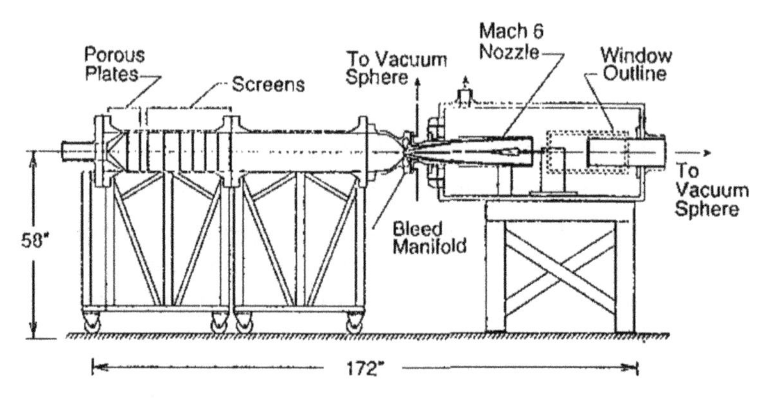


**Figure A.35.** Mach 6 Quiet Tunnel. Reproduced from Wilkinson *et al* (1992). Copyright 1992 by the American Institute of Aeronautics and Astronautics.

**Table A.39.** Operating conditions.

| Quantity | Value |
|---|---|
| $M_\infty$ | 6 |
| $p_{t_\infty}$ (MPa) | 3.28 |
| $T_{t_\infty}$ (K) | 450 |

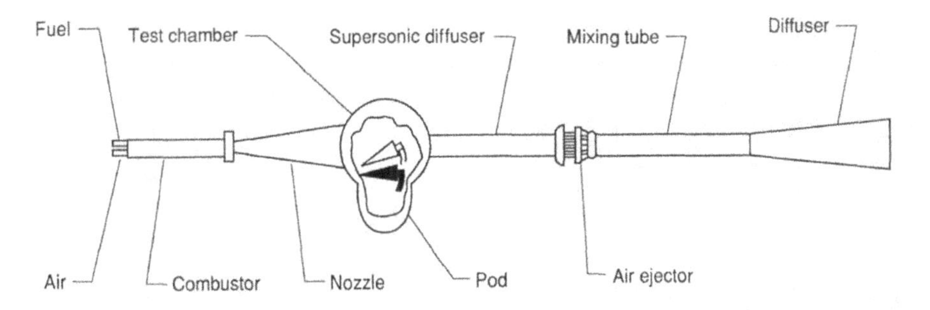


**Figure A.36.** NASA Langley 8 Foot High Temperature Tunnel. Reproduced from Parrott *et al* (1989). Copyright US Government. Image stated to be in public domain.

**Table A.40.** Operating conditions.

| Quantity | Value |
|---|---|
| $M_\infty$ | 6.8 |
| $T_{t_\infty}$ (K) | 1300 to 2000 |
| $q_\infty$ (kPa) | 11.7 to 86.2 |
| $Re_\infty \times 10^{-6}$ ($m^{-1}$) | 0.9 to 9.9 |
| Test time (s) | 120[a] |

[a] Maximum.

### A.22.3 NASA Langley 8 Foot High Temperature Tunnel

Parrott *et al* (1989) describe the NASA Langley 8 foot High Temperature Tunnel (HTT). The facility is a hypersonic blowdown wind tunnel. The configuration is shown in figure A.36. High stagnation temperatures are achieved through combustion of a mixture of methane and air under high pressure in a combustor. The combustion products are expanded through an axisymmetric, conical contoured nozzle with an exit diameter of 2.4 m (8 ft). The test section is a free jet, and the gas stream exits to a supersonic diffuser where it is pumped to atmospheric conditions by a single-stage, annular ejector. The operating conditions are listed in table A.40.

### A.22.4 NASA Langley 11 inch Hypersonic Tunnel

Schaefer (1965) describes the operational characteristics of the NASA Langley 11 inch Hypersonic Tunnel. A schematic of the facility is shown in figure A.37. The purpose of the tunnel was investigation of pressure, heat transfer and force measurement. The tunnel was configured with four interchangable nozzles providing a Mach number range from 6.8 to 18. The operating conditions are listed in table A.41.

### A.22.5 NASA Langley 1 Foot Hypersonic Arc Tunnel

Schaefer (1965) describes the operating conditions of the NASA Langley 1 Foot Hypersonic Arc Tunnel. A schematic of the facility is presented in figure A.38. Air is heated by a 1.5 MW DC arc heater and exhausts through a conical nozzle and steam ejector into the atmosphere. The test section diameter is 30.48 cm (1 foot). The operating conditions are listed in table A.42.

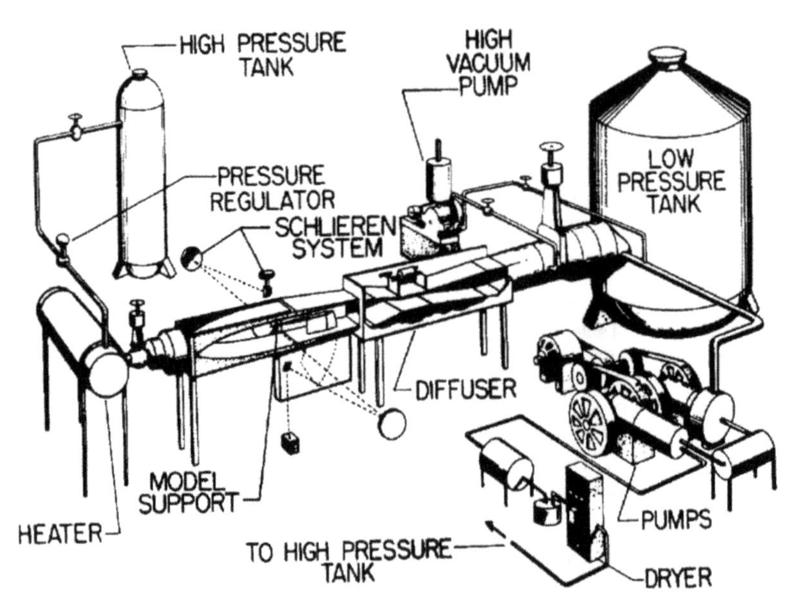


**Figure A.37.** NASA Langley 11 inch Hypersonic Tunnel. Reproduced from Schaefer, (1965). Copyright US Government. Image stated to be in public domain.

**Table A.41.** Operating conditions.

| Quantity<br>Gas | $M_\infty = 6.8$<br>Air | $M_\infty = 9.6$<br>Air | $M_\infty = 10.5$<br>He | $M_\infty = 18.0$<br>He |
|---|---|---|---|---|
| $p_{t_\infty}$ (MPa) | 0.48 to 3.72 | 1.52 to 4.76 | 1.38 to 5.52 | 2.76 to 11.03 |
| $Re_\infty \times 10^{-6}$ (m$^{-1}$) | 1.64 to 13.1 | 0.98 to 3.28 | 8.86 to 32.2 | 3.94 to 32.8 |
| Running time (s) | 70 to 100 | 100 | 14 | 10 |

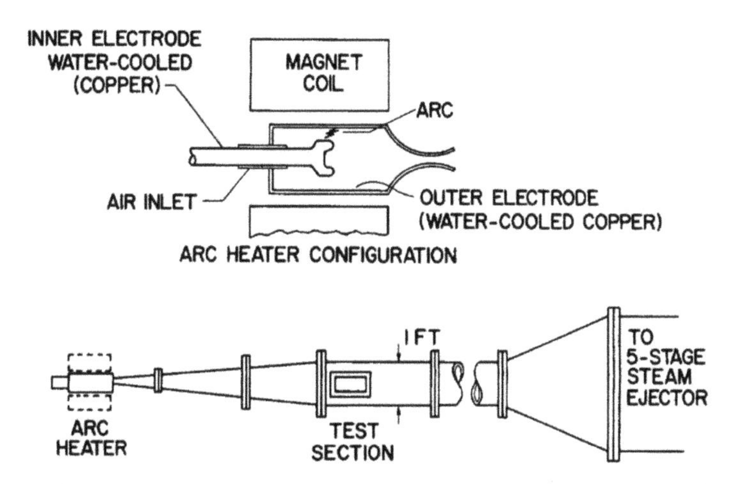


**Figure A.38.** NASA Langley 1 Foot Hypersonic Tunnel. Reproduced from Schaefer, (1965). Copyright US Government. Image stated to be in public domain.

**Table A.42.** Operating conditions.

| Quantity | Min | Max |
|---|---|---|
| $M_\infty$ | 12 | 12 |
| $H_\infty$ (MJ kg$^{-1}$) | 3.5 | 7.0 |
| $p_{t_\infty}$ (MPa) | 1.52 | 2.76 |
| $Re_\infty \times 10^{-6}$ (m$^{-1}$) | 0.033 | 0.066 |
| Running time (s) | 60 | 900 |

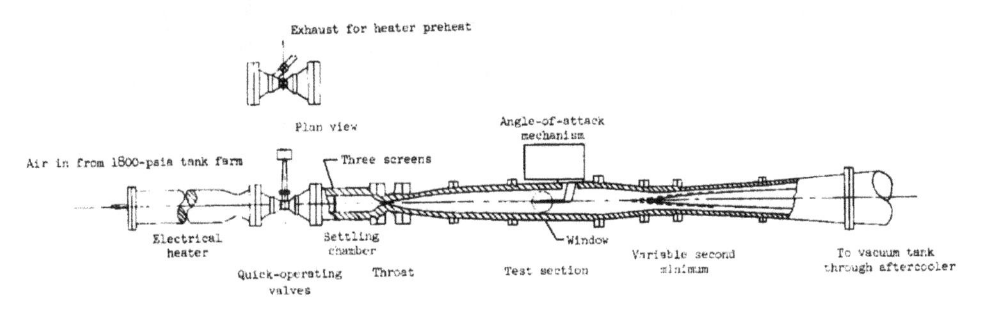


**Figure A.39.** NASA Langley Hypersonic Flow Apparatus. Reproduced from Schaefer, (1965). Copyright US Government. Image stated to be in public domain.

## A.22.6 NASA Langley Hypersonic Flow Apparatus

Schaefer (1965) details the operating characteristics of the NASA Langley Hypersonic Flow Apparatus. A schematic of the facility is shown in figure A.39. The tunnel is used for force, pressure, heat transfer and flutter measurements. Air is

**Table A.43.** Operating conditions.

| Quantity | Min | Max |
|---|---|---|
| $M_\infty$ | 10.03 | |
| $H_\infty$ (MJ kg$^{-1}$) | 1.0 | 1.0 |
| $p_{t_\infty}$ (MPa) | 5.52 | 8.27 |
| $T_{t_\infty}$ (K) | 833 | 978 |
| $Re_\infty \times 10^{-6}$ (m$^{-1}$) | 4.27 | 6.56 |
| Running time (s) | 180 | |

**Figure A.40.** NASA Langley Mach 8 Variable Density Hypersonic Tunnel. Reproduced from Schaefer, (1965). Copyright US Government. Image stated to be in public domain.

heated by a 2000 kVA DC resistance heat exchanger. The test section is 38.1 cm in diameter. The heated air exhausts through an aftercooler to a vacuum tank. The operating conditions are listed in table A.43.

### A.22.7 NASA Langley Mach 8 Variable Density Hypersonic Tunnel

Schaefer (1965) describes the operating characteristics of the NASA Langley Mach 8 Variable Density Hypersonic Tunnel. A photograph of the facility is shown in figure A.40. Air is heated by a combination of Dowtherm and electrical resistance. The axially symmetric contoured nozzle has an exit diameter of 45.7 cm. The air exhausts to a vacuum tank or the atmosphere. The operating conditions are listed in table A.44.

**Table A.44.** Operating conditions.

| Quantity | Min | Max |
|---|---|---|
| $M_\infty$ | 7.5 | 8.0 |
| $p_{t_\infty}$ (MPa) | 0.10 | 20.2 |
| $T_{t_\infty}$ (K) | 644 | 839 |
| $Re_\infty \times 10^{-6}$ ($\mathrm{m}^{-1}$) | 0.33 | 39.4 |
| Running time (s) | 90 | 600 |

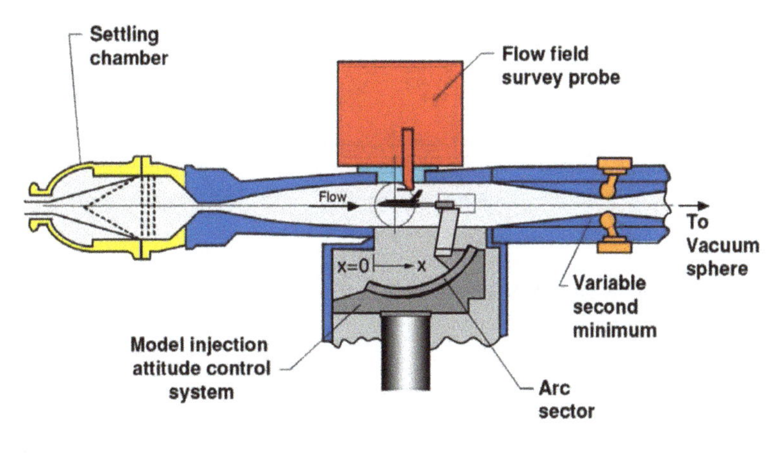


**Figure A.41.** NASA Langley 20 inch Mach 6 Tunnel. Reproduced from Rufer and Berridge (2012). Copyright American Institute of Aeronautics and Astronautics.

### A.22.8 NASA Langley 20 inch Mach 6 Tunnel

Schaefer (1965), Berry *et al* (1999) and Rufer and Berridge (2012) describe the operating characteristics of the NASA Langley 20 inch Mach 6 Tunnel. A schematic of the facility is shown in figure A.41. The air is heated with a Dowtherm heat exchanger. The nozzle is two dimensional and contoured with a test section 50.8 cm × 50.8 cm. The heated air exhausts through a movable second throat into the atmosphere using a annular ejector. The operating conditions are shown in table A.45.

### A.22.9 NASA Langley 20 inch Mach 8.5 Tunnel

Schaefer (1965) describes the operating characteristics of the NASA Langley 20 inch Mach 8.5 Tunnel. A schematic of the facility is shown in figure A.42. Air is heated using an electrical resistance heater. The axially symmetric contoured nozzle has an exit

**Table A.45.** Operating conditions[a].

| Quantity | Min | Max |
|---|---|---|
| $M_\infty$ | 6 | 6 |
| $p_{t_\infty}$ (MPa) | 0.14 | 3.28 |
| $T_{t_\infty}$ (K) | 447 | 533 |
| $Re_\infty \times 10^{-6}$ (m$^{-1}$) | 1.1 | 27.9 |
| Run time (min) | 1 | 20 |

[a] Rufer and Berridge (2012).

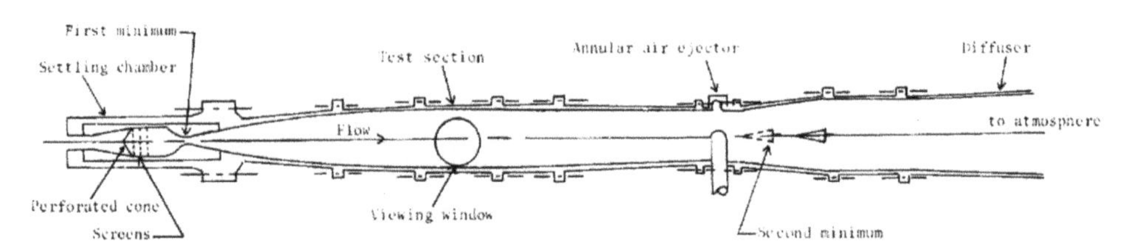


**Figure A.42.** NASA Langley 20 inch Mach 8.5 Tunnel. Reproduced from Schaefer, (1965). Copyright US Government. Image stated to be in public domain.

**Table A.46.** Operating conditions.

| Quantity | Value |
|---|---|
| $M_\infty$ | 8.5 |
| $p_{t_\infty}$ (MPa) | 11.0 to 17.2 |
| $T_{t_\infty}$ (K) | 839 |
| $Re_\infty \times 10^{-6}$ (m$^{-1}$) | 15.7 to 24.6 |
| Running time (min) | 7[a] |

[a] Maximum.

diameter of 53.3 cm. The heated air exhausts through a second throat to the atmosphere using an annular ejector. The operating conditions are listed in table A.46.

### A.22.10 NASA Langley 20 inch Hypersonic Arc Heated Tunnel

Schaefer (1965) describes the operating characteristics of the NASA Langley 20 inch Hypersonic Arc Heated Tunnel. A schematic is presented in figure A.43. Test gases are air and nitrogen and heated by an alectric arc. The tunnel is a free jet with a conical nozzle. The operating conditions are shown in table A.47.

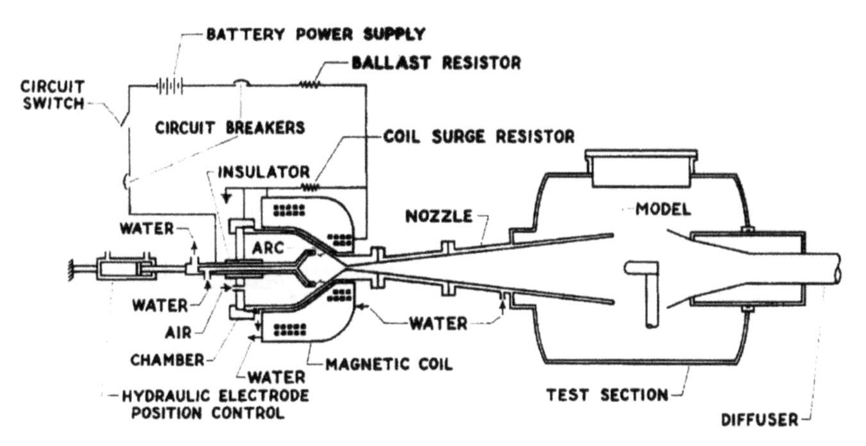


**Figure A.43.** NASA Langley 20 Inch Hypersonic Arc Heated Tunnel. Reproduced from Schaefer, (1965). Copyright US Government. Image stated to be in public domain.

**Table A.47.** Operating conditions.

| Quantity | Arc power (MW) | | |
|---|---|---|---|
| | 1.8 | 1.95 | 2.06 |
| $M_\infty$ | 6 or 10 | 6 or 10 | 6 or 10 |
| $H_\infty$ (MJ kg$^{-1}$) | 10.9 | 6.4 | 3.7 |
| $p_{t_\infty}$ (MPa) | 0.52 | 1.08 | 2.68 |
| Run time (s) | 60[a] and 180[b] | | |

[a] At Mach 10.
[b] At Mach 6.

## A.22.11 NASA Langley 22 inch Helium Tunnel

Schaefer (1965) describes the operating conditions of the NASA Langley 22 inch Helium Tunnel. The purified helium is supplied at ambient temperature or heated using electrical resistance heater. The contoured nozzles exhaust into the test section whose diameter is 57.2 cm. An image of the tunnel is shown in figure A.44. The gas exhausts to a vacuum tank. The operating conditions are shown in table A.48.

## A.22.12 NASA Langley 12 inch Hypersonic Ceramic-Heated Tunnel

Clark (1965) details the design and operation of the NASA Langley 12 inch Hypersonic Ceramic-Heated Tunnel (HCHT). A schematic of the facility is shown in figure A.45. High stagnation temperature is achieved using a pebble bed heater 20.32 cm in diameter and 2.34 m in height containing 9.53 mm diameter zirconia ($ZrO_2$) pebbles and heated using air and propane, or air, propane and oxygen. The heated air exits the pebble bed and expands through an 16° total angle conical nozzle with diameter 30.5 cm to the free jet test section. The air exhausts through the diffuser and

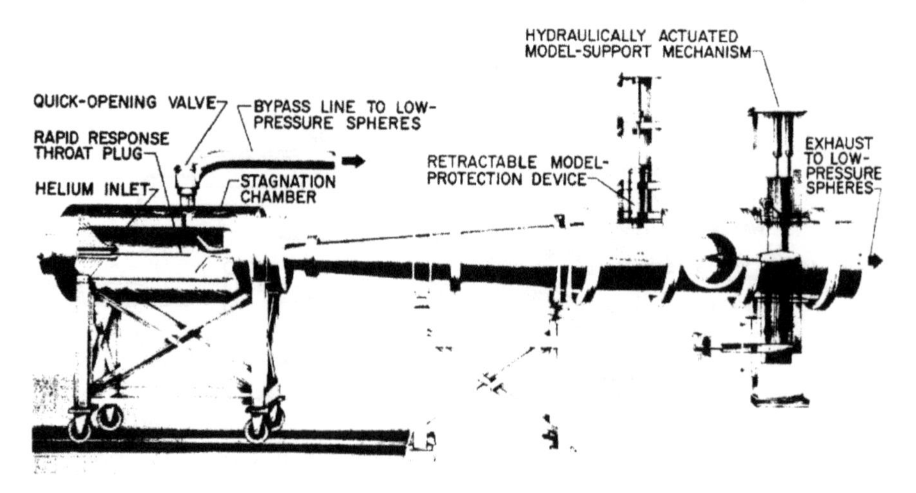


**Figure A.44.** NASA Langley 22 Inch Helium Tunnel. Reproduced from Schaefer, (1965). Copyright US Government. Image stated to be in public domain.

**Table A.48.** Operating conditions.

| Quantity | Min | Max |
|---|---|---|
| $M_\infty$ | 18 | 24 |
| $p_{t_\infty}$ (MPa) | 3.44 | 24.8 |
| $T_{t_\infty}$ | 300 | 478 |
| $Re_\infty \times 10^{-6}$ ($m^{-1}$) | 9.8 | 49.2 |
| Run time (s) | 40 | 60 |

aftercooler to the vacuum sphere. The operating conditions are shown in table A.49. A similar ceramic-heated tunnel for Mach 4 and Mach 6 is described in Trout (1963).

## A.22.13 NASA Langley Mach 10 and Mach 20 helium tunnels

Watson and Bushnell (1971) describes the design and operational characteristics of the NASA Langley Mach 10 and Mach 20 high Reynolds number helium tunnels. A schematic of the facility is shown in figure A.46. Helium is stored in two pressure vessels with individual volumes of 28.3 $m^3$ at pressures up to 45.5 MPa. Each test leg comprises a stagnation chamber, nozzle and test section. The gas exhausts to two vacuum spheres of volume 3209 $m^3$ each. The nozzle diameters are 91.4 cm (Mach 10) and 152.4 cm (Mach 20). The operating conditions are listed in table A.50.

## A.22.14 NASA Langley Hypersonic Nitrogen Tunnel

Schaefer (1965) details the operating characteristics of the NASA Langley Hypersonic Nitrogen Tunnel. A schematic of the facility is shown in figure A.47. The tunnel has an

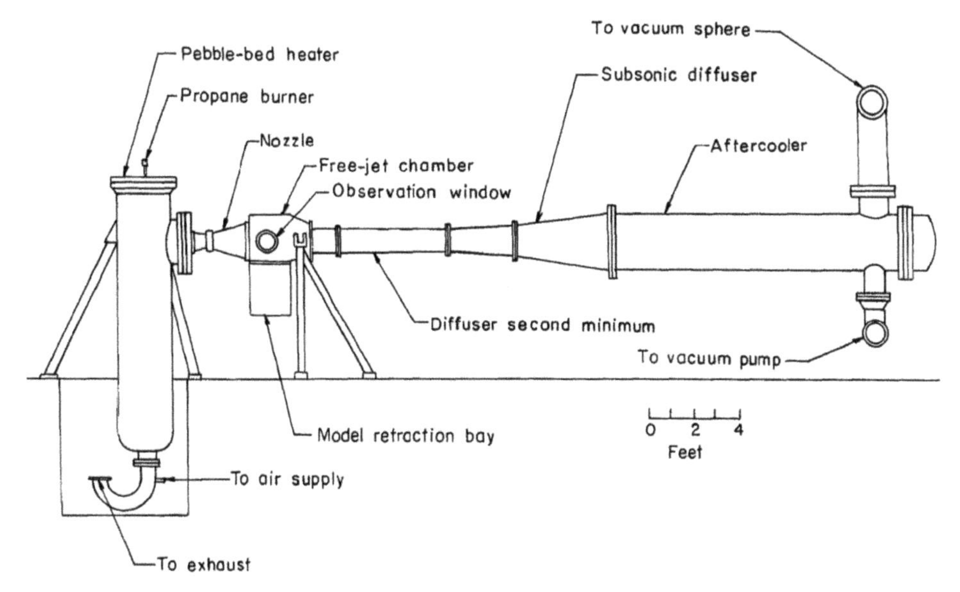


**Figure A.45.** NASA Langley 12 Inch Hypersonic Ceramic-Heated Tunnel. Reproduced from Clark (1965). US Government. Image stated to be in public domain.

**Table A.49.** Operating conditions.

| Quantity | Min | Max |
|---|---|---|
| $M_\infty$ | 12.93[a] | 13.62[b] |
| $p_{t_\infty}$ (MPa) | 0.41 | 4.24 |
| $T_{t_\infty}$ (K) | 1367 | 2083 |
| Running time (s) | 30[c] | 60[c] |
| $Re_\infty \times 10^{-6}$ ($m^{-1}$) | 0.056[c] | 0.75[c] |

[a] Equilibrium flow.
[b] Frozen flow.
[c] Schaefer (1965).

axisymmetric contoured nozzle with diameter of 43.2 cm. The test gas is nitrogen and is heated by a tungsten grid resistance heater. The gas exhausts through an afterooler into a vacuum sphere. The operating conditions are shown in table A.51.

## A.22.15 NASA Langley 2 Foot Hypersonic Facility

Schaefer (1965) describes the operating characteristics of the NASA Langley 2 Foot Hypersonic Facility. A schematic is shown in figure A.48. The facility is a closed circuit continuous operating tunnel. Air is heated by an electrical resistance heater. The test section is 60.96 cm ×60.96 cm in cross section and 1.37 m in length. The tunnel also operates at Mach 3 and 4.5. The operating conditions at Mach 6 are shown in table A.52.

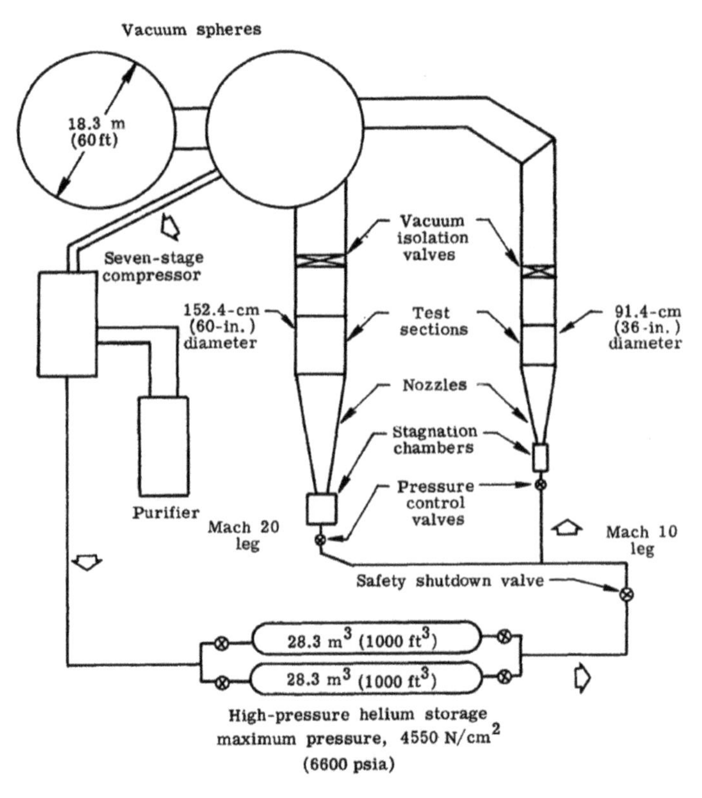


**Figure A.46.** NASA Langley Mach 10 and Mach 20 High Reynolds Number Helium Tunnels. Reproduced from Watson and Bushnell (1971). Copyright US Government. Image stated to be in public domain.

**Table A.50.** Operating conditions.

| Quantity | $M = 10$ | $M = 20$ |
|---|---|---|
| Gas | He | He |
| $p_{t_\infty}$ (MPa) | na | 2.1 to 13.8[a] |
| $T_{t_\infty}$ (K) | 300 | 300 |
| $Re_\infty \times 10^{-6}$ ($m^{-1}$) | na | 96.4[b] |
| Test time (s) | na | 5 |

[a] Watson and Bushnell (1971).
[b] Maximum (Watson and Bushnell 1971).

### A.22.16 NASA Langley Hotshot Tunnel

Schaefer (1965) details the operating characteristics of the NASA Langley Hotshot Tunnel. A schematic is presented in figure A.49. The test gas (air, helium or nitrogen) is initially stored in a pressure vessel. A bank of 720 capacitors are charged to 7.5 kV and discharged across a pair of electrodes to add up to 2.0 MJ. The heated

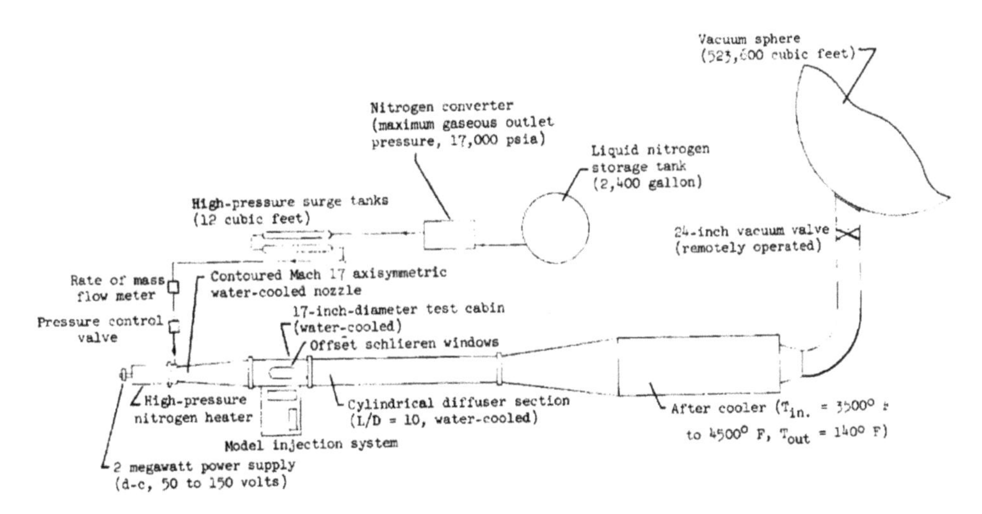


**Figure A.47.** NASA Langley Hypersonic Nitrogen Tunnel. Reproduced from Schaefer, (1965). Copyright US Government. Image stated to be in public domain.

**Table A.51.** Operating conditions.

| Quantity | Value |
|---|---|
| Gas | $N_2$ |
| $M_\infty$ | 18 |
| $p_{t_\infty}$ (MPa) | 103.4[a] |
| $T_{t_\infty}$ (K) | 2478 |
| $Re_\infty \times 10^{-6}$ ($m^{-1}$ | 0.51 to 2.57 |
| Running time (s) | >30 |

[a]Maximum.

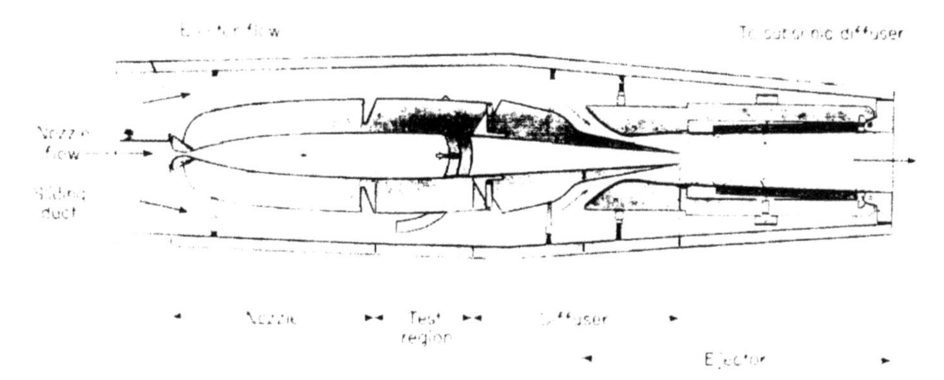

**Figure A.48.** NASA Langley 2 Foot Hypersonic Facility. Reproduced from Schaefer, (1965). Copyright US Government. Image stated to be in public domain.

**Table A.52.** Operating conditions.

| Quantity | Min | Max |
|---|---|---|
| $M_\infty$ | 6 | 6 |
| $p_{t_\infty}$ (MPa) | 0.20 | 0.38 |
| $T_{t_\infty}$ (K) | 422 | 533 |
| $Re_\infty \times 10^{-6}$ (m$^{-1}$) | 1.64 | 3.94 |
| Runtime | continuous | |

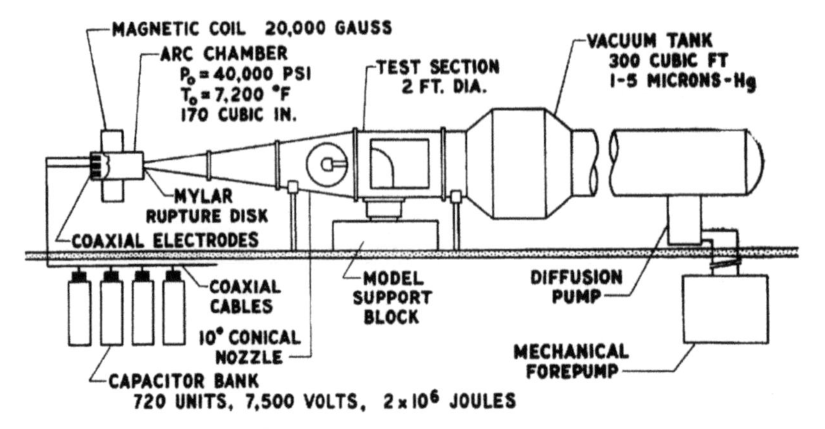


**Figure A.49.** NASA Langley Hotshot Tunnel. Reproduced from Schaefer, (1965). Copyright US Government. Image stated to be in public domain.

**Table A.53.** Operating conditions[a].

| Quantity | Min | Max |
|---|---|---|
| $M_\infty$ | 12 | 28 |
| $p_{t_\infty}$ (MPa) | 103.4 | 103.4 |
| $T_{t_\infty}$ (K) | 3000 | 3000 |
| $Re_\infty \times 10^{-6}$ (m$^{-1}$) | 0.033 | 3.28 |
| Running time (s) | 40 | 120 |

[a] Using $N_2$.

gas expands through a 10° total angle conical nozzle to the test chamber with diameter 60.96 cm and exhausts to a vacuum chamber. The operating conditions for nitrogen are listed in table A.53.

### A.22.17 NASA Langley Continuous Flow Hypersonic Tunnel

Schaefer (1965) describes the operating characteristics of the NASA Langley Continuous Flow Hypersonic Tunnel. A schematic of the facility is shown in figure A.50. Air is heated by an electrical resistance heater. Two interchangable test sections with cross section 78.74 cm ×78.74 cm are capable of Mach 10 and 12. The tunnel is a closed circuit facility using a vacuum tank for starting. The operating conditions are listed in table A.54.

### A.22.18 NASA Langley Pilot Model Expansion Tube

Schaefer (1965) describes the operating characteristics of the NASA Langley Pilot Model Expansion Tube. A schematic of the facility is shown in figure A.51. A variety of test gases can be used. Energy is provided by a 2.5 MJ capacitor bank charged to 12 kV. The tube diameter is 10.16 cm. The facility can be operated in three modes, namely, expansion tube, shock tube and shock tunnel. The operating conditions are shown in table A.55.

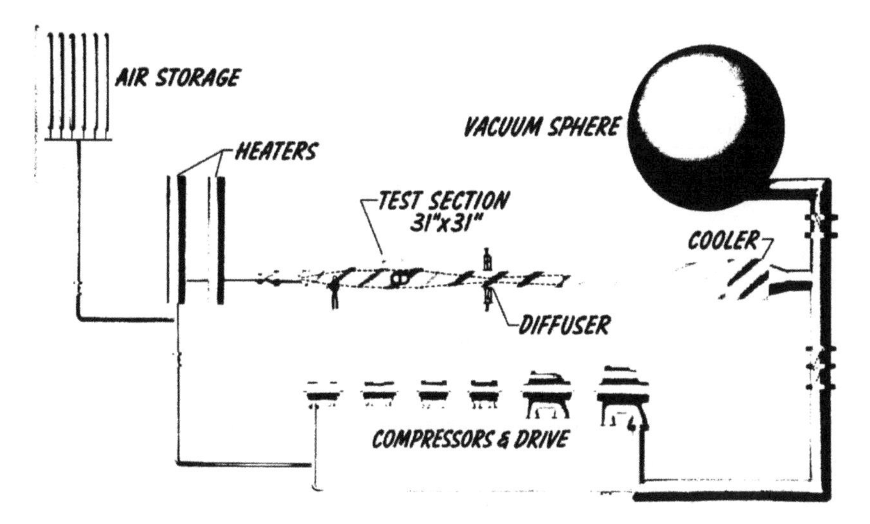


**Figure A.50.** NASA Langley Continuous Flow Hypersonic Tunnel. Reproduced from Schaefer, (1965). Copyright US Government. Image stated to be in public domain.

**Table A.54.** Operating conditions.

| Quantity | $M_\infty = 10$ | $M_\infty = 12$ |
| --- | --- | --- |
| $p_{t_\infty}$ (MPa) | 1.52 to 10.1 | 6.08 |
| $T_{t_\infty}$ (K) | 1033 | 1311 |
| $Re_\infty \times 10^{-6}$ (m$^{-1}$) | 1.64 to 5.58 | 2.0 |

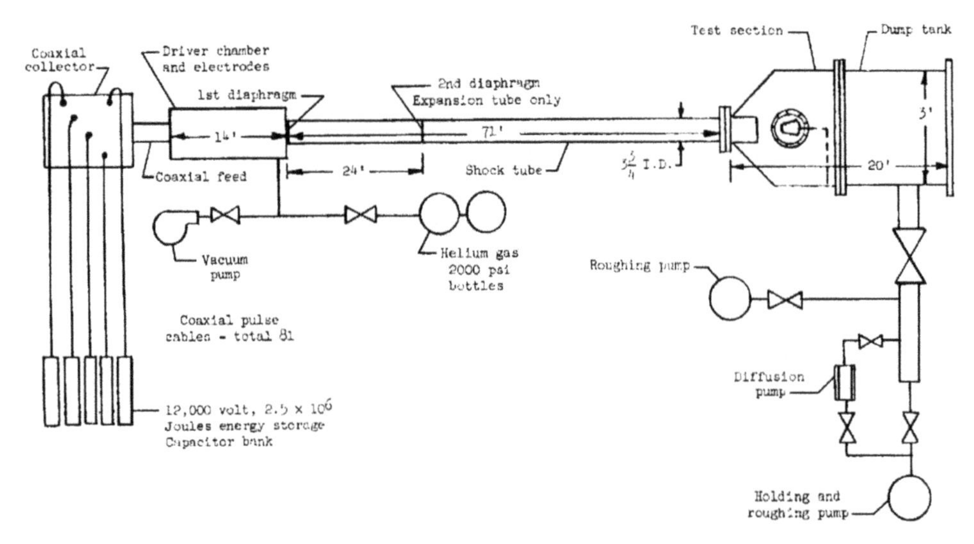


**Figure A.51.** NASA Langley Pilot Model Expansion Tube. Reproduced from Schaefer, (1965). Copyright US Government. Image stated to be in public domain.

**Table A.55.** Operating conditions.

| Quantity | Expansion Tube | Shock Tube | Shock Tunnel |
|---|---|---|---|
| $M_\infty$ | 15 to 30 | 2 to 3 | 7 to 16 |
| $H_\infty$ (MJ kg$^{-1}$) | 10.5 to 41.9 | 4.0 to 20.9 | 4.0 to 20.9 |
| $Re_\infty \times 10^{-6}$ (m$^{-1}$) | 0.33 to 3.3 | 0.033 to 16.4 | $3.3 \times 10^{-4}$ to $9.8 \times 10^{-2}$ |
| Running time ($\mu$s) | 50 to 400 | 200 to 500 | 2000 to 5000 |

### A.22.19 NASA Langley Hypersonic Aerothermal Dynamics Facility

Schaefer (1965) describes the NASA Langley Hypersonic Aerothermal Dynamics Facility, also known as the NASA Langley 4 foot Hypersonic Arc Tunnel. A schematic of the test sections is shown in figure A.52. Air is heated by a 10 MW to 20 MW DC arc. Test section diameters of 60.96 cm and 1.22 m. The nozzle is conical. The air exhausts into a 30.48 m diameter vacuum sphere. The operating conditions are indicated in table A.56.

### A.22.20 NASA Langley—other hypersonic tunnels

Additional hypersonic tunnels at NASA Langley are listed in table A.57 and together with their principal research use in table A.58.

## A.23 Naval Ordnance Laboratory

### A.23.1 NOL Tunnel No. 4

The US Naval Ordnance Laboratory 12 cm ×12 cm Hyperballistic Tunnel No. 4 began operation in May 1950 at the Naval Ordnance Laboratory in White Oak,

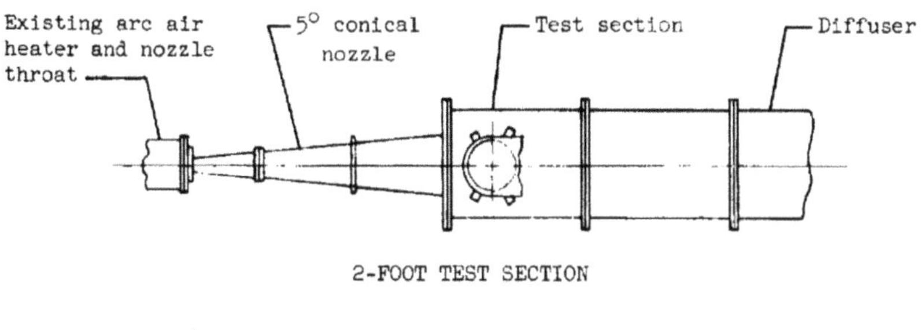


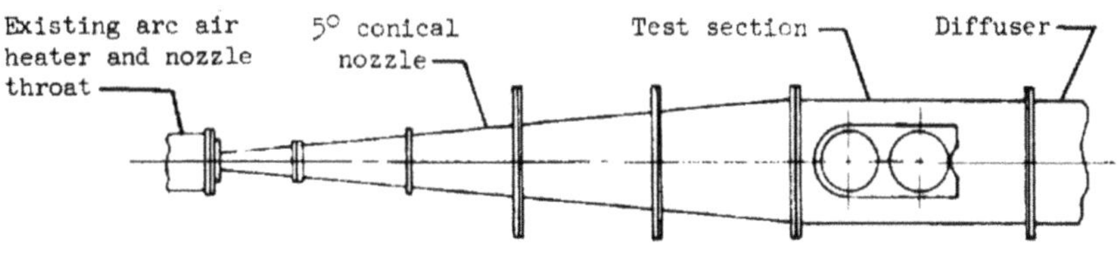


**Figure A.52.** NASA Langley Hypersonic Aerothermal Dynamics Facility. Reproduced from Schaefer, (1965). Copyright US Government. Image stated to be in public domain.

**Table A.56.** Operating conditions.

| Quantity | Min | Max |
|---|---|---|
| $M_\infty$ | 8 | 18 |
| $H_\infty$ (MJ kg$^{-1}$) | 3.5 | 14.0 |
| $p_{t_\infty}$ (MPa) | 1.72 | 10.34 |
| $Re_\infty \times 10^{-6}$ (m$^{-1}$) | $3.3 \times 10^{-3}$ | 0.33 |

**Table A.57.** NASA Langley—other hypersonic tunnels.

| Tunnel | $M_\infty$ |
|---|---|
| 10 MW arc tunnel | 5.1 to 6.85 |
| Mach 6 high Reynolds Number | 6 |
| 7 inch Mach 7 Pilot Tunnel | 6.5 to 7.7 |
| 11 inch ceramic-heated tunnel | 2 to 6 |
| Hypersonic aeroelastic tunnels | 10 and 20 |
| 8 Foot high temperature structures tunnel | 6.8 to 7.7 |
| Hypersonic aeroelastic tunnels | 10 and 20 |

Maryland. An artist's rendition of the tunnel is shown in figure A.53 and details are presented in Wegener (1951) and Wegener and Lobb (1953). The wind tunnel was a continuous blowdown air facility with the test section connected to a 2000 m$^3$ vacuum sphere. The nozzle has an arbitrary curvature at the throat followed by two 50 cm long

**Table A.58.** NASA Langley—other hypersonic tunnels.

| Tunnel | Purpose |
|---|---|
| 10 MW arc tunnel | Materials and structures testing |
| Mach 6 High Reynolds Number | Fluid dynamics |
| 7 inch Mach 7 pilot tunnel | Heat transfer, ablation and technique development |
| 11 inch ceramic-heated tunnel | High temperature materials |
| Hypersonic aeroelastic tunnels | Aeroelastic, thermal and dynamics |
| 8 foot high temperature structures tunnel | Structures and thermal protection |
| Hypersonic aeroelastic tunnels | Aeroelastic, thermal and dynamics |

**Figure A.53.** Artist's rendition of NOL 12 cm ×12 cm hyperballistic wind tunnel. Reproduced from Wegener, (1951). Copyright United States Naval Ordnance Laboratory.

plane diverging walls, followed by a square test section 12 cm × 12 cm. The operating conditions are listed in table A.59. The facility was dismantled in 1968 and the original nozzle forwarded to the von Kármán Institute for Fluid Dynamics in Belgium.

## A.23.2 NOL Tunnel No. 7

The US Naval Ordnance Laboratory Boundary Layer Channel Tunnel No. 7 was located at the Naval Ordnance Laboratory in White Oak, Maryland. The nozzle

**Table A.59.** Operating conditions.

| Quantity | Value | |
|---|---|---|
| | Min | Max |
| $M_\infty$ | 4 | $11^a$ |
| $H_\infty$ (deg K) | 288 | 773 |
| $p_{t_\infty}$ (MPa) | $0.1^b$ | $1.0^{b,c}$ |

[a] Later upgraded to $M = 18$.
[b] Cited in Lobb *et al* (1955).
[c] Cited in Wegener *et al* (1953).

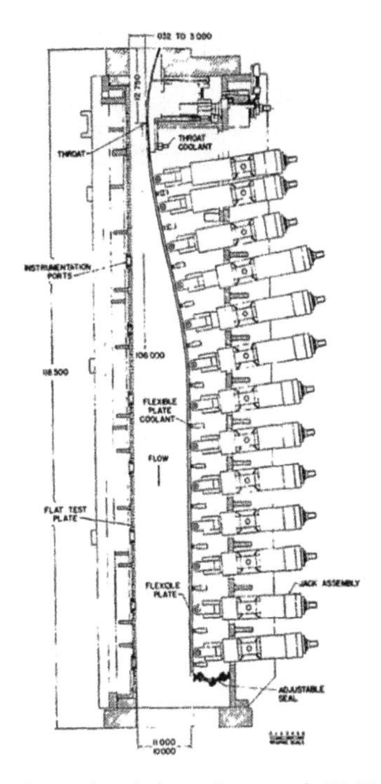


**Figure A.54.** Schematic of nozzle. Reproduced from Lee *et al* (1969). Copyright United States Naval Ordnance Laboratory.

geometry is a half-symmetric shape comprising a flat plate extending from the nozzle throat for a length of 2.54 m and an adjustable flexible curved plate as illustrated in figure A.54 and shown in figure A.55 (Lee *et al* 1966). The flat plate diverges from 30.48 cm at the throat to 34.29 cm at the nozzle exit to accommodate the boundary layer growth on the sidewalls. A propane-fired heater was used to heat the supply air. The vacuum system is the same as used by the Naval Ordnance Laboratory Tunnel No. 4. The flat and curved plates were actively cooled. The operating conditions of the tunnel are shown in table A.60. The tunnel was deactivated in the 1970s.

**Figure A.55.** Photo of nozzle. Reproduced from Lee *et al* (1969). Copyright United States Naval Ordnance Laboratory.

**Table A.60.** Operating conditions.

| | Value | |
|---|---|---|
| Quantity | Min | Max |
| $M_\infty$ | 3 | 7 |
| $T_{t_\infty}$ (deg K) | 300 | 811 |
| $p_{t_\infty}$ (MPa) | 0.010 | 1.01 |

# A.24 NYU hypersonic wind tunnels

The New York University (NYU) Aerospace Facility was established in 1964 based upon a proposal to the National Aeronautics and Space Administration submitted by Dr. Lee Arnold, chairman of the NYU Department of Aeronautical Engineering and Dr. Antonio Ferri, director of the NYU Aerospace Laboratory (McCormick *et al* 2004). The facility was located in the Bronx, New York City, adjacent to the Harlem River and hence known as the NYU Harlem River Laboratory. The operating characteristics of two of the facilities are shown in table A.61.

# A.25 ONERA

## A.25.1 F4

Sagnier and Vérant (1998) and Masson *et al* (2002) discuss the design and operation of the Office National d'Etudes et de Recherches Aérospatiales (ONERA) F4 high

**Table A.61.** Operating conditions.

| Quantity | Tunnel 1 | Tunnel 2 |
|---|---|---|
| $M_\infty$ | 6 | 14 |
| $p_{t_\infty}$ (MPa) | 13.2[a] | 203[a] |
| $T_{t_\infty}$ (K) | 1033[a] | 1389[a] |

[a] Maximum (McCormick *et al* 2004).

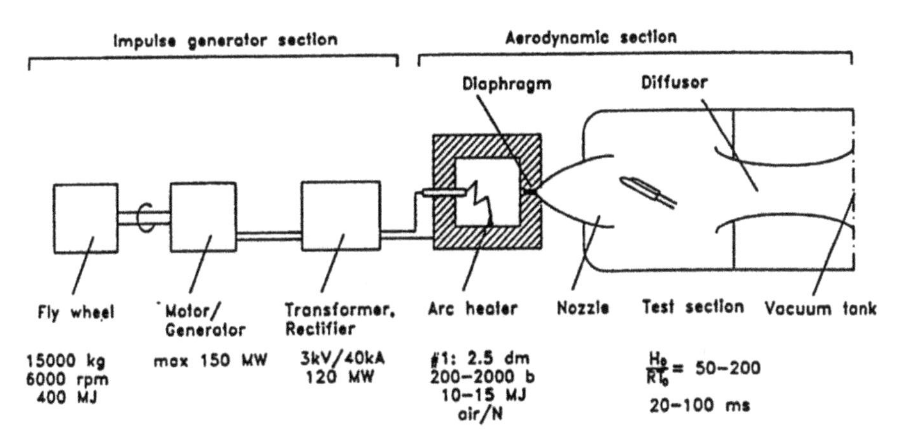


**Figure A.56.** ONERA F4. Reproduced from Grönig (1995) with permission of Springer.

enthalpy wind tunnel. The facility is an arc heated, blow down, free jet hypersonic wind tunnel also known as a 'hot shot' tunnel located in Le Fauga-Mauzac Center near Toulouse, France. The facility was built for the Hermes space program and inaugurated in June 1992. A sketch of the facility is presented in figure A.56. The test gas is heated by an electric arc in a chamber of 10 L to 15 L volume that is initially pressurized and at ambient temperture. The electric arc is generated by a flywheel driving a motor-generator connected to a transformer and rectifier (Grönig 1992). The chamber is cylindrical in shape and the electrodes are located on the lateral sides. The shape of the electrodes induces a rotation of the arc thereby improving efficiency of the heating process. The electrode energy is provided by an impulse generator at a power of 150 MW for several tens of milliseconds. The operating conditions are listed in table A.62.

### A.25.2 R1Ch, R2Ch, R3Ch and R5Ch

Morzenski (1986), Chanetz *et al* (1998) and Chanetz *et al* (2020) describe the hypersonic blow down tunnels R1Ch, R2Ch, R3Ch and R5Ch located in Chalais-Meudon, France. A schematic of R3Ch is shown in figure A.57. The air is supplied by a shared 500 m$^3$ reservoir at 25 MPa. The R1Ch and R2Ch air is heated up to 700 K using a heat exchanger at a mass flow rate of 40 kg s$^{-1}$. The Reynolds number

**Table A.62.** Operating conditions.

| Quantity | Min | Max |
|---|---|---|
| $M_\infty$ | 7.2[a] | 13[a] |
| $p_{t_\infty}$ (MPa) | 20.2 | 760 |
| $T_{t_\infty}$ (K) | 2000 | 6000 |
| Test time (ms) | <400[b] | |

[a] Sagnier and Vérant (1998).
[b] Desse *et al* (2014).

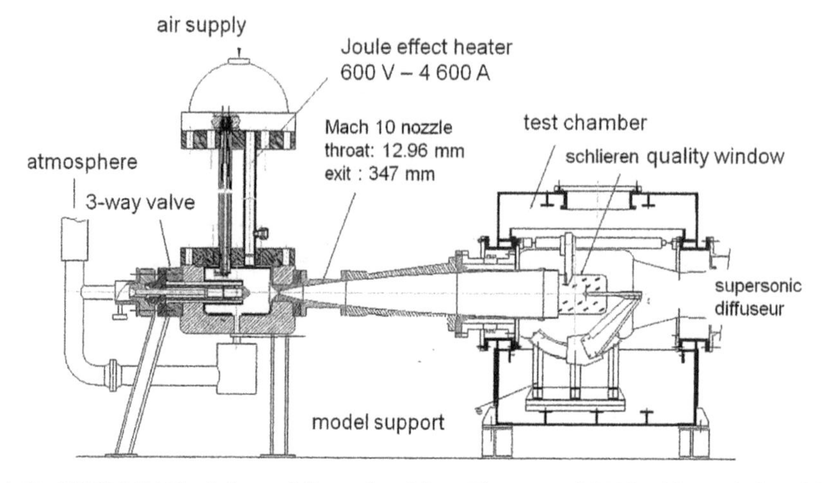


**Figure A.57.** ONERA R3Ch wind tunnel. Reproduced from Chanetz *et al* (2020) with permission of Springer.

**Table A.63.** Operating conditions.

| Quantity | R1Ch | R2Ch | R3Ch | R5Ch |
|---|---|---|---|---|
| $M_\infty$ | 3 or 5 | 3 to 7 | 10 | 10 |
| $p_{t_\infty}$ (MPa) | 1.5[a] | 8.0[a] | 1.2 to 12 | 0.25 |
| $T_{t_\infty}$ (K) | 400[a] | 700[a] | 1100 | 1050 |

[a] Maximum (Chanetz *et al* 2020).

range for these two tunnels is $2.1 \times 10^6$ m$^{-1}$ to $5.1 \times 10^6$ m$^{-1}$ achieved by varying the stagnation pressure between 0.05 MPa to 8.0 MPa (Chanetz *et al* 2020). The R3Ch facility utilizes a Joule instant heater to increase the air temperature to 1100 K. The nozzle exit diameter for the Mach 5 to 7 nozzles in R2Ch is 0.326 m. The R3Ch nozzle has an exit diameter of 0.35 m. Test times in R1Ch and R2Ch range from several minutes (when exhausted to atmosphere) to approximately 30 seconds (when exhausted to vacuum sphere of 500 m$^3$). The operating conditions are listed in table A.63. R3Ch and R5Ch are no longer in operation (Chanetz 2022).

### A.25.3 S4MA

Laverre (1987) and Vennemann (1999) describe the hypersonic blow down tunnel S4MA located in Modane-Avrieux, France. A schematic of the facility is shown in figure A.58. Air is stored in a series of tanks with total volume of 29 $m^3$ at pressures up to 40 MPa. A cylindrical pebble bed heater with diameter of 2 m and height of 10 m contains 12 tons of alumina pebbles that can be heated to a maximum temperature of 1850 K by propane combustion. The heated air exhausts through a contoured nozzle into the test cell. Three nozzles are available with corresponding test section Mach numbers of 6.4, 10 and 12. The Mach 6.4 nozzle has an exit diameter of 685 m and a length of 3.6 m. The Mach 10 and 12 nozzles have an exit diameter of 994 mm and total length approximately 7 m each. The test section is a cube with side lengths of 3 m. The flow exits the test cell to the atmosphere, or a vacuum chamber ranging in size from 3000 $m^3$ to 4000 $m^3$, depending on the test conditions. The operating conditions are listed in table A.64.

## A.26 Oxford University

### A.26.1 Gun Tunnel

McGilvray *et al* (2015) describe the Oxford University Gun Tunnel. The Oxford Gun Tunnel was designed and manufactured by Bristol Sidley Engines. The tunnel was constructed in 1965. The facility was moved to Oxford University in the mid 1970s. A schematic of the tunnel is shown in figure A.59. The tunnel comprised an

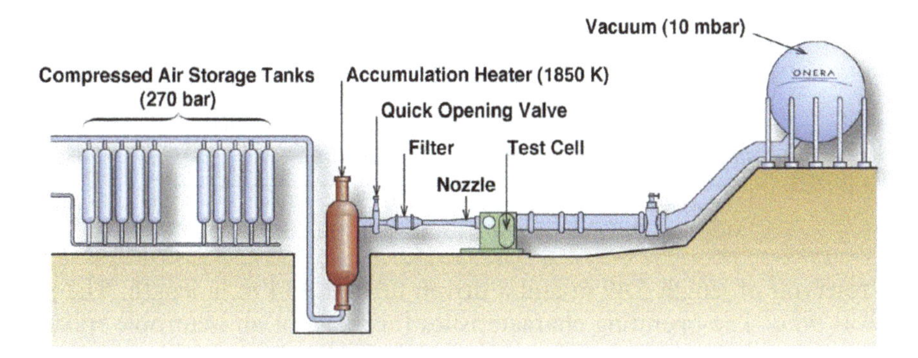


**Figure A.58.** ONERA S4. Copyright Bruno Chanetz.

**Table A.64.** Operating conditions.

| Quantity | Min | Max |
|---|---|---|
| $M_\infty$ | 6.4 | 12 |
| $Re_\infty \times 10^{-6}$ ($m^{-1}$) | 3.2 | 28 |
| Test time (s) | 25 | 90 |

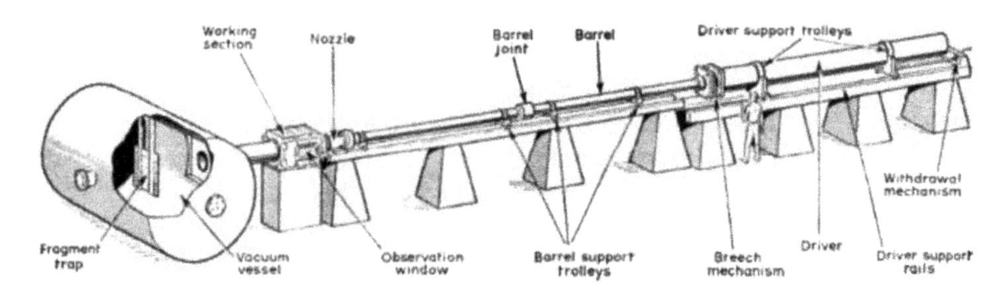


**Figure A.59.** Oxford University Gun Tunnel. Copyright 2015 by Matthew McGilvray, Luke Doherty, Richard Morgan, David Gildfind.

**Table A.65.** Operating conditions.

| Quantity | Value |
|---|---|
| $M_\infty$ | 7.1 |
| $p_{t_\infty}$ (MPa) | 6.34 |
| $T_{t_\infty}$ (K) | 720 |
| $T_w/T_{t_\infty}$ | $\approx 0.47$ |

**Table A.66.** Operating conditions[a].

| Quantity | $M_\infty = 6$ | $M_\infty = 8$ | $M_\infty = 9$ |
|---|---|---|---|
| $p_{t_\infty}$ (MPa) | 13.2[a] | 13.2 | 13.2 |
| $T_{t_\infty}$ (K) | 1300[b] | 1300 | 1300 |
| $Re_\infty \times 10^{-6}$ ($\text{m}^{-1}$) | 39.4 | 19.8 | 8.2 |

[a] Anonymous (1990).
[b] Maximum (Anonymous 1990), 720 K in LICH mode.

inline reservoir of length 5.49 m and a driven tube of 9.14 m in length. The piston mass was 96 g. The operating characteristics for a test of an isentropic spike inlet (Matthews *et al* 2005) are shown in table A.65. The overall operating conditions are shown in table A.66. See also Buttsworth (2016).

### A.26.2 T6 Stalker Tunnel

Collen *et al* (2021) describe the development and operation of the Oxford University T6 Stalker Tunnel. The facility can operate in four distinct modes, namely, Reflected Shock Tunnel (RST), Expansion Tube (ExT) and two different shock tube (ST) modes. All four modes use the same driver section. A schematic of the facility is shown in figure A.60. The free piston driver was originally part of the T3 tunnel at the Australian National University. The driver section in 9 m in length. A piston of either 36 kg or 89 kg is used with piston velocities up to 150 m s$^{-1}$. The driven tube

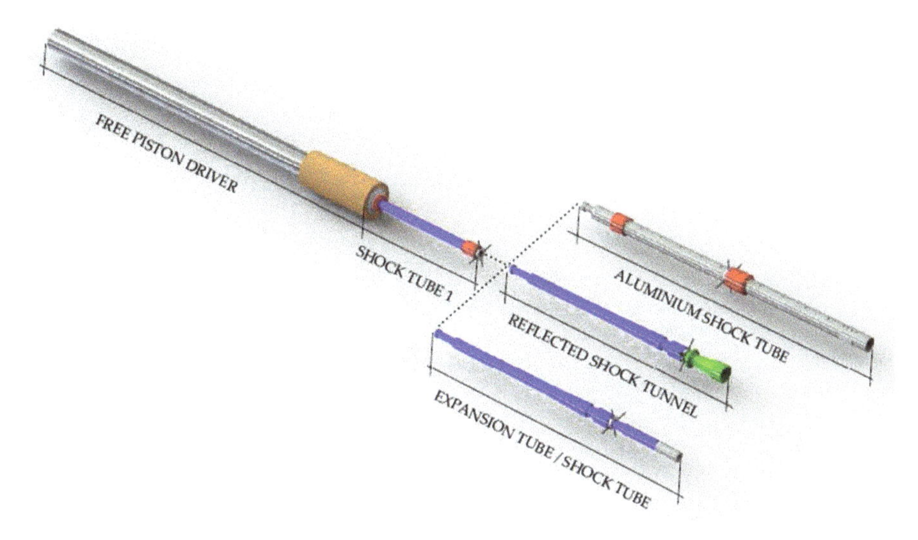


**Figure A.60.** Oxford University T6. Reproduced from Collen *et al* (2021) with permission of Springer.

**Table A.67.** Operating conditions.

| Quantity | Reflected Shock Tunnel | Expansion Tunnel | Shock Tube |
|---|---|---|---|
| $U_\infty$ (km s$^{-1}$) | 6.5[a] | 15[a] | 18[a] |
| Flow core diameter (mm) | 150–200 | 50–120 | 80 or 200 |
| Test time ($\mu$s) | 1000–3000 | 50–500 | 2–50 |

[a] Maximum (McGilvray *et al* 2015).

(barrel), nozzle and test section were formerly part of the Oxford Gun Tunnel (section A.26.1). The test section is 1.1 × 0.9.9 m. The RST mode is capable of Mach 6, 7 and 8. The operating conditions are listed in table A.67.

# A.27 People's Republic of China

## A.27.1 Chinese Academy of Aerospace Aerodynamics FD-21

Li *et al* (2018) and Zhixian *et al* (2018) describe the design and operation of the free piston shock tunnel FD21 at the Chinese Academy of Aerospace Aerodynamics (CAAA) in Peking. The facility was constructed in 2014–2017. A schematic is shown in figure A.61. The tunnel comprises a high pressure air reservoir (24 m$^3$), driver tube (compression tube), driven tube (shock tube), nozzle, test section and vacuum tank. The driver tube gas is helium or a mixture of helium and argon, and the driven tube gas is air. The driver tube length is 75 m and diameter is 0.668 m. The driven tube is 34 m in length and 0.29 m in diameter. The nozzle exit diameter is 2 m. The tunnel is designed for pistons from 120 kg to 600 kg. The operating conditions are listed in table A.68.

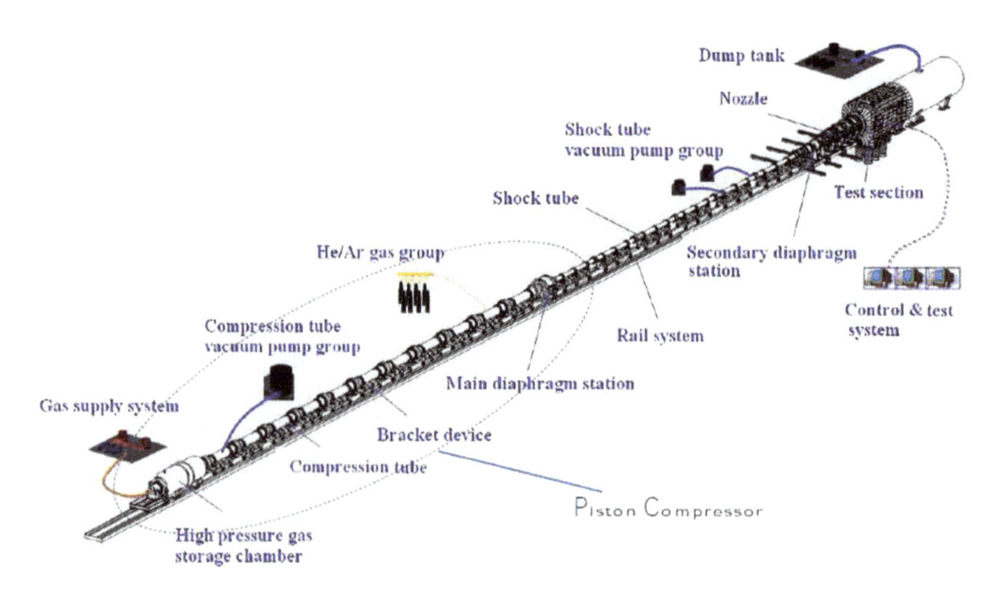


**Figure A.61.** FD21. Copyright Universität der Bundeswehr München.

**Table A.68.** Operating conditions.

| Quantity | Min | Max |
|---|---|---|
| $H_\infty$ (MJ kg$^{-1}$) | 5 | 28 |
| $U_\infty$ (km s$^{-1}$) | 3 | 7 |

## A.27.2 Chinese Academy of Sciences Shock Tunnel JF12

Jiang and Yu (2014) and Wang *et al* (2016) discuss the design and operational characteristics of the JF12 ('Hyper-Dragon') detonation driven shock tunnel. The tunnel is also known as the Long-test-duration Hypervelocity Detonation-driven Shock Tunnel (LHDst) and located at the State Key Laboratory of High Temperature Gas Dynamics, Institute of Mechanics, Chinese Academy of Sciences, Peking. A schematic of the facility is shown in figure A.62. The damping section at the far left is 19 m in length with a diameter of 400 mm and dissipates the wave system generated by the detonation. The driver section operates in the backwards propagating mode whereby the detonation is initiated at the diaphragm initially separating the driver and a transition section. The driver section is 99 m in length and 400 mm in diameter. A transition connects the driver and driven sections which gradually increases the diameter from 400 mm to 720 mm. The driven section is 89 m in length with a diameter of 720 mm. Two different contoured nozzles are available, one with length of 8 m and exit diameter of 1.5 m for Mach numbers of 5–7, and other with length of 15 m and exit diameter of 2.5 m for Mach numbers of 7–9. The test section is connected to a E-shaped vacuum tank with volume 600 m$^3$ and

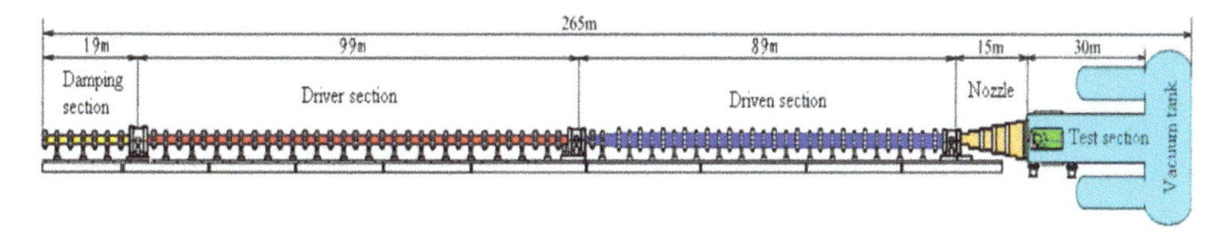


**Figure A.62.** JF12 detonation driven shock tunnel. From Jiang and Yu (2014); reprinted by permission of the American Institute of Aeronautics and Astronautics, Inc.

**Table A.69.** Operating conditions.

| Quantity | Value |
|---|---|
| $M_\infty$ | 5 to 9 |
| Flight altitude (km) | 25 to 50 |
| Test time (ms) | >100 |

**Figure A.63.** National University of Defense Technology Hypersonic Quiet Tunnel. From Dundian *et al* (2017); reprinted by permission of the American Institute of Aeronautics and Astronautics, Inc.

length of 50 m. Test times longer than 100 ms are obtained. The facility can operate at the tailored interface condition (Wittliff *et al* 1959). The operating conditions are noted in table A.69. The next generation tunnel JF-22 is in progress with design maximum velocity of 10 km $s^{-1}$ (Post 2022).

### A.27.3 NUDT Hypersonic Quiet Tunnel

Dundian *et al* (2017) describe the design and operational characteristics of the Hypersonic Quiet Tunnel at the National University of Defense Technology (NUDT), Changsha, Hunan, People's Republic of China. A photograph of the facility is shown in figure A.63. The tunnel comprises the settling chamber, nozzle

**Table A.70.** Operating conditions.

| Quantity | Value |
|---|---|
| $M_\infty$ | 6 |
| $p_{t_\infty}$ (MPa) | 0.50 |
| $T_{t_\infty}$ (K) | 600[a] |
| $Re_\infty \times 10^{-6}$ (m$^{-1}$) | 3.1[b] |
| Test time (s) | >15 |

[a] Maximum.
[b] At $p_t$ = 0.50 MPa and $T_t$ = 600 K.

and test section. High pressure air is heated up to 650 K and delivered to the settling chamber. Upstream of the converging-diverging nozzle the boundary layer on the settling chamber wall is removed through an annular bleed slot. The axisymmeric nozzle exit diameter is 300 mm. The vacuum tank volume is 650 m$^3$. Quiet flow (defined as rms pitot pressure fluctuations below 0.1% of the mean pitot pressure) is achieved at a total pressure up to 0.5 MPa. The operating conditions are listed in table A.70.

### A.27.4 Peking University Mach 6 Quiet Tunnel

Zhang *et al* (2013), Zhang (2014) and Zhang (2020) describe the Mach 6 Quiet Tunnel (M6QT) at Peking University Department of Aeronautics and Astronautics. The facility is a blow down tunnel. A short axisymmetric nozzle with exit diameter 120 mm exits into an open jet test section. The facility is capable of continuous operation with run time up to 30 s. Zhang *et al* (2013) indicate quiet tunnel operation with rms pitot pressure fluctuations equal to 0.10% of mean pitot pressure are achieved at a total pressure $p_t$ = 0.34 MPa and total temperature $T_t$ = 410 K.

## A.28 Princeton University

### A.28.1 Mach 8 Hypersonic Boundary Layer Tunnel

The Princeton University Mach 8 Hypersonic Boundary Layer Tunnel is described in Baumgartner (1997) and Baumgartner *et al* (1997). A schematic of the facility is shown in figure A.64. High pressure air is supplied by a 63 m$^3$ tank system to the heater assembly where the temperature is increased to approximately 800 K to avoid condensation in the test section. The air expands through a converging-diverging nozzle to the test section, and exhausts through the diffuser and cooler to the atmosphere using a two stage ejector system. The overall length of the wind tunnel is approximately 9 m excluding the ejector system and heater. The operating conditions are indicated in table A.71.

### A.28.2 Helium Hypersonic Tunnel

Bogdonoff and Hammitt (1954), Hammitt and Bogdonoff (1956) and Bogdonoff and Vas (1959) describe the design and operation of the Princeton University blow

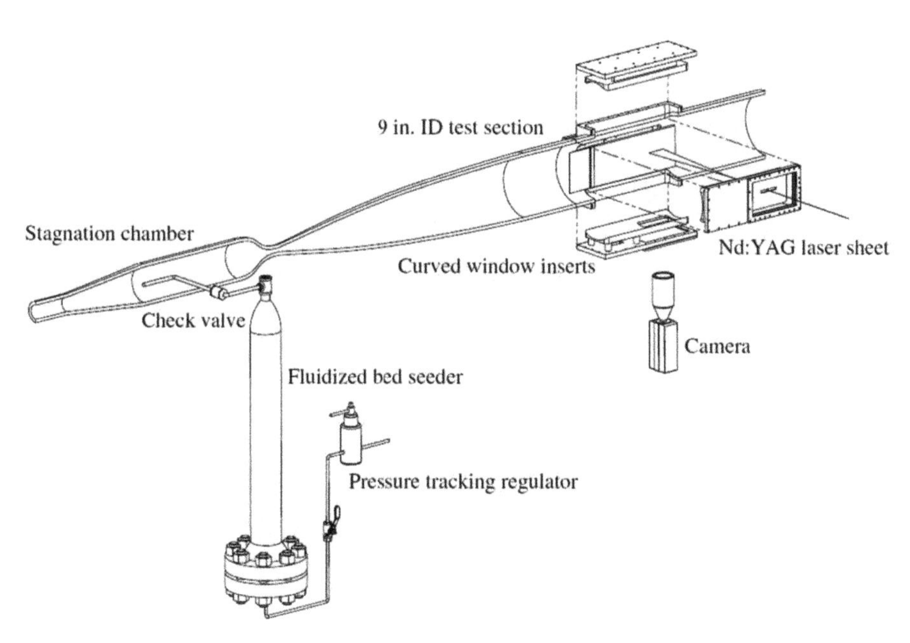


**Figure A.64.** Princeton Mach 8 Hypersonic Boundary Layer Tunnel. Reprinted with permission. Reproduced from Williams *et al* (2018). Copyright 2017 Cambridge University Press.

**Table A.71.** Operating conditions.

| Quantity | Min | Max |
|---|---|---|
| $M_\infty$ | 8 | 8 |
| $p_{t_\infty}$ (MPa) | 1.7 | 10 |
| $T_{t_\infty}$ (K) | 840[a] | 840 |
| $Re_\infty \times 10^{-6}$ ($m^{-1}$) | 3 | 20 |
| Test time (s) | 240[a] | |

[a]Maximum.

down Helium Hypersonic Tunnel. A photograph of the facility, located at the Forrestal Research Center of Princeton University, is shown in figure A.65 from Bogdonoff and Hammitt (1954). The gas is helium to avoid condensation when expanded to high Mach numbers in the test section, and was supplied from pressurized semi-trailers at 17 MPa. The maximum settling (stagnation) chamber pressure was 10.3 MPa. The helium exhausted through a converging diverging nozzle into the test section. The diverging section of the nozzle was a 5.5° conical section joined to the 8.26 cm diameter cylindrical test section. The gas exited to the atmosphere through an air ejector system. The operating conditions are listed in table A.72. The large run time was effectively determined by the helium capacity in the semi-trailer.

**Figure A.65.** Settling chamber, nozzle and test section. Reproduced from Bogdonoff and Hammitt (1954). Copyright US Government. Image stated to be in public domain.

**Table A.72.** Operating conditions.

| Quantity | Min | Max |
|---|---|---|
| $M_\infty$ | 12.7 | 14 |
| $p_{t_\infty}$ (MPa) | 5.5[a] | 9.7[a] |
| Test time (min) | 10[b] | |

[a] Hammitt and Bogdonoff (1956).
[b] Maximum (Bogdonoff and Hammitt 1954).

## A.29 Purdue University Mach 6 Quiet Tunnel

Schneider (2008) and Chynoweth *et al* (2017) describe the design and operation of the Boeing/AFOSR Mach 6 Quiet Tunnel (BAM6QT) at Purdue University. Figure A.66 is a schematic of the tunnel. The facility is a Ludwieg tube and was designed by Prof. Steven Schneider for research on hypersonic transition to turbulence. The driver tube is 37.3 m in length with a 0.44 m diameter. The converging diverging nozzle expands the flow to Mach 6 with a nozzle exit diameter of 0.24 m. The test flow exhausts to a 113.3 $m^3$ vacuum tank. Approximately every 200 ms the expansion fan in the driver tube reflects from the contraction causing a drop in stagnation pressure. During the ensuing 200 ms period, the flow in the test section is quasi-static. The shape of the divergent section of the nozzle was designed to minimize Görtler instability, and the surface was polished to a mirror finish to eliminate any roughness that would stimulate transition on the nozzle wall boundary

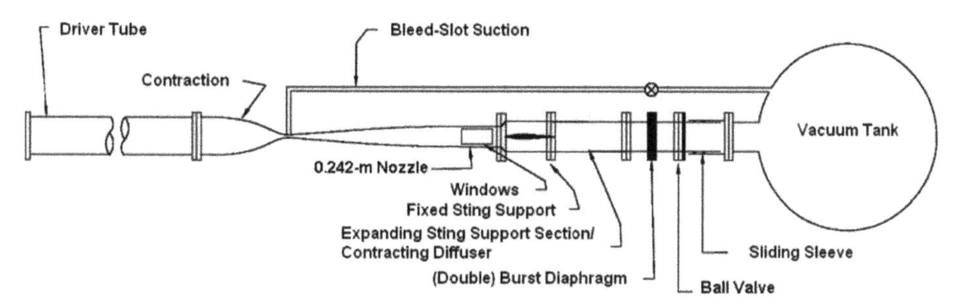


**Figure A.66.** Boeing/AFOSR Mach 6 Quiet Tunnel. Reproduced from Chynoweth *et al* (2017). Copyright Dr Steven P Schneider.

**Table A.73.** Operating conditions.

| Quantity | Value |
|---|---|
| $M_\infty$ | 6 |
| $p_{t_\infty}$ (MPa) | 2.2[a] |
| $T_{t_\infty}$ (K) | 473[a] |
| $Re_\infty \times 10^{-6}$ ($m^{-1}$) | 12.5[b] |
| Test time (s) | 5[c] |

[a] Maximum.
[b] Maximum for quiet flow.
[c] Typical.

layer and thus generate Mach wave disturbances within the test section. A unique feature is an axisymmetric suction slot located upstream of the nozzle throat that removes the boundary layer formed on the surface of the driver tube, and thus a new laminar boundary layer forms on the nozzle surface at the throat. With the bleed slot operating the freestream pitot pressure fluctuations within the test section are less than 0.02% of the mean (Steen 2010). The operating conditions are indicated in table A.73. Quiet flow conditions in the test section are achieved at stagnation pressures up to 0.93 MPa (Chynoweth *et al* 2017).

## A.30 RWTH Aachen Shock Tunnel TH2

Olivier *et al* (2002) and Olivier (2016) present the design and operational characteristics of the Rheinisch-Westfälische Technische Hochschule (RWTH) Aachen Shock Tunnel TH2. The facility is a dual model shock tunnel that can operate as a conventional shock tunnel using typically helium as the driven gas, or as a detonation shock tunnel using a stoichiometric mixture of oxygen and hydrogen with varying admixtures of helium and argon depending on the desired test conditions. The driver and driven tubes inner diameter is 140 mm and the tube wall thickness is 80 mm. The conventional shock tunnel comprises a 6 m driver tube and 15.4 m driven tube, with the driver and driven sections initially separated by a

**Table A.74.** Operating conditions.

| | Value | |
|---|---|---|
| Quantity | Min | Max |
| $M_\infty$ | 5.3 | 12.1 |
| $H_\infty$ (MJ kg$^{-1}$) | 1.7 | 14.4 |
| $p_{t_\infty}$ (MPa) | 7.0 | 56.0 |
| $T_{t_\infty}$ (K) | 1520 | 7400 |
| $Re_\infty \times 10^{-6}$ (m$^{-1}$) | 0.8 | 4.1 |
| Test time (ms) | 2 | 10 |

diaphram consisting of two 10 mm thick stainless steel plates (at maximum driver pressure of 150 MPa) and driven section initially terminated by brass or copper diaphragm. The driven section connects to a 5.8° conical nozzle with a exit diameter of 586 mm. The detonation shock tunnel comprises a 6.4 m damping section, a 9.4 m detonation section and a 15.4 m driver section. The operational characteristics are listed in table A.74.

## A.31 Sandia National Laboratories

### A.31.1 Hypersonic Wind Tunnel (HWT)

Beresh *et al* (2015) summarize the design and operating characteristics of the Sandia National Laboratory's Hypersonic Wind Tunnel (HWT). The facility began operation in 1962 and was further modernized in 1977 and subsequent to 2015. The schematic of the wind tunnel is shown in figure A.67 from Beresh *et al* (2015). The facility is a blow down wind tunnel operating at Mach 5, 8 or 14 using air ($M = 5$) or nitrogen ($M = 8$ and 14). The gas is heated using a 3 MW electrical resistance heater that replaced an earlier pebble bed heater using propane heated alumina pebbles. The typical run time is 30 s to 40 s with the gas exhausting to vacuum spheres with a total volume of 850 m$^3$. The operating conditions are listed in table A.75.

### A.31.2 Hypersonic Shock Tunnel (HST)

Lynch *et al* (2022) discuss the Hypersonic Shock Tunnel (HST) at Sandia National Laboratories. The facility is a free-piston-driven reflected shock tunnel. A schematic of the facility, commissioned in 2021, is presented in figure A.68 from Lynch *et al* (2022). The driver tube is 5.9 m in length with a diameter of 0.27 m, and the driven tube is 7.3 m in length with a diameter of 0.082 m. The current configuration uses a 11.9 kg piston. The 7.9° conical nozzle with exit diameter 0.36 m discharges into a 0.5 m diameter test section of length 1.4 m. The air exhausts into a dump tank of 0.98 m diameter and 2.7 m height. The operating conditions are listed in table A.76

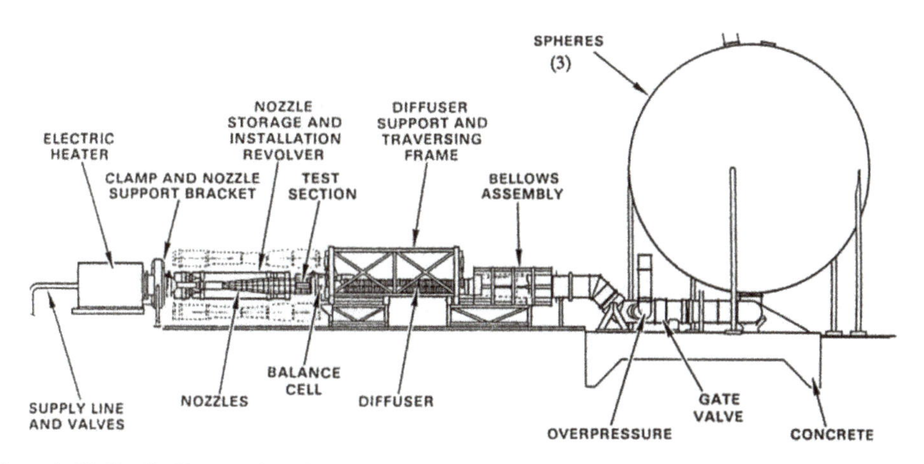


**Figure A.67.** Sandia Hypersonic Wind Tunnel. Reproduced from Beresh *et al* (2015). Copyright Steven J Beresh.

**Table A.75.** Operating conditions.

| Quantity | Mach 5 | Mach 8 | Mach 14 |
|---|---|---|---|
| Gas | Air | $N_2$ | $N_2$ |
| $p_{t_\infty}$ (MPa) | 1.38[a] | 6.89[a] | 20.68[a] |
| $T_{t_\infty}$ (K) | 900[a] | 900[a] | 1400[a] |
| Test section diameter (m) | 0.46 | 0.36 | 0.46 |
| $Re_\infty \times 10^{-6}$ ($m^{-1}$) | 3 to 30 | 3 to 20 | 1 to 4 |
| Test time (s) | 60[a] | | |

[a] Maximum.

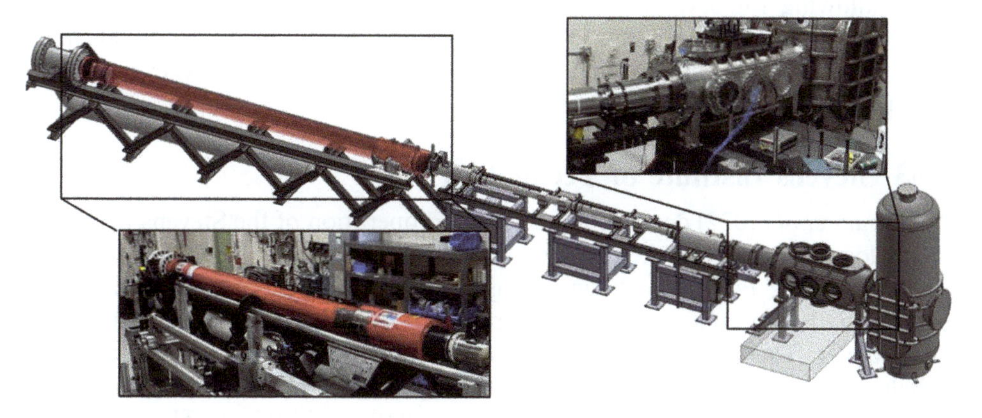

**Figure A.68.** Sandia Hypersonic Shock Tunnel. Copyright Kyle Lynch.

**Table A.76.** Operating conditions.

| Quantity | Value |
|---|---|
| $M_\infty$ | 9 to 12 |
| $H_\infty$ (MJ kg$^{-1}$) | 4.5 and 9 |
| $p_{t_\infty}$ (MPa) | 12 to 18 |
| $T_{t_\infty}$ (K) | 3000 to 6000 |
| Test time (ms) | 1 |

(Lynch 2022). Upgrades to the piston and driver tube are underway to increase run time and stagnation conditions (Lynch 2022).

## A.32 Stanford University

### A.32.1 Stanford 3.5 inch Expansion Tube

Ben-Yakar and Hanson (2002) described the design and operation of the Stanford University 3.5 inch (89 mm) Expansion Tube. A schematic of the facility is shown in figure A.69. The facility is 12 m in length including the dump tank. The driver and expansion gases are helium, with the driver tube diameter of 101.6 mm, while the driven and expansion tube diameters are 88.9 mm. The square test section is 27 cm ×27 cm. Operating conditions are shown in table A.77.

### A.32.2 Stanford Six Inch Expansion Tube

Heltsley *et al* (2006) describe the characteristics and performance of the Stanford University Six Inch Expansion Tube. The driver, driven and expansion tubes are each comprised of multiple individual sections of length 0.9 m (driver), 1.2 m (driven) and 2.7 m (expansion). The total overall length of the combined driver, driven and expansion tubes is 12 m. The interchangable sections are mounted on rollers allowing for several hundred different possible configurations. The inner diameter is 14 cm. The test section is connected to a dump tank of radius 30 cm and height of 122 cm. The facility began operation in August 2005 (Heltsley *et al* 2006). The operating conditions are listed in table A.78.

## A.33 Stevens Institute of Technology

Shekhtman *et al* (2022) describe the design and operation of the Stevens Institute of Technology Shock Tunnel. The facility is located in Hoboken, New Jersey. A schematic of the facility is shown in figure A.70. The tunnel comprises a driver section (with optional extension), driven section, nozzle, test section and dump tanks. The facility is designed to operate at Mach 6 with a stagnation enthalpy of 1.5 MJ kg$^{-1}$. The driver section length is 5.0 m (with an additional 5.0 m optional extension), and the driven section length is 11.07 m. The test section inner diameter is 0.610 m. The test section rhombus is approximately 0.33 m high

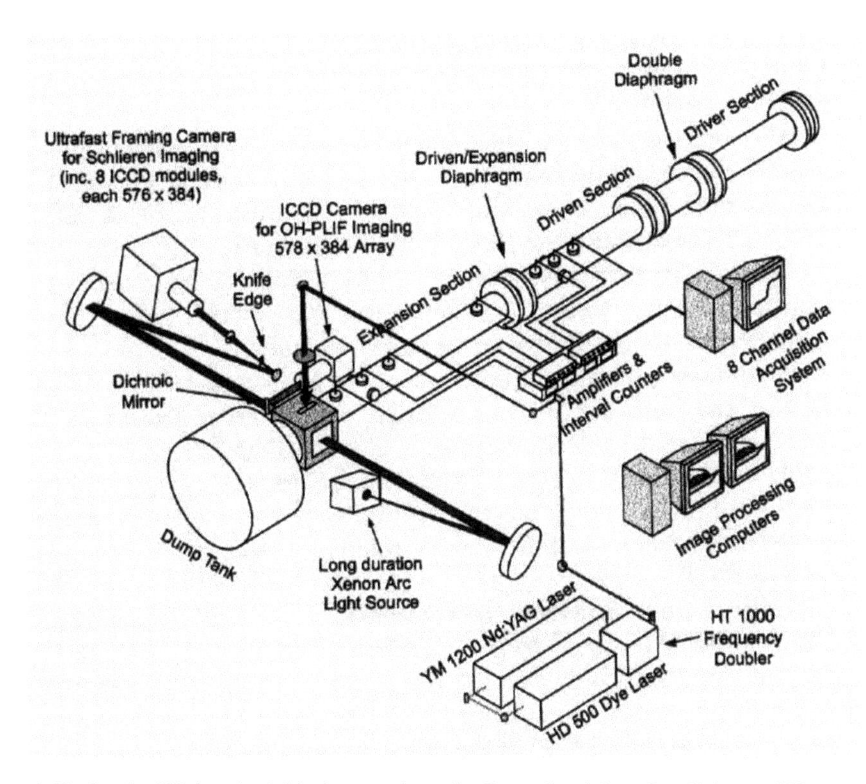


**Figure A.69.** Stanford University 3.5 inch expansion tube. Reproduced from Ben-Yakar and Hanson (2002). Copyright Ben-Yakar and Hanson.

**Table A.77.** Operating conditions.

| Quantity | Value |
|---|---|
| $M_\infty$ | 8 to 12 |
| $H_\infty$ (MJ kg$^{-1}$) | 3 to 6 |
| $p_{t_\infty}$ (MPa) | 3.3[a] |
| $T_{t_\infty}$ (K) | 4188[a] |
| Test time ($\mu$s) | 170 to 400[b] |

[a] Maximum (Gu and Olivier 2020).
[b] Gu and Olivier (2020).

and 1.5 m long. The combined dump tank volume is 8.78 m$^3$. Operational characteristics are indicated in table A.79.

## A.34 TUB hypersonic Ludwieg tube (HLB)

Radespiel *et al* (2016) describe the hypersonic Ludwieg tube facility at Technische Universität Braunschweig (TUB). See also Estorf *et al* (2005), Wolf *et al* (2007) and

**Table A.78.** Operating conditions.

| Quantity | Min | Max |
| --- | --- | --- |
| $M_\infty$ | >1 | >8 |
| $H_\infty$ (MJ kg$^{-1}$) | na | 8 |
| $T_{t_\infty}$ (K) | na | na |
| $p_{t_\infty}$ (MPa) | na | na |

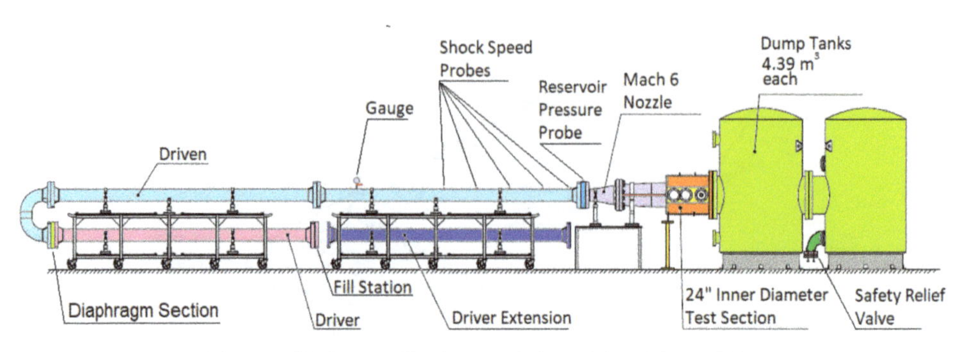


**Figure A.70.** Stevens Shock Tunnel. Copyright Nick Parziale.

**Table A.79.** Operating conditions.

| Quantity | Value |
| --- | --- |
| $M_\infty$ | 6 |
| $H_\infty$ (MJ kg$^{-1}$) | 1.5 |
| $p_{t_\infty}$ (MPa) | 0.345 to 10.3 |
| $Re_\infty \times 10^{-6}$ (m$^{-1}$) | 0.35 to 8.1 |
| Test time (ms) | 4 |

Marineau *et al* (2019). A photo of the wind tunnel is shown in figure A.71. The high pressure section is 17 m length with a 3 m heated section. The test section diameter is 496 mm and the length is 940 mm. The air exhausts to a 6 m$^3$ dump tank. The facility was extended to allow experiments of a propulsive jet, and also extended by a tandem nozzle to provide Mach 3 capability (Stephan *et al* 2015, Wu *et al* 2015, Stephan and Radespiel (2017). The operating conditions are listed in table A.80 (Radespiel 2022).

**Figure A.71.** Photo of HLB. Copyright Rolf Radespiel.

**Table A.80.** Operating conditions.

| Quantity | Value | |
|---|---|---|
| | Min | Max |
| $M_\infty$ | 6 | 6 |
| $p_{t_\infty}$ (MPa) | 0.35 | 2.84 |
| $T_{t_\infty}$ (K) | 420 | 485 |
| $Re_\infty \times 10^{-6}$ ($m^{-1}$) | 3 | 30 |
| Run time (ms) | | 80 |
| Test time (ms) | | 60 |

# A.35 Texas A&M University

## A.35.1 Mach 6 Quiet Tunnel (M6QT)

Hofferth *et al* (2010) describe the design and operational characteristics of the Texas A&M University National Aerothermochemistry and Hypersonics Laboratory Mach 6 Quiet Tunnel (M6QT). The blow down facility is the former Mach 6 Quiet Tunnel at NASA Langley Research Center (section A.22.2) and was moved to Texas A&M in 2005. A schematic (notional scale) of the test section is shown in figure A.72. A laminar boundary layer is maintained on the inner surface of the nozzle for a sufficient distance to achieve a test rhombus free of disturbances arising

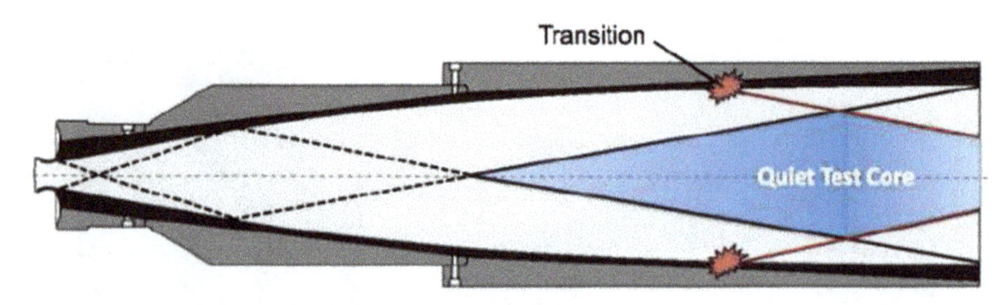


**Figure A.72.** Mach 6 Quiet Tunnel. From Hofferth *et al* (2010); reprinted by permission of the American Institute of Aeronautics and Astronautics, Inc.

**Table A.81.** Operating conditions.

| Quantity | Min | Max |
|---|---|---|
| $M_\infty$ | 6 | 6 |
| $T_{t_\infty}$ (K) | 450[a] | 450 |
| $Re_\infty \times 10^{-6}$ ($m^{-1}$) | 3.0[b] | 11.0[b] |
| Runtime (s) | 40 | |

[a] Typical (Hofferth *et al* 2010).
[b] Quiet flow (Bowersox 2022c).

from boundary layer transition of the nozzle wall boundary layer. The nozzle exit diameter is 19 cm (Hofferth *et al* 2010). The air exhausts through a two stage air ejector system. The operational characteristics are listed in table A.81.

### A.35.2 Hypervelocity Expansion Tunnel (HXT)

Dean *et al* (2022) describe the design and operating characteristics of the Hypervelocity Expansion Tunnel (HXT) at the Texas A&M National Aerothermochemistry and Hypersonics Laboratory. A schematic of the facility is shown in figure A.73. The overall length of the tunnel is 30.5 m and the diameter is 48 cm. The driver section length is 1.5 m and can be pressurized to 13.8 MPa using either helium or air. The driven section length is 4.6 m, and the acceleration length is 9.1 m in length. The diverging nozzle is 6.1 m long with its diameter expanding from 48 cm to 91 cm. The test section length is 1.8 m with a square cross section of 1.5 m ×1.5 m. The dump tank is 1.06 m in diameter and 6.1 m long. The tunnel can also be reconfigured as a reflected shock tunnel. The operating conditions are listed in table A.82. The tunnel has operated at Mach numbers up to 23, and upgrades in progress will increase maximum stagnation enthalpy over 30 MJ $kg^{-1}$.

### A.35.3 Actively Controlled Expansion Tunnel (ACE)

Semper *et al* (2009), Tichenor *et al* (2010), Semper *et al* (2012) and Leidy *et al* (2020) describe the design and operational characteristics of the Actively Controlled

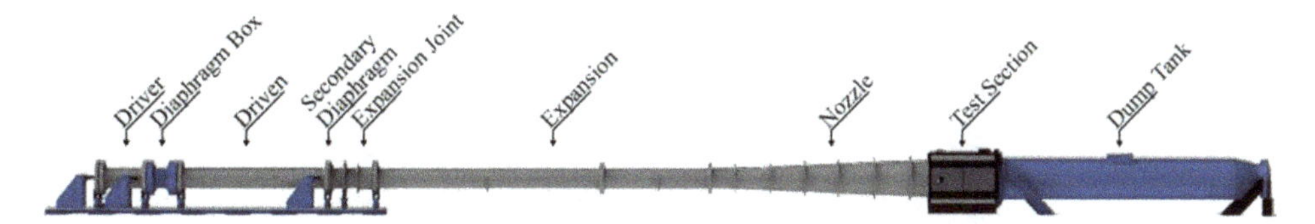


**Figure A.73.** Hypervelocity Expansion Tunnel (HXT). From Dean *et al* (2022); reprinted by permission of the American Institute of Aeronautics and Astronautics, Inc.

**Table A.82.** Operating conditions.

| Quantity | Min | Max |
|---|---|---|
| $M_\infty$ | 3[a] | 23[b] |
| $H_\infty$ (MJ kg$^{-1}$) | 0.4[c] | 33[d] |

[a] Effective range from CFD (Dean *et al* (2022).
[b] Bowersox (2022a).
[c] Dean *et al* (2022).
[d] Bowersox (2022b).

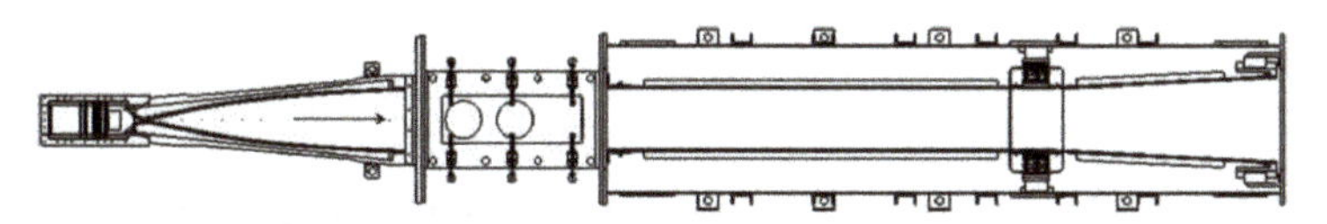

**Figure A.74.** Actively Controlled Expansion (ACE) wind tunnel. Reproduced from Leidy *et al* (2020). Copyright 2020 by Andrew Leidy and Rodney Bowersox.

Expansion (ACE) hypersonic wind tunnel at the Texas A&M University National Aerothermochemistry and Hypersonics Laboratory. A schematic of the test section is shown in figure A.74. A unique feature of the facility is the capability to actively control the Mach number during operation of the tunnel. The test cross section is 22.86 cm × 35.56 cm. Change in the test section Mach number is achieved by varying the nozzle throat height using a flexure facility designed and fabricated at Texas A&M. The operating conditions are listed in table A.83.

## A.36 United States Air Force Academy Mach 6 Ludwieg Tube M6LT

Cummings and McLaughlin (2012) and Decker *et al* (2015) describe the design and operation of the United States Air Force Academy (USAFA) Mach 6 Ludwieg Tube

**Table A.83.** Operating conditions.

| Quantity | Min | Max |
|---|---|---|
| $M_\infty$ | 5 | 8[a] |
| $p_{t_\infty}$ (MPa) | 0.14 | 1.03 |
| $T_{t_\infty}$ (K) | 300 | 533 |
| $Re_\infty \times 10^{-6}$ ($m^{-1}$) | 0.5[b] | 10.0[c] |
| Runtime (s) | 50 | |

[a] Tichenor *et al* (2010).
[b] Bowersox (2022b).
[c] Bowersox (2022a).

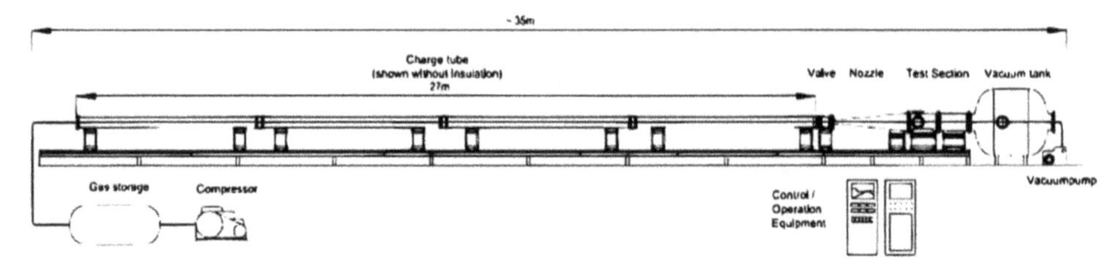


**Figure A.75.** USAFA Mach 6 Ludwieg Tube M6LT. Reproduced from Decker *et al* (2015). Copyright US Government. Image stated to be in public domain.

(M6LT) at Colorado Springs, Colorado. A schematic of the facility is shown in figure A.75 and a photograph in figure A.76. The facility comprises a compressor, gas storage, charge tube, nozzle, test section and vacuum tank. There is also a dryer between the compressor and gas storage. The charge tube is 27 m in length and can be heated up to 673 K. The test section is 0.496 m in diameter. A fast acting valve at the nozzle throat initiates the flow. The operating conditions are shown in table A.84.

## A.37 University of Arizona Mach 5 Ludwieg Tube

Bearden *et al* (2022) describe the design and operation of the Mach 5 Ludwieg Tube (LT5) at the University of Arizona. A schematic of the facility is presented in figure A.77. The facility components include a compressor, driver tube, nozzle, test section, diffuser and vacuum chamber. The pressure and vacuum infrastructure is shared with the Mach 4 Quiet Ludwieg Tube (QLT4). The driver tube length is 25.4 m with a 0.4 m inner diameter. The nozzle diameter is 0.381 m with a length of 3.5 m. The driver tube can be heated up to 453 K. The operating conditions are indicated in table A.85.

**Figure A.76.** Air Force Cadet 2nd Class Eric Hembling examining the USAFA M6LT. Reproduced from Armstrong (2019). Copyright US Government. Image stated to be in public domain.

**Table A.84.** Operating conditions.

| Quantity | Min | Max |
|---|---|---|
| $M_\infty$ | 6 | 6 |
| $p_{t_\infty}$ (MPa) | 0.5 | 4.0 |
| $T_{t_\infty}$ (K) | 300 | 673 |
| $Re_\infty \times 10^{-6}$ (m$^{-1}$) | 2.5 | 30 |
| Run time (ms) | 100 | 150 |

## A.38 University of Delft Hypersonic Ludwieg tube

Schrijer and Bannink (2008) discusses the design and operation of the Hypersonic Test Facility Delft (HTFD) Ludwieg tube. A photograph of the facility is shown in figure A.78. The facility comprises a storage tube, fast acting valve, nozzle, test section and vacuum tank. The storage tube is a reflexed type comprising two sections joined by a 180° bend. The total length of the storage tube is 29 m comprising a cold section (23 m) and hot section (6 m). Only the portion of the tube corresponding to the mass flow into the test section during the test is heated. The diameters of the cold and hot tube are 59 mm and 49.25 mm, respectively. The nozzle is conical with a 7° half angle. The test section diameter is 30 cm, with a usable section 20 cm ×20 cm. Operating conditions are listed in table A.86.

## A.39 Universität der Bundeswehr München HELM

Mundt (2016) describes the design and operation of the Universität der Bundeswehr München (University of Federal Armed Forces München) High Enthalpy

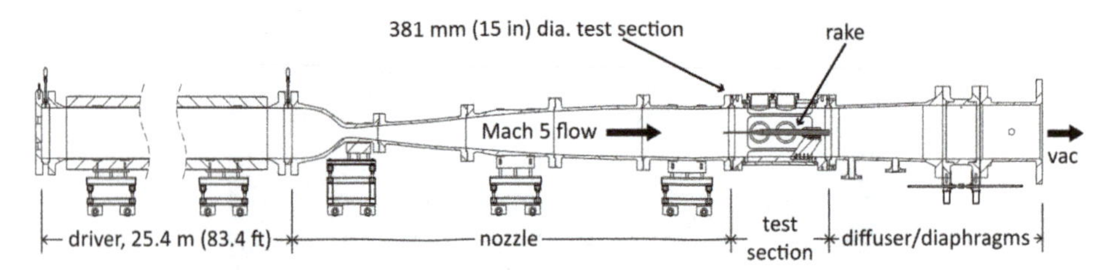


**Figure A.77.** University of Arizona Mach 5 Ludwieg Tube (Bearden *et al* 2022). Copyright A Craig.

**Table A.85.** Operating conditions.

| Quantity | Min | Max |
|---|---|---|
| $M_\infty$ | 5 | 5 |
| $p_{t_\infty}$ (MPa) | 0.34 | 2.07 |
| $T_{t_\infty}$ (K) | 385 | 450 |
| $Re_\infty \times 10^{-6}$ (m$^{-1}$) | 5.2 | 39.9 |
| Test time (ms) | 100 | |

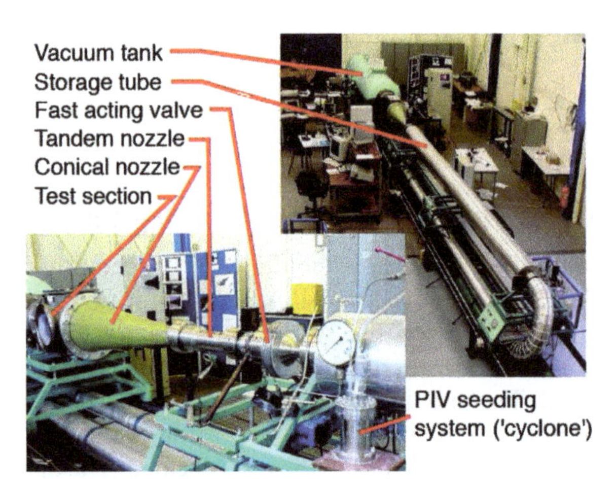


**Figure A.78.** University of Delft HTFD Ludwieg tube. From Schrijer and Bannink (2008); reprinted by permission of the American Institute of Aeronautics and Astronautics, Inc.

Laboratory Munich (HELM) free piston shock tunnel. The facility began operation in 2010. A photograph of the driver tube is shown in figure A.79. The compression tube length is 21 m with a diameter of 286 mm. Variable shock tube lengths from 8 m to 11 m are available with diameter of 95 mm. The sections are mounted on

**Table A.86.** Operating conditions.

| Quantity | Min | Max |
|---|---|---|
| $M_\infty$ | 6 | 11 |
| $p_{t_\infty}$ (MPa) | 0.28 | 8.92 |
| $T_{t_\infty}$ (K) | 300 | 585 |
| $Re_\infty \times 10^{-6}$ ($m^{-1}$) | 1.6 | 19.7 |
| Runtime (ms) | ⩽200 | |

movable chariots to facilitate interchangability. The nozzle exit diameter is 684 mm. The operating conditions are listed in table A.87.

## A.40 University of Notre Dame

### A.40.1 AFOSR-Notre Dame Large Mach 6 Quiet Tunnel

Lakebrink *et al* (2018) and Lax and Leonov (2022) describe the design of the current AFOSR-Notre Dame Large Mach 6 Quiet Tunnel and the Mach 10 Quiet Tunnel under development at the University of Notre Dame. A schematic of the Mach 6 Quiet Tunnel is shown in figure A.80. The facility is a Ludwieg tube. The driver tube is 60 m in length with a diameter of 60 cm. The converging diverging nozzle has an exit diameter of 60 cm. The air exhausts into an open test section which is connected to a dump tank with diameter 3.7 m and length of 16 m. Opening the fast acting shutter value at the contraction exit initiates the flow into the nozzle and test section. The design operating conditions are shown in table A.88.

### A.40.2 Arc heated hypersonic tunnel

Baccarella *et al* (2016) and Hoberg *et al* (2019) describe the design and operational characteristics of the Arc-heated Combustion Test-rig (ACT-1) at the University of Notre Dame. Hedlund *et al* (2018) utilized this facility for a hypersonic transition and shock wave boundary layer interaction study. Panoramic views of ACT-1 is shown in figure A.81(a) and the arc heater in figure A.81(b). A 260 kW DC power supply supplies the pulsed electric arc heater with discharge currents up to 500 A. The heated gas is mixed with oxygen or other gases as required and flows through a constant area duct (plenum) to the converging diverging nozzle. The facility is equipped with contoured axisymmetric nozzles for Mach 4.5, 6 and 9 (air). The operating conditions are listed in table A.89. In 2021-2022 this facility underwent significant modification including increasing the stagnation pressure up to 3 MPa, stagnation temperature up to 6000 K due to replacement of the gas heater with the Huels type arc heater. The facility was rebranded ND ArcJet.

**Figure A.79.** HELM. Copyright Christian Mundt.

**Table A.87.** Operating conditions.

| Quantity | Value |
| --- | --- |
| $U_\infty$ (km s$^{-1}$) | 5 to 7 |
| $H_\infty$ (MJ kg$^{-1}$) | 20 to 25 |
| $p_{t_\infty}$ (MPa) | 100 |
| Test time (ms) | <2 |

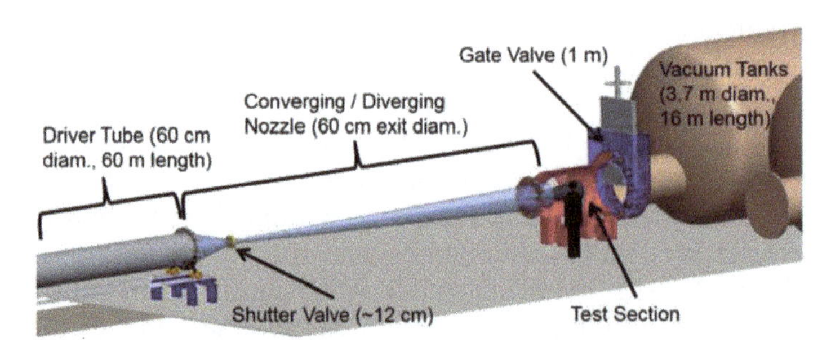


**Figure A.80.** AFOSR-Notre Dame Large Mach 6 Quiet Tunnel. Reprinted with permission. Reproduced from Lakebrink *et al* (2018). Copyright 2017 by The Boeing Company. Published by the American Institute of Aeronautics and Astronautics, Inc., with permission.

**Table A.88.** Operating conditions.

| Quantity | Value |
|---|---|
| $M_\infty$ | 6 |
| $p_{t_\infty}$ (MPa) | 1[a] |
| $T_{t_\infty}$ (K) | 430[b] |
| $Re_\infty \times 10^{-6}$ (m$^{-1}$) | 11[a] |

[a]Nominal (Lakebrink *et al* 2018).
[b]Minimum (Lakebrink *et al* 2018).

(a) ACT-1 (ND ArcJet)

(b) ACT-1 (ND ArcJet)

**Figure A.81.** ACT-1 (ND ArcJet). Photos courtesy of Dr Sergey Leonov.

## A.41 University of Texas at Austin

Erengil and Dolling (1993), Brusniak and Dolling (1994), Unalmis *et al* (2001) and Murphree *et al* (2021) describe the the University of Texas at Austin Mach 5 Hypersonic Tunnel. The tunnel is a blow down facility. Two banks of nichrome wires each dissipating 400 kW heat the air in the stagnation chamber. The heated air exhausts through a converging-diverging nozzle to the rectangular test section which is 17.78 cm in height and 15.24 cm in width (Schmisseur and Dolling 1992). The operating conditions are listed in table A.90.

## A.42 University of Texas at San Antonio Mach 7 Hypersonic Wind Tunnel

Bashor *et al* (2019), Hoffman *et al* (2020, 2021) and Hoffman *et al* (2022) describe the design and operation of the University of Texas at San Antonio Mach 7 Hypersonic Wind Tunnel. A schematic of the facility is shown in figure A.82. The

**Table A.89.** Operating conditions.

| Quantity | $M = 4.5$ | $M = 6$ | $M = 9$ |
|---|---|---|---|
| $p_{t_\infty}$ (MPa) | 0.05 to 3.0[a] | 0.05 to 3.0 | 0.05 to 3.0 |
| $T_{t_\infty}$ (K) | 1000 to 6000[a] | 1000 to 6000 | 1000 to 6000 |
| Test time (s) | 1[b] | 1[b] | 0.4[b] |
| Exit diameter (mm) | 60.5 | 67.6 | 181 |

[a] Leonov (2022).
[b] Maximum (Baccarella *et al* 2016).

**Table A.90.** Operating conditions[a].

| Quantity | Min | Max |
|---|---|---|
| $M_\infty$ | 4.92 | 4.95 |
| $p_{t_\infty}$ (MPa) | 2.4 | 2.6 |
| $T_{t_\infty}$ (K) | 335 | 365 |
| $Re_\infty \times 10^{-6}$ (m$^{-1}$) | 52 | 62 |
| Runtime (s) | 10 | 90 |

[a] Clemens (2022).

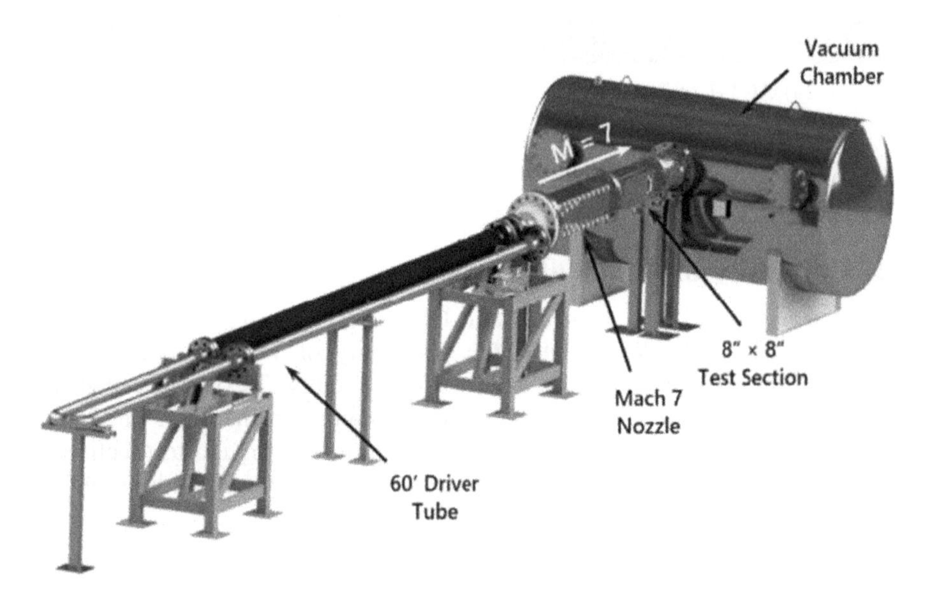


**Figure A.82.** University of Texas at San Antonio Mach 7 Ludwieg Tube. Copyright Christopher Combs.

facility is a Ludwieg tube. The driver tube diameter is 10.2 cm and the length is 18.3 m. A two-dimensional converging-diverging nozzle accelerates the gas to Mach 7.2 in a test section whose cross section is 20.32 cm ×20.32 cm and length is 97.5 cm. The vacuum tank capacity is 6.06 m$^3$. The test gas is typically air, although it is

**Table A.91.** Operating conditions.

| Quantity | Min | Max |
|---|---|---|
| $M_\infty$ | 7.2 | 7.2 |
| $p_{t_\infty}$ (MPa) | 0.34 | 13.8 |
| $T_{t_\infty}$ (K) | 700[a] | 700 |
| $Re_\infty \times 10^{-6}$ (m$^{-1}$) | 0.5 | 196.9 |
| Test time[b] (ms) | 65 | 100 |

[a] Maximum.
[b] Steady state.

possible to test with nitrogen and other gases. The operating conditions are described in table A.91.

## A.43 University of Southern Queensland

Buttsworth (2009) describes the design and operation of the free-piston compression-heated Ludwieg tube at the University of Southern Queensland, Toowoomba, Australia (TUSQ). The facility can be operated as either a Ludwieg tube (with or without free piston compression) or an atmosphere pressure blow down tunnel. A schematic of the Ludwieg tube is shown in figure A.83. The Ludwieg tube comprises a 16 m long tube. Piston masses of 80 gram and higher are used. In the Ludwieg tunnel operational mode, the pressure in the tube is slowly increased until the diaphragm at the entrance to the nozzle is ruptured and the gas exhausts through the converging-diverging nozzle. In the free-piston-driven Ludwieg tunnel operational mode, the piston is initially positioned downstream of the primary valve separating the high pressure air reservoir and the Ludwieg tube. Upon opening the valve, the piston is driven down the tunnel, the test gas is compressed, the diaphragm at the entrance to the nozzle bursts and the test gas exhausts through the converging-diverging nozzle. Different contoured nozzles are used depending on the mode of operation and test section Mach number (table A.92). The test section is 60 cm in diameter and 83 cm in length. The Mach 6 operating conditions are listed in table A.93.

## A.44 University of Queensland

### A.44.1 T4

Hannemann *et al* (2016) describe the design and operational characterisics of the University of Queensland free piston shock tunnel T4, also known as the T4 Stalker tube. The facility was designed for testing of scramjets and was commissioned in 1987. A schematic of the tunnel is shown in figure A.84. The compression tube is 26 m in length with an inner diameter of 229 mm. The shock tube is 10 m in length with an inner diameter of 79 mm. The air reservoir that drives the piston (92 kg) surrounds the compression tube. The converging-diverging nozzles vary in exit

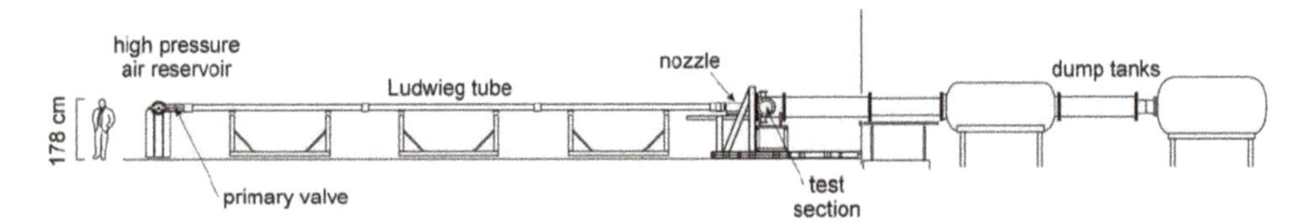


**Figure A.83.** University of Southern Queensland Ludwieg Tube. Copyright David Buttsworth.

**Table A.92.** Nozzle configurations.

| Operational mode | $M_\infty$ | $D$ (mm) |
|---|---|---|
| Blowdown | 2 | 81.3 |
| | 3 | 225.6 |
| | 4.5 | 215.9 |
| Ludwieg | 2 | 41.2 |
| | 4 | 128.6 |
| | 6 | 217.6 |
| | 7 | 217.7 |

**Table A.93.** ROC[a] for Mach 6 nozzle.

| Quantity | Value |
|---|---|
| $M_\infty$ | 5.95 |
| $p_{t_\infty}$ (MPa) | 0.25 to 5.0 |
| $T_{t_\infty}$ (K) | 565 |
| $Re_\infty \times 10^{-6}$ (m$^{-1}$) | 1.75 to 35.1 |
| Test time (ms) | >100 |

[a] Rated operating conditions.

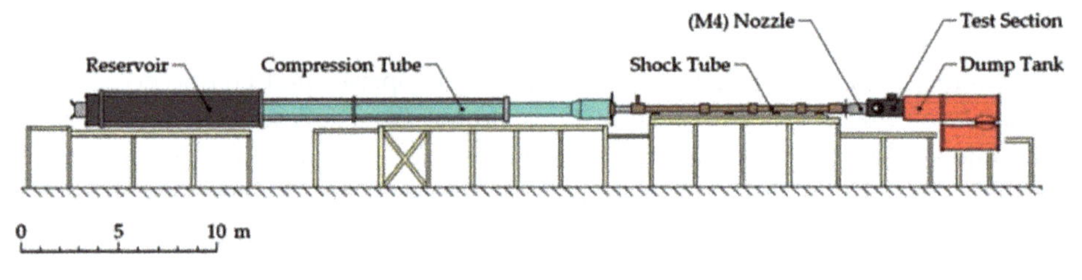


**Figure A.84.** T4 Stalker tube. Reporduced from Hannemann *et al* (2016). With permission of Springer.

**Table A.94.** Operating conditions.

| Quantity | Min | Max |
|---|---|---|
| $M_\infty$ | 4 | 10 |
| $H_\infty$ (MJ kg$^{-1}$) | 2.5 | 12 |
| $p_{t_\infty}$ (MPa) | 50[a] | 90 |

[a]Typical.

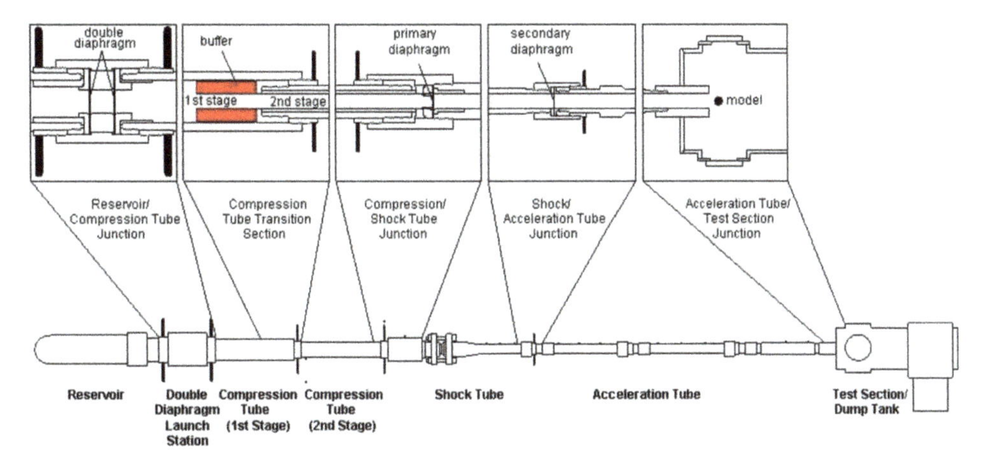


**Figure A.85.** X2 with compound piston. From Scott *et al* (2005); reprinted by permission of the American Institute of Aeronautics and Astronautics, Inc.

diameter from 135 mm at Mach 4 to 375 mm at Mach 10. The operating conditions are shown in table A.94. See also Paull *et al* (1995) and Stalker (2006).

### A.44.2 X2

Gildfind *et al* (2016) describe the University of Queensland X2 Expansion Tunnel. The expansion tunnel was commissioned in 1995 (Scott *et al* 2005) following earlier developments with small expansion tunnels TQ and X1. The total length is 25 m and the driven tube diameter is 85 mm. The original compound piston driver (figure A.85) was replaced by a 35 kg single piston (figure A.86), and further changed to a 10.5 kg single piston. The total length of the facility is 17 m. A contoured nozzle provides a design exit $M = 10$ for an inflow $M = 7.3$. The facility is capable of reproducing freestream conditions for Earth reentry at velocities up to 15 km s$^{-1}$ (Lefevre *et al* 2021). The facility can also be operated as non-reflected shock tube with driven sections of 85 mm and 150 mm diameter (Jacobs *et al* 2015), Morgan (2022). Sample operating conditions are shown in table A.95.

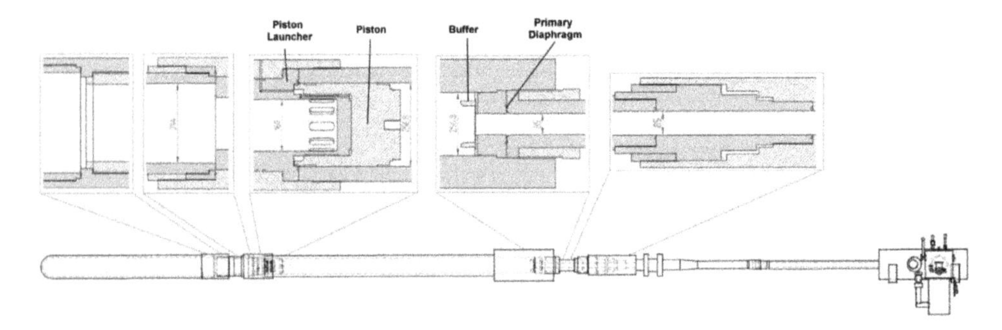


**Figure A.86.** X2 with single piston. From Scott *et al* (2005); reprinted by permission of the American Institute of Aeronautics and Astronautics, Inc.

**Table A.95.** Operating conditions[a].

| Quantity | Min | Max |
|---|---|---|
| $M_\infty$ | 5 | 14 |
| $p_{t_\infty}$ (MPa) | 5 | $10^4$ |
| $T_{t_\infty}$ (K) | 2000 | 20 000 |
| Test time ($\mu$s) | 50 | 500 |

[a] Morgan (2022).

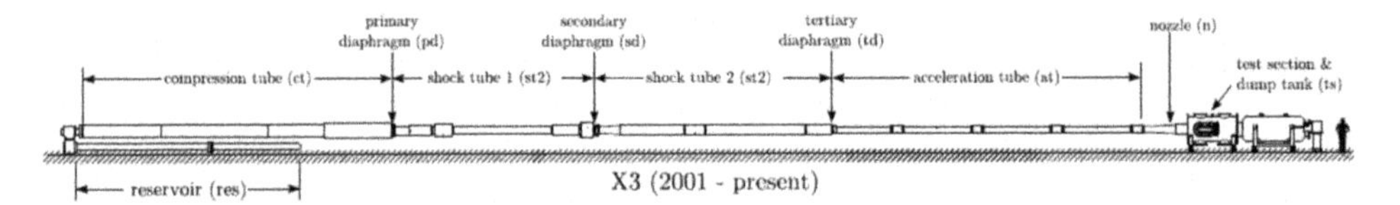


**Figure A.87.** X3. Reproduced from Gildfind *et al* (2016) with permission of Springer.

### A.44.3 X3

Gildfind *et al* (2016) describe the University of Queensland X3 Expansion Tunnel. A scaled schematic of the facility is shown in figure A.87, and an unscaled general layout is presented in figure A.88 showing the overall longitudinal size. The tunnel was commmissioned in 2001. The contoured Mach 10 nozzle has an exit diameter of 440 mm, and a Mach 12 nozzle has an exit diameter of 600 mm (Morgan 2022). A 200 kg piston was installed in 2011. The facility can also be operated as a non-reflected shock tube (Morgan *et al* 2006). The operating conditions are listed in table A.96.

### A.44.4 X3R

Stennett *et al* (2018, 2019) and Stennett (2020) describe the X3R reflected shock tunnel at the University of Queensland. The X3R is an alternate operating mode of

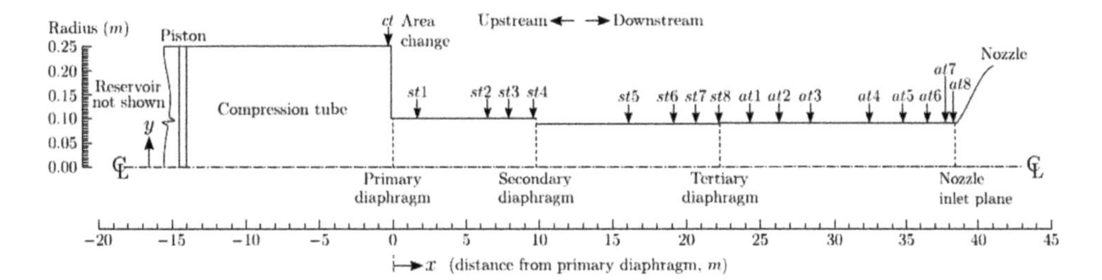


**Figure A.88.** X3 general layout. Reproduced from Gildfind *et al* (2016) with permission of Springer.

**Table A.96.** Operating conditions[a].

| Quantity | Min | Max |
|---|---|---|
| $M_\infty$ | 5 | 15 |
| $H_\infty$ (MJ kg$^{-1}$) | >100 | |
| $p_{t_\infty}$ (MPa) | 5 | 1000 |
| $T_{t_\infty}$ (K) | 2000 | 15 000 |
| Test time (ms) | 0.2 | 1.0 |

[a] Morgan (2022).

the existing X3 tunnel, and was designed for testing of scramjet engines. A schematic of the facility is presented in figure A.89. The driver tube is 14 m in length with an inner diameter of 500 mm, and the shock tube is 22 m in length with a variable inner diameter of 180 mm to 200 mm (Stennett *et al* 2019). Pistons with mass of 100 kg, 200 kg, 280 kg and 560 kg are used to optimize test time at different freestream conditions. The Mach 7 nozzle has an exit diameter of 757.2 m. The operating conditions are listed in table A.97.

## A.45 University of Toronto Institute for Aerospace Studies

The University of Toronto Institute for Aerospace Studies 27.94 cm × 38.1 cm (11 inch × 15 inch) hypersonic shock tunnel is described in Chan *et al* (1965). The tunnel comprises two parts, namely, (1) the shock tube driver and driven section, and (2) the wind tunnel section (figure A.90). The wind tunnel section comprises the expansion nozzle, test section and receiving tanks. The combustion driver of the shock tube, the driven section and test section are 2.13 m, 4.85 m and 6.71 m in length, respectively. A normal shock is generated in the driver tube by ignition of a hydrogen–oxygen mixture which bursts a diaphragm. The shock wave propagates into the driven section and reflects from the nozzle generating a region of high temperature gas that expands into the test section to the design Mach number. The operating conditions are shown in table A.98 (Wittliff 1987).

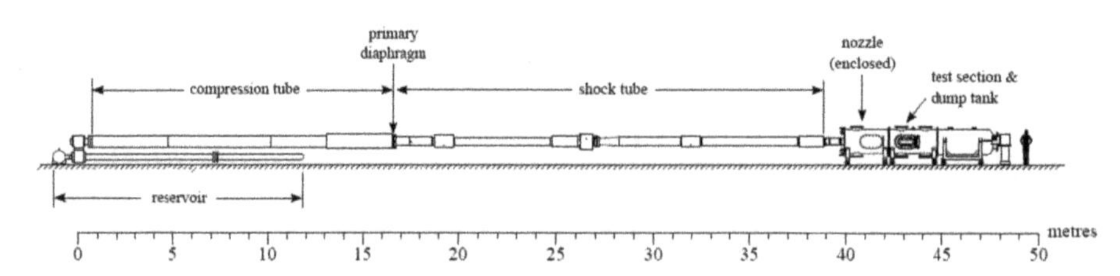


**Figure A.89.** X3R. Copyright Richard Morgan.

**Table A.97.** Operating conditions[a].

| Quantity | Value |
|---|---|
| $M_\infty$ | 7 |
| $p_{t_\infty}$ (MPa) | 13 |
| $T_{t_\infty}$ (K) | 6000 |
| Test time (ms) | 10 |

[a] Morgan (2022).

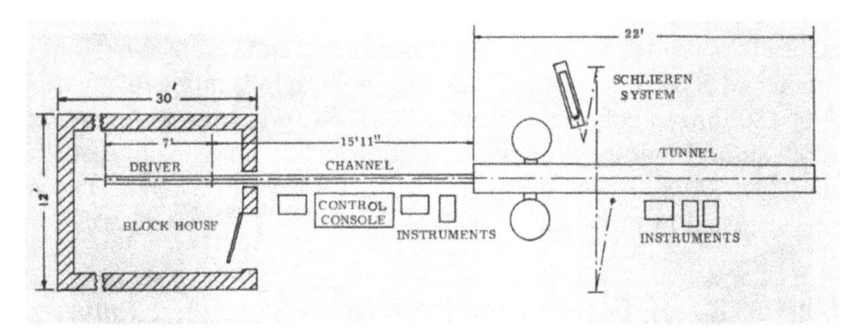


**Figure A.90.** UTIAS hypersonic shock tunnel. Reproduced from Chan *et al* (1965). Copyright US Government. Image stated to be in public domain.

## A.46 Virginia Tech hypersonic wind tunnel

Virginia Tech Hypersonic Wind Tunnel (2022) describes the design and operation of the Virginia Tech open jet blow down hypersonic wind tunnel located at Blacksburg, Virginia. A schematic of the facility is shown in figure A.91. The total length is 4.06 m. Test gases include air, nitrogen, argon, helium and other safe gases. An electric heater (220/380 V, 15-20 kW) increases the gas stagnation temperature up to 800 K for Mach 4 and above. The nozzle exit diameter is 100 mm. The test chamber dimensions are 360 mm × 226 mm × 200 mm. Operating conditions are indicated in table A.99.

**Table A.98.** Operating conditions.

| Quantity | Value |
|---|---|
| $M_\infty$ | 8.3 to 12.5 |
| $p_{t_\infty}$ (MPa) | 35.9[a] |
| $T_{t_\infty}$ (K) | 2111 |
| $Re_\infty \times 10^{-6}$ ($m^{-1}$) | 39.4[a] |

[a] Maximum.

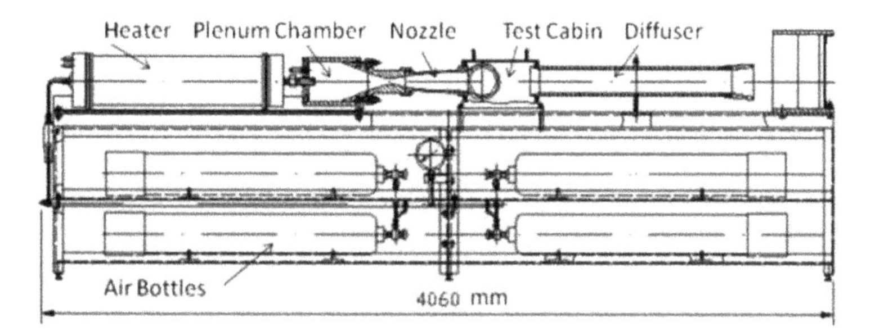


**Figure A.91.** Schematic of Virginia Tech hypersonic wind tunnel. Copyright Joseph Schetz.

**Table A.99.** Operating conditions.

| Quantity | Min | Max |
|---|---|---|
| $M_\infty$ | 2 | 7 |
| $p_{t_\infty}$ (MPa) | 0.25 | 12.7 |
| $T_{t_\infty}$ (K) | 290 | 645 |
| $Re_\infty \times 10^{-6}$ ($m^{-1}$) | 29.6 | 56.0 |
| Runtime (s) | 3 | 4 |

## A.47 von Kármán Institute for Fluid Dynamics

### A.47.1 Longshot tunnel

Richards and Enkenhus (1970), Simeonides (1990), Grossir *et al* (2014, 2015) and Grossir and Ilich (2018) describe the design and operation of the von Kármán Institute for Fluid Dynamics (VKI) Longshot Hypersonic Wind Tunnel. A schematic of the facility is shown in figure A.92. The facility was established in the 1960s and subsequently upgraded. The VKI Longshot is a gun tunnel. The facility comprises a driver tube of length 6.1 m, driven tube of length 27.5 m, reservoir, contoured or conical nozzle and open jet test section with volume 17 $m^3$.

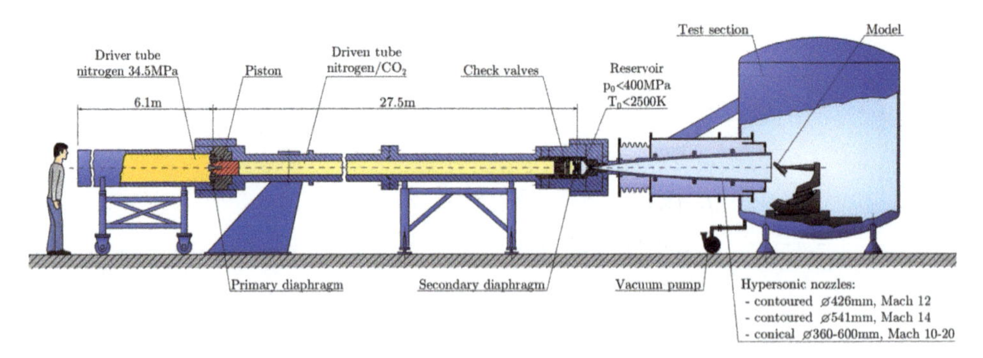


**Figure A.92.** VKI Longshot wind tunnel. Copyright Guillaume Grossir.

**Table A.100.** Operating conditions.

| Quantity | Value |
|---|---|
| $M_\infty$ | 12 to 24[a] |
| $p_{t_\infty}$ (MPa) | 400[b] |
| $T_{t_\infty}$ (K) | 2800[b] |
| $Re_\infty \times 10^{-6}$ ($m^{-1}$) | 3.5[c] to 12[c] |
| Running time (ms) | 20[c] |

[a] Richards and Enkenhus (1970).
[b] Maximum (Grossir 2022).
[c] Grossir *et al* (2015).

Two contoured nozzles of diameters 426 mm (Mach 12) and 541 mm (Mach 14) and 6° half-angle conical nozzles of diameters 360 mm to 600 mm (Mach 10 to 20) are used. The driver gas is nitrogen, and the driven gas can be nitrogen, carbon dioxide or helium. The piston mass can vary from 1.5 kg to over 4.5 kg (Grossir and Ilich 2018). The operating conditions are indicated in table A.100. See also Saavedra *et al* (2022).

### A.47.2 H3 Hypersonic Wind Tunnel

Vanhée (1989), Simeonides (1990) and Grossir *et al* (2015) describe the design and operation of the von Kármán Institute for Fluid Dynamics (VKI) H3 Hypersonic Wind Tunnel. A schematic of the facility is presented in figure A.93. The facility is a blowdown to vacuum wind tunnel. Air is heated to 500 K and expanded through a converging diverging nozzle to an open jet test section. The air exhausts through a supersonic ejector to atmosphere. A Mach 5 nozzle is under construction and is expected to be commissioned in 2023 (Grossir 2022). The operating conditions are indicated in table A.101.

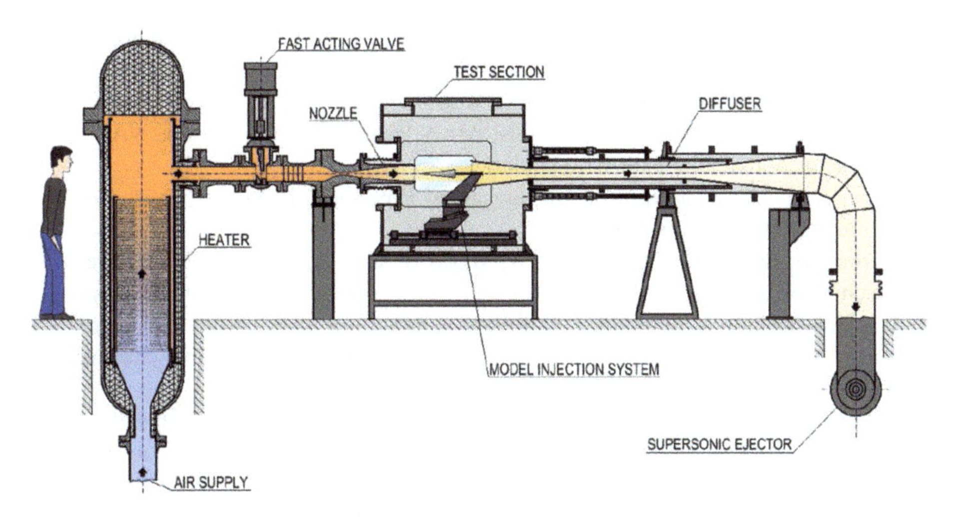


**Figure A.93.** VKI H3 Hypersonic Wind Tunnel. Reproduced from Grossir *et al* (2015). Copyright by G Grossir, D Masutti, O Chazot and the von Karman Institute for Fluid Dynamics.

**Table A.101.** Operating conditions.

| Quantity | Value |
|---|---|
| $M_\infty$ | 6 |
| $p_{t_\infty}$ (MPa) | 0.61 to 3.55 |
| $T_{t_\infty}$ (K) | 450 to 550 |
| $Re_\infty \times 10^{-6}$ ($m^{-1}$) | 6 to 30 |
| Running time (s) | 30 |

# References

Adamov N, Vasenev L, Zvegintsev V, Mazhul I, Nalivaichenko D, Novikov A and Kharitonov A 2006 Characteristics of the AT-303 hypersonic wind tunnel. Part 2: aerodynamics of the HB-2 reference model *Thermophys. Aeromech.* **13** 157–71

Albrechcinski T, Boyer D, Chadwick K and Lordi J 1995 Calspan's upgraded 96" hypersonic shock tunnel: its development and application in the performance of research and testing at higher enthalpies AIAA Paper 95-0236 (Reston, VA: American Institute of Aeronautics and Astronautics)

Anonymous 1990 Aerospace technology—technical data and information on foreign test facilities Report GAO/NSIAD-90-71-71FS, United States General Accounting Office, Washington, DC

Armstrong J 2019 Photo of Ludwieg tube https://www.defense.gov/Multimedia/Photos/igphoto/2002171170/

Baccarella D, Liu Q, Passaro A, Lee T and Do H 2016 Development and testing of the ACT-1 experimental facility for hypersonic combustion research *Meas. Sci. Technol.* **27** 045902

Baloga P and Nagamatsu H 1955 Instrumentation of GALCIT hypersonic wind tunnels *GALCIT Memorandum 29* California Institute of Technology, Pasadena, CA

Bashor I, Hoffman E, Gonzalez G and Combs C 2019 Design and preliminary calibration of the UTSA Mach 7 hypersonic Ludwieg tube AIAA Paper 2019-2859 (Reston, VA: American Institute of Aeronautics and Astronautics)

Baumgartner M 1997 Turbulence structure in a hypersonic boundary layer *PhD Thesis* (Princeton, NJ: Princeton University)

Baumgartner M, Erbland P, Etz M, Yalin A, Muzas B, Smits A, Lempert W and Miles R 1997 Structure of a Mach 8 turbulent boundary layer AIAA Paper 97-0765 (Reston, VA: American Institute of Aeronautics and Astronautics)

Bearden K, Padilla V, Taubert L and Craig S 2022 Calibration and performance characterization of a Mach 5 Ludwieg tube *Rev. Sci. Instrum.* **93** 085104

Beck W, Eitelberg G, McIntyre T, Baird J, Lacey J and Simon H 1995 The high enthalpy shock tunnel in Göttingen (HEG) *Proc. 18th Int. Symp. on Shock Waves*; K Takayama (Berlin: Springer) pp 677–82

Beckwith I, Chen F, Wilkinson S, Malik M and Tuttle D 1990 Design and operational features of low disturbance wind tunnels at NASA langley for Mach Numbers from 3.5 to 18 AIAA Paper 1990-1391 (Reston, VA: American Institute of Aeronautics and Astronautics)

Ben-Yakar A and Hanson R 2002 Characterization of expansion tube flows for hypervelocity combustion studies *J. Propuls. Power* **18** 943–52

Beresh S, Casper K, Wagner J, Henfling J, Spillers R and Pruett 2015 Modernization of Sandia's Hypersonic Wind Tunnel AIAA Paper 2015-1338 (Reston, VA: American Institute of Aeronautics and Astronautics)

Berry S, Horvath T, Kowalkowski M and Liechty D 1999 X-33 (Rev-F) aeroheating results of test 6770 in NASA Langley 20-inch Mach 6 air tunnel *Technical Memorandum 1999-209122* (Washington, DC: National Aeronautics and Space Administration)

Bogdonoff S and Hammitt A 1954 The Princeton helium hypersonic tunnel and preliminary results above $M = 11$ *WADC Technical Report 54-124* Aeronautical Research Laboratory, Wright Air Development Center, Air Research and Development Command, United States Air Force, Wright-Patterson AFB, OH

Bogdonoff S and Vas I 1959 Preliminary investigation of spiked bodies at hypersonic speeds *J. Aerosp. Sci.* **26** 65–74

Borovoy V, Egorov I, Skuratov A and Struminskaya I 2013 Two dimensional shock wave boundary layer interaction in the presence of an entropy layer *AIAA J.* **31** 80–93

Bowersox R 2022a private communication

Bowersox R 2022b The Hypervelocity Expansion Tunnel (HXT) https://nal.tamu.edu/facilities/

Bowersox R 2022c The Mach 6 quiet tunnel (M6QT) https://nal.tamu.edu/facilities/

Boyd C and Ragsdale W 1991 Hypervelocity wind tunnel 9 facility handbook, 3rd edn *Technical Report NAVSWC TR 91-616* (White Oak, Silver Spring, MD: United States Naval Ordnance Laboratory)

Brusniak L and Dolling D 1994 Physics of unsteady blunt-fin-induced shock wave turbulent boundary layer interactions *J. Fluid Mech.* **273** 375–409

Buttsworth D 2009 Ludwieg tunnel facility with free piston compression heating for supersonic and hypersonic testing *Proc. 9th Australian Space Science Conf.* ed W Short and I Cairns (Sydney: National Space Society of Australia Ltd) pp 153–62

Buttsworth D 2016 Gun tunnels *Experimental Methods of Shock Wave Research* ed O Igra and F Seiler (Berlin: Springer) pp 381–96

Chan Y, Mason R and Reddy N 1965 Instrumentation and calibration of UTIAS 11 in ×15 in hypersonic shock tunnel *Technical Note 91* (Toronto, Canada: University of Toronto)

Chandel D, Nompelis I and Candler G 2018 Computations of high enthalpy shock propagation in electric arc shock tube (EAST) at NASA Ames AIAA Paper 2018-1722 (Reston, VA: American Institute of Aeronautics and Astronautics)

Chanetz B 2022 private communication

Chanetz B, Benay R, Bousquet J-M, Bur R, Pot T, Grasso F and Moss J 1998 Experimental and numerical study of the laminar separation in hypersonic flow *Aerosp. Sci. Technol.* **2** 205–18

Chanetz B, Délery J, Gilliéron P, Gnemmi P, Gowree E and Perrier P 2020 *Experimental Aerodynamics* (Berlin: Springer)

Chue R, Tsai C-Y, Bakos R, Erdos J and Rogers R 2002 NASA's HYPULSE facility at GASL—a dual mode, dual driver reflected shock/expansion tunnel *Advanced Hypersonic Test Facilities* ed D Marren and F Lu (Reston, VA: American Institute of Aeronautics and Astronautics) pp 29–71

Chynoweth B, Edelman J, Gray K, McKiernan G and Schneider S 2017 Measurements in the Boeing/AFOSR Mach 6 quiet tunnel on hypersonic boundary layer transition AIAA Paper 2017-3632 (Reston, VA: American Institute of Aeronautics and Astronautics)

Clark L 1965 Description and initial calibration of the langley 12 inch hypersonic ceramic-heated tunnel *Technical Note D-2703* (Washington, DC: National Aeronautics and Space Administration)

Clemens N 2022 private communication

Collen P, Doherty L and Subiah S *et al* 2021 Development and commissioning of the T6 stalker tunnel *Exp. Fluids* **72**

Cornell Bulletin 1945 Curtiss-Wright Corp. Gives Lab to Cornell; To Aid Air Research 1945 *The Cornell Bulletin*

Cox R 1962 Experimental facilities for hypersonic research *Prog. Aerosp. Sci.* **3** 137–78

Cummings J, Foster T and Lockman W 1965 Heat transfer test of a 0.006 scale thin skin thermocouple space shuttle model (50-0, 41-T) in the NASA Ames Research Center 3.5 foot hypersonic wind tunnel at Mach 5.3 (IH28) *Contractor Report 147616* (Washington, DC: National Aeronautics and Space Administration)

Cummings R and McLaughlin T 2012 Hypersonic Ludwieg tube design and future usage at the US Air Force Academy AIAA Paper 2012-0734 (Reston, VA: American Institute of Aeronautics and Astronautics)

Dean T, Blair T, Roberts M, Limbach C and Bowersox R 2022 On the initial characterization of a large scale hypervelocity expansion tunnel AIAA Paper 2022-1602 (Reston, VA: American Institute of Aeronautics and Astronautics)

Decker R, Semper M, Anthony J and Cummings R 2015 Starting characteristics of the US Air Force Academy Mach 6 Ludwieg tube AIAA Paper 2015-3338 (Reston, VA: American Institute of Aeronautics and Astronautics)

Desse J, Soutade J, Viguier P, Picart P, Ferrier M and Serre L 2014 Digital holographic interferometry at F4-ONERA hypersonic wind tunnel *16th Int. Symp. on Flow Visualization*

Dufrene A 2006 Design and characterization of the hypervelocity expansion tube *Master's Thesis* University of Illinois at Urbana-Champaign, Urbana, IL

Dufrene A, MacLean M, Parker R, Wadhams T, Mundy E and Holden M 2010 Characterication of the new LENS expansion tunnel facility AIAA Paper 2010-1564 (Reston, VA: American Institute of Aeronautics and Astronautics)

Dufrene A, Sharma M and Austin J 2007 Design and characterizaation of a hypervelocity expansion tube facility *J. Propuls. Power* **23** 1185–93

Dundian G, Shihe Y and Xiaoge L 2017 Design and performance of a hypersonic quiet wind tunnel at NUDT AIAA Paper 2017-2305 (Reston, VA: American Institute of Aeronautics and Astronautics)

Eggers A, Hansen C and Cunningham B 1955 Theoretical and experimental investigation of the effect of yaw on heat transfer to circular cylinders in hypersonic flow *Research Memorandum RM A55E02* (Washington, DC: National Aeronautics and Space Administration)

Eitelberg G 1994 First results of calibration and use of the HEG AIAA Paper 94-2525 (Reston, VA: American Institute of Aeronautics and Astronautics)

Erdos J, Bakos R, Castrogiovanni A and Rogers R 1997 Dual mode shock-expansion/reflected-shock tunnel AIAA Paper 1997-560 (Reston, VA: American Institute of Aeronautics and Astronautics

Erdos J, Calleja J and Tamagno J 1994 Increases in the hypervelocity test envelope of the HYPULSE shock-expansion tube AIAA Paper 94-2524 AIAA Paper 1997-560 (Reston, VA: American Institute of Aeronautics and Astronautics)

Erengil M and Dolling D 1993 Effects of sweepback on unsteady separation in Mach 5 compression ramp interactions *AIAA J.* **31** 302–11

Erickson W 1963 An extension of estimated hypersonic flow parameters for helium as a real gas *Technical Note D-1632* (Washington, DC: National Aeronautics and Space Administration)

Estorf M, Wolf T and Radespiel R 2005 Experimental and numerical investigation on the operation of the hypersonic Ludwieg tube braunschweig *Proc. 5th European Symp. on Aerothermodynamics for Space Vehicles* ed D Danesy (Noordwijk: European Space Agency) pp 579–86

Faro I, Small T and Hill F 1951 Hypersonic flow at a Mach number of 10 *J. Appl. Phys.* **22** 220–2

Felgueiras L D 2018 Performance design of hypervelocity shock tube facilities *Master's Thesis* Instituto Superior Tecnico, Lisbon, Portugal

Fiore A and Law C 1975 Aerodynamic calibration of the aerospace research laboratories $M = 6$ high Reynolds number facility *Technical Report ARL TR 75-0028* Aerospace Research Laboratories, Wright-Patterson AFB, OH

Fletcher D 2005 Fundamentals of hypersonic flow—aerothermodynamics *Critical Technologies for Hypersonic Vehicle Development* ed D Gaitonde and D Fletcher (NATO Research and Technology Office) pp 3-1–47

Friehmelt H, Koppenwallner G and Müller-Eigner R 1993 Calibration and first results of a redesigned Ludwieg expansion tube AIAA Paper 93-5001 (Reston, VA: American Institute of Aeronautics and Astronautics)

Gildfind D, Morgan R and Jacobs P 2016 Expansion tubes in Australia *Experimental Methods of Shock Wave Research* ed O Igra and F Seiler (Berlin: Springer) pp 399–431

Gnemmi P, Srulijes J and Seiler F *et al* 2016 Shock Tunnels at ISL *Experimental Methods of Shock Wave Research* ed O Igra and F Seiler (Berlin: Springer) pp 131–80

Goldfeld M, Maslov A, Starov A, Shuminskii V and Yaroslavtsev M 2016 IT-302M hotshot wind tunnel as a tool for the development of hypersonic technologies *Proc. Int. Conf. on the*

*Methods of Aerophysical Research (ICMAR 2016),* Khristianovich Institute of Theoretical and Applied Mechanics (New York: AIP Publishing) pp 030020-1–8
Gregorek G 1962 Initial calibrations and performance of the ARL twenty inch hypersonic wind tunnel *ARL 62-393* Aeronautical Research Laboratories, Office of Aerospace Research, United States Air Force, Wright-Patterson AFB, OH
Gregorek G and Lee J 1962 Design performance and operational characteristics of the ARL twenty inch hypersonic wind tunnel *ARL 62-392* Aeronautical Research Laboratories, Office of Aerospace Research, United States Air Force, Wright-Patterson AFB, OH
Gromyko Y, Polivanov P, Sidorenko A, Buntin D and Maslov A 2013 An experimental study of the natural noise in the transit-M hypersonic wind tunnel *Thermophys. Aeromech.* **20** 481–93
Grönig H 1992 Shock tube application: high enthalpy european wind tunnels *Proc. 18th Int. Symp. on Shock Waves* ed K Takayama (Berlin: Springer) pp 3–16
Grossir G 2022 private communication
Grossir G and Ilich Z 2018 Numerical modeling of the VKI longshot compression process *Flow characterization and modeling of hypersonic wind tunnels vol VKI Lecture Series 2018/19* ed O Chazot and G Grossir (Sint-Genesius-Rode: von Karman Institute for Fluid Dynamics) ch 2
Grossir G, Masutti D and Chazot O 2015 Flow characterization and boundary layer transition studies in VKI hypersonic facilities AIAA Paper 2015-0578 (Reston, VA: American Institute of Aeronautics and Astronautics)
Grossir G, Paris S, Rambaud P and Hove B V 2014 Design of static pressure probes for improved freestream characterization in hypersonic wind tunnels AIAA Paper 2014-1410 (Reston, VA: American Institute of Aeronautics and Astronautics)
Gu S and Olivier H 2020 Capabilities and limitations of existing hypersonic facilities *Prog. Aerosp. Sci.* **113** 1–27
Hammitt A and Bogdonoff S 1956 Hypersonic studies of the leading edge effect on the flow over a flat plate *Jet Propul.* **26** 241–6 and 250
Hannemann K and Beck W 2002 Aerothermodynamics research in the DLR high enthalpy shock tunnel HEG *Advanced Hypersonic Test Facilities* ed D Marren and F Lu (Reston, VA: American Institute of Aeronautics and Astronautics) pp 205–37
Hannemann K, Itoh K, Mee D and Hornung H 2016 Free piston shock tunnels HEG, HIEST, T4 and T5 *Experimental Methods of Shock Wave Research* ed Shock Wave Research (Berlin: Springer) pp 181–263
Hannemann K, Schramm J, Wagner A and Camillo G 2018 The high enthalpy shock tunnel Göttingen of the German Aerospace Center (DLR *J. Large-Scale Res. Facil.* **4** A133
Hedlund B, Houpt A, Gordeyev S and Leonov S 2018 Measurement of flow perturbation spectra in Mach 4.5 corner separation zone *AIAA J.* **56** 2699–711
Heltsley W, Snyder J, Houle A, Davidson D, Mungal M and Hanson R 2006 Design and characterization of the Stanford 6 inch expansion tube AIAA Paper 2006-4443 (Reston, VA: American Institute of Aeronautics and Astronautics)
Hill F 1956 Boundary-layer measurements in hypersonic flow *J. Aeronaut. Sci.* **23** 35–42
Hill F 1959 Turbulent boundary layer measurements at Mach numbers from 8 to 10 *Phys. Fluids* **2** 668–80
Hill J, Reeder M and Thomas L *et al* 2022 Hypersonic laminar, transition and turbulent boundary layer measurements using FLEET velocimetry AIAA Paper 2022-1704 (Reston, VA: American Institute of Aeronautics and Astronautics)

Hoberg E, Huffman C, Sanchez-Plesha N, Running C, Kato N, Im S-K and Juliano T 2019 Flow characterization of a supersonic and hypersonic arc-heated wind tunnel *Flow Characterization of a Supersonic and Hypersonic Arc-Heated Wind Tunnel* (Reston, VA: American Institute of Aeronautics and Astronautics)

Hofferth J, Bowersox R and Saric W 2010 The Mach 6 quiet tunnel at Texas A&M: quiet flow performance AIAA Paper 2010-4794 (Reston, VA: American Institute of Aeronautics and Astronautics)

Hoffman E, Bashor I and Combs C 2020 Construction of a Mach 7 Ludwieg Tube at UTSA AIAA Paper 2020-2998 (Reston, VA: American Institute of Aeronautics and Astronautics)

Hoffman E, LaLonde E, Garcia M, Elizondo V D, Chen I, Bilbo H and Combs C 2022 Characterization of the UTSA Mach 7 Ludwieg tube AIAA Paper 2022-1600 (Reston, VA: American Institute of Aeronautics and Astronautics)

Hoffman E, Rodriguez J, Garcia M, Elizondo V D, LaLonde E and Combs C 2021 Preliminary testing of the UTSA Mach 7 Ludwieg tube AIAA Paper 2021-2979 (Reston, VA: American Institute of Aeronautics and Astronautics)

Holden M 1975 Shock wave—turbulent boundary layer interaction in high speed flow *Technical Report 75-0204* Aerospace Research Laboratories, US Air Force Systems Command, Wright-Patterson AFB, OH

Holden M 1990 Large energy national shock tunnel (LENS) *Technical Report* CALSPAN-UB Research Center, Buffalo, NY

Holden M 2010 The LENS facilities and experimental studies to evaluate the modeling of boundary layer transition, shock/boundary layer interaction, real gas, radiation and plasma phenomena in contemporary CFD codes *Aerothermodynamic Design, Review on Ground Testing and CFD* ; O Chazot and K Bensassi NATO RTO-EN-AVT-186 pp 2-1–2-64

Holden M and Parker R 2002 LENS hypervelocity tunnels and application to vehicle testing at duplicated flight conditions *Advanced Hypersonic Test Facilities* ed D Marren and F Lu (Reston, VA: American Institute of Aeronautics and Astronautics) pp 73–110

Holden M, Wadhams T, Smolinski G, MacLean M, Harvey J and Walker B 2006 Experimental and numerical studies on hypersonic vehicle performance in the LENS shock and expansion tunnels AIAA Paper 2006-0125 (Reston, VA: American Institute of Aeronautics and Astronautics)

Hornung H 1992 Performance data of the new free-piston shock tunnel at GALCIT AIAA Paper 92-3943 (Reston, VA: American Institute of Aeronautics and Astronautics)

Hornung H, Sturtevant B, Belanger J, Anderson S, Brouillette M and Jenkins M 1995 Perforamance data of the new free-piston tunnel T5 at GALCIT *Proc. 18th Int. Symp. on Shock Waves*; K Takayama (Berlin: Springer) pp 603–10

IT-302M Hypersonic Hot-Shot Wind Tunnel 2022 http://www.itam.nsc.ru/en/science/experimental_facilities/it-302m_hypersonic-hot-shot_wind_tunnel_.html

Itoh I, Komuro T, Sato K, Ueda S, Tanno H and Takahashi M 2002 Characteristics of free-piston shock tunnel HIEST: 1st Report, tuned operation of free-piston driver *Trans. Japan Soc. Mech. Eng.* B **68** 2968–75 (in Japanese)

Itoh K 2002 Characteristics of the HIEST and its applicability for hypersonic aerothermodynamic and scramjet research *Advanced Hypersonic Test Facilities* ed D Marren and F Lu (Reston, VA: American Institute of Aeronautics and Astronautics) pp 239–53

Itoh K, Ueda S, Komuro T, Saito K, Takahashi M, Miyajima H and Koga K 1997 Design and construction of HIEST (high enthalpy shock tunnel) *Proc. Int. Conf. on Fluid Engineering* vol 1 (Tokyo: JSME Press) pp 353–8

Itoh K, Ueda S, Komuro T, Sato K, Takahashi M, Miyajima H, Tanno H and Muramoto H 1998 Improvement of a free piston driver for a high-enthalpy shock tunnel *Shock Waves* **8** 215–33

Itoh K, Ueda S, Tanno H, Komuro T and Sato K 2002 Hypersonic aerothermodynamic and scramjet research using high enthalpy shock tunnel *Shock Waves* **12** 93–8

Jacobs C, McIntyre T, Morgan R, Brandis A and Laux C 2015 Radiative heat transfer measurements in low density titan atmospheres *J. Thermophys. Heat Transfer* **29** 835–4

Jiang Z and Yu H 2014 Experiments and development of long test duration hypervelocity detonation driven shock tube (LHKst) AIAA Paper 2014-1012 (Reston, VA: American Institute of Aeronautics and Astronautics)

Kemp J 1970 The Ames M-50 helium tunnel *Technical Note TN D-5788* (Washington, DC: National Aeronautics and Space Administration)

Kharitonov A, Zvegintsev V, Vasenev I, Kuraeva A, Nalivaichenko D, Novikov A, Paikova M, Chirkashenko V, Shakhmatova N and Shpak S 2006 Characteristics of the AT-303 hypersonic wind tunnel, part 1: velocity fields *Thermophys. Aeromech.* **13** 1–16

Kimmel R, Borg M, Jewell J, Lam K, Bowersox R, Srinivasan R and Fuchs S 2017 AFRL Ludwieg Tube Initial Performance AIAA Paper 2017-0102 (Reston, VA: American Institute of Aeronautics and Astronautics)

Kumar D 1995 Hypersonic control effectiveness *PhD Thesis* College of Aeronautics, Cranfield University, Cranfield, UK

Lakebrink M, Bowcutt K, Winfree T, Huffman C and Juliano T 2018 Optimization of a Mach 6 quiet wind tunnel nozzle *J. Spacecr. Rockets* **55** 315–21

Laumann E and Herrera J 1967 Jet propulsion laboratory wind tunnel facilities *Technical Memorandum 33-3335* Jet Propulsion Laboratory, California Institute of Technology, Pasadena, CA

Laverre J 1987 Amelioration de la soufflerie S4MA pour les essaies Hermès *Technical Paper 1987-60* Office National d'Etudes et de Recherches Aérospatiales (ONERA)

Lax P and Leonov S 2022 Condensation-limited operational maps of Notre Dame large quiet tunnels AIAA Paper 2022-1719 (Reston, VA: American Institute of Aeronautics and Astronautics)

Lee R, Yanta W, Leonas A and Carner J 1966 The NOL boundary layer channel *Technical Report NOLTR 66-185* (White Oak, Silver Spring, MD: United States Naval Ordnance Laboratory)

Lefevre A, Gildfind D, Gollan R, Jacobs P, McIntyre T and James C 2021 Expansion tube experiments of magnetohydrodynamic aerobraking for superorbital Earth reentry *AIAA J.* **59** 3228–40

Leidy A, Neel I, Tichenor N and Bowersox R 2020 High-speed schlieren imaging of cylinder-induced hypersonic-shock-wave-boundarylayer interactions *AIAA J.* **58** 3090–9

Leonov S 2022 private communication

Li C, Sun R, Wang Y, Chen X and Bi Z 2018 Reliability improvement of the piston compressor in FD-21 free-piston shock tunnel *Technical Report, 5th Int. Conf. on Experimental Fluid Mechanics, Münich*

Liepmann H and Roshko A 1957 *Elements of Gas Dynamics* (New York: Wiley)

Lino da Silva M, Carvalho B, Smith A and Marraffa L 2016 High-pressure H2/He/O2 combustion experiments for the design of the ESTHER shock tube driver AIAA Paper 2016-4156 (Reston, VA: American Institute of Aeronautics and Astronautics)

Lino da Silva M *et al* 2020 European shock tube for high enthalpy research: design and instrumentation, manufacturing and acceptance testing AIAA Paper 2020-0624 (Reston, VA: American Institute of Aeronautics and Astronautics)

Lobb R, Winkler E and Persh J 1955 NOL hypersonic tunnel No. 4 results VII: experimental investigation of turbulent boundary layers in hypersonic flight *Technical Report NAVORD Report 3880* (White Oak, Silver Spring, MD: United States Naval Ordnance Laboratory)

Ludwieg H 1955 Der Rohrwindkanal *Z. Flugwiss.* **3** 206–16

Ludwieg H 1957 The tube wind tunnel—a special type of blowdown tunnel *Report 143* Advisory Group for Aerospace Research and Development

Lukasiewicz J 1973 *Experimental Methods in Hypersonics* (New York: Marcel Dekker)

Lynch K 2022 private communication

Lynch K, Grasser T, Farias P, Daniel K, Spillers R, Downing C and Wagner J 2022 Design and characterization of the sandia free piston reflected shock tunnel AIAA Paper 2022-0968 (Reston, VA: American Institute of Aeronautics and Astronautics)

Marineau E, Grossir G, Wagner A, Leinemann M, Radespiel R, Tanno H, Chynoweth B, Schneider S, Wagnild R and Casper K 2019 Analysis of second mode amplitudes on sharp cones in hypersonic wind tunnels *J. Spacecr. Rockets* **56** 307–18

Marineau E and Hornung H 2009 Heat flux calibration of T5 hypervelocity shock tunnel conical nozzle in air AIAA Paper 2009-1158 (Reston, VA: American Institute of Aeronautics and Astronautics)

Marren D and Lafferty J 2002 The AEDC hypervelocity wind tunnel 9 *Advanced Hypersonic Test Facilities* ed D Marren and F Lu (Reston, VA: American Institute of Aeronautics and Astronautics) pp 467–78

Maslov A 2003 Experimental and theoretical study of hypersonic laminar flow control using ultrasonically absorptive coatings (UAC) *Technical Report ISTC 01-7005* Khristianovich Institute of Theoretical and Applied Mechanics, European Office of Aerospace Research and Development (EOARD). ADA423851

Maslov A, Shumsky V and Yaroslavtsev M 2013 High enthalpy hot shot wind tunnel with combined heating and stabilization of parameters *Thermophys. Aeromech.* **20** 527–38

Masson A, Sagnier P and Mohamed A 2002 The ONERA F4 high enthalpy wind tunnel *Advanced Hypersonic Test Facilities* ed D Marren and F Lu (Reston, VA: American Institute of Aeronautics and Astronautics) pp 441–66

Matthews A, Jones T and Cain T 2005 Design and test of a hypersonic isentropic-spike intake with aligned cowl *J. Propuls. Power* **21** 838–43

McCormick B, Jumper E and Newberry C 2004 *Aerospace Engineering Education During the First Century of Flight* (Reston, VA: American Institute of Aeronautics and Astronautics)

McGilvray M, Doherty L, Morgan R and Gildfind D 2015 T6: The Oxford University Stalker Tunnel AIAA Paper 2015-3545 (Reston, VA: American Institute of Aeronautics and Astronautics)

McLellan C, Williams T and Bertram M 1950 Investigation of a two step nozzle in the langley 11 inch hypersonic tunnel *Technical Note 2171* (Washington, DC: National Advisory Committee on Aeronautics)

Miller C and Jones J 1983 Development and performance of the nasa langley research center expansion tube/tunnel, a hypersonic-hypervelocity real-gas facility *14th Int. Symp. on Shock Tubes and Shock Waves* ed B Milton and R Archer (Sydney: Sydney Shock Tube Symposium Publishers, University of New South Wales Press)

Mills M 2015 Arnold engineering development complex Von Karman facility hypersonic wind tunnels AIAA Paper 2015-1336 (Reston, VA: American Institute of Aeronautics and Astronautics)

Morgan R 2022 private communication

Morgan R, McIntyre T, Gollan R, Jacobs P, Brandis A, McBilvray M, van Diem D, Gnoffo P, Pulsonnetti M and Wright M 2006 Radiation measurements in nonreflected shock tunnels AIAA Paper 2006-2958 (Reston, VA: American Institute of Aeronautics and Astronautics)

Morzenski L 1986 Moyens d'essais et de mesures de souffleries à rafales R2 et R3 de Chalais-Meudon *Note Technique 97/1865 AN* Office National d'Etudes et de Recherches Aérospatiales (ONERA)

Mundt C 2016 Development of the new piston-driven shock-tunnel HELM *Experimental Methods of Shock Wave Research* ed O Igra and F Seiler (Berlin: Springer) pp 265–83

Murphree Z, Combs C, Yu W, Dolling D and Clemens N 2021 Physics of unsteady cylinder-induced shock wave transitional boundary layer interactions *J. Fluid Mech.* **918** 1–27

Oliveira B 2021 High pressure He/H2/O2 mixtures combustion on the ESTHER driver: experiment and modeling *Master's Thesis* Ténico Lisboa, Lisbon, Portugal

Olivier H 2016 The aachen shock tunnel TH2 with dual driver mode operation *Experimental Methods of Shock Wave Research* ed O Igra and F Seiler (Berlin: Springer) pp 111–29

Olivier H, Zonglin J, Yu H and Lu F 2002 Detonation-driven shock tubes and tunnels *Advanced Hypersonic Test Facilities)* ed F Lu and D Marren (Reston, VA: American Institute of Aeronautics and Astronautics) pp 135–203

Parobek D, Boyer D and Clinehens G 1985 Development and compatibility of flow seeding techniques for LV measurements in a diversity of research test flows *Technical Memorandum AFWAL-TM-85-204-FIMN* Aeronautical Laboratories, Flight Dynamics Laboratory, Aeromechanics Division, Experimental Engineering Branch, Wright-Patterson AFB, OH

Parrott T, Jones M and Albertson C 1989 Fluctuating pressures measured beneath a high temperature, turbulent boundary layer on a flat plate at a Mach Number of 5 Technical Paper 2947 (Washington, DC: National Aeronautics and Space Administration)

Paull A, Stalker R and Mee D 1995 Experiments on supersonic combustion ramjet propulsion in a shock tunnel *J. Fluid Mech.* **296** 159–83

Penaranda F and Freda M 1985 Aeronautical facilities catalogue, volume 1—wind tunnels *NASA RP 1132* (Washington, DC: National Aeronautics and Space Administration)

Pirrello C, Hardin R, Heckart M and Brown K 1971 An inventory of aeronautical ground research facilities, volume 1: wind tunnels *NASA CR 1874* (Washington, DC: National Aeronautics and Space Administration)

Post T D 2022 China to Build Mach 30 Wind Tunnel by 2022 https://www.thedefensepost.com/2021/08/25/china-mach-30-wind-tunnel/

Prince S 2022 private communication

Radespiel R 2022 private communication

Radespiel R, Estorf M, Heitmann D and Munoz F 2016 Hypersonic Ludwieg tube *Experimental Methods of Shock Wave Research* ed O Igra and F Seiler (Berlin: Springer) pp 433–58

Raman K 1974 A study of surface pressure fluctuations in hypersonic turbulent boundary layers *Contractor Report CR-2386* (Washington, DC: National Aeronautics and Space Administration)

Resler E and Bloxsom D 1952 Very high Mach number flows by unsteady flow principles Limited distribution monograph, Cornell University Graduate School of Aeronautical Engineering, Ithaca, NY

Richards B and Enkenhus K 1970 Hypersonic testing in the VKI longshot free-piston tunnel *AIAA J.* **8** 1020–5

Rufer S and Berridge D 2012 Pressure fluctuation measurements in the NASA langley 20 inch Mach 6 wind tunnel AIAA Paper 2012-3262 (Reston, VA: American Institute of Aeronautics and Astronautics)

Running C, Juliano T, Jewell J, Borg M and Kimmel R 2019 Hypersonic shock wave boundary layer interactions on a cone/flare *Exp. Therm. Fluid Sci.* **109** 109911

Saavedra J, Grossir G, Chazot O and Paniagua G 2022 Start-up analysis of a hypersonic short duration facility *AIAA J.* **60** 2060–74

Sagnier P and Vérant J-L 1998 Flow characterization in the ONERA F4 high-enthalpy wind tunnel *AIAA J.* **36** 522–31

Scaggs N, Burggraf W and Gregorek G 1963 The ARL thirty-inch hypersonic wind tunnel initial calibration and performance ARL 63-223, Aeronautical Research Laboratories, Office of Aerospace Research, United States Air Force, Wright-Patterson AFB, OH

Schaefer W 1965 Characteristics of major active wind tunnels at the NASA langley research center *Technical Memorandum X-1130* (Washington, DC: National Aeronautics and Space Administration)

Schmisseur J and Dolling D 1992 Unsteady separation in sharp fin-induced shock wave turbulent boundary layer interactions at Mach 5 AIAA Paper 92-0748 (Reston, VA: American Institute of Aeronautics and Astronautics)

Schneider S 2008 Development of quiet hypersonic tunnels *J. Spacecr. Rockets* **45** 641–64

Schrijer F and Bannink W 2008 Description and flow assessment of the delft hypersonic Ludwieg tube AIAA Paper 2008-3943 (Reston, VA: American Institute of Aeronautics and Astronautics)

Schülein E 2003 Optical skin friction measurements in the short duration Ludwieg tube facility *20th Int. Congress on Instrumentation in Aerospace Simulation Facilities (ICIASF 2003)* pp 157–68

Schülein E 2022 Concave bump for impinging shock control in supersonic flow *AIAA J.* **60** 2749–66

Schülein E and Zheltovodov A 1998 Development of experimental methods for the hypersonic flows studies in Ludwieg tube *Int. Conf. on Methods of Aerophysical Research (ICMAR 1998), Part I* ed V Fomin and A Kharitonov 191–9

Scott M, Morgan R and Jacobs P 2005 A new single stage driver for the X2 expansion tube AIAA Paper 2005-6970 (Reston, VA: American Institute of Aeronautics and Astronautics)

Semper M, Pruski B and Bowersox R 2012 Freestream turbulence measurements in a continuously variable hypersonic wind tunnel AIAA Paper 2012-0732 (Reston, VA: American Institute of Aeronautics and Astronautics)

Semper M, Tichenor N, Bowersox R, Srinvasan R and Worth S 2009 On the design and calibration of an actively controlled expansion hypersonic wind tunnel AIAA Paper 2009-799 (Reston, VA: American Institute of Aeronautics and Astronautics)

Shekhtman D, Hameed A, Segall B, Dworzanczyk A and Parziale N 2022 Initial shakedown testing of the stevens shock tunnel AIAA Paper 2022-1402 (Reston, VA: American Institute of Aeronautics and Astronautics)

Simeonides G 1990 The VKI hypersonic wind tunnels and associated measurements techniques *Technical Memorandum 46* von Kármán Institute for Fluid Dynamics

Smelt R 1946 A critical review of German research on high speed flow *J. R. Aeronaut. Soc.* **50** 899–934

Smith C 1966 The starting process in a hypersonic nozzle *J. Fluid Mech.* **24** 625–40

Stalker R 1960 Isentropic compression of shock tube driver gas *ARS J.* **30** 564

Stalker R 1967 A study of the free-piston shock tunnel *AIAA J.* **5** 2160–5

Stalker R 1972 Development of a hypervelocity wind tunnel *Aeronaut. J.* **76** 374–84

Stalker R 2006 Modern developments in hypersonic wind tunnels *Aeronaut. J.* **110** 21–39

Steen L 2010 Characterization and development of nozzles for a hypersonic quiet wind tunnel *Master's Thesis* Purdue University, West Lafayette, IN

Stennett S 2020 Development of an extended test time operating mode for a large reflected shock tunnel facility *PhD Thesis* Centre for Hypersonics, School of Mechanical and Mining Engineering, The University of Queensland, Brisbane, Australia

Stennett S, Gildfind D and Jacobs P 2019 Optimising the X3R reflected shock tunnel free-piston driver for long duration test times *Proc. 31st Int. Symp. on Shock Waves—vol 2,* International Shock Wave Institute; A Sasoh, T Aoki and M Katayama (Berlin: Springer) pp 189–96

Stennett S, Gildfind D, Jacobs P and Morgan R 2018 Performance optimization of X3R: a new reflected shock tunnel mode for the X3 expansion tube AIAA Paper 2018-3563 (Reston, VA: American Institute of Aeronautics and Astronautics)

Stephan S and Radespiel R 2017 Propulsive jet simulation with air and helium in launcher wake flows *CEAS Space J.* **9** 195–209

Stephan S, Wu J and Radespiel R 2015 Propulsive jet influence on generic rocket launcher base flow *CEAS Space J.* **7** 453–73

Stetson K 1980 Hypersonic boundary layer transition experiments *Technical Report AFWAL-TR-80-3062* Flight Dynamics Laboratory, Air Force Wright Aeronautical Laboratories, Air Force Systems Command, Wright-Patterson AFB, OH

Stollery J, Maull D and Belcher B 1960 The Imperial College hypersonic gun tunnel August 1958 *J. R. Aeronaut. Soc.* **64** 24–32

T-313 Hypersonic Wind Tunnel 2022 http://www.itam.nsc.ru/en/science/experimental_facilities/t-313_wind_tunnel.html.

Taylor O 2019 Effects of surface roughness on boundary layer transition in a hypersonic flow *PhD Thesis* Imperial College, London, UK

Thomas S, Lee J, Stephens J, Hostler R and Kamp W V 2010 The Mothball, Sustainment, and Proposed Reactivation of the Hypersonic Tunnel Facility (HTF) at NASA Glenn Research Center Plum Brook Station AIAA Paper 2010-4533 (Reston, VA: American Institute of Aeronautics and Astronautics)

Tichenor N, Semper M, Bowersox R, Srinvasan R and North S 2010 Calibration of an actively controlled expansion hypersonic wind tunnel AIAA Paper 2010-4793 (Reston, VA: American Institute of Aeronautics and Astronautics)

Trimpi R 1962 A preliminary theoretical study of the expansion tube, a new device for producing high enthalpy short duration hypersonic gas flows *Technical Report R-133* (Washington, DC: National Aeronautics and Space Administration)

Trimpi R 1966 A theoretical investigation of simulation in expansion tubes and tunnels *Technical Report R-243* (Washington, DC: National Aeronautics and Space Administration)

Trimpi R and Callis L 1965 A perfect-gas analysis of the expansion tunnel, a modification to the expansion tube *Technical Report TR-R-133* National Aeronautics and Space Administration

Trout O 1963 Design, operation, and testing capabilities of the langley ll-inch ceramic-heated tunnel *Technical Note D-1598* (Washington, DC: National Aeronautics and Space Administration)

TsAGI T-116 Wind Tunnel 2022 http://tsagi.com/experimental_base/t-116/.

TsAGI T-117 Wind Tunnel 2022 http://tsagi.com/experimental_base/t-117/.

TsAGI T-132 Wind Tunnel Complex 2022 http://tsagi.com/experimental_base/t-131/.

Unalmis O, Clemens N and Dolling D 2001 Experimental study of shear layer acoustics coupling in Mach 5 cavity flow *AIAA J.* **39** 242–52

Vaganov A, Neyland V, Noev A, Skuratov A, Kashin V, Lifits A and Nemykin V 2014 Laminar-turbulent transition on body of revolution with surface breaks *Proc. 17th Int. Conf. on Methods of Aerophysical Research* Khristianovich Institute of Theoretical and Applied Mechanics

Vanhée J 1989 The H3 hypersonic wind tunnel: new implementation and calibration *Technical Report VKI-IN-86* von Kármán Institute for Fluid Dynamics

Vennemann D 1999 Hypersonic test facilities available in western europe for aerodynamic/aerothermal and structure/material investigations *Phil. Trans. R. Soc.* A **357** 2227–48

Vincente F and Foy N 1963 Hypersonic wind tunnels in the United States *Report No. TOR-169 (3305)-1* Revision 1, Aerospace Corporation

Virginia Tech Hypersonic Wind Tunnel 2022, https://www.aoe.vt.edu/research/facilities/hyperson.html

Wang Y, Hu Z, Liu Y and Jiang Z 2016 Starting process in a large scale shock tunnel *AIAA J.* **54** 1240–9

Watson R and Bushnell D 1971 Calibration of the Langley Mach 20 high Reynolds number helium tunnel including diffuser measurements *NASA TM X-2353* (Washington, DC: National Aeronautics and Space Administration)

Wegener P 1951 Summary of recent experimental investigations in the N.O.L. hyperballistics wind tunnel *J. Aeronaut. Sci.* **18** 665–70

Wegener P and Lobb R 1953 An experimental study of a hypersonic wind-tunnel diffuser *J. Aeronaut. Sci.* **20** 105–10

Wegener P, Winkler E and Sibulkin M 1953 A Measurement of turbulent boundary-layer profiles and heat-transfer coefficient at $M = 7$ *J. Aeronaut. Sci.* **20** 221–2

Wendt J 1987 European hypersonic wind tunnls *AGARD CP 428, Aerodynamics of Hypersonic Lifting Vehicles* Advisory Group for Aerospace Research and Development pp 2-1–24

Wilkinson S, Anders S, Chen F-J and Beckwith I 1992 Supersonic and hypersonic quiet tunnel technology at NASA langley AIAA Paper 92-3908 (Reston, VA: American Institute of Aeronautics and Astronautics)

Willems S, Gülhan A and Steelant J 2015 Experiments on the effect of laminar-turbulent transition on the SWBLI in H2K at Mach 6 *Exp. Fluids* **56**

Williams F 1985 *Combustion Theory* (Cambridge, MA: Perseus Books)

Williams O J H, Sahoo D, Baumgartner M L and Smits A J 2018 *J. Fluid Mech.* **834** 237–70

Wittliff C 1987 A survey of existing hypersonic ground test facilities—North America *AGARD CP 428, Aerodynamics of Hypersonic Lifting Vehicles* pp 1-1–17

Wittliff C, Wilson M and Hertzberg A 1959 The tailored-interface hypersonic shock tunnel *J. Aerosp. Sci.* **26** 219–28

Woike M and Willis B 2002 NASA Glenn Research Center's hypersonic tunnel facility *Advanced Hypersonic Test Facilities* ed F Lu and D Marren (Reston, VA: American Institute of Aeronautics and Astronautics) pp 427–39

Wolf T, Estorf M and Radespiel R 2007 Investigation of the starting process in a Ludwieg tube *Theor. Comput. Fluid Dyn.* **21** 81–98

Wu Y, Yi S, He L, Chen Z and Zhu Y 2015 Flow visualization of Mach 3 compression ramp with different upstream boundary layers *J. Vis.* **18** 631–44

Yu H-R, Esser B, Lenartz M and Grönig H 1992 Gaseous detonation driver for a shock tunnel *Shock Waves* **2** 245–54

Zhang C-H 2014 The development of hypersonic quiet wind tunnel and experimental investigation of hypersonic boundary layer transition on a flared cone (in Chinese) *PhD Thesis* Peking University, Peking, People's Republic of China

Zhang C-H 2020 Evolution of the second mode in a hypersonic boundary layer *Phys. Fluids* **32** 171206

Zhang C-H, Tang Q and Lee C-B 2013 Hypersonic boundary layer transition on a flared cone *Acta Mech. Sin.* **29** 48–53

Zhixian B, Bingbing Z, Hao Z and Xing C 2018 Experiments and computations on the compression process in the free piston shock tunnel FD21 *Technical Report, 5th Int. Conf. on Experimental Fluid Mechanics, Münich*

Milton Keynes UK
Ingram Content Group UK Ltd.
UKHW051932160524
442820UK00002B/9

9 780750 35000